R. Stein, R. W. Macdonald

The Organic Carbon Cycle in the Arctic Ocean

Springer-Verlag Berlin Heidelberg GmbH

Ruediger Stein · Robie W. Macdonald (Eds.)

The Organic Carbon Cycle in the Arctic Ocean

With 194 Figures, 22 in colour, and 62 Tables

Springer

Editors

PROFESSOR DR. RUEDIGER STEIN

Institute for Polar and Marine Research
27568 Bremerhaven
Germany
Email: rstein@awi-bremerhaven.de

DR. ROBIE W. MACDONALD

Institute of Ocean Sciences
Sidney B.C. V8L 4B2
Canada
Email: macdonaldrob@dfo-mpo.gc.ca

Cover photo (Research Icebreaker "Polarstern"): H. Grobe, AWI Bremerhaven

ISBN 978-3-642-62351-6

Library of Congress Cataloging-in-Publication Data

The organic carbon cycle in the Arctic Ocean / Ruediger Stein, Robie W. Macdonald (eds.).
p. cm.
Includes bibliographical references.

1. Chemical oceanography. 2. Carbon cycle (Biogeochemistry) 3.
Biogeochemistry--Arctic Ocean. I. Stein, Ruediger. II. MacDonald, Robin W., 1931-

ISBN 978-3-642-62351-6 ISBN 978-3-642-18912-8(eBook)
DOI 10.1007/978-3-642-18912-8
http://www.springer.de

Softcover reprint of the hardcover 1st edition 2004

Typesetting: Fotosatz-Service Köhler GmbH, Würzburg
Cover design: E. Kirchner, Heidelberg

Printed on acid-free paper 32/3141/as 5 4 3 2 1 0

Preface

The flux, preservation, and accumulation of organic carbon in marine systems are controlled by various mechanisms including primary production of the surface water, supply of terrigenous organic matter from the surrounding continents, biogeochemical processes in the water column and at the seafloor, and sedimentation rate. For the world's oceans, phytoplankton productivity is by far the largest organic carbon source, estimated to be about 30 to 50 Gt (10^9 tonnes) per year (Berger et al. 1989; Hedges and Keil 1995). By comparison, rivers contribute about 0.15 to 0.23 Gt y^{-1} of particulate organic carbon (Ittekkot 1988; Hedges et al. 1997). On average, only about 0.1–0.2 Gt y^{-1} of the organic carbon supplied to the world's oceans has been buried in Holocene sediments, with perhaps 85–90% of that being deposited in deltaic, shelf and upper slope sediments (Romankevich 1984; Berner 1989; Hedges and Keil 1995). To refine the global ocean carbon budget in general and, specifically, to quantify organic carbon burial because it is so important for climatic change, we need detailed measurements of the organic carbon flux in all ocean environments, especially for continental margins (e.g., Liu et al., 2000). With the notable exception of the Arctic Ocean, data have been available for some time on organic carbon sources, pathways and burial in a variety of ocean environments. Even though the Arctic Ocean is (1) surrounded by the world's largest shelf seas, (2) influenced by large river (organic matter) discharge, (3) highly sensitive to climate change and (4) an important feedback component of the global climate system, a comprehensive compilation of data on processes controlling organic carbon flux, accumulation, and its temporal and spatial variability has not been available. A recent monograph published in Russian (Romankevich and Vetrov; 2001) provides a restricted view of the organic carbon cycle addressing only the Russian arctic shelf seas and excluding sediment core data.

With this book, we present an overview of the sources, pathways, and burial of organic carbon in the circum-Arctic continental margin and deep-sea areas. To understand the Arctic Ocean's organic carbon cycle, important system components are addressed here in topical chapters dealing with:

- processes controlling the terrigenous organic carbon input,
- primary productivity,
- the role of dissolved organic carbon,
- organic carbon flux through the water column,
- biological and geochemical processes at the seafloor, and
- the burial of organic carbon and its variability in time and space.

The book serves a common theme – organic carbon. Accordingly, we have attempted both to collate all available data from each shelf and the basins of the Arctic Ocean and to present these data in common or parallel formats. A coherent picture of sediment and organic carbon inputs from the ocean's margins is beginning to emerge as a consequence of reasonably complete coverage for major arctic rivers. In contrast, we lack crucial measurements of, for example, coastal erosion, particle flux, sediment core data and primary production for many regions of the Arctic. Future programs focussed on various aspects of arctic biogeochemistry should help to fill these gaps.

We and the authors of the different chapters of this book are grateful to numerous colleagues for their efforts in offering many suggestions for improvement of the manuscripts, for providing unpublished data, for discussions as well as for technical support: R. Benner, G. Budeus, S. Claesson, K. Dittmers, E. Fahrbach, H. Grobe, G. Harrison, C. Hartmann, J. Hatzky, B. Hollmann, T. Korotkova, H. Köhler, A.P. Lisitzin, W. Luttmer, M. Maksimova, L. Mayer, B. Meon, F. Niessen, M. Reigstad, S. Reiser, G. Rohard, U. Schauer, H.-W. Schenke, S. Solomon, A. Spitzy, L. Tranvik, and D. Yashin. Support for various components of the study has been provided by the Department of Fisheries and Oceans (DFO) Canada; Department of Natural Resources of Russian Federation; European Union; German Ministry of Education and Science (BMBF); National Science Foundation (NSF); National Oceanic & Atmospheric Administration (NOAA); Office of Naval Research (ONR); Norwegian Research Council; Nordic Council of Ministers; Nordic Arctic Research Programme (NARP); Russian Foundation of Basic Research; Ministry of Industry, Science and Technology of Russian Federation; U.S. Geological Survey; Exxon Exploration Company. We thank Marty Bergmann for financial support and encouragement to produce this book. RS thanks the Alfred Wegener Institute for supporting a fourmonths stay at the Institute of Ocean Sciences Sidney, where a major part of this book has been compiled.

February 2003

R. Stein and
R.W. Macdonald

Contents

List of Authors

R. M. W. Amon
Alfred Wegener Institute for Polar and Marine Research,
27568 Bremerhaven, Germany, ramon@awi-bremerhaven.de
New Address: Departments of Marine Sciences and Oceanography,
Texas A&M University, Fort Crockett Campus, 5007 Ave. U,
Galveston, TX 77551, USA, *amonr@tamug.edu*

E. Bauerfeind
Institute for Baltic Sea Research, Seestrasse 15, D-18119 Rostock-Warnemünde, Germany, *eduard.bauerfeind@io-warnemuende.de*

A. Boetius
Alfred Wegener Institute for Polar and Marine Research,
27568 Bremerhaven, and International University Bremen, Campus-ring 1, 28759 Bremen, Germany, *aboetius@awi-bremerhaven.de*

G. I. Batova
All-Russia Research Institute for Geology and Mineral Resources of the World Ocean, 1, Angliysky ave., 190121, St.Petersburg, Russia, *vniio@g-ocean.spb.su*

J. Christensen
Bigelow Laboratory for Ocean Sciences, West Boothbay Harbor,
ME 04575, USA, *jchristensen@bigelow.org*

L. W. Cooper
Department of Ecology and Evolutionary Biology, 10515 Research Drive, suite 100, Bldg. A, University of Tennessee, Knoxville,
TN 37932, USA

D. Birgel
Alfred Wegener Institute for Polar and Marine Research,
27568 Bremerhaven, Germany, *dbirgel@awi-bremerhaven.de*

H. Deubel
Alfred Wegener Institute for Polar and Marine Research,
27568 Bremerhaven, Germany, *hdeubel@awi-bremerhaven.de*

H. Eicken
Geophysical Institute, University of Alaska Fairbanks, Fairbanks,
AK 99775-7320, USA, *hajo.eicken@gi.alaska.edu*

K. Fahl
Alfred Wegener Institute for Polar and Marine Research,
27568 Bremerhaven, Germany, *kfahl@awi-bremerhaven.de*

M. Fortier
GIROQ, Département de biologie, Université Laval, Québec, Québec, G1K 7P4, Canada, *Martin.Fortier@giroq.ulaval.ca*

M. Fukuchi
Center for Antarctic Environment Monitoring, National Institute of Polar Research, 9-10, Kaga 1-chome, Itabashi-ku, Tokyo 173-8515, Japan, *fukuchi@nipr.ac.jp*

C. Gobeil
Institut national de la recherche scientifique, Université de Quebec, CP 7500, Sainte-Foy, Québec, Canada, G1V 4C7, *charles_gobeil@inrs-ete.uquebec.ca*

V. V. Gordeev
P.P. Shirshov Institute of Oceanology, The Russian Academy of Sciences, 36 Nakhimovsky prospect, 117997 Moscow, Russia, *gordeev@geo.sio.rssi.ru*

J. M. Grebmeier
Department of Ecology and Evolutionary Biology, 10515 Research Drive, suite 100, Bldg. A, University of Tennessee, Knoxville, TN 37932, USA, *jgreb@utkux.utk.edu*

M. N. Grigoriev
Permafrost Institute RAS, 677010 Yakutsk, Russia, *grigoriev@mpi.ysn.ru*

A. Grantz
Department of Geological and Environmental Sciences, Stanford University, USA, *agrantz@usgs.gov*

M.J. Hameedi
NOAA, SC1, 1305 East-West Highway, Silverspring, MD 20910, USA, *Jawed.Hameedi@noaa.gov*

B. Hargrave
Department of Fisheries and Oceans, Marine Environmental Science Division, Bedford Institute of Oceanography, Dartmouth, Nova Scotia, NS B2Y 4A2, Canada, *HargraveB@mar.dfo-mpo.gc.ca*

H. R. Harvey
Chesapeake Biological Laboratory, University of Maryland, PO Box 38, Solomons, MD 20688, USA. *Harvey@cbl.umces.edu*

H.-W. Hubberten
Alfred Wegener Institute, Research Unit Potsdam, Telegrafenberg A 43, 14473 Potsdam, Germany, *hubbert@awi-potsdam.de*

M. Jakobsson
Center for Coastal and Ocean Mapping/ Joint Hydrographic Center, University of New Hampshire, USA, *martin.jakobsson@unh.edu*

W. Jokat
Alfred Wegener Institute for Polar and Marine Research, 27568 Bremerhaven, Germany, *wjokat@awi-bremerhaven.de*

M. Klages
Alfred Wegener Institute for Polar and Marine Research,
27568 Bremerhaven, Germany, *mklages@awi-bremerhaven.de*

Y. Kristoffersen
Institute of Solid Earth Physics, University of Bergen, Norway,
yngve.kristoffersen@ifjf.uib.no

A.V. Kursheva
All-Russia Research Institute for Geology and Mineral Resources
of the World Ocean, 1, Angliysky ave., 190121, St.Petersburg, Russia,
a.kursheva@mail.ru

A.P. Lisitzin
P.P. Shirshov Institute of Oceanology, The Russian Academy
of Sciences, 36 Nakhimovsky prospect, 117997 Moscow, Russia,
Russia, *lisitzin@geo.sio.rssi.ru*

R.W. Macdonald
Institute of Ocean Sciences, PO Box 6000, Sidney, B.C. V8L 4B2,
Canada, *MacDonaldRob@pac.dfo-mpo.ga.ca*

R. Macnab
Geological Survey of Canada (retired), Canada,
ron.macnab@ns.sympatico.ca

E.V. Narkevskiy
All-Russia Research Institute for Geology and Mineral Resources
of the World Ocean, 1, Angliysky ave., 190121, St.Petersburg, Russia,
egor_narkevsky@mail.ru

A.S. Naidu
Institute of Marine Sciences, University of Alaska Fairbanks,
P.O. Box 757220, Fairbanks, Alaska 99775-1080, USA,
Ffsan@ims.uaf.edu

T. Noji
U.S. National Marine Fisheries Service, Ecosystem Processes Division,
Northeast Fisheries Science Center, Howard Laboratory, Sandy Hook,
New Jersey, USA, *thomas.noji@noaa.gov*

E.-M. Nöthig
Alfred Wegener Institute for Polar and Marine Research,
27568 Bremerhaven, Germany, *enoethig@awi-bremerhaven.de*

K. Olli
Institute of Botany and Ecology, University of Tartu, Lai 40, 51005
Tartu, Estonia, *olli@ut.ee*

R. Peinert
Institute for Baltic Sea Research, Seestrasse 15, D-18119 Rostock-
Warnemünde, Germany, *rolf.peinert@io-warnemuende.de*

V.I. Petrova
All-Russia Research Institute for Geology and Mineral Resources of
the World Ocean, 1, Angliysky ave., 190121, St.Petersburg, Russia;
vera@vitali.spb.ru

D. Piepenburg
Academy of Sciences, Humanities and Literature Mainz, c/o Institute for Polar Ecology, Kiel University, Wischhofstr. 1-3, 24148 Kiel, Germany, *dpiepenburg@ipoe.uni-kiel.de*

V. Rachold
Alfred Wegener Institute for Polar and Marine Research, 27568 Bremerhaven, Germany, *vrachold@awi-bremerhaven.de*

E. A. Romankevich
P.P. Shirshov Institute of Oceanology, The Russian Academy of Sciences, 36 Nakhimovsky prospect, 117997 Moscow, Russia, *biogeoch@sio.rssi.ru*

E. Sakshaug
The Biological Station, Norwegian University of Science and Technology, N-7491 Trondheim, Norway, *Egil.sakshaug@vm.ntnu.no*

H. Sasaki
Department of Biotechnology, Senshu University of Ishinomaki, Ishinomaki, Miyagi 986, Japan, *sasaki@isenshu-u.ac.jp*

I. Schewe
Alfred Wegener Institute for Polar and Marine Research, 27568 Bremerhaven, Germany, *ischewe@awi-bremerhaven.de*

L. Schirrmeister
Alfred Wegener Institute, Research Unit Potsdam, Telegrafenberg A 43, 14473 Potsdam, Germany, *lschirrmeister@awi-potsdam.de*

C. J. Schubert
EAWAG, Limnological Research Center, -6047 Kastanienbaum, Switzerland, *carsten.schubert@eawag.ch*

V. P. Shevchenko
P.P. Shirshov Institute of Oceanology, The Russian Academy of Sciences, 36 Nakhimovsky prospect, 117997 Moscow, Russia, *vshevch@geo.sio.rssi.ru*

T. Soltwedel
Alfred Wegener Institute for Polar and Marine Research, 27568 Bremerhaven, Germany, *tsoltwedel@awi-bremerhaven.de*

R. Stein
Alfred Wegener Institute for Polar and Marine Research, 27568 Bremerhaven, Germany, *rstein@awi-bremerhaven.de*

A. A. Vetrov
P.P. Shirshov Institute of Oceanology, The Russian Academy of Sciences, 36 Nakhimovsky prospect, 117997 Moscow, Russia, *vetrov@sio.rssi.ru*

P. Wassmann
Norwegian College of Fishery Science, University of Tromsø, N-9037 Tromsø, Norway, *paulw@nfh.uit.no*

D. Weiel
Alfred Wegener Institute for Polar and Marine Research,
27568 Bremerhaven, Germany, *dweiel@awi-bremerhaven.de*

T. E. Whitledge
Institute of Marine Sciences, University of Alaska Fairbanks, P. O. Box
757220, Fairbanks, Alaska 99775-1080, USA

M. B. Yunker
7137 Wallace Dr., Brentwood Bay, BC, Canada,
V8M 1G9, *mark-yunker@telus.net*

A. G. Zinchenko
All-Russia Research Institute for Geology and Mineral Resources
of the World Ocean, 1, Angliysky ave., 190121, St.Petersburg, Russia,
annazinchenco@mail.ru

Chapter 1

The Arctic Ocean: Boundary Conditions and Background Information

1.1 Bathymetry and Physiography of the Arctic Ocean and its Constituent Seas

M. Jakobsson, A. Grantz, Y. Kristoffersen, R. Macnab

1.1.1 Introduction

A key to many scientific studies of ocean and sea floor processes is an accurate description of the sea floor morphology. For Arctic scientists this has posed a particular problem because less is known about the Arctic Ocean bathymetry and physiography than for the other oceans. For nearly 60 years after Nansen's collection of deep bathymetric soundings during his epic expedition with the vessel *Fram* drifting in the Arctic pack ice from the New Siberian Islands to the Fram Strait (Nansen 1902), it was a common belief that the entire central Arctic Ocean consisted of one deep basin. The perennial sea ice cover, within which Fram drifted, has severely hampered the systematic collection of bathymetric information from surface vessels. Following World War II the former Soviet Union started a sparse, but systematic, collection of bathymetric soundings in the central Arctic from ice stations established at great risk using airplanes to land scientists on the pack ice in order to carry out soundings through the ice. This exploration constituted a major breakthrough in Arctic Ocean mapping and revealed that this ocean was underlain by a morphologically complex sea floor consisting of abyssal plains interspersed between large submarine ridges (see e.g., Weber 1983).

The latest contribution to Arctic Ocean bathymetric portrayals is the International Bathymetric Chart of the Arctic Ocean (IBCAO). This is a digital 2.5×2.5-km grid-model on a polar stereographic projection that provides a detailed and coherent description of the morphology of the Arctic Ocean seafloor north of 64°N (Jakobsson et al., 2000). IBCAO provides a significant improvement over previous recent bathymetric portrayals (e.g., Canadian Hydrographic Service 1979; Perry et al. 1985) because the compilers were able to use legacy and modern data sets that were previously unavailable. These data included echo soundings from United States and British navy nuclear submarine cruises between 1958–1988, echo soundings from nuclear submarine cruises from the SCICEX program 1993–1999, and echo soundings from icebreaker and research vessel cruises carried out by Canada, Germany, Norway, Russia, Sweden, and the United States. In addition, the IBCAO compilers used published bathymetric contour maps in areas where original depth measurements were sparse or not available. These include the recently published Russian 1:5,000,000 scale bathymetric contour map "Bottom Relief of the Arctic Ocean" (VNIIOkeangeologia 1999).

The IBCAO grid-model has recently been used to estimate water mass volumes and areas of the Arctic Ocean and its constituent seas (Jakobsson et al. 2002) as well as to classify and calculate the areas of the first order physiographic provinces of the Arctic Ocean (Jakobsson et al. 2003). This chapter aims to summarize the current view of Arctic Ocean bathymetry and physiography including a summary of the recently published results on water mass volumes and the areas of physiographic provinces. It is not possible to cover all previous valuable contributions to our current knowledge of Arctic Ocean bathymetry and physiography in this short chapter. For this, the reader is referred to more extensive review articles, e.g., Weber (1983), Weber and Roots (1990), and Johnson et al. (1990).

1.1.2 Definition of the Arctic Ocean and its Constituent Seas

In the scientific literature, the Arctic Ocean is commonly referred to as a nearly landlocked ocean consisting of a deep central basin surrounded by seven epicontinental seas, i.e. the Barents, Kara, Laptev, East Siberian, Chukchi, Beaufort, and Lincoln Seas (e.g. Menard and Smith 1966; Weber 1989). How-

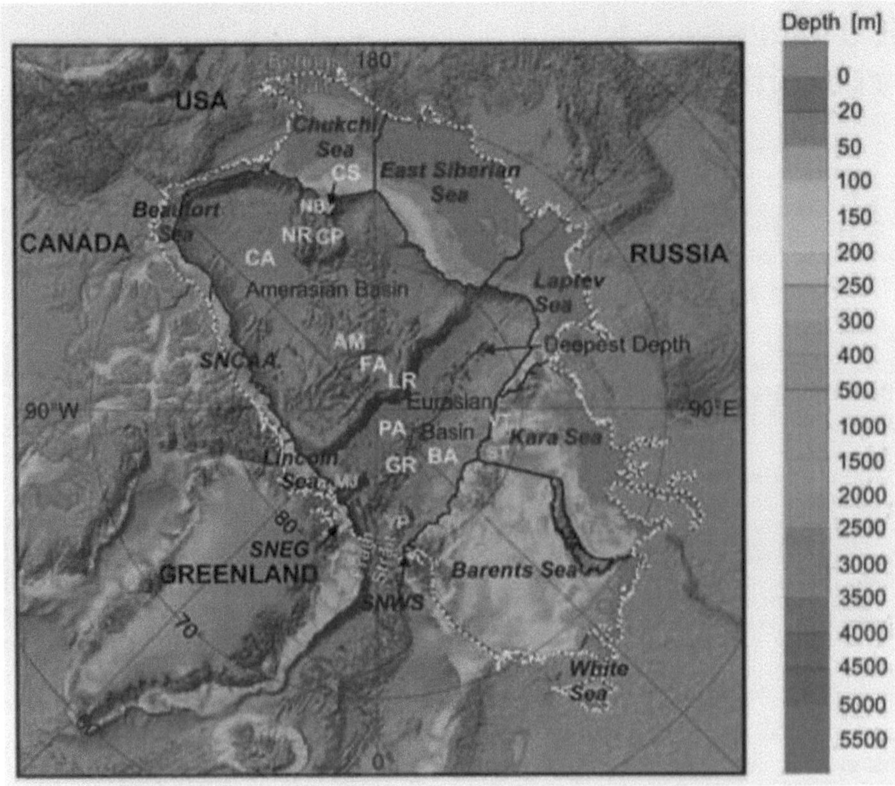

ever, the limits of these epicontinental seas have not, in most cases, been consistently or precisely defined. In order to perform quantitative analyses, such as computations of areas and water volumes, a precise definition is required that allows the construction of polygons enclosing regions of interest. This information is available from the International Hydrographic Organization (IHO), which formally defined the world's oceans and seas in Publication S-23 "Limits of Oceans and Seas" (International Hydrographic Organization 2001). In this publication, the geographic limits of the world's oceans and seas are defined by geographic positions of "fix-points" that are connected by either a rhumb line or a coastline.

The IHO-defined Arctic Ocean spans a much larger region than the one commonly used in the scientific literature. In addition to the seven epicontinental seas mentioned above, IHO included the following in their definition of the Arctic Ocean: the Greenland, Norwegian, Iceland, and White Seas; Baffin and Hudson Bays; Davis and Hudson Straits; and the waterways of the Canadian Arctic Archipelago. If one wishes to use an internationally established definition of the Arctic Ocean and its constituent seas, IHO S-23 is a useful guide. The IHO-defined Arctic epicontinental seas do not strictly conform to the shallow continental shelves, since their outer seaward limits are constructed primarily by rhumb lines drawn between fixed geographic points that are arbitrarily determined. For this reason, and to conform to the definition commonly used in scientific literature, Jakobsson (2002) defined the Arctic Ocean using only portions of the S-23 document. This proposed definition confines the Arctic Ocean to: the oceanic deep Arctic Ocean Basin; the broad continental shelves of the Barents, Kara, Laptev, East Siberian and Chukchi Seas; the

Fig. 1.1.1. Overview map showing the extent of the Arctic Ocean and its constituent seas as defined by Jakobsson (2002) to confine the oceanic deep Arctic Ocean Basin; the broad continental shelves of the Barents, Kara, Laptev, East Siberian and Chukchi Seas; the White Sea and the narrow continental shelves of the Beaufort Sea; and the Arctic continental margins of the Canadian Arctic Archipelago and Northern Greenland, including the Lincoln Sea (see text). The shaded relief portraying the bathymetry is derived from the IBCAO model. Area, water volume, and mean depth estimations for the Arctic Ocean and its constituent seas, as defined in this Figure, are shown in Table 1.1.2. Abbreviations used: AM: Alpha-Mendeleev Ridge complex; BA: Barents Abyssal Plain; CA: Canada Abyssal Plain; CP: Chukchi Plateau; CS: Chukchi Spur; FA: Fletcher Abyssal Plain; GR: Gakkel Ridge; LR: Lomonosov Ridge; MJ: Morris Jesup Rise; NB: Northwind Basin; NR: Northwind Ridge; PA: Pole Abyssal Plain; SNCAA: Shelf of Northern Canadian Arctic Archipelago; SNEG: Shelf of Northeastern Greenland; SNWS: Shelf of Northwestern Svalbard; ST: Svytaya Anna Troughs; YP: Yermak Plateau; VT: Voronin

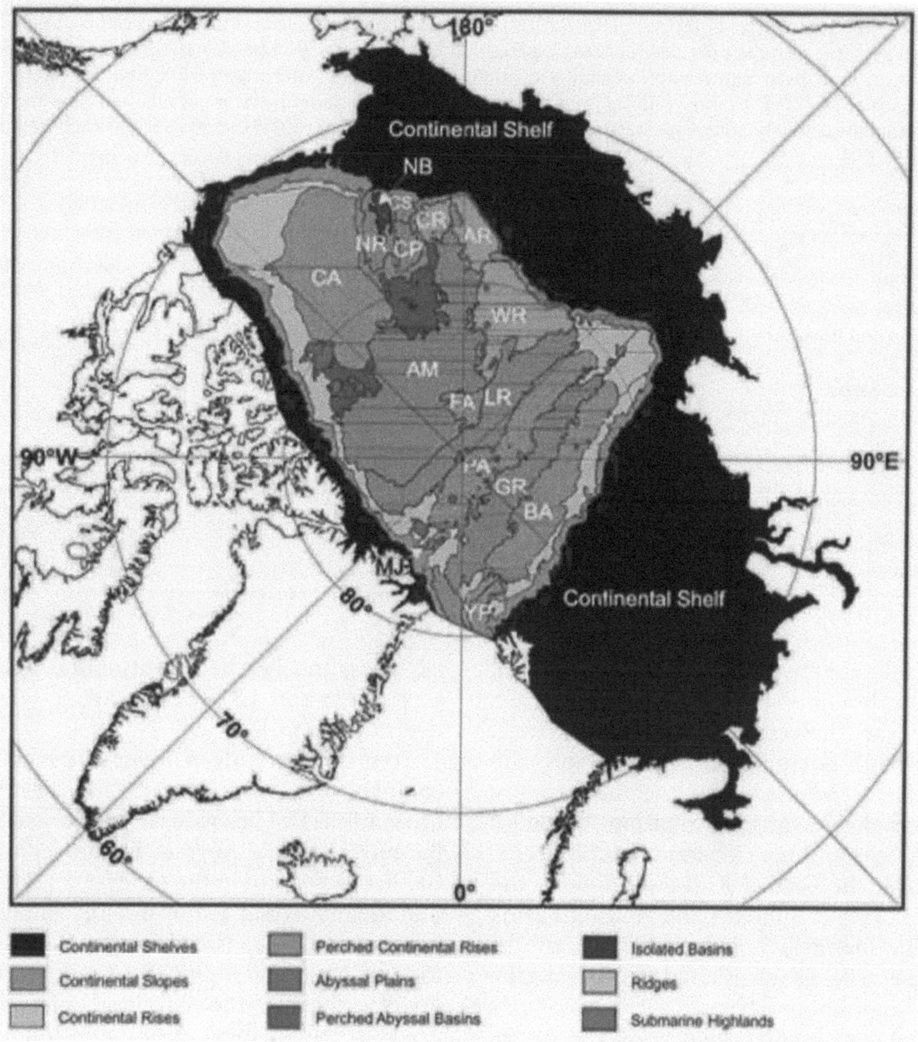

Fig. 1.1.2. Physiographic provinces of the Arctic Ocean as classified by Jakobsson et al (2003). The calculated areas for these physiographic provinces are shown in Table 1.1.1. Abbreviations used: AM: Alpha-Mendeleev Ridge complex; AR: Arlis Perched Rise; BA: Barents Abyssal Plain; CA: Canada Abyssal Plain; CP: Chukchi Plateau; CR: Chukchi Perched Rise; CS: Chukchi Spur; FA: Fletcher Abyssal Plain; GR: Gakkel Ridge; LR: Lomonosov Ridge; MJ: Morris Jesup Rise; NB: Northwind Basin; NR: Northwind Ridge; PA: Pole Abyssal Plain; YP: Yermak Plateau; WR: Wrangel Perched Rise

White Sea and the narrow continental shelves of the Beaufort Sea; and the Arctic continental margins of the Canadian Arctic Archipelago and Northern Greenland, including the Lincoln Sea (Fig. 1.1.1). The lateral boundaries between the epicontinental seas are taken from the S-23 document, but their outer limits that front upon the deep central Arctic Ocean basin have been modified to conform with the shelf edge that is described in the IBCAO model. There are, however, three areas of continental shelf, two of them previously belonging to the IHO-defined Greenland Sea, that are not included in any of the epicontinental seas listed above, but which nevertheless are situated inside the nearly landlocked area defined as the Arctic Ocean. These small continental shelf areas are herein referred to as the shelves of Northwestern Svalbard, Northeastern Greenland and Northern Canadian Arctic Archipelago (Fig. 1.1.1).

1.1.3
Bathymetry and Physiography

The deep central basin of the Arctic Ocean is nearly surrounded by continental massifs and the shallow continental shelves of Russia, Alaska, Canada, and Greenland. The only real break in this ring of continental crust is the Fram Strait between northeastern Greenland and northwestern Svalbard (Figs. 1.1.1 and 1.1.2). The Arctic Ocean basin is com-

Table 1.1.1. Physiographic provinces of the Arctic Ocean and their calculated areas (Summarized from Jakobsson et al. in prep). Note that the area presented for the province "Submarine Highlands" (written within brackets) has been included into the physiographic provinces that these features are located within. Furthermore, the total Area of Continental Shelves are given to ~5025×10^3 km², which is ~ 27×10^3 km² smaller than what one gets summarizing the areas of all Shelf Seas in Table 1.1.2. The reason for this is that the Chukchi Spur was classified as a Ridge in Jakobsson et al (2003) whereas it was included in the Chukchi Sea in Jakobsson (2002)

Physiographic Province	Area (1000 km²)	Relative Area (%)
1. Continental Shelves	5025	52.7
2. Continental Slopes	541	5.7
3. Continental Rises	733	7.7
4. Perched Continental Rises	362	3.8
5. Abyssal Plain	1122	11.8
6. Perched Abyssal Basins	222	2.3
7. Isolated Basins	23	0.2
8. Ridges	1506	15.8
9. Submarine highlands (65) (0.7)		
Total (Rounded from sum of non rounded original calculations)	**9534**	**100**

monly divided into two major sub-basins: the Eurasian Basin, bounded by the Lomonosov Ridge and the shallow shelves of Barents, Kara, and Laptev Seas and northern Greenland and the Amerasian Basin bounded by the Lomonosov Ridge and the shelves of East Siberian, Chukchi, and Beaufort Seas and the Canadian Arctic Archipelago. In turn, the Gakkel Ridge subdivides the Eurasian Basin into Amundsen and Nansen Basins and the Alpha-Mendeleev Ridge complex subdivides Amerasian Basin into Canada and Makarov Basins.

The shallow continental shelves of the Arctic Ocean, defined as the area from the coastline out to the shelf break, make up as much as 52.7% of the total area of the Arctic Ocean (Table 1.1.1) (For details of the area calculations see Jakobsson et al. 2003). This is in sharp contrast to the rest of the world's oceans, where the combined area of continental shelves and slopes have been estimated to range from only 9.1% and 17.7% (Menard and Smith 1966). Several large submarine canyons, of which the Voronin and Svytaya Anna Troughs are the largest, indent the Arctic continental shelves. Beyond the shelf breaks are the relatively steep continental slopes followed by the more gently sloping continental rises, which extend from the foot of the slopes to the flat abyssal plains (Fig. 1.1.2). Three of the continental rises, the Chukchi, Arlis and Wrangel continental rises (Fig. 1.1.2), have been classified as "Perched Continental Rises" by Jakobsson et al. (2003) since they extend only from the continental slope to sediment dams that lie one to one and one half km above the abyssal plain that lies down slope. The Continental Slopes, Continental Rises, and Perched Continental Rises make up 5.7%, 7.7% and 3.8% of the Arctic Ocean, respectively (Table 1.1.1).

Four primary deep-water areas of low relief comprise the Arctic Ocean's Abyssal Plains (Figs. 1.1.1 and 1.1.2). These are in the Eurasian Basin, the Barents and Pole Abyssal Plains separated by the Gakkel Ridge, and in the Amerasian Basin, the large Canadian Abyssal Plain and the smaller Fletcher Abyssal Plain separated by the Alpha-Mendeleev Ridge. The sediment thickness reaches several kilometers beneath the Barents, Pole and Fletcher Abyssal Plains (Jokat et al. 1992; Jokat et al. 1995) and more than 6 km beneath the Cananda Abyssal Plain (Grantz, et al. 1990; Jackson and Oakey 1990). The Pole Abyssal Plain is generally the deepest abyssal plain in the Arctic Ocean with depths exceeding 4000 m over large areas. The fact that the neighboring Barents Abyssal Plain is shallower has been ascribed to the large input of terrestrial sediment from the large rivers that drain into the adjacent Kara and Laptev Seas (Johnson 1969) and input detritus from glacial erosion on the Barents-Kara shelf during the Quaternary (Elverhøi et al. 1998). The two sub-circular basins associated with the northwestern and northeastern corner of the Canada Abyssal Plain are classified by Jakobsson et al. (2003) as "Perched Abyssal Basins" since they lie "perched" at ca. 100–500 m above the Abyssal Plain. The physiographic provinces Abyssal Plain and Perched Abyssal Plain, which are estimated to underlie ca 11.8% and 2.3% of the Arctic Ocean, respectively (Table 1.1.1), include irregular areas of smaller Submarine Highlands (Fig. 1.1.2) (for details see Jakobsson et al. 2003).

Submarine ridges make up the largest physiographic province of the Arctic Ocean (Fig. 1.1.2 and Table 1.1.1). As much as 15.8% is estimated to be underlain by the large ridges of the Arctic Ocean as well as smaller ridges with flat to gently rounded crests that sometimes have been referred to as Marginal Plateaus. The North Atlantic Mid-Ocean Ridge system enters the Arctic Ocean through the Fram Strait where the Spitzbergen Fracture Zone and Lena Trough links Knipovich Ridge in the Greenland Sea with Gakkel Ridge, which bisects the Eurasian Basin. The existence of a spreading ridge in the Arctic Ocean was first proposed on the basis of the pattern of earthquake epicenters and a small number of soundings (Heezen and Ewing 1961; Gakkel 1962). The deepest area of the Arctic Ocean is found in the axial valley of the Gakkel Ridge near 81°20′ N 120° 45′ W close to the Laptev Sea margin where the IBCAO grid indicates 5243 m (Jakobsson 2002). However, this depth represents an interpolation over a grid-cell of 2.5×2.5 km. The map published by VNIIOkeangeologia (1999) shows a depth of 5260 m at this location.

The Lomonosov Ridge divides the Arctic Ocean into the Eurasian and Amerasian Basins. It follows the same directional trend as the Gakkel Ridge, crossing from the Northern Greenland margin, passing close to the North Pole, and joining the shelf off the New Siberian Islands. This submarine ridge reaches depths shallower than 700 m, and it plays an important role in the Arctic Ocean oceanographic circulation (e.g. Anderson et al. 1994; Rudels et al. 1994). It is considered to be a continental sliver that rifted from the Barents-Kara Continental Shelf during the propagation of the North Atlantic Mid-Ocean Ridge into the Arctic Ocean (e.g. Karasik 1968, 1974; Wilson 1963; Vogt et al. 1979; Vink et al. 1984). The ridge has a flat crest that is thought to be inherited from a prominent unconformity, formed from sub-aerial and shallow marine erosion when the ridge first subsided below sea level about 50 million years ago (Jokat et al. 1995). In addition, it was recently discovered that areas of the ridge crest that lie above ca. 1000 m of water depth have been affected by extensive erosion (Jakobsson et al. 1999) that has been attributed to ice-grounding (Polyak et al. 2000) or possible ice-grounding in combination with strong bottom currents during the Late Pleistocene (Jakobsson et al. 2001).

The largest submarine ridge system in Arctic Ocean is the Alpha-Mendeleev Ridge, which lies in the Amerasian Basin. It was discovered by scientists on the drifting ice station ALPHA during the International Geophysical Year in 1957–58 (Hunkins 1961). This complex physiographic feature is morphologically very different from the Lomonosov Ridge, with numerous valleys and elevations and generally less steep slopes.

In addition to large ridge systems, the Arctic Ocean contains a number of smaller systems: the Amerasian Basin contains the Northwind Ridge and the Chukchi Spur-Chukchi Plateau composite ridge of the Chukchi Borderland; the Eurasian Basin contains the Yermak Plateau and Morris Jessup Rise (Figs. 1.1.1 and 1.1.2). All of these smaller ridges are connected to their adjacent continental shelves.

The north-trending ridges comprising the Chukchi Continental Borderland enclose an extensional basin with a generally flat sea floor at depths around 2000 m. This is the Northwind Basin that due to its unique physiography led Jakobsson et al. (in prep) to create the physiographic province "Isolated Basins", which underlay 0.2% of the Arctic Ocean (Fig. 1.1.2 and Table 1.1.1).

1.1.4 Volumes, Areas and Mean Depths of the Arctic Ocean and its Constituent Seas

Volume estimations of water masses and areal extents of seas may play an important role in large-scale oceanography studies such as basin-wide circulation patterns or chemical budgets. This section briefly summarizes the volume and area calculations of the Arctic Ocean and its constituent seas made by Jakobsson (2002) using the IBCAO grid-model. For details the reader is referred to that study, in which the seafloor area and volume were calculated for different depths starting from the present sea level and progressing in increments of 10 m to a depth of 500 m and in increments of 50 m from 550 m down to the deepest depth within each of the analyzed seas (using both IHO's originally defined limits and redefined limits of the Arctic Ocean seas). Summarized here are only the area and water volumes calculated for the redefined Arctic Ocean and its constituent seas that conform more closely to what commonly is used in scientific literature.

Table 1.1.2 summarizes the volumes, areas and mean depths of the seas of the Arctic Ocean. The entire Arctic Ocean comprises 2.6% of the total area of the world's oceans, but less than 1% of the volume. The large shallow continental shelf provinces that make up ~52.7% of the Arctic Ocean, as mentioned above, leads to a shallow mean depth of 1361 m and a relatively small total volume of 12990×10^3 km^3 for the entire Arctic Ocean. Of the epicontinental seas, the Barents Sea, with a mean depth of 200 m, has the largest area and volume.

Table 1.1.2. Area, volume and mean depth in the Arctic Ocean calculated with redefined limits for the constituent seas as shown in Figure 1.1.1 (modified from Jakobsson 2002). The presented total area of the Arctic Ocean in Table 1.1.1 is ~0.07 % smaller than the presented area in this Table. This is due to the different calculation techniques used in Jakobsson (2002) vs Jakobsson et al. (in prep). Abbreviations used in this Table: SNCAA: Shelf of Northern Canadian Arctic Archipelago; SNEG: Shelf of Northeastern Greenland; SNWS: Shelf of Northwestern Svalbard

Seas	Area (10^3 km^2)	Volume (10^3 km^3)	Mean depth (m)
Barents Sea	1512	302	200
White Sea	85	5	56
Kara Sea	926	121	131
Laptev Sea	498	24	48
East Siberian Sea	987	57	58
Chukchi Sea	620	50	80
Beaufort Sea	178	22	124
Lincoln Sea	64	16	257
Central Arctic Ocean Basin	4489	12339	2748
SNCAA	146	49	338
SNEG	30	4	119
SNWS	6	1	93
Totals and mean depth	9541	12990	1361

The Laptev and East Siberian Seas constitute the shallowest portion of the shelves with mean depths of only 48 m and 58 m, respectively, and with generally low bathymetric relief (Fig. 1.1.1 and Table 1.1.2). Together with the Chukchi Sea the Laptev and East Siberian Seas form a large shallow province of the continental margin that represent as much as 22% of the total area of the Arctic Ocean, but only 1% of the water mass volume. The small Lincoln Sea and the adjacent area of the Northern Canadian Arctic Archipelago have the largest mean depths, 257 m and 338 m respectively, of the continental shelves. Given the properties of the large shallow continental shelves, it seems reasonable that the oceanographic circulation in the Arctic Ocean ought to be much more sensitive to eustatic sea level changes than the rest of the world's oceans. An analysis of IBCAO bathymetry by Jakobsson (2002) shows that the average depth for all the Arctic Ocean shelves, except the continental shelf occupied by the Barents and Kara Seas and the very small Lincoln Sea and the adjacent Northern Canadian Arctic Archipelago, is less than the estimated eustatic sea level drop of 121 m during LGM (Fairbanks 1989). From this it follows that most, if not all, of the Arctic Ocean shelf region could not play a role in the ocean circulation during LGM nor during the Early- and Mid Weichselian glaciations, nor the Saalian glaciation.

1.2 The Arctic Ocean: Modern Status and Recent Climate Change

R.W. MACDONALD, E. SAKSHAUG, R. STEIN

1.2.1 Modern Hydrography and Sea-ice Cover of the Arctic Ocean

The oceanography of the Arctic Ocean is unique by virtue of six dominant features including: 1) Strong stratification supported primarily by large river inflows, 2) Limited connections with the Pacific and Atlantic Oceans to the south making the Arctic Ocean a "mediterranean" sea; 3) A high Coriolis parameter favouring boundary currents; 4) Enormous continental shelves; 5) Sea-ice formation and melting and; 6) Strong seasonal forcing (runoff, ice formation, sunlight). These factors together provide the foundation for the Arctic Ocean's biogeochemical cycles and transport of properties including sediments and organic carbon.

The Arctic Ocean owes much of its stratification to the approximately 4000 km^3 y^{-1} of freshwater inflow (Aagaard and Carmack 1989; Lewis et al. 2000; Stein 2000) which sustains a low-density surface layer (Figs. 1.2.1 and 1.2.2). All of the deep water is supplied by dense, saline water from the Atlantic Ocean. The Pacific water, which enters through the very shallow Bering Strait (~50 m sill depth), is less saline than the Atlantic water (<33 compared to ~34.8), therefore occupying the upper 300 m or so of the water column, and is confined mostly to the

Fig. 1.2.1. A schematic diagram showing surface circulation of the Arctic Ocean including seawater exchange with the Pacific and Atlantic, ice drift and runoff from the arctic drainage basins (modified from Stein et al. 1994; AMAP 1998)

Canadian Basin (McLaughlin et al. 1996; Morison et al. 1998).

Water in these vertical layers is modified within the Arctic, particularly over the shelves, by sea-ice formation/melt and runoff. The surface layer, or polar mixed layer (PML, ~0–50 m), cycles seasonally; brine produced by ice formation in winter tends to destabilize the water column allowing it to mix while in summer, melting ice and freshwater runoff produces stratification with a fresher surface layer (5–10 m). Most of the biological production occurs in this upper water layer and in the sea ice. Beneath the PML is found a region of cold water with increasing salinity (the *halocline*); an important consequence of the halocline is that exchange is limited between the deep ocean and the surface ocean. Within the Canada Basin the halocline is dominated by water from the Pacific Ocean which has been modified as it passes over the Chukchi Shelf. At the top of the Atlantic layer there is a complex region thought to be supported by insertion of dense water produced on the shelves (Fig. 1.2.2, Melling and Lewis 1982; Aagaard et al. 1985). This region has been called the lower halocline to distinguish it from the Pacific layer (upper halocline), but it is probably a composite water mass produced by brine drainage from the Eurasian shelves, the

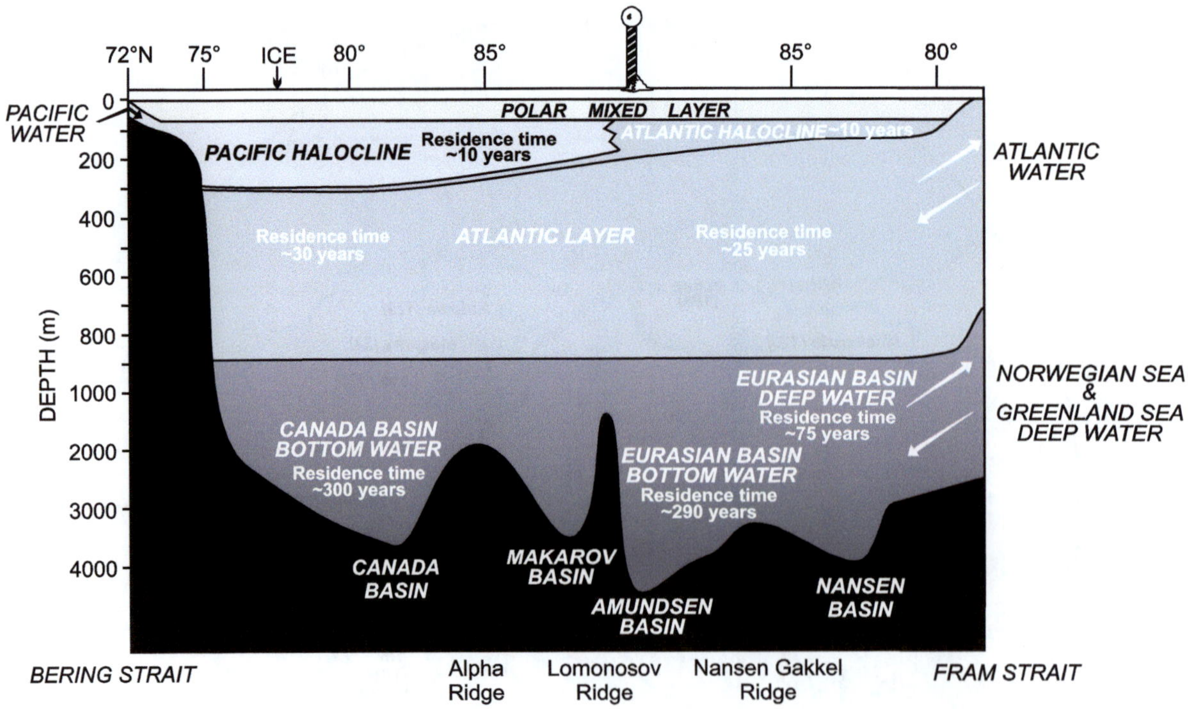

Fig. 1.2.2. A schematic diagram showing the stratification of the Arctic Ocean (modified from Macdonald and Bewers 1996)

Barents Sea shelf being a key one (Schlosser et al. 1995).

The time scale of transport or replacement of water masses varies among the layers: the surface layer has a residence time of about 10 years in the Arctic (Östlund 1982) The haloclines also have residence times of this order, but as we proceed deeper into the Atlantic layer, the residence time increases to perhaps 30 years. The basin waters have the longest time scale, measured in centuries (Fig. 1.2.2, Schlosser et al. 1997).

Horizontally, the Arctic Ocean can be subdivided into two main basins separated by the Lomonosov Ridge (Fig. 1.2.2). The basins are to some degree decoupled from one another exhibiting distinct differences in freshwater content (Aagaard and Carmack 1989), water-mass structure (McLaughlin et al. 1996), current structure (Rudels et al. 1994) and probably particulate transport (Matthiessen et al. 2000; Gobeil et al. 2001).

Exchange between the Arctic Ocean and oceans to the south (Figs. 1.2.1 and 1.2.2) is organized to a great degree by the stratification within the Arctic. The largest exchanges occur with the North Atlantic where re-circulation makes it a particularly difficult task to estimate net flows (e.g. see discussions in Woodgate et al. 1999; Fahrbach et al. 2001). Rudels and Friedrich (2000) summed up the exchanges as follows: about 1–1.5 Sv of water enters the Arctic in the West Spitsbergen current through Fram Strait (but may be as high as 3–5 Sv according to Woodgate et al. (1998)) and a further ~2 Sv enter directly into the Barents Sea (and see Ingvaldsen et al. 2002). On the west side of Fram Strait, 3– 3.5 Sv of water is exported from the Arctic in the East Greenland Current (but may be as high as 9 ± 2 Sv (Woodgate et al. 1999)). Atlantic water dominates the Barents Sea and Eurasian Basin, top to bottom. Water also enters the Arctic from the Pacific Ocean through Bering Strait, estimated at a flow of 0.83 Sv (Roach et al. 1995). This nutrient and silicate-rich water dominates surface waters for much of the Canadian Basin and also provides an important upstream supply for the water leaving the Arctic Ocean through the Archipelago. This latter transport is uncertainly estimated at somewhere between 0.7 and 2.7 Sv (Melling 2000).

Circulation within the Arctic reflects the high Coriolis parameter in that strong cyclonic currents are associated with basin boundaries (Fig. 1.2.3, Aagaard 1984; Aagaard 1989; Rudels et al. 1994; McLaughlin et al. 1996; Jones 2001; Woodgate et al. 2001; Schauer et al. 2002). These currents provide a relatively rapid pathway for the transport of conservative water properties, perhaps the best example being the radionuclide tracers of the northern European reprocessing plants (Smith et al. 1998). It seems likely that these boundary currents may also

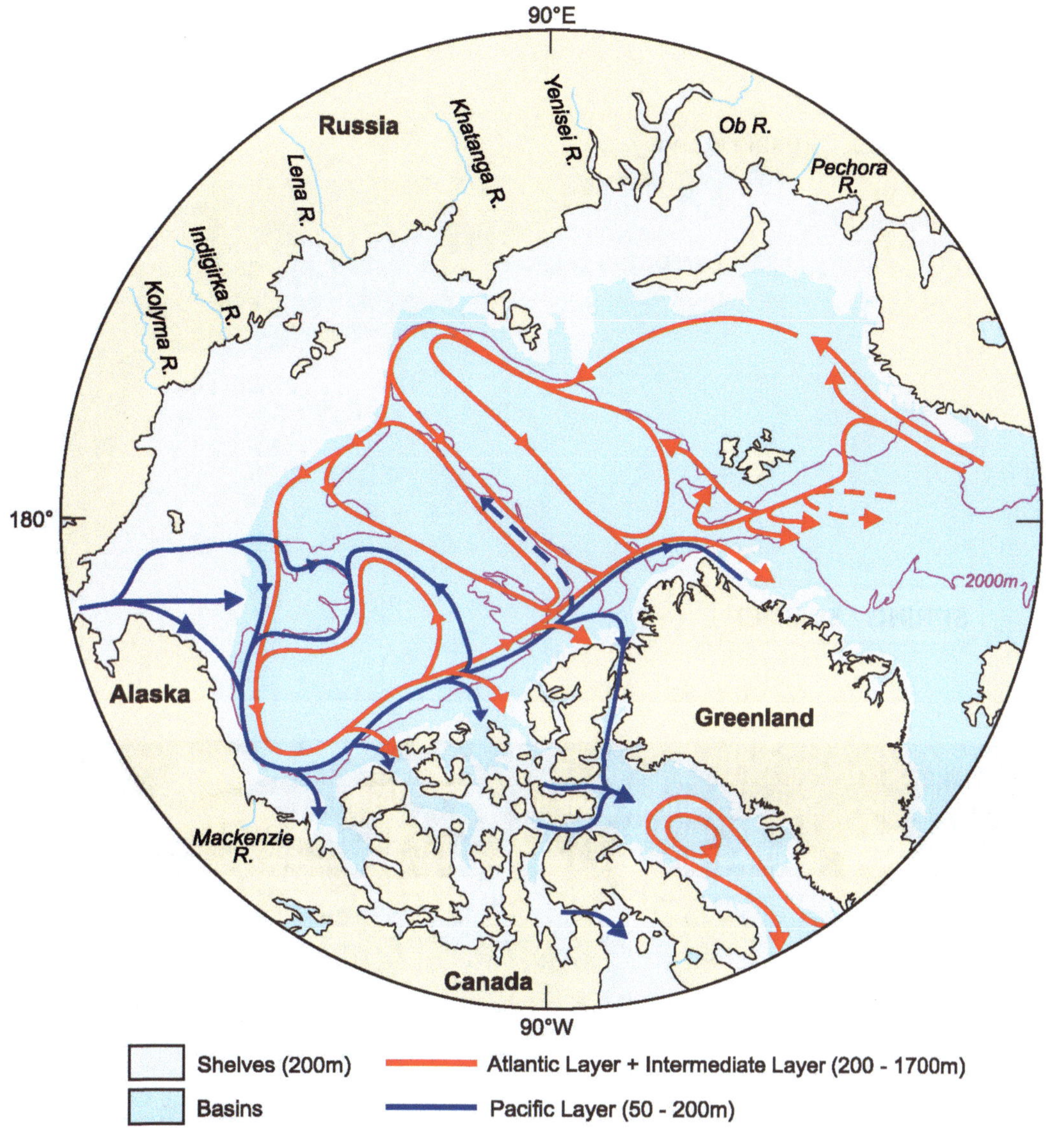

Fig. 1.2.3. A schematic diagram showing the locations of boundary currents in the Arctic Ocean (modified from Rudels et al. 1994; McLaughlin et al. 1996; Jones 2001)

be important for the transport of suspended material (e.g., see Gobeil et al. 2001).

Of all oceans, the Arctic contains the largest shelves (over 50% of the Arctic Ocean area (Fig. 1.2.3)). These shelves are important regions for sea-ice production (see Chapter 2.4, Fig. 2.6) and for the storage, trapping and processing of freshwater and dissolved and particulate inputs from land (Fig. 1.2.4). How effectively these regions trap or process material depends on the residence time of shelf waters, which have been estimated to vary from as little as 0.2–1.5 years for the Beaufort and Chukchi Shelves to over 3 years for the enormous East Siberian shelf (Macdonald 2000).

Large seasonal cycles leave their greatest imprint on the shelves (compare Figs. 1.2.4a and 1.2.4b). In spring and summer, rivers discharge most of their sediment and freshwater loads, large areas of the shelf are usually clear of ice, and virtually all of the annual biological production occurs (Fig. 1.2.4a). In winter, freshwater inputs are vastly reduced and constrained by ice to the nearshore (Fig. 1.2.4b). On the middle and outer shelf, brine produced by the rapid formation of ice in flaw leads and polynyas destabilizes the water column leading to mixing or, under the right circumstances, the production of dense water which flows along the shelf bottom to enter the ocean interior (Melling and Lewis 1982) transporting with it regenerated nutrients and resuspended bot-

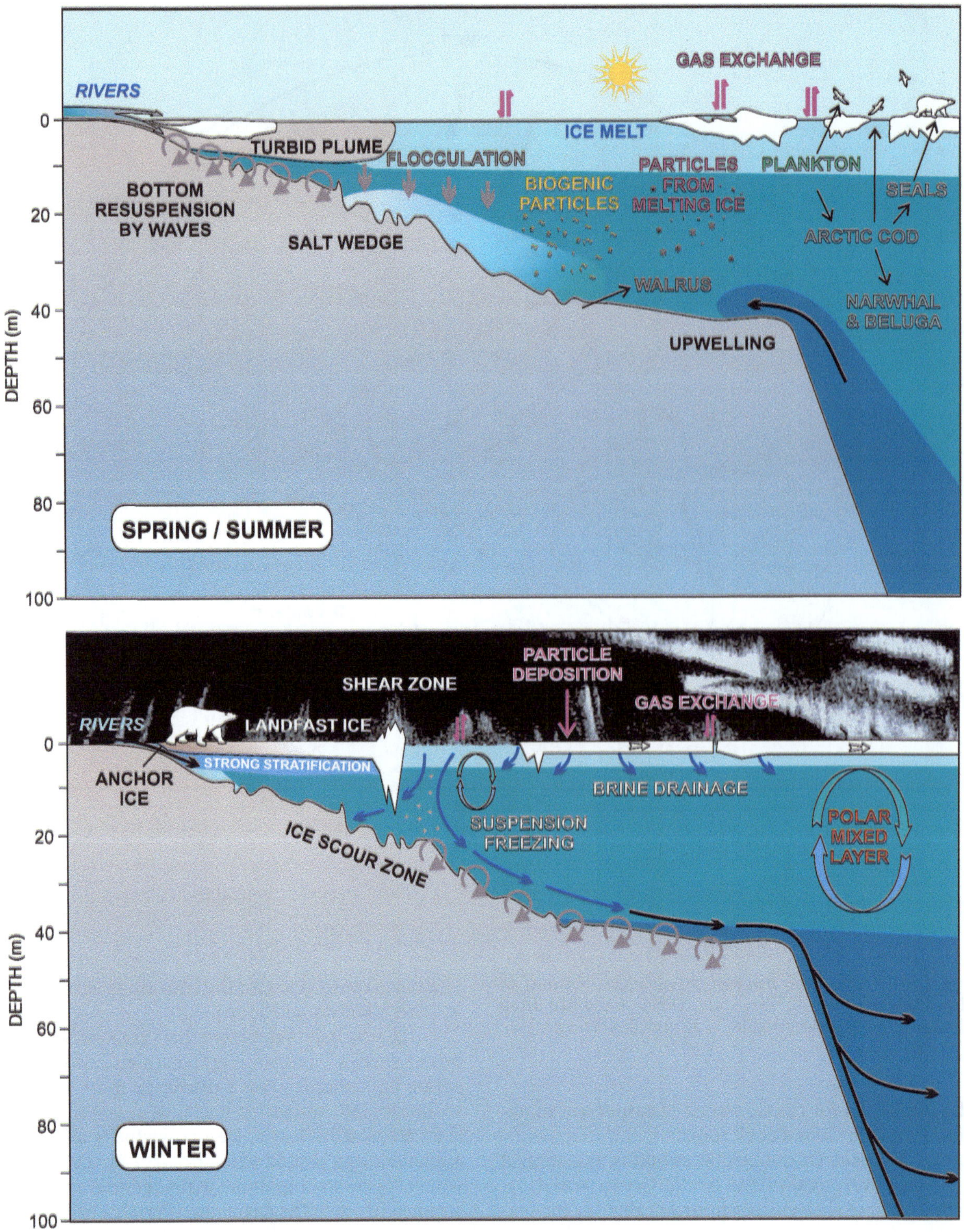

Fig. 1.2.4. A schematic diagram showing shelf processes affecting the transport and production of organic carbon during a) Summer and b) Winter (modified from Macdonald and Bewers 1996)

tom sediments. As much as 78,000 $km^3\ y^{-1}$ (~2.5 Sv) of near-freezing water may be supplied to the halocline by this process alone (Aagaard et al. 1981).

The most intense ice production during winter occurs in flaw leads and polynyas which are located over the shelves (Fig. 1.2.1). Ice production under open-water conditions can lead to suspension freezing during which suspended sediments are vigorously incorporated into sea ice which is then exported from the shelf to melt and deposit its load elsewhere (see Chapter 2.4, Fig. 2.6, and Nürnberg et al. 1994; Eicken et al. 1997, 2000). Sea ice controls the exchange of heat and other properties between atmosphere and ocean and, together with snow cover, determines the penetration of light into the sea. Ice also provides a surface for particle and snow deposition, a biological habitat above, beneath and within the ice and, when it melts in summer, stratification of the upper ocean.

Ice cover in the Arctic Ocean comprises a permanent pack in the ocean's interior with marginal seas exhibiting a seasonal ice cover from approximately October to June. The area of open water over shelves during summer varies widely between years (Serreze et al. 1995; Parkinson et al. 1999). Landfast ice, which grows to about 2 m in thickness by the end of winter, forms inshore of the 20 m isobath. The discontinuity between the landfast ice and offshore ice, which drifts with the general circulation, forms a flaw lead system where open water may be observed intermittently throughout winter (see Fig. 1.2.1 and Reimnitz et al. 1994; Macdonald et al. 1995). Most shelves export ice (see Chapter 2.4, Fig. 2.6) with the notable exception of the East Siberian Sea which may import large amounts of ice from the Beaufort Sea. The Arctic Ocean itself exports on average about 2800 $km^3\ y^{-1}$ of ice through Fram Strait with a large year-to-year variability (Dickson et al. 2000).

Arctic Ocean surface circulation, as represented by the ice drift, is probably better understood than for most oceans due to data collected from drifting buoys (Rigor 1992; Rigor et al. 2002). There are two domains of ice drift. In the interior western Arctic Ocean, ice circulates in the large clockwise Beaufort Gyre centered in the vicinity of 80°N 150°W (Fig. 1.2.1) but the size and center of the gyre vary on the decadal scale with atmospheric pressure fields. Along the margin of the East Siberian Sea and across the Eurasian Basin, the Transpolar Drift is the major organizing feature (Fig. 1.2.1). Ice in the Transpolar Drift may traverse from the Laptev Sea into the Greenland Sea in a couple of years whereas ice in the Beaufort Gyre may take six or more years before eventually exiting the Arctic (AMAP 1998; Rigor et al. 2002).

1.2.2 The Arctic Ocean and Global Change

Despite early evidence of cyclical change in northern ice conditions and in the Arctic's ecology (see for example, Gudkovich 1961; Vibe 1967; Bockstoce 1986), the general view among many western physical scientists throughout the 1960s–1980s was that the Arctic Ocean was a relatively stable place (Macdonald 1996). This view has been replaced by one of an ocean where major shifts can occur in a very short time, forced by natural variation in the atmospheric pressure field associated with the so-called "Arctic Oscillation." (Serreze et al. 2000; Wallace and Thompson 2002). These recent rapid changes in the Arctic, however, must be viewed in the context of changes that extend back millennia, some of them surprisingly abrupt (Alley et al. 2002).

1.2.2.1 *The distant past*

During the last 400,000 years, the earth has experienced four ice ages which have written records in glacial ice accumulating in Antarctica (Petit et al. 1999) and for the past 250,000 years or so, in Greenland (Dansgaard et al. 1993; Sowers and Bender 1995). The overall surface air-temperature change between glacial and interglacial periods is thought to have been about 12 °C, but perhaps more significant than temperature variation have been the accompanying changes in continental ice masses, sea-ice climate and global ecosystems. In particular, sea-ice cover has proven to be a master variable in the equation of change. During the last glacial maximum, sea ice was locked within the Arctic and seasonal or perennial sea ice extended well south into the North Atlantic Ocean (de Vernal et al. 1993; Darby et al. 1997). The change from glacial to deglacial to interglacial can be seen widely in arctic sediments both in sedimentation rate and in the amounts and sources of organic material deriving from primary production or land (see for example, Stein et al. 1994c; Phillips and Grantz 1997; Nørgaard-Petersen et al. 1998; Darby et al. 2001; Stein et al. 2001) (see Tables 7.8.2 and 7.9.2 in Chapters 7.8 and 7.9, respectively).

During the last glacial maximum (LGM), about 20 Cal. kyrs. BP, sea level dropped by approximately 120 m (Fairbanks 1989). This exposed much of the Arctic Ocean's enormous continental shelves (Fig. 1.2.5) with the consequence that inflowing rivers discharged sediment and water directly towards the continental slope and into the interior basins. Sediment and organic carbon input by turbidity cur-

Fig. 1.2.5. A simplified schematic diagram showing the distribution of the ice sheet (white shading) around the Arctic Ocean during the Last Glacial Maximum (adapted from Arkhipov et al. 1986; Fulton 1989; Svendsen et al. 1999; Spielhagen 2001; Mangerud et al. 2002; http://instaar.Colorado.EDU/QGISL/ak_paleoglacier_atlas/). Question marks and the black line outline the proposed distribution of an ice sheet in the easternmost Kara Sea. Black arrows indicate glacial ice and sediment input to the ocean. The loss of continental shelves due to the sea-level drop is approximated by using the 100 m isobath as the shoreline

rents probably was much more important than it is at present (e.g., Stein et al. 1999a, and see Chapter 7.5). Furthermore, with a drawdown of up to 120 m, Bering Strait became dry land (presently the sill depth is about 50 m) preventing the relatively fresh, nutrient- and silicate-rich waters of the North Pacific Ocean from entering the Arctic Ocean, and sea level dropped below the sill depth for most of the channels of the Canadian Archipelago, forcing the Arctic Ocean to communicate with the Atlantic predominantly through Fram Strait and secondarily through Nares Strait (between Ellesmere Island and Greenland). The fundamental geographic changes in the size of the shelves and in the connections with the Atlantic and Pacific Oceans mean that during the LGM, the Arctic Ocean was not at all the same as the present one.

During the LGM, large parts of northern Canada, Greenland and Eurasia were covered by major ice sheets (Fig. 1.2.5, Arkhipov et al. 1986; Fulton 1989; Svendsen et al. 1999; Spielhagen 2001; Mangerud et al. 2002). The extent of the Eurasian ice sheet remains under discussion. Grosswald and Hughes (1999) propose a huge LGM Eurasian ice sheet covering the entire area from Scandinavia to eastern Siberia and Alaska. Many of the glacial features in northern Siberia, however, have turned out to be much older, and there is evidence of a more restricted Eurasian ice sheet during the LGM, with its eastern limit in the western Kara Sea (Svendsen et al. 1999; Mangerud et al. 2002). It has also not yet been determined whether a (local?)

LGM ice sheet covered the easternmost Kara Sea and, if so, whether this ice sheet was connected with the main Barents-Kara ice sheet (e.g., Svendsen et al. 1999; Polyak et al. 2002; Stein et al. 2002). The possible existence of an extensive LGM ice shelf on Chukchi Rise is discussed by Polyak et al. (2001). Coarse-grained terrigenous material (including organic matter) incorporated into such ice sheets can be transported onto the shelves during glacial periods. After breakup of the glacial Arctic Ocean sea ice, following the glacial maximum, icebergs were free to circulate into the central Arctic Ocean within the major current systems, progressively releasing their sediment load along the transport pathway as they melted (Darby et al. 1997; Phillips and Grantz 2001). From the composition and distribution of ice-rafted glacial erratics in late Quaternary sediments from the central Arctic Ocean, specific continental source areas such as the Canadian Arctic Islands and parts of Eurasia can be identified, thus allowing paleo-current systems to be reconstructed (e.g., Phillips and Grantz 2001).

With sea-level rise, about 15,000 years ago the Bering land bridge was flooded (Hopkins 1979) and then gradually submerged (Dyke et al. 1996) allowing the Pacific Ocean access to the Arctic Ocean. This sequence together with inundation of the continental shelves had enormous consequences for the oceanography and regional biogeography for the western Arctic and the Archipelago (Dunton 1992; Héquette et al. 1995; Dyke et al. 1996). Nutrients, organic carbon and North Pacific biota could enter the Arctic through Bering Strait and shelves once again could play an important role in the sequestering of sediments and storage of fresh water.

Although climate has been described as "exceptionally stable" during the past 10,000 years (Dansgaard et al. 1993), it has actually continued to undergo substantial fluctuations. Indeed, it seems that very small shifts in temperature, perhaps only a degree or two, account for the so-called Medieval Warm Period (1100–1400 AD) and Little Ice Age (1450–1850 AD) that followed it (for the relevance of these terms see Bradley and Jones 1993; Crowley and Lowery 2000). Both of these minor and sporadic deviations in the temperature record had dramatic consequences for the Arctic Ocean's oceanography and ecology (McGhee 1996; Ogilvie and Junsson 2000; Alley et al. 2002). During the past two centuries, small changes in ice and watermass distribution have continued to impact Arctic Ocean ecosystems, sometimes leading to population migrations but certainly requiring adaptation (Vibe 1967; Miller et al. 2001).

1.2.2.2
Recent change and the Arctic Oscillation

During the 1990s, it became widely recognized that the Arctic was undergoing dramatic change (Dickson 1999; Macdonald 1996; Morison et al. 2000; Serreze et al. 2000) and many of the observed changes have been attributed to the extremely high

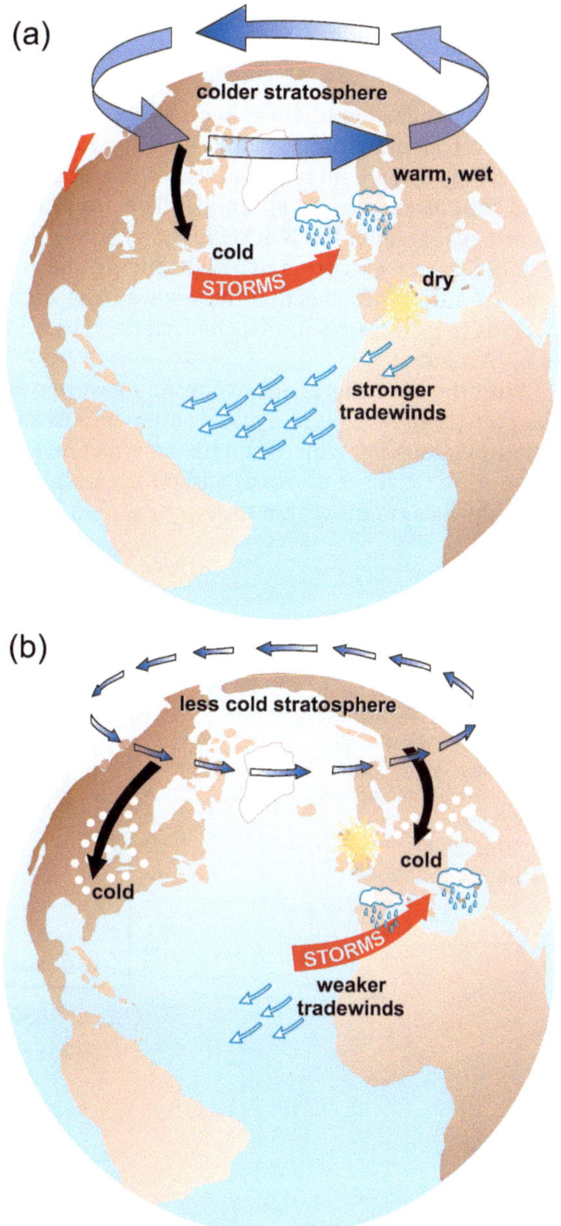

Fig. 1.2.6. A schematic diagram of the polar vortex and North Atlantic storm tracks showing a) The effect of the positive phase of the Arctic Oscillation and b) The effect of the negative phase of the Arctic Oscillation. (Figure courtesy of J. Wallace, University of Washington, http://www-nsidc.colorado.edu/arcticmet/patterns/arctic_oscillation.html)

Arctic Oscillation (AO) index in the late 1980s to early 1990s. This index has recently returned to more normal conditions (Johnson et al. 1999; Björk et al. 2002; Boyd et al. 2002). A high AO index corresponds to a strengthening of the polar vortex (Fig. 1.2.6) which is manifested at the sea surface by a relatively low atmospheric pressure over the north pole and a change in the location of storm tracks into and through the Nordic seas (Thompson and Wallace 1998; Wallace and Thompson 2002). The AO correlates strongly (85–95%) with the more commonly used indicator of large-scale wind forcing, the North Atlantic Oscillation (NAO) (the NAO is the normalized gradient in sea-level air pressure between Iceland and the Azores – see for example, Hurrell 1995; McCartney 1997; Deser 2000; Dickson et al. 2000; Serreze et al. 2000). The long-term record available for the NAO (Fig. 1.2.7) shows that this index was exceptionally high during the late 1980s to mid 1990s, but that there have been other periods (e.g., the early 1900s) characterized by high NAO/AO index.

With the lowering of atmospheric pressure over the pole (high AO index), cyclonic atmospheric circulation intensifies which then has an effect on sea-ice motion (Rigor et al. 2002) and on the pathways of freshwater runoff within the Arctic Ocean (Guay et al. 2001; Macdonald et al. 2002). Although it is not yet clear whether part of the change is an early manifestation of warming forced by greenhouse gas emissions, the contrast between high and low AO index provides an opportunity to investigate how the Arctic might respond to climate change because climate-change projections and change associated with the AO are similar. It is not our intent to discuss all of the changes reviewed in the above-cited literature. Rather, we highlight here those changes that appear to have the greatest potential to affect organic carbon cycling in the Arctic Ocean.

Variation in sea-level atmospheric pressure as recorded by the AO index demonstrates that the Arctic Ocean exhibits at least two modes of behaviour (Proshutinsky and Johnson 1997) and that these modes cascade from primary forcing (air pressure) into wind fields (Serreze et al. 2000), ice drift patterns (Rigor et al. 2002), watermass distributions (McLaughlin et al. 1996; Morison et al. 1998), freshwater pathways in the ocean (Steele and Boyd 1998; Macdonald et al. 1999, 2002; Ekwurzel et al. 2001; Guay et al. 2001; Johnson and Polyakov 2001), and ice cover (Serreze et al. 1995; Maslanik et al. 1996; Maslanik et al. 1999; Dickson et al. 2000). From the perspective of organic carbon, the most

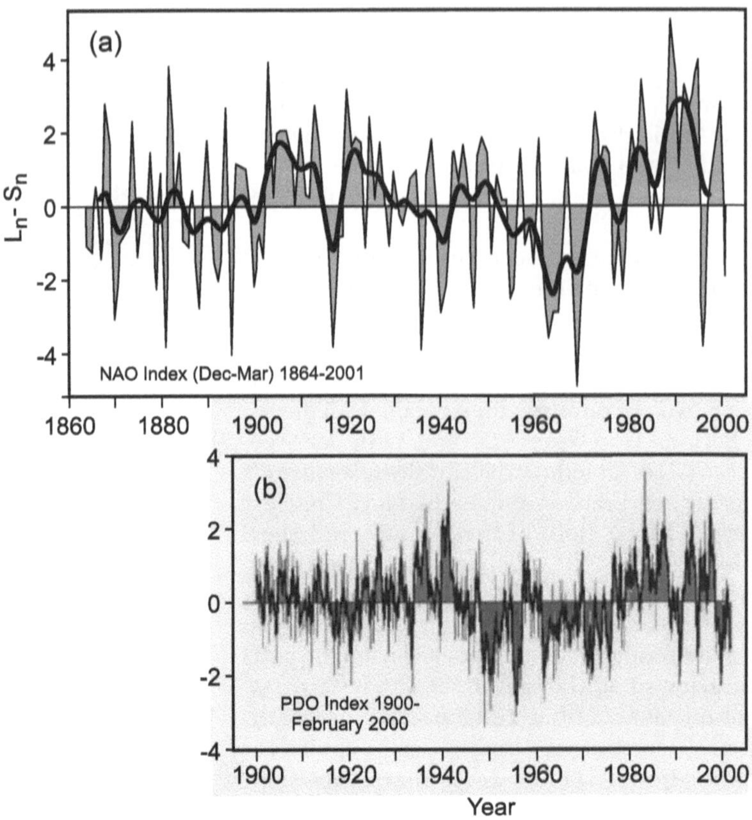

Fig. 1.2.7. a) The North Atlantic Oscillation (NAO) Index from 1860 to 2000 (source Hurrell 2002) and b) The Pacific Decadal Oscillation (PDO) winter index (November-March) since 1900 (source University of Washington: http://tao.atmos.washington.edu/pdo/)

important changes will be associated with ice cover, ice drift, surface currents and drainage basin hydrology. These changes then impact marine organisms depending on their habitat and response time to change. Whereas micro-organisms can respond to change within a few days, mesozooplankton take weeks-to-months, and larger animals take years to decades.

Effects on sea- ice cover and drift

There has been an overall decrease in the sea-ice cover in the Arctic during the past 120 years. Since the spring of 1881, when sea ice extended to the south of Iceland, the areal ice cover in the Atlantic sector has decreased by 50 % between 1966 and 1980 and by >50 % since 1990 (Vinje et al. 1996). However, multi-decadal, decadal and shorter quasi-periodic cycles have been a noted feature of Arctic climate patterns in this sector extending back for centuries.

During the 1990s, the science community recognized with some alarm that arctic sea ice had been undergoing retreat over the past three decades. Observed changes included: a reduction in area covered by sea ice (Maslanik et al. 1996; Johannessen et al. 1999; Parkinson et al. 1999; Vinnikov et al. 1999; Levi 2000), an increase in the length of the ice melt season (Smith 1998; Stabeno and Overland 2001; Rigor et al. 2002), a loss of multiyear ice (Johannessen and Miles 2001), a general decrease in the thickness of ice over the central Arctic Ocean (Rothrock et al. 1999) and an increase of ice melt in the Beaufort Sea (McPhee et al. 1998; Macdonald et al. 1999).

Changes to the ice cover like those listed above clearly have important consequences for organic carbon cycling, especially over the shelves but probably in the basins as well (see, for example, Gobeil et al. 2001). Open water during late summer leaves coastlines vulnerable to long-fetch storms and hence accelerated coastal erosion. Furthermore, open water over the shelf alters mixing, upwelling, river plume distribution, and light climate – all of which impact primary production – and it enhances bottom resuspension through wave orbital velocities, and alters the exchange characteristics between shelf and interior ocean. Model projections that suggest the seasonal demise of ice for the entire Arctic Ocean by 2100 (Vinnikov et al. 1999; Flato and Boer 2001), if realized, imply a fundamentally different Arctic Ocean carbon cycle.

Analyses of satellite data from 1978–94 reveal an average loss of 3 % (0.3×10^6 km^2) per decade of ice cover (Cavalieri et al. 1997; Parkinson et al. 1999). Shelves of the Eastern Arctic contribute significantly to the estimated ice losses. Multi-year ice is apparently being lost at an even greater rate, estimated at 7% per decade, partly replaced by first-year ice (Johannessen and Miles 2001).

The large seasonal range in area covered by ice (Fig. 1.2.1) makes it difficult to assess temporal trends. Nevertheless, the satellite data available since the late 1970s clearly imply a reduction of 2 % per decade of total ice area in winter (Johannessen et al. 1999) and a significant shift in the marginal seas toward first-year ice which is easier to melt than multiyear ice because it is thinner and saltier. The total area of arctic sea ice including the marginal seas, which varies from about 13×10^6 km^2 in winter to 5×10^6 km^2 in summer, has shrunk by about 0.6×10^6 km^2 between 1978 and 1997 (Johannessen and Miles 2001). The Arctic Ocean began to exhibit a much stronger seasonal modulation in ice cover in about 1989 (Fig. 1.2.8) with the

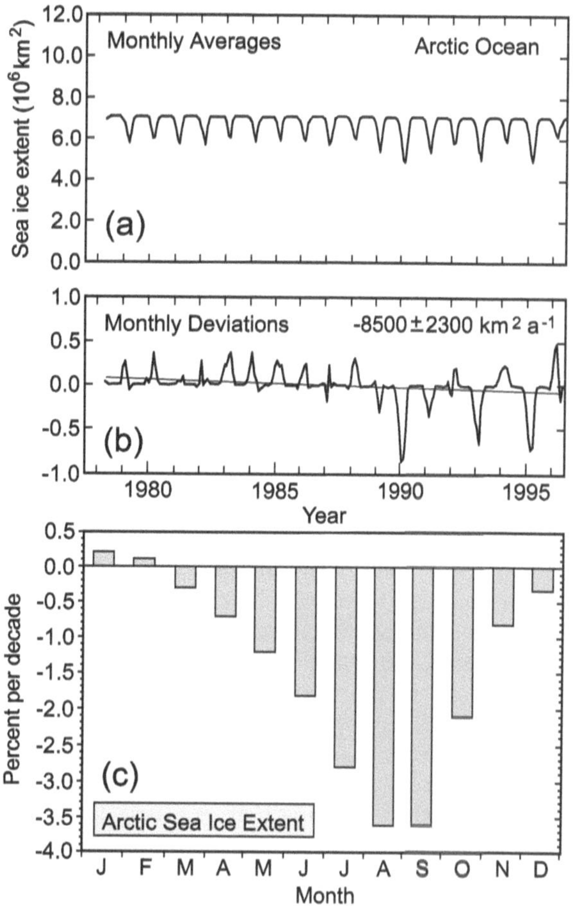

Fig. 1.2.8. Ice cover in the Arctic Ocean showing a) Monthly averages in ice cover for the period from 1979 to 1996 for the Arctic Ocean (after Parkinson et al. 1999)), b) The monthly deviations in ice cover for the same area showing the transition in 1990 to seasonally clear shelves (after Parkinson et al. 1999) and c) The change in sea ice extent in % per decade (1979–1995) showing the ice loss to be predominantly a late winter – summer phenomenon (after Serreze et al. 2000)

East Siberian and Beaufort Seas experiencing anomalous areas of open water in late summer at various times during the 1990s (Serreze et al. 1995; Maslanik et al. 1999; Parkinson et al. 1999; Rigor et al. 2002). The greatest loss of sea-ice cover occurs during the period from June to October (Fig. 1.2.8 c, Serreze et al. 2000).

The trends in ice cover with time (Figs. 1.2.8 a and 1.2.8 b) suggest that the wholesale clearing of ice from shelves during summer is a phenomenon of the 1990s, timed with (Russian Shelves) or slightly delayed from (Beaufort Shelves) the shift to strong positive AO index in 1989. Maslanik et al. (1996) draw the connection between increased cyclones and increased poleward transport of heat, which are observed during high index conditions, and the absence of ice in late summer over the Siberian shelves. Based on results of a coupled sea/ice/ocean model, Zhang et al. (2000) suggest that there is a strong correlation between sea-ice thinning and the AO (~80%) due to dynamical effects, and that the Eurasian and Canada Basins respond differently to AO forcing. This point is important in light of the findings from repeat submarine surveys that ice thickness has decreased over the central Arctic by about 1.3 m between 1958–1976 and the 1990s (Wadhams 1997, 2000; Rothrock et al. 1999). According to several models (Polyakov and Johnson 2000; Zhang et al. 2000; Holloway and Sou 2001), the submarine observations may have been conducted primarily in that part of the ocean that underwent thinning in response to a shift to high AO index. The conclusion of reduction of ice thickness, while valid for the domain of submarine measurements, is not necessarily true for the whole Arctic Ocean and an alternative hypothesis that ice thickness *distribution* changed but ice *volume* may not have changed in response to the AO needs to be carefully evaluated. The loss of ice cover between low and high states of the AO (or NAO) index conditions is estimated at 590,000 km^2 in the Barents and Greenland Seas (Dickson et al. 2000) and, if the remarkably open ice in the East Siberian Sea in 1990 and the Beaufort Sea in 1998 is a product of the high AO index of the early 1990s, then perhaps half as much again ice loss occurred over the Russian and North American Shelves.

From data collected between 1979 and 1997, Rigor et al. (2000) determined that sea-ice melt begins in the marginal seas by the first week of June and advances rapidly to the pole in two weeks. Freezing begins at the pole on 16 August, returning to the marginal seas by late September for a total melt season length of about 58 days at the pole and 100 days toward the margin. Based on satellite data (SSMR and SSM/I) predominantly from the Beaufort Sea, Smith (1998) estimated that the length of the melting season has been increasing by about 5.3 days per decade during 1979–1996. In contrast, Rigor et al. (2000) found a shortening of the melt season in the western Arctic of 0.4 days per decade and an increase of about 2.6 days per decade in the Eastern Arctic.

Although it has long been recognized that large-scale ice-drift patterns in the Arctic undergo change (Gudkovich 1961), it was not until the International Arctic Buoy Programme (IABP) that sufficient data became available to map the ice drift in detail and thereby directly evaluate the role of the AO in changing ice drift trajectories. Two characteristic modes of arctic ice motion are associated with the AO index (Figs. 1.2.9 a and 1.2.9 b; Proshutinsky and Johnson 1997; Rigor et al. 2002). The ice-motion scheme shown by drifting buoys is reasonably well corroborated by models investigating the influence of the atmospheric variability inherent in the AO (Maslowski et al. 2000; Polyakov and Johnson 2000). There are two overarching differences between the ice circulation modes; 1) during low index conditions (Fig. 1.2.9 a), ice in the Transpolar Drift tends to move directly from the Laptev Sea across the Eurasian Basin and out into the Greenland Sea whereas during high index conditions (Fig. 1.2.9 b), ice in the Transpolar Drift takes a strong cyclonic diversion across the Lomonosov Ridge and into the Canada Basin (Mysak 2001) and 2) during high index conditions (Fig. 1.2.9 b), the Beaufort Gyre shrinks back into the Beaufort Sea and becomes more disconnected from the rest of the Arctic Ocean exporting less ice to the East Siberian Sea and importing little ice from the region to the north of the Archipelago – a region known to contain the Arctic's thickest multi-year ice (Bourke and Garrett 1987).

There are also changes in the time required for ice to transit the ocean (Figs. 1.2.9 c and 1.2.9 d; Rigor et al. 2002) and for the destinations of ice exported from shelves. During winter under high index conditions, there is an increase in ice advection away from the East Siberian and Laptev Sea coasts leading to the production of more new, thin ice in the coastal flaw leads (Polyakov and Johnson 2000; Rigor et al. 2002), a decrease in the advection of ice from the western Arctic into the eastern Arctic, possibly an increased advection of ice from the Arctic Ocean to the Barents Sea through the Spitsbergen-Franz Josef Land passage (Polyakov and Johnson 2000), and an increase in the 900,000 km^2 of ice advection out of the Arctic at Fram Strait (Morison et al. 2000; Rigor et al. 2002). Increased ice export through Fram Strait can be produced by

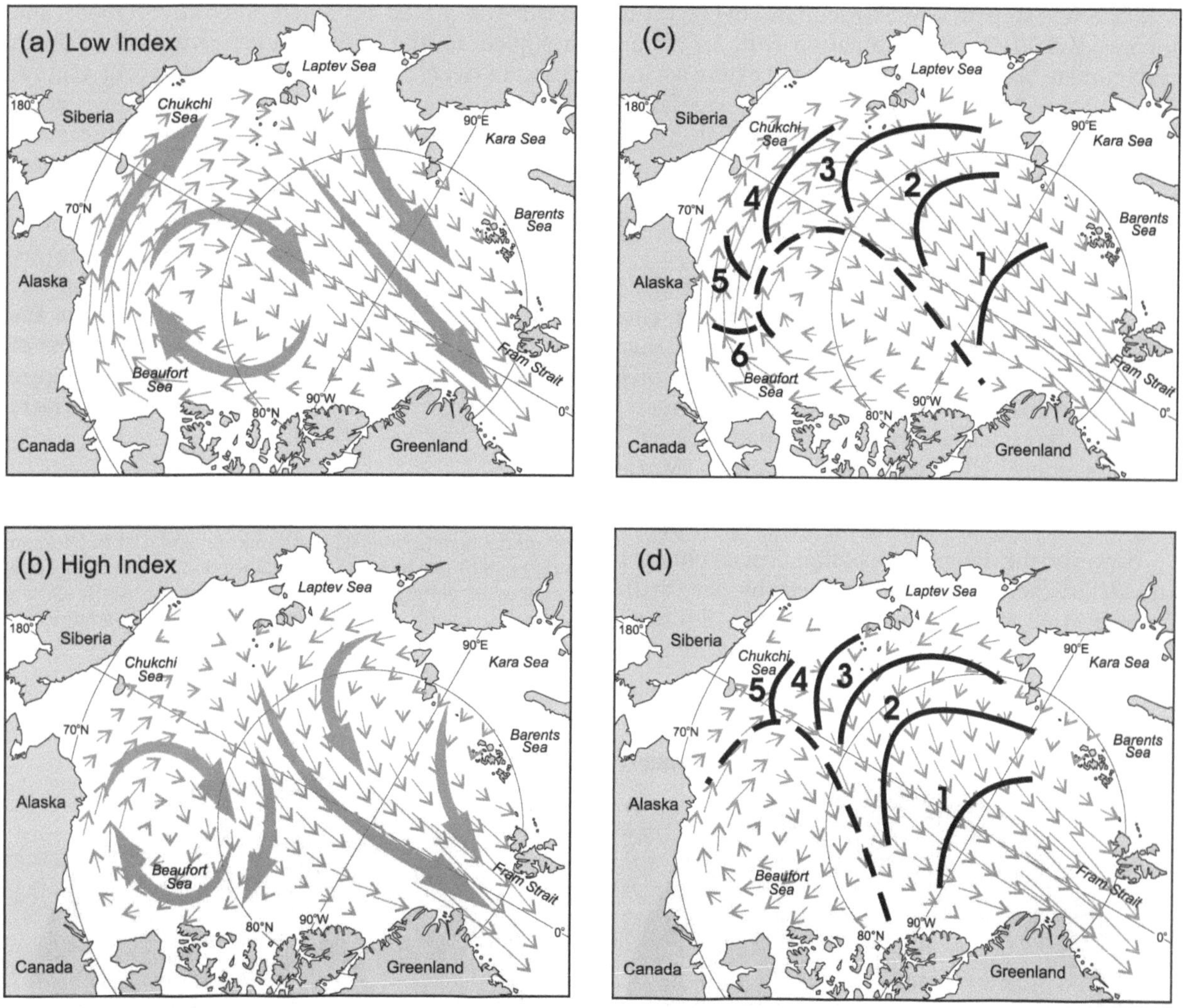

Fig. 1.2.9. Ice drift patterns for a) Years with low AO^- index (anticyclonic conditions) and b) High AO^+ index (cyclonic conditions) (after Maslowski et al. 2000; Polyakov and Johnson 2000; Rigor et al. 2002). Time taken in years for sea ice to reach Fram Strait is shown for c) low AO^- index conditions and d) high AO^+ index conditions (after Rigor et al. 2002)

shifts to both negative and positive AO phases (Dickson et al. 2000).

The response in ice-drift trajectories to change in the AO index (Rigor et al. 2002) together with alteration in the length of ice-melt season therefore have important implications for the sources, amounts, sinks and efficiency of transport of material by arctic sea-ice.

Although the AO/NAO indices can be used to explain much of the recent variability in the Arctic Ocean and Nordic Seas, they are not applicable to the subarctic Pacific (specifically, the Bering Sea). There, the Pacific Decadal Oscillation (PDO; Fig. 1.2.7b), the Northern Oscillation (NO) and the El Niño – Southern Oscillation (ENSO) have been applied to questions of climate cycles and climate change (see for example, Trenberth and Hurrell 1994; Francis et al. 1998; Hare and Mantua 2000; Stabeno and Overland 2001). Both the PDO and NO indexes indicate substantial climate shifts in the North Pacific and the Bering Sea at roughly 14-year intervals with the last great shift occurring in 1976 (Francis et al. 1998). A positive PDO index (co-varying with negative NO) implies warm sea surface temperature (SST) anomalies in the Northeast Pacific and cold anomalies in the Northwest Pacific.

After 1976, a predominantly positive PDO has been associated with El Niños, which have occurred 3.5 times more frequently than during the preceding period, with warm winter anomalies in the southeast Bering Sea and with a deepening of the Aleutian Low. In consequence, there has been reduced ice cover with sea ice largely absent from the Middle Bering Shelf in 1979–1981 (Overland et al. 2001; Stabeno and Overland 2001). In 1984–1991, a temporary shift to a weak Aleutian Low (typical for negative PDO) in combination with La Niña-type climate brought cold winters and extensive ice cover to the Bering Sea and ice cover was fairly large also in 1989–1998 (Niebauer et al. 1999; Stabeno and Overland 2001).

Effects on water currents and watermass distribution

Changes in shallow ocean currents are most important in their potential to alter particulate and organic carbon processes, because it is here that primary production occurs and it is here that runoff and coastal erosion deposit their products. Over the large scale, surface water pathways will to some extent reflect ice-drift trajectories (Morison et al. 2000; Jones 2001) and therefore respond in like manner to the state of the AO (Fig. 1.2.9).

However, the AO produces other crucial changes in surface water not represented by ice drift. Salinification of surface water in the Eurasian Basin was noted by Steele and Boyd (1998) and assigned to the diversion of river inflow from Russian rivers toward the east. Models (Fig. 1.2.10 a; Dickson 1999; Johnson and Polyakov 2001; Maslowski et al. 1998) and geochemical measurements (Figure 10b, Macdonald et al. 1999, 2002; Ekwurzel et al. 2001; Guay et al. 2001; Schlosser et al. 2002) clearly show that with the high index conditions of the late 1980s, river water entering the Laptev and Kara shelves was forced to the east rather than directly off the shelf and into the Transpolar Drift. Under high index conditions of the early 1990s, perhaps 1000 km^3 y^{-1} or more of runoff from the Lena, Ob and Yenesei rivers

Fig. 1.2.10. The change in river inflow pathways between low AO$^-$ index (1979) and high AO$^+$ index year(1990–94) (based on model results from W. Maslowski reproduced in Dickson 1999) and b) A schematic diagram showing the amounts and changes in pathways for river inflow to the Arctic Ocean under AO$^-$ conditions (blue) and AO$^+$ conditions (red)

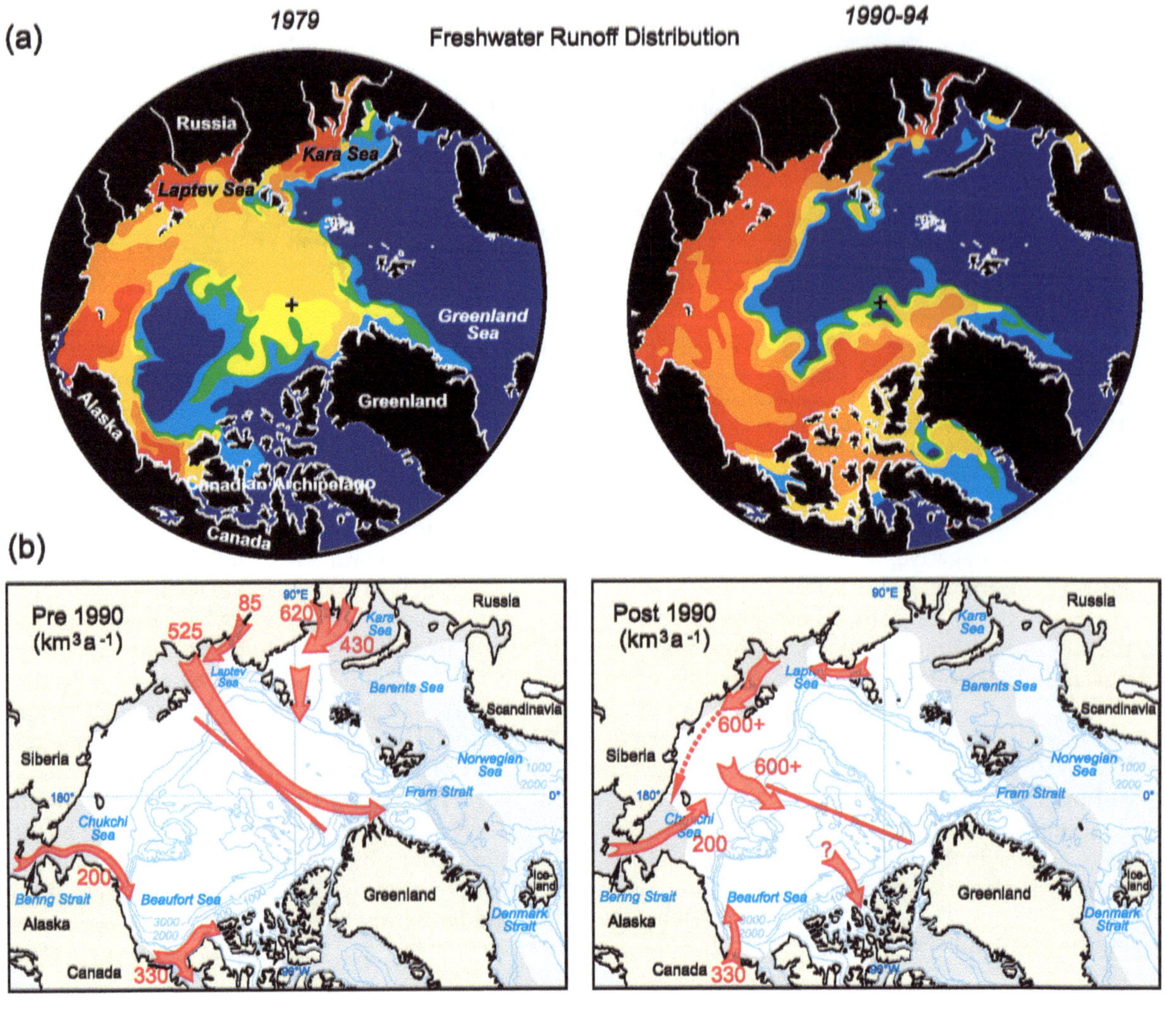

stopped entering the Eurasian Basin and entered, instead, the East Siberian Shelf and thence the Canadian Basin possibly to exit the Arctic Ocean via the Archipelago (Fig. 1.2.10) (Morison et al. 2000). These changes suggest fundamental alteration in plume dynamics, arctic coastal currents and shelf-basin exchanges. Changes like these have the potential to be reflected in arctic sediments as changes in the sources of sediment supply (Darby et al. 2001).

Effects on hydrology

Given the changes in winds and temperature that occur with shifts in the AO, it is to be expected that precipitation and evaporation within the Arctic will also be affected both in amount and seasonality (Serreze et al. 2000) and that these will subsequently be reflected in river flow (Semiletov et al. 2000). Enhanced, or episodic, river flow together with permafrost degradation are likely to lead to enhanced particle erosion and supply of relict organic carbon to coastal oceans through rivers. Due to sparse networks and short time series, it is difficult to assess with confidence the spatial or temporal variation of precipitation within the Arctic. Nevertheless, records suggest that net precipitation (precipitation minus evaporation) north of 70°N is 36% higher during periods of high index compared to low index (Serreze et al. 1995).

Precipitation minus evaporation (P–E) integrated over a drainage basin should be equivalent to river discharge for the basin. Arctic rivers exhibit large interannual variation (Semiletov et al. 2000; Shiklomanov et al. 2000; Stewart 2000) making it difficult to link river flow to precipitation or temperature trends or to climate variables like the AO. For example, Shiklomanov et al. (2000) suggested little change in mean annual discharge for arctic rivers between the 1920s and 1990s whereas Semiletov et al. (2000) found recent increases for several Eurasian rivers and Lammers et al. (2001) found evidence of increased winter discharge from rivers in Siberia and Alaska in the 1980s compared to the 1960s–70s. Within Canada, the Mackenzie Basin has undergone exceptional warming between 1961 and 1990 (Stewart et al. 1998); nevertheless, increased basin temperatures are not obviously recorded in this river's hydrology (Fig. 1.2.11a; Stewart 2000) or in other arctic rivers (Shiklomanov et al. 2000). Instead, one sees evidence of 3- to 4-year periodicity in peak flow and alterations in the seasonal shape of the hydrograph with higher flows delayed well into August suggesting changes in both total annual discharge and its seasonality. Such patterns appear to be only partially related to the AO as

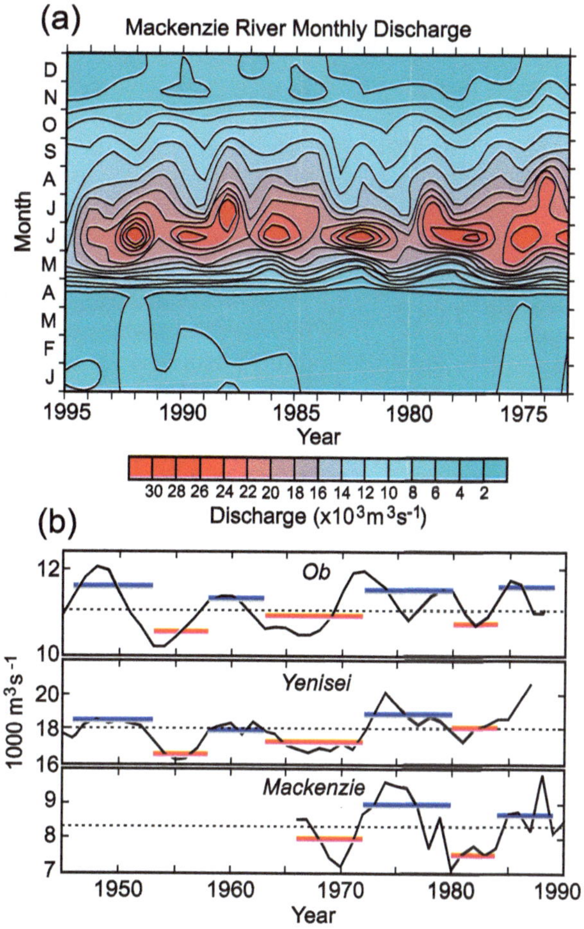

Fig. 1.2.11. a) The monthly river discharge for the Mackenzie River from 1973 to 1995 inferred from observations at Arctic Red River and the Peel River (after Stewart 2000) and b) the relationship between the cyclonic (AO^+)/anticyclonic (AO^-) states and the three-year smoothed monthly river discharge for the Ob, Yenisei and Mackenzie Rivers (after Johnson et al. 1999)

evidenced by significant correlations between runoff and precipitation for the Mackenzie Basin and variation in North Pacific storm tracks (Bjornsson et al. 1995).

The river discharges for the Ob, Yenisei and Mackenzie Rivers appear to show a positive relationship with the North Pole pressure anomaly with a lag in discharge of about 0.5–0.7 years (Fig. 1.2.11b; Johnson et al. 1999) but such a relationship runs counter to the enhanced precipitation observed during high index conditions. Even if all variation in arctic river discharge at the 4–5 year time scale is assigned to shifts in AO/NAO index, the maximum effect on annual flow would be about 5–15% which is within the present range of interannual variability (for example see Johnson et al. 1999; Semiletov et al. 2000).

Effects on marine ecosystems

Changes in marine ecosystem distributions can often be directly linked to large-scale change associated with shifts in atmospheric forcing (i.e., the AO, NAO or PDO). In the Barents Sea, primary production was 30% higher in the warm period between 1989 and 1995 than in the cold 1960s, mainly due to a 4- to 5-fold increase of primary production in the additional 650 000 km² of ice-free water (Slagstad and Wassmann 1996). The Barents Sea stock of *Calanus finmarchicus* presumably was up to 10 times larger during this warm period, mainly due to large import from the Norwegian Sea (Skjoldal and Rey 1989). After the warm period, conditions reversed: ice cover increased in the Barents Sea, the wind-mixed layer became shallower, blooms appeared earlier and were more rapid, and nitrate depletion became more pronounced. The resulting mismatch between phytoplankton blooms and grazers lowered zooplankton production (Olsen et al. 2002).

In the Beaufort Sea, a comparison of data collected during 1997–1998 with older data from 1979–1980 revealed an impoverished sea-ice fauna (Melnikov et al. 2002). As a consequence of the high AO index of the early 1990s, increased melting of sea ice altered its physical properties, and fresh water from runoff stratified the upper ocean with the consequence that diatoms became scarcer and were replaced by freshwater green algae. Nematodes, copepods, amphipods, and turbellarians all vanished.

A great deal of evidence has been assembled to show that the North Pacific Ocean can undergo regime shifts which pervade physical and biological systems (Hare and Mantua 2000). Recent – and dramatic – change in the Bering Sea ecosystem has revealed how physical change can completely disrupt ecosystem structure from top to bottom (Niebauer 1998; Springer 1998; Wyllie-Echeverria and Wooster 1998; Brodeur et al. 1999; Hunt et al. 1999; Rugh et al. 1999; Saar 2000; Stockwell et al. 2000; Macklin 2001; Stabeno and Overland 2001).

1.2.2.3
The Future

All climate scenarios predict considerable warming in the Arctic during the coming 100 years (ACIA 2002a, 2002b). With such warming, there will be less sea ice, an increase in the area of permanently ice-free waters well into the Siberian Shelf Seas and, ultimately, the sea ice in the deep Arctic Ocean may vanish in summer (Flato and Boer 2001).

Some scenarios cautiously predict that the polar vortex will strengthen and conditions will, as a result, become more like those observed during strong positive phases of the AO/NAO index. Accordingly, there would be more intense and/or higher frequency of winter storms, and more rapid and early melting of sea ice especially around the Arctic Ocean margins. On the other hand, it is also possible that positive AO, NAO and PDO conditions are not sustainable as they bear the seeds that trigger the negative phases. For example, in the Atlantic sector, the warm period 1989–1995 brought vast amounts of rain into the Nordic Seas and the resulting freshening is today evident in the North Atlantic Deep Water flow, which now has reached the American seaboard (Dickson et al. 2002). There are also signs that the deep-water production in the Nordic Seas is slowing (Hansen et al. 2001) which may then reduce the import of Atlantic Water into the Nordic Seas enhancing the probability of negative AO and NAO conditions. In the North Pacific, the PDO shifted toward negative values in 1999, possibly leading to a reduction in El Niños and greater ice cover in the Bering Sea. Climate oscillations are likely to continue, as they always have, but there will presumably be superimposed a long-term trend of warming.

Observing how components of the Arctic's organic carbon cycle respond to the extreme positive or negative phases of climate oscillations provides a way to project the consequences of climate change. However, it seems likely that climate change in the Arctic will impact the marine and terrigenous components of the carbon cycle differently.

For the marine system, large areas of the Arctic Ocean becoming free of ice will admit more light and allow more mixing, with the consequence that net primary production will increase. If warming is most pronounced in spring, leading to early melting, the consequent extension of the growth season will presumably also increase annual primary production. These kind of general changes will, however, be experienced differently by the various shelves and basins of the Arctic Ocean and it will be instructive to seek evidence for specific changes in phytoplankton communities and timing of blooms as discussed in Chapter 3.9. The above simplistic predictions must be qualified because upwelling, wind-driven vertical mixing, and freshwater supply from sea ice and rivers – all of which will change under future climate scenarios – will also play a large role. A second, very important change in marine productivity will have to do with the loss of ice-algal communities to be replaced by pelagic communities which then alters vertical flux of organic carbon and coupling between the upper ocean and benthos (Walsh 1989). Such change could potentially occur over much of the Arctic

Ocean to become widely recorded in sediment geochemistry (Gobeil et al. 2001) and in the preservation of carbon in shelf and basin sediments.

The terrigenous component of the Arctic's carbon cycle could be altered at many locations including basin headwaters, rivers, estuaries, shelves, the interior ocean and at the exits of the Arctic Ocean. Within the drainage basins, the melting of permafrost (Weller and Lange 1999; IPCC 2001; Vörösmarty et al. 2001) and subsequent release of sediments and relict organic matter offers an immediate way to enhance the loading of these to coastal seas. Such inputs would be further augmented by accelerated coastal erosion brought on by thermal destruction of ice-bonded coasts, sea-level rise, and enhanced fetch due to loss of sea-ice cover over the shelves. Some of this organic matter would become metabolized and some would be buried, but much of the DOC component would remain in the upper ocean to be transported in the surface current systems (Opsahl et al. 1999). If the diversion of river inflow eastward over the Russian shelves becomes a prominent feature of the Arctic Ocean in the future, then DOC (and part of the POC) will likewise be diverted from the Eurasian Basin into the Canada Basin. Not only would this alter the DOC pathway through the Arctic, but it would also change the storage of DOC within the Arctic because the residence time of surface waters in the Canada Basin (~10 y) is longer than the ~2-year transit time for water in the Transpolar Drift (for discussion of recent changes in hydrological pathways in the Arctic Ocean and their consequences see, Steele and Boyd 1998; Ekwurzel et al. 2001; Guay et al. 2001; Johnson and Polyakov 2001; Macdonald et al. 2002). The loss of ice cover together with prospect of increased incident UV radiation (Weatherhead and Morseth 1998) and increased DOC residence time in the Arctic Ocean makes for an interesting, and difficult to predict, feedback loop wherein more of the terrestrial and marine DOC may become photo-oxidized, but the increased terrestrial DOC might also serve to protect the upper ocean ecosystem from incident UV.

Finally, because ice provides such an important medium of transport for particulate matter (see Chapter 2.4, Fig. 2.6), change in ice drift pathways, in the strength and location of ice formation and in the location of ice melting all promise to alter particle transport pathways. Although details here are difficult to predict, it seems likely that a retreat of the ice edge back into the Arctic would shorten the ice transport route from the Russian shelves to the Greenland Sea and, accordingly, would reduce particle export from the Arctic Ocean by this route. Furthermore, an alteration in ice composition toward more first-year ice would imply a greater loss of particulates during transport because first year ice will be more prone to melting and fracturing en-route. It remains difficult to predict how projected change will impact crucial processes like suspension freezing in shelf flaw leads and ice export from shelves during winter (Macdonald 2000).

1.3 Tectonic Evolution of the Arctic Ocean – Overview and Perspectives

W. Jokat

At the end of the 19th century most of the world's surface and oceans had been discovered with only a few remaining uncharted regions in the far north and south. At that time the polar regions were investigated only along their margins. However, there was a continuing interest in charting these areas in order to find shorter routes to the east along the northern sea route. Investigations in the Arctic pack ice and under a hostile climate were extremely difficult given the existing ships and new strategies were required to complete the explorations. During a three year expedition (1893–96) Fridjof Nansen opened the door to Arctic research. The famous drift of the Fram in the pack ice was both the first scientific expedition to the Arctic Ocean and a technical break-through. For the first time, the drift allowed the gathering of reliable scientific data from the interior ice covered Arctic Ocean. Nansen's bathymetric map of the Arctic Ocean revealed a deep ocean basin surrounded by the large land masses of Eurasia, North America and Greenland.

After reaching the north pole, interest in Arctic research dwindled in the western countries and, until the 50's of the last century, only Russian scientists equipped ice islands on a regular basis to gather basic oceanographic and meteorological information (Kristoffersen 1990). Ice islands are large, thick ice floes that have calved off the ice shelves in the Arctic. They drift with the pack ice. During the Cold War only few results of these investigations were available to western scientists with, for example, the discoveries of the bathymetry of the Arctic Ocean remaining secret. Until the geophysical year in 1957, the understanding of the morphology and the evolution of the Arctic Ocean was not very different from that of Nansen's map. During the next two decades US and Canadian Institutions mounted projects on several ice islands (Table 1.3.1) to gather new geoscientific data which were subse-

Table 1.3.1. List of research activities on ice islands

Year of expedition	Expedition
since 1937	numerous Soviet drifting stations
1952–1974	Fletcher's Ice Island
1957–1958	Alpha
1959–1960	Charlie
1961–1965	Arlis
1971, 1972, 1975–1976	Aidjex
1979	Lorex
1979–1982	Fram
1983	Cesar
1984–??	Hobson's Choice

quently made available to the public. Geophysical investigations were part of these research programmes. In particular, seismic instruments operated through holes in the ice floes provided the first information on the general structure of the seafloor and the sediment distribution. However, the tectonic evolution of the Arctic Ocean remained speculative until the plate tectonic revolution began in the early 70's.

The interior Arctic Ocean, which was not accessible for surface ships until the 1980s, was surveyed by long-range aircraft of Russian and US institutions (Karasik, 1968; Vogt et al. 1979, 1982) to acquire aeromagnetic data (Fig. 1.3.1). A general model, which is still valid, was established in the late 1970s and early 1980s. The magnetic data together with sparse bathymetric data from numerous Russian and joint US/Canadian drift stations (Table 1.3.1) and the few unclassified submarine data, allowed the identification of the large-scale structures in

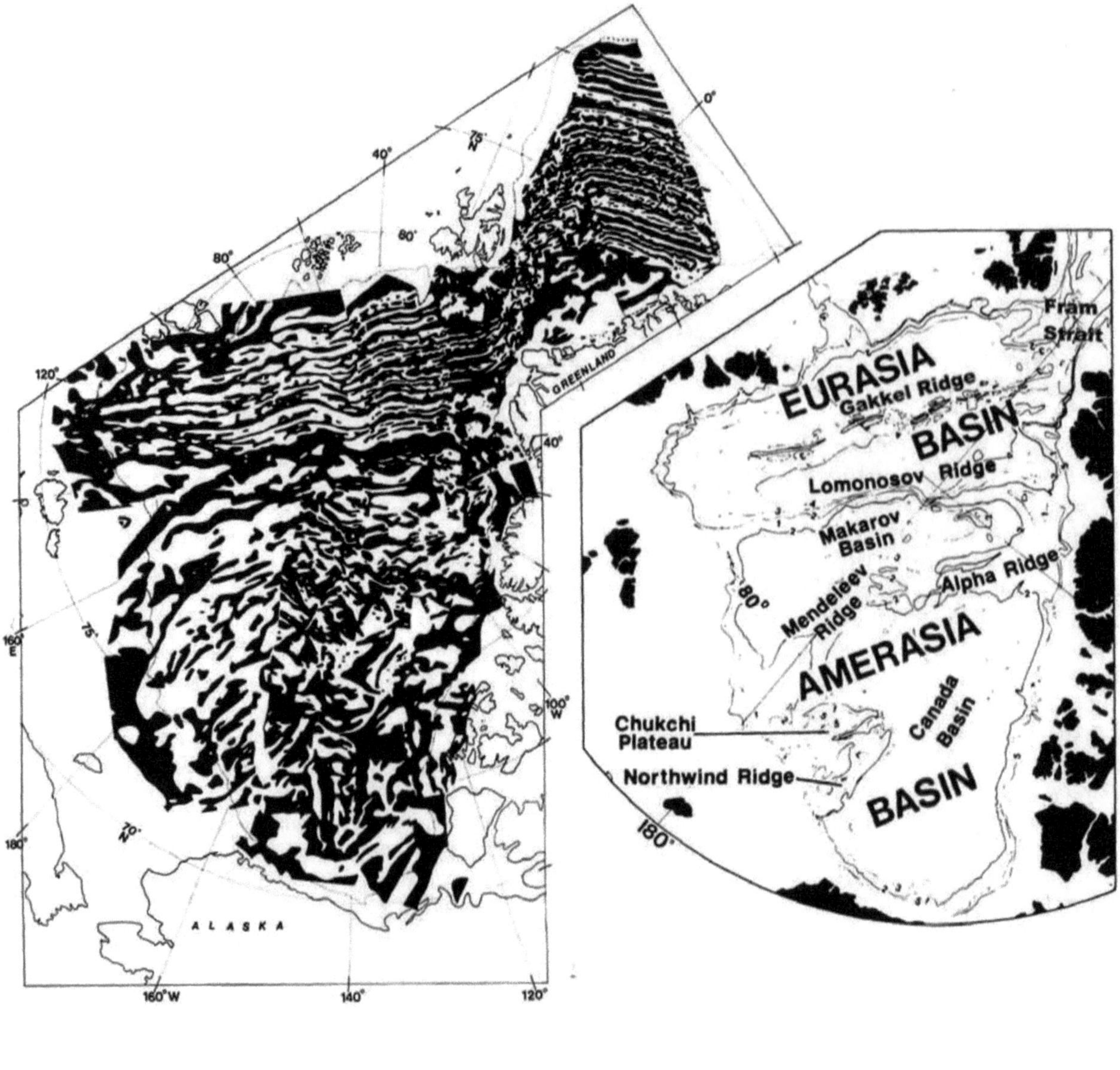

Fig. 1.3.1. Residual magnetic anomaly map for the Arctic based on results from Russian and American surveys. The right map shows the geological features with a simplified bathymetry of the Arctic basins and ridges (Modified from Vogt et al. 1982 and Coles and Taylor 1990)

the Arctic. Both geophysical data sets provided the basis for a first-order tectonic interpretation of the ridges and basins.

Eurasian Basin – Gakkel Ridge/Amundsen and Nansen basins/Lomonosov Ridge

Regional aeromagnetic data (flight spacing > 20 km) indicated the presence of seafloor spreading anomalies in the basins north and south of Gakkel Ridge (Fig. 1.3.1). The Gakkel Ridge, located in the middle of the Eurasian Basin, is an active spreading centre. According to the tectonic evolution of the North Atlantic, the oldest seafloor spreading anomaly was dated to an age of 58 Myr. The interpretation of the spreading history showed that the Gakkel Ridge is presently the slowest spreading ridge (full rate 0.6–1.2 cm y^{-1}) of the world-wide ridge system. Here, in the rift valley, the greatest water depths of the Arctic Ocean are to be found (5400 m). The Eurasian Basin is bordered to the north by the Lomonosov Ridge. The origin of this ridge, which is more than 1800 km long, was unknown before the magnetic data became available. The continuous seafloor spreading anomalies allowed only one interpretation: in Early Cenozoic times (>58 Ma) the Lomonosov Ridge was attached to the Barents and Siberian shelves (Wilson 1963). With the onset of seafloor spreading in the Eurasian Basin the Lomonosov Ridge split off these shelves. It is surprising that the Lomonosov Ridge survived the separation as a structural unit, perhaps as a consequence of the slow spreading rate along the Gakkel Ridge. In general, the ridge has subsided until today it is submerged to water depths of 1000–1500 m in the east and to more than 2000 m on the Greenland part (Jakobsson et al. 2000). North of Greenland and Svalbard the aeromagnetic and bathymetric data indicate the presence of two plateaux, the Morris Jesup Rise and Yermak Plateau. The entire Morris Jesup Rise and the northern Yermak Plateau show strong magnetic anomalies, typical of extensive magmatic activity during the separation of continents. Both features formed a unit before chron C13 (36 Ma) and were later separated by the ongoing seafloor spreading processes. As a consequence of the constant spreading in the Eurasia Basin and the relative movements between Greenland and Svalbard, Fram Strait began to open. From 10 Myr onwards these movements created the only deep water connection of the Arctic Ocean (Lawver et al. 1990) to the world ocean. This tectonic event had, therefore, a profound influence on the earth's climate, perhaps favouring the occurrence of numerous glacial/interglacial cycles in the northern hemisphere.

Amerasia Basin – Canada Basin/Chukchi Plateau/Alpha-Mendeleev Ridge/Makarov Basin

The aeromagnetic data indicate a weak pattern of seafloor spreading anomalies in the Canada Basin (Vogt et al. 1982). Their dating is quite speculative since only few anomalies are visible. Here, the combination of the aeromagnetic and onshore geological data from Alaska and northern Canada provided constraints for the tectonic model. Most of the models suggest that the Canada Basin was formed by seafloor spreading in the Cretaceous around 130 Myr. At that time Alaska was attached to the Canadian Arctic Islands. With the onset of seafloor spreading Alaska rotated counter-clockwise into its present-day position. In this model the Chukchi Plateau is interpreted to be stretched continental crust once attached to the Canadian Arctic (Grantz et al. 1998). The Alpha-Mendeleev Ridge, discovered in 1956/57 (Hunkins 1961), remains the most problematic feature created during the Mesozoic history of the Arctic Ocean. The ridge, which is 250–800 km wide, strikes more or less parallel to the Lomonosov Ridge. In bathymetric cross sections it is roughly symmetrical with greatest elevation at the centre. From the bathymetry it is not clear if the Alpha-Mendeleev Ridge has been formed through the same processes as the Lomonosov Ridge. The magnetic field is highly variable and no seafloor spreading anomalies are visible. Thus, it has been suggested that the Alpha-Mendeleev complex as well as the Makarov Basin were formed during or after the opening of the Canada Basin within the Cretaceous Quiet period. Investigators have suggested that this blocky ridge may be: (1) of continental origin; (2) a former spreading centre; (3) a result of "hot spot" activity; and (4) a former region of subduction or compression. Currently, models in which the Alpha Ridge represents a former spreading centre or "hot spot" trail are favoured (Forsyth et al., 1986; Lawver and Mueller, 1994).

Numerous short cores of Quaternary age were recovered from the ice islands. However, only at three locations have Cretaceous and Early Cenozoic sediments been recovered from Alpha Ridge (Kitchell and Clark 1982; Bukry 1984). Dredged material recovered by the CESAR expedition from elevated basement provided undatable, fragmented and weathered alkaline volcanic rock (Van Wagoner and Robinson 1985).

During the late 80's, a new era of geoscientific Arctic research began with the availability of ice-capable research vessels which could enter the Arctic pack ice. Research from drifting ice islands was very expensive, since all equipment had to be transported by aircraft onto the floe and support was

Table 1.3.2. List of geoscientific icebreaker cruises to the Central Arctic Ocean

Year of expedition and area	Ship and country
1987 Nansen Basin, Gakkel Ridge	Polarstern (Germany)
1988 Chukchi Plateau	Polar Star (USA)
1991 Nansen/Amundsen Basin/Lomonosov Ridge	Oden/Polarstern (Sweden/Germany)
1993 Canada Basin/Chukchi Plateau	Polar Sea (USA)
1994 Canada Basin/Lomonosov Ridge	Polar Sea/Louis S. St. Laurent (USA/Canada)
1995 Amundsen/Makarov Basin/Lomonosov Ridge	Polarstern (Germany)
1996 Lomonosov Ridge/Makarov Basin	Oden (Sweden)
1998 Alpha Ridge/Lomonosov Ridge	Polarstern/Arktika (Germany/Russia)
2000 Mendeleev Ridge	Akademic Fedorov (Russia)
2001 Nansen Basin/Gakkel Ridge	USCGC Healy/Polarstern (USA/Germany)
2001 Nansen Basin/Lomonosov Ridge	Oden (Sweden)

necessary continuously during the drift. In contrast to icebreaker expeditions, the drift of the floe was impossible to control and difficult to predict. The benefit in using drifting ice islands was that areas that were out of reach for any surface ship at that time could be investigated. In contrast to ice islands, ships can carry heavier gear, more personnel and can steam at least to an approximate location within a certain time. Several expeditions organised by American and European institutions (Table 1.3.2) have provided a vast amount of high quality and more detailed data on the ridges and basins of the High Arctic

- Bathymetry: multi-beam/swath bathymetry instead of single beam data
- Coring: 6–12 m instead of 1–3 m
- Dredging: detailed sampling of the mid-ocean Gakkel Ridge
- Geophysics: Continuous profiles to the acoustic basement instead of short, shallow profiles

The following is a brief summary of what these ship-based expeditions have found. Seismic data indicate that the Lomonosov Ridge is covered by almost 500 m of Cenozoic sediments resting on an erosional unconformity which formed during the subsidence of the ridge below sea level after separation from the Barents and Siberian shelves. The new geophysical and geological information (Jokat et al. 1992) confirmed the continental origin of the Lomonosov Ridge. The sediment thicknesses in the Nansen and Amundsen Basins as well as the crustal thickness beneath the rift valley of the Gakkel Ridge were measured by recent expeditions and the first visit to the Alpha Ridge by surface ships during 1998 (Jokat et al. 1999) recovered basement samples. Dating of these samples indicate a Cretaceous age of the central Alpha Ridge. The crustal thickness across the Mendeleev Ridge has been determined and the sediment distribution of the Canada Basin/Chukchi Plateau was investigated.

Despite the very successful expeditions one basic problem remained. Only few cores containing information on the Cenozoic/Mesozoic environment of the Arctic Ocean have been recovered (e.g., Jackson et al. 1985; Clark et al. 1986). These cores do not provide a sufficient basis to extend the well-documented Quaternary geological history back in time in order to refine the general geophysical tectonic models for the Arctic. Thus, it remains unknown whether or not present Arctic pack ice is the exception in geological history or the rule. From onshore investigations, there is evidence that the Arctic Ocean was warm some 90 Ma ago (Tarduno et al. 1998). Was the Arctic a tropical paradise during the Cretaceous and most of the Cenozoic? To answer this question, long cores are urgently needed. The Ocean Drilling Programme has identified the Arctic Ocean to be an important area for deciphering the earth's climate history (Hovland 2001). Long cores of some hundred metres in length, likely to be drilled in a few years, may radically change our view of the tectonic and climatic evolution of this remote ocean and its global importance.

1.4 Geochemical Proxies used for Organic Carbon Source Identification in Arctic Ocean Sediments

R. STEIN, R.W. MACDONALD

1.4.1 Introduction

In order to interpret the organic carbon data in relation to transport pathways, depositional environment and (paleo-)climatic/environmental changes as well as for the calculation of flux, remineralization and burial rates of marine and terrigenous organic carbon, information about the origin of the

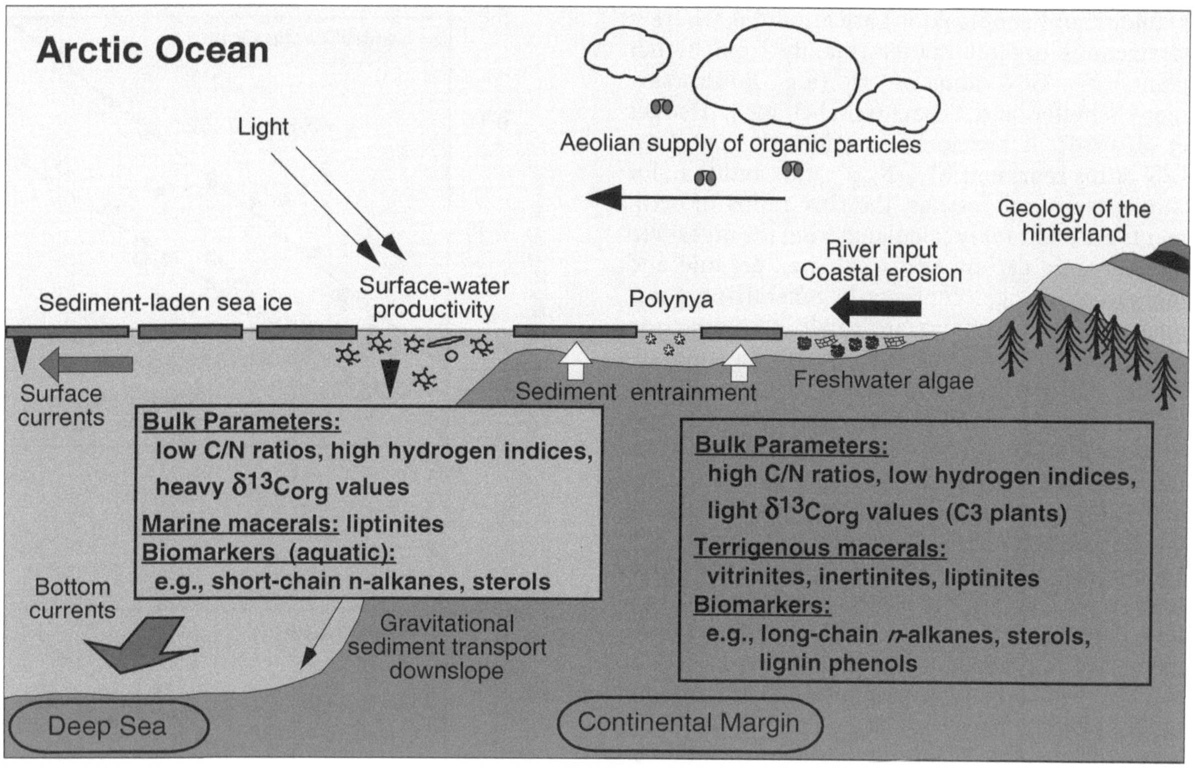

Fig. 1.4.1. Schematic diagram summarizing the factors controlling organic carbon fluxes in the Arctic Ocean. Organic-geochemical proxies used for the identification of aquatic (marine/freshwater) and terrigenous (higher plant) organic matter are shown

organic carbon preserved in the sediments is required. This information, i.e., estimates of the marine and terrigenous proportion of the organic matter, can be obtained by means of organic-geochemical proxies which include organic-geochemical bulk parameters (C/N ratios, $\delta^{13}C_{org}$ values, Rock-Eval parameters), maceral composition and specific biomarker distributions. Here, a short summary about the use and background of the different proxies will be given, emphasizing parameters for which a broader data base is available for Arctic Ocean sediments and which are used in this book (Fig. 1.4.1; for further general information on other organic-geochemical proxies see e.g., Tissot and Welte 1984; Hollerbach 1985; Stein 1991; Peters and Moldowan 1993; Meyers 1997; Rullkötter 2000, and further references therein; for the application of biomarkers and proxies to questions of sources of organic matter in the Arctic see e.g., Yunker et al. 1995; Fahl and Stein 1997; 1999; Goñi et al. 2000; Schubert and Calvert 2001).

When using organic-geochemical proxies as indicators of spatial and temporal changes in organic-carbon sources in high-latitude ice-covered regions, such as the Arctic Ocean with its marginal seas, it is important to consider the complex factors controlling the organic-carbon flux in these areas (Fig. 1.4.1; see Fahl and Stein 1999). First, the strong fluvial supply in the marginal seas may provide freshwater (aquatic) organic matter as well as terrestrial (higher plant) organic matter. Second, due to the more or less closed sea-ice cover, phytoplankton productivity in the Arctic Ocean is very low and may represent a mixed signal of surface-water productivity and productivity of sea-ice communities. Third, the organic matter deposited on the shelves, which is a mixture of terrigenous higher plant material, freshwater organic material and, to a limited extent, marine organic matter, may be incorporated into the sea ice and further transported offshore. Thus, terrigenous organic matter may be deposited far away from where it enters the ocean.

1.4.2 Organic geochemical bulk parameters

Carbon/nitrogen (C/N) ratios

Carbon/nitrogen (C/N) ratios measured by elemental analysis are often used to distinguish between marine (algal) and terrigenous (higher plant) organic matter in marine sediments. C/N ratios of marine organic matter (mainly phytoplankton and zooplankton) are around 6, whereas terrigenous organic matter (mainly from higher plants) has C/N ratios of >15 (e.g., Bordowskiy 1965; Scheffer and Schachtschabel 1984; Hedges et al. 1986). It has to be considered, that these C/N ratios represent "C_{org}/N_{org}" ratios and that, for simple technical reasons, the C/N ratios of sediments are commonly calculated from the measured total organic carbon and total (i.e., organic and inorganic) nitrogen contents. In general, inorganic nitrogen concentrations are small in comparison to those of organic nitrogen, justifying this approach. If the organic carbon content is low (<0.5%), however, the inorganic nitrogen may become a major proportion of the total nitrogen. Furthermore, inorganic nitrogen is bound as ammonium in clay minerals (N_{bou}), especially in the clay mineral illite (Stevenson and Cheng 1972; Müller 1977). In the Laptev Sea, especially in its eastern part, and in the central Arctic Ocean, high illite contents of about 45 to 65% were determined in surface sediments (Stein, et al. 1994a; Rossak et al. 1999), which may result in a significant amount of N_{bou} being fixed in the sediments. In these cases, the absolute C/N ratio may be misleading as an organic carbon source indicator (although relative changes in the ratios may still be used) (see Chapters 7.5 and 7.9).

A total organic carbon (OC) versus total nitrogen (N_{tot}) diagram may help to estimate the amount of N_{bou} from the positive N_{tot} intercept at OC=0 (see Ruttenberg and Goñi 1997; Schubert and Calvert 2001), as shown in examples from the central Arctic Ocean and the Laptev and Kara seas (Fig. 1.4.2). For the sediments from the central Arctic Ocean (Fig. 1.4.2a), there is an intercept of 0.04% N_{tot} at 0% OC, suggesting the presence of a significant proportion of N_{bou} in the samples. N_{bou} values were determined on the same set of samples by direct measurements (Schubert and Calvert 2001). These values vary between 0.027 and 0.066% and support the estimated average of 0.04% N_{tot}. In the Laptev Sea, the proportion of N_{bou} is probably even higher, as suggested from the intercept of 0.05 and 0.10% N_{tot} at 0% OC (Fig. 1.4.2b). The data points with the

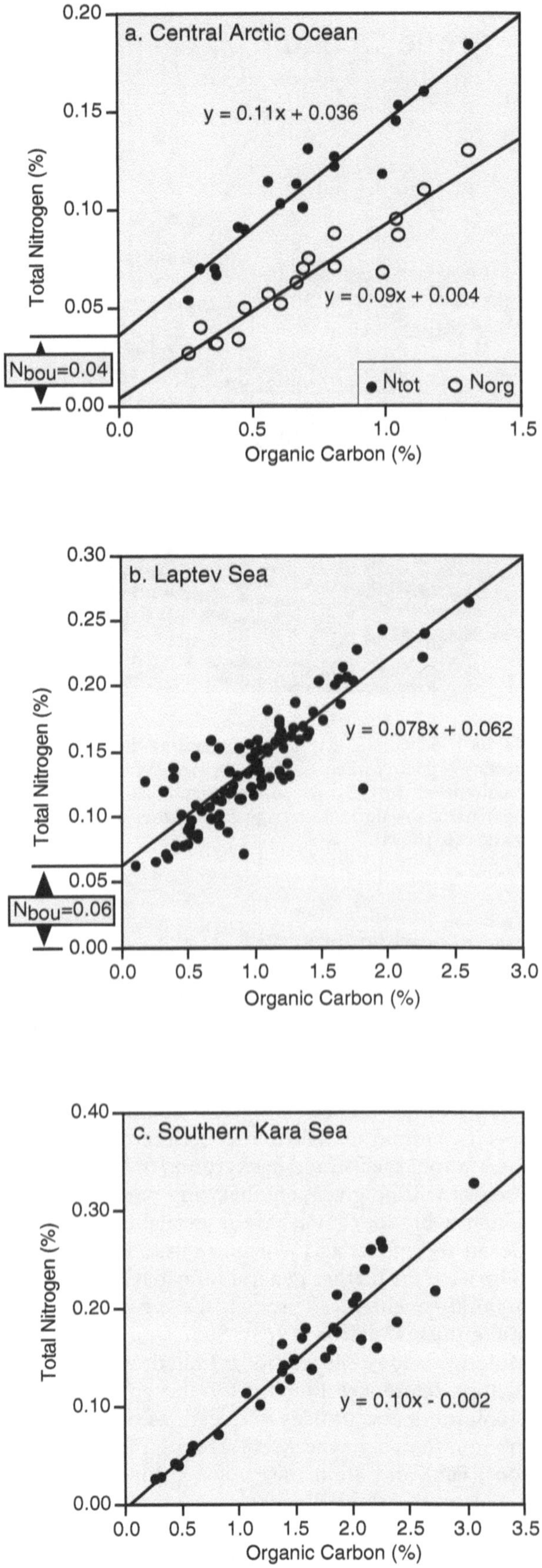

Fig. 1.4.2. Total organic carbon (TOC) versus total nitrogen (N_{tot}) correlation in Arctic surface sediments. The estimated amount of inorganic nitrogen (N_{bou}) is indicated. For the central Arctic Ocean sediments, the correlation between TOC and organic nitrogen (N_{org}) is also shown. See text for further explanation ▶

a. Central Arctic Ocean (data from Schubert and Calvert 2001)

b. Laptev Sea

c. Southern Kara Sea

higher estimated N_{bou} value of 0.10% derive mainly from the eastern Laptev Sea which is characterized by maximum illite contents (Rossak et al. 1999). From this it is obvious that the calculated C/N ratios of the central Arctic Ocean and Laptev Sea surface sediments may be at least partly artifactually depressed leading to an underestimate of the terrigenous proportion of organic matter. After a correction for the proportion of N_{bou} (using $N_{org} = N_{tot} - N_{bou}$), these C/N ratios significantly increase (see Fig. 2a) and indicate a more mixed terrigenous/marine to terrigenous source (for further discussion see Chapters 7.5 and 7.9).

According to the OC versus N_{tot} diagram, the proportion of N_{bou} is probably insignificant in the surface sediments from the Kara Sea (Fig. 1.4.2c; for discussion see Chapter 7.6). A very similar situation is recorded in the Chukchi Sea (see Chapter 7.3).

Carbon stable isotope composition of the organic matter

A common approach for estimating the relative proportions of marine and terrigenous organic carbon in marine sediments is to assume two end members of different isotopic composition (e.g., Hedges et al. 1988; Prahl et al. 1994; Goñi et al. 2000). Assuming a linear mixing of terrigenous and marine organic matter, TerrOC can be calculated using the following equation:

$$\text{TerrOC}\ (\%) = \frac{\delta^{13}C_{sample} - \delta^{13}C_{mar}}{\delta^{13}C_{terr} - \delta^{13}C_{mar}} \times 100$$

where $\delta^{13}C_{terr}$ and $\delta^{13}C_{mar}$ are the isotope values of the terrigenous and marine organic carbon endmembers.

The terrigenous endmember can be constrained with confidence. Light isotopic $\delta^{13}C_{org}$ values of −26 to −28‰ are typical of terrigenous organic matter (i.e., land plants using the C_3 pathway of photosynthesis). Land plants using the C_4 pathway for carbon fixation and, thus, characterized by heavy $\delta^{13}C_{org}$ values of about −14‰ (O'Leary 1988; Meyers 1994), are of very minor importance in high latitude areas (Teeri and Stowe 1976). The $\delta^{13}C_{org}$ values determined in particulate organic matter from the Lena River, which dominates the riverine input into the Laptev Sea, for example, vary between −25.7 and −28.8‰ with an average of −27.1‰ (Rachold and Hubberten 1999). For Ob and Yenisei draining into the Kara Sea, $\delta^{13}C_{org}$ values of −28 to −28.7‰ and -26.5‰, respectively, were measured (Fernandes and Sicre 2000; Krishnamurthy et al. 2001; Fahl et al. 2002). For the Mackenzie/Beaufort area, −26.5‰ to −27‰ has typically been used for the terrigenous endmember (Goñi et al. 2000; Naidu et al. 2000).

The marine endmember is much less well constrained in the Arctic Ocean environment. In low to mid latitudes, heavy $\delta^{13}C_{org}$ values of −20 to −22‰ are given as characteristic for marine organic matter (marine algae) (Meyers 1994 and further references therein). In high latitudes, however, $\delta^{13}C_{org}$ values of marine phytoplankton may approach terrestrial values, ranging between −16.7 and −30.4‰ (e.g., Rau et al. 1982, 1989, 1991; Goericke and Fry 1994). This high variability can be related to several factors such as higher concentration of dissolved CO_2 at low surface water temperatures, cell growth rate, cell size, and cell membrane CO_2 permeability (e.g., Laws et al. 1995; Goñi et al. 1997; Rau et al. 1997). Furthermore, $\delta^{13}C_{org}$ values in sea-ice algae may reach heavy values of −15 to −8‰ (Gibson et al. 1999). Schubert and Calvert (2001) measured a strong enrichment in ^{13}C in phytoplankton samples recovered from meltwater ponds in the central Arctic Ocean ($\delta^{13}C_{org}$ values of −15.6 to −18.6‰). This high variability in the marine endmember often makes it difficult to determine organic carbon sources in marine sediments from sedimentary $\delta^{13}C_{org}$ values. When using $\delta^{13}C_{org}$, one should be aware of these problems and interpret the data with caution, or better still, apply the isotope approach together with other proxies for organic carbon source (Goñi et al. 2000; Schubert and Calvert 2001). The application of $\delta^{13}C_{org}$ values as proxies for the origin of organic carbon are presented and discussed specifically for the Beaufort, Chukchi, Laptev and Kara seas in Chapters 7.2, 7.3, 7.5, and 7.6, respectively.

Rock Eval Pyrolysis

In organic-carbon-rich (OC >0.5%), immature sediments, Rock-Eval pyrolysis parameters (S2 peak, hydrogen and oxygen indices) are also useful indicators for the characterization of the composition of the organic-carbon fraction (i.e., to estimate the amount of terrigenous and marine proportions) (e.g., Tissot and Welte 1984; Peters 1986; Stein 1991). The pyrolysis is conducted on bulk sediment samples to determine the amount of hydrocarbons (HC) already present in the sample (S1 peak in mg hydrocarbons per gram sediment), the amount of hydrocarbons generated by pyrolytic degradation of the kerogen during heating up to 550°C (S2 peak in mg hydrocarbon per gram sediment), the amount of carbon dioxide generated during heating up to 390 °C (S3 peak in mg carbon dioxide per gram sediment), and the temperature of maximum pyrolysis yield (Tmax value in °C) (Espitalié et al. 1977). The hydrogen index (HI) corresponds to the quantity of pyrolyzable hydrocarbons per gram OC (mgHC/gOC), the oxygen index (OI) corresponds

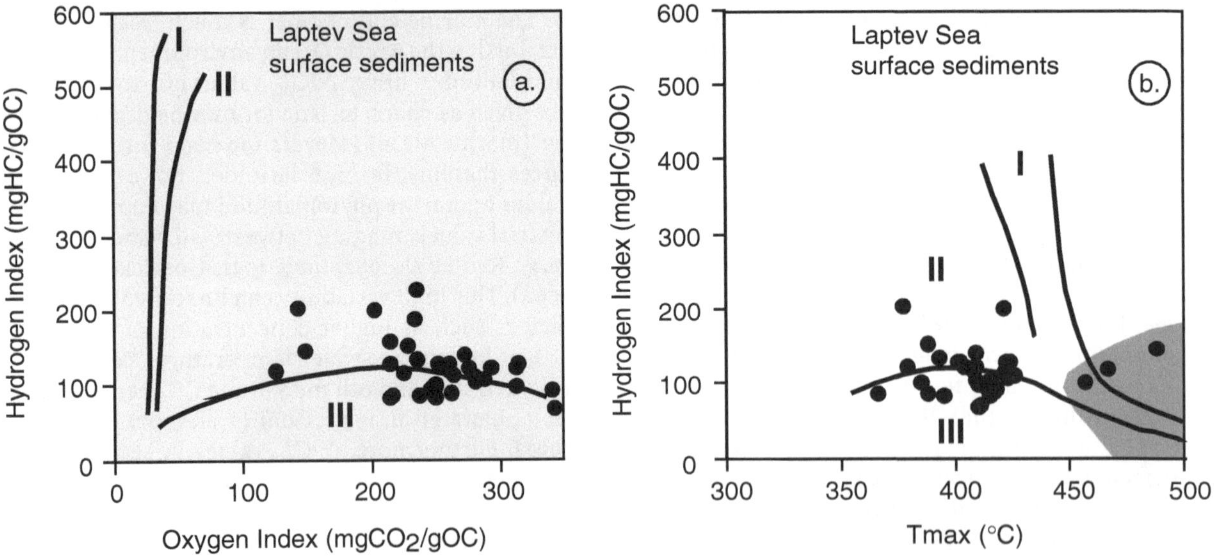

Fig. 1.4.3. Results of Rock-Eval measurements performed on surface sediments from the Laptev Sea.
a. Hydrogen versus oxygen index diagram
b. Hydrogen index versus Tmax diagram. The gray area marks samples containing more mature/reworked organic carbon

to the quantity of carbon dioxide per gram OC ($mgCO_2/gOC$). In immature sediments, HI values of < 100 mgHC/gOC are typical of terrigenous organic matter (kerogen type III), whereas HI values of 300 to 800 mgHC/gOC are typical of marine organic matter (kerogen types I and II) (Fig. 1.4.3a; Tissot and Welte 1984). Tmax values < 425 °C are indicative for immature organic matter. In combination with HI values, Tmax values may give further information about the composition of the organic matter (Fig. 1.4.3b).

When using hydrgen index values it has to be considered that this proxy only allows a very general classification of the origin of the organic matter and does not allow to give more precise estimates of the proportions of terrigenous and marine organic matter in percentages. Based on the comparison of hydrogen index values and maceral composition in immature sediments obtained within the Deep Sea Drilling Project (DSDP), for example, Stein et al. (1986) showed that in samples with hydrogen index values around 100 mgHC/gOC the relative proportions of terrigenous organic matter vary between 60 and 100 %. Furthermore, bulk sediment samples with low OC contents (< 0.5 %) may yield spuriously low hydrogen index values due to adsorption of hydrocarbons onto clay mineral surfaces ("mineral matrix effect"; Espitalié et al. 1984), overestimating the terrigenous proportion of OC in marine sediments.

The use of Rock-Eval results to determine organic carbon composition is presented and discussed in Chapters 7.5 and 7.6.

1.4.3
Maceral composition

Macerals, organic particles determined by microscopy under incident and fluorescent light, allow a precise and quantitative distinction of marine and terrigenous organic matter in sediment samples (for method and classification see Taylor et al. 1998; Wagner and Dupont 1999; Boucsein and Stein 2000). Vitrinite/huminite, inertinite, detritus (vitrinite/huminite and inertinite with grains <10 μm) and specific liptinites (sporinite, cutinie, and freshwater alginite) are counted as terrigenous particles, whereas marine particles are represented mainly by lamalginite, dinoflagellate cysts, and marine liptodetrinite. Particles <10 μm showing fluorescence are counted as "terrigenous-marine liptodetrinite" because a precise source indication is not possible due to the small particle size. As an example, the maceral composition of surface sediments from a transect across the Laptev Sea continental margin is shown in Figure 1.4.4, indicating a clear decrease in relative proportions of terrigenous organic carbon from the river mouth towards the continental slope. Furthermore, the proportion of small-sized detritus increases in the offshore direction (Fig. 1.4.4), indicating a sorting and decrease in grain size of terrigenous organic particles during long-distance transport. Further details on maceral distribution in Laptev Sea sediments have been published by Boucsein and Stein (2000) and Boucsein et al. (2000). In this book, the maceral approach has been used to estimate the marine and terrige-

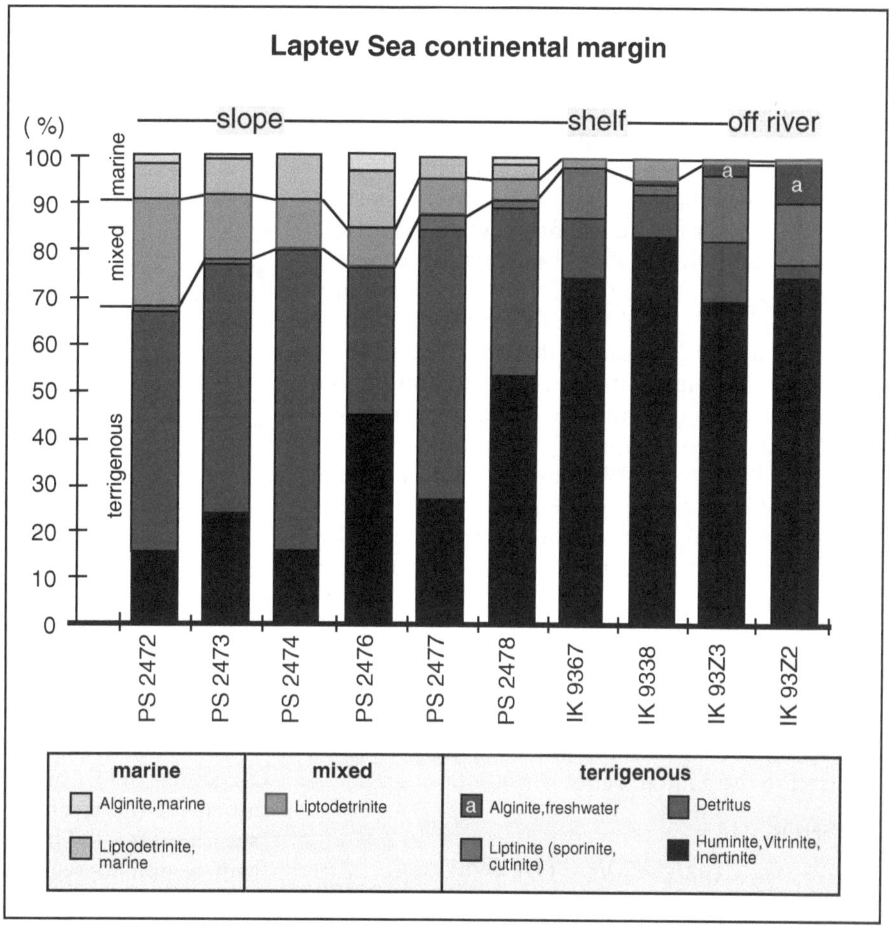

Fig. 1.4.4. Maceral composition in surface sediments from a transect across the Laptev Sea continental margin (from Boucsein and Stein 2000). For the location of transect see Figure 7.5.1 (Chapter 7.5)

nous proportions of organic matter in sediments from the Laptev and Kara seas (see Chapter 7.5 and 7.6).

1.4.4 Biomarker composition

Whereas the organic-geochemical bulk parameters such as C/N ratios, $\delta^{13}C_{org}$ values and HI values are important proxies for identifying general origins of organic matter (e.g., Stein 1991; Meyers 1997), specific biomarkers often provide more precise information about the different marine and terrigenous proportions of the organic-carbon fraction in marine sediments, (e.g., Volkmann et al. 1980, 1993, 1998; Brassell et al. 1986; Prahl and Muehlhausen 1989; Sikes et al. 1991; Yunker et al. 1995; Fahl and Stein 1999; Goñi et al. 2000). The distribution of long-chain C_{25}, C_{27}, C_{29}, and C_{31} *n*-alkanes may incisively point to land-derived vascular plant material (e.g., Prahl and Muehlhausen 1989; Yunker et al. 1995). Other biomarker proxies for the input of higher land plants are specific sterols (campesterol and β-sitosterol) (e.g., Huang and Meinschein 1976; Volkman 1986). One of the best proxies for vascular plant organic carbon input is lignin (e.g., Hedges and Mann 1979; Goñi and Hedges 1992; Goñi et al. 2000). From the specific lignin phenols, an even more detailed identification of different terrigenous sources is possible (e.g., Kattner et al. 1999).

Biomarkers often used as tracers of algal-derived organic matter are short-chain *n*-alkanes (C_{17} and C_{19} *n*-alkanes) (e.g., Blumer et al. 1971; Prahl and Muehlhausen 1989; Yunker et al. 1995) as well as specific sterols (e.g., dinosterol and brassicasterol) and short-chain fatty acids (e.g., de Leeuw et al. 1983; Nichols et al. 1984; Volkmann et al. 1993; Fahl and Stein 1997). When using these biomarkers as indicators of organic carbon source, it has to be considered that in environments strongly influenced by fluvial discharge (such as the Beaufort, Laptev and Kara sea shelves), these biomarkers may derive from both aquatic (freshwater) algae as well as from marine phyto- and zooplankton. Consequently a pronounced algal signature is not necessarily marine in origin, and all data have to be interpreted cautiously. More detailed discussions on the use of biomarkers as organic carbon source indicators in Arctic Ocean sediments are published in Yunker et al. (1995), Fahl and Stein (1997 1999), and Goñi et al. (2000).

One approach to estimate the proportion of marine and terrigenous organic matter is based on long-chain *n*-alkanes. Prahl and Carpenter (1984) examined long-chain *n*-alkane distributions in surface sediments from the Pacific coast off Washington State, and found that the *n*-alkane concentrations (C_{25} – C_{31}) in µg/gOC were fairly constant despite the large variability in OC contents (ratio = 277 +/– 87 µg/gOC). Based on this observation, they developed a simple binary model assuming (1) long-chain *n*-alkanes and terrestrial organic carbon (ALKOC) are exported at the mouth of the Columbia River to the Washington coast in a fixed ratio and (2) a change in the ALKOC ratio determined in coastal surface sediments is caused by dilution of terrestrial organic carbon with marine organic carbon that does not contain long-chain *n*-alkanes (Prahl and Carpenter 1984; Prahl and Muehlhausen 1989; Prahl et al. 1994). Accordingly, the percentage of terrigenous organic carbon (TerrOC) at each station can be calculated from the concentration of long-chain *n*-alkanes in µg/gOC normalized to the ALKOC ratio:

$$\text{TerrOC (\%)} = \frac{(C_{25-31}\ (\mu g/g\text{Sed.})/\text{OC}\ (g/g\ \text{Sed.}))_{\text{sample}}}{\text{ALKOC}} \times 100$$

Very recently, this approach has been successfully applied for the Kara Sea. Fernandes and Sicre (2000) found that the Ob and Yenisei rivers also supply long-chain *n*-alkanes into the Kara Sea in a relatively constant ratio to the organic carbon content (ALKOC ratio = 451) (Fig. 1.4.5), as indicated by the following equation:

$$\text{Sum } C_{25-31} = 451 \times \text{OC} + 0.117\ (r = 0.95)$$

where the sum of odd *n*-alkanes C_{25} – C_{31} is given in µg/g Sed. and OC in g/g Sed. The correlation between the long-chain *n*-alkanes and OC content of the surface sediments from the rivers and estuaries indicates that in the absence of long-chain *n*-alkanes, OC contents reach values of about zero which implies that the organic matter is almost entirely terrigenous (Fig. 1.4.5). Using the equation for TerrOC and an ALKOC ratio of 451, TerrOC can then be calculated (see Chapter 7.6 for further discussion).

Concerning the "*n*-alkane approach" to estimate the proportions of terrigenous and marine organic carbon it is important to note that the ALKOC ratio of the Ob and Yenisei rivers (451) is higher than that determined for the Columbia River (Prahl and

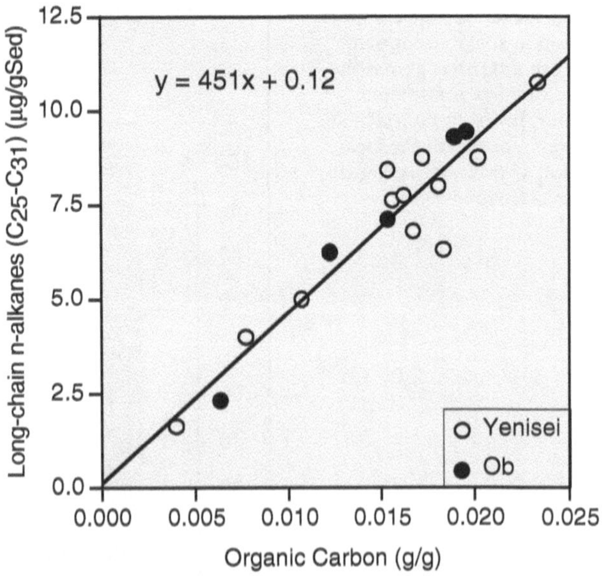

Fig. 1.4.5. Correlation between odd long-chain *n*-alkanes (C_{25}-C_{31}) and total organic carbon in surface sediments from the Ob and Yenisei rivers and estuaries (from Fernandes and Sicre 2000)

Carpenter 1984), but it is lower than that determined for the rivers draining into the Laptev Sea (see Chapter 7.5). Thus, the ALKOC ratio seems vary from region to region and has to be empirically evaluated for each coastal system. Furthermore, this implies that the ALKOC ratio cannot simply be used for estimating the relative proportion of terrigenous organic matter in the central Arctic Ocean where the importance of terrigenous organic carbon input from the different marginal seas (characterized by different ALKOC ratios) is generally not known and may vary from area to area.

1.4.5 The application of redox markers to organic carbon sediment geochemistry

The metabolism of labile organic carbon in sediments draws down dissolved oxygen with the consequence that redox conditions vary with depth in sediments and between sediment locations. The withdrawal of dissolved oxygen within sediments, which is controlled by the intensity of organic carbon supply and its rate of metabolism, sets up a redox gradient that progresses from oxic near the surface to suboxic (no dissolved oxygen but the presence of oxygen containing molecules like nitrate and sulphate) to anoxic at depth (Bender et al. 1989). In the case of anoxic ocean basins, for which there are presently no recorded examples within the Arctic Ocean, bottom water may, itself, become anoxic along with the sediments below. Anoxic water or sediment conditions may then feed back into the carbon metabolism process having consequences for organic carbon preservation (Hulthe et al. 1998). Because the organic carbon delivery to sediment surfaces varies with time often arriving in pulses (Lampitt 1985), sediment redox conditions likewise may vary with time responding on one hand to variable metabolic rates and on the other hand to re-supply of oxygen by diffusion into sediments from bottom water (Gobeil et al. 2001).

The supply of organic carbon, its composition and metabolism are frequently evaluated in sediments through bulk analyses of organic carbon and general composition as represented by, for example, C/N and $\delta^{13}C_{org}$ values. Aside from difficulties in assigning unique source compositions, this approach reveals the properties of more-or-less recalcitrant organic carbon remaining after metabolism has occurred in the more labile fraction. Because most of the marine organic carbon is labile (e.g., unsaturated fatty acids, sterols, alkanes), much of it is lost on the way to the sediments or recycled within the sediments (Belicka et al. 2002). As a consequence, organic carbon measured in sediments can be considered as the recalcitrant organic carbon "ashes" remaining after a substantial portion of the original carbon has been transformed to metabolic products including CO_2.

In contrast, the redox state in sediments (oxic vs anoxic), which is controlled by the rate of carbon metabolism, reflects to some degree the intensity and variability of the "flames" of labile organic carbon burndown. As a consequence, elements that respond in a well-defined way to redox conditions in sediments provide a complementary view of organic carbon to that derived by measuring the remaining organic carbon.

Although redox tracers have been used to infer organic carbon fluxes qualitatively for some time (e.g., Pedersen et al. 1989) it is only recently that their full potential is being recognized (Thomson et al. 2001). For the Arctic Ocean there appear to be very few studies that have used redox elements to infer organic carbon mediated processes in sediments (Aldahan et al. 2000; Danyushevskaya et al. 1980; Gobeil 1997; Gobeil et al. 2001).

Redox tracers that have been applied to sediment diagenesis problems include manganese, arsenic, rhenium, cadmium, iron, silver, phosphorus and acid-volatile sulphide (AVS) (Gobeil 1997; Gobeil et al. 2001; Luther et al. 1997; Sullivan and Aller 1996; Sundby et al. 1992). In all cases, these redox indicators are produced, accumulated, removed or recycled in sediments by varying redox conditions and by redox gradients. Manganese has long been recognized as a redox-sensitive element due to its often-observed enrichment in marine surface sediments. This element enters sediments in solid form as oxides or hydroxyoxides. As soon as oxygen becomes depleted in sediments, the oxidized forms of manganese become reduced to the soluble divalent form which then diffuses upward in the sediment to re-precipitate when it meets dissolved oxygen (Luther et al. 1997). Accordingly, the Mn distribution in sediment reflects the penetration of oxygen into sediments and the approximate location of the suboxic region where Mn oxides become reduced. On the other hand, AVS is produced under sulphidic conditions when sulphate reduction occurs. AVS therefore provides a sensitive marker of the redox boundary where sulphide is present in pore water. Cadmium precipitates with sulphide and therefore its natural enrichment in sediments reflects sulphidic conditions and may under the right circumstances provide an indication of the intensity of organic carbon metabolism (Gobeil 1997; Pedersen et al. 1989). Rhenium precipitates under suboxic conditions but quickly re-dissolves in the presence of oxygen; accordingly, the absence of Re reflects the presence of oxygen.

Although each of these markers by itself provides evidence of the activity of the organic carbon cycle in sediments, their greatest – and, as yet, barely tapped – potential appears to lie in using them in combination to infer the dynamic processes accompanying organic carbon fluxes and sediment accumulation. For example, Gobeil et al. (1997) concluded from the vertical offset observed between Mn and Cd enrichments in sediments from the Chukchi Sea that the redox boundary in these sediments must fluctuate widely implying that organic carbon supply must likewise fluctuate (Fig. 1.4.6a). More recently, Gobeil et al. (2001) measured vertical profiles for organic carbon, AVS, Mn and Re in widely-distributed Arctic basin sediments. Based on the vertical distributions of these properties and on diffusion length scales, these authors concluded that the sediments were not at steady state but rather appeared to have been subjected to a recent (<50 y) increase in labile organic carbon loading which was especially evident in the Re profile (Fig. 1.4.6b). Such an increase over large length scales suggests similar large-scale change in the organic carbon flux to Arctic basin sediments. These preliminary studies suggest that further application of redox markers together with detailed studies of organic carbon composition should provide new insights on the preservation and remineralization of organic carbon in Arctic Ocean sediments.

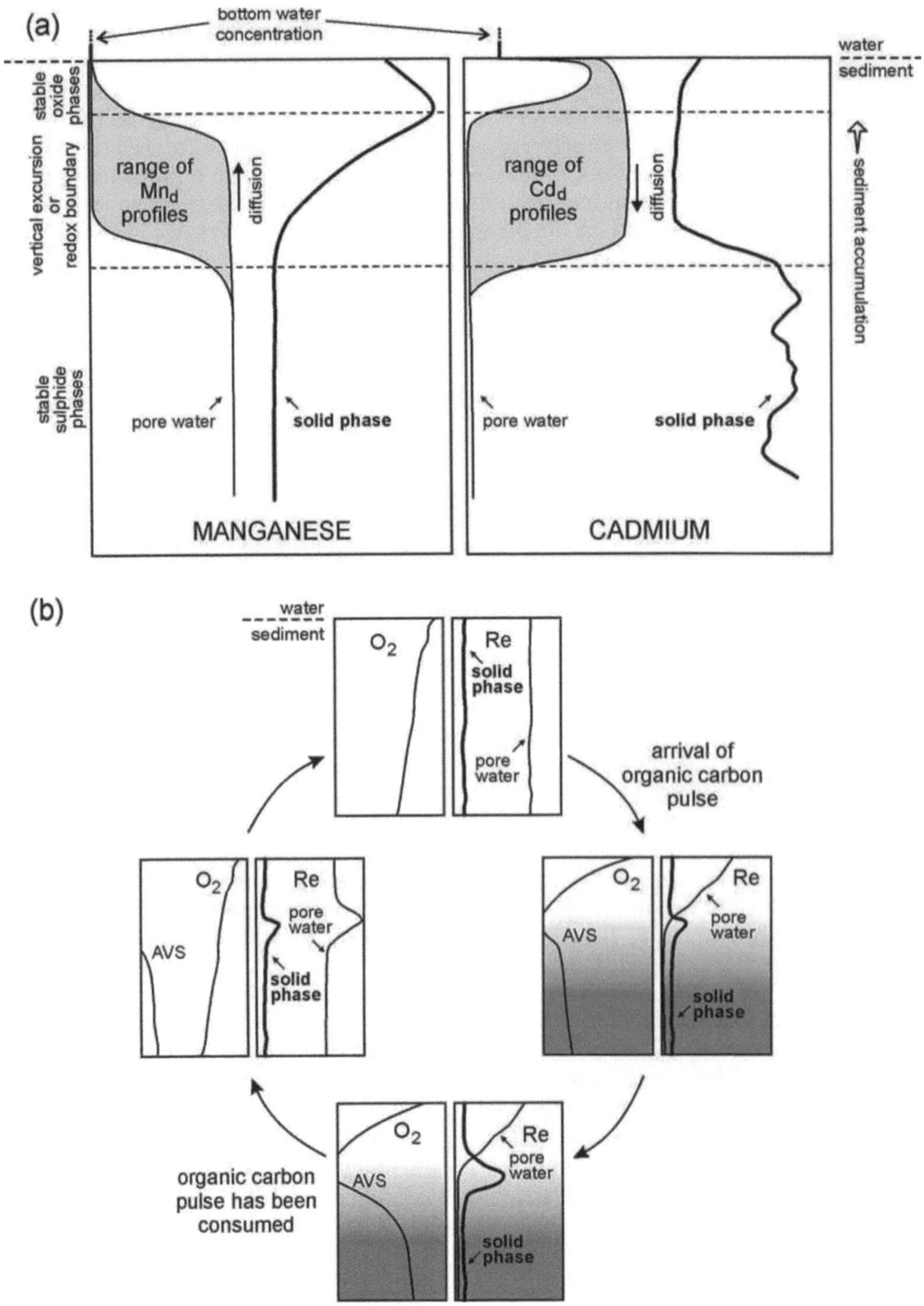

Fig. 1.4.6. A schematic diagram showing a) how manganese and cadmium distributions in sediments are impacted by pulsed inputs of organic carbon and b) how such oscillations may be recorded by rhenium. In Figure a, the shaded areas show the range of porewater profiles that results from a vertically migrating redox boundary. Solid-phase manganese reflects this process as a smooth profile with near-surface enrichments whereas solid-phase cadmium records variability in the strength of sulfide precipitation below the permanent anoxic boundary. Basically, the oscillation of the redox boundary acts to separate solid-phase manganese and cadmium in the sediment. In Figure b (top), oxic sediment conditions are poor for capturing Re. Once a pulse of carbon arrives at the sediment surface, anoxia occurs at some depth in the sediments. Re then begins to precipitate at the oxic-anoxic boundary below which acid-volatile sulfide (AVS) also accumulates. As the organic carbon becomes consumed, oxygen then re-penetrates deeply into the sediments remobilizing some (or all) of the Re. The differing time scales between oxygen diffusion and organic matter metabolism allow for discord between these various markers when the carbon arrives in pulses. (Figure modified from Gobeil 1997; Gobeil et al. 2001)

Chapter 2

Modern Terrigenous Organic Carbon Input to the Arctic Ocean

V. Rachold, H. Eicken, V.V. Gordeev, M.N. Grigoriev, H.-W. Hubberten, A.P. Lisitzin, V.P. Shevchenko, L. Schirrmeister,

2.1 General Introduction

The main objective of this chapter is to assess recent fluxes of terrigenous dissolved and particulate organic carbon into the Arctic Ocean. The most important terrigenous sources of organic matter (OM) in the ocean are (1) river and groundwater discharge, (2) coastal erosion, (3) sea-ice input and (4) aeolian material fluxes. The organic carbon fluxes of each of these pathways will be considered separately and a evaluation will be made of their roles in the total balance of OM input to the ocean.

2.2 River Input

V.V. Gordeev, V. Rachold

2.2.1. Introduction

Before addressing the question of river discharge and, in particular, its contribution to terrigenous organic carbon input to the ocean, it is necessary first to define clearly what we mean by the Arctic Ocean and its drainage basin. Prowse and Flegg (2000) showed that in the literature evaluations of river water discharges to the Arctic Ocean vary widely. These authors conclude that this variability derives simply from differences in the how the Arctic Ocean has been defined geographically. At the widest scale, Shiklomanov et al. (2000) defined "All Arctic Regions" (AAR) to include "river basins in North America that drain into the Hudson Bay, and the Yukon River (Alaska) and the Anadyr River (Russia) that drain into the north Pacific south of the Bering Strait". Greenland, the eastern edge of Norway and the Canadian Arctic Archipelago are also included in this definition of the Arctic. The total contributing area is 23.7×10^6 km² and the total river discharge amounts to 5250 km³ y⁻¹.

From our point of view it seems more reasonable to follow the definition that was adopted at the NATO Research Workshop of the Arctic Ocean Freshwater Budget – the so-called "Arctic Ocean River Basins – AORB" (Lewis 2000). There, it states that the "Arctic Ocean is defined as being bounded by: the Russian main land, a line across Bering Strait, the north coast of Alaska and the northernmost limit of the islands in the Canadian Arctic Archipelago, then across Kennedy Channel to Peary Land, across Svalbard, down to the Nordkapp of Norway and back to the Russian coast". This definition comprises only the coastal basins that drain directly towards the Arctic Ocean including the northern edges of the Canadian islands and the northern tip of the Greenland ice cap. The total contributing area for the AORB definition is 15.5×10^6 km² and the total river discharge amounts to 3299 km³ y⁻¹, with a range from 3043 to 3546 km³ y⁻¹ (Prowse and Flegg 2000).

2.2.2 River water and suspended matter

River water discharge is particularly important for the Arctic Ocean because, although the this ocean contains only 1.0% of the world ocean water, it receives 11% of the global runoff (Shiklomanov 1998). In Table 2.1 the average multiannual discharges of freshwater, suspended matter (SM), and dissolved organic carbon (DOC), particulate organic carbon (POC) and total organic carbon (TOC) into the Arctic Ocean are summarized. The largest Arctic rivers in terms of water discharge (in km³ y⁻¹) are: Yenisei – 620, Lena – 525, Ob – 404, Mackenzie – 330, Pechora – 131 and Kolyma – 122. The western rivers (Sev. Dvina, Onega, Mezen) show a maximum discharge in May (Fig. 2.1) whereas the the Siberian rivers, Ob, Yenisei, Lena, Indigirka etc., exhibit their highest discharge in June.

The concentration of SM in the Arctic rivers is low ranging from 8 mg l⁻¹ in the Yenisei to 207 mg l⁻¹ in the Indigirka. The average SM concentration is 36 mg l⁻¹ for the Russian Arctic rivers and 63 mg l⁻¹ for all Arctic rivers. The first assessments of SM

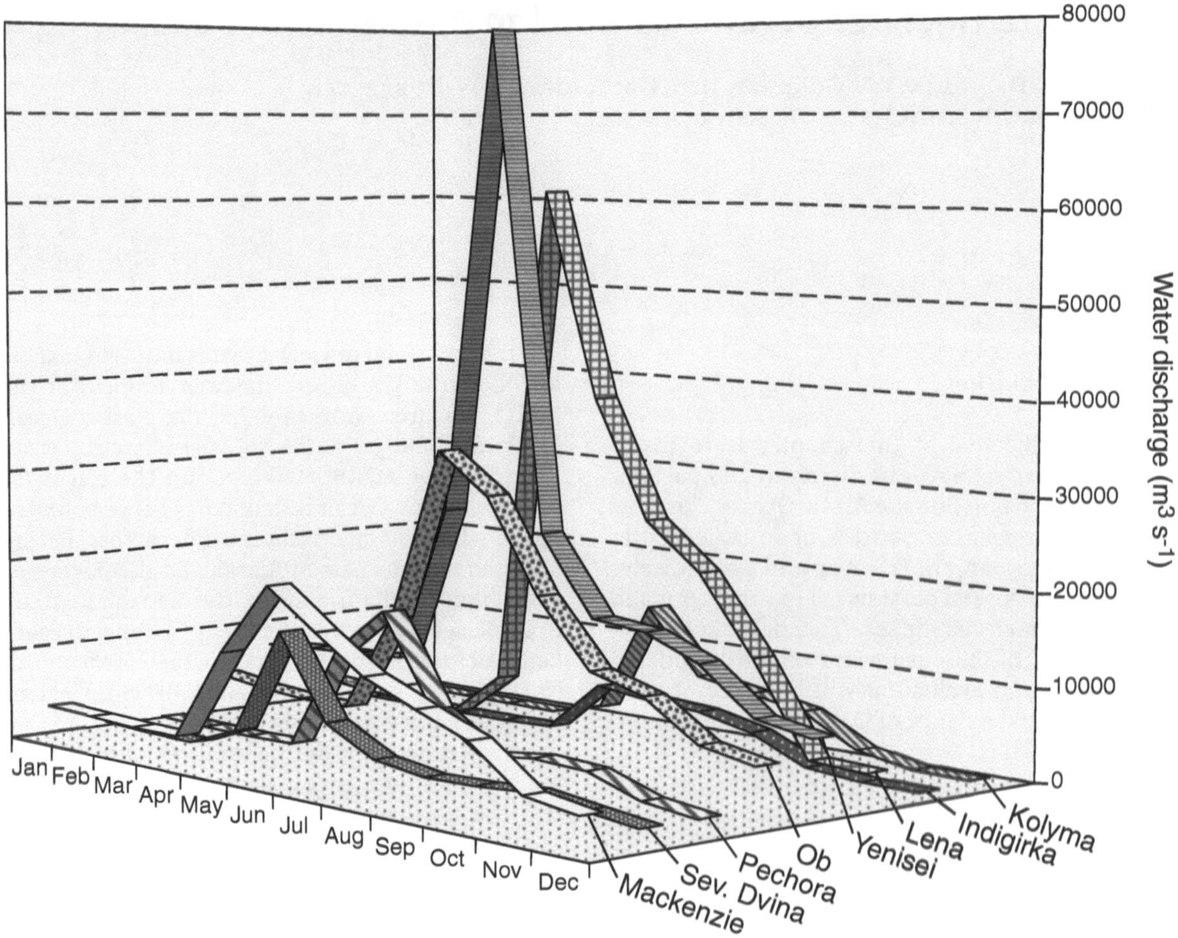

Fig. 2.1. Seasonal variations in water discharge of the largest Arctic rivers. The monthly discharge data are taken from the Regional, Electronic, Hydrographic Data Network For the Arctic Region (http://www.R-ArcticNET.sr.unh.edu/) and represent the average discharges for the years 1980–1993

fluxes of the Russian Arctic rivers appeared at the end of the 1940s (Shamov 1949; Lopatin 1952), based on the Roshydromet data from the 1920s. In North America regular measurements of SM discharge began in the 1970s (US Geological Survey and Water Survey of Environment Canada). Since that time many evaluations of SM fluxes have been published culminating in a pan-Arctic review article synthesizing the sediment fluxes of 8 large Arctic rivers to the Arctic Ocean and coastal seas (Holmes et al., 2002). The best estimates of the contemporary average annual sediment flux in these rivers were provided and are included in Table 2.1. Taking into account corrections made by Holmes et al. (2002), the total SM flux from the Eurasian territory to the Arctic Ocean was reduced from 115×10^6 t y^{-1} (Gordeev et al. 1996) to 102.2×10^6 t y^{-1} (Table 2.1). The Mackenzie River, which is the fourth largest Arctic river in terms of water discharge (330 km^3 y^{-1}; Macdonald et al. 1998) is the largest river in terms of sediment discharge (124×10^6 t y^{-1} of sediment; Holmes et al. 2002). Thus, using the Arctic Ocean River Basins definition (see above), the Mackenzie River is absolutely dominating in the Canadian Arctic, providing 90% of the water and 99% of the SM discharge for that region (Table 2.1).

It is interesting to note that the East Siberian rivers in comparison to the western Russian Arctic rivers (west of the Lena River, which is located at the boundary between the Eurasian and North American tectonic plates) are characterized by lower runoff, higher SM concentration and by significantly lower water mineralization, organic matter and nutrient concentrations. The rivers eastward of the Lena River resemble more closely the North American rivers than the western Russian Arctic rivers (Gordeev et al. 1996).

Table 2.1. Average multiannual riverine water and suspended matter discharges and fluxes of DOC, POC and TOC to the Arctic Ocean. The majority of data in this table is taken from Gordeev et al. 1996 and Gordeev 2000. Note that sometimes POC + DOC is not equal to TOC because of different sources of data. For the content of TOC, "total" refers to "average concentrations". This average calculated for the large rivers was adopted to "other areas" where no concentration data are available. [1] Holmes et al. (2002); [2] Smirnov et al. 1988; [3] Romankevich and Vetrov 2001; [4] Lobbes et al. 2000; [5] Artemiev 1997; [6] Nesterova 1960; [7] Maltseva et al. 1987; [8] Magritsky 2001; [9] Rachold and Hubberten 1999; [10] Cauwet and Sidorov 1996; [11] Mulholland and Watts 1982; [12] AMAP Report 1998; [13] Macdonald et al. 1998; [14] Spitzy and Leenheer 1991; [15] Degens et al. 1991; [16] based on [6]; [17] based on [4]

River	Area	Water disch.	Tot. susp. matter		Content			Flux			TOC	POC in dry SPM	DOC×100 / DOC+POC
					DOC	POC	TOC	DOC	POC	TOC			
	$10^3 km^2$	$km^3 y^{-1}$	$g m^{-3}$	$10^6 t y^{-1}$		$mg l^{-1}$			$10^6 t y^{-1}$		$t km^{-2} y^{-1}$	wt.%	
White and Barents Seas													
Onega	57	15.9	18	0.30	–	–	20.7[2]	–	–	0.33	5.8	–	–
N. Dvina	357	110	37	4.1[1]	11.6	2.6	15.3[3]	1.28	0.28	1.68	4.7	6.8	76
Mezen	78	27.2	33	0.6[1]	12.1[4]	1.8[4]	7.0[2]	0.25[4]	0.04[4]	0.19	2.4	1.1	–
Pechora	324	131	72	9.4[1]	12.7	0.3	13.0[5]	1.66	0.04	1.70	5.2	0.43	97
Other area	570	179	19	3.5	–	–	13.7	–	–	2.45	4.3	–	–
Total	1386	463	39	17.9[1]	–	–	13.7	4.18[3]	0.45[3]	6.35	4.4	–	–
Kara Sea													
Ob	2545	404	37	15.5[1]	9.1[6]	0.9[6]	7.1[2]	3.68[16]	0.36[16]	2.87	1.1	–	91
Nadym	64	18	22	0.4	–	–	5.0[2]	–	–	0.09	1.4	–	–
Pur	112	34.3	18	0.7[1]	–	–	6.7[1]	–	–	0.23	2.0	–	–
Taz	150	44.3	21	0.7[1]	–	–	–	–	–	–	–	–	–
Yenisei	2594	620	8	4.7[1]	8.5[4]	0.3[4]	7.4[7]	4.86[4]	0.17[4]	4.59	1.8	–	96
Pyasina	182	86	39	3.4	–	–	–	–	–	–	–	–	–
Other area	867	275	20	5.5	–	–	7.2	–	–	1.98	–	–	–
Total	6589	1480	21	30.9[1]	–	–	7.2	–	–	10.6	1.6	–	–
Laptev Sea													
Khatanga	364	85.3	20	1.7	–	–	6.3	–	0.04[9]	0.54	1.5	2.3	–
Anabar	100	17.3	24	0.4	–	–	5.1	–	–	0.09	0.9	–	93
Olenjok	218	32.8	38	1.1[1]	10.2[4]	0.83[4]	7.2	0.32[4]	0.026[4]	0.24	1.1	–	92
Lena	2448	523	39	20.7[1]	6.6[10]	1.1[10]	7.7	3.6[9]	1.2[9]	4.8[9]	1.9	5.8	86
Omoloy	39	7	18	0.04[1]	2.8[4]	0.3[4]	–	0.003[4]	0.001[4]	–	–	5.7	75
Yana	225	31.9	130	4.0[1]	2.8[4]	1.6[4]	6.7	0.085[4]	0.05[4]	0.21	0.9	1.5	64
Other area	197	40.3	16	0.65	–	–	9.2	–	–	0.37	1.9	–	–
Total	3597	738	39	28.6[1]	–	–	9.2	–	–	6.8	1.9	–	–

Table 2.1 (continued)

River	Area	Water disch.	Tot. susp. matter		Content			Flux			TOC	POC in dry SPM	DOC×100 / DOC+POC
					DOC	POC	TOC	DOC	POC	TOC			
	$10^3 km^2$	$km^3 y^{-1}$	$g m^{-3}$	$10^6 t y^{-1}$		$mg l^{-1}$			$10^6 t y^{-1}$		$t km^{-2} y^{-1}$	wt.%	
East Siberian Sea													
Indigirka	360	54.2	207	11.1 [1]	4.8 [4]	3.5 [4]	7.7	0.24 [4]	0.17 [4]	0.42	1.2	1.6	58
Alazeya	68	8.8	80	0.1 [1]	–	–	–	–	–	–	–	–	–
Kolyma	647	122	83	10.1 [1]	4.6 [4]	3.1 [4]	8.1	0.46 [4]	0.31 [4]	0.99	1.5	3.0	60
Other area	252	48.2	80	3.85	–	–	8.0	–	–	0.38	1.5	–	–
Total	1327	233	110	25.15 [1]	–	–	8.0	–	–	1.86	1.4	–	–
Chukchi Sea (excluding Alaska)													
Amguema	29.6	9.2	6	0.05	–	–	6.7	–	–	0.06	2.0	–	–
Other area	64.6	11.2	58	0.65	–	–	6.7	–	–	0.07	1.1	–	–
Total	94.2	20.4	34	0.7	–	–	6.7	–	–	0.13	1.4	–	–
Total Eurasian Arctic													
	12987	2932	36	102.2 [1]	6.6 [4]	1.3 [4]	8.8	19.4 [17]	3.81 [17]	25.7	2.0	–	–
Chukchi Sea (Alaska) and Beaufort Sea													
Kobuk	24.7 [11]	–	–	–	–	–	–	–	–	0.04 [12]	–	–	–
Kuparuk	8.11 [11]	–	–	–	–	–	–	–	–	0.014 [12]	–	–	–
Mackenzie	1787 [12]	330 [13]	168	124 [1]	5.2 [14]	7.2 [15]	12.5 [15]	1.3 [13]	2.1 [13]	4.1	2.1	3.3	62
Other area	726	37	–	1.1	–	–	–	0.19	0.055	0.24	–	–	–
Total Canadian Arctic													
	2513	367	–	125.1 [1]	5.1	5.8	11.6	1.9	2.15	4.3	1.7	2.1	48
Total Arctic													
	15500	3299	63	227.3 [1]	–	–	9.1	–	–	30.0	1.9	–	–

2.2.3 Fluxes of organic carbon

The estimation of riverine DOC and POC fluxes into the Arctic Ocean has long attracted the attention of many investigators. In the Russian Arctic rivers systematic measurements of organic matter (OM) were started in the mid-1930s (Maltseva et al. 1978). Most of the data, which were collected by the Roshydromet System, were obtained on unfiltered water samples using permanganate oxidation in an acidic medium and by dichromate oxidation in an acidic medium followed by recalculation into TOC concentrations based on correction coefficients that were specific for the various geographic zones and seasons (Sawyer and Semenov 1971; Semenov 1977). Only a few hundred determinations of DOC and POC were made after filtration of river water through GF/F and GF/C filters by means of wet combustion with colorimetric recording of the liberated CO_2, or by means of photochemical oxidation with infrared spectroscopic recording of CO_2 (Romankevich and Artemiev 1985; Artemiev 1997).

The first assessments of TOC discharge for the Russian Arctic rivers were obtained by Skopintsev and Krylova (1955) and Alekin and Brazhnikova (1964). A summary of the Roshydromet data for the period from 1936 to 1965 and later up to 1980 (Table 2.2) was provided by Maltseva et al. (1978, 1987), Maltseva (1980) and Smirnov et al. (1978, 1988). Two decades later new estimates of OM discharge from the Russian territory to the Arctic Ocean appeared in Artemiev (1997), Gordeev et al. (1996), Gordeev and Tsirkunov (1998) and Gordeev (2000). The most recent work by Romankevich and Vetrov (2001) is an important summary of all available information on the carbon cycle in the Russian Arctic Seas including the riverine discharge of OM.

Several articles published in English are available on the riverine OM input to the World Ocean, in which Arctic rivers are considered to some extent (Duce and Duursma 1977; Schlesinger and Melack 1981; Meybeck 1982, 1993; Michaelis et al. 1986; Degens et al. 1991). The transport of organic carbon to the oceans by the North American rivers was summarized in the work of Mulholland and Watts (1982). During the last few years several new studies on the topics of discharge and origin of riverine OM and its distribution in the Arctic Ocean have been presented (Ittekkot 1988; Telang et al. 1991; Cauwet and Sidorov 1996; Macdonald et al. 1998; Lara et al. 1998; Opsahl et al. 1999; Rachold and Hubberten 1999; Kattner et al. 1999; Lobbes et al. 2000).

Based on a review of the existing information, Table 2.1 presents our best estimates of DOC, POC and TOC fluxes from rivers of the Eurasian and Canadian Arctic. It has to be noted that in this section we evaluate only the riverine OM discharge to the Arctic Ocean or so-called gross river flux, i.e. the amounts of substances transported by the river to the land/sea boundary. In the river/sea mixing zone riverine material, including OM, is subject to intensive change in quality and quantity. However, it is not the objective of this section to consider the behavior of riverine OM at the river/sea boundary which is discussed in detail in Chapter 4.

At present, reliable TOC data are available for most rivers whereas only major rivers have been studied separately for DOC and POC concentrations. Table 2.1 shows the existing data and the fluxes calculated on the basis of these analytical data where, if possible, the DOC and POC components are separated. Unfortunately, the existing database is not sufficient to distinguish between DOC and POC for each shelf area and in some cases only TOC fluxes can be reported. POC fluxes have been published by Romankevich et al. (2000). However, it has to be noted that their flux data are not based on analytical POC values but, rather, were quantified based on the assumption that the organic carbon concentration comprises 2% of the total suspended matter for each river.

We estimate that total TOC discharge to the Eurasian Arctic Ocean is 25.7×10^6 tC y^{-1}. The implied average TOC concentration of 9.1 mg l^{-1} in the Arctic rivers is similar to the world-wide average of 9.9 mg l^{-1} (Meybeck 1993), and the TOC flux to the

Table 2.2. TOC fluxes of the former USSR territory (1936–1980) (Maltseva et al. 1987) (* OM was recalculated to TOC as 2:1)

Sea	Area 10^3 km^2	Water discharge km^3 y^{-1}	TOC flux 10^6 tC y^{-1}	Specific TOC flux t km^{-2} y^{-1}
White and Barents	1250	418	5.52	4.4
Kara	6200	1337	12.52	2.0
Laptev	3670	799	8.52	2.3
East Siberian	1390	238	1.70	1.2
Chukchi	102	28.6	0.90	1.0
Total	12612	2808	28.36	2.2

Arctic accounts for ca. 8.1% of the global TOC flux (370×10^6 tC y^{-1}, Meybeck 1993). A recent estimate of the TOC flux to the seas of the Russian Arctic (Romankevich and Vetrov 2001) is 23.5×10^6 tC y^{-1}, which is 10% lower than our value of 25.7×10^6 tC y^{-1}. However, both studies report practically the same TOC concentration in river water of the Russian Arctic (8.6 and 8.8 mg l^{-1}, respectively) and the difference between the two evaluations is mainly explained by the difference in total water discharges: 2730 km^3 y^{-1} estimated by Romankevich and Vetrov (2001) and 2932 km^3 y^{-1} used in this study.

What are the main sources of OM in the rivers? Riverine OM is generally derived from allochthonous sources (eroded soil and plant material) with a much smaller component from autochtonous sources (freshwater aquatic production) (Meybeck 1982; Ittekkot 1988; Lobbes et al. 2000). C/N ratios and stable carbon isotope ratios are useful in distinguishing between autochthonous and allochthonous sources, and studies in the Lena River indicate that POC is formed mainly from allochthonous detrital organic material (Lara et al. 1998; Rachold and Hubberten 1999; Lobbes et al. 2000), which is in general agreement with the small phytoplankton biomass in the Lena River (Sorokin and Sorokin 1996) and the relatively old average radiocarbon age of OM of the Laptev Sea bottom sediments (Kuptsov and Lisitzin 1996). Figure 2.2 shows the relationship between $\delta^{13}C$ values of POC and reciprocal POC concentrations in the water volume (POC [mg l^{-1}] = SPM [mg l^{-1}] · TOC [%] · 10^{-2}) of the Lena, Yana, and Khatanga basins. The linear correlations suggest two end-member mixing with the two hypothetical end-members being detrital and autochthonous organic matter (Rachold and Hubberten 1999). In boreal forests like the Arctic taiga about 84% of the terrestrial carbon is stored in soils and 16% in living vegetation (Dixon et al. 1994). It can be expected that the thawing of permafrost due to global warming will elevate the amount of soil-derived terrigenous DOM discharged into the Arctic Ocean (Opsahl et al. 1999).

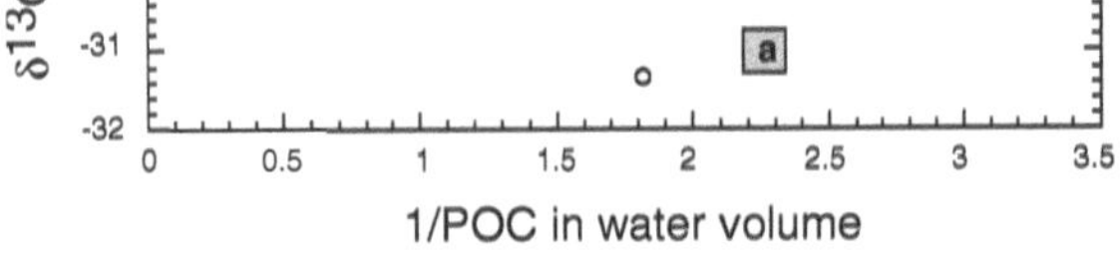

Fig. 2.2. $\delta^{13}C$ values of POC vs. 1/POC concentrations in the water volume (POC [mg l^{-1}] = SPM [mg l^{-1}] · TOC [%] · 10^{-2}) of the Lena, Yana, and Khatanga basins. The linear correlations suggest two-endmember mixing. Hypothetical detrital [d] and autochthonous [a] endmembers are presented (Rachold and Hubberten 1999)

It has to be noted that POC (and TOC) fluxes (Table. 2.1) refer to the suspended matter only and do not include the river bedload. The transport of bedload and the relationship between the amount of SM and bedload in the Arctic rivers is still poorly understood. Lopatin (1952) noted that field measurements of bedload by "device-catchers", especially in large rivers during freshet, did not provide adequate results for volume and regime of bedload transport. In agreement with observations in the Mackenzie River (Carson et al. 1998), in the first approximation it is assumed that in the plain rivers the bedload does not exceed 5–10% of the SM flux. Taking into account that the bedload consists mainly of sand with an organic carbon content of 0.5–1.0%, a rough evaluation of the bedload carbon input to the Arctic Ocean is $0.05\text{-}0.2 \times 10^6$ tC y^{-1} (average: 0.13×10^6 tC y^{-1}). This value is negligible compared with the total TOC flux.

There are very few data available on seasonal variations of DOC, POC and TOC concentrations and fluxes in Arctic rivers. A review of Roshydromet data on seasonal variation of TOC discharge into the Arctic Seas, which was published by Maltseva et al. (1987), indicates that 58 to 78% of the total TOC is discharged during freshet (Table 2.3) and that the portion of TOC transported to the Arctic seas during this period increases from west to east. Let us consider a more recent example of the seasonal variations in TOC concentrations and fluxes in the lower reaches of the Lena River (Fig. 2.3; data taken from Cauwet and Sidorov 1996). Minimum TOC concentrations are observed in winter (3.1–4.8 mgC l^{-1}) while maximum concentrations occur during freshet in June–July (9.6–

Table 2.3. TOC flux in different seasons, in % of annual volume (Maltseva et al. 1987)

Sea	Spring flood	Summer-autumn	Winter
White and Barents	58	34	8
Kara	63	26	11
Laptev	75	22	3
East Siberian	73	24	3
Chukchi	78	21	1

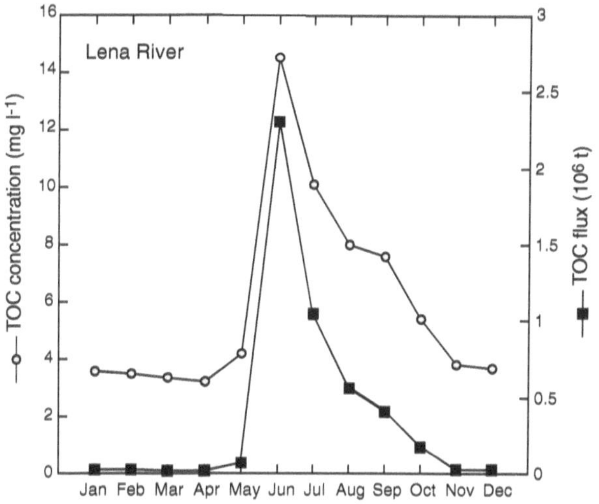

Fig. 2.3. Seasonal variations of TOC concentrations (data from Cauwet and Sidorov 1996) and TOC fluxes in the lower reaches of the Lena River. The TOC fluxes are based on the TOC concentrations given by Cauwet and Sidorov (1996) and the water discharge data for the period of 1980–1993 taken from the Regional, Electronic, Hydrographic Data Network For the Arctic Region (http://www.R-ArcticNET.sr.unh.edu/)

14.4 mgC l^{-1}). This pattern is even more pronounced for the TOC fluxes (TOC concentration multiplied by water discharge), which reveals that during winter the TOC flux of the Lena River is almost negligible (Fig. 2.3). Consequently, the average seasonal TOC concentrations and fluxes to the Laptev Sea (Table 2.4) show that freshet accounts for more than 50% of the annual TOC discharge of the Lena River while the total winter TOC export does not exceed 4% of the annual value. A similar pattern has recently been observed for the DOC transport of the Yenisei and Ob Rivers (Fig. 2.4; Köhler et al. 2003). The comparison with the Lena River (Fig. 2.3) clearly shows that the proportion of OM transported during freshet increases from west to east (Ob < Yenisei < Lena) as previously suggested by Matseva et al. 1987 (see above). Note that for the Lena TOC is shown whereas for the Ob and Yenisei DOC are displayed.

Groundwater discharge

The flux of dissolved OM with groundwater constitutes one of the pathways of terrigenous organics to the Arctic Ocean. A review of the chemical composition of groundwaters of the upper hypergenesis zone of various areas of the Earth has shown

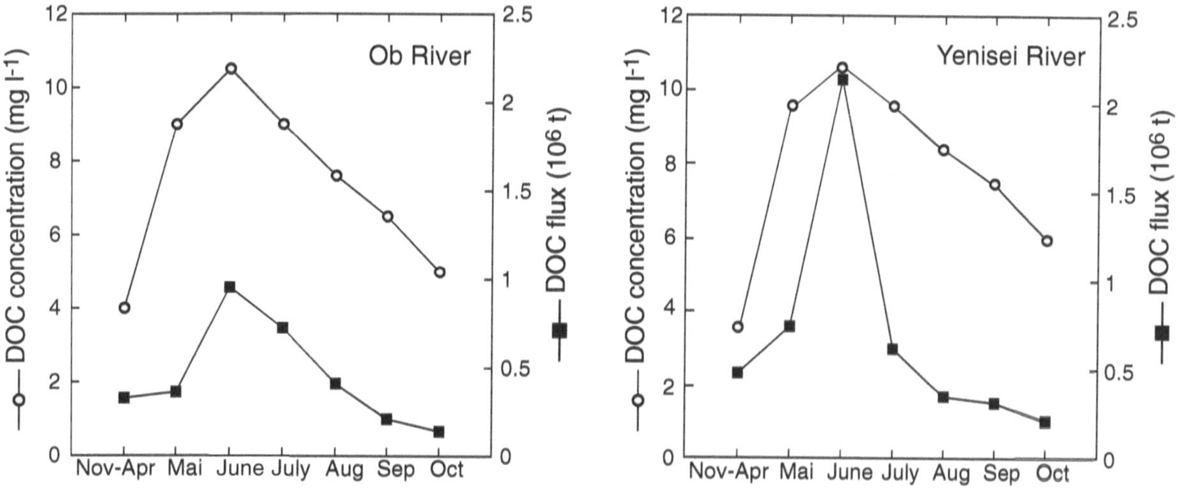

Fig. 2.4. Seasonal variations of DOC concentrations and fluxes of the Ob and Yenisei Rivers (data from Köhler et al. 2003)

Table 2.4. Average concentrations and fluxes of TOC to the Laptev Sea (Cauwet and Sidorov 1996). 1- flood; 2- summer and autumn; 3- winter; 4- average (sum for fluxes)

	Head of the delta				Mouth of the delta			
	1	2	3	4	1	2	3	4
Concentration, mgC l^{-1}	12.6	7.4	3.6	10.2	11.8	8.4	3.7	10.0
Flux, 10^3 tC yr^{-1}	3930	1300	130	5360	3680	1480	140	530

Table 2.5. River and groundwater discharges to the Arctic Seas (in $km^3 y^{-1}$) according to Gordeev et al. (1999)

Sea basin River	River water discharge (RWD)	Groundwater discharge (GWD)	GWD/RWD %
Barents and White Sea			
Northern Dvina	110	27.8	25.3
Onega	15.9	3.5	22.0
Mezen	27.2	3.74	13.75
Pechora	131	5.0	38.5
Other area	179	26.8	15.0
Total	463	68	14.7
Kara Sea			
Ob	429	76	17.7
Pur	34.3	1.7	5.0
Taz	44.3	1.6	3.6
Yenisei	620	39	6.3
Other area	333	33	9.9
Total	1478	152	10.3
Laptev Sea			
Anabar	17.3	0.91	5.3
Olenyok	36	3.1	8.6
Lena	525	37	7.0
Yana	34	3.3	9.7
Other area	133	12.7	9.5
Total	745	57	7.65
East Siberian Sea			
Indigirka	61	4.6	7.5
Kolyma	132	8.5	6.4
Other area	57	4.4	7.7
Total	250	16.9	6.8
Total Eurasian Arctic (excluding Chukchi Sea)	2936	294	10.0

that the DOC concentration averages ca. 5.9 mg l^{-1} (Shvets 1973). From the available data on DOC in river waters of the Arctic for the period of winter low flow, Romankevich and Vetrov (2001) accepted a figure of 6.5 mgC l^{-1} as an average concentration.

Gordeev et al. (1999) sought a relationship between the river discharge and groundwater discharge and their associated transport of dissolved inorganic nutrients to the Arctic Seas. The groundwater discharge for each watershed was evaluated by taking the groundwater discharge module from the map (Kudelin 1975). The small area between the hydrological station located closest to the shore line and the shore line itself were not included due to the absence of data. The calculations show that the ratio between river and underground water discharge averages 10:1. For selected rivers and seas the following proportions of river vs. underground water discharges were quantified (in $km^3 y^{-1}$) (Table 2.5): Sev. Dvina – 110:27.8, Pechora – 130:5, White and Barents seas – 463:68 (14%); Ob – 429:76, Yenisei – 620:39, Kara Sea – 1478:152 (10%); Lena – 525:37, Laptev Sea – 745:57 (7.6%), Indigirka – 61: 4.6, Kolyma – 132:8.5, East Siberian Sea – 250:16.9 (6.7%), all Eurasian Arctic (without Chukchi Sea) – 2936:294 (10%). The numbers in brackets correspond to the percentages of groundwater discharge relative to river discharge. During winter, river discharge is supported mainly by underground water input to the main river stream (Gordeev and Sidorov 1993). For this reason, DOC and nutrient concentrations in groundwaters were assumed to be equal to their concentrations in river water during very low runoff in winter (Gordeev et al. 1999). Based upon these assumptions the DOC underground flux to the Eurasian part of the Arctic Ocean is evaluated as 1.9×10^6 tC y^{-1}. Considering the proportion between the Eurasian and Canadian

Arctic ($12987:2513\times10^3$ km^2) the groundwater discharge from the Canadian Arctic amounts to ca. 48 km^3 yr^{-1} corresponding to a TOC flux of 0.37×10^6 tC y^{-1}. This very rough estimate yields a total Arctic groundwater DOC flux of about 2.3×10^6 tC y^{-1}. The POC flux from groundwater is assumed to be negligible.

2.3 Organic Carbon Input to the Arctic Seas Through Coastal Erosion

M.N. Grigoriev, V. Rachold,
H.-W. Hubberten, L. Schirrmeister

2.3.1 Introduction

Shore dynamics directly reflecting complicated land-ocean interactions play an important role in the balance of sediments, organic carbon and nutrients in the Arctic basin. Nevertheless, the contribution of coastal erosion to the material budget of the Arctic Seas has often been underestimated. In recent years, however, several studies have underlined the importance of coastal erosion for the sediment budget of the Arctic Seas. Reimnitz et al. (1988a) made calculations for 344 km of Alaska coast in the Colville River area, finding that coastal erosion supplied 7 times more sediments to the Alaskan Beaufort Sea than did rivers. Are (1999) suggested that the amount of sediment supplied to the Laptev Sea by rivers and shores are at least of the same order of magnitude but that the coastal erosion input is probably much larger than the input of the rivers. This finding was supported by Rachold et al. (2000), who concluded that the sediment flux to the Laptev Sea through coastal erosion is two times larger than the river input. In the Canadian Beaufort Sea on the other hand, the Mackenzie River input is the dominant source of sediments and coastal erosion is much less important (Macdonald et al. 1998), which indicates that pronounced regional differences in the ratio between riverine and coastal erosion sediment input have to be considered.

Numerous studies addressing coastal erosion in various Arctic Seas have been published in the literature (Table 2.6). However, most of these papers deal only with coastal retreat rates and sediment input and publications considering the organic carbon flux are limited to Macdonald et al. (1998), Yunker et al. (1991, 1993) for the Canadian Beaufort Sea; Stein and Fahl (2000) for the Laptev Sea; Semiletov (1999a, 1999b, 2000) for the Laptev, East Siberian and Chukchi Seas; and Lisitzin (1990), Ronov (1993) in general. Recently a review of the organic carbon fluxes to the Russian Arctic Seas has been presented in Romankevich and Vetrov (2001).

In the following we present a quantitative assessment of the organic carbon input to the Arctic Seas through coastal erosion. It must be cautioned that these are the best available estimates of the contribution of coastal erosion to sediment and organic carbon input and may contain considerable error. The evaluation is based upon a combination of data for coastal erosion sediment input and organic carbon concentrations of the coastal sections. Emphasis will be laid on the Laptev Sea and East Siberian Seas, where our own field studies have been performed from 1998 to 2000 (Rachold 1999, 2000; Grigoriev and Kunitsky, 2000; Rachold and Grigoriev 2001). Based on published information listed above, quantification will be extended to cover all Arctic Seas.

Table 2.6. Published information on coastal erosion

Region	References
Canadian Beaufort Sea	Mackay (1963), Harper et al. (1985), Harper (1990), Hill et al. (1986, 1991), Dallimore et al. (1996), Macdonald et al. (1998), Wolfe et al. (1998)
Alaskan Beaufort Sea	Hume et al. (1972), Reimnitz and Barnes (1982), Naidu et al. (1984), Reimnitz et al. (1988a)
White Sea and Barents Sea	Zenkovich (1962), Suzdal'sky (1974), Medvedev (1972), Velikotsky (1998)
Kara Sea	Popov et al. (1988), Vasiliev (1995), Sovershaev (1996), Koreisha et al. (1997)
Laptev Sea	Toll (1897), Gakkel (1957, 1958), Grigoriev (1966), Kluyev (1970), Are (1980, 1985, 1987, 1999), Grigoriev (1993, 1996), Grigoriev and Kunitsky (2000), Rachold et al. (2000, 2002)
East Siberian Sea	Pavlidis et al. (1988), Grigoriev and Kunitsky (2000), Razumov (2000)
Chukchi Sea	Shuisky and Ogorodnikov (1981), Shuisky (1983, 1986), Pavlidis et al. (1988)
Russian Arctic (in general)	Zenkovich (1962), Kaplin et al. (1971), Arkhikov et al. (1982), Budyko and Izrael (1987), Lisitzin (1990), Kaplin and Selivanov (1999), Lopatin (1999)

It should be noted that our assessment is based on total organic carbon (TOC) concentrations of the bulk, ice-rich, coastal permafrost sediments and does not distinguish between dissolved and particulate organic carbon species. Furthermore, we presently have no information about the fate of the organic matter derived from coastal erosion once it enters the shelf, i.e. whether the organic carbon will be transported in particulate form or whether it will be partly transferred into dissolved species. Therefore, it is possible to provide information only on the TOC fluxes, which correspond to the gross coastal organic carbon input, i. e. the amount of organic carbon entering the sea as a consequence of coastal erosion.

2.3.2 Data sources and methods

Generally, the methodology to quantify TOC input through coastal erosions involves the following steps:

1. The quantification of coastal retreat rates of representative key sections, which can be done by long term field measurements, the comparison of different-time topographic maps, satellite images and aerial photographs or a combination of both. A detailed description of this method is given in Rachold et al. (2000).
2. The determination of the coastal morphology (mainly cliff height and slope) to quantify the volume of material supplied (m^3 y^{-1}) based on linear coastal retreat rates (m y^{-1}).
3. The analysis of the composition of the coastal sediments, which includes the determination of ice content and specific density of the sediments. These data are essential to quantify the sediment mass input through coastal erosion.
4. The determination of the TOC concentration of the coastal sediments at the key sites to estimate the TOC flux.
5. Finally, the data obtained for key sections must be extrapolated to the entire coastline. This can be done by segmenting the coast into homogeneous segments depending on the locations of substantial changes in sediment texture, ice content, or TOC concentration.

Laptev and East Siberian Sea

Recently, several key sections of the Laptev Sea coast have been evaluated in detail for coastal retreat rates and composition of the coastal sediments during the Russian-German expeditions within the Laptev Sea 2000 project from 1998 to 2000 (Rachold 1999, 2000; Rachold and Grigoriev 2001). Using methodology described above, the coast was divided into segments and the parameters needed to quantify the sediment input through coastal erosion were assigned to each segment as was average TOC concentration for the coastal sediments.

The Laptev Sea coastline is characterized by the frequent occurrence of ice-rich deposits with about 30% of the coast consisting of the Ice Complex with an ice content of up to 80%. Our studies clearly show that erosion of the Ice Complex is of major importance for the sediment budget of the Laptev Sea, contributing about 76% of the coastal sediment flux. A similar situation can be observed in the East Siberian Sea. At many coastal sections the Ice Complex is intensively eroded because wave-cut notches are formed along the bases during storms, which subsequently results in block disintegration of the coast. The still-frozen blocks are washed away by seawater and the dominant erosion rates are in the range of 2–6 m y^{-1}. Due to thermal denudation of the upper and middle parts of the ice-rich cliffs, coastal retreat rates can remain high even during temporary attenuation of the cliff foot abrasion. We estimate the average retreat rate of coastlines consisting of Ice Complex and Lake-Thermokarst deposits to be 2.5 m y^{-1}.

Table 2.7 shows average TOC concentrations of thermokarst sediment and the Ice Complex of the Laptev Sea region, which were determined by CNS-elemental analyzer after removal of carbonate carbon. As seen from the table, the TOC values are highly variable ranging from <1 to >30% and we will use 4% as an average TOC content for the Ice Complex and thermokarst sediments of the Laptev Sea coast in the following. Other types of coasts must be considered such as ice-poor coasts formed from Pleistocene-Holocene sediments and rocky or other non-icy coasts. The ice-poor and non-icy coasts are characterized by significantly lower TOC concentrations averaging 1% and 0.3%, respectively.

Table 2.8 summarizes the characteristics of the main types of coasts in the Laptev Sea region.

Based on the length of each coastal type and the parameters given in the table, we quantify the TOC flux to the Laptev Sea through coastal erosion as 1.8×10^6 tC y^{-1}. The major portion enters the eastern Laptev Sea (0.9×10^6 tC y^{-1}), whereas the TOC flux to the central and western Laptev Sea amount to 0.56×10^6 tC y^{-1} and 0.34×10^6 tC yr^{-1}, respectively.

Based on the studies of Grigoriev and Kunitsky (2000) and Razumov (2000), we are able to extend these calculations to the East Siberian Sea (Table 2.9) where ca. 2.2×10^6 tC y^{-1} are exported to the shelf by coastal erosion.

Table 2.7. TOC concentrations of coastal sediments of the Laptev Sea region (in %)

Location	TOC max	TOC min	TOC stn. dev.	TOC avg.
Big Lyakhovsky Island				
Ice Complex (right section)	17.5	0.9	3.3	**3.2**
Ice Complex (left section)	38.2	3.1	8.4	**11.9**
Alas deposits (right section)	3.1	1.2	0.5	**2.0**
Alas deposits (left section)	7.1	1.5	2.1	**3.8**
Bykovsky Peninsula				
Alas deposits	16.5	0.8	4.3	**6.3**
Ice Complex	21.5	1.3	3.4	**4.1**
Olenyok Channel (Lena Delta), Nagym Section				
Ice Complex	21.5	0.2	6.4	**4.3**
Olenyok Channel (Lena Delta), Buor-Khaya Section				
Ice Complex	21.6	1.1	6.8	**9.2**
Average ice-rich coast				**4.0**

Table 2.8. Coastal types of the Laptev Sea and their characteristics in regard to TOC flux associated with coastal erosion

	Ice Complex and thermokarst deposits	Ice-poor Pleistocene-Holocene coasts	Rocky and other types of non-icy coasts
Total length of the coast (km)	2400	1600	3200
Average retreat rate (m y^{-1})	2.5	1	0.0001–0.1 (avg. 0.05)
Average ice content (%)	50	10–30 (avg. 20)	0–4 (avg. 2)
Average cliff height (m)	10	5	10–30 (avg. 20)
Average TOC concentration (%)	4	1	0–1 (avg. 0.3)
Coastal sediment flux (10^6 t y^{-1})	44.4	9.5	4.5
Coastal TOC flux (10^6 t y^{-1})	1.78	0.1	0.015

Table 2.9. Coastal types of the East Siberian Sea and their characteristics in regard of TOC flux associated with coastal erosion

	Ice Complex and thermokarst deposits	Ice-poor Pleistocene-Holocene coasts	Rocky and other types of non-icy coasts
Total length of the coast (km)	2400	1600	1900
Average retreat rate (m y^{-1})	3	1	0.0001–0.1 (avg. 0.05)
Average ice content (%)	50	10–30 (avg. 20)	0–4 (avg. 2)
Average cliff height (m)	10	5	10–30 (avg. 20)
Average TOC concentration (%)	4	1	0–1 (avg. 0.3)
Coastal sediment flux (10^6 t y^{-1})	52.5	10.9	3.1
Coastal TOC flux (10^6 t y^{-1})	2.1	0.1	0.01

White, Barents, Kara and Chukchi Seas

Unfortunately, the other Russian Arctic Seas are have not been studied in such detail. Publications on coastal retreat erosion, where they exist, do not follow the methodology given above. The best estimates are summarized by Romankevich and Vetrov (2001), who used all available sources to evaluate the TOC flux. Although some of their estimates, especially those of the White, Barents and Kara Seas, seem too high, in the following we will use their TOC input estimates because more accurate data are not available at present.

Note that the Chukchi Sea data given by Romankevich and Vetrov (2001), which are in fact taken from Lisitzin (1990), refer to the Asian Sector only. To roughly quantify the sediment and TOC flux to the entire Chukchi Sea we multiply their value by a factor of 2.

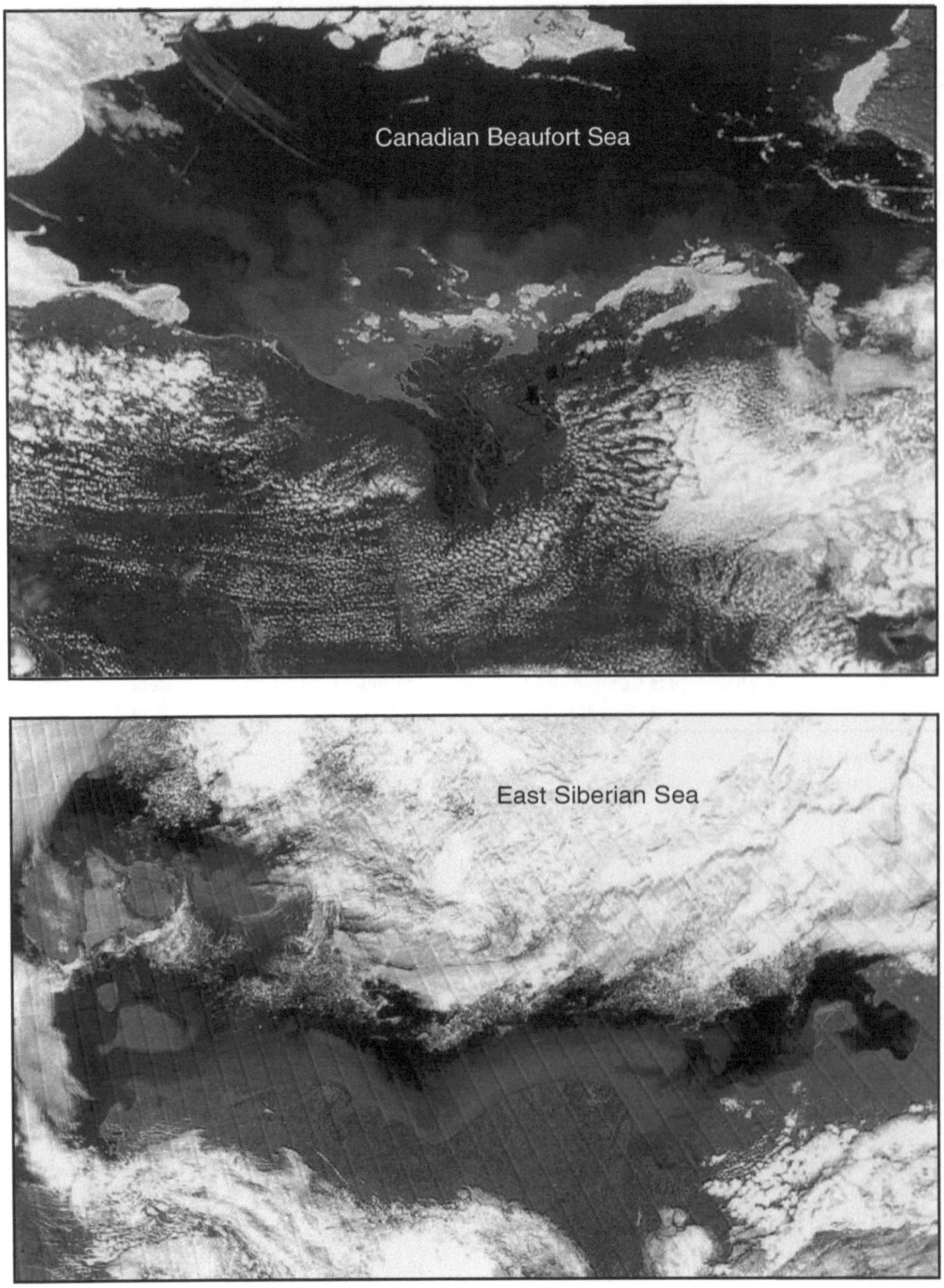

Fig. 2.5. Satellite images of the East Siberian and Beaufort Sea showing the distribution of suspended sediments. The strong river plume of the Mackenzie River is clearly seen in the Beaufort Sea, whereas in the East Siberian Sea very high turbidities are observed along the coastline. Images are taken from http://www.visibleearth.nasa.gov

Beaufort Sea

As discussed in Rachold et al. (2000) the best estimate of the coastal erosion sediment input to the Canadian Beaufort Sea (5.6×10^6 t y^{-1}) was published by Hill et al. (1991). Based on this value and on an average TOC concentration of the coastal sediments of up to 5% Macdonald et al. (1998) estimate a maximum coastal TOC input into the Canadian Beaufort Sea of 0.3×10^6 tC y^{-1}. In their further calculations Macdonald et al. (1998) accept an average of 0.06×10^6 tC y^{-1} for the coastal TOC input, which will be applied in our calculations.

The coastal sediment flux into the Alaskan Beaufort was quantified by Reimnitz et al. (1988a)

as 2.3×10^6 t y^{-1}. Organic carbon concentrations of shallow marine Alaskan shelf sediments have been published by Naidu et al. (1985). In general, the concentrations are low (less than 1–1.4%). Based on these data we will use a maximum TOC input of 0.03×10^6 tC y^{-1} for the Alaskan Beaufort Sea.

2.3.3 Coastal organic carbon input

In total 431×10^6 t yr^{-1} of sediment and 6.7×10^6 tC y^{-1} enter the Arctic Ocean (Table 2.10). Approximately 60% of the total TOC flux originates in the Laptev and East Siberian Seas. The predominant sources are Ice Complex deposits, which are widespread in Northeast Siberia. The highest coastal TOC flux is observed in the East Siberian Sea, even though the Laptev Sea coastline is considerably longer. This is due to the dominance of the Ice Complex along the coastline of the East Siberian Sea. Satellite images of the East Siberian and the Beaufort Sea (Fig. 2.5) clearly show the major sources of sediment: the strong river plume of the Mackenzie River is visible in the Beaufort Sea, whereas in the East Siberian Sea high turbidities, which are related to coastal sediment input, are observed along the coastline.

Semiletov (1999a, 1999b) stated that in the Russian Arctic marginal seas, the transport of organics (in the form of particulate matter) induced by coastal erosion has a value similar to that of the DOC transport of the rivers. Semiletov´s evaluation shows an annual flux of particulate organic carbon along the Yakutian coast (from eastern Taymyr to Chukotka) of $3.5–7.0 \times 10^6$ tC y^{-1} (Semiletov, 1999b, 2000). Based on our estimation given in Table 2.10, the sum of the TOC input of the Laptev Sea, East Siberian Sea and the Asian sector of the Chukchi Seas is quantified as 4.4×10^6 tC y^{-1}, which is in general agreement with Semiletov´s data and underscores the importance of coastal TOC input in the Siberian Arctic.

Table 2.10. Sediment and TOC flux to the Arctic Ocean through coastal erosion. [1] Romankevich and Vetrov (2001), [2] this study, [3] refers to the entire Chukchi Sea (twice the value of Romankovich and Vetrov (2001) given for the Asian sector of the Chukchi Sea only; [4] sum of Canadian Beaufort Sea (according to Macdonald et al. 1988) and Alaskan Beaufort Sea (based on Reimnitz et al. 1988a and Naidu et al. 1985)

	Sediment flux (10^6 t y^{-1})	TOC flux (10^6 t y^{-1})
White Sea [1]	60	0.3
Barents Sea [1]	59	0.5
Kara Sea [1]	109	1
Laptev Sea [2]	58.4	1.8
East Siberian Sea [2]	66.5	2.2
Chukchi Sea [3]	70	0.8
Beaufort Sea [4]	7.9	0.09
Total	**430.8**	**6.69**

2.4 The Role of Arctic Sea Ice in Transporting and Cycling Terrigenous Organic Matter

H. Eicken

2.4.1 Introduction

The role of sea ice in the Arctic Ocean's budget of terrigenous organic carbon (tOC) is presently not well understood. While it is obviously not a source of tOC as such, the ice cover can potentially be of great importance in providing the fastest, most effective means for basin-wide transport as well as export of tOC from the Arctic into the European Nordic (in particular Greenland and Barents) Seas. In contrast with other ocean basins, Arctic surface waters exhibit unusually high concentrations of terrigenous dissolved organic carbon (tDOC), such that the basin-scale transport of tOC is dominated by the dispersal of riverine tDOC (Opsahl et al. 1999; Lobbes et al. 2000). Whereas sea ice may have slightly elevated levels of tDOC (Thomas et al. 1995; Opsahl et al. 1999), and has been attributed with a significant fraction of the flux of tDOC out of the Arctic Basin through Fram Strait, it is of much greater importance for the transport of particulate organic matter.

Evidence has accumulated over the past two decades demonstrating that entrainment of sediments into sea ice (at particulate concentrations of tens to hundreds of mg l^{-1}) is a common phenomenon in the Arctic and particularly widespread over the broad, shallow Eurasian shelves (e.g., Reimnitz et al. 1994; Pfirman et al. 1997; Eicken et al. 2000). Terrigenous particulate organic matter (tPOM) deposited on the shelves may thus potentially be removed and transported over large distances with sea ice moving with the Transpolar Drift away from the Siberian Arctic across the Pole and into the Greenland Sea (Pfirman et al. 1997). Given that the flux of tPOM into the central Arctic and the Nordic Seas is greatly limited due to the inefficiency of transport in the water column and the bathymetric isolation of the central basins from shelf-slope transport, sea ice input could thus even come to dominate the total flux of tPOM in the inner Arctic and Greenland and Barents Seas. Studies of modern sediment transport (Pfirman et al. 1990; Hebbeln and Wefer 1991; Reimnitz et al. 1993b; Eicken et al. 2000) together with interpretation of the sediment record (Bischof and Darby 1997; Nørgaard-Pedersen et al. 1998), suggest that sediments released from sea ice account for a major fraction of the Holocene sedimentation in the deep Arctic Ocean and the Greenland Sea. Hence, in the context of the (terrestrially-derived) carbon budget of the Arctic Ocean, sea ice has to be considered as an efficient means of redistributing tPOM over large distances in short periods of time. For the regional budgets of tPOC in different sectors of the Arctic, the ice cover can hence serve as a potentially important, yet mostly overlooked, sink or source function. This problem is highlighted by a recent study of the carbon budget of the Mackenzie shelf by Macdonald et al. (1998), which identified the sea-ice associated carbon and particulate fluxes as one of the key unknowns in the system. In this contribution, we will briefly review the magnitude of sea-ice transport of dissolved and particulate terrigenous organic matter and attempt an assessment of its importance as a redistributor of tOC.

2.4.2 Methods

While measurements of organic carbon parameters reported below have generally been obtained using standard methods (for details, see Naidu et al. 2000, for stable-isotope measurements, Reimnitz et al. 1993a, for sedimentological analysis, Stein et al. 1999, and Lindemann 1999, for kerogen microscopy and pyrolysis, and Wheeler et al. 1997, for measurements of dissolved organic carbon), the sampling process differs substantially from standard oceanographic or sediment-core analysis. With the exception of surface aggregations of sediments, typically occurring as a result of melt (Nürnberg et al. 1994), sampling of the dissolved and particulate load of sea ice requires coring with specialized equipment (see Eicken et al. 1995, for detailed description of methodology and overview of standard ice-core parameters). Measurements of DOC on melted cores require special precautions to minimize contamination (Thomas et al. 1995). The particulate matter is typically concentrated through filtration, preferably in a laminar-flow, clean-bench (Reimnitz et al. 1993a; Lindemann 1999). Further sample processing can then proceed as with other standard sediment samples.

Assessments of transport of materials by ice also require estimates of the ice volume fluxes which are mostly determined through one or a combination of the following: drifting buoy data (Rigor and Colony 1997; Pfirman et al. 1997), satellite remote-sensing data (Thomas and Rothrock 1993; Alexandrov et al. 2000), and modelling (Alexandrov et al. 2000). Recently, progress has also been made in identifying and mapping the distribution of sediment-laden ice using satellite remote-sensing techniques (for details see, Reimnitz et al. 1993b; Kolatschek 1998; Eicken et al. 2000).

2.4.3 Sea ice transport in the Arctic Ocean and entrainment of particulate matter

Summarizing direct measurements, remote sensing and modelling, Figure 2.6 shows the net export of sea ice from the marginal seas into the Arctic Ocean and the export through Fram Strait into the Greenland Sea. With the wind-driven Beaufort Gyre dominating circulation in the North American Arctic as well as the Chukchi and eastern East Siberian Seas (Proshutinsky and Johnson 1997), the latter regions contribute on average only small amounts of exported first-year ice to the ice budget of the Arctic Ocean. In contrast, the other main wind-driven circulation feature of the Arctic Ocean, the Transpolar Drift, results in an average export of more than 1000 km^3 of first-year ice from the Laptev, western and central East Siberian and Kara Seas (Proshutinsky and Johnson 1997, Alexandrov et al. 2000).

A major fraction of this ice forms during fall and the early stages of winter over the broad Siberian shelves in comparatively shallow water of mean depths around 50 m in the Laptev and East Siberian Seas (Fig. 2.7; Timokhov 1994). Tidal and wind mixing during fall freeze-up as well as thermohaline mixing associated with rapid ice growth and brine rejection promote the resuspension of sediments (Sherwood 2000) from the seafloor and furthermore enhance the rate of formation of frazil ice in the water column (Smith et al. 1990). The combination of these processes results in the entrainment of substantial amounts of resuspended particulate matter into the newly formed ice (Figs. 2.7 and 2.8; Pfirman et al. 1990; Reimnitz et al. 1993b, Eicken et al. 2000). While other processes such as anchor ice formation, deposition of river sediments onto a flooded fast ice cover and aeolian deposition may also contribute to the particulate load, the process of suspension freezing, though not at all well understood, appears to be the most significant contributor to the phenomenon of sediment-laden sea ice in the North American and Siberian Arctic (Osterkamp and Gosink 1984; Reimnitz et al. 1993a;

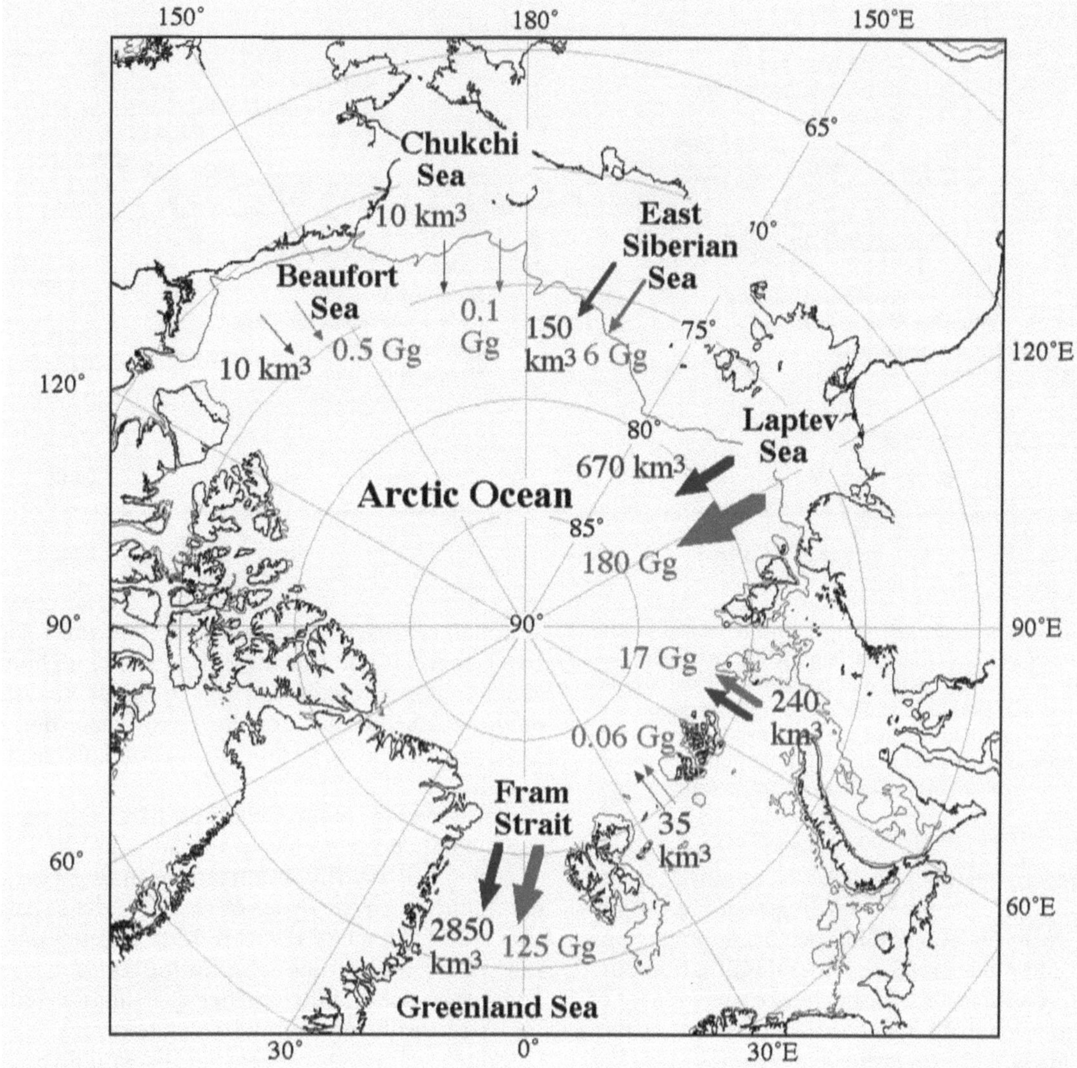

Fig. 2.6. Net annual export of sea ice from the marginal seas into the Arctic Basin and through Fram Strait into the Greenland Sea (black; unit in km³) and associated transport of terrigenous particulate organic carbon (grey; unit in Gg = 10^3 t). Ice-transport data compiled from Timokhov (1994) for the Chukchi, East Siberian and Kara Seas, Alexandrov et al. (2000) for the Laptev Sea (based on an average ice thickness of 1.5 m), and Vinje (1987) for the Barents Sea. Given the lack of data for the Beaufort Sea and Canadian Archipelago, the upper maximum limit for export to the Arctic has been assessed as 10 km3, with a remote-sensing based ice-mass balance model suggesting no net import into the Arctic Ocean from this region (Thomas and Rothrock 1993). Ice-associated transport of tPOC is based on data from Macdonald et al. (1998), Naidu et al. (2000), Romankevich et al. (2000), Stein et al. (1999), Lindemann (1999); Stein (1996), Hulth et al. (1996), and Stein and Fahl (Chapter 7.6) as outlined in Table 2.12. The 100-m isobath is shown in grey

Dethleff et al. 1993; Eicken et al. 2000). Shipboard observations and satellite remote sensing have established in recent years that the entrainment of particulates into sea ice is a widespread phenomenon in the Arctic and that ice-rafting of particulates figures prominently in the overall sediment budget of the Arctic Ocean. While a detailed analysis of the processes controlling entrainment and export of particulates is beyond the scope of this contribution, we will examine in more detail how sea-ice transport contributes to the redistribution and regional supply of tOC, both, in dissolved and particulate form.

2.4.4
Dissolved organic carbon in sea ice

Most ionic species dissolved in seawater are too large to be incorporated into the crystal lattice of ice (Ih) as a solid solution (Weeks and Ackley 1986). Thus, as a result of solute segregation at the ice-water interface, only between 15 and 30 % of the salt

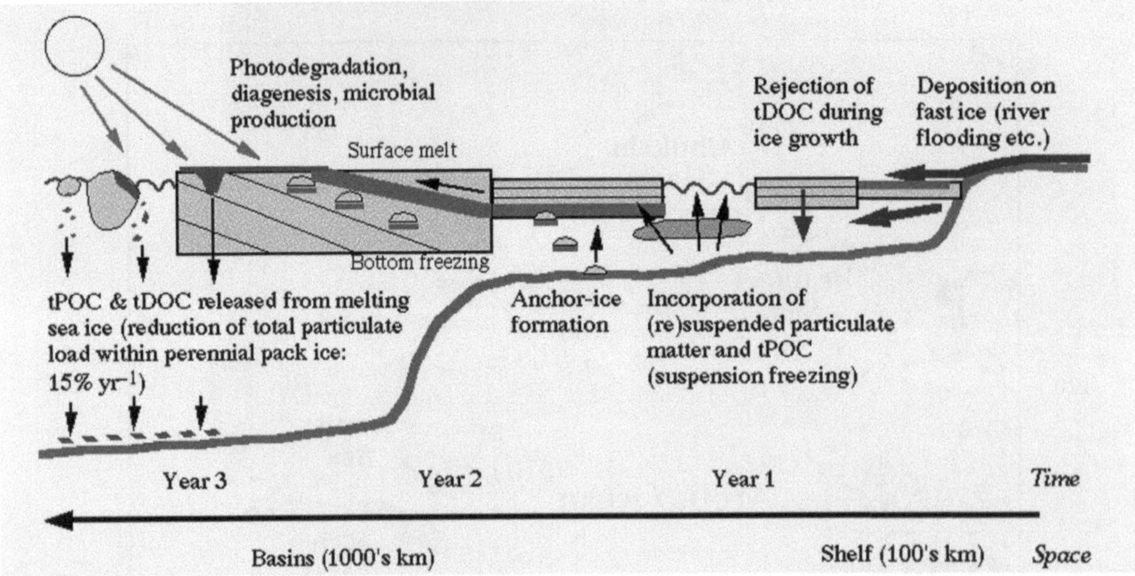

Fig. 2.7. Schematic depiction of entrainment, transport and loss of terrigenous organic carbon (particulate and dissolved) by sea ice (processes and features depicted are not drawn to scale). Note that while anchor ice formation and direct deposition onto fast ice are responsible for entrainment of tPOC, they are not deemed as quantitatively important for basin-scale transport as suspension freezing

contained in seawater is actually retained in the corresponding volume of newly grown ice, mostly in the form of sub-millimeter brine inclusions (Weeks and Ackley 1986). Further desalination during ice growth and melt can reduce the salinity of sea ice to values well below 10% of that of seawater. While one would expect the same to hold true for dissolved organic molecules which are subject to similar constraints with regards to size and charge balance, little work has been done on the segregation of DOM during ice growth. Studies of the partitioning of glucose between growing ice and the parent solution indicate that solute concentrations in the ice range between <1 and 20% of that in the liquid (Miyawaki et al. 1998). Experiments by Amon (Chapter 4) and Giannelli et al. (2001) indicate that the bulk of the DOM fraction is expulsed from the ice along with the salt. Thus, it appears reasonable to assume that sea ice is not a particularly effective transport medium for (t)DOC. If the entrained DOC fraction in sea ice was to scale with the inorganic ionic fraction (which is reduced to mean salinities of about 2, Eicken et al. 1995), the concentrations of entrained DOC in sea ice of the Arctic Basin should be lower by about a factor of 15 than those in the source areas. The few data available on concentrations of DOC in sea ice suggest that this holds true (Thomas et al. 1995; Opsahl et al. 1999), although sea-ice biota may accumulate and release DOC, thereby complicating the picture (Thomas et al. 1995). Nevertheless, this is likely to affect only the marine and not the terrestrially derived DOC pool in the ice (Amon et al. 2001, Lobbes et al. 2000).

With average concentrations of tDOC of approximately 60 to 70 μM in the Siberian shelf seas (Kattner et al. 1999), concentrations in first-year sea ice would be on the order of 12 μM, if the segregation and drainage of tDOC were to parallel that of salt (with first-year ice salinities of about 6, Weeks and Ackley 1986). With further desalination reducing salinities to on average 2 in the central Arctic (Eicken et al. 1995), concentrations would drop to 4 μM. This corresponds reasonably well with a measurement on multiyear ice sampled in Fram Strait which contained between 5.4 and 7.5 μM tDOC (Opsahl et al. 1999). Given that maximum tDOC concentrations of 550 μM are found near the estuaries and deltas (Lobbes et al. 2000), the fate of ice forming in waters affected by river plumes of, e.g., Lena, Ob, Yenisei or the Mackenzie is of particular importance. A major fraction of the coastal sea ice over the Siberian and Beaufort Sea shelves becomes landlocked in fall. After the ice breaks up in summer it melts in place in most years, rather than being exported into the interior Arctic (Dean et al. 1994, Rigor and Colony 1997). Hence, the ice with the highest concentrations of tDOC does not contribute substantially to export from the shelf.

Based on the ice volume fluxes summarized in Figure 2.6 and average tDOC concentrations of 12 and 4 μM in first- and multi-year sea ice, respectively, the sea-ice flux of tDOC has been estimated for the different sectors of the Arctic Ocean and

Fig. 2.8. Aerial (top, several hundred meters across) and ship-based photograph (bottom, several tens of meters across) of sediment-laden sea ice in the northeastern Laptev Sea (August 1995). Note the patchy distribution of darker, sediment-laden ice and the uniformly dark appearance of surface melt ponds due to sediment layers at the bottom. The white, linear features visible in the upper panel correspond to leads that opened and refroze after sediment concentrations in the water column had subsided, resulting in clean ice

summarized in Table 2.11. It is assumed that export through Fram Strait balances import from the shelves and that the reduction in tDOC concentrations in the ice is adequately described by the reduction in bulk ice salinity. In comparison with other terms in the Arctic Ocean's DOC budget, such as the total annual river input of 23 Tg yr^{-1} (Wheeler et al. 1997), sea-ice transport is of little to no significance.

Table 2.11. Sea–ice associated fluxes of terrigenous dissolved organic carbon (tDOC) into the Arctic Ocean region (1 Tg = 10^6 t)

Region	Ice volume flux, $km^3\ y^{-1}$	Flux of tDOC, Tg y^{-1}
Export to interior Arctic Ocean by first-year ice from shelves		
Chukchi & Beaufort Sea	20	0.003
East Siberian Sea	150	0.019
Laptev Sea	670	0.087
Kara Sea	240	0.031
Barents Sea	35	0.004
Export with multi-year ice through Fram Strait		
Arctic Ocean export	–2850	–0.12
Net balance (input to Arctic Ocean)		
Arctic Ocean input		0.024

While of little apparent quantitative importance, sea ice may still play an important role in the cycling of tDOC in the Arctic Ocean. Dissolved and particulate impurities entrained in first-year ice tend to be concentrated in the surface layers of the ice during subsequent years as a result of surface melt (Weeks and Ackley 1986; Nürnberg et al. 1994; Eicken et al. 2002; Fig. 2.7). Hence, tDOC entrained into sea ice is subjected to much higher time-integrated levels of shortwave radiation (incl. UV) and potentially also photochemically produced oxidants than the same compounds transported in the water column. While little is known about photochemical degradation of DOM under such conditions in the ice cover (Sydnes 1991, Amon, this volume), sea-ice entrainment may thus provide a unique reactive pathway for the breakdown and modification of DOM in the marine system. Moreover, such photodegradation is also likely to affect the bioavailability of DOM to sea-ice organisms (Thomas et al. 1995, Bussmann 1999; see also discussion further below).

2.4.5 Particulate organic carbon in sea ice

As described above, ice growth in well-mixed shallow waters, in particular over the broad Siberian shelves, can result in the entrainment of high concentrations of suspended particulate matter (SPM) into the ice cover (Figs. 2.7 and 2.8). Here, we will assess whether the transfer of organic carbon of terrigenous origin associated with these ice-entrained sediments represents a significant contribution to the tOC flux through the Arctic Ocean. While marine POC may also be present in sea ice in higher concentrations (both from ice biota as well as POC associated with marine sediments), we will exclusively focus on the tOC component associated with the particulate load.

Quantifying the sediment transport by sea ice presents a considerable challenge, not least due to the patchy distribution of sediments (Fig. 2.8), our lack of understanding of entrainment processes and the overall difficulty in obtaining data on the areal distribution and SPM concentrations characteristic of sediment-laden ice (Pfirman et al. 1990; Reimnitz et al. 1993b; Eicken et al. 2000). Data on sediment concentrations in sea ice require direct sampling of the ice and have shown concentrations to range between <10 to >1000 mg l^{-1}. Heavily sediment-laden ice exhibiting a strong discoloration typically contains SPM at concentrations of several hundreds of mg l^{-1}. Studies in the Siberian Arctic indicate that even visually "clean" ice grown in shallower water (<30 m depth) exhibits SPM background concentrations of around 2 to 20 mg l^{-1} (Dethleff et al. 1993; Eicken et al. 1997), at least half of which can be considered to represent lithogenic sediments rather than particulates of biogenic origin (Nürnberg et al. 1994). Average SPM concentrations in visibly discolored, sediment-laden ice sampled in different regions range between 40 mg l^{-1} in the Beaufort Gyre region (Reimnitz et al. 1993a), 192 mg l^{-1} in the coastal Beaufort Sea (Osterkamp and Gosink 1984), 212 mg l^{-1} in ice grown from slush in the Beaufort Sea (Kempema et al. 1989), 70 mg l^{-1} in first-year ice in the Laptev Sea and 381 mg l^{-1} in ice from the eastern Laptev and Western East Siberian Sea (Eicken et al. 2000) to 662 mg l^{-1} in the upper layers of multi-year ice in the central Arctic (Nürnberg et al. 1994).

Few data are available on organic carbon concentrations in ice-rafted sediments. Lindemann (1999) completed an extensive survey of samples collected in the Laptev Sea, with a median POC fraction of 2.3% for 71 samples from the northern and 3.6% for 40 samples from the southern Laptev Sea. POC fractions in sediments sampled in multi-year ice from the central Arctic amounted to 1.8 wt-% (8 samples, Reimnitz, unpublished). Wollenburg (1993) found between 0.6 and 6.8 wt-% POC (averaging 2.1 wt-%) in sediments sampled from sea ice in the Eurasian Arctic. These values correspond well with organic carbon contents of surface shelf sediments, typically ranging between 1 and 2% in the Laptev Sea (Stein et al. 1999) and 1 and 1.5% over the Canadian Beaufort Shelf (Macdonald et al. 1998). An enrichment of organic carbon in ice-entrained sediments is attributed to selective incorporation of the clay and silt fraction, exhibiting higher concentrations of POC (Reimnitz et al. 1998, Lindemann 1999).

The fraction of organic carbon of terrigenous origin associated with sediments transported by sea ice is less well established. Based on kerogen microscopy and pyrolysis data, Lindemann (1999) estimated the median tPOC in sediments sampled in sea ice of the southern and northern Laptev Sea as 76 and 83%, respectively. This corresponds reasonably well with surface sediment data (Stein et al. 1999). Isotopic analysis ($\delta^{13}C$) of a small set of ice-rafted sediments from the Beaufort Sea is less conclusive but most likely shifted by in-situ primary production (Naidu et al. 2000). Surface sediments in the Beaufort Sea exhibit terrigenous fractions between 50 and 85%, while the Chukchi and East Siberian Seas range between 50 and <10% (Naidu et al. 2000) based on isotopic evidence. Given potentially large variations among terrigenous and marine endmembers in $\delta^{13}C$, in particular when considering the overall comparatively narrow range

Table 2.12. Fluxes of terrigenous particulate organic carbon (tPOC) associated with sediment-laden sea ice in the Arctic Ocean region (1 Tg = 10^6 t)

Region	Ice volume flux $km^3\ yr^{-1}$	POC_{SPM} %	$tPOC_{SPM}$ %	SPM_{ice} $mg\ l^{-1}$	f_{vol} %	Flux of tPOC $Tg\ y^{-1}$	SPM_{ice} flux $Tg\ y^{-1}$
Import with first-year ice from shelves							
Beaufort Sea	10	1.3[a]	60[b]	150	5	0.5×10^{-3}	0.075
Chukchi Sea	10	1.5[b]	10[b]	10	5	0.007×10^{-3}	0.005
East Siberian Sea	150	1.5[c]	30[b]	100	10	6×10^{-3}	1.5
Laptev Sea	670	2.5[d]	80[e]	150	10	180×10^{-3}	10
Kara Sea	240	1.6[f]	50[f]	100	10	17×10^{-3}	2.4
Barents Sea	35	1.3[f]	30[f,g]	10	5	0.06×10^{-3}	0.018
Export with multi-year ice through Fram Strait							
Arctic Ocean export						-125×10^{-3}	-8.6
Net balance (input to Arctic Ocean)							
Arctic Ocean input						55×10^{-3}	5.4

Data sources – a: Macdonald et al. 1998; b: Naidu et al. 2000; c: Romankevich et al. 2000; d: Stein et al. 1999; e: Lindemann 1999; f: Stein and Fahl, Chapter 7.6; g: Hulth et al. 1996; for other data see explanations in text

of values, future studies of ice-entrained sediments may benefit greatly from determinations of more discriminating tracers or biomarkers such as the hydrocarbons singled out by Yunker et al. (1995).

To arrive at the total fluxes of tPOC in the Arctic Ocean, we have estimated average SPM concentrations in sediment-laden ice (SPM_{ice}) based on the data presented above (Table 2.12). With only very limited data available for the Chukchi and Barents Seas, SPM concentrations at the background level (10 mg l^{-1}) have been assumed here. However, recent evidence from the Shelf-Basin Interaction (SBI) study indicates that both in the Chukchi and Beaufort Seas higher sediment loads may be quite common (Eicken and Gradinger, unpubl. data, 2002). The volume fraction (f_{vol}) of sediment-laden ice is conservatively estimated based on the areal distribution of shallow-water entrainment regions (water depth < 20 m, Sherwood 2000) and the likelihood and extent of past entrainment events observed in these regions. The total export and net input into the Arctic Ocean have been estimated assuming a loss of 15% of the total sediment load due to melt-out every summer based on an average residence time of three years in the Arctic Ocean (Fig. 2.7; Freitag 1999; Eicken et al. 2000). Given the substantial errors associated in particular with the estimation of SPM_{ice} and f_{vol}, the total export from the different regions are at best order-of-magnitude estimates.

More detailed data sets available from individual entrainment and export events can help constrain these estimates. For example, an entrainment event in the eastern Laptev/western East Siberian Sea in 1994/95 has been studied in detail by combining ground measurements, remote sensing and modelling (Kolatschek 1998; Lindemann 1999; Eicken et al. 2000). It could be established that the roughly 100,000 km^2 of highly sediment-laden ice produced in the vicinity of the New Siberian Islands resulted in the export of 18.5 Tg of sediment by sea ice. Based on the extensive measurements of POC and tPOC employing kerogen microscopy and pyrolytic analysis on 70 of the samples collected from the area (Lindemann 1999), we can estimate the total export of terrigenous carbon to be on the order of 0.35 Tg (0.83×2.3% tPOC fraction of total SPM). Upon melting of the ice, most of which was exported into the Greenland Sea within less than three years (Eicken et al. 2000), the ensuing flux of released sediments can represent a significant contribution to the regional carbon budget (Fig. 2.7), with as much as 0.7 g C m^{-2} d^{-1} in the seasonally ice-covered Greenland Sea. This compares with typical OC sedimentation fluxes of 0.1 g C m^{-2} d^{-1} in the Arctic Ocean and 1.5 g C m^{-2} d^{-1} in the Greenland Sea as based on the compilation of data by Hargrave et al. (1994).

Thus, despite the substantial uncertainties involved in these assessments, it is clear that sea-ice export of tPOC can play an important role in the carbon budget of the Arctic Ocean. This is also borne out by a comparison with the magnitude of other terms in the Arctic Ocean's carbon budget (see Chapters 4, 5 and 7). For example, sea-ice export from the Laptev Sea accounts for between one fifth and one half of the annual riverine input of POC, based on different input estimates for the area (Gordeev et al. 1996; Romankevich et al. 2000). While the corresponding fraction of total ice export of POC drops to below 1% of the total riverine input for the entire Arctic Ocean (Gordeev et al. 1996), release of POC from melting sea ice still constitutes an important term in the carbon budget of deep basins and remote seas that receive little direct input. Nevertheless, given the large uncertainty in several of the terms listed in Table 2.12, more research is needed to establish better estimates of ice-associated transport and in particular how this flux varies regionally and temporally. Most important, it needs to be investigated whether the combination of high ice-production rates, high concentrations of SPM in newly-grown sea ice and overall high fractions of tOC in resuspended and entrained surface sediments do in fact single out the Laptev Sea as contributing an order of magnitude more tOC to the Arctic Ocean via the sea-ice conveyor than all other sources combined.

2.4.6 Conclusions

The major conclusions from this survey are that (1) transport of terrigenous dissolved organic carbon (tDOC) by sea ice constitutes a minor to negligible component of the Arctic Ocean's tDOC budget, (2) entrainment and export of terrigenous particulate organic carbon (tPOC) into sea ice from the shallow shelves to the deep basins and the European Nordic Seas represents an important term in the Arctic Ocean's carbon budget, (3) tPOC and tDOC entrained into sea ice are potentially important for ice-associated primary (and secondary) production, and (4) due to the rapidity and effectiveness of ice transport as well as the potential for substantial alterations in its composition, sea-ice transport provides for a unique pathway of terrigenous organic carbon through the Arctic system, with implications for the marine sediment record.

Quantitatively, sea-ice transport is of particular regional importance in the Laptev Sea, where the average export into the Arctic Ocean has been estimated as 0.18 Tg (of C) y^{-1} (note, however, that given the considerable uncertainties in some of the source terms, this is only an order-of-magnitude estimate) and accounts for between one fifth and one half of the total riverine input of tPOC. An individual entrainment event has been shown to remove as much as 0.35 Tg (of C) of tPOC. Sea ice is furthermore of great significance as a source of tPOC released from melting ice (mostly during the summer months) in the central Arctic, where other modes of transport are not as efficient, and in the Greenland Sea, where high net melt rates (up to 10 m of ice y^{-1}) result in significant ice-derived particulate fluxes. Sediment trap data (Hebbeln et al. 1991) and tracking of sediment-laden sea-ice fields (Eicken et al. 2000) indicate that in the Greenland Sea and parts of the Eurasian central Arctic, the bulk of the lithogenous particle flux to the seafloor can be derived from the ice cover. In comparison with source sediments, ice-entrained material is sorted towards the fine-grained (clay and silt) fractions (Reimnitz et al. 1998; Lindemann 1999), which in turn are associated with higher concentrations of organic carbon (Mayer 1994). Given the substantial errors associated with these first assessments, future work needs to focus on more widespread measurements of the tPOC fraction in ice-rafted sediments as well as on the temporal and spatial variability of entrainment and export by sea ice.

Apart from the present-day quantitative importance of sea-ice rafting of organic matter, two further issues can be of substantial importance for the Arctic's carbon budget and need to be investigated in more detail. First, suspended particulates in sea ice tend to aggregate near the surface of the ice as a result of surface melting and bottom ice accretion (Fig. 2.7; Pfirman et al. 1990; Nürnberg et al. 1994; Freitag 1999) and are thus subjected to arguably the most extreme environmental conditions at the low-temperature end in the marine realm with temperatures below −20 °C, salinities above 150, pH below 5 and high concentrations of oxidants due to high UV fluxes. At the same time, particulates in sea ice are associated with high bacterial biomass (Junge et al. 2001), which can result in significant respiration and turnover of organic carbon. Incubation experiments by Bussmann (1999) and tracer studies by Schell (1983) indicate that tOC may in fact enter the marine food web. Hence, it needs to be investigated to what extent diagenetic modification and remineralisation of the tOC fraction of ice-rafted material can significantly affect the quantity and composition of organic matter delivered to the seafloor after release from the ice.

This aspect ties in with the second important question related to variability in time and space of ice-transport of organic carbon. As evident from Table 2.12 and Fig. 2.6, both sea-ice and associated tPOC fluxes can vary regionally by several orders of magnitude. Similar variability can be expected for the same location on geological time scales, depending on sea level, riverine supply of particulates and freshwater and other factors. This has potentially important implications for the transfer of organic carbon from the shelves into the deeper Arctic Ocean and the European Nordic Seas. Currently, conditions in the Laptev Sea are such that it represents by far the single largest source term in the ice-transported flux of tOC. However, as conditions for entrainment and export (mostly driven by atmospheric and to a lesser extent oceanic parameters, see Eicken et al. 2000) change, other source areas may come to dominate the ice-associated flux of tOC. Since the latter constitutes a dominant term in parts of the Arctic under present-day conditions, the close linkage between sea-ice and tOC fluxes in geological present and past warrants a closer look at paleoceanographic sediment records to separate genuine changes in regional carbon budgets from those associated with changes in sea-ice entrainment and export.

2.5 Aeolian Input

V. P. Shevchenko, A. P. Lisitzin

Numerous studies have shown that aerosols are of importance for atmospheric chemistry and climate of the Arctic (Pacyna 1991; Barrie 1996; Leck et al. 1996) and other parts of the World Ocean (Duce et al. 1991; Lisitzin 1996). There are only a few regional estimates of the importance of aeolian material fluxes onto the Arctic surface (Mullen et al. 1972; Macdonald et al. 1998; Zdanowicz et al. 1998) and even fewer consider organic carbon fluxes (Macdonald et al. 1998; Romankevich and Vetrov 2001). Macdonald et al. (1998) estimated an aeolian input of particulate organic carbon to the Canadian Beaufort Shelf (area of about 0.06×10^6 km^2) of $<0.001\times10^6$ t y^{-1}, Romankevich and Vetrov (2001) calculated that the aeolian input of particulate organic carbon to the Russian Arctic Seas (area of about 4.477×10^6 km^2) is about 0.387×10^6 t y^{-1}.

The estimates presented here are based on results of aerosol research in the marine boundary layer over the seas of the Russian Arctic and the Central Arctic in 1991-1999 (Shevchenko et al. 1999 a, b, c, 2000). During August-September the mass concentration of the insoluble coarse fraction varies from 0.02 to 1.38 µg m^{-3} (0.23 µg m^{-3} on average, n = 55 samples). In most samples organic matter (vegetation fibers, pollen, diatoms etc.) is the main component by mass. The content of the sum of organic carbon and soot carbon varies from 7.54 to 48.9% (30.2% on average, n = 18 samples). Taking into account that the organic carbon percentage in Arctic aerosols is about 76% of the total non-carbonate carbon and that soot carbon percentage is 24% (Rosen et al. 1981; Shevchenko et al. 1999a), we suggest that the organic carbon content of insoluble Arctic aerosols is about 23% and soot carbon content is about 7.2%.

The vertical fluxes of aerosol deposited on the sea or land surface can be estimated from data on aerosol concentrations and size distribution. Away from the coastal areas the aerosol particle size distribution has been shown to vary only slightly (Kuusk et al. 1996; Smirnov et al. 1996; Shevchenko et al. 2000), which simplifies the estimation of vertical fluxes. The above data on mineral and chemical composition and pollution of aerosols can be recalculated both as values for total vertical fluxes of aerosol and for fluxes of various components (for example, carbon). This method, which involves some assumptions, gives only rough values of the fluxes and, when necessary, can be supplemented and checked by direct determination of the fluxes by measuring the snow depth (estimation of fluxes for the period of snow cover), ice and bottom sediments, including measurements using various precipitation gauges.

In the following we present the dry deposition fluxes of aeolian material to the sea surface in the Russian Arctic. These are based on the assumption that the sedimentation rate of insoluble coarse (>1 µm) particles, which form the major portion of aerosols by mass (Shevchenko et al. 2000), is equal to 2 cm s^{-1} (Duce et al. 1991). Accordingly, the fluxes for the Arctic range from 0.03 to 1.68 mg m^{-2} d^{-1} (on the average 0.39 mg m^{-2} d^{-1} or 142 mg m^{-2} y^{-1}). To quantify the total fluxes of insoluble aerosols per year we assume that the dry deposition is 25% of the total deposition. Furthermore, we extend our summer data on aerosol content to the whole year, by considering that the aerosol concentration in summer is several times lower than in winter and spring, but the amount of aerosol deposition is three to four times larger in summer than in spring. The flux of the aeolian suspension calculated under these assumptions is equal to 570 mg m-2 y^{-1} (1.56 mg m^{-2} d^{-1}).

The flux of the aeolian material to the drifting ice surface in the Arctic can be determined independently based on the mean content of suspended particulates in fresh snow, which averages 2.19 mg l^{-1} (the mean value calculated from data presented by Mullen et al. 1972; Darby et al. 1974; Pfirman et al. 1989; Dibb 1996; Dethleff et al. 1998; Shevchenko et al. 1999a; n = 87 cases), and the mean precipitation over the Arctic Ocean which averages 285 mm y^{-1} (Voskresenskii and Petrov 1985). This direct method produces a value of 624 mg m^{-2} d^{-1} (1.71 mg m^{-2} d^{-1}) for the vertical flux of aerosols. The two independant methods produce an average value of vertical aerosol flux of about 597 mg m^{-2} y^{-1}, which is far higher than traditionally accepted.

Considering the area of the Arctic and the carbon concentrations presented above, the aeolian

Table 2.13. Aeolian input of particulate carbon (organic + soot) in the Arctic Ocean. For the calculation, we use following average values: total (dry and wet) aeolian flux of insoluble particles is 597 mg m^{-2} y^{-1}; particulate carbon (organic and soot) content is 30.2 % (for further explanation see text)

	Area (10^6 km^2)	Total Sediment Flux (10^6 t y^{-1})	Total TOC flux (10^6 t y^{-1})
Barents Sea	1.512	0.904	0.273
Laptev Sea	0.498	0.298	0.090
Kara Sea	0.926	0.553	0.167
East Siberian Sea	0.987	0.589	0.178
Chukchi Sea	0.62	0.371	0.112
Beaufort Sea	0.178	0.106	0.032
Central Arctic	4.489	2.68	0.809
Total Arctic	**9.541**	**5.695**	**1.720**

fluxes of organic carbon (incl. soot carbon) to the Arctic can be quantified as about 1.7×10^6 t y^{-1} (Table 2.13). Of course, these values require further refinement. The above estimates imply that the role of aerosols in the Arctic has been underestimated and that aeolian input of particulate organic and black carbon should not be neglected as a source of particulate carbon in the Arctic.

2.6 Summary and Concluding Remarks

- To evaluate the importance of the different sources of terrigenous organic carbon in the Arctic Ocean, we present a brief summary of the TOC fluxes (Table 2.14), which have been quantified in the preceding sub-chapters. Our evaluation shows that 80 % of the total terrigenous TOC flux to the Arctic Ocean (40.8×10^6 t y^{-1}) is attributed to dissolved and particulate river discharge (30×10^6 t y^{-1}). Coastal erosion contributes the second largest proportion (ca. 15 %, corresponding to ca. 6.7×10^6 t y^{-1}), whereas other sources, i.e. river bedload, underground water discharge, sea ice input and aeolian deposition, are of lesser importance (<7 %).
- Despite the overall dominance of river discharge, TOC input from coastal erosion exceeds the river TOC flux in some regions. The distribution of river and coastal TOC fluxes in various Arctic Seas (Fig. 2.9) illustrates that proportions vary regionally. In contrast to all other Arctic Seas, coastal erosion in the East Siberian Sea supplies more TOC than rivers, which is partly due to the absence of large rivers, and partly due to the dominance of easily eroded, high organic containing Ice Complex deposits along the coastline. In the Laptev Sea, where Ice Complex deposits are widespread as well, the coastal erosion TOC input is comparable to that of the East Siberian Sea. However, the strong inflow of the Lena River produces a considerably larger TOC flux.
- The highest TOC input, seen in the Kara Sea, is related to the river discharge of the Yenisei and Ob Rivers, which are the largest and third largest rivers in the Arctic in terms of water discharge.
- The most interesting conclusion implied by our calculations concerns the overall sediment budget of the Arctic Ocean: the total coastal erosion component of ca. 430×10^6 t y^{-1} (Table 2.10) is almost twice as large as the total sediment input from rivers of ca. 230×10^6 t y^{-1} (Table 2.1). Although the coastal erosion estimates are subject to considerable uncertainty (except for the Laptev and East Siberian Seas, which are reasonably well studied), this finding indicates that coastal erosion is likely to be the main process supporting modern sedimentation in the Arctic Ocean.

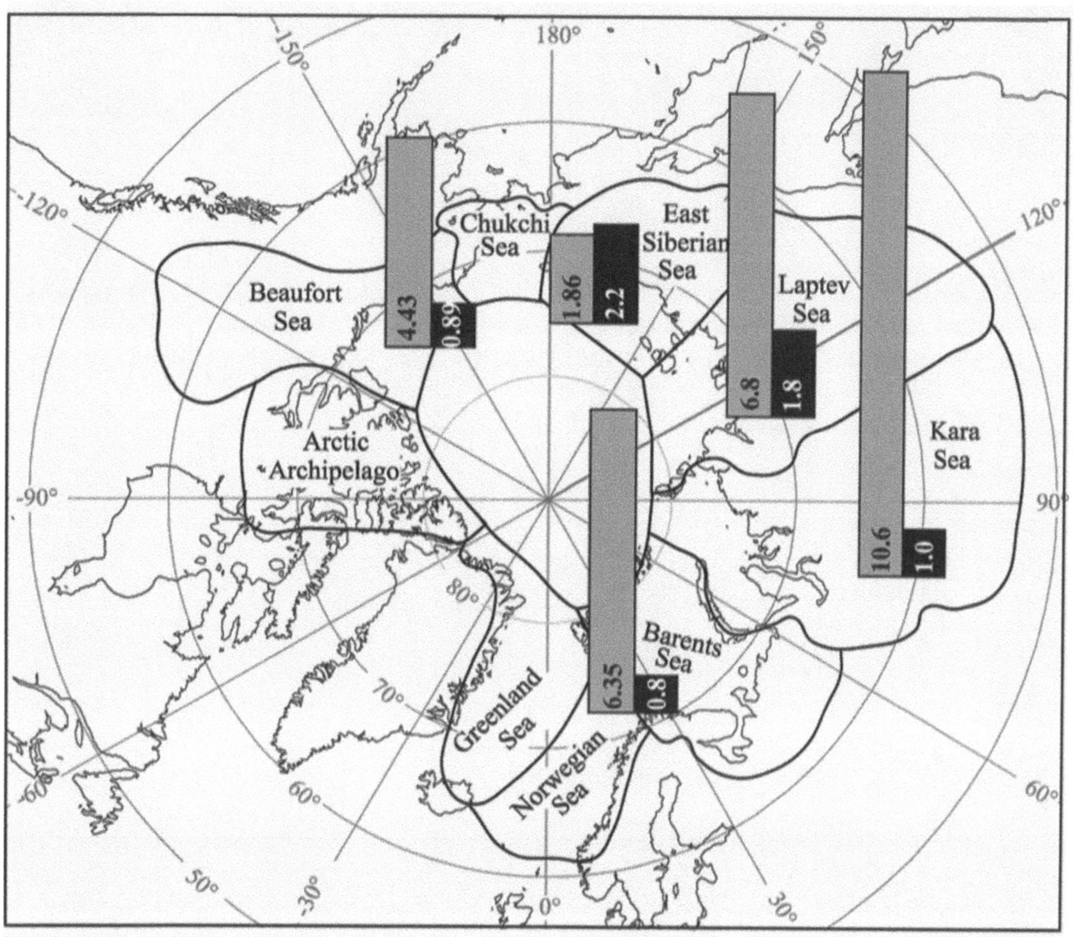

Fig. 2.9. Riverine and coastal TOC fluxes to the Arctic Ocean (10^6 t C y^{-1}). Grey bars refer to river input and black bars to coastal input. Note that the sum is shown for Beaufort and Chukchi Sea and that Barents Sea input data include White Sea. The drainage systems are taken from http://www.R-ArcticNET.sr.unh.edu/

Table 2.14. TOC flux to the Arctic Ocean

Source	DOC flux (10^6 t yr^{-1})	POC flux (10^6 t yr^{-1})	TOC flux (10^6 t yr^{-1})
River discharge	–	–	30
River bedload	–	0.13	0.13
Underground water discharge	2.3	–	2.3
Coastal erosion input	–	–	6.69
Aeolian deposition*	–	1.72	1.72
Total input			**40.8**

* Organic carbon + soot carbon.

Chapter 3

Primary and Secondary Production in the Arctic Seas

E. Sakshaug

3.1 Introduction

Marine bio-production in the Arctic and Subarctic is based primarily on microscopic unicellular algae (phytoplankton) in the water column and microalgae associated with ice. Algae utilise light to produce organic matter by photosynthesis, thereby reducing CO_2 while releasing O_2 and producing carbohydrates which, according to the needs of the algae, can be converted into essential compounds such as proteins and nucleic acids by incorporating nitrogen, phosphorus, sulphur, and other elements.

The produced organic matter is eaten by herbivorous grazers, mainly zooplankton, which in turn may be eaten by fish, ultimately ending in top carnivores, with a loss of 75–80 % of the organic matter from one trophic level to the next. The main losses are associated with respiration within the organisms themselves, microbial degradation of dissolved organic matter, sinking cellular remains and faecal pellets, thereby releasing CO_2 and nutrients. Thus only a small fraction of the organic matter reaches the seabed – the deeper the water column, the smaller the fraction.

This chapter reviews phytoplankton dynamics and the primary production of the arctic seas. Much information has been compiled in the last 30 years (Zenkevitch 1963; Herman 1989; Hood and Calder 1981; Barnes et al. 1984; Smith 1990; Sakshaug et al. 1991; Sakshaug et al. 1994; Kassens et al.1999; Loughlin and Ohtani 1999). However, recent investigations have shown that the primary production in the deep Arctic Ocean is far higher than assumed earlier. Moreover, the production and stock of dissolved organic carbon (DOC) is about as large as it is in other seas (Gosselin et al. 1997; Wheeler et al. 1997; Gradinger 1999; Agatova et al. 1999; Bussmann and Kattner 2000). On the other hand, knowledge of winter survival strategies of microalgae remains limited. There is also increasing recognition that arctic and subarctic ecosystems and population variability are closely associated with marine climate variability, making the term "ecological balance" not very meaningful in a polar context (Melnikov 1997; Niebauer et al. 1999, Ottersen et al. 2000; Stabeno et al. 2001; Olsen et al. 2003).

3.2 Major Algal Groups and Their Distribution

Phytoplankton and ice algae are unicellular and predominantly multiply by binary fission. Diatoms, chrysophytes, dinophytes (dinoflagellates), prymnesiophytes, and green flagellates are most common in arctic and subarctic seas. Picoplankton (<2 nm), among them cyanobacteria, seem to be important only in the southern fringes of the Subarctic, for instance in the oceanic Bering Sea (Sukhanova et al. 1999). Most phytoplankton are nanoplankton (5–20 µm) yet many diatom and dinophyte species are microplankton (>20 µm), with *Ceratium* spp reaching 500–750 µm. All algae possess chlorophyll *a*, which is supplemented by group-specific auxiliary pigments. These are organically soluble (chlorophylls *b*, *c*, *d* and carotenoids) or water-soluble (phycobilins in some cryptophytes, cyanobacteria and red algae). Whereas Chl*a* has absorption maxima in blue (435 nm) and red light (675 nm), carotenoids absorb bluegreen-green light and phycobilins, green-yellow light (Jeffrey et al. 1997).

Diatoms (Chl*c*; size range 5–1000 µm) possess, like kelp, the golden brown carotenoid fuxoxanthin. They need silicate because the cell is surrounded by a non-growing and sculptured silicious cell wall that is constructed as a box with a "lid" and a slightly smaller "bottom" half. Diatoms exist as centric or pennate forms that have radial and bilateral symmetry, respectively. During division, half the daughter cells inherit the "bottom" half of the siliceous wall, thus becoming a little smaller than the parent cells. When the minimum cell size is reached, maximum size is restored by sexual reproduction.

The majority of quantitatively important marine planktonic diatoms are centric (*Chaetoceros* spp and *Thalassiosira* spp); however, some pennates are also common (*Fragilariopsis*, *Navicula*, *Pseudo-nitzschia*). Pennates such as *Fragilariopsis*, *Navicula* and *Nitzschia frigida* are predominant in sea ice, especially in spring (Syvertsen 1991; Booth and Horner 1997; von Quillfeldt 1997), yet on the underside of pack ice the centric diatom, *Melosira arctica*, can establish mats of >0.5 m thickness, unique to the Arctic (Gradinger 1999). Dense ice-algal stocks may reach 1 g Chl*a* m^{-3} (Thomas and Dieckmann 2002).

Chrysophytes are flagellates with the same pigments as diatoms and are found chiefly in waters with lowered salinity; for example, the colony-forming *Dinobryon* spp.Various green flagellates (Chl*b*) and cryptophytes are at times important in sea ice, in its meltponds, and in the surrounding low-salinity waters (Horner 1989; Gradinger 1996; Sukhanova et al.1999).

Dinophytes (Chl*c*) form a heterogeneous group of flagellated autotrophic and heterotrophic species, some of which combine both modes of living (mixotrophs; Jones et al. 1994). Thecate species are covered with cellulosic platelets (theca), examples being the autotrophic *Ceratium arcticum* and *C. longipes* and the heterotrophic *Protoperidinium* spp, which are common in the Arctic and Subarctic. The latter may ingest bacteria, dead microscopic matter and DOC. Naked (athecate) dinophytes are important in waters influenced by sea ice (Braarud 1935; Gosselin et al. 1997). Whereas the thecate autotrophs have the mahogany-red carotenoid pigment peridinin, most of the naked autotrophs have 19′-hexanoyloxyfucoxanthin, a close relative of fucoxanthin.

Prymnesiophytes (Chl*c*) are autotrophs with 19′-hexanoyloxyfucoxanthin as the most important carotenoid yet mixotrophy is known (Jones et al. 1994). The group includes the genus *Phaeocystis*, which in the Arctic and Subarctic presumably is represented by *P. pouchetii* (Baumann et al. 1994; Zingone et al. 1999). *Phaeocystis* blooms are made up by non-motile cells that form 1–2 mm large lobe-formed colonies, which are assemblages of cells and seawater enclosed by a semi-permeable and strong yet plastic membrane. This membrane probably protects against infections and, to some extent, grazing (Hamm et al. 1999). Coccolithophorids, which represent another group of prymnesiophytes, are covered by sculptured calcareous platelets (coccoliths). Among them the cosmopolitan and non-flagellated species, *Emiliania huxleyi*, is by far the most important species, both in the Arctic and the Subarctic (Medlin et al. 2000).

3.2.1 Distribution of species

Because vast areas are under-sampled and many species, especially those without a hard cell wall, disintegrate or become misshapen when preserved, making identification impossible (Sournìa 1978), the number of species of phytoplankton and ice algae in arctic seas is presumably significantly higher than the number that has been recorded.

Since 1902, the record of arctic and subarctic phytoplankton species has increased from 115 to about 300, among which diatoms and dinophytes comprise at least 160 and 35 species, respectively (Gran 1902; Sukhanova et al. 1999). The western Barents Sea and the Sea of Okhotsk are the most species-rich areas in terms of phytoplankton (Loughlin et al. 1999) and the deep Arctic Ocean the poorest (Horner 1984; Melnikov 1997). Most of the phytoplankton species are known from all arctic seas albeit strains are presumably different physiologically. However, the boreal diatom, *Neodenticula seminae*, is restricted to the Pacific side (Loughlin et al. 1999).

In sea ice, >300 species of diatoms and flagellates have been recorded although some of them exist also as phytoplankton (Booth and Horner 1997). The number of heterotrophic species is presumably large, comprising about 35% of the flagellate taxa (Ikävalko and Gradinger 1997).

Among species that form phytoplankton blooms, *Chaetoceros* and *Thalassiosira* species are ubiquitous in the Arctic and Subarctic (Horner 1984; Booth and Horner 1997; Hasle and Heimdal 1998; Sukhanova et al. 1999). They, as well as *Phaeocystis*, can form blooms of several billion cells m^{-3}. Except for the Arctic Ocean, *Phaeocystis* is abundant in all subarctic and arctic seas, including fjords and estuaries (Sakshaug 1972; Barnard et al. 1984; Eilertsen and Taasen 1984; Smith et al.1991; Sandvik et al. 1996; Gosselin et al. 1997; Hasle and Heimdal 1998; Sukhanova et al. 1999; Wassmann 1999; Stuart et al. 2000).

Emiliania huxleyi frequently forms blooms in the North Sea, south of Iceland (Braarud et al. 1953; Holligan et al. 1993), and in Norwegian coastal waters and fjords, where it seems to be introduced from North Sea stocks every summer (Birkenes and Braarud 1952; Berge 1962; Johnsen and Sakshaug 2000). It is also common in the Norwegian Sea and the Subarctic Pacific (Paasche 1960; Sukhanova et al. 1999). It formed blooms in the summers of 1997, 1998 and 2000, on the north-east Bering shelf – the farthest north in the Pacific it has been observed (Stockwell et al. 2001; Murata and Takizawa 2002). *Coccolithus pelagicus* and many other coccolithophorid species are common in the North Atlantic Current and in the oceanic Bering Sea (Sakshaug et al. 1981; Sukhanova et al. 1999).

Most of the southern imports are ephemeral at high latitudes. In 1927, during a highly positive phase of the North Atlantic Oscillation (NAO), the temperate diatom *Asterionella japonica* formed a bloom at the Norwegian coast at 62.5°N (Gran 1929). In the early 1990s, also coinciding with strongly positive NAO, the toxic prymnesiophyte, *Chrysochromulina leadbeateri*, and some toxic dinophytes, formed blooms along the Norwegian coast, some of which reached the Kola Peninsula (Johnsen et al. 1997; 1999).

3.2.2 Nutritional and chemical properties

Because protein represents 30–50% of a healthy algal cell and carbohydrate 40–60%, phytoplankton resemble legumes in nutritional value. Fat content is <10% except for some large-celled diatoms. A significant fraction of the fat is made up by polyunsaturated fatty acids (PUFAs; Metz et al. 2001), which are transferred upwards in the food web by grazing (Falk-Petersen et al 1998; Henderson et al. 1998). PUFAs are particularly abundant in shade-adapted ice algae (B. Kroon, personal communication). In nutrient-starved cells the protein fraction is small (Sakshaug and Holm-Hansen 1977; Sakshaug et al. 1983).

Among phytoplankton carbohydrates, glucans (polymers of glucose) are predominant. Cellulose (β-1.4 glucan) forms the wall material of thecate dinoflagellates, and starches (α-1.4 glucan) serve as storage material. Diatoms, chrysophytes, prymnesiophytes and some athecate dinophytes produce the water-soluble β-1,3 glucan (chrysolaminarin), which can be abundant in starved cells because photosynthesis may proceed for a while whereas cell division and protein synthesis are quickly arrested. β-1,3 glucan is released both from decaying cells and from living cells of some species into the water, and from some of these species profusely so if the cells are nutrient deficient (Myklestad and Haug 1972; Myklestad 1974).

β-1,3 glucan is the most abundant carbohydrate in the oceans (Myklestad 1974) contributing significantly to the pool of biodegradable DOC. In strongly-stratified and highly-productive surface layers up to half of the DOC pool may be biodegradable. The DOC pool in the surface layer presumably ranges from 0.6–1 g C m^{-3} in the central Arctic Ocean (Wheeler et al. 1997; Guay et al. 1999), 0.85–1.8 g C m^{-3} in the Norwegian Sea (Børsheim and Myklestad1997), and 1.5–6 g m^{-3} along the Siberian part of the Bering shelf (Walsh 1995; Agatova et al. 1999). Judging from the occurrence of a late winter minimum, biodegradable DOC is likely to be metabolised by bacteria within a few months (Børsheim et al. 1999; Børsheim 2000).

In the Greenland Sea, about 35 g DOC m^{-3} is accumulated in the upper 200 m in summer, almost on par with new production (Børsheim 2000). In the bottom layer of fast ice, DOC can increase from 3 to 25 g m^{-3} as the ice biota grow to around 60 g C m^{-3} (Smith et al. 1997). In drifting multiyear ice, however, DOC averages only 3.8 g m^{-3} (Bunch and Harland 1990; Thomas et al. 1995; Melnikov 1997).

Chrysophytes, prymnesiophytes and dinoflagellates, especially *Phaeocystis, Emiliania,* and ice algae, produce copious amounts of dimethylsulphopropionate (DMSP) and dimethyl sulphide (DMS), which can act as osmolytes, cryoprotectants and antioxidants (Kirst and Wiencke 1995; Sunda et al. 2002). DMSP is converted to DMS, yielding acrylic acid as a byproduct (Turner et al. 1988; Matrai et al. 1996; Schultes et al. 2000; Simó et al. 2000). In seawater, this conversion is presumably mediated by bacterial grazing (Kiene 1992) and, possibly, by decay of algal blooms due to viral attack, as indicated for *Emiliania* blooms (Bratbak et al. 1995). In sediment pore-water, concentrations of acrylic acid can exceed 50 µmol m^{-3}, retarding the respiration rate of organic matter (Vairavamurthy et al. 1986).

DMS evades from the ocean into the lower atmosphere where it may decompose and play a role in generating sulphuric aerosols. By acting as condensation nuclei, these sulphuric aerosols may in turn support low cloud formation that leads to cooling (Charlson et al. 1987). This proposed mechanism for DMS production to affect climate, however, has not yet been unequivocally substantiated (Groene 1995).

3.3 Limitation and Control of Primary Production

The maximum growth rate of phytoplankton is determined by the genetic constitution of a species (or strain) and temperature. The growth rate and production can be suppressed if a nutrient occurs in such low concentration that uptake of it by the algae is retarded. Usually, this limiting factor is nitrogen or phosphorus (for diatoms, also silicate limitation occurs; Droop 1974) unless light limits photosynthesis. The degree of limitation is affected by controlling factors that modify the nutrient supply and light, such as vertical mixing and grazing (Thingstad and Sakshaug 1990). In the northeast Pacific iron generally controls nitrate uptake (Frost and Kishi 1999) and this can happen temporarily in other subarctic regions. It has, for instance, been observed in April in the Trondheimsfjord, Norway, in the transient oligotrophic situation in April between the spring bloom in March and the entrainment-driven bloom in May (Öztürk et al. 2002).

The concept of one limiting factor is further complicated in natural communities by the presence of many species with different light and nutrient strategies. An oligotrophic community is by definition nutrient-limited because addition of the limiting nutrient enhances the growth rate (thus community biomass and production). Paradoxically, such communities are usually characterised by cells that are not very nutrient-deficient, indicating that the nutrient requirement of the predominant species may fit the nutrient flux. Increasing the supply of the limiting nutrient will force a shift towards predominance of other species that fit the new and higher flux of nutrients (thus increasing growth rate and community biomass further). Near steady state, neither of these species show distinct signs of starvation. Consequently, strongly starved populations are in the main restricted to transient situations, such as when the nutrient flux changes from high to low relative to community needs during the conclusion of a phytoplankton bloom. Presumably cells that are starved beyond a certain extent are not viable.

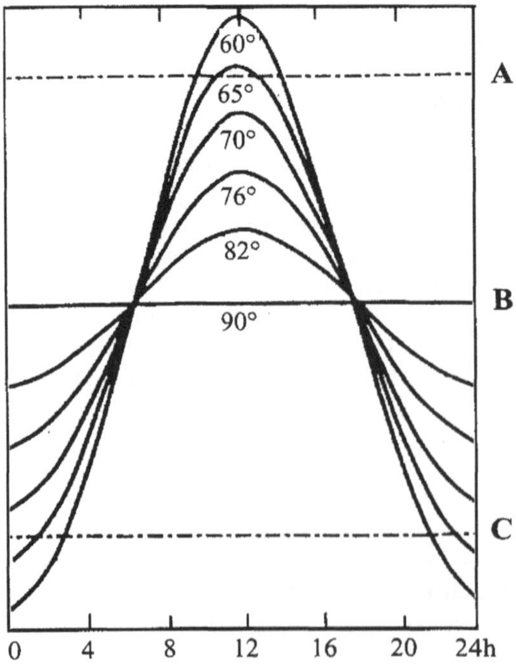

Fig. 3.1. The daily solar orbit as seen from different latitudes at the Earth's surface. The orbit can be regarded as fixed for a given latitude whereas the seasons arise from vertical declination-dependent movements of the horizon (horizontal line), which represents winter solstice in its uppermost position (A), summer solstice in its lowermost position (C) which is 47.1° lower, and the vernal and autumnal equinoxes in its mid position (B), when the sun is in zenith and night and day are equally long over the whole Earth. (from Sakshaug and Slagstad 1991)

3.3.1 Light

The arctic light regime offers extreme variations, from midnight sun to winter darkness. In addition to this extreme seasonal range in light intensity, algae are subject to a vertical light gradient in the water column, which is strongly amplified by the presence of an ice cover. Moreover, solar elevation is low and the light angle at the ocean's surface is "flatter". During summer solstice, the solar elevation at the North Pole is only 23.45° (Fig. 3.1). Thus the maximum surface irradiance (clear midsummer day) is low, theoretically ranging from 1200 µmol m^{-2} s^{-1} over 400–700 nm wave length (known as Photosynthetically Available Radiation or PAR) at the North Pole to 1700 µmol m^{-2} s^{-1} (PAR) at 60°N (Fig. 3.2).

Scattered clouds not covering the sun can increase the clear-sky irradiance by 5–10%. However, a thin but continuous layer of cirrus clouds or haze may attenuate irradiance by 20–30%, and a 20–100 m thick layer of polar fog, typical over ice-free areas in summer, by up to 60–70% (English 1961; Apollonio 1980; Bischof et al. 1998). Low clouds associated with passing low-pressure fronts can attenuate light by 80–90%, averaging 70% for several solar passes and the intervals between them (Sakshaug and Slagstad 1991).

At solar elevations >10°, glassy open waters reflect 3–10% of the down-welling irradiance, increasing towards 100% near sunset. In wavy and foaming sea, reflection losses are about 30% (Kirk 1994). Dry snow under clear skies reflects 80–97% of the light (Andersen 1989; Winther et al. 2001), decreasing to 45–70% during melting (Langleben 1969). Sea-ice reflects 40–80%, but surface ponds may lower this to <20% (Rothrock 1978). The average reflection loss for the whole Arctic Ocean in midsummer is about 46%, assuming 85% ice cover (Andersen 1989), but may be considerably lower if there is less ice cover. In an ice-free Arctic Ocean, average reflection losses would be about 10–20%.

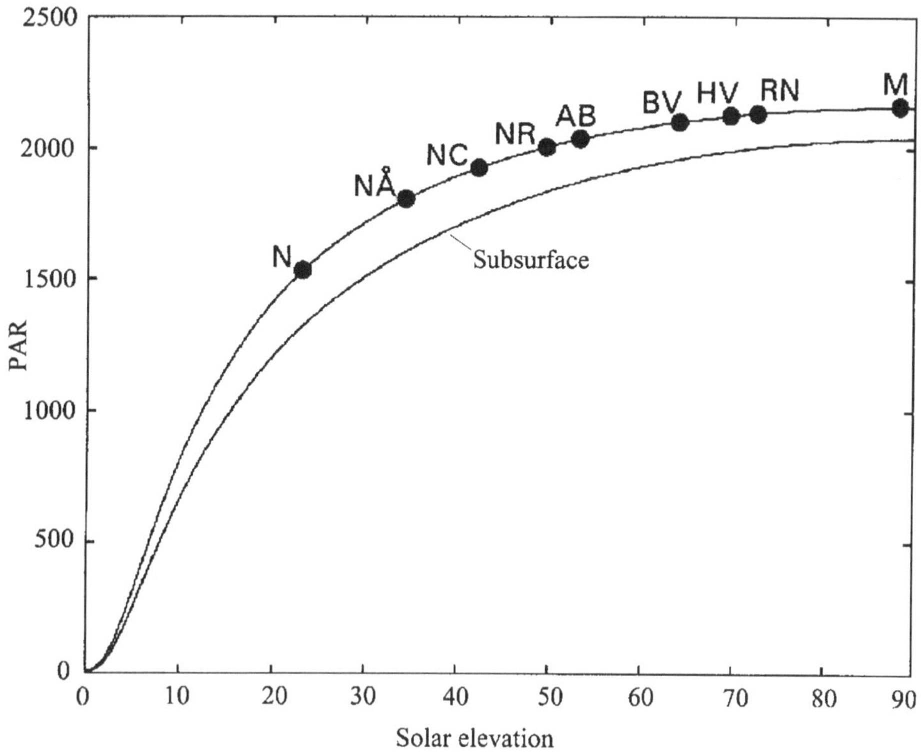

Fig. 3.2. Clear-sky irradiance (PAR = Photosynthetically Available Radiation = total irradiance over 400–700 nm; μmol m^{-2} s^{-1}) at different solar elevations assuming a maritime atmosphere and maximum irradiance at true noon on summer solstice in different places. Upper line: irradiance at the sea surface; lower line: irradiance just below the sea surface. Spectral distribution varies little for elevations >10°. (Adapted from Bird 1984). N: North Pole; NÅ: Ny Ålesund, Svalbard; NC: North Cape, Point Barrow; NR: Nome, Reykjavik, Trondheim; AB: Anchorage, Bergen, Oslo, Stockholm, Helsinki, St. Petersburg; Brussels, BV: Vancouver; HV: Halifax, Vladivostok; RN: Rome, New York; M: Miami, Gran Canaria

Attenuation in water and ice

The clearest (pre-bloom) arctic and subarctic waters have a scalar vertical attenuation coefficient (k_s PAR) of 0.07 m^{-1} in the upper 50 m, implying that 1% of the surface light reaches a depth of 66 m and 0.1% of the surface light at least 130 m (Dalløkken et al. 1994). Common arctic and subarctic Chl*a* concentrations, 0.5–10 mg m^{-3}, increases k_s(PAR) from 0.081 to 0.3 m^{-1}, yielding a 1% light depth of 57–15 m. Extreme Chl*a* concentrations, >60 mg m^{-3}, common in highly-productive parts of the Bering shelf, can increase k_s (PAR) to >1.3 m^{-1} and lower the 1% light depth to <3.5 m.

Silt causes the water to take on an icy greenish or, in extreme cases such as glacier-fed estuaries, a milky-white colour due to strong scattering. For example, in Kongsfjorden, Svalbard, the 1% depth is only 6 m in the zone of freshwater influence, in contrast to 30–35 m near the fjord mouth (Halldal and Halldal 1973; Bischof et al. 1998). The densest silt suspensions may cause darkness at depths <0.1 m (Eilertsen et al. 1989). Scattering in coccoliths generates a silt-like signal that makes *Emiliania* blooms more "visible" than other blooms. Thus the bloom on the Bering shelf in 1997 raised k_s from 0.1 m^{-1} to 0.7 m^{-1} for 4 mg Chl*a* m^{-3}, which is 2.5 times the expected k_d for a diatom bloom with the same Chl*a* concentration (Stockwell et al. 2001).

Yellow matter (Gilvin), which is formed by the Maillard reaction between free amino groups and carbohydrates in decaying organisms, is omnipresent in the sea (Painter 1983; McCarthy et al. 1996) but little studied in the Arctic. It absorbs mainly blue and UV light. The allochthonous supply from rivers is vast, and although much of it sediments near the freshwater/seawater interface, a fraction is transported long distances, for instance from the Baltic Sea along the Norwegian Coast Sea and to the Barents Sea (Aas et al. 2002).

Snow has extremely high attenuation coefficients, from 16 to 45 m^{-1} (Weller and Parker 1972; Maykut and Grenfell 1975), implying that a mere 15 cm of snow on top of the ice can attenuate irradiance to 10% of the surface value. Moreover, the high attenuation coefficient of the sea ice alone, 1.3–2.5 m^{-1}, can reduce irradiance immediately below the ice and its algal layers to 0.2–5% of surface PAR

Table 3.1. Irradiance immediately below different types of ice and their algal layers in the Barents Sea. The lowest values pertain to ice with a thin layer of snow and a layer of ice algae underneath. Based on measured measurement of E_s(PAR; (Sakshaug et al. 1994)

	Ice thickness (m)	Light under ice (% of surface PAR)	
		sunny	cloudy
One-year ice	0.8–1.5	1–0.25	3.5–1
Two-year ice	1–3	3.5–1.5	5–0.5
Ice >2 years	2.4–3.5	2.5–2	0.5–0.2

Table 3.2. The 1% light depth (m) for various chlorophyll *a* concentrations (mg^{-3}) in a homogeneous water column of "clearest Nordic Seas water" ($k_d = 0.07\ m^{-1}$), at 440 nm, 550 nm and PAR, during a diatom bloom with absorption characteristics 0.024, 0.006, and 0.04 m^2 (mg Chl*a*)$^{-1}$ in the three light fields, respectively. Surface irradiance is set at unity in each case. Based on measurements of E_d

Chl*a* mg m^{-3}	440 nm	550 nm PAR	400–700 nm
	blue	green	
0	230	61	66
1	104	57	49
2	68	53	39
5	33	44	24
10	18	34	15
20	9.2	24	8.5
40	4.7	15	4.5

(Table 3.1). Consequently, thawing and the establishment of meltponds can enhance irradiance under the ice by >10-fold. Essentially, sea ice and the snow on top of it form a thin yet optically dense layer that transforms the light field of the water immediately beneath the ice into a light field that in ice-free waters would be typical for 30–100 m depth.

Spectral effects

Pure seawater is transparent mainly to blue light (clearest at 475 nm), followed by green light, and is nearly opaque to red light and UVB (Fig. 3.3). Pure sea ice and snow are almost identical to pure water with respect to spectral discrimination. The 1% depth for blue light can reach 200 m (three times the 1% PAR depth) in pre-bloom waters, penetrating to the bottom of most of the arctic shelf seas (Table 3.2). In contrast, the average 10% light depth for UVB (310 nm) in pre-bloom waters averages 0.9–3.8 m in Kattegat and Norwegian coastal waters, where yellow matter is abundant, 4–9 m off Svalbard, and 14 m in the Norwegian Atlantic Current (Aas and Højerslev 2001). Because of the efficient attenuation of red light by seawater, marine photosynthesis is dependent mainly on the distribution of blue-bluegreen light.

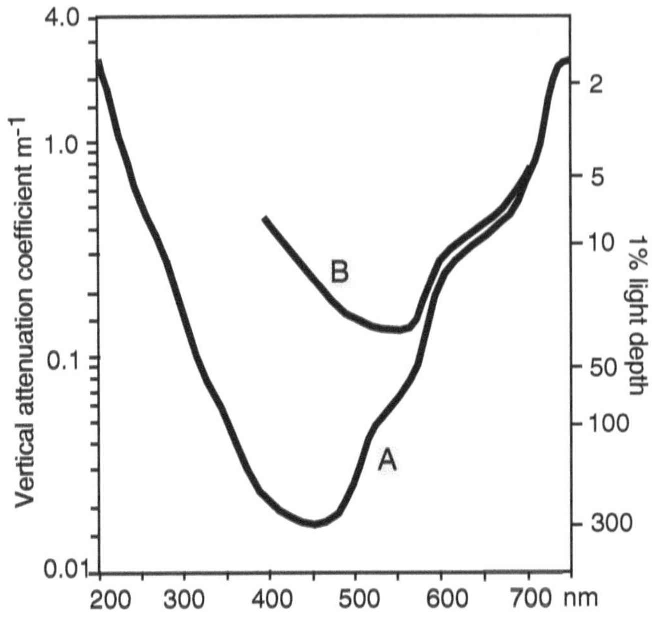

Fig. 3.3. Spectrally dependent diffuse vertical attenuation coefficient and 1% light depth for "clearest seawater" (lower curve marked A) and for pre-bloom green (Gilvin-infested or so-called Case II) waters from the Trondheimsfjord (upper curve marked B). Data taken from Smith and Baker (1981) and G. Johnsen (pers. comm.)

Algae, in contrast to pure seawater, absorb mainly blue-bluegreen and red light while being virtually green-transparent. Thus, when a phytoplankton bloom develops, the 1% light depth decreases more rapidly for blue light than for green light. Self-shading by phytoplankton in the water column is therefore considerably more pronounced than apparent from the commonly measured PAR depths, which are more similar to light depths for green light (Table 3.2).

3.3.2 Nutrients

The maximum (winter) surface-layer concentration of nutrients is dependent on the extent of admixture of deep water and its nutrient concentration. The latter is low in the Atlantic Arctic and the Atlantic Subarctic because these areas receive north-flowing upper water that is "biologically spent". However, after having sunk in the Greenland and the Labrador Seas and moved southwards as a deep current for a century or two towards the Southern Ocean, the water becomes enriched with nutrients that originate from sinking and decaying organic matter. By the time such water ascends in the Southern Ocean, it is 3–5 times richer in nutrients. A fraction of this water in turn spreads to the world's oceans and enters the Pacific Ocean as nutrient-rich Antarctic Bottom Water and Pacific Intermediate Water. Thus the world's oldest and most nutrient-rich deep water is found in the North Pacific, where deep-water production is hindered by low surface-layer salinity (Schlosser et al. 2001). Consequently, at depths >300 m in the subarctic Pacific and the oceanic Bering Sea, nutrient concentrations reach a phenomenal 45 mmol m^{-3} of nitrate, 3.5 mmol m^{-3} of phosphate and 100–300 mmol m^{-3} of silicate (Coachman et al. 1999; Shiomoto 1999).

These high concentrations to some extent are reflected in the surface winter nutrient maximum in the Bering Sea. However, restricted vertical mixing and a small, albeit significant, primary production in winter force surface-layer concentrations to be lower than in the Southern Ocean yet higher than in the Atlantic Arctic (Table 3.3). Moreover, the Bering shelf exhibits a strong horizontal gradient. There is about 25 mmol nitrate m^{-3} along the shelfbreak in winter, 15–20 mmol m^{-3} on the Middle and Outer Shelves, and much less along the north-east coastal Alaskan Shelf due to the long distance from the deep waters and nutrient-poor freshwater supply (Coachman and Walsh 1981). Nutrient transport from the Pacific Ocean can be traced in the Amerasian Basin at 200–300 m depth but not in shelf waters except for part of the Chukchi Sea (Carmack et al. 1997).

Except for the iron-limited eastern oceanic Bering Sea and the eastern subarctic Pacific, surface concentrations of nitrate, phosphate and silicate approach the detection limit following the conclusion of the spring bloom. Even on the Bering shelf, nitrate concentration is usually <1 mmol m^{-3} in summer, demonstrating efficient utilisation of the winter nutrients. Ammonium concentrations in the arctic seas are usually <1 mmol m^{-3} but can, due to high biological activity, reach 3–6 mmol m^{-3} occasionally in the surface layer of the Bering shelf and up to 10–15 mmol m^{-3} near the bottom of the Middle Bering shelf. This is high enough to inhibit the uptake of nitrate by phytoplankton (Whitledge and Luchin 1999).

The nitrate:phosphate ratio of arctic waters ranges from 11–16 (mol mol^{-1}). Because silicate ranges from 6–8 mmol m^{-3} in Atlantic Water to 60 mmol m^{-3} in Pacific waters, silicate:nitrate ratios range from 2.36 in the Chukchi Sea to 1.87 in the Eastern Canadian Arctic, 0.57 in the Labrador Sea, and 0.31 in Eurasian waters (Codispoti 1979; Anderson and Dyrssen 1981; Harrison and Cota 1991). The highly species-dependent N:P ratio of averages 16, thus N:P ratios <16 are presumably nitrogen limiting. Because arctic rivers are rich in available nitrogen (and silicate) but poor in phosphate (Macdonald et al. 1987; Gordeev et al. 1997), P-limitation of phytoplankton growth is likely in brackish water with salinity <25–26 (Sakshaug and Myklestad 1973; Sakshaug et al. 1983).

Table 3.3. Winter surface nutrients, mmol m^{-3} in various arctic and antarctic seas

	Barents Sea (Atlantic)	Bering Sea (Pacific)	Ross Sea (Antarctic)	Scotia Sea (Antarctic)
Nitrate	12	10–30	25	30
Phosphate	0.85	1.0–2.0	2	2
Silicate	6–8	25–60	50–60	100
Ammonium	2	2.8	4	4

3.4 Primary Production and Growth Rate

Whereas photosynthesis is restricted to daytime, loss of organic matter from the cells (respiration) takes place both day and night. The difference between photosynthesis and loss over 24 h (except for release of DOC) is known as primary production, originally defined as plant matter which is available for herbivores (Lindemann 1942). The photosynthetic rate may reach about half its daytime value in midsummer nights at latitudes higher than 80°N (Eilertsen et al. 1989). Excretion of DOC represents a loss from the cells and production that provides a growth substrate for microheterotrophs. Production of DOC, which may reach >30 % of total primary production in extreme cases (Myklestad 1974), is not included in most arctic data sets before the mid 1990s.

The majority of photosynthetic rate measurements (P) in arctic seas are derived from incubations in artificial light or on shipboard in natural sunlight which is screened to obtain different irradiances (so-called P vs. E measurements; for definitions and references, see Sakshaug et al. 1997). It is now possible to carry out *in situ* incubations of ice algae (Mock and Gradinger 1999); however, older data are derived from ice that has been melted.

3.4.1 New vs. regenerative primary production

The concept of new vs. regenerative production is appropriate for timescales <1 year. New production is based on nutrients that are supplied from outside the euphotic zone, by upwelling or admixture of nutrient-rich deep water and, near land, also input from rivers (fixation of molecular nitrogen by planktonic cyanobacteria is presumably negligible in the Arctic Ocean considering their low abundance). Regenerative production is based on almost immediate uptake of nutrients released by the organisms that live inside the euphotic zone (closed system).

Phytoplankton blooms are characterised by a high incidence of new production (>90 %; Kristiansen and Lund 1989) whereas oligotrophic communities are largely regenerative (<30 % new), with predominance of species that have high affinity for nutrients (Kristiansen et al. 1994; Cota et al. 1996). Both sedimentation (vertical flux out of the euphotic zone) and the amount of biomass that can be harvested are ultimately limited by the magnitude of new primary production. Obviously, the incidence of new production is highest in areas with almost continual upwelling, for instance along shelfbreaks and where wind-driven vertical mixing is sufficiently deep, and smallest where vertical mixing does not reach the nutrient-rich deep layer even in winter, such as in the deep Arctic Ocean.

Away from river influence, uptake of nitrate can be used as a "tracer" for new primary production because ammonia, urea and several species of organic nitrogen represent recent biological release. Because waters rich in zooplankton release much urea (Conover and Gustavson 1999), the ratio between new and regenerative production can be overestimated by measuring only nitrate and ammonium uptake. Although rivers by definition supply new nitrogen, nitrate cannot be used as a tracer for new production because of the large riverine supply of other nitrogen species.

3.4.2 Chla:C ratio, light saturation index, photoacclimation

Because algal carbon is difficult to estimate in natural communities due to the omnipresence of other organic matter, the Chl*a* concentration is conveniently used to express algal biomass (Tables 3.4 to 3.6). The Chl*a*:C ratio of algae, however, is acclimation-dependent and therefore highly variable. The Chl*a* concentration is therefore not a good indicator of carbon biomass. In fact, the Chl*a*:C ratio ranges from 0.003–0.08 (w:w) in arctic and subarctic waters, with the majority of data within 0.01–0.04 w:w (Platt et al. 1982; Sakshaug 1989; Johnsen and Hegseth 1991; Hegseth 1993; Booth et al. 1993; Suzuki et al. 1997; Buck et al. 1998). The average Chl*a*:C ratio for the Arctic Ocean in the growth season is 0.036 w:w (Booth and Horner 1997), and for the iron-limited eastern subarctic Pacific, 0.020 w:w (Booth et al. 1993).

The Chl*a*:C ratio is lowest in algae that are acclimated to strong light and/or low nutrient supply, and highest for algae grown in weak light and/or large nutrient supply. Thus low ratios are typical for stable, shallow surface layers and meltponds in summer, and high ratios for algae growing at the underside of the sea ice. The Chl*a*:C ratio increases roughly inversely with the logarithm of the growth irradiance up to the point where irradiance

Table 3.4. Integrated Chl*a* concentration (mg m^{-2}) in different types and layers of pack ice (Gradinger 1999)

Canadian and Alaskan Coastal:	26–300	bottom 3–6 cm
Central Arctic	0.1–0.14	bottom 2–4 cm
Central Arctic	<0.01–7.1	total core
Central Arctic median values	0.02–1.6	total core

Table 3.5. Phytoplankton characteristics for the Northeast Polynya in 1992–1993 (Booth and Smith 1997).

	Average	Minimum	Maximum	SD
Chl*a* mg m^{-3}	1.38	0.04	7.38	1.19
Autotroph C, mg m^{-3}	79.2	5.95	417	82.6
Autotroph Chl*a*:C	0.016	0.094	0.0038	0.02
PP, mg m^{-3} h^{-1}	0.86	0.05	2.57	0.63
μ (d^{-1})*	0.60	0.035	1.78	0.44
Flagellate-C of aut-C	0.23	0.003	0.91	0.25

* *based on ^{14}C uptake.*

Table 3.6. Average biomass in the oceanic Bering Sea and the waters south of the Aleutian Arc (average for Western Gyre, Alaskan Gyre and the area between). Based on data from Taniguchi (1999)

	Oceanic Bering Sea	Subarctic Pacific South of Central Bering
Chl*a* (mg m^{-3})	0.9	0.4
Phytopl. (mg C m^{-3})	45	18
Hetero. micropl. (mg C m^{-3})	8.4	10
Mesozoopl. (mg C m^{-3})	22.5	10

becomes inhibiting (Hegseth 1989). Variation in the Chl*a*:C ratio of microalgae is included in recent models of the phytoplankton growth rate (Geider et al. 1996).

Diatoms consistently seem to have Chl*a*:C ratios that are 1.5–2.5 times higher than those of prymnesiophytes grown in similar growth regimes (Verity et al. 1991; Johnsen et al. 1992). Thus light-acclimated diatoms may have a Chl*a*:C ratio of 0.015–0.020 (w/w), compared to 0.005–0.011 for light-acclimated *Phaeocystis* and *Emiliania* (Sakshaug 1989; Booth et al. 1993; Johnsen et al. 1992). Accordingly, a diatom bloom of 12–25 mg Chl*a* m^{-3} equals a *Phaeocystis* bloom of 8–10 mg Chl*a* m^{-3} in terms of carbon biomass. Correspondingly, the photosynthetic parameters Pm (the maximum photosynthetic rate) and α (the maximum light utilisation coefficient or "photosynthetic efficiency"), which are similar when normalised to carbon biomass, are 1.5–2.5 times higher in prymnesiophytes than in diatoms when normalised to Chl*a* (Verity et al. 1991; Stuart et al. 2000).

The light saturation index E_k (= P_m: α) also expresses the acclimation state. Low E_k indicates shade-acclimation whereas high E_k indicates acclimation to strong light (Escoubas et al. 1995). In arctic and subarctic waters, E_k for shade-acclimated microalgae ranges from 0.5–60 µmol m^{-2} s^{-1} and is lowest in algae growing at the underside of the ice. E_k can reach up to 700 µmol m^{-2} s^{-1} in light-acclimated microalgae (Andersen 1989; Smith and Sakshaug 1990). E_k averages 163 µmol m^{-2} s^{-1} in western arctic summer phytoplankton (Harrison and Cota 1991), 94 µmol m^{-2} s^{-1} in West Greenland coastal summer plankton (Jensen et al. 1999), and 43 µmol m^{-2} s^{-1} for summer phytoplankton in the Atlantic Domain of the Barents Sea (Rey 1991). The latter perhaps reflects stronger shade acclimation in the more deeply mixed waters of the Atlantic Barents Sea.

3.4.3 Growth rate

The algal growth rate is the net carbon turnover rate of the cells over 24 h and equals the daily primary production (not including DOC) divided by carbon biomass. For cultures of arctic and subarctic diatoms grown at −0.5 °C in continuous light, the maximum growth rate is 0.4–0.6 d^{-1}, averaging 0.5 d^{-1} (Gilstad and Sakshaug 1990). Studies of shipboard cultures grown at <5 °C indicate a species-dependent ceiling of 0.35–0.9 d^{-1} for arctic algae (Table 3.5; Kristiansen and Lund 1989; Sakshaug 1989; Hegseth 1993).

The average growth rate of arctic diatoms forms saturation functions with irradiance and day length (Sakshaug and Slagstad 1991). A wide optimum range, 50–400 µmol m^{-2} s^{-1}, and weak inhibition at >500 µmol m^{-2} s^{-1}, reveals the "essence" of photoacclimation, namely minimisation of the change in the growth rate for a change in the light regime (Sakshaug and Holm-Hansen 1986). Yet for irradiances decreasing below 50 µmol m^{-2} s^{-1}, the growth rate inevitably decreases roughly with the logarithm of the irradiance (Sakshaug and Andresen 1986). Even though irradiance can be <7.5 µmol m^{-2} s^{-1} in the permanently shaded algal layer at the bottom of the ice, growth rates can be 0.12–0.16 d^{-1} (Suzuki et al. 1997).

Introducing variable temperature using Eppley's (1972) equation, the growth rate data for arctic diatoms can be used to construct a guideline for the growth rate under light and temperature control, assuming nutrient surplus:

$$\mu(d^{-1}) = 0.9e^{0.0633T}[D/(12+D)](1 - e^{-0.35E/Ek})$$

$$E_k = 140 \sin(0.493d) + 10 \text{ (positive values)}$$

where T is centigrade, D is day length (h), and d is the day of the year since winter solstice (21 December). E_k varies with the time of year, from 10 μmol m^{-2} s^{-1} in winter to 150 μmol m^{-2} s^{-1} in summer. The equation may be relevant for phytoplankton in the wind-mixed layer, especially if acclimation from shade to strong light is faster than acclimation in the opposite direction such that phytoplankton are acclimated to the light regime of the upper part of the water column. The equations predict a maximum growth rate of 0.54 d^{-1} at −1.9 °C and 0.85 d^{-1} at 6 °C.

Arctic diatoms do not seem to grow faster than their temperate cousins at low temperature. However, arctic microalgae are not likely to grow at temperatures >12–15 °C, and many species have optima <5 °C (Stapleford and Smith 1996). On the other hand, whereas arctic microalgae obviously grow well in strong light at <5 °C, 20 common temperate species from the Trondheimsfjord could not at irradiances >500 μmol m^{-2} s^{-1}. Yet they grew well at those temperatures in weak light (K. Tangen and E. Sakshaug, unpublished data). Perhaps some temperate species have an oversummering problem in the high Arctic?

3.4.4 Growth strategies

It is theoretically impossible to predict which species will become predominant in communities of more than 3–5 microalgal species because of inherent chaos in the models (Huisman and Weissing 2001). Yet it can be assumed that the community will be dominated by fast-growing, bloom-forming diatom species unless growth conditions for diatoms are unfavourable (Brand 1991). The latter obviously happens during silicate limitation and apparently also during iron limitation. The latter is known to particularly affect large bloom-forming diatom species that can sink rapidly (Muggli et al. 1996; Muggli and Harrison 1997). By suppressing the diatom component of the community, insufficient supplies of silicate and iron can lower export production by up to 30 % compared to blooms of the same magnitude in which diatoms are predominant (Tréguer and Pondaven 2000).

Patches of ammonium-rich water may inhibit nitrate uptake or protein synthesis (Kristiansen and Lund 1989; Taguchi and Smith 1997; Varela and Harrison 1999), favouring regenerative production at the expense of new production, also lowering export production.

Because of strong self-shading by diatoms in the water column, diatom blooms are associated with shallow mixing, <40 m (Sakshaug and Holm-Hansen 1984). In contrast, the Chla-poor *Phaeocystis* is often associated with depths of mixing of 60–80 m (Goffart et al. 2001; Olsen et al. 2003). This species can also thrive in strong light because of its extremely plastic cycle of photo-protective pigments (Moisan et al. 1998).

Emiliania huxleyi blooms are associated with stratified or shallowly mixed (<30 m) layers and strong light. Some clones tolerate 1000–1600 μmol m^{-2} s^{-1} without showing signs of photoinhibition (Balch et al. 1992; Nanninga and Tyrrell 1996). Moreover, this species, in contrast to large diatoms, has small requirements for iron and other micronutrients (Brand 1991) and seems to be favoured at low phosphate concentration (Tyrrell and Taylor 1996). Although a protective function of the coccoliths against strong light has not been observed, the very weak light below the highly light-scattering *Emiliania* blooms may enhance its competitiveness.

Coccoliths constitute up to 40–50 % of the carbon in the cells (Balch et al. 1996) and are produced from bicarbonate, mediated by the enzyme carbonic anhydrase. They may therefore provide an inexpensive way to keep the CO_2 concentration inside the chloroplasts high, reducing the energy expenditure of CO_2 uptake (Anning et al. 1996). Like *Emiliania*, many micro-algal species, including some large diatoms, have the enzyme carbonic anhydrase and therefore can generate intracellular CO_2 from bicarbonate (Sültemeyer 1998; Goldman 1999). It has been suggested that some diatoms may shift to C_4 metabolism when zinc deficiency inhibits carbonic anhydrase activity (Reinfelder et al. 2000). Some species are possibly carbon-limited at normal concentrations of CO_2 (Riebesell et al. 1993).

3.5 Seasonality

In most arctic and subarctic waters, the growth season begins with a bloom that feeds on the winter nutrients and is triggered by the seasonal increase in light and stabilisation of the upper layers. The growth season ends, depending on latitude, in late summer or autumn when the critical depth for algal growth is reduced to <20 m and/or when formation of sea ice begins (Hegseth 1997). In the eastern

Subarctic Pacific and in the shelfbreak area near the Atlantic Current north of Svalbard, grazing can suppress the phytoplankton biomass, effectively hindering a bloom from developing (Banse and English 1999; Wassmann 2001).

3.5.1 Pre-bloom, winter and survival

Few species seem to have designated winter survival features such as the heavily silicified resting spores known from a few diatom species, and the cysts of some dinophytes. A heterotrophic mode of winter survival has not been found important in arctic diatoms (Horner and Alexander 1972) but cannot be excluded in mixotrophic species considering the large pool of bio-available DOC both in the water and the sea ice.

Many microalgae are presumably shade-acclimated through the autumn, yielding a gradual lowering of the growth and respiration rates (Sakshaug and Andresen 1986), and in the limit becoming functionally resting stages that should have little problem surviving winter darkness for several months (Antia 1976). Such stages can look like normal cells, perhaps except for being more densely packed with organic matter and pigments (Doucette and Fryxell 1983). Such cells of the centric diatom *Skeletonema costatum* are known from the Murman littoral (Kashkin 1964).

The re-emergence of phytoplankton in spring presumably implies a reversal of the gradual autumn acclimation for many species. According to numerical plankton models, the size of the winter phytoplankton biomass may have a noticeable impact on the development of the bloom and its magnitude (Sakshaug and Slagstad 1991). However, diatom species that produce resting spores profusely, such as *Chaetoceros socialis* under N and P limitation (not under Si limitation), arrive in abundance "suddenly" and fairly independently of the light regime in the preceding weeks. The germination of the spores requires a critical day length of perhaps 11–12 h and blue light perhaps penetrating >200 m, both of which can happen in pre-bloom water from early March and onwards (Eilertsen et al. 1993; Chautón 1997).

3.5.2 Spring blooms, vertical mixing and ice-edge blooms

A phytoplankton bloom begins when the integrated gross photosynthetic rate in the mixed part of the water column becomes larger than the losses (Sverdrup 1953). The losses include respiration, sinking, grazing and release of DOC. The depth above which this happens, known as the critical depth, is itself a variable that decreases with increasing attenuation coefficient of the water and increases with the surface irradiance. At a depth of mixing of 20 m, a bloom can in principle begin 4–6 weeks earlier than at a depth of mixing of 100 m (Sakshaug and Slagstad 1991).

In the permanently ice-free waters of the Bering Sea and the Atlantic Barents Sea, stability is thermally induced, creating a surface wind-mixed layer of 20–50 m thickness in late May - early June (Stabeno et al. 1999; Olsen et al. 2003). In the Atlantic Barents Sea, the Norwegian Sea, the Icelandic Sea and the Irminger Sea, the bloom peaks in late May - early June (Steemann-Nielsen 1935; Halldal 1953; Paasche 1960). In the subarctic waters north of Iceland, however, the spring bloom culminates in late March - late April, as in Norwegian fjords and permanently ice-free subarctic coastal waters (Gislason and Astthorsson 1998).

In areas with first-year ice, the growth season begins with an ice-edge bloom that forms a 20–100 km wide belt off the ice edge (Gran 1931; Braarud 1935; Alexander and Niebauer 1981; Sakshaug and Skjoldal 1989; Head et al. 2000). Such blooms develop rapidly when the ice breaks up because nutrient-rich water is exposed to light while the water from the melting ice establishes an upper mixed layer of only 15-35 m depth, separated from the underlying waters by a strong pycnocline.

In a spring following a particularly cold winter, the ice cover may reach south to the northern part of the "warm" waters of the Atlantic Barents Sea and south of the Bering shelfbreak. Because of early melting from below, the bloom will begin earlier after a cold winter than after a "warm" winter when the ice cover does not spread this far south. Consequently the timing of the bloom in these areas is extremely sensitive to the distribution of sea ice (Sakshaug and Skjoldal 1989; Skjoldal and Rey 1989; Stabeno et al. 2001).

Strong stratification in Arctic Water causes rapid bloom development and nutrient depletion. However, as long as the sea ice continues breaking up, exposing nutrient-rich water, the bloom will follow the retracting ice edge. Thus an ice-edge bloom, in reality a spring bloom "on the move," is established. Although at times poorly defined, like the ice edge itself because of local weather and currents (Niebauer 1991; Guest et al. 1995; Fennel and Johannessen 1998), the ice-edge bloom is in principle young near the ice edge and in its concluding phase on the far side. In late summer and autumn, ice-edge blooms do not form because the remnants of the first-year ice drift over biologically "spent" water. Because of the ice-edge bloom, the ice-edge zone is the most important habitat for life in the seasonally ice-covered Arctic.

Because it is dependent on the timing of the melting of sea ice, the ice-edge bloom begins in mid April – early May at the southernmost fringes of the first-year ice. Close to the multiyear ice, it is likely to begin in July (Zenkevich 1963; Alexander and Niebauer 1981; Sakshaug and Slagstad 1991; Eilertsen et al. 1993; Strass and Nöthig 1996). Consequently the growth season lasts for 180–200 days in the southernmost ice-covered parts and for only 70–100 days close to the multiyear ice (Strass and Nöthig 1996; Niebauer et al. 1999). In the Greenland Sea, however, southward transport of ice and late melting causes a late ice edge bloom, in May, as far south as Denmark Strait (Braarud 1935).

In the Atlantic sector and parts of the Arctic Ocean that are not influenced by Pacific water, diatom blooms can reach 8–30 mg Chl*a* m^{-3} (Gosselin et al. 1997; Grahl et al. 1999; Olsen et al. 2003). On the middle and outer Bering shelves and in the Chukchi Sea, 16–36 mg m^{-3}, <16 mg m^{-3}, and >100 mg m^{-3} Chl*a* m^{-3}, respectively, have been observed (Alexander and Niebauer 1981; Sambrotto et al. 1986; Walsh 1995. The spring bloom presumably is responsible for 50–65% of the annual new production in the Barents Sea and on the Bering shelf. This leaves 50–35% of the production for summer and autumn (Sambrotto and Goering 1983; Sakshaug and Slagstad 1992; Sakshaug et al. 1995; Niebauer et al. 1995).

Beneath the drifting multiyear ice in the Arctic Ocean, phytoplankton blooms are dependent on the melting of the snow on top of the sea ice and the partial melting of the ice itself. Thus the growth season may last for only 6 weeks (English 1961; Zenkevich 1963). In polynyas, however, the growth season is extended by 2–4 months because of the early melting of the ice (Smith et al. 1997; Suzuki et al. 1997). In estuaries in Svalbard, the growth season lasts for 4 months or less. It can, however, be restricted to <6 weeks where large input of silt-laden glacial water restricts the growth season to the window of time between the melting of the sea ice and the accelerated melting of snow and glacial ice.

3.5.3 The post bloom

Except where iron limitation and/or strong grazing keeps the nutrient requirements of the algal biomass small, the post-bloom phase is low in surface nutrients. This is particularly true for Arctic Water, which is separated from the underlying water by a marked pycnocline. The regenerative euphotic zone however, deepens to 30–35 m depth in the course of the summer. Only a Chl*a* maximum layer near the pycnocline contributes to the new production.

In less stratified waters, secondary small blooms can arise, depending on episodic supply of nutrients. Such pulsed supply can occur more or less regularly, for instance in connection with the passing of low-pressure atmospheric systems. The fronts in principle supply new nutrients by mixing whereas the calmer and less cloudy days between passing fronts ensure good conditions for growth (Fig. 3.4; Sakshaug and Slagstad 1992; Sakshaug et al. 1995). Only winds exceeding 8 m s^{-1} in the Barents Sea and 9–12 m s^{-1} in the Bering Sea (due to lower salinity) have a marked impact on primary production through enhanced vertical mixing and Ekman suction – the latter can reach 25–30 m d^{-1} under the most intense lows. Because summer winds are usually stronger in the Bering Sea than in the Barents Sea, the impact on annual primary production may not be very different (Slagstad and Wassmann 1996; Overland et al. 1999).

In some areas, hydrodynamics in combination with topography keeps primary production high by bringing nutrients to the euphotic zone more or less continually. Examples are tidal mixing and turbulence generated by shear forces where currents meet, the turbulent wake behind banks and islands in the path of a current (island and bank effect), and upwelling in straits, gyres and along shelf-breaks. Thus primary production in summer is particularly large near Svalbard, Bjørnøya and the Barents Sea, the Bering shelfbreak, and along the Aleutian Arc. Entrainment by river water is important in the Eurasian Arctic and in estuaries, and can "lift" 5–20 fold as much deep water to the surface as the freshwater flux itself. However, entrainment does not bring nutrient-rich water of the deep Arctic Ocean to the surface.

In Atlantic Water, the surface layer post-bloom community typically contains 0.6–0.8 mg Chl*a* m^{-3} in July, decreasing to <0.3 mg m^{-3} in early September when the depth of mixing is larger. Arctic Water typically has 0.2–0.5 mg Chl*a* m^{-3} in summer, and a 3–5 m thick Chl*a* maximum layer near the pycnocline with >20 mg m^{-3} in exceptional cases (Nielsen and Hansen 1995; Heiskanen and Keck 1996; Luchetta et al. 2000). On the Bering shelf, surface layer concentrations of 2–4 mg Chl*a* m^{-3} are common in summer (Sukhanova et al. 1999).

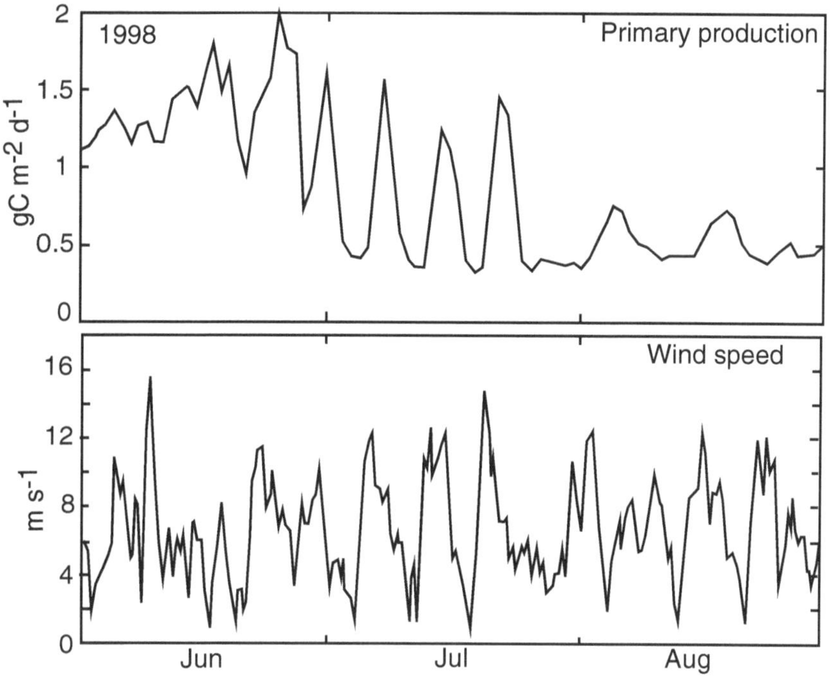

Fig. 3.4. Modelled primary production (upper panel) and wind-speed in Atlantic Water of the Barents Sea, summer 1998 (lower panel). Wind data and the hind-cast model are courtesy of the Norwegian Meteorological Institute; production model, D. Slagstad)

3.6 Distribution of Primary Production

Being highly dynamic, phytoplankton communities can change 10-fold in biomass within a week. In a high-speed movie, the Chl*a* distribution in ice-free arctic seas would resemble the flickering of the northern lights. Accurate estimation of primary productivity therefore requires a very fine-meshed spatial and temporal grid of measurements, well beyond what is possible today. Areal and temporal estimates of primary production are instead based on extrapolation and interpolation using relatively simple models and knowledge of algal physiology, and are therefore not very accurate. Luckily, light and nutrient characteristics of arctic and subarctic seas offer certain constraints on primary production estimates, but the sparse knowledge of grazing rates and the sinking characteristics of individual species of microalgae still represent a problem.

The arctic and subarctic seas are conventionally grouped here into the deep Arctic Ocean, polynyas, the arctic shelf seas, the Atlantic sector (the Nordic Seas, comprising the Greenland, Icelandic and Norwegian Seas, and the Baffin Bay, Hudson Bay, and the Labrador Sea), the Bering shelf, the oceanic Bering Sea, and the Okhotsk Sea (Table 3.7). Unless specified otherwise, primary production is given as total particulate production.

3.6.1 The deep Arctic Ocean Basin

The deep Arctic Ocean Basin is characterised by surface waters <0 °C and 3–5 m thick multiyear ice (Carmack et al. 1997). In summer it is strewn with leads that move about at the whims of wind and currents. The areal multiyear ice-cover in summer is 80–90% of the winter ice-cover (AMAP 1998). If a lead lasts for a week or more, productivity will be enhanced both in the lead itself and in its immediate surroundings.

Diatom blooms develop in the upper 10 m of the water column in July in years with early melting of the snow cover, and secondary peaks have been recorded in early October (Gran 1904; Usachev 1961; Kawamura 1967; Booth and Horner 1997; Fortier et al. 2002). In the poorest plankton communities, at >85°N, flagellates, especially athecate heterotrophic dinophytes, are predominant (Gosselin et al. 1997; Booth and Horner 1997).

According to transect data from August 1994 (Wheeler 1997), integrated phytoplankton in the mixed layer is only 1.8–25 mg Chl*a* m^{-2} (<1 mg Chl*a* m^{-3}), and the median value for ice algae is only <1.6 mg Chl*a* m^{-2} (Gosselin et al. 1997). Ice patches with up to 40 mg Chl*a* m^{-2}, however, do occur (Sherr

Table 3.7. Approximate annual particulate primary production (new and total in g C m^{-2}); DOC not included) and total areal primary production in Tg (10^6 t C) in different arctic seas. Data for the Nordic Seas, the Barents Sea, the Bering Sea, and the Chukchi Sea, are based on measurement and modelling of nitrate in the water column, yielding new production, and model estimates for total production on basis on f-ratios and ^{14}C, where available. New production in the deep Arctic Ocean is based on phosphate fluxes. Other areas are represented by ^{14}C measurements of total production. For Siberian shelf seas except the Barents and Chukchi Seas, new production is assumed to be 25% of total production. The database for the arctic shelves and the oceanic Bering Sea is sparse, and figures for the Kara and East Siberian Seas are calculated by extrapolation of data from neighbouring seas. Production in sea ice is included. Based on data from Platt et al. (1982), Wassmann and Slagstad (1993), Walsh and Dieterle (1994), Vedernikov et al. (1995), Gordeev et al. (1996), Springer et al. (1996), Gosselin et al. (1997), Macdonald et al. (1997), Smith et al. (1997), Zheng et al. (1997), Boetius and Damm (1998), Maita et al. (1999), Sorokin and Sorokin (1999), Noji et al. (2000), Klein et al. (2002), Tremblay et al. (2002), Anderson et al. (2003), Olsen et al. (2003), Vetrov and Romankevic (Chapter 7.7), Harrison, G. (pers. comm.)

	Area 10^3 km²	Primary Prod. (gC m^{-2} y^{-1})		Ave. areal tot.PP, Tg · y^{-1}
		Total	New	
Central Deep Arctic	**4.489**	**>11[1]**	**<1**	**>50**
Nansen Basin[2]	–	5–30	–	–
Arctic shelves	**5,052**	**32**	**8**	**279**
Barents Sea	1,512	<20–200[3]	<8–100	136
Barents north slope	–	35	16	–
White Sea	90	25	6	2
Kara Sea	926	30–50	7–12	37
Laptev Sea	498	25–40	6–10	16
East Siberian Sea	987	25–40	6–10	30
Chukchi Sea	620	20–>400	5–>160	42
Beaufort Sea	178	30–70	7–17	8
Lincoln Sea	64	20–40	5–10	3
Other (Can. Arctic)	182	20–40	5–10	5
NE Water Polynya	<50	20–50	13–32	–
North Water Polynya[7]	<31	150	90	–
Total High Arctic	**9,541**	**>26**	**<5**	**>329**
Atlantic sector	**5,000**	**97**	**50**	**483**
Baffin Bay	690	60–120	25–50	62
Hudson Bay	820	50–70	25–35	49
Greenland Sea	600	70	40	42
Labrador Sea	1,090	100	45	110
Norwegian Sea	1,400	80–150	35–65	160
Icelandic Sea	400	100–200	45–90	60
West Spitzbergen	–	120	55	–
Bering Shelf	**1,300**	**>230**	–	**>300**
Alaskan coastal	–	50–75[4]	<20	–
Siberian Coastal	–	>400[5]	>160	–
Middle, Outer Shelf	–	150–175	30–50	–
Shelf Break	–	450–900	170–360	–
Bering oceanic[6]	**1,000**	**60–180**	–	**155**
Okhotsk Sea	**1,600**	**100–200**	–	**240**

[1] DOC represents additional 6 g C m^{-2} y^{-1}.
[2] lowest production in the northern part.
[3] values >60 g C m^{-2} y^{-1} in southwestern Atlantic part; total Barents Sea average 90 g C m^{-2} y^{-1}.
[4] 35% of primary production enters detritus pool.
[5] >75% of primary production enters detritus pool.
[6] lowest production in eastern part.
[7] pertains to northern part; total area up to 80,000 km².

et al. 1997). In coastal multiyear ice, the Chl*a* concentration is two orders of magnitude larger than in drifting multiyear ice (Table 3.4), possibly because of larger nutrient supply. However, difficulties in estimating the substantial contribution by flagellates in drifting multiyear ice may have led to estimates that are too small (Gradinger 1996). Multiyear ice supports a functional microbial web (Sime-Ngando et al. 1997; Vézina et al. 1997; Gradinger 1999).

Chl*a* concentrations >10 mg m^{-3} are common only in the lowermost 10 cm of the drifting multiyear ice (Gradinger 1996), especially in so-called "dirty ice" (Juterzenka and Knickmeier (1999). Extreme concentrations can reach 1500 mg Chl*a* m^{-3} (Andersen 1989). Commonly, 4–62% of the algal biomass resides in the 20–40 cm above the bottom 10 cm of the ice (Gradinger 1999), implying gross underestimation of algal biomass if only the lowermost 10 cm is sampled. Although usually ignored, meltponds can yield concentrations around 30 mg Chl*a* m^{-3} (Gradinger 1996).

The earliest estimates of the primary production of the deep Arctic Ocean suggested <1 g C m^{-2} y^{-1} (Apollonio 1959; English 1961), the lowest for any sea. However, production in the multiyear ice was ignored, like it also was in a later estimate (Subba Rao and Platt 1984). The production of DOC, later suggested at 0.6 g C m^{-2} y^{-1} for multiyear ice (Legendre et al. 1992), also was not included.

In contrast, the 1994 transect (Gosselin et al. 1997) indicates that because of significant primary production in the multiyear ice and leads, the annual primary production of the deep Arctic Ocean is >11 g C m^{-2} y^{-1} (Table 3.7), assuming a 120-day growth season. To this comes 6 g DOC m^{-2}. Thus the annual primary production in the deep basins is >15 g C m^{-2} y^{-1}. Recent studies in the Canadian Basin (DOC included; Cota et al. 1996) indicate a similar value, 15 g C m^{-2} y^{-1}. Studies of O_2 draw-down and in-growth of nitrate in the deep water of the same basin indicate primary productivity close to 30 g C m^{-2} y^{-1} (Macdonald and Carmack 1991; Macdonald et al. 1993), which is comparable to 19–38 g C m^{-2} y^{-1} suggested for the Arctic Ocean and its shelves on basis of oxygen distribution (Pomeroy 1997).

The 1994 transect indicates that algae in multiyear ice contribute about 60% of the primary production of the whole water column in the deep Arctic Ocean, compared to 3% in the first-year ice of the surrounding regions. Although adjusted upwards by an order of magnitude in the last 10 years, the deep Arctic Ocean remains the least productive of all oceans. The permanent ice cover and the strong stratification efficiently prevents large supply of new nutrients, so new production is extremely low, presumably <1 g C m^{-2} y^{-1} (Table 3.7; Anderson et al. 2003).

3.6.2 Polynyas

Polynyas are open waters surrounded by sea ice. Many recur year after year and are, like the ice edge, noted for their abundance of seals and whales (Stirling 1997). Robert Peary described the North Water Polynya as an "arctic oasis" (Dunbar 1969). Arctic polynyas are of the "latent heat" type, characterised by prolific production of ice in intensely cold weather and persistent wind, which transports the newly formed sea ice away. Come milder weather in spring, production of ice stops and the surface ice thaws, resulting in an extended growth season locally.

Polynyas are common between fast and drifting ice along the coasts of the Arctic Ocean (coastal polynyas or flaw leads), in the Canadian Archipelago, the northern Greenland Sea, and on the Bering shelf (Fig. 3.5; AMAP 1998; Niebauer et al. 1999). Because they are not easily accessible, few large-scale biological investigations have been undertaken until the 1990s. The greatest efforts have been concentrated on the Northeast Polynya on the Northeast Greenland coast, the North Water Polynya in the Canadian Archipelago, and the Saroma-ko Lagoon on the northeast tip of Hokkaido.

The Northeast Water polynya is situated off the northernmost part of northeast Greenland. Weak wind and low albedo in May expands the open area and keeps it open so that ultimately the open surface stretches across 500–1,000 km (Böhm et al. 1997). Two fast-ice shelves, "Norske Øer" and the "Ob' Bank", prevent invasion of ice from the outside (Minnett et al. 1997). Findings of Inuit artefacts indicate that the Northeast Polynya has existed for a long time.

The transit time of water along ice-free stretches in the Northeast Water Polynya is 1.5–3 months, sufficient for phytoplankton blooms to develop (Böhm al. 1997). The Chl*a* concentration averages 1.38 mg m^{-3} (Table 3.5) but may reach 7.5 mg m^{-3} during the spring bloom in mid June-August. Diatoms make up about 70 mg C m^{-3}, which is about 85% of the phytoplankton biomass. Of this, *Chaetoceros socialis* alone contributes 52–100%. Autotrophic flagellates, including small photosynthetic dinophytes, make up 11.5% and *Phaeocystis*, 3.3% (Booth and Smith 1997). Nitrate and, to some extent, silicate, are depleted during the diatom bloom, making ammonium and silicate controlling (Kattner and Budéus 1997).

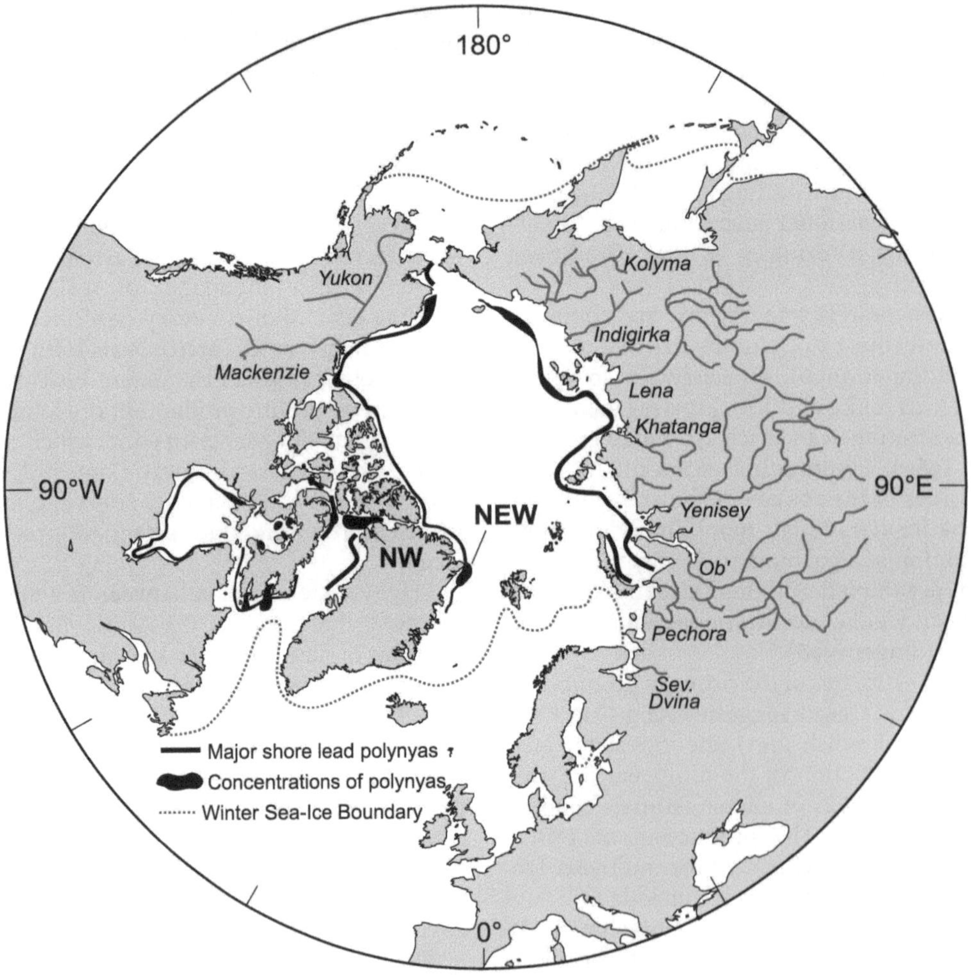

Fig. 3.5. Polynyas and major shore (lead) polynyas in the Arctic Ocean. Dotted line: Average maximum ice extent. NW: North Water Polynya; NEW: Northeast Water Polynya (adapted from AMAP 1998)

Daily primary production averages about 0.56 g C m^{-2} from late May to mid August, of which new production contributes <65%. Because the growth season is at most 90 days, annual total primary production is presumably 20–50 g C m^{-2} and new production, 13–32 g C m^{-2} (Table 3.7; Smith et al. 1997).

The North Water (NOW) Polynya is surrounded by Thule, Grisefjord and, during maximum expansion in July, also by Pond Island, Nanisivik and Arctic Bay on Baffin Island, and Resolute Bay. It is almost completely covered with ice in winter, 50% of it <30 cm. The ice-free surface begins to expand in late March or early April, first on the Greenland coast and up to 45 days later along the Ellesmere coast, reaching a total of about 80,000 km^2 in July (Barber et al. 2001; Yackel et al. 2001). Although of the "latent heat" type, cells of upwelling warm water where the polynya spreads along the Greenland coast indicate that parts of it represent the "sensible heat" type (Lewis et al. 1996).

The North Water Polynya exhibits a phytoplankton bloom in late May - early June with its duration depending on the frequency of wind-generated nutrient pulses. The bloom is concluded by strong stratification. Along the Ellesmere coast deep mixing prevents blooms from forming whereas in the north, advection of arctic water through Smith Sound and tidal mixing and local upwelling along the north Greenland coast create blooms that last until the end of August. Near the Carey islands, the island effect generates blooms, and near the northern part of Ellesmere Island, a late-summer bloom is triggered by the melting of sea ice. Thus the North Water Polynya, through its large size and diverse topography, offers a textbook case that represents almost every process that can influence primary production in the Arctic.

Data for the northern part of the NOW polynya, 31,000 km^2, show that annual new and total primary production is very high, 90 and 150 g C m^{-2}, respectively, thanks to efficient upwelling (Klein et al. 2002; Tremblay et al. 2002). This is on par with primary production in the Atlantic Barents Sea and several-fold higher than in the Northeast Water Polynya.

The Saroma-ko lagoon, which has a depth <10 m and covers 200 km^2 on the north-eastern tip of Hokkaido, is the southernmost of all seasonally ice-covered areas (44 °N) and therefore has a distinct day-night cycle throughout the year (Fukuchi et al. 1997). Ice is formed in January and reaches a maximum towards the end of that month. The ice cover, reaching a thickness of <0.5 m, lasts only for 3 months (Kudoh 1997). Because of the strong surface irradiance, the algae begin to grow in the bottom 3–4 cm of the ice at 0.5–2.8% surface light in February, during which Chl*a* increases from 2 mg m^{-2} to 38 mg m^{-2}, peaking at about 120 mg m^{-2} in March. The photosynthetic ciliate *Mesodinium rubrum* is predominant together with heterotrophic flagellates, mainly choanoflagellates <8 μm size (Sime-Ngando et al. 1997). Diatoms are less prominent (Robineau et al. 1997), presumably because they are limited by silicate (Michel et al. 1997).

3.6.3 Arctic Shelf Seas

The Eurasian shelves contribute most of the area of the arctic shelf seas. They include in addition to the Siberian shelf seas also the Barents Sea, which is distinguished from the others by its proximity to the Atlantic Ocean and the large transport of Atlantic Water (average 3 Sv) across it.

The Atlantic surface flow makes the southwest part of the Barents Sea permanently ice-free and functionally an extension of the Norwegian Sea in terms of primary and secondary production. In winter, the wind-mixed layer reaches the bottom (<250 m; Olsen et al. 2003), and the bloom fully develops when thermal stratification reduces the depth of mixing to 20–80 m in the first 1–2 weeks of June. Because of the larger depth of mixing, the bloom as a rule lasts longer than in Arctic Water, and it also fixes more carbon. Based on modelling of nitrate consumption in the water column, the average annual new primary production in the Atlantic Barents Sea is 40–70 g C m^{-2} (Slagstad and Støle-Hansen 1991). Where local currents and topography favour continual upwelling, such as at the Svalbard Bank, new production can reach 100 g C m^{-2} or more. Total annual primary production averages 80–150 g C m^{-2}, reaching >200 g C m^{-2} y^{-1} locally (Fig. 3.6).

The other Eurasian shelves, including the northern Barents Sea, but not the Chukchi Sea, are characterised by a stratified layer of Arctic Water, which is poor in nutrients because the ice-cover effectively prevents upwelling of the deep water at the shelf edge. Consequently, the nutrient-rich water near the halocline at 200–300 m depth (salinity about 33), which is of Bering shelf origin but has been reworked over the Chukchi Shelf, is not accessible for phytoplankton growth in the western Arctic. This water traditionally has extended to the Lomonosov Ridge but was in the early 1990s pushed back to the Alpha Mendeleev Ridge (McLaughlin et al. 1996; Morison et al. 1998).

Annual total primary production in these shelf seas presumably ranges from from 15–20 g C m^{-2} in their northernmost parts close to the multiyear ice, to 70 g C m^{-2} where the growth season is long due to early melting of the sea ice. In these stratified waters, which have no contact with nutrient-rich Atlantic Water, new production presumably does not reach higher than 25% of the total primary production. According to Barents Sea models (Slagstad and Støle Hansen 1991), it is likely that ice-poor years exhibit higher (30% or more) total primary production than ice-rich years due to longer growth season and deeper yet not-too-deep mixing.

Although data are sparse, it can be assumed that the ice edge bloom is the major event of the year in the arctic shelf seas, furnishing roughly half of the annual new production. The Chl*a* maximum layer near the pycnocline can contribute from <5 to 15 g C m^{-2} y^{-1}, mainly as new production, depending on the length of the growth season (Luchetta et al. 2000).

Primary production in the Chukchi Sea in part reflects the very high nutrient fluxes in a plume of Bering Sea water shortly north of the Bering Strait, where annual primary production can reach >400 g C m^{-2} y^{-1} (Walsh and Dieterle 1994). On the Mackenzie Shelf on the Canadian side, in contrast, total primary production is only 30–70 g C m^{-2} y^{-1}, averaging 50 g C m^{-2} y^{-1} (Macdonald et al. 1987). New productivity is 15–25 g C m^{-2} y^{-1}, averaging 20 g C m^{-2} y^{-1}. The Bering summer water nutrient signal is weak, represented by a layer at 60–80 m depth, and of little significance for the primary production because vertical mixing does not reach deeper than 60 m. Likewise, high concentrations of nutrients in the halocline (~150–200 m) are unavailable. Vertical mixing is shallower in ice-rich than in ice-poor years so, similar to the Eurasian shelf seas, production in ice-rich years is about 30% smaller than in ice-poor years.

Daily primary production in first-year ice of the Bering shelf and the Barents and Greenland Seas contributes <1 g C m^{-2}, of which about half takes place in the lowermost 5 cm of the ice, which is only 1–3% of total primary production (Sukhanova et al.

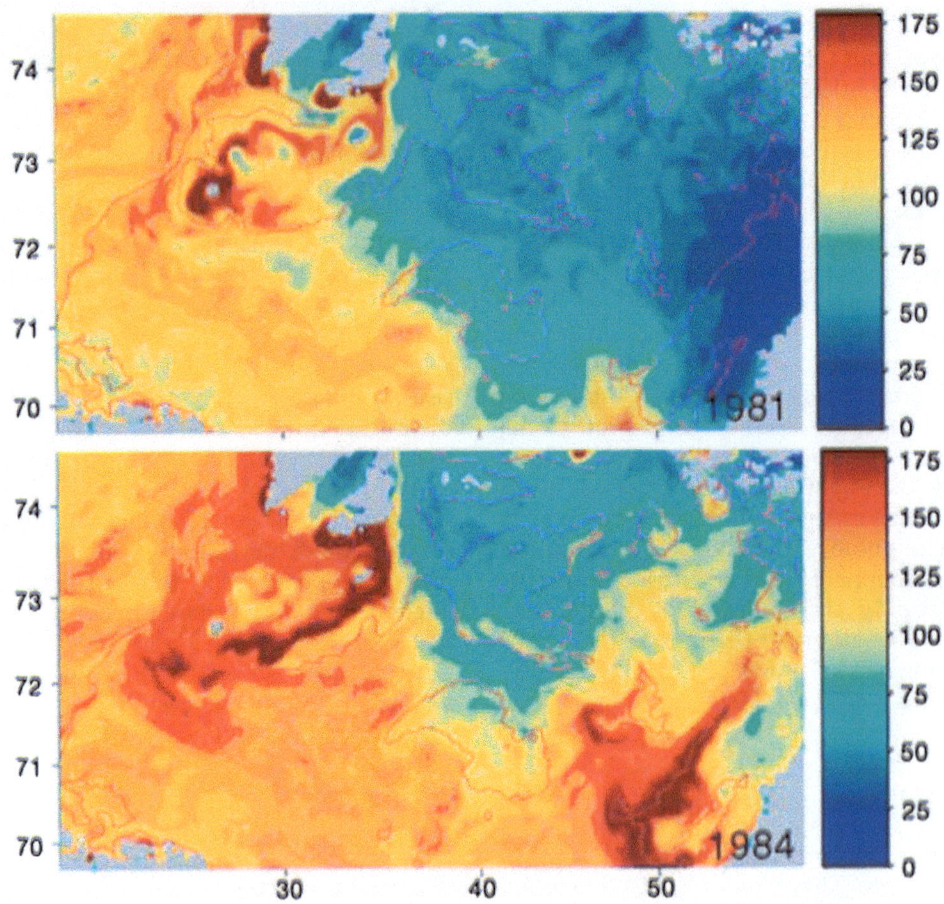

Fig. 3.6. Modelled annual total primary production (g C m^{-2}) in the Barents Sea in a "cold year" (1981; large areal ice cover) and a "warm year" (1984; small areal ice cover), adapted from Slagstad and Wassmann (1996). Latitude (vertical scale) and longitude (horizontal scale) are given

1999). In the Laptev and East Siberian Seas, however, algae in a mixture of first-year and multi-year ice contribute about 30% (Romankevich and Vetrov 2001). On the Mackenzie Shelf, ice algae likely contribute 5–10% (Horner and Schrader 1982; Macdonald et al. 1987). In the southern Hudson Bay, annual primary production in first-year ice is presumably <5 g C m^{-2} y^{-1} (Runge et al. 1991). First-year ice does not have a well-developed microbial ecosystem (Mock and Gradinger 1999).

In arctic estuaries, primary production depends on the onset of the spring bloom (early March to July; Hasle and Heimdal 1998) and the optical properties in the water. Annual primary production can reach 5 g C m^{-2} in areas with extremely short growth season and large supply of silt (Rysgaard et al. 1996), and 40–150 g C m^{-2} in fjords where this does not happen (Hop et al. 2002). The higher range is similar to that in Norwegian fjords (Parsons et al. 1988; Eilertsen et al. 1989; Parsons et al. 1989; Matthews and Heimdal 1980).

3.6.4
The Atlantic sector: The Nordic Seas, Baffin Bay, Hudson Bay and Labrador Sea

Whereas the Norwegian, Icelandic and Labrador seas are as a rule permanently ice-free, the Greenland Sea and the Baffin and Hudson Bays are covered by seasonal ice. In the Greenland Sea, the ice-covered area has ranged from <3,000–27,000 km^2 in the last 20 years (Toudal and Valeur 1996). In the parts that are affected by melting of sea ice, the phytoplankton dynamics resemble those of the arctic shelf seas, and the bloom may take place as late as May. Annual primary production averages 70 g C m^{-2} and new production, 40 g C m^{-2} (Noji et al. 2001). In the Baffin Bay, annual primary production is 60–120 g C m^{-2} (Platt et al. 1982; Jensen et al. 1999) and in Hudson Bay, assuming more pronounced stratification and a larger incidence of P-limitation, 50–70 g C m^{-2}.

The Norwegian Sea is similar to the southwest Barents Sea with respect to conditions for phytoplankton growth even though surface temperature is higher, up to 10–12 °C in summer in the southern fringes, compared to <6 °C in the Atlantic Barents Sea. Presumably the annual primary production is about the same, from 80–150 g C m^{-2}, of which 40–70 g m^{-2} is new. In the Icelandic Sea, annual primary production is presumably in the order of 100–200 g C m^{-2} (Stefansson et al. 1987; Gislason and Astthorsson 1998), of which a little less than half is likely to be new. In the Labrador Sea, where the productive season begins in early April and lasts to mid November, annual primary production is around 100 g C m^{-2} (G. Harrison, pers. comm.)

3.6.5 Bering Shelf

The fabulously high primary production of the Bering shelf is localised to the so-called "greenbelt", where daily productivity may commonly reach 5 g C m^{-2} (Fig. 3.7; Springer et al. 1996). On the northwest (Siberian) coastal shelf, where wind and topography generate upwelling (Nihoul et al. 1993), the annual production is >400 g C m^{-2}. Of this, new production makes up 40 % (Walsh et al. 1989; Hansell et al. 1993). On the nutrient-poor Alaskan side, however, annual primary production is only 50–75 g C m^{-2}, of which <20 g C m^{-2} is new (Coachman and Walsh 1981; Whitledge et al. 1986).

North of St. Lawrence Island, annual primary production may reach >800 g C m^{-2} (Springer et al. 1996), making this area among the most productive ecosystems in the ocean. Other greenbelt areas are the shelfbreak and the waters off the Kamchatka Peninsula, where the annual primary production is >300–900 g C m^{-2} (Sapozhnikov et al. 1993), and the waters along the Aleutian Arc, with 200–300 g C m^{-2} (McRoy et al. 1972; Schell and Saupe 1989). On the outer Bering shelf, the annual primary production is >200 g C m^{-2} and on the middle shelf, 150–175 g C m^{-2}, of which 30–50 g C m^{-2} is new (Sambrotto et al. 1986; Walsh and McRoy 1986; Whitledge et al. 1986; Walsh and Dieterle 1994).

The *Emiliania* bloom in 1997 covered 210,000 km^2 (700 km × 300 km), mainly on the Northeast shelf. It succeeded a diatom bloom that had concluded on the middle shelf in April. In June and July, a high pressure over the Bering Sea caused the weakest winds and smallest cloudiness on record (Overland et al. 1999), establishing a shallow mixed layer that was fresher and warmer than normal, efficiently reducing the nutrient supply (Stabeno et al. 2001). Thus stratification, nutrient depletion and unusually strong light triggered the *Emiliania* bloom.

Fig. 3.7. Schematic distribution of total (new and regenerative) primary production in the Bering Sea (Springer et al. 1996)

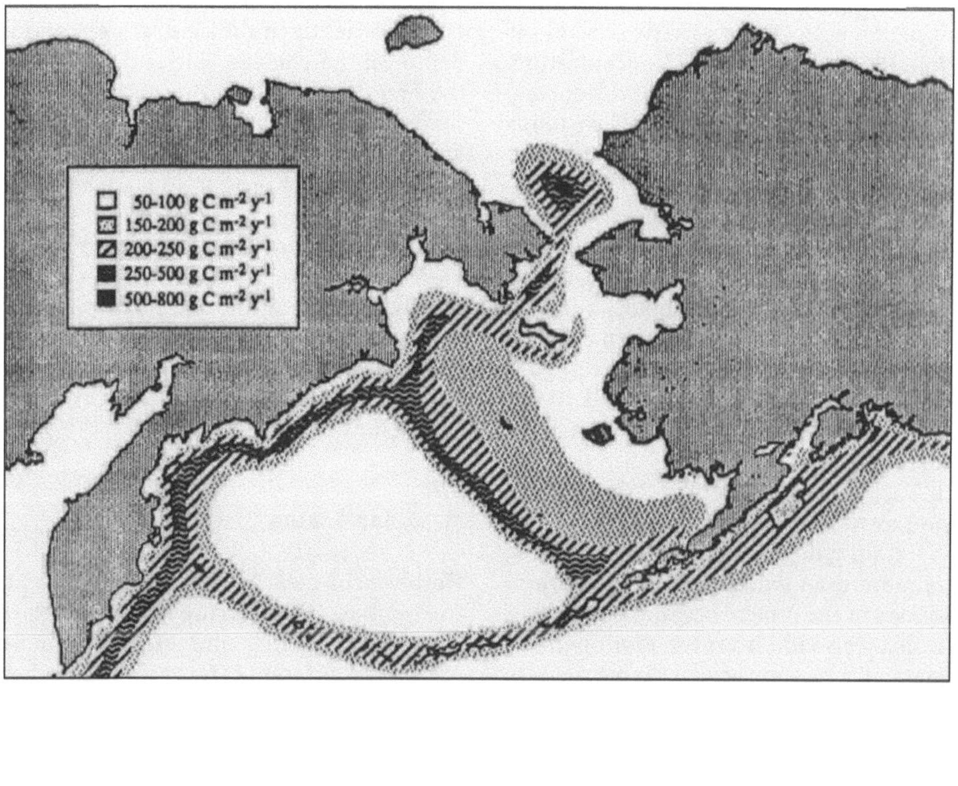

The *Emiliania* bloom was observed in the warm depleted waters in early July and persisted through the autumn, reaching 4.5 billion cells m^{-3} by September, and 4 mg Chl*a* m^{-3} and daily primary production up to 925 mg C m^{-2} (Stockwell et al. 2001). Incidentally, copepod biomass reached several times its average value. Although present at average abundance, krill did not swarm in the surface, causing food shortage and mass mortality of short-tailed shearwaters (*Puffinus tenuirostris*). From 1998 and onwards, the water column has freshened over the middle shelf because of more sea ice, lessening the transport of nutrient-rich water from shelf break onto the shelf.

3.6.6 Oceanic Bering Sea

Because of the largely wind-driven and efficient exchange of water with the Subarctic Pacific through several straits in the Aleutian Arc (average flow into the Bering Sea of 12.5 million m^3 s^{-1} vs outflow of 11.7 million m^3 s^{-1}; Stabeno et al. 1999), the oceanic Bering Sea appears as an extension of the Subarctic Pacific. Thus, the eastern part of the oceanic Bering Sea resembles the low-productive Gulf of Alaska whereas the western oceanic Bering Sea resembles the richer West Subarctic Gyre and the Oyashio region east of the Kuriles (Whitledge and Luchin 1999; Koike et al. 2001).

Like the eastern Gulf of Alaska, the eastern oceanic Bering Sea is a High-Nitrate-Low-Chlorophyll (HNLC) regime (Whitledge and Luchin 1999; Wong and Matear 1999). As a result of iron limitation in combination with strong grazing control by microplankton, the surface nutrient concentration in summer may be higher than that observed during winter in Atlantic Water (Banse and English 1999; Maldonado and Price 1999). As a result, there is hardly any seasonal variation in phytoplankton biomass despite seasonal differences in primary production (Frost and Kishi 1999). The western oceanic Bering Sea, like the western Subarctic Pacific, is less limited by iron if at all limited, presumably because of its proximity to Asian sources of aeolian dust. In spite of intensive grazing (Taniguchi 1999), a phytoplankton bloom occurs in May (Fujishima et al. 2001).

In addition to the east-west gradient, iron limitation is weaker in the oceanic Bering Sea than in the Subarctic Pacific, and the average Chl*a* concentration in the central oceanic Bering Sea is higher, 0.9 mg m^{-3} vs. 0.4 mg m^{-3} (Taniguchi 1999). Moreover, there is more than twice as much mesozooplankton biomass in the central oceanic Bering Sea than there is farther south; however, microheterotrophic biomass may be about equal (Table 3.6).

In contrast to the Gulf of Alaska, the oceanic Bering Sea exhibits a diatom bloom in April–June, peaking at 6–10 mg Chl*a* m^{-3}. There are also 2–3 modest autumn peaks, possibly caused by relaxation of the grazing control due to downward migration of the mesozooplankton and/or larger aeolian iron flux from Asia as the low pressures intensify from summer to autumn (Banse and English 1999). Because of the weaker iron limitation, diatoms make up a higher fraction of phytoplankton biomass in the oceanic Bering Sea than in the Gulf of Alaska (Koike et al. 2001; Nagata et al. 2001).

Annual primary production in the central oceanic Bering Sea is poorly known but is presumably about 250 g C m^{-2} (DOC included; Maita et al. 1999). In the eastern oceanic Bering Sea, annual productivity is presumably 60–180 g C m^{-2}, of which 30–90 g C m^{-2} is new (Springer et al. 1996).

3.6.7 Sea of Okhotsk

The Sea of Okhotsk is topographically a sibling of the Bering Sea, but is characterised by thick sea ice from December until June. Melting of the sea ice creates a vast freshwater supply that establishes a shallow pycnocline that is distinct within the euphotic zone through the whole summer and autumn, especially in the north-eastern and northern parts. Summer salinity is 28–33 in the upper 30 m and summer temperature is 6–10 °C (Zenkevich 1963).

Phytoplankton biomass is typically concentrated right above the pycnocline, which is situated at 30–50 m depth in the central part and at 12–25 m depth on the shelves. The spring bloom begins in the first days of May, culminating in early July, and there is a secondary peak in September (Sorokin and Sorokin 1999).

Daily primary production reaches up to 3 g C m^{-2} (in some patches >5 g C m^{-2}) during the spring bloom and about 1 g C m^{-2}during the secondary bloom. Whereas large diatoms are predominant during the spring bloom, phytoplankton <10 µm size are predominant in late autumn and early winter (Shiomoto 1997). Annual primary production is presumably 100–200 g C m^{-2}, with lowest values for the sea's interior.

3.7 Mesozooplankton

Herbivorous mesozooplankton remove phytoplankton biomass, transferring it up through the grazing food chain to fish and higher animals. Because mesozooplankton can graze microheterotrophs

>20 µm size such as ciliates, they also tap into the microbial food web. Typical for the marine environment, there is on average more mesozooplankton than phytoplankton biomass, in the Barents Sea, for instance, >30 vs 20 g C m^{-2} on average (Sakshaug et al. 1994).

Around 260 mesozooplankton species are known from subarctic and arctic seas, from 37 in the East Siberian Sea to 131 in the Barents Sea (Zenkevich 1963). In the Laptev Sea and the Nansen Basin, 53 species have been registered, of which 44 are calanoids that occur with decreasing number of species and biomass towards shallower waters (Kosobokova et al. 1998). Some of these species prefer low salinity and are therefore important in the shelf seas.

Among the "large" copepods (1–5 mm adult size), the subarctic species, *Calanus finmarchicus*, and *Oithona atlantica* are common in Atlantic Water and adjacent coastal waters. They are to some extent exported into Arctic Water. The arctic species, *Calanus hyperboreus* and *Metridia longa*, are common both in Atlantic and arctic waters whereas the arctic species, *Calanus glacialis* and *O. similis* are common in Arctic Water (Falk-Petersen et al. 1999). In the Pacific Arctic, *Neocalanus* spp, *Eucalanus bungii* and *Metridia pacifica* are common in the oceanic Bering Sea whereas *Calanus marshallae* and various small copepods such as *Pseudocalanus* spp are common on the Middle and Coastal Bering shelves (Smith and Schnack-Schiel 1990; Loughlin et al. 1999; Stockwell et al. 2001).

C. finmarchicus overwinters at depths >400 m until early March, well below the sill depth of the Barents Sea, so it cannot be advected into the Barents Sea in winter (Østvedt 1955; Diel and Tande 1992). This makes it strongly dependent on the flow of the major branches of the Atlantic current (Skjoldal et al. 1987). *C. finmarchicus* produces 1-3 generations per year along the Norwegian coast and <1 generation per year north of 78°N (Pedersen et al. 2001). Spawning typically begins in late March and takes >1 month (Tande and Hopkins 1981). A new generation is developed from May–July (Head et al. 2000; Tande and Slagstad 1990).

Calanus glacialis has a biannual cycle in the northern Barents Sea and an annual cycle in the Atlantic Barents Sea, in the Davis Strait, the Fram Strait, in the Godthåb Fjord and off Nova Scotia (Tande 1991 and references therein). Its spawning behaviour is extremely flexible through rapid mobilisation of ovaries and high rates of egg production as soon as the food supply is ample (Hirche 1989). Cohorts that spawn in April favour annual generations whereas those that spawn in May and June favour biennial generations (Tande and Slagstad 1990).

Calanus hyperboreus has been found as far south as the western coast of Norway albeit not reproducing there (Matthews et al. 1978). In the central Greenland gyre it descends in July to 500–1000 m depth, spending the winter there, and ascends in April (Hirche 1997). Moulting into adults begins in November and is completed by March-April, before the ascent (Hirche and Niehoff 1996).

Neocalanus flemingeri spawns in January–February at >250 m depth in the Oyashi region and the western Subarctic Pacific and reaches copepodite stage I in March and most of copepodite V by early June. It breeds annually, but a few overwintering copepodite IV stages have a biennial cycle (Kobari and Ikeda 2001). *N. plumchrus* and *N. flemingeri* form a biomass peak in spring in the Gulf of Alaska. There and farther west they subsequently descend in late spring and early summer and are succeeded by small copepods and non-copepods (Goldblatt et al. 1999).

Among herbivorous krill, *Thysanoessa inermis* and to some extent, *T. raschii*, are important in Atlantic Water (Dalpadado and Skjoldal 1991; 1996), the latter is also common on the Middle and Coastal Bering shelf (Loughlin et al 1999). In the Subarctic Atlantic, *T. longicauda* can be important and in the Pacific sector, *T. longipes* and *Euphausia pacifica* (Smith 1991; Mackas and Tsuda 1999). *T. raschii* is entirely herbivorous and is known to graze *Phaeocystis* efficiently (Hamm et al. 2001), whereas *T. longicaudata* and *T. inermis* may occasionally eat copepods (Båmstedt and Karlson 1998).

Four large amphipods, *Apherusa glacialis*, *Onisimus* spp and *Gammarus wilkitzkii*, are common in the multiyear ice, where they live in cavities, brine channels and on the underside of the ice. *G. wilkitzkii*, which is the largest (3–4 cm), constitutes at times >90% of the amphipod biomass in the multiyear ice. *Apherusa* can be common also in first-year ice, and *Onisimus* in fast ice (Hop et al. 2000). Both *G. wilkitzkii* and *A. glacialis* have high fecundity, which counteracts large losses due to advective transport (Melnikov 1997; Poltermann et al. 2000). A fourth occasionally important amphipod species, *Themisto libellula*, lives in ice-filled waters and is not dependent on ice.

3.7.1 Mesozooplankton biomass

The "large" herbivorous copepods can comprise up to 70–90% of the mesozooplankton biomass in the arctic and subarctic seas. *C. finmarchicus* individuals contain on average 0.13–0.22 mg C and

C. glacialis individuals, 0.22–0.36 mg C. *C. hyperboreus* individuals are biggest, containing 1–1.8 mg C. Copepods are most carbon-rich in autumn when they are loaded with wax esters. Adult individuals of *T. inermis* and *T. raschii*, also richest in carbon in the autumn, contain about 5 and 3.5 mg C, respectively (Smith and Schnack-Schiel 1990).

In the upper 500 m of the Amundsen, Makarov and northern Nansen Basin, mesozooplankton biomass averages 0.8 g C m^{-2} (*C. finmarchicus* and/or *C. hyperboreus* representing 30–45% of the biomass). In the southern Nansen Basin, the average is 2.3 g m^{-2} (*C. finmarchicus* 35%), and in the Greenland Sea, 4.2 g m^{-2} (noncalanoids and non-*Metridia* 55%) (Mumm et al. 1998). On the Laptev Shelf, the mesoplankton biomass in summer is 0.05–0.8 g C m^{-2} and in the adjacent waters of the Nansen Basin, 2.3–4.0 g C m^{-2} (*C. finmarchicus* and *C. glacialis* predominant; Kosobokova et al. 1998; Abramova 1999).

In the Northeast Water Polynya, mesozooplankton biomass ranges from 0.07–5.3 g C m^{-2}, of which *C. glacialis* makes up 76% (Hirche and Kwasniewski 1997). *C. finmarchicus* is scarce because advective import with Atlantic Water is small (Ashijan et al. 1997).

In the Atlantic Barents Sea, *C. finmarchicus* biomass peaks in June at about 8 g C m^{-2} whereas *C. hyperboreus* biomass increases with latitude, peaking at around 1.5 mg C m^{-2} in late July or August (Tande 1991). *C. glacialis* biomass in arctic waters presumably peaks in August at about 5 g C m^{-2} (Tande and Slagstad 1990). In subarctic waters north of Iceland, mesozooplankton biomass (60% *C. finmarchicus)* peaks in late May (Gislason and Astthorsson 1998).

At the Bering shelfbreak and on the outer Bering shelf, *N. plumchrus* represents from 0.6 g C m^{-2} in the beginning of April to 5.6 g C m^{-2} at the end of May (33% and 94% of mesozooplankton biomass, respectively). *C. marshallae* is predominant in June with 0.8 g C m^{-2} of a total mesozooplankton biomass of 1.4 g C m^{-2} (Vidal and Smith 1986).

Krill biomass represents <0.15 g C m^{-2} (*T. inermis*) and <0.05 g m^{-2} (*T. longicaudata*) in the Atlantic Barents Sea (Dalpadado and Skjoldal 1991). Both species spawn in June, matching the spring bloom. In the Bering shelfbreak and on the outer shelf, *Thysanoessa* spp represent up 0.5 g C m^{-2}, with small temporal variation from March to early June (Vidal and Smith 1986).

Mesofauna biomass associated with the underside of the pack ice averages 1.9 g C m^{-2}, of which small copepods are predominant. The amphipod meiofauna represents <0.1-7.4 g C m^{-2} (median 3 mg C m^{-2}), which is considerably higher than the mesoplankton biomass in the water column (Werner and Arbizu 1999).

3.7.2
Grazing and mesozooplankton production

Only adults and copepodite V of *Calanus* spp stocks attain large enough biomass to graze phytoplankton efficiently. These stages, and perhaps also those of *Neocalanus* spp, ingest 7–18% of their body carbon daily (Smith and Schnack-Schiel 1990; Båmstedt al. 1991). For *Thysanoessa inermis* and *T. raschii*, 3% d^{-1} has been indicated (Vidal and Whitledge 1982).

In the Norwegian Sea, the grazing rate of *C. finmarchicus* is 2% of the body carbon per day during the pre-bloom phase, reaching 30% during the bloom and 14% during the post-bloom stage (Irigoyen et al. 1998). Large mesozooplankton stocks can, therefore, decimate phytoplankton stocks within few days during the bloom and the post-bloom phase. Moreover, because they are as abundant under the ice as in the water (0.45–0.55 g C m^{-2}), copepods in the arctic part of the Barents Sea exert a significant grazing impact even if their biomass corresponds to only 10–20% of the microheterotrophic biomass (Hansen et al. 1996). Egg production is particularly demanding, requiring a daily intake of up to 30% of the phytoplankton stock in *Calanus* spp (Booth et al. 1993; Hirche and Kwasniewski 1997).

In Hudson Bay, copepods graze ice algae efficiently in the interface between ice and water (Runge et al. 1991). In Resolute Bay, where small *Pseudocalanus* spp dominate the mesoplankton, 4.2% d^{-1} of the algal biomass is grazed in the ice and 36.8% d^{-1} in the upper water column (Hattori and Saito 1997). Intense grazing of algae released from the ice matrix by *Pseudocalanus acuspes* has been observed under the ice in Barrow Strait in the Canadian Arctic (Fortier et al. 2002). In the Saroma-ko lagoon, on the other hand, the small mesozooplankton biomass grazes only <1% of the phytoplankton biomass daily (Saito and Hattori 1997).

In the multiyear ice, microheterotrophs are responsible for 35% of the daily grazing, and turbellarians, nematodes and crustaceans roughly 20% each. The median total grazing rate is around 1.8 mg C m^{-2} d^{-1}, which corresponds to an annual total of about 1 g C m^{-2}. This corresponds to <10% of the annual primary production in the multiyear ice (Gradinger 1999). *Apherusa glacialis* feeds mainly on ice algae and phytodetritus whereas *Gammarus wilkitzkii* and *Onisimus* spp feed both on calanoids and algae (Scott et al. 1999).

Variations in phytoplankton biomass have strong impact on the ingestion rate for copepods, especially the fast-growing stages (Hygum et al. 2000). Nauplii and early-stage copepodites of *C. finmarchicus* are food-limited at <0.5–0.7 mg Chl*a* m^{-3}, which is a common concentration during the post-bloom phase (Campbell et al. 2001). Also the reproductive capacity of copepods is strongly affected by fluctuations in the quality and abundance of food. Thus egg production in *C. finmarchicus* can be stopped by starvation, but may resume again if the food supply is increased (Carlotti and Hirche 1997). Moreover, egg mortality seems to increase with the abundance of females and stage V copepodites (Ohman and Hirche 2001).

According to experiments with 16 copepod species in 15 laboratories world-wide, it seems that diatoms can reduce the egg production rates or egg viability of copepods (Ban et al. 1997), as observed in other studies (Uye 1996). Possibly, antimitotic compounds in diatoms are ubiquitous and hazardous to egg production. Yet *C. finmarchicus* females in the Norwegian Sea on average produce 44 eggs d^{-1} during the diatom bloom compared to 8 eggs d^{-1} during the pre-bloom phase (Niehoff et al. 1999). And in the Hudson Bay, production of copepod eggs is negligible until after the ice-algal bloom in mid June (Runge et al. 1991).

The diet affects the sinking rate of faecal pellets. Because *Emiliania* coccoliths make the pellets heavy, the sedimentation rate inside a bloom patch of *Emiliania* can be 70 times larger than outside it at comparable grazing pressure (van der Wal et al. 1995). The small copepod *Acartia tonsa* produces heavy faecal pellets when fed ciliates or diatoms, and light pellets when fed heterotrophic flagellates, including dinophytes, representing a 40% difference in the sinking rate (Feinberg and Dam 1998).

The few estimates of annual mesozooplankton grazing rates are based on models. Estimates are 1 g C m^{-2} for multiyear ice, 5–10 g C m^{-2} for Eurasian shelves and Atlantic Water, and >18 g C m^{-2} for the most high-productive parts of the Bering shelf and the Chukchi Sea (Walsh et al. 1989; Tande and Slagstad 1990; 1992; Gradinger 1999). In the shallowest parts of the Bering shelf (<60 m), most of the primary production enters the detritus pool, which supports considerable production of zoobenthos (Walsh et al. 1989).

3.7.3 Match-mismatch

The grazing food web and sedimentation are in principle competing processes. The more phytoplankton is grazed, the less of the primary production is left for sedimentation and *vice versa*. When phytoplankton blooms and large zooplankton stocks coincide in space and time, the grazing efficiency is large (match) and sedimentation small. Conversely, when phytoplankton blooms and large zooplankton stocks are separated in time and space, grazing efficiency is small (mismatch) and sedimentation large.

In Atlantic Water, where the depth of mixing is relatively large, blooms begin later and last longer than the intense bursts in Arctic Water. This matches well with late development of copepodite stage V of *Calanus finmarchicus* (Skjoldal et al. 1987; Dalpadado and Skjoldal 1991; Kosobokova 1999). On the other hand, this species will be mismatched to early blooms, which presumably is the main reason why *C. finmarchicus* is an expatriate in the arctic part of the Barents Sea (Melle and Skjoldal 1998). The flexible *C. glacialis* presumably avoids serious mismatch by adjusting the egg production to the development of the phytoplankton bloom whether it begins early or late.

The mismatch is most pronounced in years when the spring follows a particularly cold winter, which brings sea ice over relatively warm water, resulting in very early blooms. Extreme mismatch is also observed in Norwegian fjords where the freshwater supply restricts vertical mixing most of the winter and spring so that the bloom occurs in February-March, depending on latitude, long before mesozooplankton stocks grow large (Wassmann 1991). Thus virtually the whole bloom sinks.

In the Atlantic Barents Sea, cold winters that caused early blooms and mismatch with *C. finmarchicus* occurred in 1979, 1981, 1992, 1996, 1998, 1999 and 2001. In these years the bloom culminated up to 4–6 weeks earlier than usual, resulting in poor development of copepodite stages I–IV in June (Olsen et al. 2003). In contrast, 1994 and 1995 exhibited late blooms and match. Although not strongly correlated, mismatch seems to be likely in Atlantic Water when the depth of mixing is <40 m and match when the depth of mixing is >40 m.

3.8 Primary Production – Impact of Climate Change

Because of the impact of areal ice cover and the depth and intensity of water-column mixing, the total marine primary production in the Arctic is highly climate-dependent. Nearly all scenarios predict warming and smaller areal ice cover in the next 100 years, and some scenarios also indicate a strengthened polar vortex – consistent with positive AO/NAO index – yet climate oscillations are also likely to exist in the future (see Chapter 1.2).

The positive phase should increase total primary production in the Arctic Ocean Periods, especially on the Eurasian side and the negative phase, *vice versa*. Earlier melting of sea ice in spring is likely to enhance annual primary production by extending the growth season. The actual outcome, however, is highly dependent on regional and local changes in upwelling, wind-driven vertical mixing, and freshwater supply from sea ice and rivers – it takes only a small decrease in salinity to counteract the impact of vertical mixing of an increase in the wind intensity.

Given long-term warming, the sea ice will disappear first in the arctic part of the Barents Sea and on the Bering shelf. This can imply, given today's wind field, that the annual particulate primary production in the arctic Barents Sea increases from <40 g C m^{-2} y^{-1} to 100–150 g C m^{-2} y^{-1}, a factor of 3.5. This is a consequence of the withdrawal of sea ice and intensified transport of Atlantic nutrients through deepened vertical mixing in a saltier water column. Along the southern fringe of the arctic Barents Sea, this can postpone the spring bloom towards late May whereas near the central Arctic Ocean the bloom will arise earlier than before, perhaps in late April. A permanently ice-free Bering shelf is likely to yield higher annual primary production than today primarily as a consequence of extended growth season with virtually continual upwelling of nutrient-rich water in the greenbelt. Moreover, if the wind field becomes more intense, primary production in the northeast Pacific and the eastern oceanic Bering Sea can increase as a consequence of enhanced supply of aeolian iron – this, however, will also depend strongly on the impact of grazing.

As long as vertical mixing in the Siberian shelf seas (except for the Barents Sea) is restricted mainly to reworking the nutrient-poor shelf waters, primary production presumably will increase little even if the ice on the shelves vanishes permanently. However, a rather rapid and significant increase can be expected when the ice edge withdraws so far into the central Arctic Ocean that wind-forced upwelling of nutrient-rich deep water can take place along the shelf break. In the central Arctic Ocean, areal primary production is likely to increase as the number and size of leads in the multiyear ice increase. Productivity, however, is not likely to exceed 20–40 g C m^{-2} y^{-1} because winter sea ice is likely to exist in the future, causing strong stratification even during an ice-free growth season.

The depth of mixing is likely to exert strong impact on phytoplankton community structure. Depths of mixing <40 m, likely where sea ice melts and in areas that are this shallow, is likely to favour diatom blooms. Mixing to 60–80 m depth is likely to favour *Phaeocystis*. Thus, unless there is an increase in freshwater input, *Phaeocystis* can become more common than today. This can happen in Atlantic Water. Moreover, if the mixing reaches >100 m depth, also possible in Atlantic Water, a low-productive community dominated by nanoflagellates may be favoured, as in the off-shelf parts of the Southern Ocean (Hewes et al. 1990). This implies little transfer of carbon to herbivores and to sediments because the grazers would largely be ciliates (Sakshaug and Walsh 2000).

Strong northward transport of water will presumably enhance the northward transport of zooplankton (Skjoldal et al. 1987; Brodeur and Ware 1992; Overland et al. 1994), for example of *C. finmarchicus* into the Arctic Ocean. This, in combination with blooms that arise earlier where waters become permanently ice-free, is likely to cause gross mismatch between phytoplankton blooms and *C. finmarchicus* stocks in the Siberian shelf seas. In any case, however, this species does not seem to thrive well in Arctic Water. On the other hand, many other copepod species that already exist in the Arctic Ocean are, through flexible reproduction strategies, which are demonstrated for *C. glacialis*, likely to match any phytoplankton bloom whenever it might arise. Thus the ratio of grazed to exported phytoplankton biomass in the Arctic Ocean will not necessarily change appreciably from today's level.

3.9 Summary and Concluding Remarks

- The average estimate for Arctic Ocean shelf seas, 32 g C m^{-2} y^{-1} (Table 3.7) is close to the 27 g C m^{-2} y^{-1} given for waters with <200 m depth north of 65°N by Legendre (1992). Primary production of the shelf seas contributes 279 Tg C y^{-1} (or 279×10^6 t C y^{-1}) distributed over 5.052 million km^2. The two single largest contributors are the Barents and Chukchi Seas, with 136 and 42 Tg, respectively (Romankevich and Vetrov 2001). The smallest contributions come from the west-arctic shelves, the White Sea, and the Laptev Sea, with 2–16 Tg C y^{-1} each. The deep Arctic Ocean contributes >50 Tg C y^{-1} over 4.489 million km^2, which is more than 5-fold higher than pre-1990 estimates.
- In total, primary production in the Arctic Ocean contributes >329 Tg C y^{-1} (or $>329 \times 10^6$ t C y^{-1}), which is less than the contribution from the subarctic Atlantic side and the Bering Sea (>300 Tg C y^{-1} each). Thus primary production in the vast Arctic Ocean is smaller than the primary production in its smaller (area-wise) neighbours.
- The few model-based secondary (zooplankton) production estimates range from 1 g C m^{-2} for multiyear ice, 5–10 g C m^{-2} for Eurasian shelves and Atlantic Water, and >18 g C m^{-2}, for the green-belt of the Bering shelf and the Chukchi Sea (Walsh et al. 1989; Tande and Slagstad 1990; 1992; Gradinger 1999). The corresponding grazing rate (given as 5 times the zooplankton production rate) in high-productive spots ranges from 23–67 % of the total primary production with a median of 40 %, leaving more than half the primary production for export. Because the lower per cents comprise only grazing by copepods and krill, the median export rate is likely to be lower than 60 % of the total primary production.

Chapter 4

The Role of Dissolved Organic Matter for the Organic Carbon Cycle in the Arctic Ocean

R.M.W. Amon

4.1 Introduction

Our knowledge of the Arctic Ocean as an ecosystem has increased dramatically over the last decade, particularly due to easier access to Russian literature and due to numerous icebreaking research efforts and international cooperations. Dissolved organic matter (DOM) has been recognized as an important component of the oceanic carbon cycle with a pool size of roughly $650–700 \times 10^{15}$ g C (Hansell and Carlson 2001, Hedges 2002) which matches the amount of carbon in our atmosphere. Even small changes in the oceanic dissolved organic carbon (DOC) pool could lead to significant perturbations of the global carbon cycle on time scales of 1000 to 10000 years (Hedges 1992, Hedges 2002). The biological oceanography of the Arctic Ocean is poorly understood and only very recently (the last 5 years) have we learned how the numerous unique features of the Arctic Ocean influence biological and biogeochemical processes like the DOM cycle (Wheeler et al. 1997). Three features have turned out to be of major importance for Arctic DOM biogeochemistry; 1) the large input of river water and concurrent terrestrial DOM; 2) the unique vertical stratification with cold and fresh surface water on top of warmer water supplied by the Atlantic Ocean and; 3) the extended shelf areas on the Eurasian side of the Arctic Ocean.

The Arctic Ocean is surrounded by continents with extended watersheds that are drained by numerous rivers discharging annually about 3299–4280 km^3 (Chapter 2.2; Shiklomanov et al. 2000) of water or approximately 11% of the global river discharge into this relatively small (16.7×10^6 km^3) ocean basin. Due to the extended shelf area (47% of total area) in the Arctic Ocean, coastal zones are expected to play an amplified role for the entire system influencing a variety of physical, chemical and biological processes. The vertical structure of the Arctic Ocean is heavily influenced by lateral water mass exchanges. Next to the riverine input of 0.10–0.14 Sv (Chapter 2.2; Shiklomanov 2000) the Arctic receives about 0.83 Sv of Pacific Water through Bering Strait (Anderson 2002) and about 5.0–6.4 Sv from the North Atlantic Current, either through Fram Strait or via the Barents Sea (Mauritzen 1996, Fahrbach et al. 2001, Schauer pers. comm.). Exchanges of intermediate and deep waters are limited to the Fram Strait through which about 4.7 Sv (estimated from Fahrbach et al. 2001) enter the deep Arctic basins. The large lateral exchanges in the Arctic Ocean result in relatively low residence times of Arctic water masses, typically between decades to a few hundred years (Östlund et al. 1987, Bönisch and Schlosser 1995, Schlosser et al. 1995). Major changes in Arctic Ocean hydrography have been observed in the last 12 years (Macdonald 1996, Steele and Boyd 1998, Morison et al. 2000) further challenging our ability to understand the biogeochemical cycle of organic carbon in this unique ocean basin.

The state of the art in terms of Arctic River DOM discharge is summarized in Chapter 2.2 and was also discussed in a recent review on DOM in the Arctic Ocean by Anderson (2002). To avoid too much overlap, this chapter will focus on shelf processes involving DOM, and the chemical composition of DOM and what that tells us about DOM origin, cycling, and fate. The horizontal and vertical distribution of DOM will be discussed in relation to physical processes and vertical organic carbon transport in the Arctic Mediterranean Seas (Arctic Ocean plus Greenland, Iceland, Norwegian Sea).

4.2 Riverine DOM on Arctic Shelves and Beyond

One of the big, unsolved questions in contemporary marine biogeochemistry is the fate of terrestrial-derived organic carbon in the world oceans (Hedges et al. 1997). Estimated global annual riverine DOC-fluxes (250 Tg C y^{-1} or 250×10^6 t C y^{-1}; Spitzy and Ittekkot 1991, Hedges et al. 1997) would theoretically replace the marine DOC pool in less than 3000 years. However, recent studies detect only trace amounts of terrigenous material in the oceans interior (Opsahl and Benner 1997) indicating the existence of an unknown but efficient removal process for terrigenous material within the oceanic environment. The Arctic with its pronounced river input is a prime location to address this question, particularly on the extended Eurasian shelves where the DOC concentration in rivers is typically >600 μM C (Anderson 2002; see Chapter 2.2). Two Arctic estuarine systems have been investigated with regard to estuarine mixing processes of DOM, the Lena River – Laptev Sea system (Cauwet and Sidorov 1996) and the Ob and Yenisei – Kara Sea system (Amon and Spitzy 1999, Köhler et al. 2003). The transition zone from river water to seawater, "the marginal filter" (Lisitsin 1995), has been considered a potential sink for terrigenous DOM (Sholkovich 1976, Lisitsin 1995) and is the logical first place to look for terrestrial DOM losses.

4.2.1 Estuarine Mixing

The above mentioned investigations (Cauwet and Sidorov 1996, Köhler et al. 2003) studied the distribution and chemical composition of DOM in the 3 largest rivers of the Arctic Ocean. Together these rivers have an annual DOC discharge of approximately 11.3 Tg C y^{-1} (Köhler et al. 2003, Anderson 2002) contributing about 50–70 % to the total riverine DOC discharge into the Arctic Ocean (Anderson 2002; Chapter 2.2), a representative portion for the Pan-Arctic system. The distribution of DOC along the salinity gradient was largely conservative during two independent field campaigns off the Lena Delta (Fig. 10 in Cauwet and Sidorov 1996) indicating that no major losses of DOC occurred on the Laptev shelf. During the last 4 years our lab was involved in research off the rivers Ob and Yenisei in the Kara Sea as part of the bilateral (Russian-German) project "Siberian River Run Off (SIRRO)": The nature of continental run-off from the Siberian rivers and its behavior in the adjacent "Arctic Basin" (see Stein et al. 2003b). During this time we had the chance to participate in 4 cruises to the Kara Sea during which we collected a large multiyear data set for the chemical characterization of DOM. The DOC distribution observed during the summer-fall periods in 1997, 1999, and 2000 also indicated the largely conservative behavior of DOC (Fig. 4.1) during its passage from the rivers to the open Kara Sea and towards the Arctic Ocean. However, we also noticed pronounced seasonal variations of DOC concentrations along with discharge fluctuations. DOC concentrations are highest during the onset of freshet in May and June reaching values >900 μM (Köhler et al. 2003), significantly higher than the values observed during early fall (Köhler et al. 2003). Strong seasonal variations

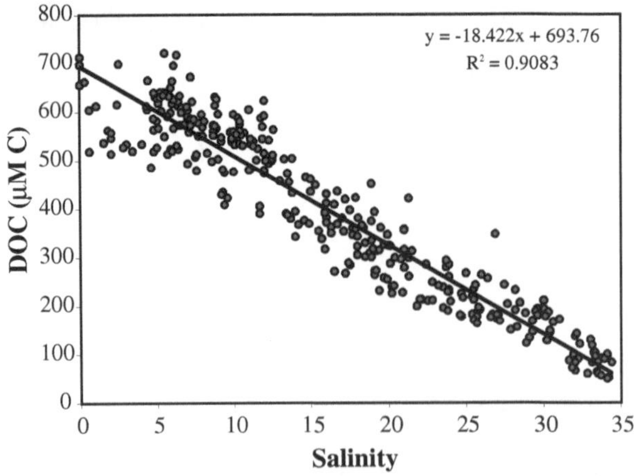

Fig. 4.1. Distribution of dissolved organic carbon (DOC) along the salinity gradient in the Kara Sea. Data are compiled from the Ob and Yenisei estuaries during the summer-fall periods of 1997, 1999, and 2000

of DOC have also been observed by Cauwet and Sidorov (1996) in the Lena river and the Laptev Sea with winter concentrations as low as 300 μM DOC and freshet values >1000 μM C. Significant changes in DOC occurred on time scales of a few weeks and this short-term variation is most likely a large component of the variation in the multiyear data set (Fig. 4.1). These variations have to be considered when estimating annual mean DOC concentrations and discharge by Arctic rivers. For more discussion of seasonal patterns of DOC concentrations and discharge see Chapter 2.2 and Köhler et al. (2003).

To confirm the conservative behavior of DOM observed in the field we conducted mixing experiments using Yenisei river water (0 ‰) and Arctic Ocean water (>33 ‰). The two endmembers were

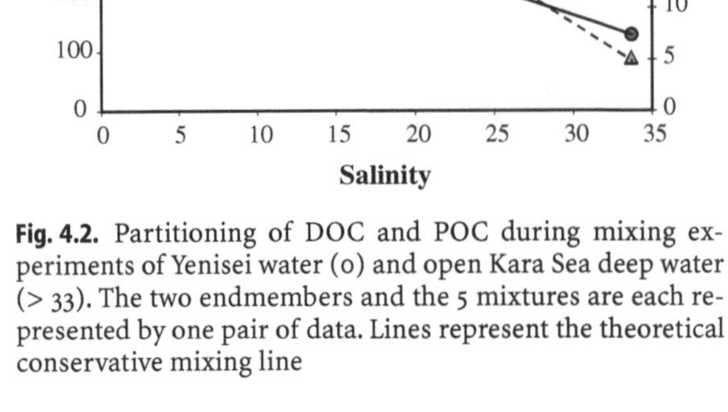

Fig. 4.2. Partitioning of DOC and POC during mixing experiments of Yenisei water (0) and open Kara Sea deep water (> 33). The two endmembers and the 5 mixtures are each represented by one pair of data. Lines represent the theoretical conservative mixing line

mixed to obtain batches of different salinity and were sampled after about 30 h for DOC and particulate organic carbon (POC) determination. Results of that experiment (Fig. 4.2) support the conservative nature of DOC, but did indicate minor DOC losses (<5%) and POC gains at low salinity (Köhler et al. 2003). From the experiments and several field observations it seems that the majority of DOC discharged into the coastal Eurasian Arctic is not affected by flocculation processes in the marginal filter and enters the outer shelf and the open Arctic Ocean.

4.2.2 Chemical characteristics and origin of DOM on the Eurasian shelf

To determine the actual sources and sinks of DOM on the shelf it is necessary to chemically characterize DOM across the salinity gradient. Fortunately, there are pronounced differences between marine and terrestrial derived organic matter that can be detected using a number of analytical methods. Again, most work was done in the rivers Ob, Yenisei, and Lena and their respective estuaries (Cauwet and Sidorov 1996, Amon and Spitzy 1999, Kattner et al. 1999, Opsahl et al. 1999, Lobbes et al. 2000, Amon and Benner 2003, Köhler et al. 2003, Amon et al. 2003). The methods used range from the determination of optical properties to elemental, isotopic, and molecular level analysis on water samples or concentrated DOM.

The large Siberian rivers Ob, Yenisei and Lena are characterized by a high content of colored dissolved organic matter (CDOM) typical for blackwater rivers rich in humic substances (Burenkov et al. 1995, Cauwet and Sidorov 1996). The high level of terrestrial CDOM plays an important role for underwater light conditions efficiently reducing the secchi-disk depth to less than 5 m in the southern Ob and Yenisei estuaries with attenuation coefficients (440 nm) between 0.7 and 15 m^{-1} (Burenkov et al. 1995). This reduction of light penetration likely effects primary production on the shelves, however, no data exist on the quantitative effect of CDOM on primary production at this point. On the other hand, CDOM has also been identified as a potential protective shield against increasing UV radiation for biota in the Arctic Ocean (Gibson and Vincent 2000) possibly counteracting the negative attenuation effects. Terrestrially-derived CDOM has a characteristic fluorescence pattern. Based on excitation/emission fluorescence spectroscopy it is possible to distinguish terrestrial from marine CDOM (Coble 1996). Samples from the Ob, Yenisei, and Mackenzie River show the typical terrestrial

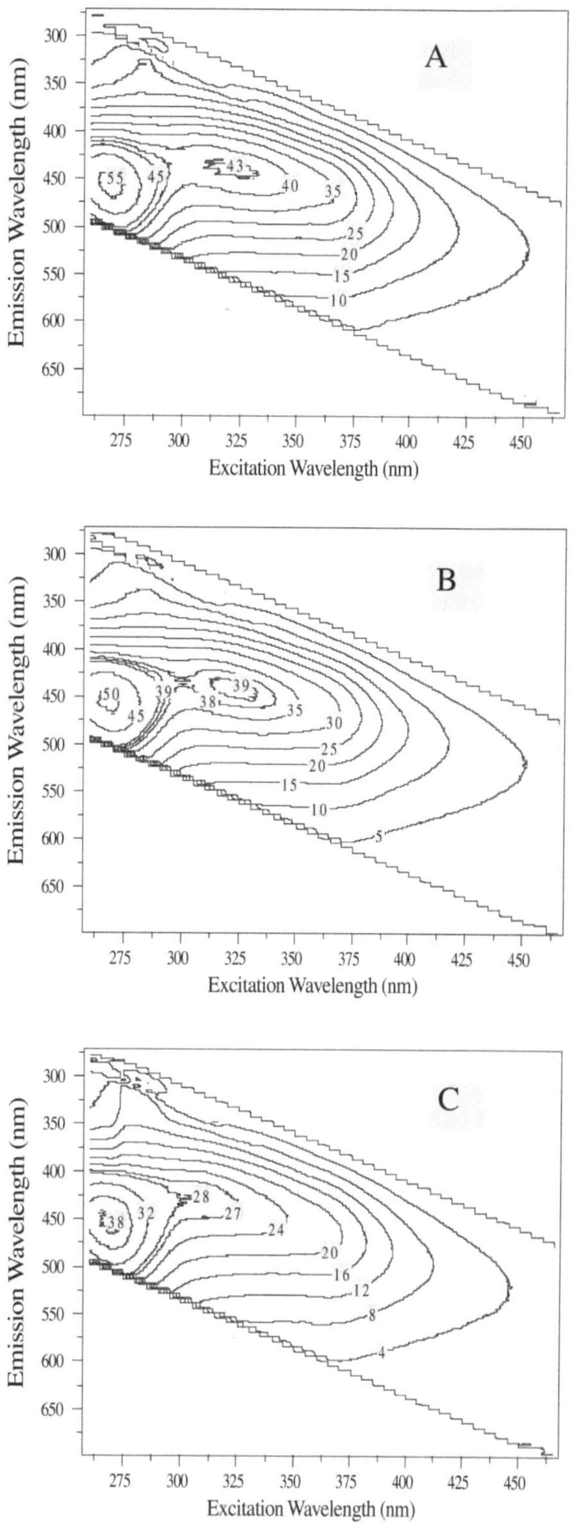

Fig. 4.3. Contour plots of fluorescence excitation/emission spectra in waters sampled from the Yenisei (A), Ob (B), and Mackenzie (C). Fluorescence signals were corrected for water Raman peaks and excitation/emission characteristics of the fluorometer. Isolines represent quinine sulfate units (1 QSU = 1 ppb quinine sulfate in 0.05 M sulfuric acid). Modified from Amon et al. (2003)

signal with two distinct fluorescence (excitation/emission) maxima at 270/455 nm and 325/440 nm in 3-D contour plots (Fig. 4.3; Amon et al. 2003). Burenkov et al. (1995) investigated the distribution of fluorescence at a similar wavelength (excitation/emision = 337/480 nm) in the Kara Sea and showed clear signs of terrigenous DOM well beyond 76°N. Fluorescence detection devices, adopted for in situ deployment, have proved to be extremely valuable tools to trace the distribution of terrigenous DOM throughout the Arctic Ocean (Guay et al. 1999, Amon et al. 2003) and will be discussed later in this chapter.

On the elemental level, C/N ratios of DOM can be used to distinguish terrestrial from marine organic matter (Thurman 1985, Benner 2002). DOM derived from phytoplankton has much smaller C/N ratios (10–20) than DOM derived from terrestrial vegetation (typically >30). Most studies conducted in and off Eurasian rivers report high C/N ratios for river DOM, typically between 40 anf 60 (Lara et al. 1998, Kattner et al. 1999, Lobbes et al. 2000, Köhler et al. 2003). However, other recent studies have reported significantly lower values (20–30) for the Lena and other Siberian rivers (Cauwet and Sidorov 1996, Wheeler et al. 1997, Gordeev et al. 1996). This discrepancy was pointed out in several studies (Kattner et al. 1999, Holmes et al. 2001, Köhler et al. 2003) suggesting that one reason for the differences might be analytical (Holmes et al. 2001). The method for DON analysis is under revision by the international community at the moment and adequate reference standards have not been used in the past making direct comparisons difficult. General analytical uncertainties with the DON measurement are reflected in a relatively low number of observations compared to DOC. Clearly we need to increase the number of observations for DON in coastal systems along with the use of international standards for quality control. One of the few studies that reported C/N values for bulk DOM across the salinity gradient in an Arctic estuary (Köhler et al. 2003) indicated a gradual change from a typical terrestrial signal (40–50) to a more marine signal (20–25) without abrupt changes due to flocculation of terrigenous DOM or DOM input by primary production (Fig. 4.4). This is consistent with the conservative mixing observed for terrestrial stable carbon isotope signatures measured in bulk DOC along a salinity gradient from the riverine endmenbers to the open Kara Sea (Spitzy et al. in prep).

More detailed chemical analysis of the isotopic and molecular composition of Arctic river and shelf DOM involved the concentration of DOM and its separation from salts. Two methods have been used for that purpose, ultrafiltration using a 1 kDa

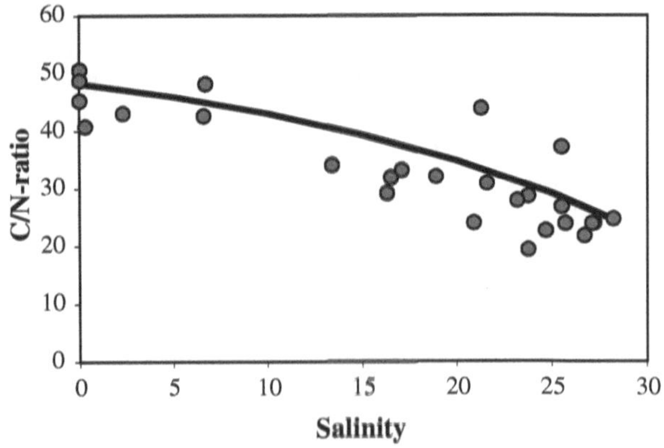

Fig. 4.4. C/N ratios in bulk DOM along the salinity gradient in the Kara Sea. The line represents the theoretical distribution of C/N values based on conservative mixing

membrane (Opsahl et al. 1999, Amon and Benner 2003) and solid state extraction using XAD resins (Kattner et al. 1999). DOM isolated by these methods from the Ob, Yenisei, and Lena estuaries has been or will be subjected to a variety of chemical analysis including elemental (C, N), isotope (^{13}C, ^{14}C, ^{15}N), liquid state NMR, lignin phenol, neutral sugar, and amino acid analysis. The ultrafiltration process typically recovered >60% of DOC from Ob and Yenisei indicating that most DOM in those rivers has a molecular weight above 1 kDa. The >1 kDa fraction includes the colloidal fraction which is often thought to be more susceptible to aggregation or flocculation than bulk DOM. However, the distribution of the >1 kDa fraction of DOM along the salinity gradient was also rather conservative and not indicative of major losses of colloids discharged by those rivers. C/N ratios determined in the >1 kDa fractions of the rivers were also high (>35), very smilar to C/N values observed in bulk DOM, and declined in a conservative fashion along the salinity gradient. Those values were determined with a CHN-analyser and lend independent support to higher C/N ratios for Eurasian river DOM as discussed above. Similar trends were observed with stable carbon isotope values of >1 kDa DOM with typical terrigenous values in the rivers (−27‰; Table 1) and steady increases across the shelf. Some of the freshwater samples in the Ob had more depleted values, below −28‰, potentially indicating the presence of DOM derived from green algae which had $\delta^{13}C$ values as low as −35‰ (Kodina 2001).

The most valuable tracer for terrigenous DOM on Arctic shelves is the concentration of lignin phenols, a compound class that is synthesized by vascular plants only. In the Laptev Sea Kattner et al. (1999) reported the presence of lignin phenols in

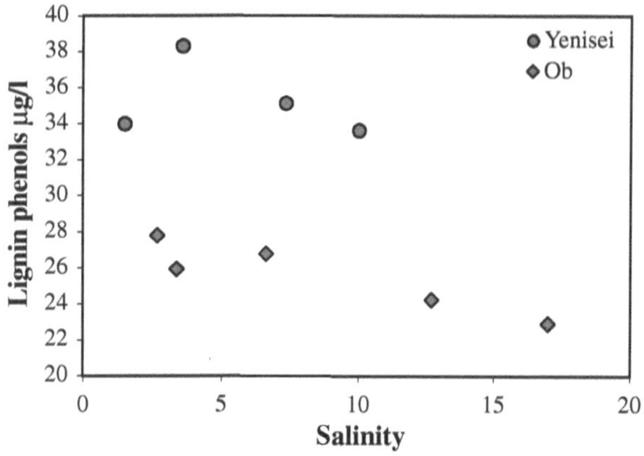

Fig. 4.5. Distribution of lignin phenols in ultrafiltered DOM (>1 kDa) along the salinity gradient in the Kara Sea. Ultrafiltered DOM represents between 50 and 70% of total DOM

XAD isolates across the shelf well into the Arctic Ocean and a largely conservative distribution. Similar results were found in the Kara Sea where Ob and Yenisei discharge DOM with very high concentrations of lignin phenols which were determined in ultrafiltered DOM (Table 4.1; Opsahl et al. 1999). The conservative distribution of lignin phenols in the >1 kDa fraction also indicated that terrestrial-derived colloids are not selectively lost during the mixing process (Fig. 4.5). Another characteristic feature of river DOM was small amounts of carbohydrate-carbon and amino acids-carbon relative to total DOC (Table 4.1) indicating that Arctic river DOM has undergone extensive degradation before entering the Kara Sea and contains little fresh phytoplankton derived DOM (Amon et al. 2001, Amon and Benner 2003). Despite the degraded nature and terrigenous origin, most of the DOM collected in Ob and Yenisei has apparently been produced within the past 50 years with ^{14}C levels well above the 1954 atmospheric value (Table 4.1; Benner et al. subm.). It seems from the above information that DOM discharged by Eurasian rivers has a dominant terrestrial component and that the vast majority of Eurasian river DOM is transported across the shelves towards the Arctic Ocean with minor losses due to estuarine mixing processes.

4.2.3 The role of bacteria and photochemical processes on the Eurasian shelf

Bacteria are the most important consumers of DOM in aquatic ecosystems, typically processing about 50% of primary production in the ocean. The role of bacteria for the carbon cycle in the Arctic ocean has been investigated by only a few studies with poor spatial and temporal resolution (Saliot et al. 1996, Rich et al. 1997). Detailed information on bacterial respiration is largely lacking for the Arctic Ocean and information on bacterial activity on the Eurasian shelf has only been addressed by Saliot et al. (1996). To determine the fate of river DOM on Arctic shelves it is necessary to determine the potential of bacteria to utilize this material on time scales typical for estuarine mixing. For that purpose we conducted a simple decomposition experiment with Yenisei river water and the natural bacterial community. We followed the bulk DOC concentration over several years and detected only minor DOC losses (<4%) during the incubation period (Köhler et al. 2003). This preliminary result confirms the largely refractory nature of DOM discharged by Eurasian rivers and supports the notion that most river DOM is transported across the shelves. More work on the bacterial production and respiration are urgently needed to better define the role of bacteria for DOM dynamics in Arctic rivers-shelf systems.

Another process that needs to be considered as a potential sink of terrigenous DOM in the coastal ocean is photooxidation (Miller and Zepp 1995,

Table 4.1. Composition of DOM in major Arctic rivers

Parameters	Ob	Yenisei	Lena+
DOC (µM)	520–960	503–722	506–1050
DON (µM)	8–12	8–15	9–28
C/N	29–44	24–51	30–58
$\delta^{13}C$ (‰)	-27.6	-27.2 to-27.1	nd
Average ^{14}C age	>1954 AD*	>1954 AD*	nd
Lignin phenols (µg/l)	21–25	31–35	14.5
Amino acid yield (%)	1.6–1.7	1.2–1.4	2.6–3.7
Neutral sugar yield (%)	2.8–3.5	1.7–2.0	1.4–1.9

DOC, DON, and C/N ratios were determined on bulk DOM. Isotopic and molecular level analysis were performed on ultrafiltered DOM (>1 kDa). *DOM contains a significant amount of bomb ^{14}C indicating that their average age is contemporary. + Data are from Cauwet and Sidorov 1996, Lara et al. 1998, Lobbes et al. 2000.

Amon and Benner 1996b, Miller et al. 2002). With growing concern about decreasing stratospheric ozone concentrations over the Arctic, UV-radiation and its damaging character to Arctic biota have received more attention (Herndl and Obernosterer 2002). An interesting recent study (Gibson and Vincent 2000) modeled the potential consequences of increasing ozone depletion versus changing concentrations of CDOM in the Arctic and found that variations in CDOM would have a stronger effect on biological UV exposure. The flip side of this of course would be that CDOM, by absorbing UV radiation, becomes photochemically oxidized and thereby chemically altered or removed from the system. The great influence of photochemical processes on DOM dynamics in the Arctic Ocean has recently been summarized by Herndl and Obernosterer (2002), but quantitative information on photochemical mineralization of terrestrial DOM in the Arctic Ocean is still lacking. In a preliminary study (using a solar simulator) that investigated the potential of photooxidation for mineralization of Arctic river DOM we found that more than 50% of DOC was lost within 72 h of irradiation. Another experiment conducted under natural conditions in the Kara Sea indicated measurable losses of oxygen and DOC due to photochemical processes (Meon pers. com.). These preliminary studies have indicated the potential role of photochemical processes but again more work is clearly needed to estimate the photochemical removal of river DOM on a pan-arctic scale. The extended ice cover and the low solar angle can be expected to put strong constraints on this process.

4.2.4 The role of sea ice formation on DOM on the Eurasian shelf

The formation of sea ice on Arctic shelves has been identified as an important factor for shelf-basin exchange processes. The density of shelf water increases due to the rejection of salty brine during formation of sea ice. This denser shelf water is thought to enter the Arctic Ocean along the continental slope potentially ventilating the deep basins (Aagaard et al. 1985, Gawarkiewicz et al. 1998, Anderson et al. 1999, Rudels et al. 2000), however, direct evidence for deep ventilation is still missing. The density of post-freezing shelf water depends on the salinity of the water prior to freezing, the higher the initial salinity the denser the post winter shelf water. Ultimately, density determines how deep shelf plumes can penetrate into the Arctic interior.

The role of sea ice formation in DOM dynamics on the Eurasian shelf has not been specifically addressed by the scientific community yet. From older chemistry literature we learned that "freezing-out" was a commonly used method to concentrate organics in solution (Shapiro 1961). With a number of simple experiments we addressed that question using Kara Sea water with different initial salinities (7, 14, 21, 27 and 31) and known initial concentrations of DOC, DON, and inorganic nutrients. We froze (−5 °C) batches of 4 liters until about 30% of the water was frozen. We then separated ice and brine water, thawed the ice, and measured salinity, DOC, DON, and nutrients in both fractions. DOM as well as nutrients behaved similar to salt resulting in enrichment of the constituents in brine water and depletion in sea ice (Fig. 4.6). Typically, only 30–50% of the dissolved species were retained in the ice and the rest remained in solution (Fig. 4.6). Under natural conditions a smaller fraction of the water body would be frozen and accumulation of organic and inorganic constituents would be less on the shelf than in the experimental enclosure. This should result in smaller (<30%) entrapment of salts and DOM in naturally-formed sea ice on the Eurasian shelves. Smaller values of entrapment of dissolved constituents in the ice-phase were indicated in a recent ice-formation study using larger enclosures (Gianelli et al. 2001).

The experiments indicated that DOM and inorganic nutrients become enriched in the water column during sea ice formation and likely move together with the dense brine water across the shelf. Assuming an average DOC concentration of 300 μM C for the Kara Sea, rejection of about 60% of DOC during sea ice formation, and an annual sea ice formation and export of 240 km^3 in the Kara Sea (see Chapter 2.4) I calculated an annual DOC "enrichment" of 0.52 Tg C for the Kara Sea. How brine formation and DOM enrichment affects the fate of shelf DOM will largely depend on the initial salinity of the water as mentioned above. Here it is important to realize that salinity and DOC concentration are inversely related in river estuaries. The same inverse relationship was found in the brine solutions of the different experiments (Fig. 4.7). The densest brine water had the lowest concentration of DOC and vice versa which demonstrates that the densest shelf water that has the greatest potential to penetrate the deep basins of the Arctic Ocean will have the least amount of terrigenous DOC (TDOC). Lower salinity shelf water will also become denser during sea ice formation in the winter, however, it is unlikely that low salinity shelf water with high terrestrial DOC concentrations will increase enough in density to penetrate through

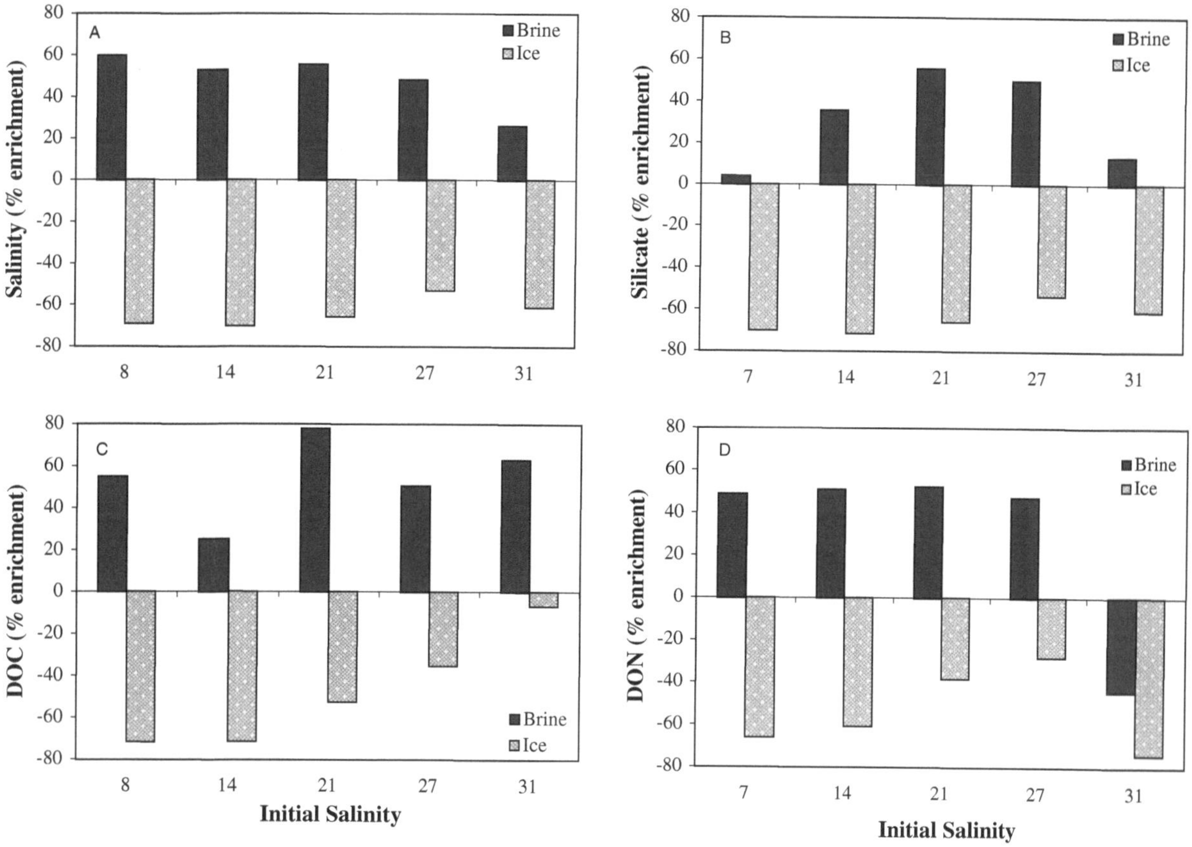

Fig. 4.6. Partitioning of sea salt (A), silicate (B), DOC (C), and DON (D), between sea ice and brine (unfrozen fraction), during freezing experiments with waters of different initial salinity from the Kara Sea. Typically, 30% of 4 l batches were frozen (−5 °C) before ice and brine water were separated for the determination of the inorganic and organic constituents in the two fractions

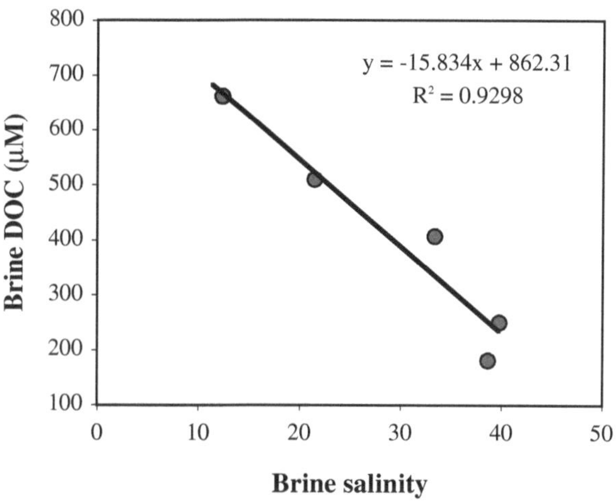

Fig. 4.7. Relationship between the concentration of salt and DOC in brine water produced during freezing experiments with Kara Sea water of varying initial salinity

the pycnocline in the Kara and Laptev Seas. This can be shown in a simple calculation assuming an average salinity of 23.4 and an average depth of 11 m for the freshwater lens in the Kara Sea (Stephantsev et al. 2000, 2001). Knowing the area (926,000 km^2) and the annual ice export (240 km^3; Chapter 2.4) of the Kara Sea one can calculate how much salt is added to the freshwater lens due to sea ice formation assuming that only 30% of the salt is removed with the ice. The resulting increase in salinity for the entire water parcel would be about 0.4, from 23.4 to 23.8, which is much lower than the salinity in the salt wedge below the freshwater lens. Schauer et al. (1997) came to a very similar conclusion for the Laptev Sea. Note that the estimate from our calculation is much less than one would predict from the experiments (33; Fig. 4.8) which were conducted in a closed system and where one third of the water was frozen. Under natural conditions the portion of water that is frozen from a given water parcel is much less than 30%, typically around 2.5%. This needs to be considered for future experiments under more realistic conditions. The main conclusion of those experiments, however, are valid and it appears that salinity and DOM will be enriched in Arctic shelf waters due to sea ice formation. Low salinity (high TDOC)

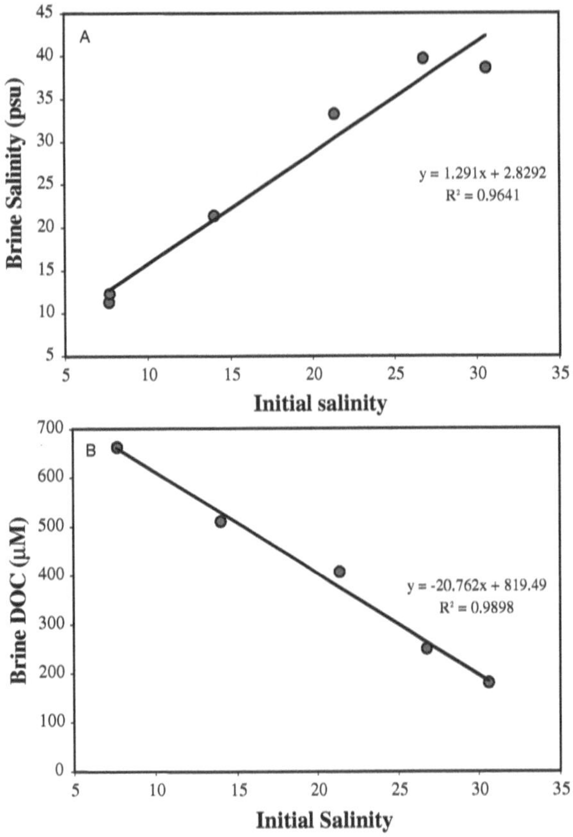

Fig. 4.8. Relationship between the initial salinity of Kara Sea water and the concentration of salt (A) and DOC (B) in brine water produced during freezing experiments

shelf water will mainly contribute to Polar Surface Waters and potentially to the Arctic halocline and not penetrate into the deep basins of the Arctic Ocean.

4.2.5 The distribution of terrestrial DOM in the central Arctic Ocean and the GIN Sea

Few studies have specifically addressed the distribution of terrigenous DOC (TDOC) in the central Arctic Ocean (Guay et al. 1999, Opsahl et al. 1999) resulting in poor horizontal and very poor vertical resolution. Estimates for the contribution of TDOC to total DOC range from 5% to peak values of 56% (Guay et al. 1999, Opsahl et al. 1999). Wheeler et al. (1997) give an average of 25% based on a mass balance estimate. The most representative data set, because it covered a wider geographical area of the Arctic Ocean, was presented by Opsahl et al. (1999) who determined the concentration of lignin phenols in DOM isolated by ultrafiltration. Opsahl et al. (1999) found 5–22% TDOC in the central Arctic and 9–27% TDOC in EGC surface waters. These estimates are lower than the general estimates by Wheeler et al. (1997) and Guay et al. (1999), however, most of the samples in the Opsahl et al. (1999) data set came from below 30 m and higher values on the very surface are possible. To gain a better understanding of the horizontal distribution of terrigenous DOM in the Arctic I compiled the available data from the literature and added some recent unpublished (Benner et al. in prep.) information (Fig. 4.9). This multi-year, multi-investigator compilation gives a general overview of the variability of DOC and TDOC in the Arctic Mediterranean Sea. Patches of high TDOC contribution were observed at different locations as far north as 86°N. Without getting too deep into this preliminary data set it is fair to state that TDOC is very abundant throughout the central Arctic Ocean but its distribution is highly variable on spatial and temporal scales. Noteworthy is the high contribution of TDOC in surface waters of the EGC while most of the GIN Sea was characterized by very low concentrations of TDOC (Fig. 4.9; Opsahl et al. 1999). Recent data confirmed the high levels of TDOC in the southward flowing EGC (12–26%) even as far south as the Denmark Strait (Fig. 4.9, Amon et al. 2003).

The export of TDOC in the EGC is an important component of the mass balance of terrigenous DOM in the Arctic Ocean. In a recent study we used in situ fluorescence measurements to trace and to better resolve the spatial distribution of TDOC in the EGC (Amon et al. 2003). Fluorescence, as mentioned before, is a simple but reliable way to trace TDOC along with other hydrographic parameters in real time. The back scatter fluorometer used by Amon et al. (2003), was specifically designed for deployment on a CTD rosette. The correlation between the fluorescence detected by the probe (ex: 330–460 nm; em: 550 nm) and TDOC (Fig. 4.10A) and the concentration of lignin phenols (Fig. 4.10B) was quite strong even at a salinity above 32 typically encountered in EGC surface waters. We used this relationship to estimate TDOC concentrations in the EGC to calculate export fluxes. As with many carbon flux estimates in the Arctic seasonal variability is not well understood due to the fact that most sampling campaigns are restricted to a few months during summer. Strong seasonal variability was observed in the volume transport through Fram Strait and in the EGC (Woodgate et al. 1999,

Fig. 4.9. Surface (< 60 m) distribution of DOC (µM) and terrestrial DOC (% TDOC) in the Arctic Mediterranean Sea. Thick gray lines across the Greenland shelf indicate the approximate positions of the fluorescence sections shown in Figure 4.11 ▶

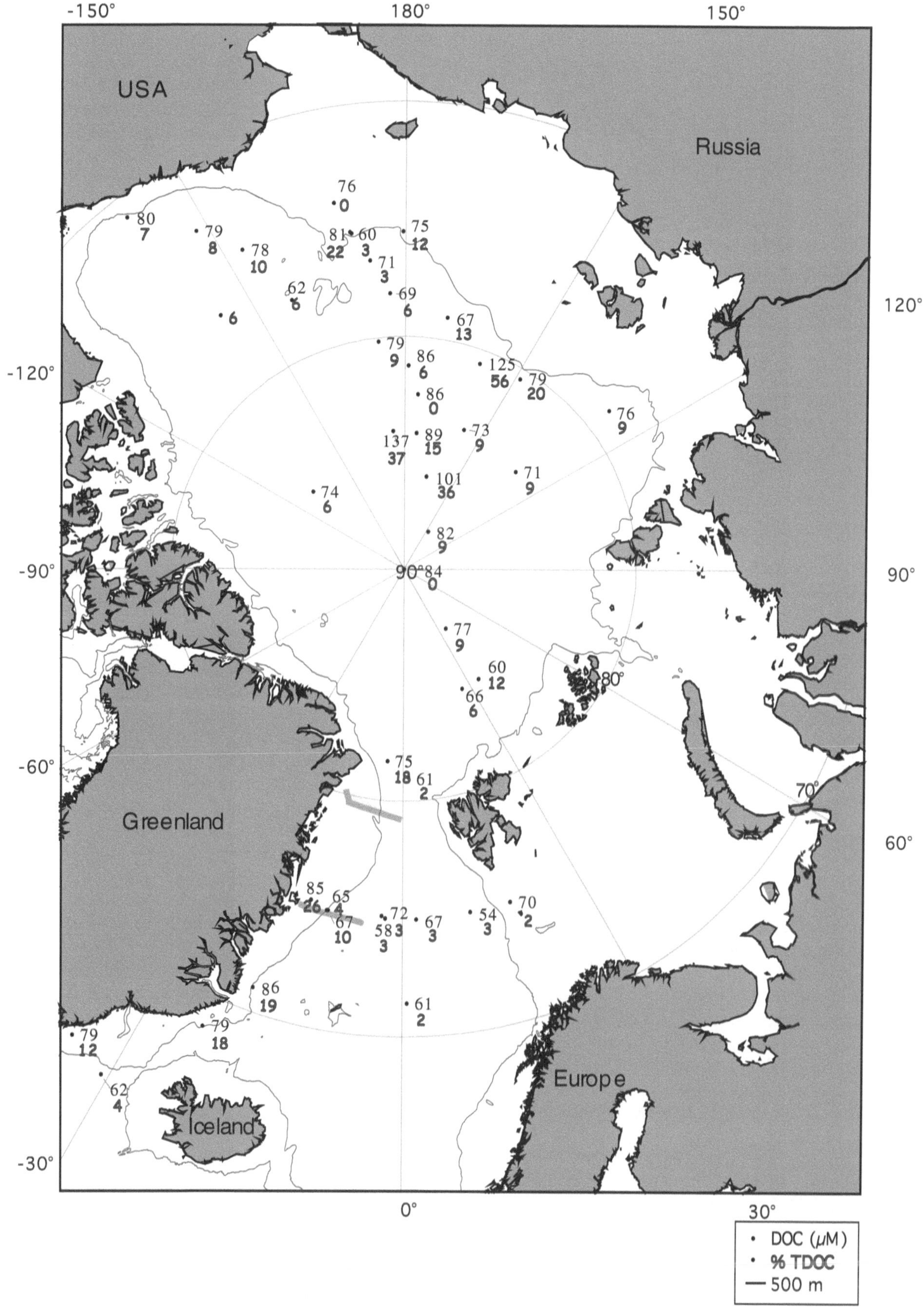
-150°
180°
150°
120°
-120°
-90°
90°
-60°
60°
-30°
0°
30°
USA
Russia
Greenland
Iceland
Europe
90°
80°
70°
• DOC (μM)
• % TDOC
— 500 m

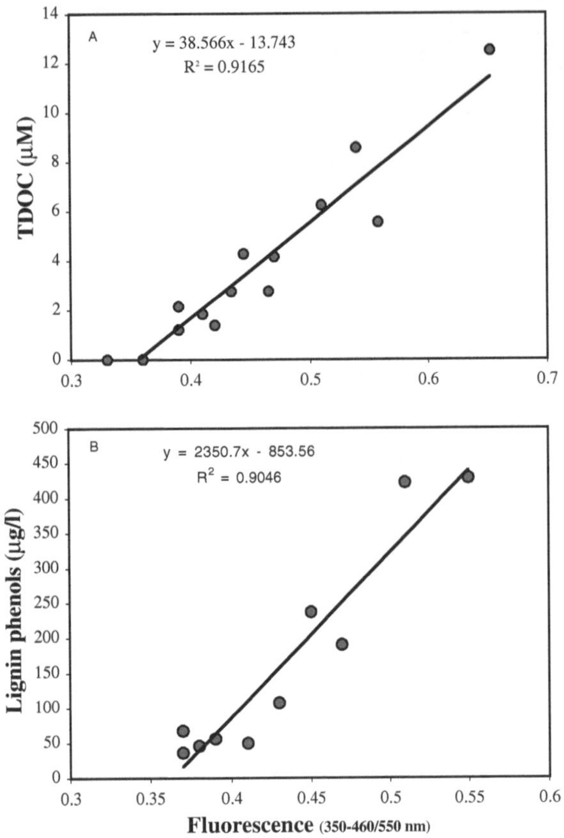

Fig. 4.10. Relationship between fluorescence (ex: 350–460 nm/em: 550 nm) and the concentration of terrestrial DOC (TDOC) during (A) a calibration experiment using water from the Ob and Yenisei as freshwater endmembers and Arctic-Atlantic water as the marine endmember and (B) the relationship between in situ fluorescence (ex: 350–460 nm/em: 550 nm) and respective in situ lignin phenol concentrations. Modified from Amon et al. (2003)

Fahrbach et al. 2001) as well as in fluorescence maxima in the EGC (Amon et al. 2003). Based on two different fluorescence cross sections at about 79°N (1993, curtesy of G. Budeus) and 75°N (1998; Figs. 4.11 A and B) and a recent flux estimate of 1.4 Sv for polar surface water in the EGC (Mauritzen 1996, Fahrbach pers. comm.) we estimated an annual TDOC export of 4.7–11.3 Tg or 20–50% of the annual river discharge to the Arctic Ocean. This is higher than our previous estimate (Opsahl et al. 1999) mainly due to the higher volume transport. Export of TDOC through the Canadian Archipelago can only be guessed at this point but elevated levels of CDOM in the North Water Polynya (Scully and Miller 2000) could be indicative for terrestrial derived DOM. With a volume transport of 1.3 Sv through the Canadian Archipelago (Lundberg and Haugan 1998) TDOC export could be of the same magnitude as in the EGC.

Up until the present we have not detected significant amounts of TDOC (lignin phenols) in Arctic or GIN Sea Deep Water (>1800 m, Opsahl et al. 1999, Benner et al. in prep.). This is consistent with our experimental observation during freezing of sea water that source water for slope transport to the Arctic Ocean interior needs to have relatively high salinity and subsequently low concentrations of TDOC (Fig. 4.8). The primary fate of terrestrially derived DOM in the Arctic ocean appears to be export towards the North Atlantic where it potentially seeds primary production due to nutrient release. Our best estimates for TDOC distribution and fluxes are still highly uncertain due to poor spatial and temporal resolution. However, the introduction of in situ fluorometers has tremendous potential to overcome this deficiency in the near future especially with recently funded international cooperations between the USA/Canada and Germany/Russia, working in the Canadian and Eurasian Arctic, respectively, to gain a pan-arctic understanding of TDOC distribution.

4.3 Distribution, chemical composition, and fluxes of marine DOM in the central Arctic Ocean

The ability to distinguish between TDOC and marine DOC is often difficult, but necessary in the Arctic Ocean in order to understand dynamics of autochthonous DOM production and utilization. Actual data on DOM production by phytoplankton and its utilization by bacteria are still very scarce in the Arctic rendering most recent estimates preliminary. Information on the chemical composition of DOM will allow us to distinguish general sources like phytoplankton and vascular plants but also to evaluate the diagenetic state and lability of marine DOM (Skoog and Benner 1997, Amon et al. 2001). We need to develop a better understanding of exchange processes of autochthonous DOM between the extended shelves, with high levels of primary production (Romankevich and Vetrov 2001; Chapter 3), and the Arctic Basins.

4.3.1 Primary production and bacterial utilization of DOM

Primary production in the central Arctic Ocean (including sea ice) and on the Eurasian shelf has recently been estimated to be around 50–75 Tg C y^{-1} (Gosselin et al. 1997: Chapter 3) and 160 Tg C y^{-1} (Romankevich and Vetrov 2001), respectively. Production estimates for the Barents and Chuckchi Seas show high variability but a conservative esti-

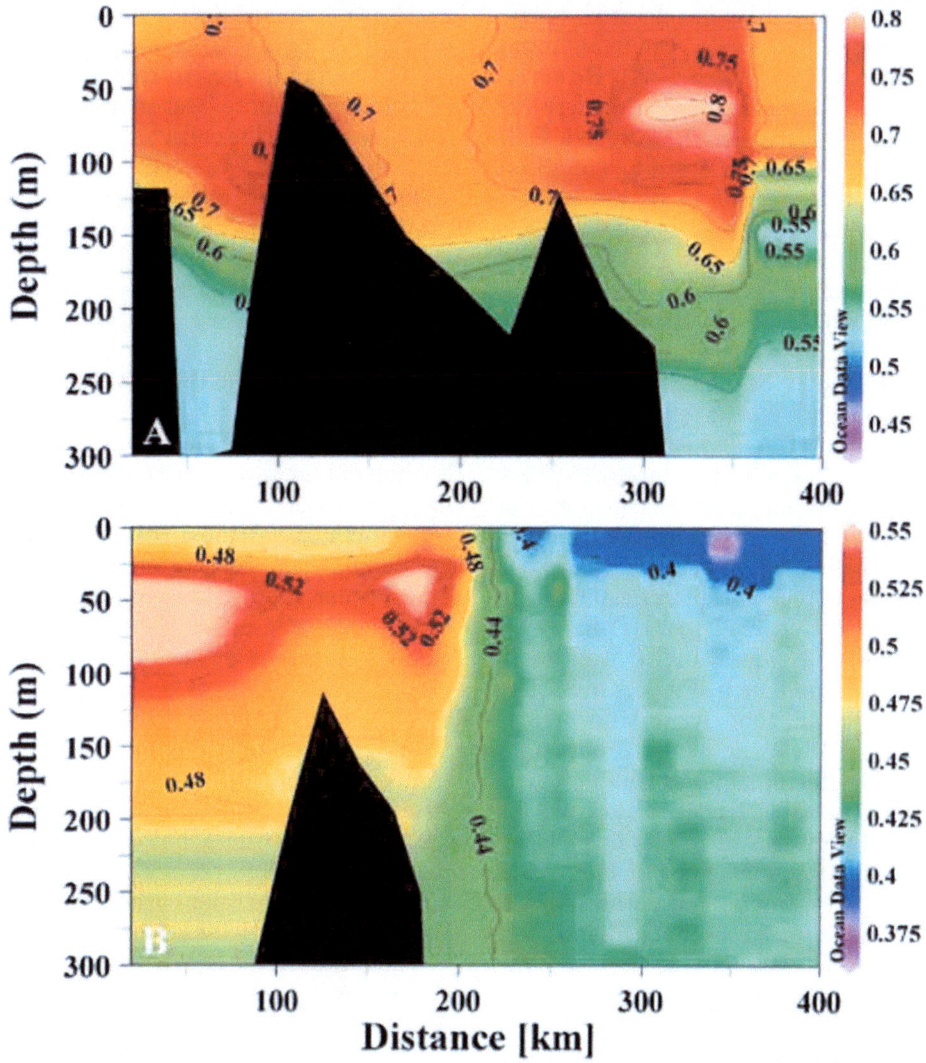

Fig. 4.11. Distribution of in-situ fluorescence (ex: 350–460 nm/em: 550 nm) in two oceanographic cross sections in (A) the Fram Strait (79°N;) and (B) across the Greenland Sea at 75°N. See also Figure 4.9. From Amon et al. (2003)

mate for primary production in the entire high Arctic is at least 329 Tg C y^{-1} (see Chapter 3). The production of DOC in the central Arctic Ocean by pelagic phytoplankton and ice algae has been determined to be around 28 Tg C y^{-1} (Gosselin et al. 1997, Wheeler et al. 1997). Gosselin et al. (1997) estimated that 20–40% of total primary production was released as DOM. Using the conservative estimate (20%) for shelf production would result in 32 Tg DOC y^{-1} produced on the shelf. Together with the estimated DOC production in the central Arctic the annual fresh phytoplankton DOM input would be about 66 Tg C y^{-1} for the total Arctic Ocean.

Bacterial mineralization processes of DOM are even less studied than primary production and represent the least studied component of the DOM cycle in the Arctic Ocean. This is particularly true for bacterial respiration. Rich et al. (1997) reported bacterial production estimates from the 1994 Arctic Ocean Section allowing a direct comparison between DOM production and bacterial DOM utilization. Based on the bacterial production data (Rich et al. 1997) and a bacterial growth efficiency of 30%, bacteria would utilize about 40 Tg C y^{-1} which is significantly more than DOM supplied by primary production during the productive season in the central Arctic Ocean (Wheeler et al. 1997). A serious mismatch between autotrophic and heterotrophic processes has been observed before in the central Arctic Ocean, based both on direct measurements (Cota et al. 1996) and on AOU estimates (Pomeroy 1997). One way to compensate for this discrepancy is through DOM supply from the shelves where primary production is thought to be much larger than in the central Arctic Ocean.

Lateral export of DOM has been suggested for the Barents Sea (Fransen et al. 2001, Anderson 2002) and for the Chukchi Sea (Walsh 1995). However, parallel estimates of bacterial production/respiration and primary production on the Eurasian shelf are very scarce. The study by Cota et al. (1996) on the Chukchi shelf, the only one to my knowledge, indicated that community respiration and primary production were fairly balanced on the shelf. Using information from Romankevich and Vetrov (2001) and Saliot et al. (1996) one can roughly estimate primary production of DOM and bacterial DOM utilization in the Laptev Sea, based on the same conversion factors and growth efficiencies used by Wheeler et al. (1997) and Rich et al. (1997). Those estimates are generally in balance with about 1.5 Tg C y^{-1} for DOM production and consumption indicating no large leftovers for export. Another argument raised against large lateral DOM exports from the shelf areas to fuel central Arctic AOU is the long transition time (Cota et al. 1996). From all we have learned about the bioavailability of DOM in the last 5 years (Amon and Benner 1996, Amon et al. 2001, Søndergaard et al. 2000, Amon and Benner 2003) we know that DOM can be partitioned into subfractions with different labilities and biological residence times. The most labile portion will be remineralized within weeks of production. However, a significant amount of semilabile DOM can survive for several month and perhaps years. Whether or not there would be enough semilabile DOM left after several years of transition from the shelf seas to the Arctic Basin (Cota et al. 1996) still needs to be investigated. From a recent study we estimated remineralization rates for semilabile DOC between 1 and 2.5 µM C y^{-1} over a 4–9 year time period in the central Greenland Sea (Amon et al. 2003) indicating the potential long term survival of DOM.

Another explanation for the observed discrepancy between production and respiration in the central Arctic Ocean might be temporal decoupling of autotrophic and DOM consumption processes which would easily be missed during short term and seasonally biased measurements (Cota et al. 1996). Both of these explanations probably contribute to account for the observed discrepancy between production and remineralization. With the available information a rough C-budget can be calculated for the central Arctic Ocean. An average of 75 Tg C y^{-1} are produced within the central Arctic Ocean according to Gosselin et al. (1997). From that, between 4 and 17 Tg POC are vertically exported (Wassmann et al., this volume), and 40 Tg C y^{-1} (Rich et al. 1997) are mineralized by bacteria. This would leave at least 18 Tg C y^{-1} for dissolution of POC to DOC to fuel the additional bacterial carbon demand (12 Tg C y^{-1}) but would not leave much to feed the rest of the heterotrophic community.

Apparently, more seasonal data on biological processes need to be collected to reproduce a realistic carbon budget. We need to determine potential sinks and sources of semilabile DOM on Arctic shelves and in the central basins and identify if significant amounts of shelf-DOM production are exported to the central basins.

4.3.2
DOM distribution and chemical composition

In general, DOC concentrations in the Arctic Ocean are higher than in any other ocean basin. In the case of surface waters this is largely a result of river inputs as demonstrated by a number of studies (Anderson et al. 1994, Opsahl et al. 1999, Bussmann and Kattner 2000, Amon and Benner 2003, Amon et al. 2003). Neglecting the terrigenous contribution, Arctic Ocean surface waters have an average DOC concentration of 71 ± 7.5 µM (Amon and Benner 2003) a concentration similar to other surface oceans (Benner 2002). Depth profiles of DOC in the different basins of the Arctic Ocean (Bussmann and Kattner 2000; Fig. 4.12) reflect the much higher surface values but also demonstrate that deep water concentrations (>1000 m) are about 50 µM DOC, higher than deep water DOC values in the Atlantic and Pacific (Hansell and Carlson 1998).

Since the Arctic Ocean is influenced by lateral exchanges of different water masses it has been most informative to look at DOC distribution with respect to typical water masses rather than only depth profiles (Bussmann and Kattner 2000, Amon et al. 2003). Typical DOC concentrations can be assigned to specific water masses (Table 4.2) for which volume transport estimates have already been established. Highest concentrations of DOC are found in Polar Surface Water (PSW) with reported average values of 78 µM DOC (Bussmann and Kattner 2000) and 76.3 ± 13.8 µM DOC (Amon and Benner 2003) and peak values up to 140 µM C (Amon and Benner 2003) and 225 µM C (Bussmann and Kattner 2000). These values are much higher than DOC values of other Arctic Mediterranean Sea basins which are typically around 60 µM DOC (Table 4.2). Inflowing Pacific water appears to have high DOC concentrations as well, averaging 71 µM C (Anderson 2002). Very consistent values, around 58 µM DOC, have been reported for Atlantic water throughout the AMS (Wheeler et al. 1997, Bussmann and Kattner 2000, Anderson 2002, Amon et al. 2003). Deep water values are typically around 50 µM DOC with lowest concentrations found in the Nor-

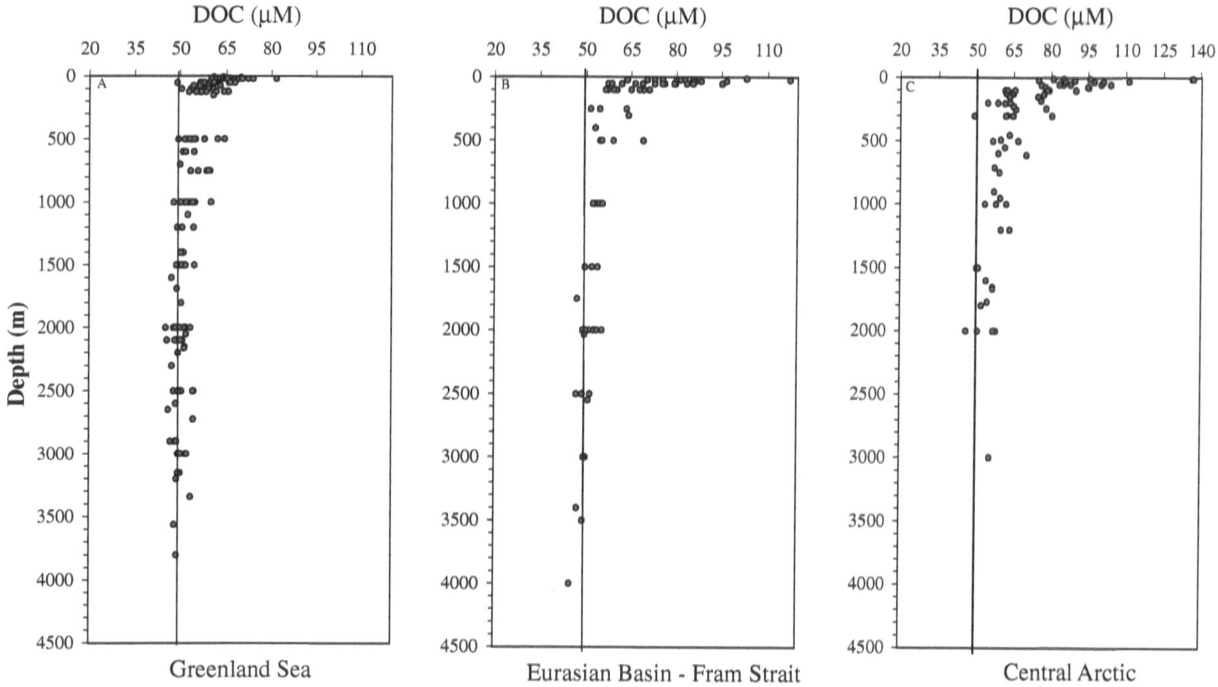

Fig. 4.12. DOC depth profiles in (A) the Greenland Sea, (B) the Eurasian Basin near Fram Strait and (C) the central Arctic Ocean

wegian Sea Deep Water and slightly higher values in deep waters of the central Arctic Ocean (Fig. 4.12, Table 4.2).

Information on the isotopic and molecular level composition of DOM in the central Arctic Ocean and GIN Sea is only beginning to emerge through international cooperations. Initial studies focused on the terrestrial component of DOM as discussed above (Opsahl et al. 1999) but ongoing studies include a variety of chemical analysis (^{13}C, ^{15}N, ^{14}C, NMR, lignin oxidation products, neutral sugars, amino acids) performed on single samples and will boost our understanding of Arctic DOM dynamics in the near future. A first data set from numerous

Table 4.2. Average DOC concentrations in water masses of the Arctic Mediterranean Sea

Water Mass	DOC (μM)	Reference
Rivers	550–630	Lobbes et al. (2000), Gordeev and Rachold (this volume)
Pacific Water	71 ± 20	Anderson (2002)
Atlantic Water	58.2 ± 4.9	Amon et al. (2003)
Polar Surface Water	76.3 ± 13.8	Amon and Benner 2003
EGC	75.8 ± 10.2	Amon et al. 2003
MAW	58.4 ± 5.1	Amon et al. 2003
GSSW	63.9 ± 8.6	Amon et al. 2003
NSSW	57.0 ± 4.7	Amon et al. 2003
ISSW	61.1 ± 4.1	Amon et al. 2003
MBDW	55	Bussmann and Kattner (2000)
ABDW	54	Bussmann and Kattner (2000)
NBDW	50	Bussmann and Kattner (2000)
EBDW	49.8 ± 2.0	Amon et al. 2003
GSDW	49.5 ± 2.0	Amon et al. 2003
NSDW	47.5 ± 1.8	Amon et al. 2003
ISDW	51.6 ± 2.7	Amon et al. 2003

EGC = East Greenland Current, MAW = Modified Atlantic Water, GSSW = Greenland Sea Surface Water, NSSW = Norwegian Sea Surface Water, ISSW = Iceland Sea Surface Water, MBDW = Makarov Basin Deep Water, ABDW = Amundsen Basin Deep Water, NBDW = Nansen Basin Deep Water, EBDW = Eurasian Basin Deep Water, GSDW = Greenland Sea Deep Water, NSDW = Norwegian Sea Deep Water, ISDW = Iceland Sea Deep Water.

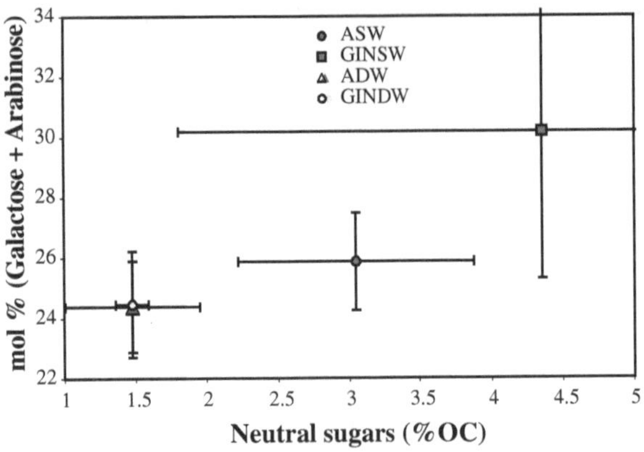

Fig. 4.13. The diagenetic state of DOM (>1 kDa) collected from surface and deep waters in the Arctic Mediterranean Sea. High neutral sugar yields and high mol% of galactose and arabinose are indicative of fresh and labile DOM. ASW – Arctic Ocean surface water, ADW – Arctic Ocean deep water, GINSW – GIN Sea surface water, GINDW – GIN Sea deep water

samples throughout the Arctic Mediterranean Sea has recently been published by Amon and Benner (2003) with a focus on the neutral sugar composition of ultrafiltered DOM. This study built on previous work (Cowie and Hedges 1992, Skoog and Benner 1997) which identified the neutral sugar yield (% DOC as neutral sugars) as an useful tool to determine the diagenetic state of DOM in the ocean. Amon and Benner (2003) showed that the neutral sugar yield can be used to predict the amount of labile and semilabile DOM and that the molecular composition of neutral sugars is also largely determined by diagenesis. Surface waters of the Arctic Mediterranean Sea contained an average of 2.8% labile DOC and their neutral sugar composition was dominated by glucose, galactose, and fucose (Amon and Benner 2003). Deep water DOM of the AMS was more degraded as indicated by a low neutral sugar yield of 1.6% and a neutral sugar composition dominated by glucose. Based on neutral sugar yield and composition, ultrafiltered DOM can be distinguished according to its diagenetic state (Fig. 4.13). GIN Sea surface waters had a higher neutral sugar yields and relatively elevated levels of galactose and arabinose (Fig. 4.13) indicating that GIN Sea DOM contained more fresh phytoplankton-derived DOM than Arctic Ocean surface water. DOM in deep waters has apparently already undergone considerable degradation with much lower neutral sugar yields and lower levels of galactose and arabinose. Surface waters of the Arctic Mediterranean Sea appear to have smaller amounts of fresh DOM than other oceanic surface waters (Amon and Benner 2003) but a significant amount of semilabile DOM which could fuel heterotrophic activity on longer time scales. The neutral sugar yield might hold the key to estimate the amount of semilabile DOM in the Arctic shelf waters in the future.

4.3.3 DOC exchanges between the Arctic Ocean and adjacent Ocean basins

Due to its relatively small volume the Arctic Ocean is heavily influenced by lateral exchanges of water masses and DOM. Volume transport estimates vary greatly in the recent literature (Mauritzen 1996, Anderson et al. 1998, Fahrbach et al. 2001) with somewhat higher estimates in the most recent assessment in the Fram Strait (Fahrbach et al. 2001) based on multiyear observations. For mass balance calculations we divided the Arctic ocean in two layers, a layer between 0 and 700 m containing water masses of Atlantic origin, and a layer below 700 m including deep water masses flowing in and out through Fram Strait (Table 4.3). The new Fram Strait flux estimates (Fahrbach et al. 2001) were combined with a new river inflow estimate (Shiklomanov et al. 2000) and existing flux estimates of surface water exchanges through the Bering Strait and the Canadian Archipelago. Recently collected information on DOC concentrations (Amon et al. 2003) were combined with existing DOC values reported for relevant water masses to estimate a mass balance for DOC sources and sinks in the central Arctic Ocean.

The largest source of DOC in the Arctic Ocean is the Norwegian Atlantic Current importing about 141 Tg DOC y^{-1}. Another large source is deep water entering the Arctic Ocean through Fram Strait importing about 88 Tg DOC y^{-1}. Other sources of DOC include the Arctic Rivers with 31 Tg DOC y^{-1}, and Pacific surface water entering through Bering Strait importing 22 Tg DOC y^{-1} (Table 4.3). The export of DOC is largest in deep waters of the East Greenland Current (EGC; 130 Tg DOC y^{-1}) and smaller DOC export fluxes of 3.8 Tg DOC y^{-1} in sea ice, and surface water export of 38 and 40 Tg DOC y^{-1} through the Canadian Archipelago and in the EGC, respectively (Table 4.3). The upper Arctic layer (<700 m) receives more DOC (194 Tg DOC) per year than is exported (134 Tg DOC). The opposite was observed for the deep layer (>700 m) which has a larger annual DOC export (130 Tg DOC) than import (88 Tg DOC). Ignoring biological processes, our estimates indicate that the Arctic Ocean is a net sink for DOC, exporting 265 Tg DOC y^{-1} and importing 282 Tg DOC y^{-1} a pattern that has also been reported by Anderson

Table 4.3. Volume and DOC fluxes in and out of the central Arctic Ocean

Water Mass	Volume transport (Sv)	DOC flux (Tg C yr^{-1})	Reference (vol. transport)
Upper Arctic: 0–700 m			
• Sea ice	- 0.09	- 3.80	Eicken (this volume)
• River inflow	+ 0.14	+ 30.8	Shiklomanov et al. 2000
• Pacific water inflow	+ 0.83	+ 22.0	Lundberg and Haugan 1996
• Atlantic water inflow	+ 6.40	+ 140.8	Fahrbach et al. 2001
• Polar Surface Water	- 1.30 (Can. Archip.)	- 37.5	Lundberg and Haugan 1996
• EGC Surface Water	- 1.40 (Fram Strait)	- 40.1	Fahrbach et al. 2001
• MAW	- 2.40	- 53.0	Fahrbach et al. 2001
Deep Arctic: >700 m			
• EGC DW	- 6.88	- 129.9	Fahrbach et al. 2001
• WSC DW	+ 4.70	+ 88.0	Fahrbach et al. 2001

The error associated with the volume tranport estimates is approximately ±9% (Fahrbach et al 2001).

(2002) if we consider the different estimates of DOC in sea ice. Contrary to Anderson's budget (Anderson 2002) our DOC budget suggests a net outflow of DOC in the deep layer, based on the new volume transport estimates through Fram Strait (Fahrbach et al. 2001). This indicates that the DOM pool within the Arctic Ocean is more dynamic than previously thought. The net input of DOC suggests efficient remineralization of DOM within the Arctic Ocean which is consistent with the observations by Rich et al. (1997) and Wheeler et al. (1997). Current estimates of primary production of DOC in the central Arctic Ocean are around 28 Tg DOC y^{-1} (Wheeler et al. 1997). Wheeler et al. (1997) estimated that bacteria utilized 12 Tg DOC more than was produced by phytoplankton during their study which would be of a similar magnitude as the net DOC import (17 Tg DOC y^{-1}) observed in our budget. On the other hand, the Arctic Ocean is thought to be a net sink for atmospheric CO_2 (Anderson et al. 1998, Bauch et al. 2000) suggesting that primary production exceeds heterotrophic respiration in the surface mixed layer. Apparently, biological processes involving DOM in the Arctic are not well resolved at the moment. More work and better spatial and temporal resolution for biological measurements are clearly needed to get to the bottom of these contradicting observations.

4.3.4 Vertical export of DOC in the Arctic Ocean

The observation that more DOC is exported than imported through the deep Fram Strait indicates that there must be considerable (42 Tg C y^{-1}) vertical transport of DOC within the central Arctic Ocean. Vertical DOC transport is also suggested by the relatively high concentrations of DOC found throughout the deep Arctic Mediterranean Sea (Fig. 4.12, Table 4.2). While the Greenland and Iceland Sea are known deep water formation sites and have been studied in recent years (Aagaard et al. 1985, Budeus et al. 1998, Anderson et al. 2000) not much is known about deep water formation in the central Arctic Ocean itself. Increasing evidence points to deep water renewal along the continental slopes of the Eurasian shelf (Anderson et al. 1999, Rudels et al. 2000) with an estimated volume transport of 1.3–1.9 Sv (Mauritzen 1996, Anderson et al. 1999). In order to account for the net outflow of 42 Tg DOC y^{-1} through the deep Fram Strait there needs to be an adequate input of semi-labile DOC to the deep Arctic Ocean. Primary production and vertical particle transport alone would not be enough to maintain the relatively high DOC concentrations in Artic deep waters. The most recent POC flux estimate for the central Arctic ranges from 4–8 Tg C y^{-1} (see Chapter 5) which would translate to an annual deep water DOC increase of 0.0025–0.05 μM C, if all POC is converted to DOC. Obviously, vertical transport of DOC during advection processes needs to be involved to sustain the observed concentrations of DOC in the deep Arctic Ocean.

From DOC depth profiles in the Greenland Gyre Amon et al. (2003) recently derived mineralization rates for semi-labile DOC on the order of 1–2.5 μM C y^{-1} based on a DOC gradient from 60 μM DOC at the surface and 50 μM DOC at 1500 m. Concentrations between 55 and 60 μM DOC are often encountered in AMS surface waters (Amon et al. 2003) during fall, well after the productive season. Assuming a concentration of 60 μM DOC as typical of source water for vertical transport and 1.3–1.9 Sv as vertical volume flow one can calculate an annual vertical DOC transport of 29–43 Tg C. This is similar to the observed net lateral export of DOC through the deep Fram Strait. Due to the presence of semi-labile components in vertically transported DOC we also need to figure in a certain loss of DOC due to bacterial utilization. The typical DOC concentrations of Arctic Ocean deep water is around 50 μM. If we assume a remineralization rate of 0.05 μM DOC y^{-1} for deep water (Hansell and Carlson 1998) we get an annual loss of 7.4 Tg C. Together with the lateral export we calculated a loss of 37–50 Tg DOC y^{-1} for the entire deep Arctic Ocean that needs to be replenished. This could be accomplished by a vertical flux of 1.6–2.2 Sv. If these assumptions are valid, vertical DOC transport would be 3–11 times higher than vertical POC transport (4–8 Tg C y^{-1}) in the central Arctic Ocean.

Similar calculations have recently been reported for the GIN Sea (Amon et al. 2003). The Greenland Gyre is a well studied location for deep water renewal with recent estimates of vertical transport and flushing times of 0.6 Sv and 20–30 years, respectively (Budeus et al. 1998). From their long term studies in the Greenland Sea Budeus et al. (1998) concluded that deep water formation in the Greenland Sea happens continuously at a rate of 150 m y^{-1}. This means that surface water with elevated DOC concentration is seasonally injected into the deep Greenland Sea. Knowing the typical pre-winter surface (<100 m) concentrations of 60.4 μM DOC and the DOC concentration of 50.3 μM C at 1500 m in the Greenland Gyre (Amon et al. 2003) we can roughly estimate the annual vertical export of DOC in the Greenland Gyre. With an approximate vertical transport of 0.6 Sv (Budeus et al. 1998) the Greenland convective gyre would produce a total annual DOC export of 13.6 Tg of DOC. If we assume that the difference between 60.4 μM and 50.3 μM DOC (about 10 μM) represents the annually produced semilabile DOC, the Greenland Gyre would annually downwell about 2.3 Tg of "new" DOC. This is about 8 times the estimated POC flux (0.3 Tg C y^{-1}; see Chapter 5) and 30% of a recent estimate for anthropogenic carbon dioxide sequestration in the Greenland Sea based on total dissolved inorganic carbon measurements (Anderson et al. 2000). Deep water formation in the Iceland Sea is still a matter of debate but a recent estimate for the formation of intermediate water in this region was reported as 0.7 Sv (Mauritzen 1996) which would represent a vertical DOC transport around 15.9 Tg DOC y^{-1}.

Altogether, the Arctic Mediterranean Sea potentially downwells up to 73 Tg DOC y^{-1} which is roughly 37% of the annual carbon burial in marine sediments or 2% of the global vertical POC export. Clearly, there remain large uncertainties associated with those back of the envelope calculations but the evidence for large vertical transport of DOC in the AMS is quite convincing. Prior studies demonstrated that high northern latitude oceans have significantly elevated (10–20 μM) pre-winter and deep-water DOC concentration compared to other ocean basins (Anderson et al. 1994, Bussmann and Kattner 2000, Amon and Benner 2003) and it seems that the higher concentrations can be sustained for decades based on residence times. We believe the reason for this is a larger standing stock of semilabile DOC within the AMS and in imported Atlantic water. Changes in DOC concentrations are especially important in areas of deep water formation where semilabile DOC can be removed from the ocean surface for centuries. It will be crucial to learn more about the mechanisms that determine the production of semilabile DOC and the source of deep water in the Arctic Ocean. The vertical export of DOC in the AMS appears to exceed the POC export by far and represents a very exciting research avenue to pursue with strong implications for the organic carbon cycle in the Arctic as well as the global oceans.

4.4
Summary and Concluding Remarks

- Dissolved organic carbon (DOC) concentrations in large Eurasian rivers (Ob, Yenisei, Lena) show a strong seasonal signal with lowest concentrations in winter (~300 µM C) and highest concentrations at the onset of freshet (>1000 µM C). DOC mixes conservatively in those rivers discharging about 11 Tg (10^6 t) DOC annually into the Arctic Ocean with minor losses (<5%) during estuarine mixing.
- The chemical composition of riverine dissolved organic matter (DOM) indicates a predominantly terrigenous origin, a modern age, and considerable degradation.
- A significant fraction (20–50%) of Arctic river DOM is exported to the North Atlantic in the East Greenland Current as indicated by organic tracers and in situ fluorescence. In situ fluorescence represents a powerful tool to trace river and shelf water in the Arctic Ocean in the future.
- The formation of sea ice on the Eurasian shelf plays an important but poorly understood role for the distribution of DOM in the Arctic Ocean.
- Biological production and consumption of DOM are still poorly resolved in the Arctic Ocean resulting in unbalanced DOC budgets. Lateral and vertical transport of DOM need to be incorporated in DOM budgets and more parallel measurements of primary production and consumption of DOM are needed.
- DOC concentrations in Arctic Mediterranean Sea (AMS) deep waters are elevated relative to other oceanic deep waters indicating frequent vertical DOC transport in this system. We estimate that about 73 Tg of DOC are downwelled annually in the AMS, about 5-times the particulate organic carbon flux.

Chapter 5

Particulate Organic Carbon Flux to the Arctic Ocean Sea Floor

P. Wassmann, E. Bauerfeind, M. Fortier, M. Fukuchi, B. Hargrave, B. Moran, T. Noji, E.-M. Nöthig, K. Olli, R. Peinert, H. Sasaki, V. Shevchenko

5.1 Introduction

Compared to other areas of the world's ocean, polar seas have a large influence on the atmosphere's CO_2 content (Anderson et al. 1990), mainly due to deep-water formation, subduction and surface water-cooling. Sequestration of atmospheric CO_2 in polar seas takes place, in part, through the action of the surface biota. One way the photosynthetically fixed carbon becomes inaccessible to the atmosphere is through gravitational sinking particulate organic carbon (POC) or vertical export with denser water of dissolved organic carbon (DOC) to intermediate depths (Fig. 5.1). The dynamics of vertical flux of biogenic matter, and to a lesser extent also the composition of exported matter has been investigated in a variety of marine ecosystems over the past two decades. Major seasonal, regional and inter-annual variations on the quantity and composition of biogenic matter exported from the upper water column have been observed. Loss of organic matter takes place mainly through settling of phytoplankton cells, detritus and zooplankton faecal pellets (Fig. 5.1). Regional, seasonal and interannual variations have thus been attributed to differences in primary production as well as retention and export food chains (e.g. Peinert et al. 1989; Wassmann 1998). In several arctic regions, river discharge and resuspension may influence primary production and vertical export. Part of the carbon supply of the Arctic Ocean is thus allochthonous: it derives from rivers and the shallow Eurasian and Chukchi Sea shelves and accumulates in the deep basins (Stein et al. 1994; Walsh 1995).

At low latitudes and in non-upwelling systems, the annual phyto- and zooplankton cycles frequently vary within narrow limits. Pelagic systems are thus potentially capable of retaining and recycling most autochthonous organic material, although, some losses due to sinking particles occur. With increasing latitude, the variability of physical forcing increases and the relationship between phyto- and zooplankton apparently becomes less coupled. This results in less predictability and a tendency towards pelagic-benthic coupling that is more episodic (Wassmann et al. 1996). Locally in arctic regions, this is reflected in decoupling between primary and secondary producers, giving rise to extensive sedimentation of ungrazed phytoplankton (e.g. Wassmann et al. 1990; Wassmann, 2001; Fortier et al. 2002; Hargrave et al. 2002). Some of the world's highest sediment carbon (Romankevich 1984) and benthic biomass concentrations (Zenkevich 1963) are found on shallow arctic shelves, which possibly results from this decoupling and a tight pelagic-benthic coupling (Grebmeier and Barry 1991).

Marginal and seasonal ice zones are predominant characteristics of arctic regions. Ice edge blooms dominated by diatoms such as *Thalassiosira antarctica*, *T. hyalina*, *Chaetoceros socialis*, *Fragilariopsis* spp. (e.g. Makarevich 1997; Wassmann et al. 1999; von Quillfeldt 2000) are believed to be the first algae to bloom. Such blooms frequently are succeeded by the colonial and/or single cell forms of the haptophyte *Phaeocystis pouchetii* (Rey and Skjoldal 1987; Skjoldal and Rey 1989; Sukhanova and Flint 2001; Ratkova and Wassmann 2002). The actual vertical flux of biogenic matter is primarily determined by an interaction between the predominant phytoplankton community and the grazer impact. It seems obvious that not all types of phytoplankton are equally favourable food items for herbivores. Diatoms are readily grazed by copepods, while the fate of *Phaeocystis* spp. is more controversial (Wassmann 1994; Makarevich 1997; Verity 2000; Sukhanova and Flint 2001), particularly in the insufficiently investigated arctic region. Flagellate blooms are either grazed or tend to decompose in the water column and fuel the microbial loop food web. Prevalence of flagellates results in recycling of the fixed carbon close to the ocean surface where it stays in exchange with the atmosphere (e.g. Anderson and Kaltin 2001).

In the Arctic, light becomes the major limiting factor for plankton growth due to the restrictions imposed by the annual light cycle and by sea ice, ice algae (see Fig. 5.1) and, locally resuspension. Areas of open water (including leads and polynyas) and reduced ice-cover (ice melt) are therefore

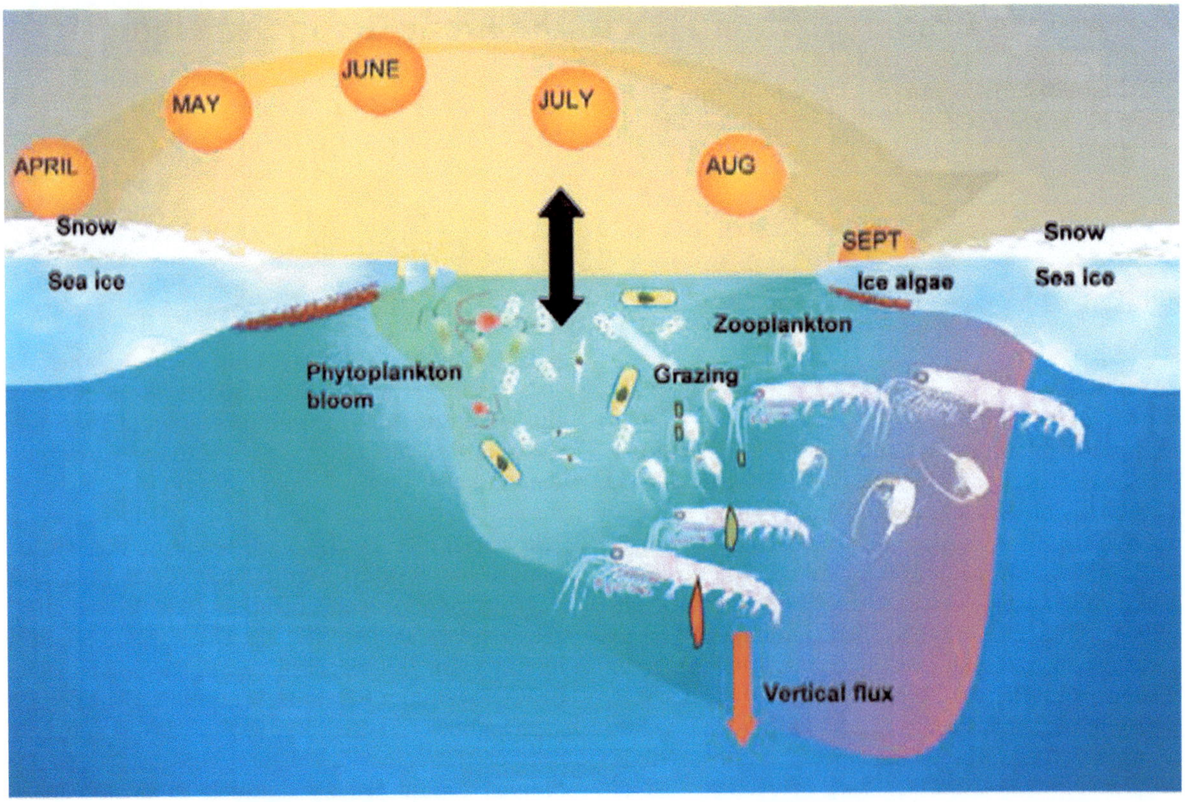

Fig. 5.1. Schematic illustration of principal processes during the productive cycle in the Arctic Ocean. The height of the sun and the declining and increasing snow and ice thickness are shown. The growth of ice algae in spring and autumn reduces the incident light. The spring bloom, the phytoplankton development, the vertical export and the dominance of phytoplankton cells in spring and faecal pellets from zooplankton grazing in summer and early autumn are illustrated. The depth of the euphotic zone is depicted. In addition, the transition from phytoplankton dominance in late spring to detritus dominance in autumn is illustrated. Redrawn after a figure provided by M. Fukuchi

expected to be regions of elevated production. In contrast, regions with continuous and thick ice-cover, estuaries, glacial-fed fjords and shallow shelves have reduced production. Thus, light is the ultimate driving force of primary production and indirectly also of the vertical flux of autochthonous carbon in these regions.

5.2 What do we Know About Vertical Carbon Flux from the Arctic Ocean?

The significance of the Arctic Ocean for global carbon budget requires understanding of processes affecting carbon cycling and vertical carbon export. Despite considerable international efforts to directly measure vertical carbon flux (e.g. JGOFS, for overviews see Berelson 2001; Ducklow et al. 2001) the current global understanding of pelagic-benthic coupling is rudimentary. Our knowledge is insufficient to understand/model dynamics of processes affecting daily and seasonal changes in vertical flux on a global scale. This is particularly true for subarctic and arctic regions with their large seasonal variation in physical forcing and poorly understood marine ecosystem dynamics. Pelagic-benthic coupling and vertical flux in many arctic regions are strongly dominated by episodic events on daily, weekly, seasonal and yearly time scales (Wassmann et al. 1996). As the physical forcing becomes more and more dominant and the response of the planktonic food web less known with increasing latitude, the episodic nature of vertical flux and its regulation gives rise to increasing uncertainty in the prediction of seasonal and annual rates.

Here we gather and present a majority of vertical flux data from the Arctic Ocean and adjacent subpolar regions with a seasonal ice-cover. We exclude investigations from arctic fjords. Annual cycles of particle fluxes have previously been studied in detail at great depths in the adjacent Nordic Seas (e.g. Honjo 1990; Hebbeln and Wefer 1991; Bodungen et al 1995; Wassmann et al. 1996;

Peinert et al. 2001a) and will not be examined hereafter. POC flux to the arctic sea floor has only been investigated in a few regions (Fig. 5.2). From the vast Arctic Ocean basin, only data from a few short-term deployments along its fringes (north of Spitsbergen, the Nansen, Amundsen and Makarov Basin) and three long-term deployments from the Lomonosov Ridge (close to the Laptev Sea), the Kara Sea and the Beaufort Sea are available (the significant data sets from the Beaufort Sea and the Kara Sea have as yet not been published and will not be presented in detail). There also exist data from the Arctic Ocean archipelago and polynyas. The highly relevant data from the SHEBA ice drift through the Chukchi and East Siberian Sea region (1997–1998) were unfortunately not available for this review.

In order to give a best possible overview, we present data from both long-term moored sediment traps in deeper regions and from short-term deployments in the upper layers and from shelves. Contrasting vertical flux of biogenic matter from various regions measured with sediment trap deployments at different depths is difficult to interpret. This dilemma, the scarcity of data, the dearth of annual investigations and the lack of concomitant investigations of primary production and plankton in the Arctic Ocean, gives rise to a rather incomplete and incoherent account. The present chapter provides an overview on existing knowledge and has a preliminary character awaiting the emergence of more adequate data. Thus, this chapter comprises modules from various investigations, followed by a short summary discussion, rather than a thorough account. The aim of the chapter is to improve our quantitative and qualitative understanding of POC flux to the sea floor of the Arctic Ocean and to provide a basis for future studies of vertical flux of biogenic matter in this region.

5.3 Case studies

The majority of the investigations summarised below have been carried out adjacent to the Arctic Ocean basin and in the Atlantic and western Eurasian sector (Fig. 5.2). With few exceptions, only investigations that have already been published are summarised here. The lack of long-term data from the central Arctic Ocean, the Canadian Basin and the central and eastern Siberian shelf is noteworthy. This reflects the relatively unconstrained access to the Arctic region through the north-eastern North Atlantic. It also reflects that considerable research has been carried out in this region, addressing the significance of the Nordic Seas for water mass formation, circulation and physical forcing in the Arctic Ocean (for recent summaries, see Anderson and Kaltin 2001; Björk et al. 2001; Fahrbach et al. 2001; Jones 2001). We present case studies from west to east, starting with the Atlantic sector.

5.3.1 North Water Polynya

B. Hargrave

The North Water Polynya (NOW) was the focus of a multidisciplinary international study in 1997–1999 [within the framework of the International Arctic Polynya Project (IAPP)]. NOW, located in northern Baffin Bay (approximately 76° to 79°N, 71° to 77°W) with a maximum area of 80,000 km^2, is the largest polynya in the western Arctic Ocean (Fig. 5.2). Arctic Ocean outflow water enters from the north through Smith Sound and combines with water from the West Greenland Current that flows northward along the western side of Greenland. Additional Arctic outflow water enters the southern part of the polynya from the east through the Canadian Archipelago via Lancaster and Jones Sounds. Ice begins to form in the polynya during October and coverage is almost complete by early December when an ice-bridge forms across Smith Sound at the southern end of Kane Basin. The ice bridge restricts movement of ice from the north during winter, but ice formed within the polynya, especially along the western side adjacent to Ellesmere Island, is often moved rapidly to the south by currents and wind. Open water may amount to 50% of the total polynya area at various times during the winter. In general, the eastern side of the polynya along the Greenland coast experiences heavier ice cover between November and April than the western side due to ice removal. The ice-bridge in Smith Sound generally breaks during late April and early May when ice rapidly leaves the polynya. Shore-fast ice disappears but some ice cover (up to 15%) remains during the summer (July–August) as floating ice pans of varying size move southward from Kane Basin through Smith Sound.

Results

Measurements of particulate organic carbon were determined in settled material collected by floating sediment traps deployed for short (<1 d) periods during cruises in September 1997, April to July 1998 and August 1999 (Michel et al. 2002). No corrections

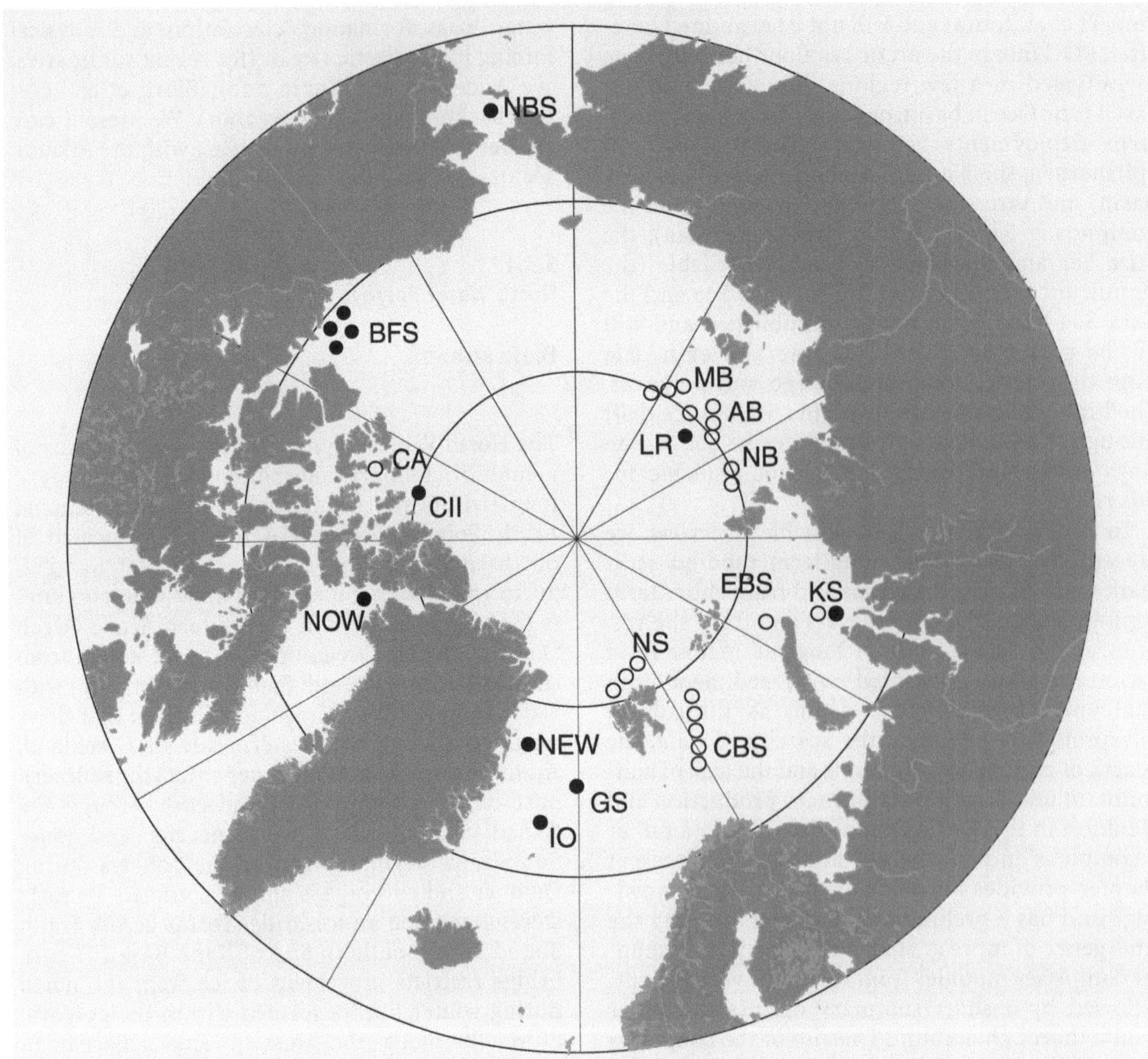

Fig. 5.2. Circumpolar view of the position of vertical export investigations in Arctic Ocean. Open circles = short-term deployments. Filled circles = long-term deployments. NOW = North Water Polynya. NEW = North East Water Polynya. IO = IS Odden. GS = Greenland Sea. CBS = Central Barents Sea. NS = Northern Spitsbergen. EBS = Eastern Barents Sea and KS = Kara Sea (see for details). NB = Nansen Basin. AB = Amundsen Basin. MB = Makarov Basin. LR = Lomonosov Ridge. BFS = Beaufort Sea. CII = Canadian Ice Island. CA = Canadian Archipelago. NBS = Northern Bering Sea

were made for dissolution of organic carbon from settled material during the collection periods. Multi-cup traps were placed at two depths >200 m on moorings at five stations over two years (September 1997–August 1999; Hargrave et al. 2002; Sampei et al, 2002). Floating sediment trap arrays consisted of cylinders (height:diameter aspect ratio >5) on drifting drogues at various depths between 50 and 100 m. Different designs of multi-cup traps (pairs of traps of similar design on each mooring) (Technicap PPS.3/3, McLean- Parflux, Oregon State-Tracer 15, Bedford Institute of Oceanography design; see Table 5.4) were used on deep moorings because of the need to combine equipment from different collaborating institutions. Some trap designs (McLean, OSU) experienced clogging during periods of high particle flux (July-August) when settled material contained gelatinous material thought to be discarded appendicularian houses. The Technicap design with the highest aspect ratio (4) did not clog. The mooring at the northern site was tilted due to high currents (velocities up to 155 cm s^{-1}) with semi-diurnal depth variations of almost 100 m. Particle collection efficiency was therefore low during these periods.

POC fluxes in surface layers (<100 m) observed with floating trap arrays in the NOW during late spring, summer and early fall months varied from 150 to 1100 mg C $m^{-2} d^{-1}$ (Michel et al. 2002; Houston and Deming 2002). Highest fluxes in upper moored traps (200 to 259 m) (maxima of 40 to 120 mg C $m^{-2} d^{-1}$) occurred between June and September at sites along the east-west transect (76°N) (Fig. 5.3). Maximum fluxes (>100 mg C $m^{-2} d^{-1}$) at S2 during June off the west Greenland coast (Fig. 5.3c) occurred about one month before the highest fluxes on the western side of the polynya (S5) (Fig. 5.3a).

Fluxes measured during summer maxima and at other times of the year at the central polynya site (S4) (Fig. 5.3b) were approximately 40% lower than rates measured to the east and west. Similar low values were measured at the northern (N2) and southern (D1) locations, although low collection efficiency (northern site) and clogging (southern site) could have been partly responsible for the reduced rates. Annual mean POC fluxes at N2, D1 and S2 (5, 2.7, and 5.2 mg C $m^{-2} d^{-1}$, respectively) were lower than yearly average values at S2 and S5 sites (14 and 27 mg C $m^{-2} d^{-1}$). The seasonal progression of increasing POC fluxes throughout the summer and fall (40 to 50% of the annual biogenic sedimentation between June and October) was similar at all locations. POC fluxes in 1998–99 were from two to five times higher than values in 1997–98 where the results from consecutive years could be compared at the same location and depth (n = 4) (Table 5.4).

Particle fluxes into moored traps 50 m above bottom were 50% higher than at shallower depths (Figs. 5.3a and 5.3c), but settled material contained more inorganic material than was present at the shallower depth. Maximum concentrations for POC (>10%) in settled material occurred in particles collected in upper moored traps at the central (S4) and eastern (S2) transect sites. C:N molar ratios in settled material were generally between 7 and 9 with exceptionally high (13–15) and low (<6) values at the northern and southern-most sites, respectively, during spring and early summer months. No carbonates were present in settled material consistent with the lack of calcareous coccolithophores.

Organic composition and elemental ratios were most similar in shallow and deep traps between late May-early June and September, periods of high flux and rapid particle settling. *Fragilariopsis* spp., *Navicula* spp., *Nitzschia*, spp., *Pauliella* spp., and *Fossula* spp. dominated the biomass of settled phytoplankton in the early summer (June) while faecal pellet numbers (cylindrical and ovoid shaped) increased in July and August. Neither *Phaeocystis* spp. nor coccolithophores were present in settled material in the NOW despite previous observations

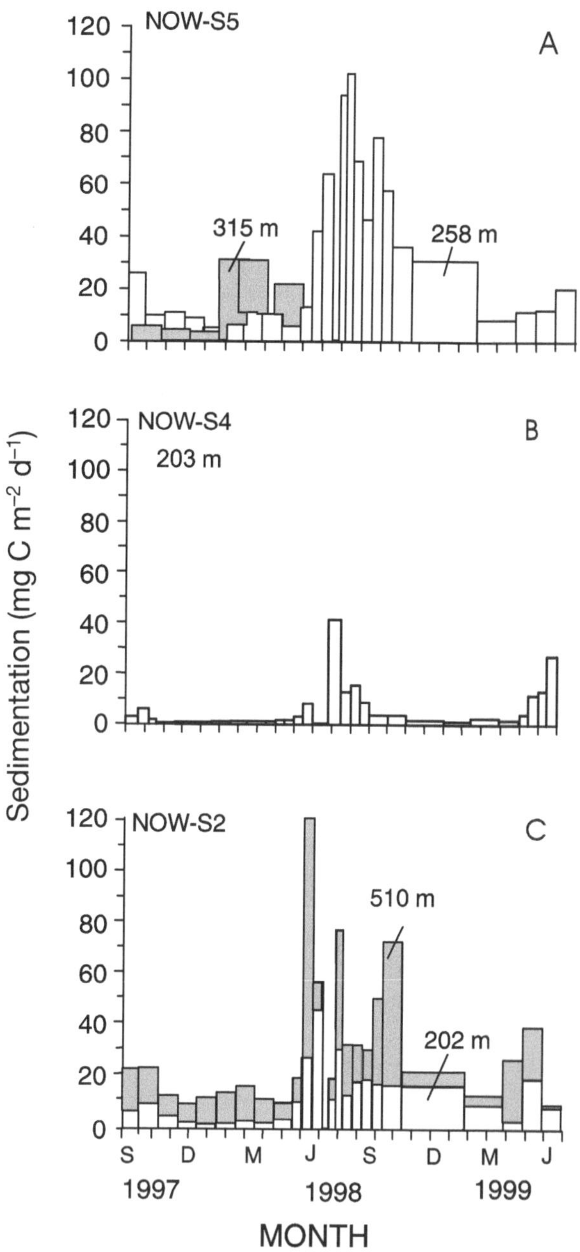

Fig. 5.3. Seasonal variation in particulate organic carbon (POC) flux into moored traps at three locations (S5, S4, S2 across 76°N latitude) in the North Water Polynya, northern Baffin Bay, between 1997 and 1999 (from Hargrave et al. 2002). Upper traps (non-shaded histograms) were moored between 202 and 258 m and lower traps (shaded histograms) were approximately 50 m above bottom (315 and 510 m). Trap collection periods varied from bi-monthly to monthly. POC sedimentation for two annual periods (1997–1998, 1998–1999; (Table 5.4) was calculated from integrated monthly fluxes. Missing data between some collection intervals was interpolated as the mean of fluxes during preceding and following sampling periods

in various subarctic regions where these species can be abundant at certain times of the year (Wassmann 1994; Sukhanova and Flint 2001).

Discussion

Published empirical models from various ocean areas predict that particle fluxes should decrease exponentially from surface layers to depths below 200 m (Wassmann et al. 1991; Berelson 2001; Hargrave et al. 2002; Sampei et al. 2002). Although floating and moored mooring configurations and different trap designs were used in the NOW, comparison of results from floating (50 to 100 m) and moored (200-300 m) collectors showed the expected decrease in flux with increasing depth. However, fluxes 50 m above bottom were usually higher than at shallower depths. The increase was predominantly due to increased mass of settled particles, not in organic matter, indicating the importance of processes such as resuspension and lateral transport of previously settled particulate matter in the lower water column of northern Baffin Bay.

Spatial and temporal changes of POC fluxes in moored traps in the NOW reflected the seasonal progression of changes in ice cover and phytoplankton biomass. SeaWiFS imagery has shown that an intense spring-summer bloom occurs in May or early June following the increase in solar radiation and opening of the polynya in April-early May (Bélanger et al. 2002). High chlorophyll in surface layers spreads from the east along the Greenland coast in April to the north in May-June and to the west off Ellesmere Island in June. The spring-summer bloom occurred earlier (mid-May to mid-June) in 1998 than in 1999 when it began in early June and continued until mid-July. The higher POC flux during the second year of the study is consistent with increased phytoplankton biomass in 1999. Higher faecal pellet fluxes observed between July and September (Sampei et al. 2002) occurred after the period of maximum phytoplankton biomass reflecting sedimentation of products of grazing by zooplankton, both gelatinous species such as appendicularians and abundant calanoid copepod species (*Calanus hyperboreus*, *C. glacialis*, and *Pseudocalanus* spp.).

Conclusions

- Year-to-year differences in melting of winter ice cover, vertical density and thermal stratification and variability in drifting pack ice may account for inter-annual variation (timing, duration and location and of blooms) of primary production by epontic and phytoplankton algal communities observed in the NOW between 1997 and 1999.
- Vertical and horizontal variations in vertical flux of POC reflect spatial and temporal patterns in the timing of settling particulate matter.
- Resuspension and lateral transport through near-bottom boundary currents are important processes for the redistribution of deposited organic matter in the NOW, especially in areas of high current velocity.
- The seasonal progression in settled material during the NOW study shows that phytoplankton-derived material predominates during and after the bloom in early summer (late May-early June) while faecal pellets and other material produced by copepods and gelatinous zooplankton communities are most abundant during late summer months (July-August).

5.3.2 North East Water Polynya

E. BAUERFEIND

The Northeast Water Polynya (NEW) was also the focus of a multidisciplinary international study in 1992–1993 [within the framework of the International Arctic Polynya Project (IAPP)]. Aside from earlier visits by aircraft, the area was studied in August 1992 and July/August 1993 and from May to July 1993. The NEW is situated at (76°–81°N and 6°–17°W) on the north-eastern Greenland shelf. It opens regularly during May, reaches its largest extent in July/August and closes in September. The formation of this polynya during summer is governed by a combination of wind, drift of ice out of the NEW with the prevailing anticyclonic movement of the water and the blocking of drifting ice by a shelf ice barrier in the southern region. To the north, shallow areas prevent ice drifting out of Fram Strait from entering the region of the polynya. The area of open water can vary to a high degree, and it ranged from 59,000 km² to 124,000 km² in the period from 1978–1993 (Böhm et al. 1997). A special feature in the NEW is the presence of a winter polynya in the northern part of the area that is generated by the prevailing strong offshore winds which break up the newly formed sea ice and advect it away from the coast (Minnet et al. 1997).

Results

During the expedition in August 1992 a series of moorings were deployed, four of which contained sediment traps. Due to technical failures some of the sediment traps sampled for a short period only. Results originating from a time series Kiel type sediment trap, installed at 130 m (200 m above bottom, 80°27′N 11°01′W) that sampled from August 1 1992 to June 28 1993, are reported here. Additionally data from a sediment trap on a short-term mooring (June 3 to July 24) are also presented. This mooring was installed at the same depth below surface 35 km from the long-term mooring during the cruise in May 1993 and sampled with a time resolution of 3 days. The traps were located below a relatively fresh and cold layer of water which is formed locally in the polynya (Budeus et al. 1997; Bignani and Hopkins 1997).

The flux of the organic carbon is referred to here as residual particulate organic carbon (rPOC) since a substantial part of the particulate organic matter can dissolve in the sampling cups of sediment traps in long-term moorings. Therefore measurements of the particulate fraction can underestimate the total organic carbon export as demonstrated in the Greenland Sea (Noji et al. 1999), the NEW (Kähler and Bauerfeind 2001) and subarctic fjords (A. Miller and P. Wassmann, unpubl. res.). This seems to be especially important when the traps are deployed at shallow depths. However, to make the results comparable to other studies in the Arctic, the uncorrected (rPOC) carbon flux data are used here, but when it comes to annual flux, both the rPOC and DOC corrected data ($POC_{corr.}$) are compiled (see Table 5.4).

Sedimentation in the NEW showed a clear pattern that was divided into three distinct phases by Bauerfeind et al. (1997). I. An increased flux period in August to October with fluxes of rPOC of up to 14 mg C m^{-2} d^{-1} (average: 6.2 mg C m^{-2} d^{-1}). II. A low-flux period after the polynya closed and ice concentrations increased to 95–100% (Fig. 4) during November to March, the average flux being 1 mg C m^{-2} d^{-1}. III. An increased-flux period from the end of March onward to the end of the deployment period (end of June) when the polynya opened again and fluxes increased to 2.2–4.7 mg Cm^{-2} d^{-1} (avg.: 2.8 mg C m^{-2} d^{-1}). For the entire mooring period, the average flux of rPOC was 2.9 mg C m^{-2} d^{-1}. Similar vertical export estimates were obtained by the short-term mooring in June/July 1993 over 41 days of operation (Fig. 5.4). The average flux rate amounted to 1.8 mg C m^{-2} d^{-1}.

The annual rPOC flux (long-term mooring plus prorated flux from the short mooring in July) amounted to 1.03 g C m^{-2} (Table 5.4). Correcting this flux for particle-associated DOC in the samples increases the carbon flux to 2.66 g C m^{-2}. Of the annual rPOC flux 57% and 29% were recorded during phases I and III, respectively.

Microscopic analysis of the sedimented material revealed that the important contributors to particulate organic carbon flux were diatoms, appendicularian houses and faecal matter, however in variable proportions. Discarded houses of appendicularians contributed most in summer and autumn 1992, when up to 37% of the rPOC could be ascribed to these particles. During the remaining time of the year appendicularian houses did not contribute significantly to vertical particle flux and were found only occasionally.

Diatoms were also very abundant in phase I (up to 10^8 cells m^{-2} d^{-1}). Of these, 50 to 70% of the cells were viable. During winter, these cells were observed only sporadically but beginning in March, the contribution by viable cells increased to 20–30%. In the period of March to June/July up to 70% of total sedimented cell were living cells. In terms of carbon, the contribution of phytoplankton (PPC) was up to 2 mg C m^{-2} d^{-1} (21%) of rPOC during the high sedimentation period in August–October. After the opening of the polynya in May/June the proportion of PPC reached a maximum of 91% (3 mg C m^{-2} d^{-1}). Species composition in the samples was dominated by diatom species of *Fragilariopsis* (*F. oceanica*, *F. cylindrus*, and *F.* sp.). To a lesser extent centric diatom species also contributed to the flux (mainly *Thalassiosira* and *Chaetoceros*). Parallel to the decline in ice concentration in June cells of the under-ice species *Melosira arctica* were observed in increasing numbers in the sediment trap samples, accompanied by *Gyrosigma* and *Pleurosigma* species. During the sampling interval in June 8-18, *Melosira arctica* dominated (5×10^8 cells m^{-2} d^{-1}) the species composition, contributing >50% to the PPC flux.

Faecal material collected was produced by appendicularians, copepods, amphipods and euphausiids. However, small and fragmented pellets were also present in samples that could not be assigned to a specific group of organisms. Sedimentation was highest in August/September (3,900 pellets m^{-2} d^{-1}) when pellets contributed between 19 and 32% to the rPOC flux. Pellets of appendicularians contributed with 50% to the pellet flux at this time, the other 50% being mainly composed of copepod pellets. The latter were present also during winter in very small quantities with 3–6% of rPOC flux, faecal material generally was less important in vertical organic carbon flux than in summer/autumn of the preceding year.

Discussion

The export of organic carbon in the NEW [1.03 g C m^{-2} y^{-1} (rPOC)] at 130 m depth, is low when compared to the flux obtained in other areas of the seasonally ice covered Greenland Sea (Peinert et al. 2001a). The DOC corrected flux of 2.66 g is slightly

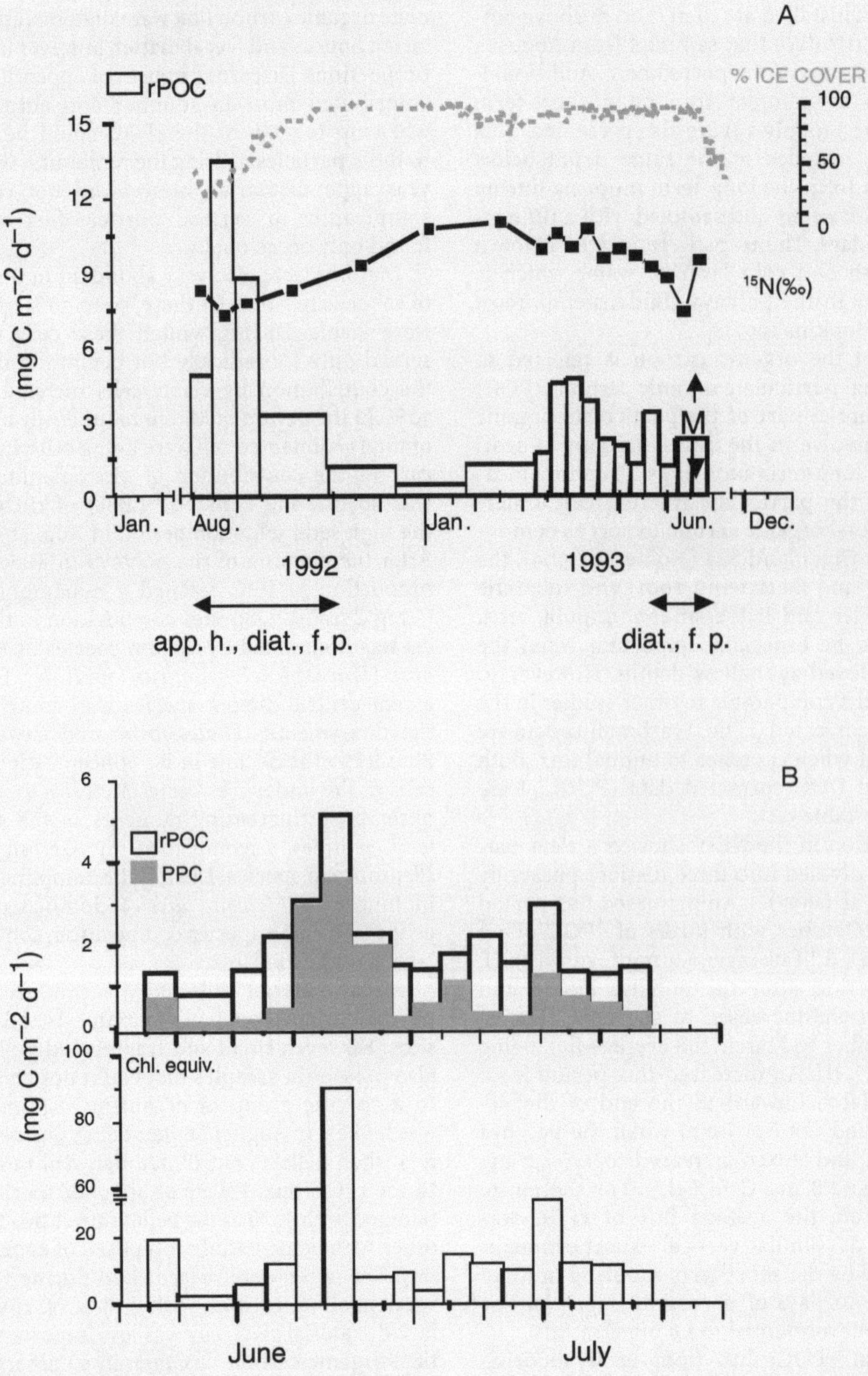

Fig. 5.4. Flux of residual particulate organic carbon (rPOC, mg C m^{-2} d^{-1}), in the NEWP. a) Obtained from the annual mooring August 1 1992 to June 28, ice concentration (%) at the mooring site and the $\delta^{15}N$ signal (‰) of the sedimented material. b) Flux of rPOC, phytoplankton carbon (PPC) and chlorophyll equivalents measured at the short term mooring (June 3 to July 24). The types of particles dominating in the sedimented particulate carbon pool in summer/autumn 1992 and in summer 1993 are indicated, App.h. = appendicularian houses, diat. = diatoms, f.p. = faecal pellets. M denotes the sedimentation event of *Melosira arctica*

higher when compared with the year round ice free Greenland Sea at 75°N, 0°W, the only region where data for DOC corrected flux data exist. Here the vertical C export was 1.9 g C m^{-2} y^{-1} and 1.44 g C m^{-2} y^{-1} for the DOC corrected and rPOC-flux, respectively (Noji et al. 2001).

Comparing the new production rates in 1993 and 1992 (Smith et al. 1997, Wallace et al. 1995) can estimate the relative significance of C export. From differences in the nitrate concentrations during the growth season a new production of 22 g C m^{-2} results for the period when the polynya was open (May-September). A similar production of 25 g m^{-2} was calculated when using nitrate removal and advection of new nitrate with the prevailing currents into the area during the existence of the polynya (Kattner et al. 1997). However, a much higher estimate is provided (75 g C m^{-2}) according to Wallace et al. (1995). Obviously, only a minor fraction (maximum ≈10%) of the produced carbon sediments to 130 m and deeper. The majority of organic carbon may be recycled in the upper water layers. This is supported by the discrepancy in faecal matter production and vertical flux during spring and summer 1993. High standing stocks of copepods (Hirche and Kwasniewski 1997) and the high faecal pellet production rates (Daly 1997) suggest high vertical export of faecal pellets during June/July. However, measured faecal matter flux was low at this time of the year, probably due to destruction and re-ingestion of the pellets by the copepods themselves. This is corroborated by Pesant et al. (1998) who also suggest a high recycling potential in the polynya in May/June. This high proportion of recycling changed during the lifetime of the polynya. Towards the end of the growth season particles of zooplankton origin again increased in importance.

Very early in the year (March/April) an increasing proportion of freshly produced material was present in the sedimented material indicated by the change in the δ^{15}N signal (Fig. 5.4). The presence of viable plankton cells indicates an early start of plankton growth that might be an effect of the winter polynya. Ice algae, of which *Melosira arctica* serves as an indicator species, are of importance to carbon export to depth. *Melosira* grows generally underneath the ice in the Arctic where it can accumulate high biomass. During the opening of the polynya and associated melting of ice, a massive sedimentation of this species was recorded (for details see Bauerfeind et al. 1997). From this, and the presence *Melosira* cells at the sediment surface in June (Ambrose et al. 1994) it becomes evident that organic carbon can be transferred rapidly to the seafloor at times in the NEW. However, this may be a localised effect, as mass sedimentation of *Melosira* was not recorded by the short-term mooring. Therefore, sedimentation of ice-associated organic matter occurs in a mosaic-like pattern in the NEW wherever ice floes melt. As ice also drifts out of the NEW into the Greenland Sea, a considerable part of the estimated new production may be laterally exported out of the area. Besides the high recycling, lateral export provides another mechanism for the observed discrepancy of calculated new production and vertical carbon export in the NEW.

Conclusions

- The export of POC at 130 m depth is ≈1 g C m^{-2} y^{-1}. Correction for DOC leakage increases the estimate by to ≈2.7 g.
- Calculations indicate that less than 10% of the new production sinks to 130 m and eventually to the bottom. The majority of organic carbon (phytoplankton and ice algae) is probably recycled in the upper water layers or advected away from the polynya.
- Sedimentation of ice-associated organic matter might occur in a mosaic-like pattern, occurring wherever ice floes melt.

5.3.3 Greenland Sea

R. Peinert, T. Noji

Particle fluxes were measured in the Greenland Sea at seasonally ice-covered stations and at a station not affected by sea ice between 1988 and 1995 (see IO and GS in Fig. 5.2). Flux patterns (seasonality, multi-annual variability, events) are related to regional ocean physics (ice regime, stratification) and biological dynamics of the pelagic system. Syntheses from this long-term interdisciplinary survey are given in Schäfer et al. (2001), with particle flux-related contributions by Noji et al. (2001), Peinert et al. (2001a, b) and Ramseier et al. (1999, 2001). Here we summarise particulate organic carbon fluxes as measured with moored, automated traps (deployment and sample treatment according to JGOFS standards as detailed in Bodungen et al. 1991) and link results to upper ocean processes.

The investigated region is part of the western Polar Province of the Nordic Seas, dominated by the southward flowing East Greenland Current that transports low-salinity water and sea ice with a polar origin (Hopkins, 1991). Ramseier et al. (2001) give an overview of the ice regime in the Greenland

Sea and its impact on particle flux. Seasonal ice cover of multi-year ice transported through Fram Strait varies spatially, although to a much lesser extent than e.g. in the Barents Sea, and the Marginal Ice Zone (MIZ) tends to follow the continental margin. A prominent feature between 71°–76°N is the Is Odden, a promontory of first-year ice related to the Jan Mayen Current in the region where deep-water formation takes place (Fig. 5.2). Is Odden is regularly observed but is highly variable in extent and form.

Stratification of the euphotic zone in the Is Odden region and MIZ is primarily controlled by ice melt. During the biologically productive spring, this region can be visualised as a north-south band running parallel to the axis of the main East Greenland Current in which phytoplankton growth conditions are patchy but favourable. Nitrate reserves available for primary production and hence potentially for export of particles are large (Noji et al. 2001; Peinert et al. 2001a).

Results and discussion

MIZ conditions in the Greenland Sea favour an early onset of phytoplankton growth due to shallow haline stratification by the injection of meltwater in early spring. The conditions create a patchy distribution containing different phytoplankton bloom stages at a scale of nautical miles (Peinert et al. 2001b). This patchiness along with the early development of phytoplankton relative to zooplankton stocks promotes a decoupling of phytoplankton growth from mineralization by herbivorous mesozooplankton. Fluxes of particles from the MIZ are thus strongly controlled by autotrophs (predominantly diatoms) and are reminiscent of particle export characteristic of spring-bloom systems.

Despite this decoupling of primary from secondary producers, the influence of mesozooplankton on particle fluxes is evidenced by the contribution of their faecal pellets to trap samples during all seasons. Important copepods in the region are large-sized *Calanus hyperboreus*, *C. glacialis* and *C. finmarchicus*. All of these produce large pellets with high sinking velocities, and a portion of this faecal matter escapes epipelagic remineralisation and sinks to depth.

At 72°N seasonal fluxes at 500 m depth in the Greenland Sea regularly increase in May (Fig. 5.5). This takes place with a time lag of weeks relative to the retreat of the ice at the mooring position and, hence, to the onset of phytoplankton growth in waters stratified by melt-water input. This increase in export is most clearly marked not in POC but in silicon and diatom cell fluxes for which winter sedimentation rates are virtually zero (Peinert et al. 2001a). Elevated POC fluxes of 10–20 mg m^{-2} d^{-1} at 500 m are observed throughout the summer until September when the growth season ends. Exceptional flux events (20–40 mg POC m^{-2} d^{-1} at 500 m depth) do take place, juxtaposed to the general pattern and can be related to specific organisms [diatoms in May 1989; radiolarians and acantharians in spring and fall 1990, respectively (Antia et al. 1993; Kohly et al. 1998; Bauerfeind et al. 1994)]. Such events can be traced to 1000 m depth almost without delay (i.e. within one 2 weeks sampling interval). At 1000 m depth fluxes in the ice-influenced region decrease due to particle degradation but the seasonal pattern remains detectable.

Differences between years are pronounced and appear to be related to exposure time of moored sediment traps to MIZ conditions. MIZ conditions favour particle export. The ice-rich years 1989 and 1990 had higher fluxes than 1991 when the impact of the ice was less (the time series stops in mid 1992, hence, this year is not considered). Fluxes in 1991 very much resemble those observed under all year ice-free conditions at 75°N (upper right in Fig. 5.5) exhibiting later increase of fluxes and lower maximal values. These fluxes represent open Greenland Sea conditions without exports from a MIZ. The strong autotrophic chlorophyll (diatom) signal in the June samples, which also have the seasonally lowest molar C:N ratio (7:1), indicate the export of freshly produced matter from the spring bloom in the central Greenland Sea. Highest POC fluxes, however, occurred in September, with a lower chlorophyll content and a higher C:N ratio compared to spring.

Noji et al. (1996) reported that annual sedimentation rates in mid-water depths (700–900 m) were lower than at 200–300 m at the open ocean station (Fig. 5.5). However deep-sea rates at 1735–2000 m (data not shown) were unusually high. The authors concluded that a major part of the spring phytoplankton biomass was consumed, oxidised and regenerated by the food web during its descent. They further concluded that an allochthonous source dominated sinking particles in the central Greenland Gyre resulting in an annual peak in sedimentation at all depths and unusually high rates of sedimentation in deep water. The material comprising the fall peak was rich in diatom valves and intact diatoms (unpublished data), the source for which could not be related to resuspension (e.g. Lampitt 1985) from the sea floor: the depth of the deepest trap was over 1.5 km above the sea floor. The origin of the material is unknown, but it is speculated to originate from the Polar Front, which may approach the trap site. Particles at the front may be transported rapidly from the surface

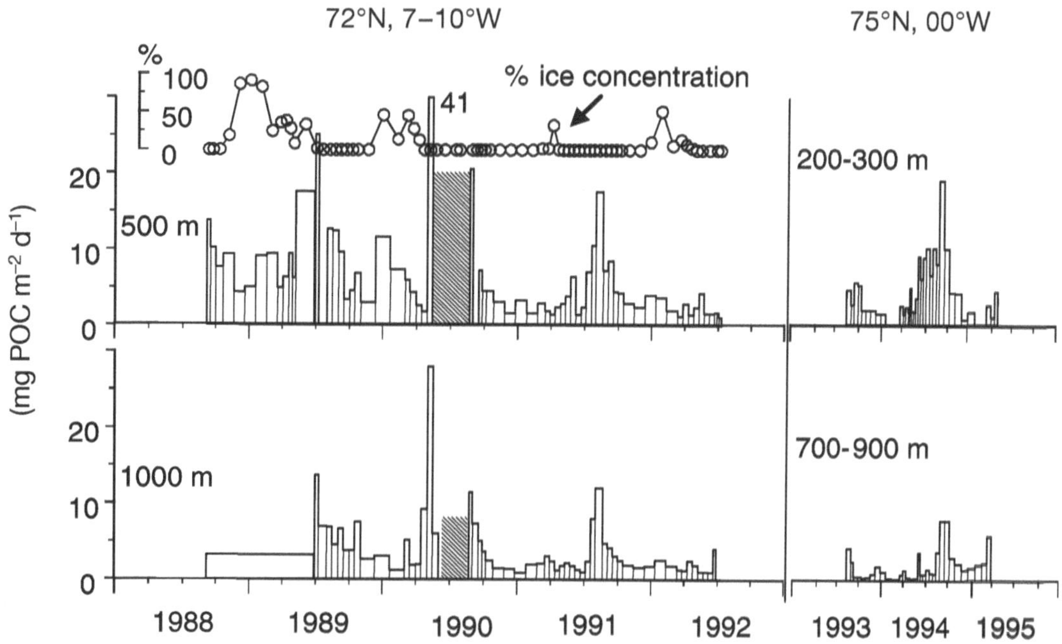

Fig. 5.5. Particulate organic carbon flux measured with moored sediment traps at 500 and 1000 m depth at 72°N (left) and at 200–300 m and 700–900 m depth at 75°N (right side of graph) in the Greenland Sea. Ice concentration (% ice cover, dots in upper left graph) as deduced from SSM/I observations are given for the 72°N station (modified from Peinert et al. 2001a). The 75°N station was ice-free during all the year

to deep water via downwelling (Blindheim and Ådlandsvik, 1995). Also, long-lived eddies at 75°N may play a role in this context (Gascard et al. 2002). If the particles sink along isopycnals extending downward into the deep central gyre (e.g. Manley and Smith, 1994), this could explain the exceptionally high sedimentation rates at depth. To our knowledge this has not been recorded previously in open-ocean systems. Although the causes for this observation are still under investigation, these findings imply that the central Greenland Sea is a focal point for the accumulation of biomass produced at the gyre margins, at least during fall.

Annually integrated fluxes at 500 m depth under the seasonal influence of sea ice amounted to 2.3 g POC $m^{-2}\,y^{-1}$ at 72°N (median value; n = 4) with a range of 1.4–3.8 g POC $m^{-2}\,y^{-1}$ (see Table 5.4). The corresponding median for 1000 m depth (n = 4) was 1.3 g POC $m^{-2}\,y^{-1}$, and the scatter is reduced (range of 1.1–1.8 g POC $m^{-2}\,y^{-1}$). 1.4 g POC $m^{-2}\,y^{-1}$ measured at 200–300 m depths at the 75°N station not affected by sea-ice fits into the lower range of above values.

Noji et al. (2001) measured particle-associated DOC in trapped particles and concludes that this significantly contributes (by up to 1/3) to total "real" particle flux. This is not easily corrected for as it is unknown which (varying) portion originates from preserved zooplankton that also enter trap cups. Although these "swimmers" are picked from samples before POC analysis, some DOC may be leached during the collection period. Rates presented in Figure 5.5 are not corrected and hence must be considered minimum POC flux estimates.

We stress that these Greenland Sea vertical carbon exports do not differ statistically from annual POC fluxes in the western Atlantic Province of the Nordic Seas, despite unmistakable differences in physical setting, biological flux control and seasonality. This emphasises the need for long multi-annual measurements to characterise biogeochemical provinces by bulk variables such as POC export.

Conclusion

- Fluxes of particles from the Greenland Sea MIZ are strongly controlled by autotrophs (predominantly diatoms). Ice-rich years have higher fluxes than ice-poor years.
- Despite an early onset of the spring bloom and apparent decoupling from mesozooplankton in the open Greenland Sea, a major part of the spring phytoplankton biomass is consumed, oxidised and regenerated by the food web in the upper layers during its descent. The organisms involved in this must be microzooplankton and microbes.

- POC fluxes at 500 m depth under the seasonal influence of sea ice amount to 2.3 g C m^{-2} y^{-1} at 72°N, with extreme years yielding 1.4 and 3.8 g C m^{-2} y^{-1}. Exceptional flux events can be related to the export of specific organisms such as diatoms, radiolarians and acantharians. Particle-associated DOC in trapped particles may contribute up to 1/3 to the total "real" particle flux.
- Vertical POC export in the Greenland Sea does not statistically differ from annual POC fluxes in the western Atlantic Province of the Nordic Seas.

5.3.4 Central Barents Sea and Northern Spitsbergen

P. WASSMANN, K. OLLI

The Barents Sea is an arctic marginal shelf sea of the north-eastern North Atlantic which is influenced by two main water masses separated by the Polar Front: Arctic Water (ArW) entering from the north-east and Atlantic Water (AW) entering from south-west. The ArW is periodically ice-covered and the ice melt during spring and summer gives rise to stratification which supports a pronounced spring bloom in the MIZ (Sakshaug and Skjoldal, 1989; Matrai and Vernet, 1997; Falk Petersen et al. 1999). In areas dominated by AW, stratification develops slowly during late spring and summer, solely supported by solar radiation, and the resulting spring bloom is less distinct (Skjoldal and Rey, 1989; Wassmann et al. 1999). The West-Spitsbergen Current transports AW to the north of Spitsbergen where it subducts under ArW and the seasonal ice cover. Stratification is weak, but increases towards the MIZ. Atlantic planktonic species are advected north of Spitsbergen, through the Barents Sea and ultimately along the Eurasian shelf break into the Arctic Ocean (Andreassen et al. 1996; Falk Petersen et al. 1999; Wassmann 2001).

The MIZ of the Barents Sea and to the east of the Svalbard Archipelago is productive (Rey and Loeng, 1985; Vernet et al. 1998). New production estimates range from about 40–60 g C m^{-2} y^{-1} (Slagstad and Wassmann, 1997; Wassmann et al. 1999; Reigstad et al. 2002). Primary production is also increased due to the long duration of the ice edge bloom as it recedes northward with the ice in late spring and summer. The early stages of the spring bloom can be found next to the ice while progressively older stages of the bloom are found southwards (Sakshaug and Skjoldal, 1989; Wassmann et al. 1999).

Here we describe the seasonal variability of vertical flux in the central Barents Sea along a transect from AW in the south to the ArW in the north (Fig. 5.2). In addition, we summarise data of the vertical export of biogenic matter in the subducting, eastbound water off the shelf of northern Spitsbergen (AW origin) and the west- and surface-bound, ArW of the Transpolar Drift. Data from north of Spitsbergen have been published by Andreassen et al. (1996). Data from the Barents Sea MIZ were collected during three cruises (I–III) corresponding to late winter (March 17–23, 1998), late spring (May 18–28, 1998) and early summer (June 30–July 9, 1999) with details reported by Olli et al. (2002).

Sediment traps were deployed from ice floes in the ice-covered regions north of Spitsbergen (NS in Fig. 5.2) at 100 m in July 1991 at 4 stations for 8–24 hrs using a TECNICAP P.P.S. 3/3 cylindrical trap (diameter of 40 cm, aspect ratio 4.8). No poison was applied. Sediment traps were deployed in the Barents Sea at 3-5 stations in March and May 1998 and July 1999 (CBS in Fig. 5.2). A drifter with an array of sediment traps at 8 depths (30, 40, 50, 60, 90, 120, 150, 200 m) was deployed for ca. 24 h. The sediment traps (KC maskiner og laboratorieudustyr, Denmark) consisted of parallel cylinders mounted in a gimballed frame equipped with a vane. The cylinders were 7.2 cm in diameter and had an aspect ration of 6.25. No poison was applied. Application of the ^{234}Thorium method (Cochran et al. 1995) at selected depths in July revealed 90–110% trapping efficiency (Coppola et al. 2002).

Results

The highest vertical export of POC north of Spitsbergen was recorded at the northernmost station (76 mg C m^{-2} d^{-1}). The other stations in this area were characterised by relatively low sedimentation rates (17–26 mg C m^{-2} d^{-1}). Viable *Melosira arctica*, both resting spores and vegetative cells (~50/50), were found in the sedimented material at all stations covered with multi-year ice, but neither ice algae nor pelagic algae were important for the total carbon flux. Faecal pellets were more important for vertical flux (2–10% of POC) than sinking algae. The pico-, nano- and microplankton carbon (PNMC) content of the vertically exported POC was generally <6%, in most cases around 1%. The molar C:N ratios in the multi-year sea ice region were variable, but generally higher than the Redfield ratio (7.7–13.0). The sedimented material at the northernmost station had a very high C:N ratio (>27).

POC sedimentation in the Barents Sea was 60–70 mg C m^{-2} d^{-1} in the AW and 20 – 50 mg C m^{-2} d^{-1}

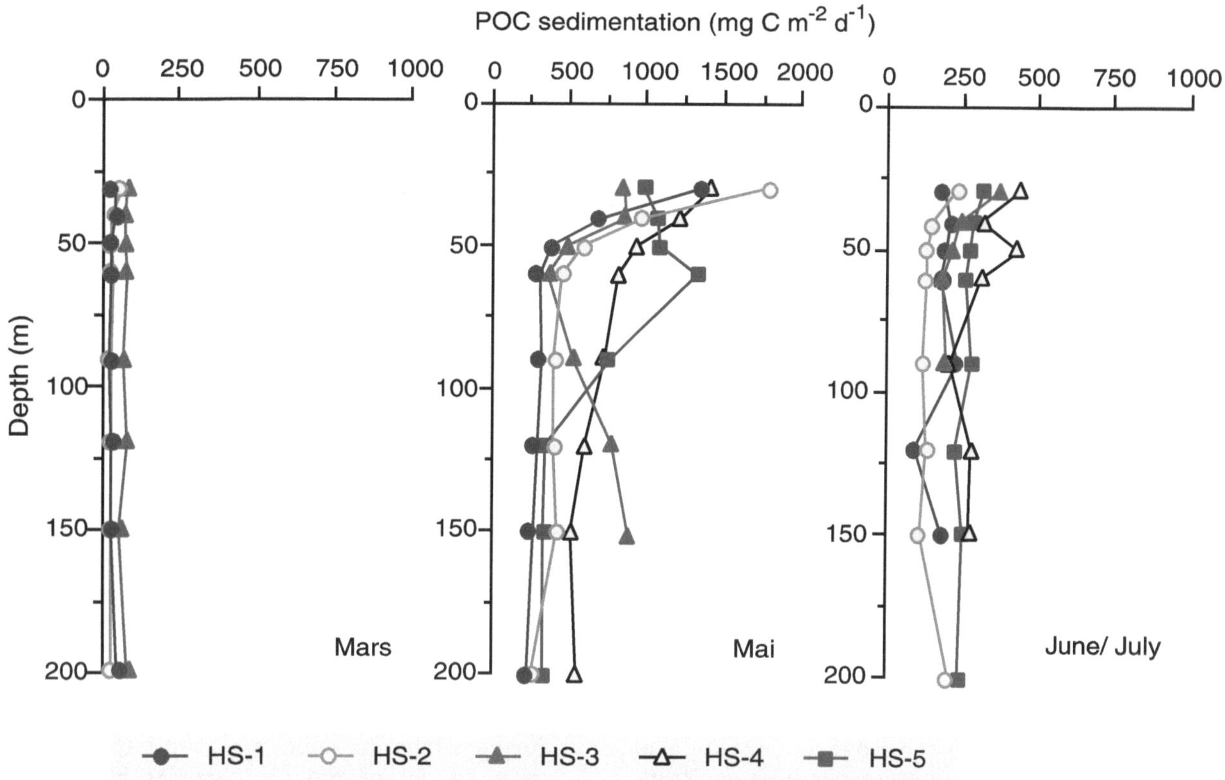

Fig. 5.6. Particulate organic carbon flux measured with drifting sediment traps during approximately 24 hours in March and May 1998 and June/July 1999 along a north-south gradient in the central Barents Sea. HS-1 in March and HS-1 to HS-3 in May and June/July were within the marginal ice zone, the remaining stations in ice-free AW

in the ArW in March (year 1998; Fig. 5.6). There was almost no decrease in vertical flux with depth in the upper 200 m. During late spring the vertical flux of POC was very high (300–1500 mg C m^{-2} d^{-1}; Fig. 5.6). Much of the vertical differences in sedimentation rates correlated with stratification. In the ArW and the MIZ the vertical flux was >500 mg C m^{-2} d^{-1} only above the main pycnocline (ca. 50 m) and decreased sharply thereafter (<500 mg C m^{-2} d^{-1}). However, in the well-mixed AW high vertical flux rates (>500 mg C m^{-2} d^{-1}) extended down to 90 m depth. Below 90 m vertical carbon flux was similar to or slightly higher (200–500 mg C m^{-2} d^{-1}) than flux in the ArW. Closer to the Polar Front (HS-III) the vertical flux was lower in the upper layer (ca. 700 mg C m^{-2} d^{-1}), but did not decrease with depth, possibly because of the specific hydrography, downward mixing and subduction associated with frontal systems. Although considerably less than during the late spring season, the vertical flux of POC in early summer was still quite high (100–400 mg C m^{-2} d^{-1}; Fig. 5.6). In summer, the vertical flux was even lower in the ArW and MIZ (generally <150 mg C m^{-2} d^{-1}) compared to the AW.

Discussion

Sea-ice in the Arctic Ocean frequently contains significant amounts of inorganic matter (Vinje and Kvambek 1991). Terrestrial matter transported by rivers to the Siberian Shelf is frozen into sea-ice (Kassens et al. 1999) and transported by the Transpolar Drift into the area north of Spitsbergen and the Fram Strait (Reimnitz et al. 1994). Melting of such ice releases the particulate matter and results in increased vertical export of terrestrial organic matter. This hypothesis is supported by observations of terrestrial material dominating the sediments below the Transpolar Drift (Stein et al. 1994). The northernmost station north of Spitsbergen revealed a relatively high POC sedimentation rate and a very high molar C:N ratio. This suggests that there was little fresh marine material in the sinking matter. The vertical flux at the northernmost station probably reflects melting ice from the Transpolar Drift and release of its particulate content derived from the Laptev Sea. The particulate content of sea ice has a patchy distribution. Consequently, the release of particulate matter from the ice will be irregular. Vertical export along the Transpolar Drift is thus characterised by variability and based predominantly on allochthonous rather than autochthonous matter.

The vertical flux in the central Barents Sea is strongly seasonal. Superimposed on this variation are differences along the north – south transect, i.e. the ArW and MIZ vs. AW. However, the spatial variation was often of qualitative rather than quantitative nature. Considering the uniformly high nutrient and very low chlorophyll concentrations in March, we interpret the vertical flux as indicative of late winter wherein vertical flux was probably fuelled by residual suspended particles (ArW), and possibly limited re-suspension (AW).

The magnitude of the vertical POC flux in late spring was somewhat unexpected. From of the upper mixed layer (50 m) the flux was approximately 500 mg C m^{-2} d^{-1} in the MIZ, which is comparable to, or higher than that observed in strong upwelling areas (Knauer et al. 1979). At shallower depths (30 and 40 m) the vertical flux of POC was up to 2000 mg C m^{-2} d^{-1}. The high input (about 250 mg C m^{-2} d^{-1}) into the deep layers (200 m) is comparable to upwelling areas (Pilskaln et al. 1996), and generally higher than in coastal and shelf areas (Peña et al. 1999; Romero et al. 2000; Roy et al. 2000). Comparable vertical export was encountered in the NOW during summer, indicating that Arctic shelves with seasonal ice zones and polynyas can be hot spots of vertical carbon export.

In late spring a relatively high proportion of the vertical flux consisted of intact phytoplankton. The variability in the vertical flux of POC along the south – north gradient during late spring was caused by differences in water column stratification. The AW was homogeneously mixed down to 90 m while a sharp halocline at ca. 30 m separated the upper and lower layers in the MIZ. Thus, the upper layer sedimentation in the AW was lower, but extended over a larger depth range than in the ArW. This probably results in a distorted picture of the true gravitational sedimentation (e.g. Gardner, 2000). In the ArW, the sediment traps were below the surface mixed layer and the sharp gradients of the vertical flux over small depth ranges reflected the impact of planktonic retention processes. This clearly demonstrates that much of the vertical flux regulation in the ocean can take place in the upper layers, within and just below the productive layer (e.g. Olli et al. 2001). When the upper part of the water column is not stratified, these regulatory processes of the vertical flux are "diluted out" over a larger depth range which must be considered when interpreting the vertical flux profiles.

The vertical flux in early summer was moderate (generally 200–400 mg C m^{-2} d^{-1} in the upper layers) and relatively higher in AW. The attenuation of the vertical flux POC with depth was only moderate, and the vertical flux at 150–200 m was quite high (120–260 mg C m^{-2} d^{-1}). In contrast to the late spring season, the vertical flux of PNMC was low in summer. It is more difficult to formulate conceptual scenarios regarding the vertical flux regulation observed during the summer cruise. Vertical flux of organic particles was higher in the AW compared to the MIZ. In contrast to the late spring cruise, this elevated flux in the AW cannot be easily ascribed to physical downward mixing, which was probably constrained by increasing stratification.

Conclusions

- In the multi-year ice-covered waters north of Spitsbergen, terrestrial material released from melting ice appears to be an important component of the vertical flux of organic matter. In addition, faecal pellets are more important in vertical carbon flux than phytoplankton.
- The average vertical POC export in the central Barents Sea and MIZ varies from ≈ 30 mg C m^{-2} d^{-1} during winter to ≈ 500 mg C m^{-2} d^{-1} in late spring. Input of organic particles below the 200 m depth is very high: 250–550 mg C m^{-2} d^{-1} during spring period and 125–260 mg C m^{-2} d^{-1} during summer.
- Despite differences in the planktonic community and the vertical mixing regime, no pronounced differences in the magnitude of vertical POC flux at 150–200 m depth occurred between AW and ArW. The spatial variability was mainly qualitative.
- The daily sedimentation loss rates of POC were ca. 1.7 times larger during spring compared to summer. This is attributed to the complexity of the planktonic community during early summer (retention food web) compared to the late spring (export food web) when only a few species dominated (*Phaeocystis pouchetii* in AW and *Thalassiosira antarctica* in ArW).

5.3.5 Eastern Barents Sea and Kara Sea

V. SHEVCHENKO

The first studies of particle fluxes in the Kara Sea and in the eastern Barents Sea were carried out in September 1993 and in August-September 1994 (Lisitzin et al. 1995; Ivanov et al. 1997; Shevchenko et al. 1998, 1999, 2000). While some reference is made

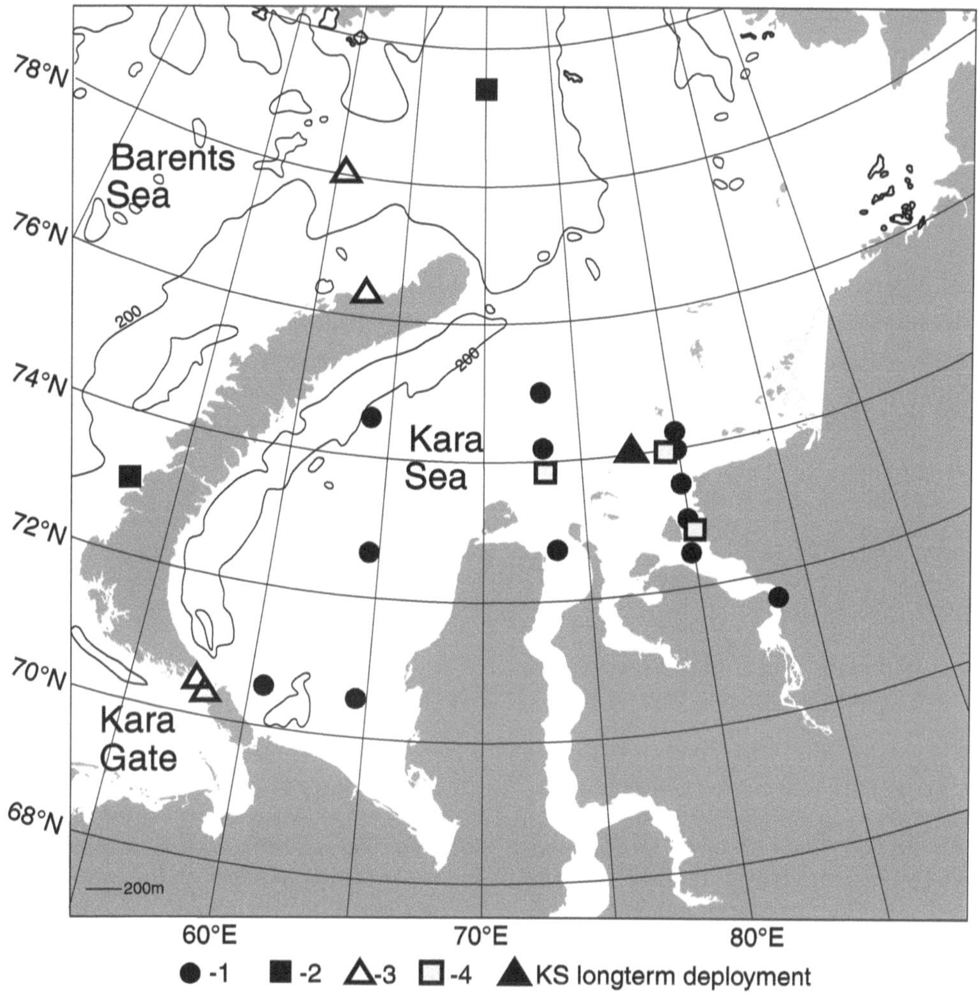

Fig. 5.7. Particulate organic carbon flux measured with moored sediment traps in the eastern Barents Sea and the Kara Sea. Filled circles = Lisitzin et al. (1995), full squares = Shevchenko et al. (1998), open triangles and open squares = this study

to these data, recent results from particle flux studies in the eastern Barents Sea and southern Kara Sea are described in detail.

Vertical particle flux was studied at 4 stations in the eastern Barents Sea in September–October 1997 (Ivanov and Aibulatov, 1998). Vertical particle flux was also studied at 3 stations in the southern Kara Sea in September 1997 (Lukashin and Rusakov, 1998) (Fig. 5.7). Small cylindrical double traps (185 mm in diameter, aspect ratio 2.6 and baffles in the trap opening) were deployed at variable depths for time intervals of approximately 1–5 days (Table 5.1). The lower part of the trap was cone-shaped with a 100 ml flask collector. The traps were poisoned with formaldehyde. For details of the analytical methods, see Lisitzin et al. (1995) and Shevchenko et al. (1999, 2000). Very recently, a long-term sediment trap has been recovered from the area north of the Kara Sea (74°'N, 80°20'E; water depth 36 m; sampling interval September 2000 to August 2001; pers. comm. B. Gaye-Haake).

Results and discussion

Station ASV-3 is located in the area of AW subduction between Franz-Josef Land and Novaya Zemlya (Fig. 5.7, Table 5.1). Vertical POC export at ASV-3 was low (< 5 mg C m^{-2} d^{-1}, Table 5.1). Primary production near ASV-3 was also very low (43 mg C m^{-2} d^{-1}; Vedernikov and Gagarin, 1998).

The material at ASV-3 consisted mainly of amorphous aggregates. These aggregates contained both colonies and individual diatom cells. At 70 m the diatoms from the genera *Thalassiosira*, *Chaetoceros* and pennates dominated. With increasing depth the proportion of pennate species (*Navicula septentrionalis*, *Nitzschia promare*) increased. However, the ratio of chloroplast-containing cells to empty

Table 5.1. Vertical POC export in the eastern Barents Sea and southern Kara Sea. For positions, see Figure 5.7. For the first annual estimate, see Table 5.3

Station	Co-ordinates		Exposure time (d)	Trap depth (m)	POC flux (mg m^{-2} d^{-1})	% POC
	Lat. (N)	Long (E)				
ASV-3	78°05′	60°00′	5	70	1.6	16.2
				130	1.0	5.5
				360	4.6	7.4
ASV-5	76°16′	62°27′	1.9	70	8.5	2.5
				85	185	2.4
ASV-7	70°18′	57°58′	3.8	80	4.2	2.0
				120	87	2.5
				165	18	0.9
ASV-8	70°17′	58°12′	2.7	55	8.3	3.3
BP-1	73°53′	73°11′	10	17	19	2.7
BP-2	73°60′	79°25′	4.5	17	0.5	8.6
BP-3	72°50′	80°10′	1.3	9	2.6	4.2

diatom valves decreased from 10.1 at 70 m to 0.85 at 360 m. A similar decrease of the proportion of chloroplast-containing cells with depth has been recorded in the Saint-Anna Trough (Shevchenko et al. 1998, 1999). This reflects a combination of organic matter degradation during descent from the euphotic zone and resuspension. Faecal pellets of crustaceans were abundant and their flux increased with depth. The increase of particle flux at 360 m is likely caused by particles derived from a well-developed benthic nepheloid layer (BNL) which is a typical feature for shelf regions and upper continental slopes (Gardner et al. 1985; Biscaye and Anderson, 1994; Aibulatov et al. 1999; Lisitzin et al. 2000).

At station ASV-5 in the outer part of the Russkaya Gavan' Bay (northern island of Novaya Zemlya) at the end of September 1997 vertical POC export at 70 m was much higher (85 mg m^{-2} d^{-1}) than at ASV-3 (Fig. 5.7, Table 5.1). Rates increased by a factor of 2 at 85 m (19 m above bottom) to 185 mg m^{-2} d^{-1} (Table 5.1). Sedimentary matter here consisted mainly of minerogenic matter. Only few empty diatom valves were found. Living conditions for phytoplankton are not optimal in this region because of a high content of suspended, minerogenic matter deriving from adjacent glaciers (Aibulatov et al. 1999).

Particle flux in the Karskie Vorota Strait was very high at station ASV-7 (Fig. 5.7, Table 5.1) where mineral particles were the main component of vertical export. Phytoplankton cells and faecal pellets were rare. Strong currents with velocities ranging up to about 65 cm s^{-1} (Pavlov et al. 1994) result in resuspension of bottom sediments.

At station BP-1 located in the outer part of the Ob estuary (Fig. 5.7, Table 5.1) vertical POC export in the middle of September was 19 mg m^{-2} d^{-1} (Table 5.1). Mineral particles were the main component of the sedimentary matter at this station. Freshwater species such as *Aulacoseira islandica* and *A. granulata*, void of chloroplasts, dominated the settled microplankton. Marine diatoms such as *Paralia sulcata* and *Thalassiosira* spp. dinoflagellates such as *Protoperidinium* spp. and silicoflagellates such as *Ebria tripartita* were less abundant. The composition of particulate matter suggests resuspended matter from a storm, which took place during the deployment.

In the open Kara Sea off the Yenisei estuary vertical POC export at station BP-2 was close to detection limit (0.5 mg C m^{-2} d^{-1}; Fig. 5.7, Table 5.1). Empty valves of the marine diatom *Thalassiosira antarctica* dominated the vertical microalgae flux, but did not contribute to the vertical C flux. Station BP-3 is located in the outer part of the Yenisey estuary in the area with an active "marginal filter" (Lisitzin, 1995). The concentration of suspended matter decreased along with the increase of salinity (Churun and Ivanov, 1999; Lukashin et al. 1999). The vertical POC flux was somewhat higher than that at station BP-2 (Table 5.1). The sedimented matter comprised mainly of clay particles. Settled microplankton was dominated by the freshwater diatoms *Aulacoseira islandica* while the contribution of marine phytoplankton (e.g. *Cylindrotheca closterium, Ebria tripartita*) was low. In addition, in September 1993, freshwater diatom species predominated in the microplankton flux in the Yenisei estuary (Lisitzin et al. 1995).

Results from the Kara Sea shelf area obtained during the short-term deployment of sediment traps by Lisitzin et al. (1995), Ivanov et al. (1997), Shevchenko et al. (1998, 1999, 2000) and Unger et al. (2000, 2001) support basically the data presented here: high vertical export of total particulate matter, but low organic carbon content. However, the first long-term sediment trap deployed in the northern Kara Sea revealed two distinct total flux maxima in June/July and September/October, related to maximum river discharge and Arctic summer enhanced primary productivity (Gebhardt et al. 2002). The annual vertical export of POC reached 9.8 g C m^{-2} (pers. comm. B. Gaye-Haake), implying an average vertical POC export of ≈ 27 mg C m^{-2} d^{-1}. This is several times higher than the majority of rates in September/October. The low vertical C export estimates in the region thus reflect primarily the low productive autumn period and that far higher export take place during spring and summer.

Conclusions

- The vertical POC export in the open Kara and eastern Barents Sea at the beginning of autumn is low, as reflected by the oligotrophic character of the basins. Amorphous aggregates dominate vertical export of particulate matter. Faecal pellets of Crustacea were abundant in a few cases.
- In the outer part of Russkaya Gavan Bay (Novaja Zemlja), in the Karskie Vorota Strait and in the outer parts of Ob and Yenisei estuaries vertical fluxes are also low, but the total flux is higher than those in the open sea because of resuspended material. Terrigenous matter apparently dominates the vertical flux.
- The high current velocities and the low aspect ratio of the sediment traps could have resulted in an underestimation of the vertical POC export.
- The first annual estimate indicates vertical C export of about 10 g C m^{-2}. The low short-term vertical C flux estimates reflect the end of the productive period, but even on an annual basis vertical export in the Kara Sea is 2–4 times lower than model simulations from the adjacent Barents Sea.

5.3.6 Laptev Sea and Lomonosov Ridge

E.-M. Nöthig, V. Shevchenko

The Laptev Sea, which is located in the centre of the Siberian shelf, is of particular interest to environmental changes in the Arctic due to its high freshwater inflow and ice production. The Lena River and other rivers are an important freshwater source characterised by a seasonally pulsed, extensive run-off in June/July and moderate run-off during May and August/September. Large amounts of sea ice are formed on the Laptev Sea shelf, the most important source for the Transpolar Drift. Extended areas are free of ice only during the months from July to September (for a general overview see Kassens et al. 1999). During the short growth season, the Laptev Sea shelf is characterised by strong surface stratification. The influence of the Lena River can be encountered far off the shelf.

From July 19–September 11 1995, fluxes of organic carbon and other particulate matter were measured at 5, 25, and 75 m depth by 11 short-term (8.5 through 25 h) deployments of sediment traps attached to ice floes at various sections of the Laptev Sea shelf (AB, NB, MB; Fig. 5.2). In addition, a long-term deployment (September 1995–August 1996) with two Aquatech multi-traps situated 150 m below surface and 150 m above bottom at 1550 m depth was moored on the western Lomonosov Ridge (Fig. 5.2). In this area ice cover is almost permanent.

Results and discussion

West of the Lomonosov Ridge in the region of the Amundsen Basin, the highest short-term flux was found. Vertical flux of POC at 25 m depth (mean 200 mg C m^{-2} d^{-1}) often exceeded the rates at 5 m depth close to the subsurface of the sea ice (mean 113 mg C m^{-2} d^{-1}; Table 5.2). Accumulation of particles was most likely favoured by the strong pycnocline at 20–30 m depth between marine water and overlying river plume. Enhanced amounts of sediment-laden detritus in the traps indicate the strong influence of the Lena outflow in this region. More than 50% of the sedimenting matter was of lithogenic origin. In the Nansen and Makarov Basins less vertical POC export was encountered in all short-term deployed traps. This is probably caused by the decreasing influence of river water with its allochthonous carbon. The percentage of lithogenic material was in most cases

Table 5.2. Vertical POC export during short-term deployments in the Laptev Sea region during late summer 1995

	Makarov Basin (mg m^{-2} d^{-1})	Amundsen Basin (mg m^{-2} d^{-1})	Nansen Basin (mg m^{-2} d^{-1})
5 m	40	113	69
25 m	38	201	56
75 m	20	20	27

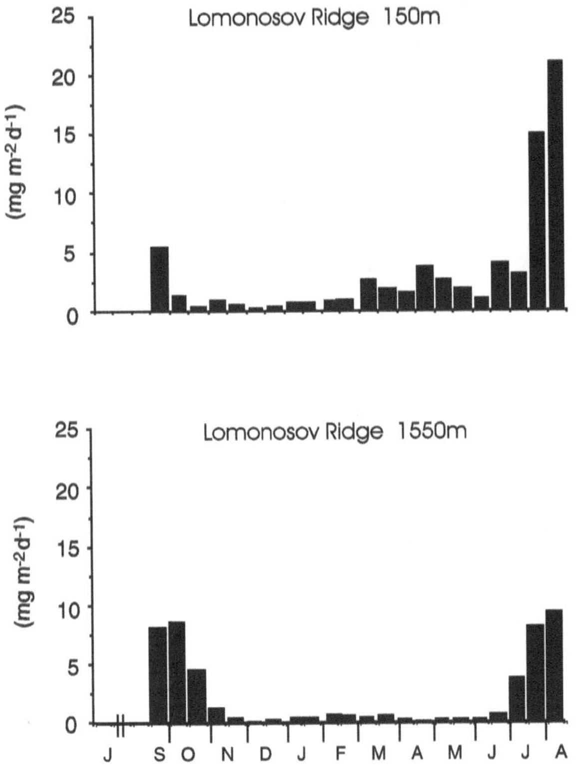

Fig. 5.8. Particulate organic carbon flux measured with moored sediment traps on the Lomonosov Ridge, Laptev Sea, at 150 and 1550 m depth from September 1995 to August 1996. Please note the difference in scales

lower than in the Amundsen Basin (~20–50%). At 75 m depth, vertical POC flux was low at all sites (mean about 20 mg C m^{-2} d^{-1}) and dominated by detrital material indicating that a high percentage of the sedimenting marine organic carbon was recycled within the upper water layers in summer (Table 5.2).

This assumption is supported by the results of Grahl et al. (1999) and by the POC flux data obtained with the long-term deployed traps (see below). The microscopic investigation indicated a strong decrease in fluxes of living particles (mainly ice algae) with depth. Besides large amounts of amorphous detritus with lithogenic particles, some faecal pellets were found.

The long-term deployed traps recorded a strong seasonal variation in vertical flux (Fig. 5.8) dominated during most of the year by the lithogenic fraction. The percentage of lithogenic material varied from ~10–75% and ~20–85% in the upper and lower trap, respectively. From July to September POC and Opal contributed a significant fraction of the total matter flux. Higher vertical POC export was encountered from mid July until the end of October (4–21 mg C m^{-2} d^{-1}). This appears to be caused by the discharge, with a maximum supply in June, and correlated well with high fractions of terrigenous material at both depths. During all other months sedimentation of POC was low (0.2–4 mg C m^{-2} d^{-1}).

The microscopic analysis indicated a strongly pronounced seasonal variability of diatom and other particulate matter flux. In the second half of March the vertical export of empty diatoms frustules increased slightly, probably due to some first ice melt. The maximum microalgae vertical export consisted of colonies in the form of mucous lumps and was recorded in July and August, with highest rates in the second half of July. From the beginning of June, the dominant species changed successively from *Nitzschia promare* to *N. frigida* and *Melosira arctica* together with *Fragilariopsis oceanica*. In September the chloroplast-containing diatom flux was still high, but decreased sharply in October. In winter, amorphous detritus and empty frustules of diatoms dominated. During the growth period, the flux of microalgae was related to the growth of ice algae, suggesting that representatives of the ice flora constitute the majority of the marine living carbon reaching the sea floor. Vertical faecal pellet export was apparently not of importance (for details, see Zernova et al. 2000).

Conclusions

- Under the permanently ice covered Laptev Sea vertical export of POC ranges between 1–1.5 g C m^{-2} y^{-1}. Besides ice algae, amorphous detritus of presumably terrigenous origin dominates the vertical export from July to October.
- Most of the autochthonous carbon is recycled in the upper 150 m.
- Regions influenced by the strong outflow of Siberian rivers, here the Lena River, exhibit relatively high vertical POC export.

5.3.7 Northern Bering Sea

H. SASAKI, M. FUKUCHI

The continental shelf of the northern Bering Sea is recognised as an area of high marine production at several trophic levels (McRoy et al. 1972; Sambrotto et al. 1984). Advection of nutrient-rich water is primarily responsible for the high primary production on the shelf (Springer and Roseneau, 1985; Grebmeier et al. 1988; Springer et al. 1989). The benthic biomass, particularly in the Chirikov Basin (northern Bering Sea), is extremely high, suggesting a close coupling between pelagic sedimentation of biogenic matter and benthic fauna (Grebmeier et al. 1989; Grebmeier and McRoy, 1989). In the seasonally ice-covered zone of this region, an interdisciplinary oceanographic study (ISHTAR: Inner Shelf Transfer And Recycling) was carried out to study fertilisation processes and pelagic-benthic coupling.

A time-series sediment trap was deployed (64° 58.62′N, 169°10.08′W) in the Chirikov Basin at the depth of 49 m (Fukuchi et al. 1988; NBS in Fig. 5.2). The trap was moored 13 m above the bottom to collect 12 consecutive samples with a collection interval of 8 days from June 20 to September 24 in 1988. Neutralised formaldehyde was used to reduce biodegradation of trap-collected particles during the deployment. For routine oceanographic observations (temperature, salinity, nutrient concentrations, chlorophyll concentrations and current velocity profiles) during the two cruises of ISHTAR (Leg. 1, June 1988) and ISHTAR (Leg 4, October 1988), see ISHTAR (1988) and Fukuchi et al. (1993).

Results

During the ice-free period the region is characterised by two distinct water masses: cold (<1.5°C) Bering Shelf Anadyr Water (BSAW) and warm (>4°C) Alaskan Coastal Water (ACW) with a well-defined frontal zone (Coachman et al. 1975; Dean et al. 1989). The Anadyr Current (AC) forms the water properties of BSAW in the western side of the region. ACW is found in the eastern side near the coast of Alaska. The Chirikov Basin with high benthic biomass is generally located below the BSAW (Grebmeier et al. 1988).

Time-series observations of water temperature and currents at 24 m depth close to the mooring site (Sasaki et al. 1993) suggest the presence of BSAW with low temperature (<2°C). A continuous, strong northward current (mean: ca. 16 cm sec^{-1}) prevailed during the first half of the period (June 20 to August 7). Water temperature increased up to about 4–5°C, and current speed (mean: 5–10 cm sec^{-1}) and directions were relatively variable during the second half of the period (August 7 to September 24). This indicates that the mooring site was located in a transition zone with alternative occurrence of cold BSAW and warm ACW. According to the temperature and current profiles, the entire observation period at the trap site can be separated into three periods: Phase 1 (June 20 to August 7), Phase 2 (August 7 to August 31) and Phase 3 (August 31 to September 24).

Nutrient concentrations (>8 μM NO_3 below 20 m depth) were generally high throughout the deployment period (Fukuchi et al. 1993). Chlorophyll *a* concentration in the upper 30 m was high (>5 μg l^{-1}) on June 21 and low (<1 μg l^{-1}) on October 6 (Fukuchi et al. 1993).

The vertical POC export varied from ca. 360–610 mg C m^{-2} d^{-1}, with a maximum between July 14 and July 22 (Fig. 5.9). The variation in the average POC flux in Phases 1–3 was small (490, 420 and 440 mg C m^{-2} d^{-1}, respectively). The integrated POC flux during the entire 96-day period is estimated as 44 g C m^{-2}. Relatively low molar C/N ratios of sinking particles, ranging from 4.9 to 7.4 (mean: 6.4), were observed during the period (Fukuchi et al. 1993).

Diatoms dominated the phytoplankton collected in the trap. The flux of intact diatom cells ranged from 2.1 to 7.8 · 10^8 cells m^{-2} d^{-1}, with the maximum flux at the beginning of Phase 1 (20–28 June) when ice floes were observed around the trap site (Sasaki et al. 1992). The diatom flux decreased at the end of Phase 1, and increased again toward the end of Phase 3. Trap-collected diatoms were classified into four groups; "ice plankton", "spring species",

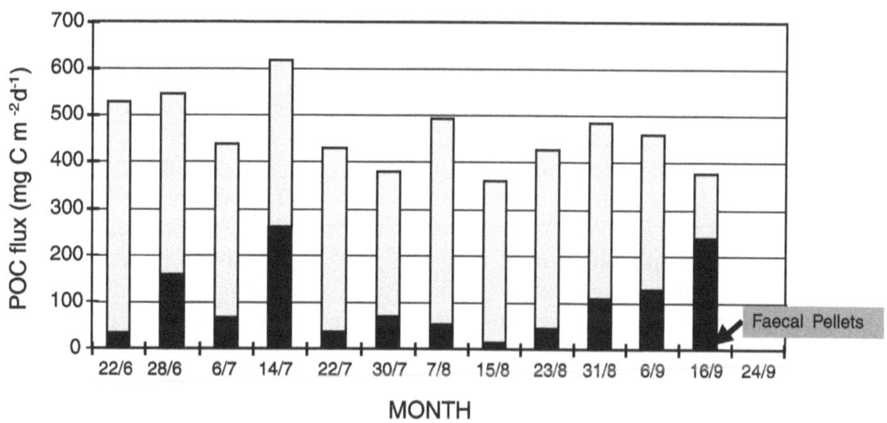

Fig. 5.9. Variability of downward POC flux at the Chirikov Basin of the northern Bering Sea (June 20 – September 24, 1988). Modified from Sasaki et al. (1993)

"summer species" and other diatoms (Saito and Taniguchi, 1978). Dominant diatoms of the "ice plankton" and "spring species" groups were *Nitzschia* spp. and *Thalassiosira* spp. respectively, among which *N. grunowii* was the most abundant. At the end of Phase 1, "summer species", such as *Chaetoceros* spp. markedly increased and replaced the two former groups. The "summer species" were probably transported with the northward flowing AC.

Cylindrical and elliptical zooplankton faecal pellets (>0.1 mm in length) were found (Sasaki et al. 1993). Possible producers of cylindrical pellets were large oceanic copepods, such as *Neocalanus flemingeri* and *N. cristatus*, but the producers of elliptical pellets were not identified. Mean total faecal pellet fluxes (cylindrical and elliptical) were high in Phase 1 (103 mg C m^{-2} d^{-1}) and 3 (158 mg C m^{-2} d^{-1}), and lowest in Phase 2 (37 mg C m^{-2} d^{-1}). The cylindrical pellets occurred mostly during Phase 1 and their contribution to vertical export of POC varied from 22% in Phase 1 to 3% in Phase 3. Elliptical pellets contributed <1% to the POC flux in Phase 1, but this markedly increased to 33% in Phase 3.

Discussion

The present results and other ISHTAR studies of the northern Bering Sea show that fairly high autochthonous production and advective inputs of allochthonous material characterised the northward flowing BSAW, particularly during early summer (Phase 1). The northward current transported abundant nutrients to support a high primary production (mean: ca. 2 g C m^{-2} d^{-1} for summer months, Walsh et al. 1989) and high zooplankton activity (Springer et al. 1989). The high vertical export of biogenic particles of autochthonous origin is represented by a high contribution of faecal pellets. These pellets with sinking rates of >60 m d^{-1} reached the sea floor (ca. 50 m) within a day (Sasaki et al. 1993), even under the strong current condition (ca. 16 m s^{-1}). The strong current also carried allochthonous organic matter, such as "summer diatoms", from the south into the shallow Chirikov Basin. The slight decrease in vertical POC export during Phase 2, accompanied by a smaller contribution of faecal pellets suggests a lower production of autochthonous particles. In Phase 3, BSAW was replaced by ACW that is relatively rich in nutrients. Consequently, the flux of total POC and faecal pellets increased again, indicating an intensified autochthonous production, probably due to a fall bloom.

The estimated carbon utilisation of benthos in the Chirikov Basin (464 mg C m^{-2} d^{-1}; Grebmeier and McRoy, 1989) was comparable to the average vertical POC export (462 mg C m^{-2} d^{-1}). The benthic communities can thus be supported by sinking organic particles, such as diatoms and zooplankton faecal pellets during summer, strongly indicating a close pelagic-benthic coupling.

Conclusions

- The vertical export of POC in the Bering Strait shelf ranges between 350–650 mg C m^{-2} d^{-1} during summer. The high sedimentation rate at 50 m depth is similar to rates encountered in the Barents Sea.
- The high vertical export directly reflects the hydrodynamic peculiarities of the Bering Strait, resulting in high nutrient supply and high primary production.
- The advection of suspended biomass and sinking particles results in a extensive import of biomass into the Chuchi Sea.

5.3.8
Canadian Ice Island

B. Hargrave

The Canadian Ice Island, a tabular iceberg approximately 5 km wide, 7 km long and 37 to 40 m thick was separated from the Ward Hunt Ice Shelf in 1983. The drift trajectory between 1983 and 1986, over the continental shelf (160 to 345 m depth) along the north coast of Ellesmere Island to the west of the Lincoln Sea, traversed one of the most permanently ice covered regions of the Arctic Ocean. Narrow leads, formed following strong easterly winds during fall and winter storms, are quickly covered by moving ice (>95% ice cover) up to 2 m thick with many ridges and thick (5–10 m) multi-year ice floes.

A permanent research camp (Canadian Ice Island) was established on the ice island in 1985. The trajectory between 1983 and 1988 followed the east to west transpolar drift along the northern margin of the Canadian Arctic Archipelago. Drift was negligible between 1986 and 1987. Maximum movement (25 km $month^{-1}$) occurred during June 1987. The ice island was semi-stationary or drifted slowly (<10 km $month^{-1}$) between 1989 and 1990 when it entered Peary Channel (depth 500 to 550 m) between Ellef Ringnes, Amund Ringnes and Axel Heiberg Islands. Current velocities measured at mid-depth when the ice island was semi-stationary were low (<5 cm s^{-1}).

The Ice Island research camp allowed mooring of multi-cup sediment traps under the ice for seasonal measurements of particle flux (79° to 81°N, 96° to 102° W) in 1986–1987 and 1989–1990 (Hargrave et al. 1989; Hargrave et al. 1994). Three traps (Kiel I – 8 cups, Kiel II – 20 cups and BIO – 14 cups) were suspended at 100 m through 1 m diameter holes melted through the ice island. The Kiel I trap was moored for nine months between September 1986 and June 1987 when water depth under the trap varied from 120 to 305 m. In September 1989, two ice holes were melted 50 m apart. The Kiel II and BIO traps, moored at 100 m on separate mooring lines, were retrieved after one year. No measurements of primary production were carried out during the trap deployments in either study but profiles of suspended chlorophyll were measured in 1986 and 1987.

Results

POC flux (September 1986 to June 1987) off Axel Heiberg Island amounted to 0.073 g C m^{-2} (Hargrave et al. 1989; Fig. 5.10a). Maximum sedimentation (0.64 mg C m^{-2} d^{-1}) occurred in February when settled particulate matter contained relatively high amounts of carbon (17%) and nitrogen (3.4%). In contrast, material collected in December–January contained less organic matter (C 3.9%. N 0.5%). C:N molar ratios varied from 6.7 to 8.0. There was a bi-modal pattern in vertical POC export with a smaller peak during October–November 1986 (0.4 mg C m^{-2} d^{-1}) followed by a larger maximum in February-March 1987 (Fig. 5.10a).

Pigment concentrations were not measured in settled material on trap retrieval in 1987, but sedimented particulate matter consisted of amorphous detritus, faecal pellets and diatoms (*Nitzschia frigida, Fragilaria oceanica*). High POC flux in March–April was correlated with high numbers of diatom frustules. Trap cups also contained numerous intact macrozooplankton (copepods, amphipods) that were manually removed before analysis of settled particulate matter. The POC content of 'swimmers' (*Calanus hyperboreus, C. glacialis, Metridia longa, Euchaeta norvegica, Xanthocalanus borealis, Oithona* spp. *Microcalanus* spp. and *Pseudocalanus* spp.), that presumably entered the trap alive and were killed of contact with $HgCl_2$ preservative, amounted to from 66 to 98% of the total POC collected.

Two traps were moored off Ellef Ringnes Island in Peary Channel in 1989–1990 (Fig. 5.10b) for a complete annual period. Annual POC fluxes in the Kiel II and BIO traps (0.093 and 0.175 g C m^{-2}, respectively) were significantly different ($p < 0.05$). Annual values for mass and inorganic matter were not significantly different between trap designs. In contrast to the 1986–1987 trap deployment where several calanoid copepod species were collected, in 1989–1990 only a single species (*Oncaea borealis*) was abundant in 'swimmers' separated from settled debris.

Maximum POC fluxes under the ice island occurred in August and September 1989 (Fig. 5.10b) when cylindrical and elliptical faecal pellets were most abundant. Using measured POC content in pellets, a maximum of 11 to 16% of POC flux would have been contained in faecal material deposited in September. Siliceous centric diatoms (*Thalassiosira* spp., *Melosira arctica*, and *Coscinodiscus marginalis, Chaetoceros* spp.), chrysophytes

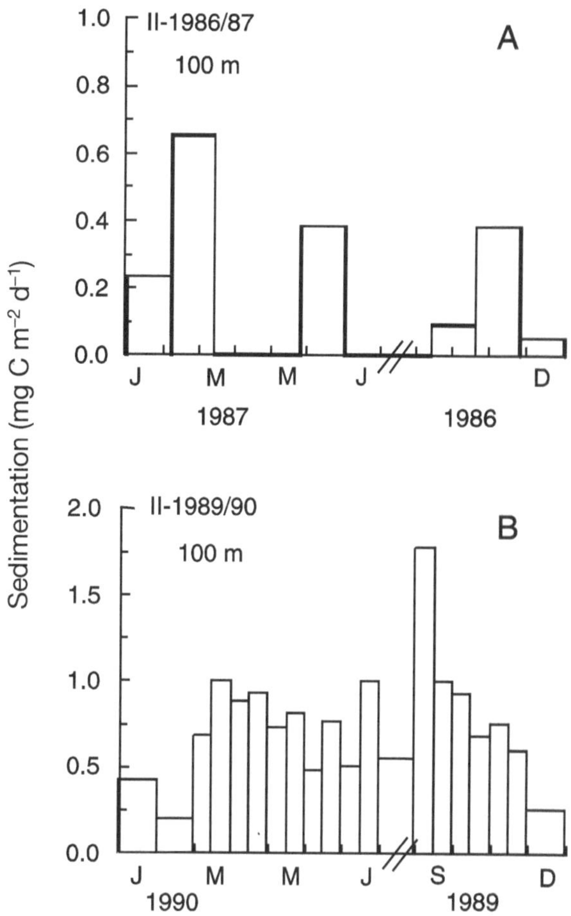

Fig. 5.10. Seasonal variations in particulate organic carbon (POC) flux in multi-cup sediment traps under the Canadian Ice Island (II) (A) Axel Heiberg Island (1986–1987), (B) Ellef Ringnes Island (1989–1990). (A) Kiel I trap (0.031 m^2, height:mouth opening aspect ratio 3.2); (B) mean POC flux for Kiel II trap (0.4059 m^{-2}, aspect ratio 1.4) and BIO Mark II trap (0.1122 m^2, aspect ratio 2.0). Trap collection periods varied from (A) 36.5 d to (B) 15 to 60 d. Data summarised from Hargrave et al. (1989) and Hargrave et al. (1994)

(*Archeomonas sphaerica* and *A. mangini* complex), silicoflagellates (*Dictyocha speculum* var. *septentrionalis*), radiolarians (*Spongotrochs glacialis* and *Amphimelissa setosa*) and sponge spicules were present in settled material at various times during the year. In general, siliceous organisms were more abundant than calcareous forms. The coccolithophore *Coccolithus pelagicus* was only present in December.

The organic carbon content in settled particles in 1989–1990 was highly variable (POC 2–40%) with annual mean values (POC 20.7%; N 4.0%) that approximated maxima observed in 1986–1987. Maximum values (POC >20%) occurred in October-November and between March–June. The annual average C:N molar ratio was 7.8 (range 4.5–14) with lowest values (4–8) between March and June.

Discussion

Hargrave et al. (1994) compared previous measurements of POC flux at polar latitudes >70°N with results from the Canadian Ice Island. Daily averages from annual integrated values in open ocean areas at latitudes from 75° to 78°N (1–13 mg C m^{-2} d^{-1}) were approximately 10-fold higher than values measured under the ice island at 79° to 81°N (0.3–0.5 mg C m^{-2} d^{-1}). An even higher mean daily rate (39 mg C m^{-2} d^{-1}) was measured in a study in a nearshore area of the Beaufort Sea (70°N) although this was in April-June, not over an annual period. The annual POC fluxes in ice-covered regions near the Canadian Arctic Archipelago in the margin of the Arctic Ocean represent the lowest published values for any arctic polar region, but share similarities with those derived from the Lomonosov Ridge (Fig. 5.8). Studies in most other regions in the Arctic presented here experience ice-free periods of varying duration, usually during summer and early autumn months and vertically export is consequently much higher.

Unlike previous studies in the arctic region, few planktonic algal species were dominant in samples of sedimented material collected under the ice island. The generalisation that calcareous planktonic organisms predominate in deposited material at polar latitudes (Honjo 1990) was not supported, neither here or in any of the other investigations. Only *Chaetoceros* spp. was abundant during August when chlorophyll and biogenic silica concentrations in settled material were high. High fluxes of chlorophyll *a* in August and September coincided with seasonal maxima in suspended chlorophyll (Hargrave et al. 1994). Maximum POC fluxes in early September (Fig. 5.10b) occurred at the time when faecal pellet fluxes were high. Much lower numbers of pellets were collected between April and July implying that zooplankton grazing increased as the summer progressed with a peak in September, following peak chlorophyll flux in August. The pattern resembles that described above for the NOW polynya in northern Baffin Bay with a delay of approximately two months (NOW – peak chlorophyll flux in June followed by faecal pellet maxima in July/August; Ice Island – peak chlorophyll flux in August with maximum pellet flux in September).

Conclusions

- Annual POC fluxes measured at 100 m under permanent ice cover over the Arctic Ocean continental shelf north of the Canadian Arctic Archipelago (<1 g C m^{-2} y^{-1}) are much lower than most measurements at polar latitudes (>70°N), but similar to estimates from the NEW and Lomonosov Ridge.
- No single planktonic algal group dominated settled material. *Chaetoceros* spp. was the most abundant species with the highest number of cells deposited during August.
- Calanoid copepod faecal pellet fluxes were greatest in September (up to 16% of total POC flux) implying increased grazing pressure following maxima in suspended chlorophyll in August.
- The dry weight of macrozooplankton (calanoid copepods, amphipods) that entered traps under the ice island as 'swimmers' was up to one-order-of magnitude higher than that of detrital material collected. The biogenic transport by mesozooplankton in the Arctic Ocean must be quantified to fully describe vertical fluxes of organic matter.

5.3.9 Canadian Archipelago: Barrow Strait

M. FORTIER

Barrow Strait (74°30′N; 96°20′W) is a shallow (150–200 m), seasonally ice-covered strait in the centre of the Parry Channel that traverses the Canadian Archipelago and links the Beaufort Sea with northern Baffin Bay (Fig. 5.2). Landfast ice consolidates in the strait between December and March with maximum ice thickness (150–240 cm) and heaviest snow cover usually occurring in late May. While the fast-ice usually breaks-up in the second half of July, a continuous eastward flow of drifting pack-ice and multiyear ice covers the strait throughout summer. Springtime, on-ice, oceanographic sampling programs have been conducted on and off in the area since the early 1980s, making Barrow Strait one of the most studied bodies of water in the Canadian Arctic Archipelago.

Before snowmelt, ice algae largely dominate springtime primary production under the snow-covered landfast ice of Barrow Strait. Maximum ice algal biomass varying from 2.5 to 25 g C m^{-2} may accumulate in the bottom ice before the initiation of the spring melt and the final release of the epontic community into the water column (Anning et al. 1988; Smith et al. 1988). Under thin snow cover, annual ice algal production may range from 5 to 23 g C m^{-2} y^{-1} for a 50-d growth season (Smith et al. 1988; Welch and Bergmann, 1989; Bergmann et al. 1991). Increased irradiance and stratification due to ice and snow melt, are the mechanisms responsible for triggering the phytoplankton bloom that normally follows the landfast ice break-up by a couple of weeks. Estimates of the contribution of ice algae to total annual primary production in the area vary from 10% (Bergman et al. 1991, Welch et al. 1992) to over 50% (Smith et al. 1988).

Estimates of the vertical export of particulate organic matter in the area have been limited to the deployment of short term sediment traps in the spring and early summer ice-covered season, with no traps moored so far during the open water season. Except for 1994, where traps where deployed until the late stages of ice ablation and break-up, most studies have extended from the start of the ice algal growth season in April until the end of the ice algal bloom which usually coincides with the period of snow melt in June or early July (Anning 1989; Michel et al. 1996; Fortier et al. 2002). Results from these studies suggest strong interannual variability in vertical POC fluxes before the ice break-up in spring and early summer (Table 5.3). Here, we

Table 5.3. Interannual variability in the range of vertical POC flux at 30 m before the ice break-up under the ice covered Barrow Strait

Collection period	POC Flux at 30 m (mg Cm^{-2} d^{-1})	Reference
25/03–29/06/1983	44–519	Anning, 1989
10/04–12/06/1985	37–328	Anning, 1989
06/04–02/06/1986	85–126	Anning, 1989
26/03–26/06/1987	5–144	Anning, 1989
02/05–25/06/1992	68–341*	Michel et al. 1996
01/05–11/07/1994	46–504	Fortier et al. 2002
01/05–11/07/1995	30–287	Fortier et al. 2002

* Collection depth at 15 m.

summarise previously published results from two contrasting years (1992 and 1994). Large interannual differences in the timing, magnitude and nature of under-ice vertical fluxes were driven by meteorological events that controlled snow cover dynamics and by interannual differences in copepod grazer assemblages.

In both years, the vertical flux of chloropigments (chlorophyll *a* (Chl *a*) + phaeopigments (Phaeo)) and POC under the ice was measured at intervals of 2 to 4 days. Sediment traps made of 1 (1992) or 2 (1994) PVC cylinders with an aspect ratio of 1:10 (10 cm diameter in 1992 and 7.56 cm diameter in 1994) were moored at 15 m below the ice bottom in 1992 and 30 and 90m below the ice bottom in 1994. The traps were filled with filtered sea water with brine added (1992) or not (1994) before deployment under the ice. Neither poison nor preservative was added (see Michel et al. 1996 and Fortier et al. 2002 for details). Daily snow depth, precipitation, air temperature and climate records for the period 1950 to 2001 were provided by the Meteorological Service of Canada in nearby Resolute Bay.

Results and discussion

The spring of 1992 was cold and experienced a suite of snow storms (early May, late May and mid-June) that resulted in a relatively thick snow cover, which persisted until the final snow melt in early July. Ice break-up did not occur until the last week of July (Fig. 5.11a). By contrast, the thin snow cover in early May 1994 was stable until a late May warm spell and associated rain-on-snow event that melted the snow cover from 20 cm to 7 cm in less than a week (Fig. 5.11a). The rain event of May 1994 was the earliest and most important recorded since 1950 (Fortier et al. 2002). The landfast ice broke up suddenly on 11 July 1994, allowing the investigation of vertical export during break-up for the first time in the region.

The persisting snow cover in 1992 limited light penetration through the ice cover and allowed a relatively high ice algae biomass to remain until the end of sampling on 24 June (Fig. 5.11b). Ice algal biomass increased until late May (average value of 75 mg m^{-2}) and then declined progressively in June where floating lumps (up to 115 mg chloropigments m^{-3}) of newly released ice algae were responsible for the high Chl *a* concentrations on 7 and 24 June (Fig. 5.11c). Concentrations of NO_3 + NO_2 in the 0–15 m layer remained high (8 to 16 μM) with no evidence of any significant under-ice phytoplankton development until the end of sampling on 24 June (Michel et al. 1996).

In 1994, the low algal biomass in the ice (max. average value of 37 mg m^{-2}) decreased rapidly after

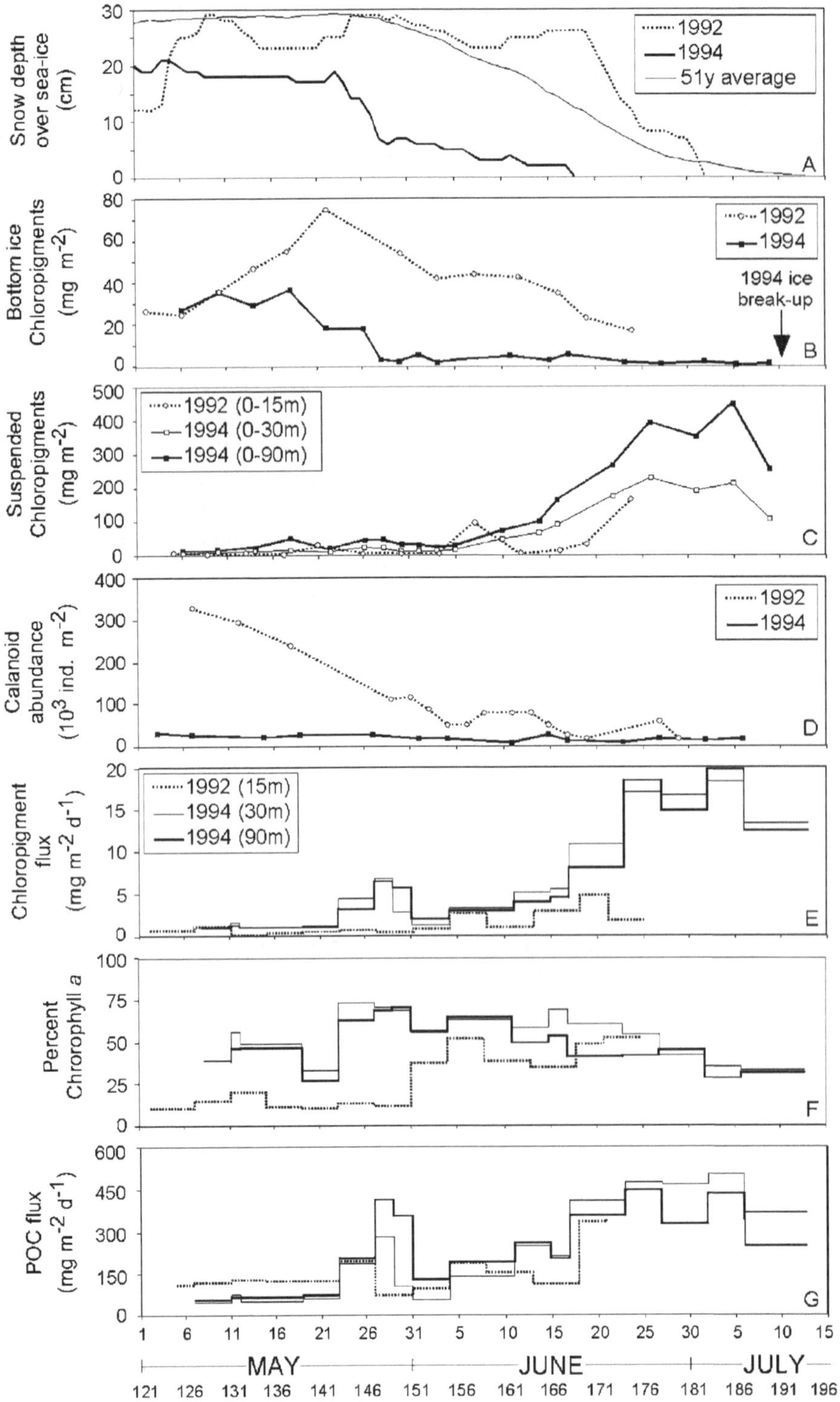

Fig. 5.11. Variation of snow depth over sea ice (A), Bottom ice chloropigments (B), suspended chloropigments (C), calanoid copepod abundance (D), vertical chloropigment flux (E), POC/chlorophyll *a* percentage in vertical export (F) and vertical POC export (G) at Barrow Strait in the Canadian Archipelago

the sudden snowmelt event in late May that allowed high irradiance to penetrate the ice cover (Fig. 5.11c). Starting in the second week of June, an intense phytoplankton bloom developed in the surface layer under the snow-free ice cover (Fig. 5.11c). Peak biomass of > 400 mg chloropigment m^{-2} in the 0-90 m layer (9 mg Chl *a* m^{-3} at 15 m) in the last week of June corresponded with decreasing concentrations of NO_2 +NO_3 from 9 µM to <2 µM (Fortier et al. 2002). In both years, the ice algal biomass was dominated by pennate diatoms of the genera *Nitzschia*, *Fragilariopsis* and *Navicula* (Michel et al. 1996; Fortier et al., unpubl. data). *Nitzschia* spp. and *Fragilariopsis* spp. prevailed during the under-ice phytoplankton bloom in 1994.

In both years, increases in the vertical flux of chloropigments and POC occurred during and immediately after the release of ice algae. In 1992, the highest POC flux (340 mg m^{-2} d^{-1}) was recorded during the final release of ice algae in the water column that coincided with the final snow melt in late June. In 1994, the abrupt release of ice algae in the last week of May resulted in an early flux (410 mg C m^{-2} d^{-1} at 90 m) of high quality particulate organic matter (C/Chloropigment: 61–70% Chl *a*) to the seafloor (Figs. 11f and 11g).

With daily vertical POC fluxes of over 300 mg C m^{-2} d^{-1} (max. of 444 mg C m^{-2} d^{-1} at 90 m) sustained for more than 3 weeks in June and early July 1994, the intense under-ice bloom that followed the early removal of the snow cover and the increase in under-ice irradiance, generated the bulk of the overall flux of particles before the ice break-up. In total 16.7 g C m^{-2} (465 mg chloropigments m^{-2}) were exported to depth (90 m) before the 11 July ice-break-up. Of this amount 82% (13.7 g C m^{-2}, 380 mg chloropigments m^{-2}) resulted primarily from the sinking of relatively fresh material with a low C/chloropigment ratio (35–50%) during the under-ice phytoplankton bloom.

Although meteorological forcing affected the timing and extent of production as well as the vertical export of particulate organic matter between the two years, the large interannual variability in the biomass of copepod grazers also affected the quality of settled material. Calanoid copepods (*Pseudocalanus acuspes*, *Calanus glacialis*, *Calanus hyperboreus*, *Metridia longa*) and the small cyclopoids (*Oithona similis*, *Oncaea borealis*) dominate the under-ice grazers in the area (Fortier, 1999). The extreme variability in the overall number and biomass of calanoid grazers under the ice reflected the exceptional abundance of *P. acuspes* in May 1992 (up to 3×10^5 m^{-2} in early May) (Fig. 5.11d). In 1994, the identifiable faecal matter collected in the sediment traps consisted primarily of intact or broken cylindrical faecal pellets of calanoid copepods. The contribution of dark, compact, ellipsoidal appendicularian pellets to the total faecal flux was small (<1%). The faecal pellet fraction of the POC flux was minimum (<5%) during the release of ice algae and the under-ice bloom, and maximum (up to 37% at 30 m) during the interval of reduced flux between the two events. This indicates that zooplankton herbivores were swamped by the abrupt flux of ice algae and the under-ice bloom (Fortier et al. 2002). Although comparable data on the vertical faecal flux are not available for 1992, several lines of evidence indicate that, in that year, pelagic grazers, dominated by *P. acuspes*, intercepted most of the biomass of ice algae released into the water column. Using a carbon budget model based on pigment concentrations in the ice, the water column and the sediment traps, Michel et al. (1996) estimated that 60–70% of the exported carbon was in the form of faecal matter in 1992.

Conclusions

- Interannual variability in the seasonal evolution of the snow cover on sea-ice (extent, thickness, duration) results in important variability in the timing and extent of under-ice export of POC by affecting the extent, timing and duration of epontic and planktonic production.
- The rapid release of ice algae in response to the early snow melt in May 1994 resulted in an important flux (>400 mg C m^{-2} d^{-1} at 90 m) of intact cells to depth (benthos), more than 2 months before the open water bloom.
- With the occurrence of an under-ice phytoplankton bloom in 1994, more than 16 g C m^{-2} of algal origin where exported to depth (90 m) before the ice break-up with daily fluxes reaching 500 mg C m^{-2} d^{-1}. This amount represents more than 25% of the annual total production estimates for Barrow Strait (60 g C m^{-2}; Welch et al. 1992).
- Calanoid grazers, dominated by *Pseudocalanus* copepodites, exploit the surface layer in spring. Their grazing is a major factor controlling vertical export of organic matter to depth before the ice break-up.

5.4 Regional Variability in POC Export Flux in the Arctic Ocean Determined Using ^{234}Th as a Tracer

S.B. MORAN

5.4.1 Introduction and Background

Over the past decade, ^{234}Th (half-life = 24.1 days) has been increasingly used to estimate the flux of POC exported from the euphotic zone in the global ocean (Moran et al. 2003). This naturally occurring, particle-reactive, radionuclide is produced continuously in seawater by alpha decay of soluble ^{238}U (half-life = 4.47×10^9 years) and has proven to be a useful tracer of upper ocean particle export on a time-scale of days to months. The use of ^{234}Th to quantify the downward flux of POC has the advantage of providing a temporally integrated record of particle export that can be determined from ships, with minimal sampling time (hours), and without the apparent biases that characterize upper ocean sediment traps. Measurements of $^{234}Th/^{238}U$ disequilibrium have been made during several recent Arctic field programs, providing valuable new insights into the regional variability in POC export fluxes in the interior Arctic, near ocean margins and in adjacent sub-polar waters.

In the surface ocean, there is typically a deficit of ^{234}Th with respect to secular equilibrium with its soluble parent ^{238}U (Fig. 5.12), which is ascribed to the uptake of ^{234}Th onto particle surfaces and subsequent sinking of particles into deeper water (Bhat et al. 1969; Matsumoto 1975; Coale and Bruland 1985). To calculate the rate at which organic carbon is transported downward using ^{234}Th as a tracer, in principle one requires only a knowledge of the ^{234}Th deficit and the POC/^{234}Th ratio of sinking particulate matter. Thus, the POC export flux is defined empirically as the product of the POC/^{234}Th ratio, which is ideally representative of sinking biogenic particles, and the depth-integrated ^{234}Th flux (Moran et al. 2003),

$$\text{POC Flux} = \left(\frac{POC}{^{234}Th}\right)_{i,w} \times \int_0^z (A_U - A_{Th})\,dz$$

where $[POC/^{234}Th]_{i,w}$ is the POC/^{234}Th ratio measured in sinking particulate matter of size i and collected at depth w, A_U is the activity of ^{238}U (dpm L^{-1}), A_{Th} is the total ^{234}Th activity (dpm L^{-1}), λ is the ^{234}Th decay constant (0.0288 d^{-1}), and z is the lower boundary of the depth-integrated ^{234}Th deficit.

Depth profiles of ^{234}Th activity are typically determined by collection of large-volume samples (several 100 L–1000 L) using in-situ pumps. Dissolved ^{234}Th is extracted *in-situ* using MnO_2 impregnated adsorber cartridges, whereas particulate ^{234}Th is collected in both suspended (*ca.* 0.4–1 µm) and large particle (>53 µm or >70 µm) size classes. ^{234}Th activities are subsequently analyzed, often

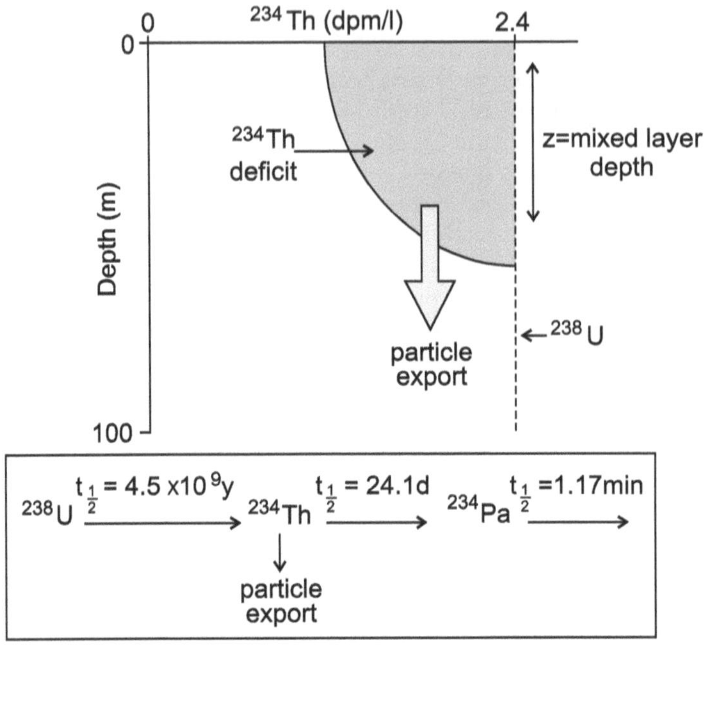

Fig. 5.12. A schematic diagram indicating a ^{234}Th deficit relative to secular equilibrium with the parent ^{238}U in Arctic surface waters, produced by particle scavenging and sinking

at-sea, using either gamma or beta counting techniques to an uncertainty of *ca.* ±5–10%. ^{238}U activities are calculated from salinity according to the relationship ^{238}U = salinity × 0.0708 @ 35 psu (dpm L^{-1}) (Chen et al. 1986). The POC/^{234}Th ratio is usually determined by analysis of material collected by filtration with *in-situ* pumps or, if available, by deployment of sediment traps.

5.4.2 Uncertainties in ^{234}Th-derived POC Export Fluxes

As recently discussed (Moran et al. 2003), it is difficult to rigorously define and measure both of the key parameters: the POC/^{234}Th ratio of sinking particles and the depth-integrated ^{234}Th flux. The largest uncertainty in estimating the export flux of POC is likely to be the POC/^{234}Th ratio of sinking particulate matter, which for the Arctic Ocean ranges by an order of magnitude; 0.06–0.80 mg C dpm^{-1} ^{234}Th (5–67 μmol C dpm^{-1} ^{234}Th) (Moran et al. 2003). Indeed, the POC/^{234}Th ratio of marine particles has been reported to vary significantly with location and time, presumably in association with changes in primary and secondary productivity, plankton community structure, export production, particle-size distribution, particle aggregation-disaggregation, and food web dynamics (Cochran et al. 1995; Moran et al. 1997; Burd et al. 2000; Moran and Smith 2000; Amiel et al. 2002; Coppola et al. 2002; Moran et al. 2003). Variability in the POC/^{234}Th ratio may also be due to differences in sampling method, which includes water bottles, large-volume pumps, and sediment traps, each of which have unique sampling biases (Amiel et al. 2002; Coppola et al. 2002; Moran et al. 2003).

Variability also exists in the integration depth (z) used in calculating the particle export flux of ^{234}Th, which in the Arctic typically ranges from *ca.* 20–200 m. In Arctic shelf environments, $^{234}Th/^{238}U$ disequilibrium frequently exists throughout the water column, presumably due to removal of ^{234}Th on abiogenic particles resuspended from the ocean bottom (Cochran et al. 1995; Moran et al. 1997; Moran and Smith 2000; Amiel et al. 2002; Coppola et al. 2002). It is therefore difficult to distinguish between biogenic and abiogenic particle export of ^{234}Th in high-energy shelf waters, which can lead to biases in calculated POC fluxes. Steady-state ($\partial^{234}Th/dt = 0$) is also usually assumed, and the advective and diffusive fluxes of ^{234}Th are assumed to be small relative to particle export of ^{234}Th.

Overall, POC export fluxes are expected to vary by at least a factor of 2–5 or more, due to variability in the POC/^{234}Th ratio and procedures used to estimate the ^{234}Th flux (Moran et al. 2003).

5.4.3 Regional Variability in Arctic POC Export Fluxes

^{234}Th-derived POC fluxes for the Arctic Ocean and the sub-polar Labrador Sea are summarized in Figure 5.13 (note that POC fluxes are reported in units of g C m^{-2} y^{-1} for consistency in this book, though the original literature data is in units of mmol C m^{-2} d^{-1}; to convert to mmol C m^{-2} d^{-1} multiply by 0.23). Clearly evident is the wide spatial variability in POC export flux. In general, shelf waters exhibit much higher fluxes compared to the central basins, which is consistent with the elevated rates of productivity in Arctic shelf-slope waters. The ^{234}Th-derived POC fluxes are also, on the whole, consistent with sediment trap observations that indicate the export of POC can be quite high in ice-free regions, particularly over the broad continental shelves. The few sediment trap data collected under the permanent ice cover indicate that particle flux is extremely low (Hargrave et al. 1994, see also Section 5.3.8), and this is generally the case for POC fluxes estimated from ^{234}Th measurements in the central Arctic (Fig. 5.13). There is evidence from sediment trap measurements for a seasonal cycle in particle flux under the permanent ice cover (Hargrave et al. 1994), and seasonal changes in POC export in the Arctic also have been reported based on ^{234}Th data (Cochran et al. 1995; Amiel et al. 2002).

The first and, to date, only upper ocean POC export fluxes reported for the central Arctic Ocean were estimated from $^{234}Th/^{238}U$ disequilibrium as part of the 1994 US-Canada Arctic Ocean Section (AOS-94) expedition (Moran et al. 1997). These results also represent the first measurements of both dissolved and particulate ^{234}Th in the central Arctic Ocean. The major features of the ^{234}Th distributions from AOS-94 indicate: 1) enhanced scavenging and particle export in the Chukchi Shelf, and; 2) particle export occurring on a time-scale of days to months at several locations in the central Arctic Ocean. These workers concluded that the observed ^{234}Th deficit in the central basins was unlikely to reflect scavenging of ^{234}Th over the shelves and lateral transport into the Arctic interior (Moran et al. 1997) but, rather, was a product of local scavenging and particle export associated with primary and secondary productivity. Moreover, they reported a covariance between the POC concentration and ^{234}Th activity of the 0.7–53 μm size particulate matter in the upper 20–30 m of the central Arctic. This was observed despite the fact that total primary production varied by over an order of magnitude along the AOS-94 section, while the production of ^{234}Th from ^{238}U is essentially constant. The implication is that ^{234}Th scavenging

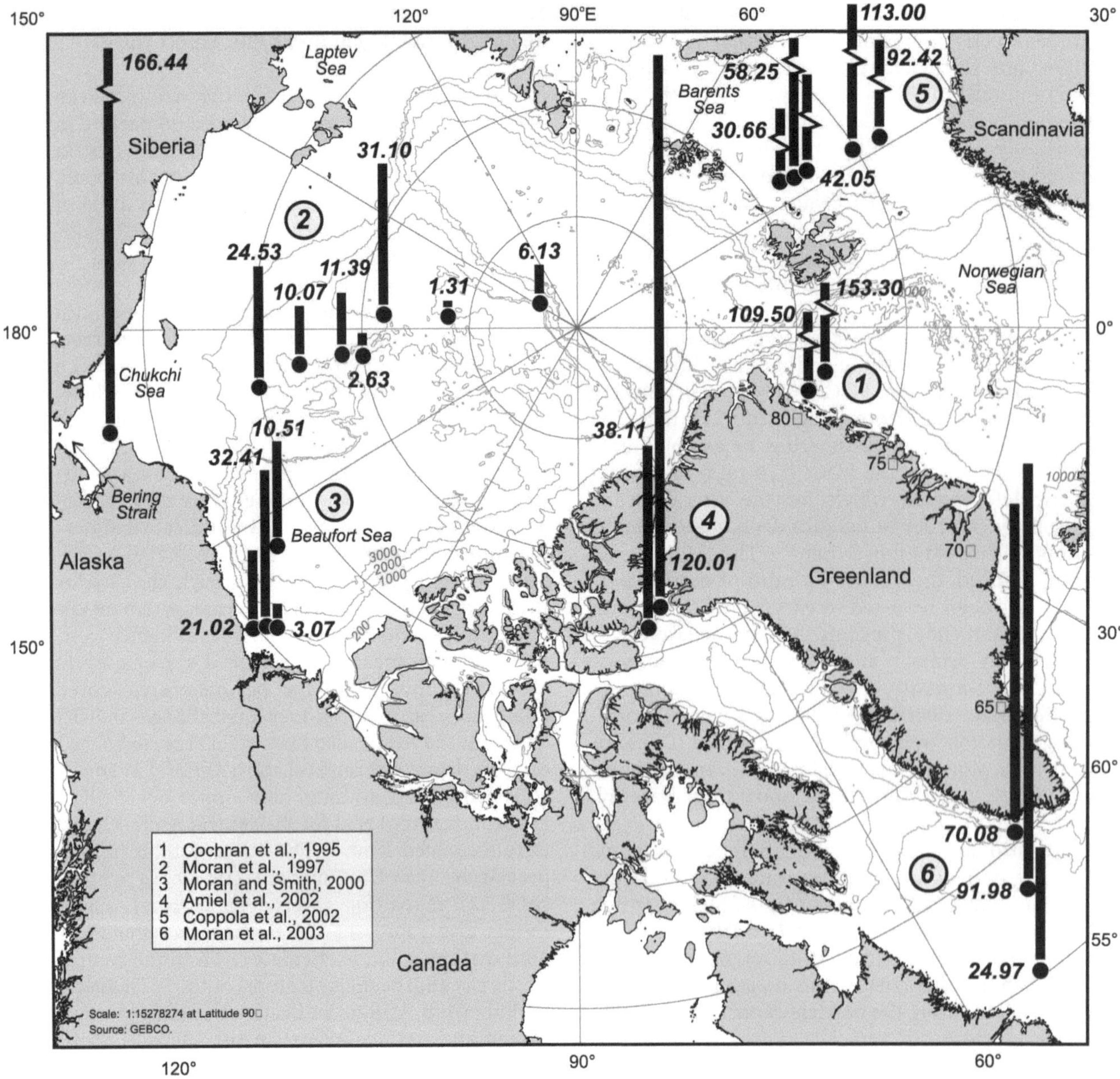

Fig. 5.13. Spatial distribution of POC export fluxes (g C m^{-2} y^{-1}) in surface water of the Arctic Ocean and the sub-polar Labrador Sea determined using $^{234}Th/^{238}U$ disequilibrium

and export from the surface waters is controlled by the formation of biogenic particulate organic carbon.

Moran et al. (1997) calculated POC fluxes using POC/^{234}Th ratios determined in 0.7–53 µm size particles and the ^{234}Th deficit in the upper 20–30 m. The highest POC export flux, 166 g C m^{-2} y^{-1}, was observed in the productive Chukchi Shelf waters (Fig. 5.13; Chapter 7.3). In the interior Arctic, POC fluxes decreased significantly, ranging from 1.3–31 g C m^{-2} y^{-1} and averaging 13 ± 9 g C m^{-2} y^{-1}. By comparison, the distribution of total primary production with latitude decreased from a maximum of 937 g C m^{-2} y^{-1} within the Chukchi Shelf to values of *ca.* 4 g C m^{-2} y^{-1} at 82°N, and then increased to *ca.* 22–123 g C m^{-2} y^{-1} from 84°N to 90°N (Gosselin et al. 1997). Total primary productivity included POC production due to water column plankton and ice algae, where a significant fraction of the total primary productivity in the ice-covered interior Arctic was due to ice algae production (Gosselin et al., 1997). Using the C:N ratio of the 0.7–53 µm size class, Moran et al. (1997) estimated the cor-

Table 5.4. Vertical POC export estimates from long-term deployments in the Arctic Ocean region. For site acronyms, see Fig. 2. NOW, left column: 1997–1998; right column: 1998–1999. NEW: rPOC; in parenthesis: POCcorr. = POC corrected for DOC leakage. IO, in parenthesis: # years and range of vertical export. KS = Kara Sea (pers. comm. B. Gaye-Haake). [BS] = simulation data from the Barents Sea, Slagstad and Wassmann (1997). N. D. = no data

Site	Position	Type of trap	Depth of trap (m)	Time (months)	Vertical flux ($g C m^{-2} y^{-1}$)	
NOW						
(N2)	78°N 75°W	BIO	203	12	1.8	nd
(S5)	76°N 77°W	Technicap PPS.3/3	258	12	8.0	11.7
(S5)	76°N 77°W	Technicap PPS.3/3	315	12	8.4	nd
(S4)	76°N 74°W	OSU Tracer 15	205	12	1.3	2.5
(S2)	76°N 72°W	Technicap PPS.3/3	198	12	3.8	6.7
(S2)	76°N 72°W	Technicap PPS.3/3	507	12	8.1	13.8
(D1)	75°N 75°W	McLean 78G-21	250	12	1.0	nd
(D2)	77°N 75°W	McLean 78G-21	406	12	1.6	nd
NEW	80°N 11°W	Kiel Trap II	130	11	1.0 (2.7)	
IO	72°N 7–10°W	Aquatec, Kiel	500	12 (4)	2.3 (1.4–3.8)	
GS	75°N 00°W	Aquatec, Kiel	200–300	12	1.4	
KS	74°N 80°W	Aquatec Kiel	36	12	9.8	
LR	81°N 138°E	Aquatec, Kiel	150	11	1.5	
CII	81°N 96°W	Kiel Trap I	100	9	0.07	
	79°N 102°W	Kiel Trap II	100	12	0.09	
	79°N 102°W	McLean MarkII	100	12	0.18	
BFS[1]						
SS1	70°N 133°W	McLean MarkII	200	11	3.7	
SS2	71°N 128W	Baker	200	11	4.6	
SS3	71°N 134°W	McLean MarkII	200	11	1.6	
SS4	70°N 136°W	Baker	200	11	6.1	
BS	73–78°N 30–40°W		75	8	[17–39]	

[1] 4 traps at the shelf edge of the Beaufort Sea were deployed for a full year, starting in April 1987 and going to March1988. They were twice deployed and poisoned with mercury. They were put on the 200 m isobath, 50 m above the bottom (R. MacDonald, unpubl. res.)

responding PON export flux to be 25 $g N m^{-2} y^{-1}$ in the Chukchi Shelf and 0.26–5.6 $g N m^{-2} y^{-1}$ (mean = 2 ± 2 $g N m^{-2} y^{-1}$) in the central Arctic.

It is important to note that the ^{234}Th-derived POC fluxes in the central Arctic Ocean are likely to be upper estimates (Moran et al. 1997) because POC/^{234}Th ratios were determined only in the suspended (0.7-53 μm) particle size class, whereas the majority of the particle flux is likely dominated by large, rapidly sinking particles (>53 μm). The POC/^{234}Th ratio often decreases with increasing particle size (Moran et al. 2003). Thus, using the POC/^{234}Th ratio of the suspended particles could result in an overestimate in POC export. Additionally, the observed ^{234}Th/^{238}U disequilibrium may not be at steady-state; rather, ^{234}Th scavenging and particle export may be the result of a phytoplankton bloom that preceded the AOS-94 sampling, and the use of a steady-state model can overestimate the particle flux during the post-bloom period (Buesseler et al. 1992).

There are very few published measurements of POC export flux in the central Arctic Ocean with which to compare the ^{234}Th-derived values. Hargrave et al. (1994) deployed sediment traps at 100 m for one year under the permanent ice cover at 79°N over the polar continental shelf off Ellef Ringes Island (*ca.* 500 m depth). Their study indicated an extremely low mean annual POC flux of 0.14 ± 0.09 $g C m^{-2} y^{-1}$. In fact, the results reported by (Hargrave et al. 1994) are lower than the more typical value of *ca.* 0.4 $g C m^{-2} y^{-1}$ recorded by sediment traps deployed in the East Greenland Sea, Fram Strait and Greenland Basin.

Export fluxes of POC determined using ^{234}Th have also been reported along a section in the Beaufort Sea, sampled in August-September, 1995, and extending from the shelf waters into the Canada Basin (Moran and Smith 2000). In this study, POC fluxes were again calculated using the steady-state assumption, measurements of POC/^{234}Th ratios in 0.7–53 μm size particles, and the observed ^{234}Th deficit in upper 50 m. POC fluxes ranged from *ca.* 17–31 g C m^{-2} y^{-1} in the shelf waters (Fig. 2), which are lower than the single value of 166 g C m^{-2} y^{-1} in the adjacent Chukchi Sea shelf waters estimated during AOS-94 (Moran et al. 1997). Fluxes of *ca.* 4–9 g C m^{-2} y^{-1} were reported for the slope waters and central Beaufort Sea, which is only slightly lower than the average value of 13 ± 9 g C m^{-2} y^{-1} reported for the central Arctic Ocean (Moran et al. 1997). These data compare with values of *ca.* 20 g C m^{-2} y^{-1} for new production based on nitrate budgets for the Beaufort Sea (Macdonald et al. 1987; Macdonald et al. 1998). Sediment trap POC export fluxes adjacent to the Beaufort Sea shelf edge at 200 m depth and 50 m above the bottom were on average 4.4 ± 1.9 g m^{-2} y^{-1} (Table 5.4).

Coppola et al. (2002) reported additional POC flux data in Arctic shelf waters as part of a study conducted in the Barents Sea (Fig. 5.13). These workers also used a steady-state model to calculate POC fluxes based on measured ^{234}Th deficits in the upper 90–200 m of the water column and POC/^{234}Th ratios determined in suspended particles (>0.6 μm) and in material collected in drifting sediment traps. The range in POC export flux of 31–113 g C m^{-2} y^{-1} calculated using the ^{234}Th model was in good agreement with values of 32–97 g C m^{-2} y^{-1} determined using the drifting sediment traps. These results are also comparable to independent simulations of the average vertical export flux of POC in this region of 73 g C m^{-2} y^{-1} (Wassmann and Slagstad 1993). The model-derived POC fluxes were 2–16 times higher, however, when calculated using POC/^{234}Th ratios in suspended particles, which were an order of magnitude higher than trap POC/^{234}Th ratios. In addition, the collection efficiency of the sediment traps was determined to be 70–100% in this study, based on a comparison of ^{234}Th fluxes determined directly using the traps and particle export of ^{234}Th calculated from the water column ^{234}Th deficit.

^{234}Th has also been used to quantify POC export fluxes in the Northeast Water (NEW) and the North Water (NOW) polynyas (Fig. 5.13). Cochran et al. (1995) used the ^{234}Th deficit in the upper 50 m of the NEW polynya and the POC/^{234}Th ratio on suspended particles to calculate POC export fluxes of 53–197 g C m^{-2} y^{-1}, with average values reported to be 110 and 153 g C m^{-2} y^{-1} in 1992 and 1993, respectively. In the North Water (NOW) polynya, average ^{234}Th-derived POC export fluxes in the upper 100 m were estimated to be 120 g C m^{-2} y^{-1} in July 1998, decreasing to 38 g C m^{-2} y^{-1} in August 1999 (Amiel et al. 2002), largely due to a decrease in the POC/^{234}Th ratio in 1999. These POC fluxes were calculated using POC/^{234}Th ratios determined in both large particles (>70 μm), which had not previously been reported for the Arctic, and in material collected using floating sediment traps. The POC fluxes determined from water column ^{234}Th profiles were within a factor of 2–5 of values determined using floating sediment traps deployed in the upper 100 m.

5.4.4 Conclusions

Several field programs conducted in the Arctic Ocean over the past decade have provided new measurements of POC export fluxes based on the determination of $^{234}Th/^{238}U$ disequilibrium in the upper waters of the permanently ice-covered interior basins, the seasonally ice-covered shelf and slope waters, and in polynyas. These studies consistently indicate particle export from the upper few hundred meters of the water column occurring on a time-scale of days to months. $^{234}Th/^{238}U$ disequilibrium observed in the interior basin and shelf-slope environments has been attributed primarily to particle settling associated with variations in local primary and secondary productivity. Over shelves, resuspension of inorganic bottom sediments also contributes to $^{234}Th/^{238}U$ disequilibrium; as is the case with sediment traps, such processes may lead to over-estimates of the vertical POC export flux.

Taken as a whole, these recent Arctic studies indicate a marked regional variability in upper ocean POC export flux, ranging from 1.3–31 g C m^{-2} y^{-1} in the central Arctic Ocean to 3.1–197 g C m^{-2} y^{-1} over shelves and within polynyas. A key issue for future studies is to determine the extent of seasonal variability in POC export, which is likely to be significant when considering the high nutrient concentrations and strong seasonal gradients in light and ice-cover that characterize the Arctic Ocean. These observations provide the basis for what is likely to be a considerable increase in ^{234}Th-derived POC fluxes in the Arctic and adjacent subpolar waters in the future, as has been the case in the global ocean (Moran et al. 2003). Indeed, major studies are currently underway in the Chukchi-Beaufort Seas (SBI, CASES) and eastern Arctic basins (ODEN cruise).

5.5 Particulate Organic Carbon Flux to the Seafloor of the Arctic Ocean: Quantity, Seasonality and Processes

P. Wassmann

5.5.1 Seasonal and Annual Estimates of Vertical Carbon Export

Most investigations show that vertical export of POC varies greatly during the productive season, with significant episodes of increased flux. The large seasonal, site- and depth-specific variability of vertical POC export makes quantitative comparison between regions and ecosystems difficult. Vertical export at a given depth is both a function of the new production and suspended biomass in the upper layers as well as the vertical flux attenuation (see below and Fig. 5.14). As primary production shows large variations (a few g to several hundred g C m^{-2} y^{-1}) between regions (for example: the permanently ice-covered regions versus NOW, Sakshaug 2002) and between years at a particular site (e.g. NOW), short-term estimates of vertical export are difficult to interpret. Further, the retention of vertically exported carbon apparently differs markedly between sites and probably also between years (see 5.5.4). Consequently, vertical POC flux attenuation varies. Thus we focus first upon integrated, long-term estimates derived from bottom-tethered sediment traps because these rates are least difficult to interpret. Estimates of the uppermost traps have been summarised in Table 5.4 and those from the Th tracer in Fig. 5.13. Later, we briefly discuss the seasonal vertical POC flux variability in the Canadian Archipelago.

The vertical carbon export in most studies with long-term moorings of sediment traps is in general underestimated because DOC in supernatant water in the collecting cups is not regularly measured. Also the Th tracer technique may underestimate vertical DOC export. As described for the NEW and Greenland Sea, this could easily increase vertical carbon export estimates by 30% or more (Bauerfeind et al. 1994, 1997; Noji et al. 1999). There is reason to believe that this is the case in all investigations where only POC is analysed. The fate of DOC in marine ecosystems is a matter of debate. It could fuel the microbial food web and be re-injected as CO_2 into the atmosphere during winter mixing. Alternatively, it could be advected below surface waters or exported to depth by aggregate and deep-water formation.

The differences in vertical carbon export between the NOW and NEW polynyas suggests that extrapolation from one polynya to the other should be treated with utmost care. The average vertical carbon export in the NOW was more than 5 times greater than that in the NEW. However, the inter-annual variability in vertical carbon export inside a polynya should also be carefully considered. The

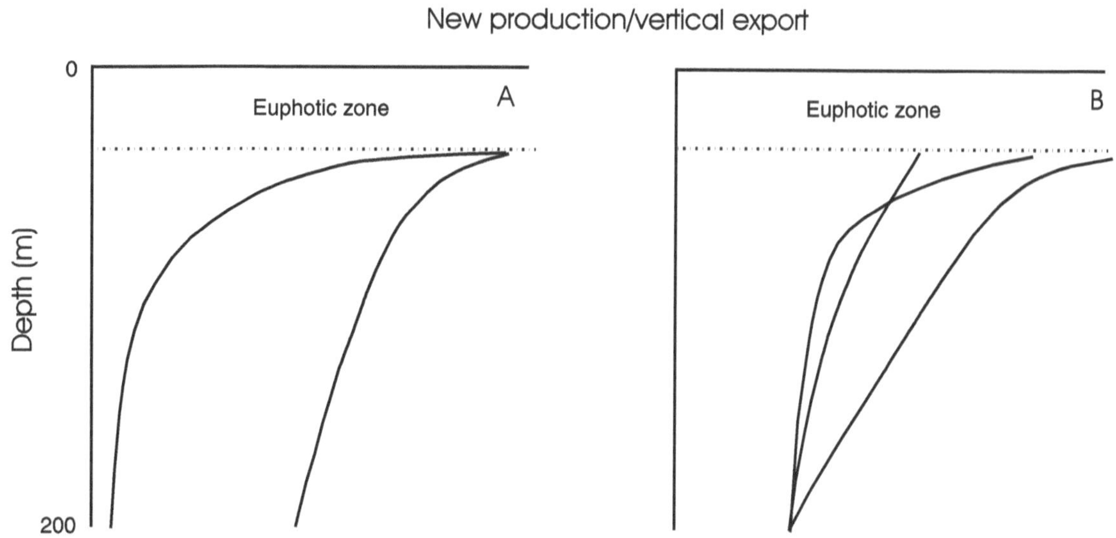

Fig. 5.14. Schematic presentation of the relationship between new production and vertical export in the upper 200 m, the twilight zone. A known new production and variable vertical carbon flux attenuation (a function of the pelagic food web recycling efficiency) will result in different vertical C export (A). From a known vertical C export at a certain depth (e.g. 200 m) the new production rate cannot be approximated unless the vertical flux attenuation is known (B)

intra-polynya variability in vertical POC export in NOW ranged between 1 to 8.4 g C m^{-2} y^{-1} (average 4.2 g C m^{-2} y^{-1} ± 77%) and 2.5 to 13.8 g C m^{-2} y^{-1} (average 8.7 g C m^{-2} y^{-1} ± 51%) for the 1997/1998 and 1998/1999 season, respectively. The interannual variability among the moorings S2, S4 and S5 at about 200 m depth was 2.9, 1.2 and 3.3 g C m^{-2} y^{-1}, respectively. Vertical carbon export was 76, 92 and 46% greater in 1997/1998 than 1998/1999, respectively. The specific production conditions and pelagic-benthic coupling scenarios of various polynyas for different years and within a single year are obviously highly variable. Without knowledge about this variability, any conclusions from comparisons between data series must remain obscure.

Also in arctic shelf regions vertical POC export can vary significantly over relatively short distances. The vertical POC export close to the Beaufort Sea shelf edge at 200 m depth and 50 m above the bottom was on average 4.4 ± 1.9 g m^{-2} y^{-1} (Table 5.4). Vertical export can thus vary over a comparatively short distance (70–71°N, 128–136°W) by about 50%. The reasons for these differences are not apparent as ice cover, currents, shelf topography etc. could be the cause. However, the average vertical POC export from the Beaufort Sea shelf is similar to that inside the NOW and even the variability inside appears to be similar in both regions.

Data from the Is Odden region the Greenland Sea with its marked variation in seasonal ice cover (Ramseier et al. 2001) also indicates a conspicuous interannual variability (Table 5.4). Vertical POC export at 500 m ranged between 1.4 to 3.8 g C m^{-2} y^{-1} with an average of 2.3 g C m^{-2} y^{-1}. The vertical POC export in the perpetually ice-free Greenland Sea was only half of this value although the trap was situated at 200 m depth. This clearly indicates that the MIZ favours accumulation of suspended biomass (stratification) and vertical export of biogenic matter (decoupling and aggregation). However, compared to the rather high primary production in the Greenland Sea (Rey et al. 2000), the vertical carbon export is astonishingly low. A similar picture is presented for the deep Arctic Ocean. Based upon phosphate deficits, Anderson et al. (2003) suggests extensive recycling of nutrients in the upper layer, resulting in an export of only 0.5 g C m^{-2} y^{-1}. Only a small fraction survives the high carbon retention in the upper layers. A comparison of the vertical C export in the Barents Sea MIZ [estimates from the model of Slagstad and Wassmann (1997), 75 m depth: 17 to 39 g C m^{-2}, March to October] with that of the Greenland Sea (Is Odden), suggests that vertical POC export in MIZs varies greatly between regions. We can only speculate why the difference is so large, but some variance may derive from the contrast between the pelagic food web structure of a wide and relatively shallow shelf compared to that of a deep, oceanic province. This suggestion is supported by the rather high vertical POC export at 38 m depth in the shallow Bering Strait where 43 g C m^{-2} was injected into the benthic boundary layer over just 96 days. For comparison, the vertical POC export in the Kara Sea was far smaller, despite the similarity in the setting. Further variance may derive from the different functioning between a MIZ where the main currents ran parallel (e.g. the western Greenland Sea) and one where perpendicular currents predominate (e.g. central Barents Sea). Although MIZs have an important role for the carbon flow in the Arctic, other differences between various MIZs, for example water column depth and the pelagic food web structure, also play a significant role for the vertical carbon export.

It could be expected that vertical carbon export under the tabular iceberg of the Canadian Ice Island is low. The estimates represent the lowest vertical POC export recorded so far (Table 5.4). The vertical carbon export in disparate ecosystems such as the NEW polynya, the central Greenland Sea and the Arctic Ocean off the Laptev Sea are quite similar (ca. 1–1.5 g C m^{-2} y^{-1}). This surprising observation suggests that carbon retention off the heavily ice-covered Laptev Sea must be modest compared to the central Greenland Sea. However, the magnitude of primary production in the Arctic Ocean is not well known (see Chapter 3) and Anderson et al. (2003) indicate that primary production (based on oxygen data) is at least an order of magnitude greater then the export production (based upon phosphate deficit). The connection between primary and export production in the Arctic Ocean region is obviously not easily established.

The large seasonal variability (1–3 months of duration) of vertical POC flux in the Canadian Archipelago (Table 5.3) illustrates how variable vertical flux can be in the Arctic. This complex investigation revealed that a suite of physical and biological processes influenced vertical POC flux: snow cover, ice-melt, ice algae versus phytoplankton growth, type of grazers, match versus mis-match of auto- and heterotrophs. The Arctic has obviously a significant potential for primary production, vertical export and pelagic-benthic coupling. Arctic ecosystems must be adapted to the tremendous interannual variability in productivity that derives from the climatologically driven changes of the physical forcing.

In summary, vertical export in the Arctic seas is highly variable. Similar to the vertical POC export estimates derived from sediment traps also the Th tracer technique (see Chapter 5.4) indicates substantial variability across the Polar Ocean (1–166 g C m^{-2}; Fig. 5.13). Hot spots with a mosaic-like distribution exist in regions adjacent to the Arctic Ocean. Shelves appear to be a precondition for high POC export. Similar ecosystems such as polynyas and MIZs can result in very different carbon export rates. Conversely, widely different ecosystems such as the ice-covered Arctic Ocean off the Laptev Sea, the permanently open central Greenland Sea and the NEW polynya can experience similar levels of carbon export. The interannual and site-specific differences in the Arctic are thus immense and extrapolations from one region to another or one year to another appear far more uncertain than in other regions of the World Ocean. No "normal" pattern of pelagic-benthic coupling in the arctic region can be derived from a few annual studies, let alone from sporadic, seasonal investigations. Future oceanographic studies in the arctic region should carefully consider the *episodic nature and mosaic-pattern* of production and pelagic-benthic coupling.

5.5.2 Ice, Light, Stratification and Vertical Carbon Export

The primary production and phytoplankton development in the Arctic is strongly influenced by light availability, stratification and grazing (see Chapter 3). In regions with weak stratification, a continuous supply of nutrients is available from below the euphotic zone due to mixing (Wassmann et al. 1999; Reigstad et al. 2002). This results in high primary (new) production (g C m^{-2}), but low concentrations of suspended biomass (g C m^{-3}). Vertical POC export is moderately high throughout the productive season in these regions. Such scenarios can be found occasionally in the NOW, and every year in the central Greenland Sea and the southern Barents Sea. Stratification in the Arctic Ocean and adjacent regions is caused either by sea ice melt or river runoff. Stratification in the MIZ supports conspicuous phytoplankton blooms, which results in large-scale export of biogenic matter from the euphotic zone, not necessarily to greater depth (e.g. >200 m). The bloom is transient due to rapidly developing nutrient limitation. Stratification in regions influenced by rivers often results in limited primary production due to high suspended loads which limit light availability on the inner arctic shelves (see 5.5.3). Autochthonous POC export is thus low. In addition, heavy zooplankton grazing can at times strongly reduce the phytoplankton standing stock, resulting in low chlorophyll/high nutrient scenarios (see 5.5.4).

The thickness and structure of the ice and snow cover, the density of ice algae and the amount of suspended matter determine the extent of light available for primary production (Sakshaug 2002). This is clearly illustrated by the investigations from the Canadian Archipelago (see 5.3.9, Fig. 5.11). Thus primary production is low under permanent ice cover, ice islands, shallow shelves (<50 m) and estuaries. Sea ice has thus three direct impacts on vertical carbon export in the Arctic. It regulates the light availability, it can result in stratification and it can contain particles which are released during ice-melt.

5.5.3 River Run-off, Resuspension and Vertical Carbon Export

The majority of previous and present investigations (n = 17) in the eastern Barents Sea, close to Novaya Zemlya, the Kara Sea and the Ob and Yenisei estuaries were carried out in relatively shallow regions (average bottom depth 55 ± 52 m). Due to heavy ice-cover access to this region with conventional vessels is possible only at the end of the productive season. 29 sediment traps have been deployed in September to October, at an average depth of 42 ± 40 m. On average, daily vertical flux of POC was about 25 mg C m^{-2} d^{-1}. Two measurements were >50 mg C m^{-2} d^{-1}. Excluding the two measurements (obvious signs of resuspension) reduced the average vertical POC flux to about 10 mg C m^{-2} d^{-1} which is about 1/3 of the minimum POC sedimentation in the central Barents Sea at the start of the productive period. The low vertical POC export in the eastern Barents and the Kara Seas reflects low productivity and the end of the productive period. It also suggests that resuspension and mineral loads of rivers and glaciers are both high in these mainly shallow regions. In the outer part of Russkaya Gavan Bay (Novaya Zemlya), in the Karskie Vorota Strait and in the outer parts of Ob and Yenisei estuaries total vertical fluxes were 2–3 orders of magnitude higher than those in the adjacent open sea. In contrast, in the central Barents Sea with its average depth of about 200 m, resuspension is of little significance. Depth and the vicinity of rivers and glaciers obviously play an important role for vertical POC export. Data from the vast, but shallow Siberian shelf with its high river discharge unfortunately do not exist. Yet, most probably the data from the eastern Barents Sea, the Kara Sea, Ob and Yenisei are characteristic for much of the Siberian shelf which represents about half of the shelves encircling the Arctic Ocean basin.

On the shallow arctic shelves autochthonous organic matter contributes relatively less to the vertical export of organic matter, and allochthonous matter can be a major source. As 10% of the global fresh-water run-off is discharged into the Arctic Ocean, a large part of it onto the Siberian shelf, the processes within and outside of estuaries have to be taken into account. The concept of the marginal filter, proposed by Lisitzin (1995), facilitates our understanding of vertical carbon export in these regions. These filters are reflected in the turbidity maximum zone of estuaries where fine suspended and dissolved matter flocculates and is injected into the benthic boundary layer. The removal of dissolved and suspended matter by the marginal filter, particularly in the inner part of estuaries, prevents a major part of the river load from being directly transported to the shelf edge and into the open ocean. The marginal filter appears to be of variable significance in different estuaries. For example, the residence time of freshwater in much shorter in the Yenisei than the Ob estuary (1 and 3 months, respectively (P. Stounzhas, pers. com.).

The marginal filter favours benthic organisms and accumulation on the shelf. A large share of terrigenous organic matter has to be resuspended and transported over the shelf before it can be deposited in deeper waters. Tidal and turbidity currents advect a mixture of terrestrial and marine matter into the Arctic Ocean basin. In regions such as the Laptev Sea where sea ice is formed in polynyas, the incorporation of the suspended load into the sea ice and its transport by the Transpolar Drift represents a significant transport mode of POC. This matter will eventually sink to the bottom along the Transpolar Drift when the ice melts (e.g. north of Spitsbergen, see 5.3.4). Most of this vertical export takes place in the Fram Strait region (Hebbeln and Wefer, 1991).

5.5.4 High Retention of Vertical Carbon Export in the Twilight Zone of the Arctic Ocean

The composition and biomass of phytoplankton is generally described as a function of physical forcing (light, nutrients and turbulence), grazing pressure, escape from grazing and competition (e.g. Margalef, 1978; Kiørboe, 1993; Wassmann, 1998). Low over-wintering populations, slow reproductive response of mesozooplankton and rapid phytoplankton growth on shelves, in some polynyas and the MIZ typically supports bloom. The deep Norwegian Sea, the northern Pacific Ocean and the Arctic Ocean, however, are important regions for over-wintering of meso- and macrozooplankton. Zooplankton is often advected from these regions onto shelves such as the Barents Sea, the Siberian shelf and the Chukchi Sea (Wassmann 2001). Signals of primary production that took place up to several years ago and *outside* the Arctic region in the form of meso- and macrozooplankton can be encountered. This life cycle strategy opens the door for match between auto- and heterotrophs throughout the Arctic. Compared to sub-polar and boreal regions, the probability of phytoplankton blooms decreases with the abundance of (multi-) annual generation of zooplankton. Several arctic zooplankton species adapted to the variable to unpredictable feeding regime imposed by the episodicity of climate and spread their life cycle over more than one year. They overwinter at depth and graze heavily and immediately on phytoplankton blooms when light permits adequate growth in the forthcoming years. Consequently, phytoplankton is often retained in arctic waters, and detritus and faecal pellets rather than fresh organic matter is injected into the benthic boundary layer.

Sedimentation of phytoplankton and ice algae is thus comparatively low in areas covered by multi-year ice. However, mosaic-like vertical export patterns for *Melosira arctica* have been suggested (see 5.3.2). Microscopic examination suggests that sedimentation of zooplankton faecal pellets is more important for vertical flux of biogenic matter than sinking algae in these regions. Faecal pellets produced by sub-ice fauna may also be a source for sinking faecal pellets (Werner 1997). Due to light-limited growth of algae, even low grazing rates may be sufficient to control phytoplankton biomass. Zooplankton and sub-ice fauna are thus important factors in regulating the vertical flux in areas covered with multi-year ice. They can easily produce high nutrients/low chlorophyll conditions. Low chlorophyll concentrations in the Arctic thus do not necessarily imply low primary production.

Except for the vigorous growth of phytoplankton in certain regions such as the NOW, the Bering Strait and the Barents Sea MIZ, the basic fate of newly produced biogenic matter in the Arctic is retention rather than export, both in absolute and relative terms. The lack of adequate investigations of the vertical export above the depth of 200–500 m, the twilight zone, where the majority of long-term sediment traps have been deployed, makes it difficult to evaluate the intensity of the vertical carbon flux as it relates to primary production. We face a lack of basic understanding on how vertical export of biogenic matter into the ocean's interior is regulated (Olli et al. 2001; Wassmann et al. 2003). With adequately vertically spaced sediment traps (e.g. those deployed in the Barents Sea, see Fig. 5.6), the vertical flux attenuation efficiency can be calculated (Wassmann et al. 2003). Unless the attenuation efficiency is known, a comparative evaluation of vertical carbon export estimates at a given depth remains obscure. A given export (new) production can result in variable vertical carbon export at a certain depth (Fig.5.14a). *Vice versa*, the carbon export at a certain depth can be the result of very different export (new) production rates (Fig. 5.14b). Data from the Greenland Sea (see 5.3.3), the NEW polynya (see 5.3.2) and the Barents Sea (Fig. 5.6) suggest that the vertical flux attenuation efficiency in the arctic regions is high. We suggest that zooplankton grazing plays an important role in constraining vertical carbon export and the pelagic-benthic coupling in these regions. Indeed, the biogenic transport or its retention by mesozooplankton in Arctic Ocean areas must be quantified to fully describe vertical fluxes of organic matter.

5.5.5 Global Warming and Vertical Carbon Export

The polar regions are predicted to incur some of the most pronounced effects of global climate change (IPCC 2001; Weller and Lange 1999). Global warming will both increase the air and sea temperatures in the region (resulting in increased stratification and decreased cooling of surface water). Increased AW advection, representing by far the largest inflow of water into the Arctic Ocean, will also take place and this influences the circulation pattern and long-term variability of the entire Arctic Ocean (Jones 2001). The effect of global warming is already revealed by the Barents Sea ice-cover that over an annual cycle varies hundreds of kilometres meridionally between the maximum and minimum extent. Considering the period 1920–1998, an overall ice extent shrinkage for April and August of about 12 and 40%, respectively, has been observed (Vinje 2001). This large difference in seasonal changes of ice extent is reflected in the net increase in the Spitsbergen seasonal temperatures over the corresponding period (winter +1, spring and summer +0.5, autumn 0.0°C; Hanssen-Bauer 2002). The large percentage difference in seasonal shrinkage indicates that global warming has a far higher impact on the disintegration than seasonal freezing rate. Connections between weather, climate, the state of the arctic ice cover and inflow of AW into the Barents Sea have been established (Grønås and Skeie 1999; Lopez et al. 2000, Ottersen et al. 2000). It is thus of fundamental importance to understand how annual, interannual and decadal variations in ice extent are related to climate forcing (e.g. NAO, AO) in the Polar Ocean.

Sea ice is obviously susceptible to changing long-term average temperatures. Preliminary forecasts estimate a reduction of the Polar ice pack of 20% during winter and 80% during summer by the end of this century (Johannessen et al. 2002). The dramatic northwards retreat of the MIZ will result in a profound widening of the MIZ and result in an extensive, stratified area that stretches from the shelves deeply into the Polar Ocean in decades to come. This situation would be a radical change from the present situation, where MIZs and polynyas support local ecosystem characteristics and processes. For example, the weak stratification characteristic for the central Barents Sea (Loeng et al. 1997) will spread northwards. Temperature will increase (Ottersen et al. 2000), Atlantic species will spread northwards and new production will probably triple (Reigstad et al. 2002). The accompanying changes in low-pressure pathways, frequency and strength will modify the annual patterns and magnitude of primary and export production (Sakshaug et al. 1995). The low productive, presently ice-covered waters fringing the Polar Ocean will turn into a stratified MIZ with nutrient depletion after a short, intense ice edge bloom. In particular, the shelf-break and adjacent fraction of the Polar Ocean basin that presently has quasi-permanent ice-cover will encounter substantial changes. Seasonally open water in this deep-water region may give rise to two scenarios. A large spring bloom off the shelf could give result in a concomitant decoupling of the pelagic and benthic systems, and rapid export of fresh particles to the deep ocean. Alternatively, the large spring bloom is matched by an adequate abundance of overwintering zooplankton that can match the bloom, with a decreased vertical export of detritus.

Global warming results in an "atlantification" of large regions of the contemporary MIZ band in the Atlantic sector of the Arctic and probably in increased export of ArW, fresh-water, ice and ice-bound detritus through the Fram Strait. In the Barents Sea, the term atlantification implies that the current MIZ will move northwards and that stratification will decrease strongly in the MIZ region. Consequently, in the future the MIZ will function like most of the AW does at present, i.e. it will be subject to significant vertical mixing.

A de-stratification of the MIZ results in significant changes in phytoplankton composition, bloom size, bloom development, vertical flux and vertical flux attenuation. More autochthonous carbon (increase in new production) and detritus (increase in grazing) will be injected into the deep ocean and BNL in the region of the present day MIZ. Global warming may also strengthen pelagic-benthic cou-

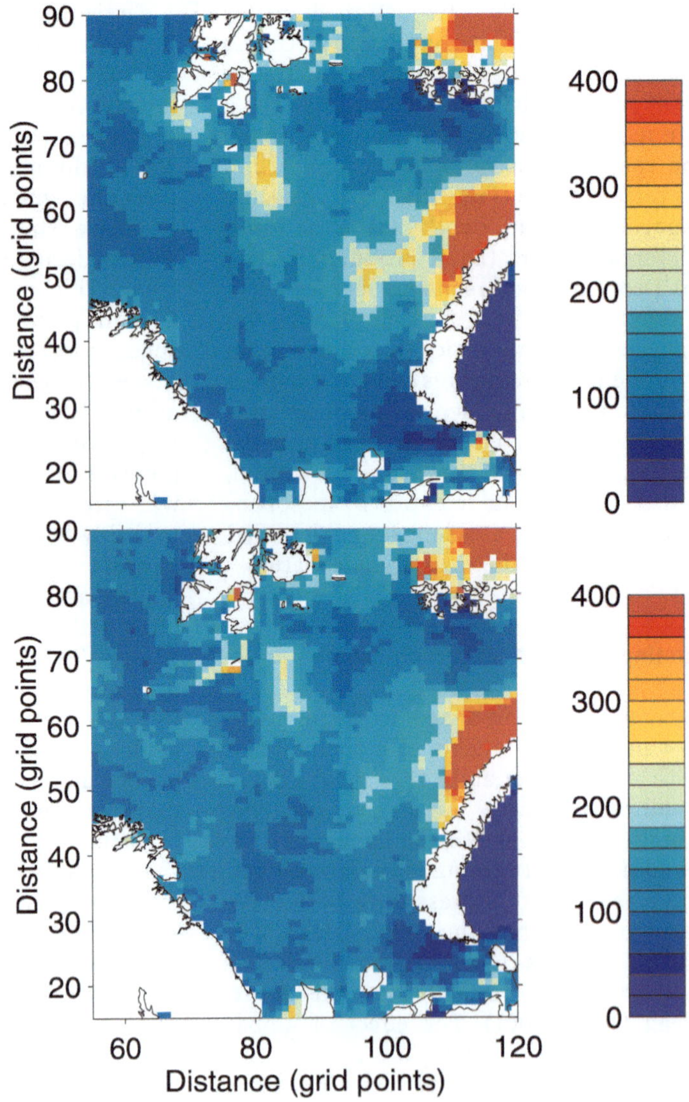

Fig. 5.15. Relative annual primary production (above) and vertical carbon flux at 75 m (below) in 1984 (warm year) in percent of 1981 (cold year). There are obvious increases in primary production and vertical carbon flux during warm years in the central Barents Sea, northwest of Novaya Zemlya and north of Franz Joseph Land, with some minor decreases in the southern and south-western part of the Barents Sea (redrawn from Slagstad and Wassmann 1997)

pling and increase the supply of fresh biogenic matter to the benthos in the periphery of water presently covered permanently by ice (which turns into a MIZ as a consequence of change). Global warming will thus result in increased primary production and vertical export of biogenic matter in this peripheral band of the contemporary permanent ice cover. In contrast, the low productivity, high retention and low biogenic export region which at present predominates the central Arctic Ocean basin will shrink.

The changes to be expected on the Siberian shelf are not self-evident. Freshwater supply may increase because of increased precipitation, but run-off will last longer or may become more continuous. Less ice-cover will increase primary production and vertical POC export. Katabatic winds during winter will have access to mix larger sea regions due to less ice cover and the incorporation of detritus into the sea-ice exported with the Transpolar Drift will increase. The consequences of climate warming will thus be of a complex nature on shelves with elevated river run-off.

The current and future lack of oceanographic data from a remote and logistically difficult region such as the Arctic Ocean represents a major challenge. We suggest that the only adequate method to address questions relating the role of the Arctic within the context of climate change is mathematical modelling based on available information on the meteorological, physical, chemical and biological variables in a region. However, few models exist which could be used to answer questions about the impact of global change on primary production and vertical export in arctic regions. In order to discuss global warming and vertical carbon export

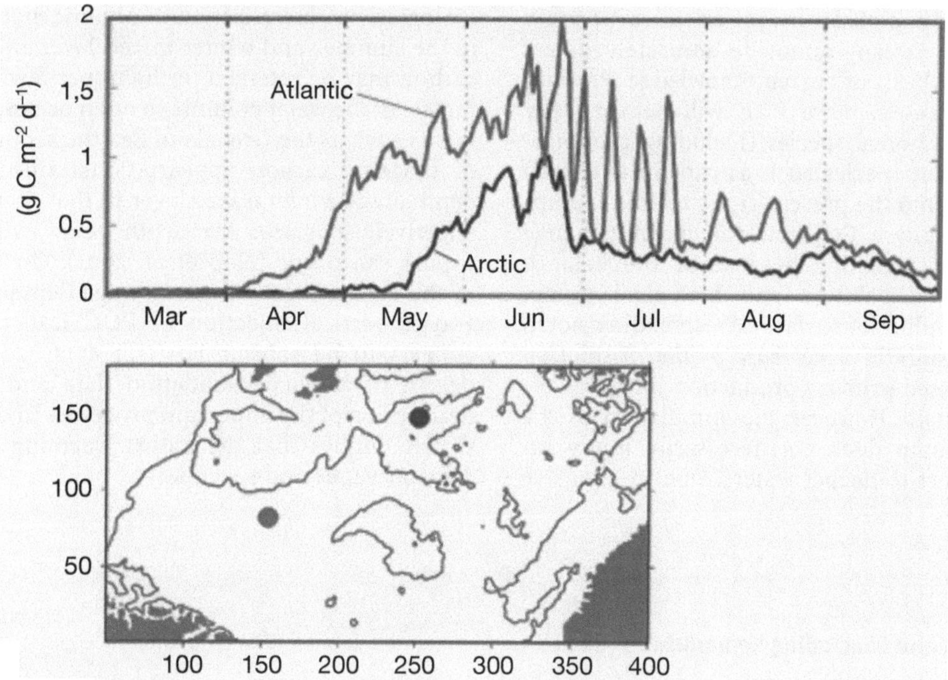

Fig. 5.16. Primary production (g C m^{-2} d^{-1}) at two different stations in the Barents Sea. The black station in the north is seasonally ice-covered and dominated by Arctic water (ArW). The gray station is permanently ice-free and dominated by Atlantic Water (AW). Primary production is higher and starts earlier at the station dominated by AW, but is lower and starts later at the ArW station (decreased light availability due to ice cover, reduced nutrient availability due to stratification). Please observe the 7 pulses of increased primary production in AW after the maximum in mid June. These are caused by low-pressure passages. The amplitude of the primary production decreases over time, most notably in August, when stratification increases and nutrient pumping is more restricted (D. Slagstad, unpubl. res.)

in more detail, we select a model example from the Barents Sea shelf (Slagstad and Wassmann, 1997). Physically-biologically coupled 3D model simulations from the Barents Sea, driven by meteorological hind-cast databases, suggest that inter-annual climatic differences are particularly reflected around the Svalbard Bank, along the Polar Front and the southern province of the MIZ (Fig. 5.15). Our best estimate of the effect of global warming on total primary and export production in the entire Barents Sea is an increase of 30–35 %, but in the MIZ interannual deviations of >200 % are encountered (Slagstad and Wassmann, 1997). This approximation, which is based on the difference between cold and warm year simulations, assumes that the ecosystem structure of the Barents Sea remains unchanged during global warming. However, there is reason to assume that changes in food web composition and ecosystem function may occur. Any attempt to predict the fate of biogenic matter in the region under altered ecological conditions such as the invasion of boreal species or changing life cycle strategies, makes such approximations non-trivial and speculative.

Ice cover and stratification determine in particular the dynamics of primary production, but a recent revision of the Barents Sea flux model (D. Slagstad, unpubl. res.) suggests that the passage of atmospheric low pressure systems may be of particular significance, primarily through nutrient pumping in weakly stratified waters (Fig. 5.16). Variations in the NAO index and the strength of the Icelandic Lows and Azores Highs will result in variable low pressure frequencies, strength, duration and pathways in the Barents Sea region. The relationship between NAO and the productivity in the non-to weakly stratified and most productive regions of the Barents Sea or any other region with similar oceanography is thus complex. Primary production and vertical C export seem to be a direct function of the local, seasonal and annual meteorology. Not only will the ice-cover (availability of PAR) co-vary with the NAO, but concomitantly, the access to nutrients will further affect the productivity in a non-linear manner. New and export production and vertical export of biogenic matter will vary accordingly.

The impact of global warming on the carbon flux in the Arctic Ocean cannot be accurately determined on the basis of current knowledge. Productivity is increased, more POC will be vertically exported and boreal species (including economically important species such as capelin and cod) may expand into the present arctic realm. The uptake of atmospheric CO_2, however, may not change during global warming because of the balance between the dissolution and biological pump. While less cooling of surface AW, spreading north and east, results in a *decrease of the dissolution pump*, increased primary production *increases the biological pump*. However, the stimulation of the biological pump does not necessarily imply increased export to deeper waters. Due to intensive biological remineralisation of organic matter within the summer and winter mixed layer, the bulk of carbon may be retained in the upper few hundred meters of the water column. In open ocean environments such as the Greenland Sea the sequestration of biogenic carbon appears constrained by the depth of the winter mixed layer, in that deep mixing effectively increases the depth below which true export can occur (Noji et al. 2001). On the shelf of the Barents Sea with an average depth of about 200 m, vertical injection of POC into the BNL, appears to be substantial (Olli et al. 2002). The dearth of adequate validation data and relevant description of the important processes in the Arctic Ocean implies that the global warming scenario remains vague and speculative.

5.6 Summary and Concluding Remarks

- Vertical export of biogenic matter is poorly resolved by measurements in the Arctic Ocean and derives primarily from the adjacent margins characterised by seasonal loss of ice cover.
- Vertical export of particulate organic carbon (POC) in the Arctic Ocean is highly variable (1.5–14 g C m^{-2} y^{-1} at roughly 200 m depths). Vertical POC export is frequently episodic depending upon a mosaic-pattern of production on the shelves fringing the Arctic Ocean and in polynyas. Similar ecosystems may support a wide range in carbon export whereas widely dissimilar ecosystems can support similar levels of carbon export.
- ^{234}Th derived POC fluxes, on the whole, are consistent with sediment trap observations. Overall, POC export fluxes are expected to vary by a factor of 2 and more due to variability in the POC/^{234}Th ratio and procedures used to estimate the ^{234}Th flux.
- Variations in sea-ice cover, light and stratification constrain the vertical export of biogenic matter, but top-down regulation through zooplankton results in extensive reduction in vertical POC export.
- The deeper regions of the Arctic Ocean are characterised by high retention of vertical carbon export in the twilight zone, but marginal ice zones (MIZ) can also show this tendency. Overwintering of large, long-lived zooplankton such as copepods (often advected from the North Atlantic) and appendicularians appear to play an important role in vertical flux attenuation.
- In the decades to come, the dramatic northwards retreat of the MIZ due to global warming will result in a profound widening of the MIZ represented by an extensive, stratified area that stretches from the shelves far into the Arctic Ocean. As a consequence, primary production and vertical export of biogenic matter will increase in the peripheral band of the contemporary permanent ice cover. In contrast, the low productivity, high retention and low biogenic export region, which at present dominates the central Arctic Ocean basin, will shrink.

Chapter 6

The Benthos of Arctic Seas and its Role for the Organic Carbon Cycle at the Seafloor

M. Klages, A. Boetius, J.P. Christensen, H. Deubel, D. Piepenburg, I. Schewe, T. Soltwedel

6.1 Introduction

The exploration of the Arctic Ocean has a long and multi-national history, but still we are only beginning to understand the role of the Arctic Ocean in the global carbon cycle. This chapter addresses the contribution of the Arctic benthos to organic carbon utilization, modification and sequestration. We review findings in benthic ecology and biogeochemistry of the Arctic Ocean from the past 15 years. Several older reviews on the Arctic benthos are available with a different focus: Zenkevitch's (1963) thorough volume covering the entire Eurasian Arctic benthos remains outstanding in its description of species composition, biomass and biogeographic aspects. Reviews by Curtis (1975), Dayton (1990), and Carey (1991) cover general aspects of the Arctic benthos as well as distribution patterns, productivity, and the feeding ecology of different species. Many recent studies have emphasized processes at the sediment-water boundary layer including respiration of sediment-inhabiting communities or single species (Piepenburg et al. 1995), calculation of carbon utilization and remineralization rates of benthic communities and their correlation to primary production (Boetius and Damm 1998), but also measuring biomass of benthic populations (Seiler 1999; Jørgensen et al. 1999; Deubel 2000). Recent discoveries of cold seeps north of the polar circle (Vogt et al. 1999) and of hydrothermal activity on the Gakkel Ridge (Thiede 2002) suggest a more diverse deep-sea benthos than previously known, but these on-going investigations cannot be reviewed herein. Advances in technology, such as reliable oxygen sensors, *in situ* micro-profilers and analysers, video-guided sampling gear, remotely operated vehicles (ROV), *in situ* lander technology, and the capability of maintaining Arctic marine invertebrates alive at low temperatures and high pressures for laboratory investigation, have allowed quantitative investigations even of the deep Arctic benthos. The availability of modern icebreaking research vessels for penetration of the high Arctic and international efforts to elucidate the role of the Arctic Ocean in the global carbon cycle have led to a greatly improved data base. However, the ability to sample at high latitudes, at great depths, or during the Arctic winter remains a challenge. To compensate for the resulting lack of important carbon cycle data such as year round primary and secondary productivity, particle export and deposition, and benthic remineralization, many investigations apply either empirical relationships from the literature or models (Cranston 1997; Seiler 1999; Deubel 2000; Klages et al. 2003). Nevertheless, large gaps remain in our current understanding of pelagic and benthic biological production of this polar sea.

In general, one of the major driving forces influencing the composition and activity of benthic assemblages is the carbon flux from the euphotic zone to the seafloor (see Chapter 5). Part of this carbon reaching the ocean floor is buried in the sediment and, therefore, sequestered from the atmosphere for millions of years. Another generally much larger part of the carbon export is remineralised by the benthos into CO_2, dissolved organic carbon (DOC) and nutrients, which are re-mixed into the hydrosphere. A third pool of carbon is dynamically stored in the biomass of the benthos. Currently it is not possible to constrain these pools of carbon and the exchange between them. The Joint Global Ocean Flux Study (JGOFS) achieved a rather comprehensive picture of the average annual exchange of CO_2 across the surface of the world ocean, but it remains open to what extent the Arctic Ocean acts as a sink or source in the global carbon cycle (Baliño et al. 2000). That the Arctic takes up large amounts of CO_2 is supported by data on the air-sea pCO_2 difference published by Takahashi et al. (1999). In the Arctic, as elsewhere in the ocean, biological processes drive seasonal and regional distributions of CO_2 fluxes. The gross primary production by ocean phytoplankton has been estimated by Bender et al. (1994) to be 103 Pg C y^{-1} (1 petagram (Pg) = 1 Gigaton = 1000 million tons). Part of this is returned to dissolved inorganic carbon (DIC) through autotrophic respiration, with the remainder being net primary production (NPP), estimated on the basis of global remote

Table 6.1. Organic carbon export relationships, as measured at 100 and 1000 m depth, for various oceanic regions of the world

Ocean region	Export at 100 m depth	Export at = 1000 m depths
Northern North Atlantic from 50–60°N[1]	–	1.9%
Northern North Atlantic from 50–80°N[2]	4.4% (at 500 m)	~2%
Southern Ocean[3]	30%	3%
Equatorial Pacific[3]	2–7%	1%
Arabian Sea[3]	20% during monsoons, otherwise 5–10%	1.7%

Data from [1] Jahnke (1996), [2] Schlüter et al. (2001), [3] Baliño et al. (2002).

sensing data to be about 45 Pg C y^{-1} (Longhurst et al. 1995; Antoine et al. 1996; Falkowski et al. 1998; Field et al. 1998; Balkanski et al. 1999). About 14 to 30% of the total NPP occurs in coastal areas (Gattuso et al. 1998), which are extensive in the Arctic. The resulting organic carbon is consumed by zooplankton (a quantitatively more important process than herbivory on land) or becomes detritus. Some organic carbon is released in dissolved form and oxidised by bacteria (Ducklow 1999) with a fraction entering the ocean reservoir as net DOC production (Hansell and Carlson 1998). Sinking of particulate organic carbon (POC) composed of dead organisms and detritus together with the vertical transfer of DOC create a downward flux of organic carbon from the upper ocean known as "export production". Recent estimates for global export production range from roughly 10 to 20 Pg C y^{-1} (Falkowski et al. 1998; Laws et al. 2000). A similar estimate for global export production of 11 Pg C y^{-1} has been derived using an inverse model of physical and chemical data from the world's oceans (Schlitzer 2000). Only a small fraction (about 0.1 Pg C) of the annual export production reaches the seafloor, mostly in the coastal areas (Gattuso et al. 1998). Heterotrophic respiration at depth converts the remaining organic carbon back to DIC. Eventually, this DIC is upwelled into the ocean's surface layer again where it may re-equilibrate with the atmospheric CO_2. These mechanisms, often referred to as the "biological pump", maintain higher DIC concentrations at depth and cause atmospheric CO_2 concentrations to be about 200 ppm lower than would be the case in the absence of such mechanisms (Sarmiento and Toggweiler 1984; Maier-Reimer et al. 1996).

For many areas of the world ocean, including the North Atlantic and the Arctic Ocean, the inverse relationship between water depth and the relative amount of organic matter that reaches the seafloor is documented and commonly accepted (Suess 1980, Hulth et al. 1994). Only a fraction of about 1 to 3 percent of the primary production reaches the seafloor at water depth below 1,000 m (Table 6.1; Baliño et al 2000; Schlüter et al. 2001). In a recursive approach, quantification of deep-sea fluxes may serve as a point of validation for conceptual and numerical models of carbon cycling in the ocean. Particularly in the Arctic Ocean with its strongly seasonal primary production regime, large and productive shelf seas and, in contrast, ice covered oligotrophic deep basins, data on fluxes and benthic standing stock may be used to test the reliability of production estimates. The enormous variability of primary production rates in the Arctic, rendering difficult a well constrained quantification of carbon cycling, is reflected, for example, by a huge range of sedimentation rates of particulate organic carbon which have been measured with sediment traps in different Arctic areas over the past 10 years (see Chapter 5). Minimum rates of only 2 to 3 mg POC m^{-2} d^{-1} have been reported from permanently ice-covered regions (Hargrave et al. 1993) suggesting low particle export. POC flux rates between 12 to 27 mg m^{-2} d^{-1} were measured in open waters indicating intermediate levels of export as reported by Honjo (1990), Bauerfeind et al. (1997) and Ramseier et al. (1997). High fluxes with more than 300 mg m^{-2} d^{-1} were found near ice edges (Hebbeln and Wefer 1991), being as high as in other regions of the world ocean. The dynamics of the vertical flux of biogenic matter through gravitational sinking from the euphotic layer to intermediate and deep water depths is described in detail by Wassmann et al. (Chapter 5).

6.2 Origin and Evolution of Arctic Habitats and Species

The origin and evolution of the Arctic biosphere was strongly influenced by tectonic processes and climate development (Menzies et al. 1973; McKenna 1980; Francis 1988). Continental plate movements at the end of the Cretaceous (80–100 million years ago) closed the previous deep-water connection between the North Pacific and primeval Arctic Ocean (Dunton 1992). The shallow Bering Strait became an effective bathymetric barrier to dispersion of, for example, bathyal Pacific-boreoarctic polychaetes, ostracods and other Pacific deep-water benthic fauna (Dunton 1992 and references cited herein). In the late Miocene the Arctic Ocean was a more temperate sea, ice-free and with open connections to both the Atlantic and Pacific Oceans. The appearance of Pacific molluscs in Iceland about 3.5 million years ago suggests that the Bering Land Bridge was breached in the late Pliocene, resulting in a migration of species through a still (at least seasonally) ice-free Arctic Ocean (Dunton 1992). This condition changed during the Quaternary, when the climate of the northern hemisphere was characterized by alternations of distinct glacial and interglacial cycles that resulted in the development of a permanently ice-covered central Arctic Ocean fringed by shelf seas with seasonally varying sea-ice cover (e.g., Thiede 1986; Dayton et al. 1994). The Quaternary was characterized by numerous glacial periods (Thiede et al. 2001) and warmer intervals of approximately equal duration of 10,000–20,000 years (Dunton 1992). The maximum extent of the last glaciation (Wisconsin or Weichselian) was reached about 20,000 Cal. yrs. BP. The changes between glacial and interglacial periods led to global sea-level variations of up to 120 m (Fairbanks 1989), resulting in the alternating loss and re-colonization of vast sublittoral habitats and impacting the diversity of benthic communities and the adaptability of benthic organisms. More recent results of sediment cores and seismo-acoustic data taken at the Lomonosov Ridge by Jakobsson et al. (2001) showed that substantial erosion caused by large ice sheets of Eurasian origin took place at the Lomonosov Ridge 1,000 m below sea level (mbsl) during the Saale glaciation (i.e. during Oxygen Isotope Stage 6 or about 140,000 Cal. yrs. BP). Comparable measurements indicating ice grounding at great water depth were recorded by Myhre et al. (1995) for the Yermak Plateau at 556 mbsl which happened 660,000 yrs. BP, and late Weichselian ice grounding was reported from the St. Anna Trough on the Kara Sea margin by Polyak et al. (1997) for 630 mbsl. It seems clear that such events have led to a massive destruction of the benthic fauna which must have escaped either to deeper waters or into areas where ice cover was not so heavy. Environmental variability has also acted on shorter time scales. The recolonization of Arctic shelves by benthic fauna and flora over the last 6,000 to 14,000 years was predominantly driven by North Atlantic emigrants with little species exchange from the North Pacific (Dunton 1992).

During the second half of the last century, significant changes in Arctic surface air temperatures, thickness of the sea ice, river discharge etc., all changes related to the North Atlantic Oscillation and Arctic Oscillation, were recorded (for details and references see Chapter 1.2). Such changes would have immediate effects on both Arctic mammals and sea-ice communities and most likely will influence both productivity and organic matter exports, hence impacting benthic communities.

6.3 Food Supply of the Arctic Benthos: Sources and Pathways

Nearly all benthic deep-sea ecosystems are heterotrophic and thus fuelled by the input of organic matter sinking either vertically down through the water column or being advected laterally from more productive regions to poorer ones (Lampitt 1985; Walsh et al. 1989, Schewe 2002). Exceptions are the chemosynthetic communities around hydrothermal vents and cold seeps, and the assemblages of microorganisms inhabiting the deep biosphere hundreds of meters below the seafloor. In the central portions of the major oceans, the vertical sinking flux of organic matter is thought to predominate over advective transport (Müller and Suess 1979; Emerson 1985; Jahnke 1989). Highest biomass and greatest productivity occurs in the euphotic zone and some fraction of this highly metabolizable material settles or is biologically transported out of this rich zone and downward through the water column (Betzer et al. 1984; Martin et al. 1987; Pace et al. 1987; Berger et al. 1988). Many deep-sea sediment trap studies have shown the strong seasonality in the sinking particle flux as driven by the overlying productivity cycle (Billett et al. 1983; Lampitt and Antia 1997; Deuser 1986; Deuser et al. 1981; see Chapter 5). However, if consumption of this incoming material by the deep-sea benthos is slow relative to the time-scale of variations in flux of the sinking material, then an approximate steady

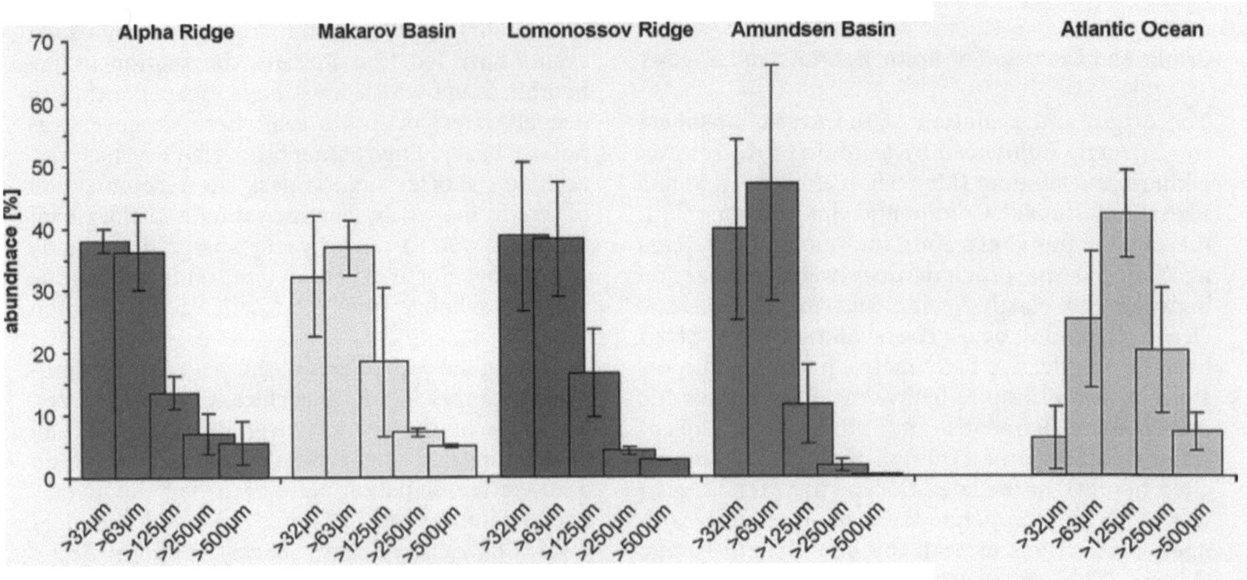

Fig. 6.1. Relative abundances of meiofauna organisms (including foraminifera) in different size-classes for various Arctic deep-sea regions as compared to the Porcupine Abyssal Plain. Data obtained at comparable water depths (from Schewe and Soltwedel 1999)

state may be considered to exist between the average populations within the sediments and the average annual euphotic zone productivity. Supporting this argument, a statistically significant relationship was found between deep-sea benthic oxygen consumption rates and annual overlying productivities (Boetius and Damm, 1998; Christensen 2000). Furthermore, the lack of dramatic seasonality in deep-sea sediment oxygen consumption rates at the Bermuda Atlantic Time-Series Site (BATS) was explained by the slow response of the benthos to seasonal pulses of sinking organic matter (Sayles et al. 1994). In contrast, if the benthos can respond quickly to an increase in the vertical rain of organic matter, then benthic responses would be expected to mirror the changes in productivity in the overlying water column; seasonality in the benthic rates would occur (Smith et al. 1987). Not all components of the benthos would be expected to respond on the same time-scales. For example, the arrival of sinking phytodetritus promoted a rapid increase in the biomass of bacteria and flagellates inhabiting surface sediments on time-scales of hours to days (Yayanos 1998). At a higher trophic level, foraminiferans and nematodes of the Northeast Atlantic Ocean show differences in their abundances which were found to be related to differences in phytodetritus input (Vanreusel et al. 1995, Smart and Gooday 1997). Also, opportunistic Arctic meiofauna organisms such as foraminiferans like *Epistominella arctica* increase in abundance soon after pulses of fresh phytodetritus (Schewe and Soltwedel 2003; see Fig. 6.3). Although the response time of deep-sea benthic communities to organic carbon fluxes is debated, an inter-annual variability in the quantity of vertically-settling utilizable organic matter reaching the sediments will likely result in concomitant variability in the benthic community and its metabolism rates (Billett et al. 1983; Rice et al. 1994; Christensen 2000). Little is known about the intra- and inter-annual seasonality in respiration of the Arctic benthos, but, on the basis of the interpretation of geochemical marker disequilibria in widely distributed basin box cores, Gobeil et al. (2001) suggest such variability on decadal or longer scales.

Primary productivity in the central Arctic is relatively low due to the reduction in light intensity by the pack-ice and its snow cover and due to the stratified and oligotrophic nutrient condition of the Arctic surface waters (Subba Rao and Platt 1984; Wallace et al. 1987; Anderson et al. 1990; and see Chapter 5.5.2). Direct measurements conducted during the Arctic Ocean Section cruise in 1994 and the Surface Heat Budget of the Arctic (SHEBA) experiment in 1997/98 as well as a reexamination of older oxygen data all suggest annual productivities of 15–30 g C m^{-2} y^{-1} (Wheeler et al. 1996; Wheeler et al. 1998; Pomeroy 1997). Actually, relatively high primary productivities for the interior ocean were first proposed by Macdonald et al. (1991; 1993) who used basin ages together with nitrate increase/oxygen decrease and a 1D model to estimate carbon flux –

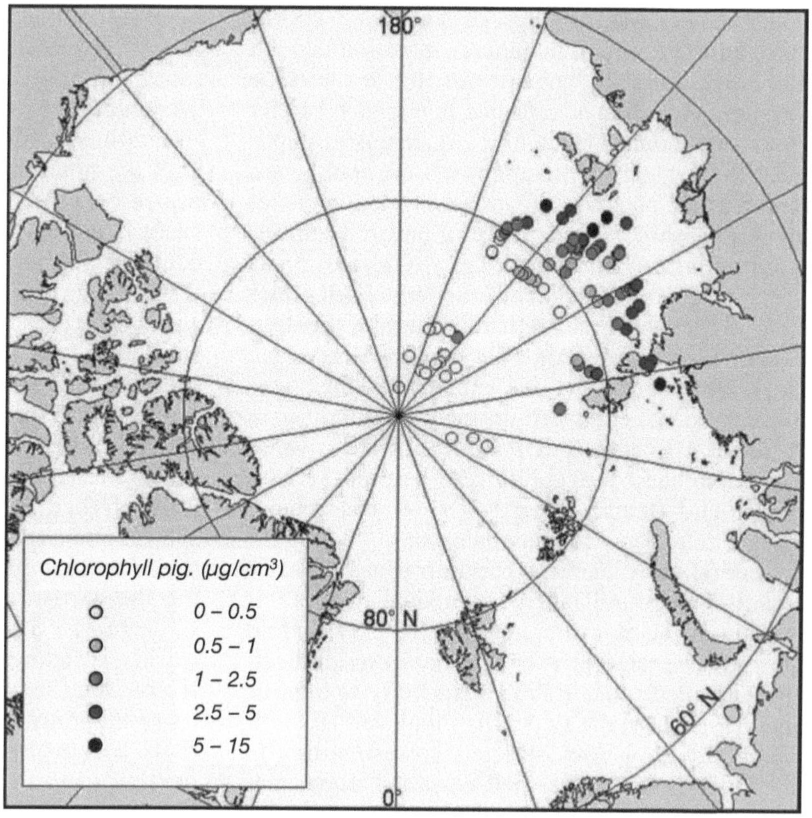

Fig. 6.2. Distribution of chlorophyll pigments in Arctic surface sediments (0–1 cm). Data extracted from PANGAEA world data center, graph kindly contributed by A. Schäfer (AWI)

they then linked this using the above cited Suess relationships to arrive at primary productivity. Hence, the carbon fluxes to the deep Arctic seafloor are expected to be lower than those found at comparable depths in other oceans (see Chapter 6.5 below). One strategy to overcome this energy limitation in deep-sea Arctic meiofauna organisms is the observed tendency towards dwarfness (see Fig. 6.1).

The few investigations of benthopelagic coupling in the Arctic Ocean showed a general relation between benthic metabolism rates and overlying primary productivity, similar to other areas of the world's ocean. Nevertheless, lateral transport of organic material may be locally significant. It is commonly accepted that among surface current regimes, the flow of water over the sea bed is an important factor influencing the distribution of sediments and food particles for benthic communities. Even the slowest near-bed current speed exceeds the fastest sinking velocities of plankton particles by at least one order of magnitude (Friedrich 1997). Flow conditions within the benthic boundary layer (BBL) not only determine particle settling and resuspension of bottom material but also influence exchange rates and processes taking place at the sediment-water interface (Thomsen et al. 2002). On a larger scale, Friedrich (1997) argues that these interactions between water flow and benthic organisms are important to the understanding of the food web, remineralization processes and organic matter cycles in the sea. Strong currents are not restricted to surface waters but may be found transiently or permanently at any depth especially near slopes or topographic features. For example, recent research from different sites along continental margin gave evidence for substantial downslope transport of organic matter to the benthic deep-sea system through canyon systems (Vetter and Dayton 1998; Rogers et al. 2002). Such investigations have not yet been carried out at the continental margin zones of the Arctic Ocean.

Limited sun light and nutrient availability, the permanent ice cover and temperatures close to the freezing point are factors reducing total productivity of the Arctic Ocean. Hence it can be assumed that the central Arctic is one of the least favorable habitats for benthic life on earth. In Figure 6.2 data on chlorophyll pigments of the upper first cm in Arctic surface sediments is given. The significance of organic matter transport from the shelves across ocean margins to the deep sea has been recognized by several oceanographic programs of the last decade (SEEP I and II: Walsh et al. 1988; Biscaye et al. 1994; ECOMARGE: Buscail et al. 1990; ACSAR: Blake and Diaz 1994). In general, it was concluded that lateral input of biogenic detritus at the slope was high enough to be reflected in enhanced activity and abundance of benthic organisms at depth (Anderson et al. 1994; Kemp 1994; Rowe et al. 1994). Hence, a key objective of several investigations of the relationship between small-organism biomass, activity and carbon turnover in Arctic sediments was to evaluate the impact of the large shelf zones on food availability at Arctic margins and the deep sea (Boetius and Damm 1998; Vanaverbeke et al. 1997). Other investigations aimed to identify the contribution of enhanced primary productivity in the cracks, fissures and polynyas of ice covered regions to food availability and benthic activity (Vetter and Deming 1994; Soltwedel and Schewe 1998; Schewe 2001, Cooper et al. 2002).

Generally, sedimentary concentrations of labile organic compounds such as particulate proteins and amino acids decrease approximately tenfold from Arctic shelves at 50 m to the deep sea, indicating a strong decline in food availability to benthic organisms down the slope. Distinct levels of food-availability have been detected, corresponding to the depth zones of the shelf edge, the upper and lower slope as well as the distant abyssal plain. Also, ice-free regions of Arctic margins are generally richer in labile organic compounds than ice-covered regions. Pigment concentrations in the sediments are highly correlated with protein and DFAA concentrations, indicating a strong relatioship between the supply with phytodetrital material and the availability of labile organic food sources to benthic organisms. Changes in the nutritious quality of organic matter down the continental slope have been estimated as the ratio of protein-carbon divided by total organic carbon (TOC). This ratio decreased in the surface sediments from around 10% at the shelf edge to 1% at the bottom of the slope (Boetius and Damm 1998).

Because the central Arctic is thought to be extremely oligotrophic, sites and regions of enhanced productivity could be important. Marginal ice zones (MIZ) and polynyas were found to be highly productive regions of ice covered polar oceans (Smith and Gordon 1997). Satellite-derived phytoplankton pigment distributions (Antoine et al. 1996; Behrenfeld and Falkowski 1997) indicate an increased productivity in the MIZ and at frontal systems in the Fram Strait. Since the MIZ is a seasonally moving feature, retreating northward as the pack ice melts in spring and summer, the impact on the underlying sediments should be transient. Schewe and Soltwedel (2003) found that in the Fram Strait, the ratio of bacterial biomass to total microbial biomass varied in relation to the position of the ice-edge, indicating a rapid response of bacteria to alterations in food availability.

Continental shelves may also be important sites with the potential for strong lateral exchanges. In this respect the Arctic is unique in that the central basin is nearly landlocked, surrounded with large continental shelves, and several of these shelves receive significant outflows from several major rivers. Whether these large shelves export significant quantities of utilizable organic carbon to the central Arctic is not well known. By examining the proportion of productivity that reached sediments of mid-latitude continental shelves, Christensen (1989) showed that wide continental shelves export little of their overall productivity while narrow shelves may export much and perhaps most of their productivity. The major Arctic shelves, including the Barents Sea, the Russian shelves, and the Chukchi Sea are the widest shelf regions in the world, exceeding 800 km from coast to shelf-edge. If large shelf regions trap nearly all their productivity so that only a few percent of their total productivity is exported, then the absolute magnitude of possible export to the central basins may be small. For the whole Arctic Ocean, Stein and Macdonald (see Chapter 8) estimated a total shelf – open ocean export of utilizable (marine) organic carbon of 14×10^6 t y^{-1} using an export rate of 5% of the primary production as given by Liu et al. (2000) for the global ocean.

For the Kara Sea Lisitzin (1995) used the term "marginal filter" describing the area at about 74°N where brackish water shifts towards fully marine, and deposition of up to 90–95% of suspended matter occurs south of this border. However, offshore export may be enhanced by additional transport mechanisms occurring in the Arctic. About 10% of the global river run-off enters the Arctic marine environment, thus leading to a strong coupling between the terrestrial and marine environment, especially in the estuaries and shallow shelves. Various national and international research programs have addressed the fate of the fluvial matter entering the shelf seas on its way to the central Arctic Ocean (for example the German interdisciplinary research project "The northern North Atlantic: A Changing Environment" (Schäfer et al. 2001), and the bilateral Russian-German programmes "System Laptev Sea" (Kassens et al. 1999) and "Siberian River Run-Off" (Stein et al. 2003b). The freshwater inflows are sufficiently large to maintain the overall estuarine circulation of the Arctic interior ocean and, consequently, would be expected to contribute to the export of organic material from the shelves

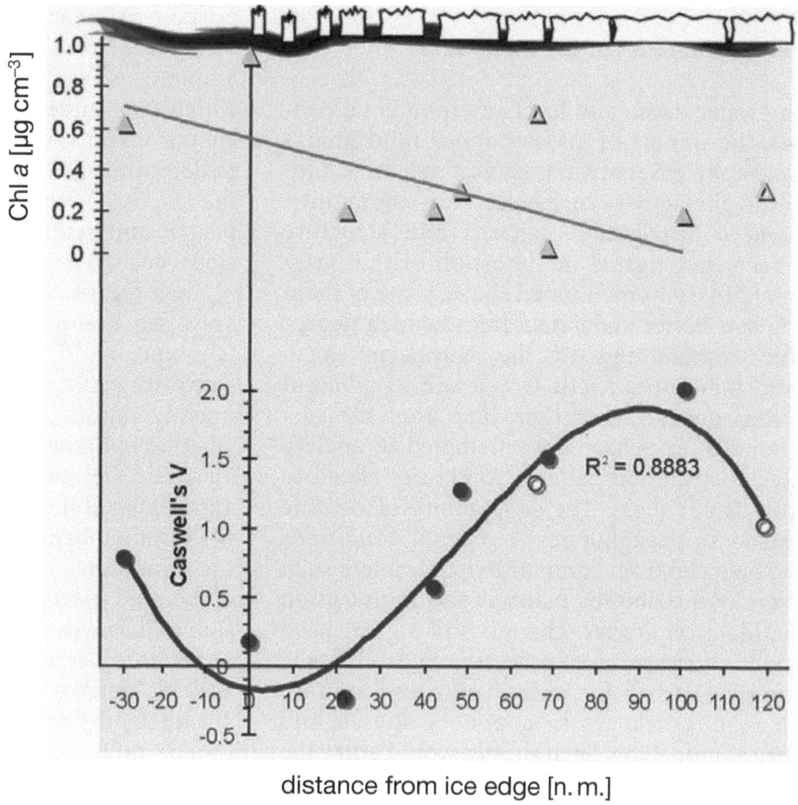

Fig. 6.3. Trends in concentrations of sediment-bound 'fresh' plant pigments (Chl *a*) and foraminiferal diversity (Caswell's V calculated here), related to the ice-coverage along a transect running orthogonally to the observed ice edge. Blank triangles and dots show results from investigations in 1997 as described in Soltwedel et al. (2000). (From Schewe and Soltwedel 2003).

to the central Arctic. Sporadic downslope turbidity flows may also move shelf and upper slope surface sediments into the central Arctic sediment column, as indicated by the frequent presence of turbidite layers in the central Arctic sediment column (Clark 1969; Stein et al. 1994a; Grantz et al. 1996; Cranston 1997; see Chapter 7.9).

Another mode of transport that may enhance the offshore export of organic material is sea-ice formation and transport (see Chapter 2.4). Ice formation in shallow water coastal environments may entrain large amounts of surface sediments and flocculant material. If this sea-ice is captured within the trans-Arctic drift, this coastal material may be released throughout the central Arctic, the Fram Strait and adjacent northern European Seas (Pfirman et al. 1989a, b; Reimnitz et al. 1993). In contrast, aeolian transport of organic material is thought to be of minor importance in the Arctic (Darby et al. 1989; see Chapters 2.5 and 8). Most deep-sea sediments from central Arctic basins contain a relatively large proportion of terrestrial material (Stein et al. 1994b; Fahl and Stein 1999; Belicka et al. 2002; see also Chapter 7.9), demonstrating the significant role of lateral transport mechanisms. However, the organic tracers employed to affirm terrestrial content represent those compounds which are highly resistant to biological utilization and biochemical destruction (Stein et al. 1994b). Because terrestrial and riverine material has already undergone significant biological degradation during its passage through the coastal seas, these materials may track processes involved in long-term burial rather than those involved in the supply of usable organic matter to active benthic communities. Results obtained from north of 85°N in permanent ice-covered regions of the central Arctic Ocean illustrate the difficulties in establishing the role of lateral transport (Soltwedel and Schewe 1998). In this study, a slight northward decrease in sediment-bound phytodetrital organic matter (chloroplastic pigments) was observed along the Lomonosov Ridge crest, suggesting northward transport from the shelf edge. However, both the enzymatic activity and biomass of bacteria did not follow the trend of northward decreasing values, confirming that the central part of the Arctic Ocean is largely decoupled from the primary production at the ice edge and on the Arctic shelves. Consequently, although there is some evidence of lateral transport of terrestrial, coastal, and shelf material to the central Arctic basins, the flow of utilizable carbon which feeds the benthic communities may remain highly dependent on the vertical sinking flux of utilizable organics from overlying productivity.

6.4 Benthic Communities of the Arctic Seas

Among water depth and local environmental conditions, the import of allochthonous food affects biogeography, diversity, population densities, biomass and the activity of the seafloor community. Zoobenthic distribution patterns and structure have been investigated in almost all of the international programs mentioned above, some of them aiming at a better understanding to which degree benthic communities of the Northern North Atlantic, the central Arctic Ocean and its adjacent shelf seas depend on carbon flux from the surface. Some studies have also attempted to analyse whether distribution patterns might be related to environmental shifts. The adaptability of benthic organisms to changing environmental conditions can be deduced from community parameters such as diversity, distribution patterns and composition of benthic assemblages. There is still a great need for quantitative studies of processes controlling the overturn of carbon in Arctic plants and animals which require field work, laboratory studies and numerical modelling. Such studies would allow the determination of reaction rates and responses of benthic organisms to natural and anthropogenic disturbances. As outlined in Chapter 1.2, there is evidence that temperature and climate of the Arctic Ocean changes, but the debate continues regarding the relative importance of natural variability (Macdonald et al. 1999; Vanegas and Mysak 2000) compared to trends produced, for example, by global warming from greenhouse gas emissions (Johannessen et al. 1996; Shindell et al. 1999). In this context, a fundamental understanding of carbon production, recycling, and preservation is crucial for the construction of an accurate carbon budget of the entire Arctic Ocean. The construction of such a budget is a prerequisite for detecting climatic changes through the response of the carbon cycle in the Arctic (Belicka et al. 2002).

6.4.1 Arctic Shelves and Margins

Data accumulated during the past 15 years on almost all aspects of benthic ecology in the Arctic Ocean (Table 6.2) shows a large variation in benthic communities of the Arctic shelves, margins and basins. This variation may result from differences in food supply, mean depth, ice cover, biogeographical history, the impact of river runoff and sediment composition. Over the shelves, runoff from rivers affects not only sea-ice formation, ocean circulation, and surface productivity, but is also an important influence on benthic assemblages. Furthermore, rivers and coastal erosion supply large amounts of suspended material (see Chapter 2) which then undergoes various biological and geochemical transformations. Clearly, it is important to determine how much suspended material enters the Arctic Ocean, its fate and its relevance for pelagic and benthic communities regionally (shelf, slope, basins) and seasonally.

Shelf regions with fine sand and mud sediments are often dominated by bivalves and polychaetes (Grebmeier and Cooper 1995, Deubel et al. 2003). In contrast, gammaridean amphipods are more frequently found in coarse-grained, sandy regimes whereas ophiuorids seem to be abundant at almost all Arctic shelves (Grebmeier et al. 1995, Piepenburg 2000). One of the most conspicuous characteristics of Arctic epibenthic communities in shelf seas is the dominance of brittle stars compared to other epibenthic species. Analysis of brittle star distribution patterns shows a non-random dispersion on a 1-100 m scale, which is most likely explained by seabed heterogeneity. At larger spatial scales, a strong depth zonation was reported by Piepenburg (2000) reflected by a pronounced decrease with water depth of two and three orders of magnitude in biomass and abundance, respectively.

Below, we provide brief descriptions of the different Arctic shelf areas, followed by a summary of current knowledge on the benthic communities inhabiting the various regions.

Northern Greenland Sea

The continental shelf of East Greenland is narrow (<100 km in off-shore extension) and rugged. Shallow banks with water depths of less than 40 m down to 150 m and troughs with water depths greater than 500 m are major features of the seabed morphology off NE Greenland between 78°N and 82°N (Piepenburg 2000).

Hydrography is strongly affected by the cold East Greenland Current which transports polar surface water and mainly multi-year sea-ice from the Arctic Ocean to the south (Coachman and Aagaard 1974, Quadfasel et al. 1987). At water depths < 100 m, temperatures are < 0 °C and salinities < 34.4. At greater depths, values are slightly higher (0.5 to 1.0 °C and 34.4 to 34.9, respectively), implying an inflow of Atlantic water masses to the shelf troughs (Budéus and Schneider 1995, Budéus et al. 1997). In general, the waters are ice-covered throughout the year (Wadhams 1981) and belong to the "permanent pack ice zone" (Hempel 1985). However, a polynya called Northeast Water (NEW) (see Chapter 5.3.2) is a recurrent annual feature south of Nordostrundingen (Schneider and Budéus

Table 6.2. Bacterial and animal densities (individuals m^{-2}) and biomass (g wet weight m^{-2}) for several benthic components from different shelf and deep sea areas of the Arctic Ocean. Bacterial results represent that within the upper 1 cm of sediments. Some macrofaunal biomass values were reported in mg of wet weight per 0.02 m^{-2} and converted to g m^{-2}. Others, where data were given in ash free dry weight for example, were not converted

Organisms and region	Range of data	Water depth (m)	Study	Comments
Bacterial numbers	**[10^8 cells cm^{-3}]**			
Central Arctic	1.54–4.85	1270–3170	Schewe (2001)	
	3.61–9.53	864–4187	Soltwedel and Schewe (1998)	
	0.36–0.88	1388–4427	Kröncke et al. (1994)	
Yermak Plateau	2.98–15.63	481–4268	Soltwedel et al. (2000)	
Svalbard	2–40	115–329	Sahm and Berninger (1998)	fjords
Barents Sea	3.03	258	Kröncke et al. (1994)	
Laptev Sea	3.5–37.1	37–3427	Boetius and Damm (1998)	
	0.1–31.5	39–3596	Grahl (1996)	
Bacterial biomass	**[µg C cm^{-3}]**			
Central Arctic	2.25–5.25	1388–4427	Kröncke et al. (1994)	
	5.86–14.64	1270–3170	Schewe (2001)	
	6.32–16.68	864–4187	Soltwedel and Schewe (1998)	
Yermak Plateau	9.09–32.16	481–4268	Soltwedel et al. (2000)	
Barents Sea	7.8	258	Kröncke et al. (1994)	
Laptev Sea	7–85	37–3427	Boetius and Damm (1998)	
	19–58	39–3596	Grahl (1996)	
Foraminifera numbers	**[ind 10 cm^{-2}]**			
Central Arctic	43–203	1270–3170	Schewe (2001)	
	60–284	864–4187	Schewe and Soltwedel (1999)	
	0–13	540–4190	Clough et al. (1997)	Foraminifera >250 µm
	1.4–36.9	1388–4427	Kröncke et al. (2000)	
	0–152.7	1051–4411	Wollenburg 1995	
NE Greenland	48–102	285–330	Newton and Rowe (1995)	Polynya; Foraminifera >150 µm
Yermak Plateau	220–2254	481–4268	Soltwedel et al. (2000)	Foraminifera >63 µm
	3.6–19.8	562–2054	Wollenburg (1995)	
Barents Sea	70–14700	240–3920	Pfannkuche and Thiel (1987)	
	35.2	258	Kröncke et al. (2000)	Foraminiferans >63 µm
	0.74–33.75	197–2994	Wollenburg (1995)	Foraminiferans >63 µm
Laptev Sea	13–4419	37–3425	Wollenburg and Kuhnt (2000)	Foraminiferans >63 µm
Foraminifera biomass	**[µg C 10 cm^{-2}]**			
Central Arctic	0–6.7	540–4190	Clough et al. (1997)	Foraminifera >250 µm
	0.088–0.4 g WW m^{-2}	1388–4427	Kröncke et al. (2000)	

Table 6.2 (continued)

Organisms and region	Range of data	Water depth (m)	Study	Comments
Metazoan meiofauna numbers	**[ind 10 cm^{-2}]**			
Central Arctic	29–94	1270–3170	Schewe (2001)	
	16–101	864–4187	Schewe and Soltwedel (1999)	
	28–254	1000–4273	Vanreusel et al. (2000)	
	1.9–154.0	1388–4427	Kröncke et al. (2000)	
NE Greenland	75–151	285–330	Newton and Rowe (1995)	Polynya
				Nematodes >63 µm ?
Morris Jesup Rise	120	1072	Vanreusel et al. (2000)	
Yermak Plateau	95–1902	481–4268	Soltwedel et al. (2000)	
Svalbard	0.8–669	10–95	Wlodarska-Kowalczuk et al. (1999)	Glacial bay
Barents Sea	348–4339	226–3920	Pfannkuche and Thiel (1987)	
	237	258	Kröncke et al. (2000)	
Laptev Sea	418–2683	65–3237	Vanaverbeke et al. (1997)	
Chukchi Sea/Bering Sea	639	540	Vanreusel et al. (2000)	
Metazoan meiofauna biomass	**[µg C 10 cm^{-2}]**			
Central Arctic	0.2–2.1	1270–3170	Schewe (2001)	only Nematoda considered
	0.3–24	1000–4273	Vanreusel et al. (2000)	only Nematoda considered
	0.27–28.23	1388–4427	Kröncke et al. (2000)	
Morris Jesup Rise	4.5	1072	Vanreusel et al. (2000)	
Yermak Plateau	10–92	481–4268	Soltwedel et al. (2000)	nematodes
Svalbard	0.2–238 [µg dw 10 cm^{-2}]	10–95	Wlodarska-Kowalczuk et al. (1999)	Glacial bay
Barents Sea	0.35–2.38	226–3920	Pfannkuche and Thiel (1987)	
	[mg AFDW 10 cm^{-2}]			
	132.3	258	Kröncke et al. (2000)	
Chukchi Sea/ Bering Sea	25.5	540	Vanreusel et al. (2000)	
Macrofauna numbers	**[ind m^{-2}]**			
Central Arctic	44–484	3743–4411	Kröncke (1998)	
	50–950	1018–4478	Kröncke (1994); Kröncke et al. (2000)	
	4–1636	780–4170	Deubel (2000)	
	95–6772	540–4190	Clough et al. (1997)	fauna >250 µm
	5–648	1000–2599	Paul and Menzies (1974)	
NE Greenland	800–8128	188–1809	Brandt and Schnack (1999)	
	1223–9815	125–515	Ambrose and Renaud (1995)	
	432–11205	183–1970	Schnack (1998)	
	424–2752	353–340	Seiler (1999)	

Table 6.2 (continued)

Organisms and region	Range of data	Water depth (m)	Study	Comments
Macrofauna numbers				
Morris Jesup Rise	132–1364	1072–3838	Kröncke (1998)	
Yermak Plateau	1056–2640	560–2531	Kröncke (1998)	
	88–1848	647–4190	Rachor and Denisenko (1997)	
Svalbard	230–7240	300–2977	Weslawski et al. unpubl.	AWI "Hausgarten"
	178–4133	30–80	Wlodarska-Kowalczuk et al. (1998)	Glacial bays of Spitzbergen
	1194–2111	10–95	Wlodarska-Kowalczuk et al. (1999)	
Barents Sea	200–5200	226–3920	Pfannkuche and Thiel (1987)	
	1276–4180	150–325	Rachor and Denisenko (1997)	
	4830	258	Kröncke et al. (2000)	
Kara Sea	112–10270	10–70	Deubel et al. (2003)	
	356–5155	17–195	Jørgensen et al. (1999)	ind 0.5 m^{-2}
Laptev Sea	84–9128	40–3344	Deubel (2000)	
	434–1545		Gukov et al. (2001)	
Chukchi Sea/ Bering Sea	330–4414	20–103	Stoker (1973)	
	270–4330	23–1700	Carey et al. (1974)	
	140–14365	19–54	Grebmeier et al. (1988)	
Macrofauna biomass	**[g WW m^{-2}]**			
Central Arctic	0.025–0.71	3743–4411	Kröncke (1998)	
	0.050–2.05	1018–4478	Kröncke (1994)	
	0.2–141	780–4170	Deubel (2000)	mg C m^{-2}
	0–270	540 – 4190	Clough et al. (1997)	mg C m^{-2}; fauna >250 µm
	0–82.7	1388–4427	Kröncke et al. (2000)	
	1–803	1000–2599	Paul and Menzies (1974)	
	0.5–1	–	George (1977) (Alpha Cordillera)	
	0.05–3.6	<100	Wacasey (1975)	(some of the references cited from
	0.2–13.5	21–2600	Carey and Ruff (1977)	AMAP Assessment Report (1998))
NE Greenland	319–9586	125–515	Ambrose and Renaud (1995)	mg AFDW m^{-2} only polychaetes
	30–756	183–1970 m	Schnack (1998)	mg C m^{-2}
	154–383	353–3400	Seiler (1999)	mg AFDW m^{-2}
Morris Jesup Rise	0.01–3.45	1072–3838	Kröncke (1998)	
Yermak Plateau	≤5.83	560–2531	Kröncke (1998)	mg WW 0.02 m^{-2}
Svalbard	4–310	30–80	Wlodarska-Kowalczuk et al. (1998)	Glacial bays of Spitzbergen
Barents Sea	0.36–25.59	226–3920	Pfannkuche and Thiel (1987)	mg AFDW 50cm^{-2}
	90.5	258	Kröncke et al. (2000)	
	13–104	80–240	Piepenburg et al. (1995)	(the latter cited from AMAP
	48–311	<100	Idelson (1934)	Assessment Report (1998))

Table 6.2 (continued)

Organisms and region	Range of data	Water depth (m)	Study	Comments
Macrofauna biomass				
Kara Sea	≤778	10–68	Klages et al. (2003)	(cited from AMAP Assessment Report (1998))
	123	< 100	Filatowa and Zenkevitch 1957	
Laptev Sea	4–2750	40–3344	Deubel (2000)	mg C m^{-2}
	16.51–77.66	22	Gukov et al. (2001)	g WW m^{-2}
Chukchi Sea/Bering Sea	1–59		Grebmeier et al. (1989)	g C m^{-2}
	17–2377		Grebmeier et al. (1988)	
	23–786	20–103	Stoker (1973)	(cited from AMAP Assessment Report (1998))
	1000–19600	18–51	Feder et al. (1994)	mg C m^{-2}
Megafauna numbers	**[ind m^{-2}]**			
NE Greenland	4–137	32–376	Starmans et al. (1999)	Key taxa: Arcturus, Ophiacantha
	32–524	80–360	Piepenburg and Schmid (1996)	NEW Polynya, only ophiuroids
Barents Sea	3–694	80–360	Piepenburg et al. (1995)	only ophiuroids
Laptev Sea	0–579	14–45	Piepenburg (2000)	only ophiuroids
Chukchi Sea/ Bering Sea	0 – 257	29–213	Ambrose et al. (2001)	only ophiuroids
Megafauna biomass	**[mg C m^{-2}]**			
NE Greenland	0.2–2.5	80–360	Piepenburg and Schmid (1996)	g AFDW m^{-2} NEW Polynya, only ophiuroids
Barents Sea	< 1–2580	80–360	Piepenburg (2000)	only ophiuroids
Laptev Sea	0–1880	14–45	Piepenburg (2000)	only ophiuroids
Chukchi Sea/ Bering Sea	0–3388	29–213	Ambrose et al. (2001)	only ophiuroids
Carbon requirements	**[mg C m^{-2} d^{-1}]**			
NE Greenland	0–54.5	718–2839	Ritzrau et al. (2001)	
	0–10	40–770 m	Piepenburg (2000)	NEW Polynya, only ophiuroids
Barents Sea	18.9–114.2	170–2577	Hulth et al. (1994)	
	17.3–121.4	226–3920	Pfannkuche and Thiel (1987)	
	< 1–22.1	80–360	Piepenburg (2000)	only ophiuroids
Kara Sea	3.5–43.2	10–68	Klages et al. (2003)	
Laptev Sea	0–11	14–45	Piepenburg (2000)	only ophiuroids
Chukchi Sea,	61.5–210.0	19–53	Grebmeier and Barry (1991)	
Bering Sea	32.9–107.0	6–161	Rowe and Phoel (1992)	

Table 6.2 (continued)

Organisms and region	Range of data	Water depth (m)	Study	Comments
Remineralization	[mg C m^{-2} d^{-1}]			
NE Greenland	7–50	150–450	Rowe et al. (1997)	NEW Polynya, Sediment Oxygen Demand
	0–5	40–770	Piepenburg (2000)	NEW Polynya, only ophiuroids
	0.25–1.4	1280–3400	Seiler (1999)	Calculated from "ship-board" incubations
	9.1–22.5 1.3–10.9	568–1950 189–390	Sauter et al. (2001) Sauter et al. (2001)	Both calculated from "ship-board" incubations and "in-situ" profiles
Yermak Plateau	6.4–47.6	550–682	Hulth et al. (1994)	Range from O_2-fluxes (profiles) and oxygen consumption (incubations) calculated
Barents Sea	5.17–24.92 7.67–36.52	405–2490 188–273	Hulth et al. (1994) Hulth et al. (1994)	Ranges calculated from O_2-fluxes (profiles) and oxygen consumptions (incubations)
	< 1–12.4	80–360	Piepenburg (2000)	only ophiuroids
Laptev Sea	0.7–14.4 0–6	14–45	Boetius and Damm (1998) Piepenburg (2000)	"ship-board" microprofiler data only ophiuroids
Chukchi Sea/Bering Sea	20.1–20.6		Grebmeier and McRoy (1989)	mmol C m^{-2} d^{-1} "ship-board" incubations
	37.2–164 ≤13.66	11–48 29–213	Devol et al. (1997) Ambrose et al. (2001)	"in-situ" flux measurements only ophiuroids

1994, 1997b, Budéus et al. 1997) which in some years extends up to 120,000 km^2 (Böhm et al. 1997). Fine-grained sediments cover the sea floor, particularly in the shelf troughs and at the continental slope (Piepenburg 1988). Coarser sediments and ice-rafted dropstones, indicating higher average bottom-water current velocities, are found mainly on the shelf banks. Seabed photographs reveal a pronounced patchiness of sea-bed facies at the 100 m scale, comprising areas of uniform soft bottom as well as gravel patches (Piepenburg 1988). For East Greenlandic fjords, Petersen and Curtis (1980) reported that the total macrofauna biomass in general is very high but decreases with distance from the coast.

In the region of the Northeast Water Polynya, sediment community biomass and respiration were measured by Rowe et al. (1997). Standing stock and faunal composition of brittle star assemblages were found to be related to meso-scale patterns in hydrography and ice cover. The latter are known to influence primary production and particle sedimentation, thus quality and quantity of food supply for the benthos (Piepenburg 2000). Starmans (1997) published results on epibenthic species composition of the coast off NE Greenland. Using a ROV at water depths between 27 and 376 m, he identified 94 species in an area of 8,250 m^2. According to the number of individuals and the dominance, echinoderms again were the most conspicuous group, as they are in most other shelf seas of the Arctic (Piepenburg et al. 2001). Species with highest abundances at different stations were the sea urchin *Strongylocentronotus pallidus*, the brittle star *Ophiocten sericeum* and the soft coral *Gersemia* cf. *fructicosa* with up to 2,643, 1,911 and 903 ind. 100 m^{-2}, respectively. Dense brittle star beds forming high standing stocks of up to several hundred ind. m^{-2} and up to several grams organic carbon m^{-2} were recorded (Piepenburg 2000). These stock figures are among the highest hitherto observed in northern seas, being in the same order of magnitude as those reported for ophiuroid mass occurrences in sublittoral and bathyal habitats of non-polar regions (Piepenburg 2000). Sub-dominant taxa, such as sea urchins, sea cucumbers and bivalves, reached only about a tenth of ophiuroid abundance and half of ophiuroid biomass.

Likewise, other photographic inventories and trawl catches clearly show that epibenthic communities at the Greenland continental margin are dominated by brittle stars (Piepenburg et al. 2001). Another conspicuous characteristic of macrofauna distribution on the shelf and upper slope off Greenland is their depth zonation. Various authors describe such patterns which more likely relate to depth-dependent ecological processes than to depth itself (Piepenburg 1988; Mayer and Piepenburg 1996; Starmans 1997; Piepenburg and Schmid 1996). Piepenburg et al. (2001) published data on macrozoobenthos distribution, abundance and biomass for the continental margin (40–3,700 m) off East Greenland between 68°N and 81°N. Macrobenthos abundance (biomass), averaged for different depth zones, was about 5,000 ind. m^{-2} (0.55 g C m^{-2}) at shelf stations at 200 to 400 m, about 2,000 ind. m^{-2} (0.20 g C m^{-2}) at mid-slope stations at 800 to 1,400 m, and about 1,800 ind. m^{-2} (0.13 g C m^{-2}) at depths $>$2,700 m (Schnack 1998). These values are in the same order of magnitude as those reported from other continental margins in comparable latitudes and depths (Dahl et al. 1976). Mean macrobenthic abundance and biomass of the entire East Greenland continental margin investigated decreased exponentially with water depth (Schnack 1998). However, not such a clear bathymetric trend was observed for macrobenthic biomass at 75°N when examined separately (Schnack 1998). At 79°N, macrofauna and polychaetes analysed from box corer samples showed a completely different depth pattern than the peracarid crustaceans in epibenthic sled catches (Brandt and Schnack 1999). Total macrofaunal abundance clearly decreased from about 7,000 ind. m^{-2} at 200 m to about 2,000 ind. m^{-2} at 800 m to about 800 ind. m^{-2} at 2,000 m depth. Ophiuroid abundances also decreased over roughly three orders of magnitude with water depth from a maximum at about 50 m to lowest values at about 800 m (Piepenburg 2000). A negative trend with depth was also obvious for biomass, albeit with a smaller decline over roughly two orders of magnitude. A similar exponential biomass gradient was reported for the invertebrate megabenthos from 500 to 4,100 m on the continental slope of the Porcupine Bight at 50 to 52°N in the north-eastern Atlantic Ocean (Lampitt et al. 1986). Polychaetes clearly dominated the shelf communities. However, their relative percentages decreased with water depth while those of Peracarida increased simultaneously. For soft-bottom communities off West Greenland, polychaetes were also found to be the most abundant and diverse taxonomic group (Andersen and Hartmann 1992).

The significant decline in biomass and abundance with water depth by as much as two and three orders of magnitude was accompanied by shifts in both faunal composition and predominance of feeding types. Food availability was found to be one of the most important factors determining the structure of macrobenthos communities. Latitude or depth alone apparently do not determine variations in species richness (Piepenburg et al. 2001). In general, some species or taxonomic groups such as sponges and brittle stars with high standing stock contribute considerably to high total benthic biomass. This implies that these groups play an especially important role in the benthic oxygen and carbon budget (see Chapter 6.5).

Surprisingly, at a first glance, organic carbon remineralization rates in soft sediments of Disko Bay, West Greenland, were found to be comparable to remineralization rates from much warmer localities, suggesting that benthic remineralization in Arctic marine environments is regulated primarily by the availability of organic matter and not by temperature (Rysgaard et al. 2000). At their shallow-water station (4 m), the sandy sediment contained a larger meiofauna population than the two deep water (150 and 300 m) stations characterised by muddy sediments. The larger number of individuals, typical representatives of interstitial communities, e.g. species belonging to Platyhelminthes, Rotifera, Gastrotricha, and Protodriloidae (Polychaeta), may in part be explained by higher food availability, as living diatoms were noted to serve as food for the meiofauna. Rysgaard et al. (2000) observed Kinorhyncha, Foraminifera, and cumacean crustaceans only at muddy stations. The bioturbation coefficient was approximately 3 times higher in the sandy sediment, corresponding to the 3 times higher number of meiofauna found in the upper 2 cm of this sediment (Rysgaard et al. 2000). Thus, they concluded that meiofauna increased the transport of solutes relative to molecular diffusion by a factor of 1.5 to 3.1, thereby stimulating microbial remineralization although grain size composition might also explain this increase.

Svalbard

The Svalbard Islands are influenced by warm Atlantic water in the south and west, and cold Arctic water in the east. Ice cover around Svalbard, and hence the position of the highly productive marginal ice zone, varies both intra- and interannually. Due to the availability of field stations on Svalbard, this area is one of the best investigated in the Arctic. Recent benthic studies in the shallow fjords and coastal areas have been concerned with growth rates and distribution of benthic megafauna (Tallqvist and Sundet 2000), the influence of particle transport and salinity gradients in tidal flats (Weslawski et al. 1999), a comparison of respiration rates in different benthic compartments (Piepenburg et al. 1995) and the effect of pollution on the benthic community (Holte et al. 1996; Alexander 1995), to give just some examples. One of the best studied areas of Svalbard is the Kongsfjord. The current interest in this fjord derives from its suitability as a site for exploring the impacts of possible climate changes because it is under the influence of Atlantic water influx and melting of tidal glaciers both of which are linked to climate variability. The pelagic ecosystem is likely to be most sensitive to the influx of Atlantic versus Arctic waters, whereas the benthic ecosystem is more affected by changes in glacial runoff and glacier retreat causing iceberg scouring and heavy mineral sedimentation (Weslawski et al. 1995, Hop et al. 2002). Most of the fjordic tidal glaciers of Spitsbergen, for example, are regressing at a maximal rate of 0.5 km per year, thus continuously exposing new areas of sea bed (Wlodarska-Kowalczuk et al. 1998). The macrofauna of these glacial bays are dominated by deposit-feeding polychaetes and bivalves but characterized by rather low wet weight biomass values ranging between 4 and 310 g m^{-2} at seven different bays studied by Wlodarska-Kowalczuk et al. (1998). Total faunal abundance was found to exceed 1,000 ind. m^{-2} at most stations with the highest being found in Templefjord (4,133 ind. m^{-2}). The number of species per sample varied from 10 in the Kongsfjord and Bettybukta up to 32 in the Ekmanfjord. In total 118 species were recorded in the study of Wlodarska-Kowalczuk et al. (1998) which is only a fraction of the entire Spitsbergen shelf benthos comprising almost 1,000 species (Gulliksen and Holte 1992). In the Skoddebukta at the southwestern tip of Spitsbergen, four distinct macrobenthic communities were identified in another study by Wlodarska-Kowalczuk et al. (1996). On shallow, stony habitats at mean water depths of 8.5 m most prominent organisms belonged to peracarid (*Mysis oculta*, *Weyprechta pinguis*, *Gammarellus homari*) and decapod crustaceans (*Eualus gaimardi*) accompanied by the bivalve *Liocyma fluctuosa*. On muddy and sandy-muddy parts of the Skoddebukta at water depth between 19 and 60 m (mean of 29.4 m) the characteristic species were bivalves (*Yoldia hyperborea*, *Macoma moesta*), the ophiuroid *Stegophiura nodosa* and the cumacean crustacean *Diastylis oxyryncha*. At greater depth (mean of 52 m) crustaceans were characteristic species (*M. oculta*, *Pagurus pubescens*, *E. gaimardi* and *Melita formosa*). The last faunal association was identified close to the river mouths and on shallow muddy-sandy bottom influenced by river inflow. Here the sipunculid *Phascolosoma* sp. and the polychaete *Spio filicornis* were abundant and dominant. Shannon-Weaver diversity indices calculated for the glacial bays were moderate and ranged between 0.383 and 2.54 (Wlodarska-Kowalczuk et al. 1996, 1998). In a recent study on the macro- and meiobenthic fauna of the Yoldiabukta bay, dominated by the bivalve *Yoldiella fraternata*, at the west coast of Spitsbergen,

Wlodarska-Kowalczuk et al. (1999) recorded maximum abundance values for nematode meiofauna of 67 ind. × 10 cm^{-2} with a biomass of 0.021 mg dry weight × 10 cm^{-2}. The maximum values measured for Harpacticoida were 13 ind. × 10 cm^{-2} and 0.036 mg dry weight × 10 cm^{-2}. The maximum total meiobenthic abundance accounted for 132 ind × 10 cm^{-2} and 0.29 mg dry weight × 10 cm^{-2}.

Svalbard is probably the only polar location where detailed studies of microbial diversity and activity have been carried out using recently developed tools of molecular ecology. A detailed investigation of benthic prokaryotic communities from 3 permanently cold stations near Svalbard (Arctic Ocean) has been conducted by Sahm and Berninger (1998). Prokaryotic cell abundances, determined by DAPI-staining, were in the range of ca 2×10^8 to 4×10^9 cells cm^{-3} wet sediment. Eubacteria always dominated the population, and the relative contribution of Archaea never exceeded 4%. These findings are similar to those for comparable studies on prokaryotic communities from temperate or warm habitats. The main interest in this and many other microbiological investigations of the Svalbard sediments is the low temperature. More than 90% of the global sea floor is colder than 4°C and it is not yet clear how microorganisms have adapted to life close to the freezing point. However, new investigations of the diversity of sulfate reducers and other bacteria in the cold sediments of Svalbard have also shown that the bacterial communities are psychrophilic and, thus, well adapted to cold temperatures and have a diversity similar to bacteria in temperate sediments (Ravenschlag et al. 2000).

Barents Sea

The Barents Sea shelf features a complex seabed morphology, comprising both shallow banks <100 m and deep troughs >300 m (Loeng 1989). Only its northern part is reckoned among the truly polar regions (Zenkevitch 1963). There, cold Arctic water masses (T <0°C) characterize the hydrography, being separated from warmer Atlantic water masses (T >2°C) in the southern Barents Sea by a sharp Polar Front at about 77°N. Arctic waters enter the Barents Sea as surface water along two main routes (through the straits between Svalbard and Franz Josef Land as well as between Franz Josef Land and Novaya Zemlya), moving in southwestern direction. This surface water inflow is probably counter-balanced by a northward transport of warmer and more saline Arctic-Atlantic bottom water, formed at the Polar Front, through the trough system of the northern Barents Sea (Loeng 1989). There is evidence that morphological features, such as shallow banks and small islands, cause mesoscale anticyclonic eddies, leading to downwelling and, hence, locally enhanced sedimentation. The northern Barents Sea, being ice-covered between September/October and May/June, belongs to the seasonal pack ice zone. The actual position of the marginal ice zone, however, oscillates considerably between years (Loeng 1989). River input and sediment load are generally lower in the Barents Sea compared with the other continental shelf seas of the Eurasian Arctic (Grebmeier and Barry 1991). Sea-bed characteristics are only scarcely known. Apparently, most of the area is covered by fine-grained brown sediment with comparatively low content of organic carbon (Zenkevitch 1963). Coarser sediments prevail on shallow shelf banks (<100 m) or in the sublittoral zone around islands (H. Grobe, pers. comm.). Sea-bed imaging reveals that stones and boulders are locally abundant (Piepenburg and Schmid 1996a). The bulk of the yearly primary production in the seasonally ice-covered northern Barents Sea is assumed to take place within the marginal ice zone which traverses the area while receding during June to August from its winter position at the Polar Front (Wassmann et al. 1991, Strass and Nöthig 1996). There is a considerable variability in composition, quality and quantity of sediment-trap contents, primarily related to ice cover, hydrography and zooplankton grazing (Andreassen et al. 1996; see Chapter 5). Surprisingly high levels of chlorophyll in sediment samples have been reported from the continental shelf and slope northeast of Svalbard (Pfannkuche and Thiel 1987), indicating a substantial sedimentation of phytodetritus to the sea bed. According to Russian benthos inventories, average benthic biomass is 5 to 10 times lower in the deeper parts of the northern Barents Sea than in the richest regions at the Polar Front and west of Novaya Zemlya (Zenkevitch 1963).

Similar to East Greenland waters, most shelf assemblages (<150 m) of the northwestern Barents Sea are strongly dominated by *Ophiocten sericeum* in terms of abundance. At several sites this species occurs in mean abundances of several hundreds ind. m^{-2}, forming very dense brittle star beds similar to those reported from several locations in other climatic zones (Aronson 1989). In these beds, *O. sericeum* could account for >99% of the ophiuroid fauna and for up to 98% of total epibenthic abundance (Piepenburg and Schmid 1996a, 1996b, 1997), and they also dominated in the trawl catches from this area (Piepenburg et al. 1996), reaching abundance shares of 65 to 98% (Schmid 1996). Other prominent taxa are sea stars (*Ctenodiscus crispatus*), sea urchins (*Strongylocentrotus pallidus*), sea lilies (*Heliometra glacialis*), and shrimps (*Pandalus borealis*). Maximum ophiuroid biomass values were about 2.5 g C m^{-2} (Piepenburg & Schmid 1996a), mainly made up by *O. sericeum*. *S. pallidus* was reported to have a maximum biomass of 1.5 g C m^{-2} (Bluhm et al. 1998). Brittle star assemblages in the shelf troughs of the Barents Sea (100 to 400 m) are characterized by intermediate densities (1 to 100 ind. m^{-2}). *O. sericeum* is less abundant or even absent (Piepenburg & Schmid 1996a, 1996b), and the communities are numerically dominated by *Ophiacantha bidentata*, the most ubiquitous species in Arctic seas (Piepenburg 2000). Ophiuroid biomass is also lower, albeit less distinctly (tens to hundreds mg C m^{-2}) because *O. bidentata* and particularly *Ophiopleura borealis* are notably larger than *O. sericeum*. On the continental slope of the western Barents Sea at 75°N in 1,340 m depth, for instance, macrobenthic abundances vary between 557 and 1,052 ind. m^{-2} (Ambrose and Renaud 1995; Brinkhurst 1991; Kendall 1996; Thomsen et al. 1995).

Kara Sea

The large shelf area of the Kara Sea (926,000 km^2) is strongly influenced by the inflow of water from the adjacent seas as well as large Siberian rivers. In the south-western area water from the Barents Sea enters the Kara Sea via the Kara Strait, whereas the eastern part of the Kara shelf is directly influenced by the large rivers Ob and Yenisei. The intense continental outflow imposes a distinct salinity gradient on the Kara shelf, which spreads northward from the river estuaries into the central Kara Sea and significantly influences the distribution of the benthic fauna and flora on this shelf sea as well as their productivity and consumption. In the western Kara Sea, Denisenko et al. (2003) observed a decreasing tendency of species numbers with depth, whereas in the eastern part of the sea an opposite trend was found (Jørgensen et al. 1999; Deubel et al. 2003). There, the macrofauna shows a remarkable increase in species numbers, biomasses and diversity towards the high salinity areas, and hence an increase with increasing distance from the coast (Deubel et al. 2003). Increasing species numbers and biomasses of macrobenthic invertebrates with increasing salinity is in accordance with results from other estuarine areas (Dahle et al. 1998; Petryashov et al. 1999).

Sirenko (2001) estimated that about 1,671 free-living invertebrate species inhabited the shelf of the Kara Sea. The macrobenthos community consists of 1,390 known species with crustaceans, polychaetes, molluscs and echinoderms as the most dominant taxa (Denisenko et al. 2003). However, on the Kara Sea shelf a strong regional variability in macrobenthic species number as well as in abundance is obvious. Species numbers vary from only 2 species m^{-2} in the southern estuaries of the Siberian rivers up to 140 species m^{-2} in the south-eastern and western part of the Kara Sea (Denisenko et al. 2003; Deubel et al. 2003). Abundances vary between 100 ind. m^{-2} and 10,000 ind. m^{-2}, whereas the highest abundances are very often directly linked to mass occurrences of single species in areas closest to the coast.

On the southern shelf of the Kara Sea, molluscs, mainly represented by the genus *Portlandia* and *Tridonta*, dominate the macrofaunal biomass (50%) and contribute 25% to the macrofaunal numerical abundance. In the central and northern part of the Kara Sea, echinoderms, especially ophiuroids are the dominant macrobenthic taxa and are distributed throughout the area. Results published by Anisimova (1989) indicate that ophiuroids occur in high species number and abundances in the Kara Sea at water depth between 150 and 200 m.

During their perennial investigations on the macrobenthic fauna in the south-eastern Kara Sea, Deubel et al. (2003) observed only slight fluctuations in macrobenthic biomass ratio over the major taxa within different years. These authors suggest that macrofaunal organisms are very tolerant to environmental impacts and integrate the large-scale seasonal variations. Jørgensen et al. (1999) and Deubel et al. (2003) demonstrated that in the southern Kara Sea high macrobenthic biomass values are located in the mixing zone of marine and riverine waters where large amounts of dissolved and particulate matter are transformed and sink to the sea floor (Lisitzin 1995). Maximum biomass values of 778 g wet weight (ww) m^{-2} (equals 57.9 g ash free dry weight (AFDW) m^{-2} by applying published conversion factors (Brey et al. 1988)) were recently reported by Klages et al. (2003), while mean values were 71.8 g ww m^{-2} (10.1 g AFDW m^{-2}). Deubel et al. (2003) give maximum and minimum abundance values of 10,280 and 196 individuals of macrozoobenthos m^{-2}, respectively (mean abundance of 1,332 m^{-2}). There was a considerable variability in species numbers m^{-2} at their sampling locations in the southern Kara Sea with minimum of 3 and maximum of 86 species (mean species number 30) resulting in low to moderate diversity indices of H' between 0.16 and 3.61 (mean value of $H' = 2.18$).

Laptev Sea

As the smallest of the Eurasian-Arctic shelf seas the Laptev Sea covers an area of only 498,000 km^2. About three quarters of the Laptev Sea belongs to the shelf with comparably low water depths (very shallow banks with less than 20 m). Similar to the

Kara Sea, the Laptev Sea is influenced by large rivers leading to reduced salinity and enhanced ice formation. Because the Laptev Sea is one of the major "ice factories" in the Arctic, this shelf sea has been the focus of several international expeditions during the past decade, leading to a considerable increase in the knowledge about the Laptev Sea system (Kassens et al. 1999). Surface sediments of the eastern Laptev Sea are dominated by silt and clay reflecting its role as a major depositional area for sediments discharged by the Lena and Yana Rivers (Lindemann 1994). The benthos of this high-Arctic shelf region has been reported to be generally poorer than in other Eurasian-Arctic seas in terms of diversity, abundance and biomass (Zenkevitch 1963, Golikov 1990). This scarcity was attributed to the severe climate in the Laptev Sea and the very low salinities due to a pronounced fluvial dilution during summer.

Benthos samples taken systematically by joint Russian-German expeditions, and the identification by specialized taxonomists has recently added another 150 species indigenous to the Laptev Sea (Petryashov et al. 1999). Over the whole shelf, >200 macrobenthic species were found (Fahl et al. 2001). Macrobenthic biomass ranged from 37 g ww m^{-2} near the mouths of the Yana and Lena Rivers and 150 g ww m^{-2} in the central western Laptev Sea (Fahl et al. 2001). The macrofaunal assemblages were dominated by brittle stars and isopods of the genus *Saduria*. Gukov et al. (2001) reported a total macrobenthos biomass in the area of the fast ice edge ranging from 16.51 to 77.66 g ww m^{-2} and abundances varying between 434 and 1,545 individuals m^{-2}. For the northern Laptev Sea Deubel (2000) estimated macrobenthic biomasses from a minimum of 0.5 to a maximum of 11.67 g AFDW m^{-2}. The number of individuals ranged from 360 to 9,128 ind. m^{-2}.

On shallow shoals of the Laptev Sea, with water depths less than 30 m, brittle star abundance and biomass is low (Piepenburg and Schmid 1997). However, on those flanks of the shelf valleys deeper than 30 m, which are not subject to reduced and fluctuating salinities and possible iceberg gouging impacts, very high densities of *Ophiocten sericeum* are recorded. These figures are similar to maximum values found for *O. sericeum* on shelf banks in the Barents Sea or the Belgica Bank off Northeast Greenland (Piepenburg and Schmid 1996b). Nor were maximum ophiuroid biomasses lower in the Laptev Sea: nearly 2 g C m^{-2} primarily made up by *Ophiura sarsi* (Piepenburg and Schmid 1997), versus about 2.5 g C m^{-2} in the Barents Sea and about 1 g C m^{-2} off Northeast Greenland, mainly contributed by *O. sericeum* (Piepenburg 2000). Except for the very shallow Laptev Sea shoals (<20 m), there is no significant difference between the Greenland, Barents and Laptev Sea shelves in terms of either ophiuroid abundance or biomass (Piepenburg 2000).

The comparison of phytoplankton, zooplankton and benthic distribution patterns suggests that different processes control the communities in the water column and at the seafloor (Fahl et al. 2001). In general, benthic distribution in the Laptev Sea is affected by the fluvial input and meso-scale current patterns rather than by large-scale biogeographical or oceanographic features. The most important factor is bottom-water salinity, which in turn is closely correlated with water depth. The influence of the sediment structure, commonly invoked as a prime benthic community determinant (Snelgrove and Butman 1994), is apparently less significant, because grain-size distribution is rather homogeneous in the Laptev Sea (Lindemann 1994). The high sediment load and freshwater inflow near the Lena River obviously limit benthic colonization. Hence, no correlation between nutrients – and related food supply to the benthos – and benthic biomass is found, even though the high input of nutrients by the Lena sustains a relatively high phytoplankton biomass near the delta (Fahl et al. 2001). This situation differs clearly from that in the Chukchi Sea where Grebmeier and Barry (1991) explained high benthic biomass by high primary productivity (300 g C m^{-2} y^{-1}) caused by upwelling and northward pumping of nutrient-rich Anadyr water through the Bering Strait (see below). In a study of the outer Laptev Sea shelf and margin sediments Boetius and Damm (1998) concluded from oxygen uptake and microbial activity that the vertical flux of POC could satisfy the benthic respiratory demand at the Laptev Sea slope and in the Arctic basin. At depths between 50–3,500 m food supply was found to be the principal factor controlling microbial activity, whereas characteristics of sediment composition, e.g. water content, grain size and TOC had a lesser impact on microbial activity. Bacterial biomass was low at the bottom of the continental slope of their area of investigation (7–26 µg C cm^{-3}), slightly higher at the ice-covered shelf edge (30–60 µg C cm^{-3}), and highest (up to 85 µg C cm^{-3}) at the ice-free shelf edge (Boetius and Damm 1998). Carbon remineralization rates calculated from total oxygen flux rates ranged from up to 22 mg C m^{-2} d^{-1} at the ice-free shelf edge, 6-11 mg C m^{-2} d^{-1} at the ice-covered shelf edge and rather low values varying between 2–5 and 0.65 mg C m^{-2} d^{-1} at the bottom of the slope of the two areas. It was concluded that advective lateral transport processes are not a major source of food to deep benthic communities. A further decline in benthic food supply and activity was found by Schewe (2001) at stations in the central Arctic Ocean, e.g. the Alpha and the Lomonosov Ridge. Chloroplastic pigment equivalents (CPE) were 0.10 and 0.17 μg cm^{-3} at the Alpha Ridge and 0.40 μg cm^{-3} at the Lomonosov Ridge, and thus an order of magnitude lower than values measured by Boetius and Damm (1998) for the Laptev Sea shelf and continental slope.

Chukchi and Bering Sea

The shallow shelf of the Chukchi Sea (620,000 km^2), ice-covered from November through June (Carey 1991), is one of the most productive areas of the Arctic shelf seas with primary productivity as high as 720 g C m^{-2} y^{-1} (see Chapter 3). This pronounced food supply fuels infauna communities with biomasses of up to 60 g C m^{-2} (Grebmeier et al. 1995). The productivity of the benthic communities is supposed to be largely a function of the high water-column productivity (Cooper et al. 2002). Results given by Ambrose et al. (2001) clearly show that echinoderms dominate the epifauna community in the area east of Wrangle Island and northwest of Alaska between 29 and 213 m water depth. The sea urchin *Echinarachnius parma* (sand dollar), and the four brittle star species *Ophiopholis aculeata*, *Ophiura maculata*, *O. sarsia*, and *Stegophiura nodosa* were most conspicuous in underwater photographs and videos taken with a ROV (Ambrose et al. 2001). Abundances of ophiuroids ranged between 0.2–257 ind. m^{-2}, and sea urchins comprised over 100 ind. m^{-2}.

Grebmeier et al. (1989) reported macrofaunal mean abundances of 641 up to 4,193 ind. m^{-2} and mean biomasses ranging between 2.0 and 15.4 g C m^{-2} for a group of stations in the northern Chukchi Sea identified by cluster analysis. However, benthic biomass was generally lower at stations influenced by the relatively warm Alaska coastal water (ACW) as compared to stations influenced by the Bering Shelf – Anadyr Water (BSAW), the latter cold and highly productive (~250–300 g C m^{-2} y^{-1} *versus* ~50 g C m^{-2} y^{-1} for ACW; see Grebmeier et al. 1989). The factors considered by Grebmeier et al. (1988, 1989) to explain benthic community structures in the northern Bering and Chukchi Seas included sediment composition and heterogeneity, benthic community composition, and feeding groups: diversity indices showed complex benthic assemblages at local spatial scales. Sediment heterogeneity and temperature, food availability and predation were found to be important factors influencing the community structures and biomass. The presence of an annually recurring polynya south of St. Lawrence Island during the ice-covered winter months might influence the re-distribution of organic carbon particles during brine formation events but Cooper et al. (2002) found little evidence that the presence of open water itself is a source of biogenic material.

Macrobenthos biomass values reported by Feder et al. (1994) from the northeastern Chukchi Sea at water depths between 18 and 51 m ranged from 1.0 to 19.6 g C m^{-2}. Here, environmental stresses (e.g. ice gouging, wave-current action, marine-mammal feeding activities) were found to be of greater importance than the factors listed by Grebmeier et al. (1989). Moreover, the findings demonstrated that there can be high standing stocks of benthos in Arctic regions with relatively low annual primary production if local carbon is augmented by POC advected from adjacent highly productive areas (Feder et al. 1994).

6.4.2 Central Arctic

Quantitative macrofaunal and megafaunal assessments in the central Arctic Ocean, which are extremely limited (Paul and Menzies 1974; Kröncke 1994, 1998; Deubel 2000), demonstrate very low species number and abundance (see Table 6.2). A comprehensive study by Deubel (2000), covering more than 60 stations, reveals decreasing species numbers, abundance and biomass both with increasing water depth as well as with increasing latitude. From this study more than 220 macrofaunal species were identified. Along a transect from the North Pole, across the Amundsen Basin and Gakkel Ridge towards the Morris Jesup Rise and Yermak Plateau, Kröncke (1998) reported macrobenthic species numbers varying from 50 to 500 per m^2 (extrapolated from the sampled area of 0.02 m^2) in the deep basins and from 176 to 324 per m^2 approaching the Yermak Plateau (see Fig. 6.4).

These values correspond well with macrofaunal species numbers given by Deubel (2000) ranging from 160 species m^{-2} on the Siberian continental slope and sporadically 4 species m^{-2} in the central part of the Arctic Ocean near the Lomonosov Ridge. Both studies showed that the macrobenthic species composition differed distinctly between different water depths. However, at similar depths, the macrobenthic assemblages in the Amundsen Basin (west of the Lomonosov Ridge) and Makarov Basin (east of the Lomonosov Ridge) did not differ statistically from each other. This observation indicates that the mid-Arctic ridge does not prevent migration of macrofauna (Deubel 2000). Both Deubel (2000) and Kröncke (1998) conclude that the low species numbers and the low abundances (median values

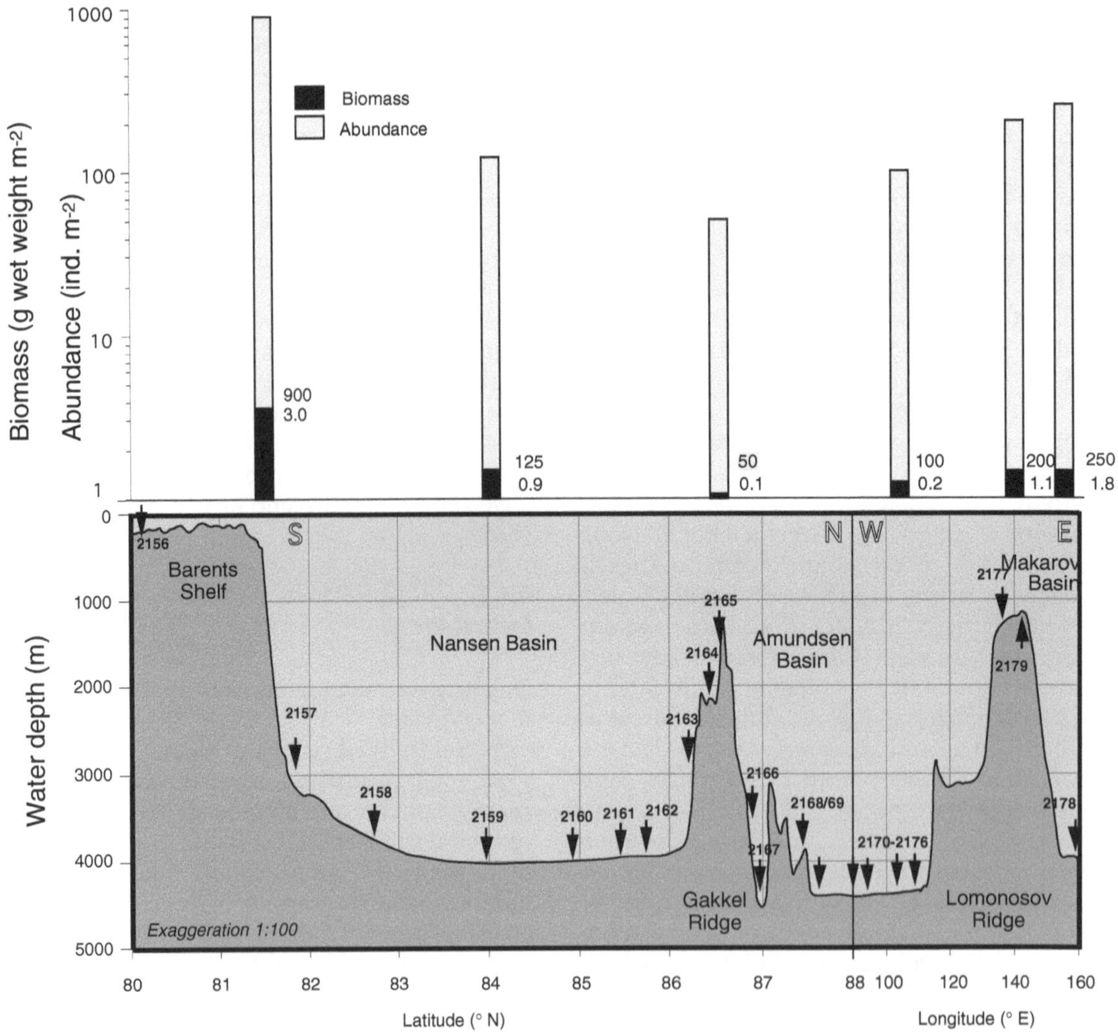

Fig. 6.4. Positions of stations sampled by Kröncke et al. (1998) along a transect from the Barents Sea shelf to the Makarov Basin during the ARCTIC '91 expedition of the RV "Polarstern". Numbers at the arrows in the lower panel refer to stations as designated during the expedition. On the upper panel average biomass and abundance values are given on a logarithmic scale at selected locations along the transect (modified after Kröncke et al. 1998)

of 8–63 ind. m^{-2} within macrobenthic assemblages northward of 83°30′N) indicate very limited organic matter influx rates to these deep-sea sediments.

Macrobenthic assemblages of the central Arctic Ocean are dominated in species number, abundance and biomass by crustaceans, echinoderms, molluscs and polychaetes, in respective order, a pattern similar to those found on the adjacent Eurasian shelf seas. Both Paul and Menzies (1974) and Kröncke (1998), noticed the increased importance of suspension feeders within the benthic assemblages and, in fact, Paul and Menzies (1974) named one of their Arctic deep-sea assemblages after the sponge, *Thenea abyssorum*, and the suspension feeding polychaete, *Spirorbis granulatus*. Within the central Arctic, Kirkegaard (1982) originally listed 32 species of polychaetes, but more recent investigations have added several new serpulid polychaete species to the list (Kupriyanova 1993a, b; Kupriyanova and Jirkov 1997). Nevertheless, most of the polychaetes belong to eurybath species and are widely distributed. Compared to other benthic taxa, the proportion of Arctic-endemic polychaetes is very low (Bilyard and Carey 1979). Echinodermata provide an essential component of macrobenthic assemblages at high latitudes. Of these, Ophiuroidea inhabits the marginal shelf areas (see previous sections), whereas Echinoidea and Holothuroidea compose regulars of the Arctic deep-sea fauna. Common inhabitants at water depths below 2,000 m include sea urchins

(*Pourtalesia jeffreysii*) and sea cucumbers (*Elpidia glacialis*). Both are not limited to the Arctic, but also inhabit abyssal depths at lower latitudes (Heding 1942; Vinogradova 1979).

In the deep basins, median biomass of the macrofauna ranges from 5 mg C m^{-2} near the North Pole to 32 mg C m^{-2} near the Siberian continental slope. The corresponding biomass values on the top of the Lomonosov Ridge range between 25 mg C m^{-2} in the north and 133 mg C m^{-2} in the south (Deubel 2000). Remarkably low macrobenthic abundance and biomasses around the Lomonosov Ridge agree reasonably well with data reported by Paul and Menzies (1974) for the Amerasian Basin and by Kröncke (1994) for the Eurasian Basin. Kröncke (1994) explained the depth-dependent distribution of macrobenthos basically as a product of decrease in food availability with increasing depth which accompanied a shift in the feeding mode of the fauna. Suspension feeders were found to be dominant on the rise and deposit feeders in the basins. The number of macrofaunal species vary from 1 to 11 per 0.02 m^2 in the basins approaching the Morris Jesup Rise and from 44 to 81 per 0.25 m^2 at the Yermak Plateau. Abundances increase from 1 to 31 per 0.02 m^2 in the basin and on the Morris Jesup Rise whereas 24–60 ind. per 0.02 m^2 have been recorded on the Yermak Plateau. As might be expected, the biomasses are low in the Amundsen and Nansen basins (0.7–14 mg 0.02 m^{-2}) and at the Morris Jesup Rise (0.5–69 mg 0.02 m^{-2}) but increase to 117 mg 0.02 m^{-2} at the Yermak Plateau (Kröncke 1998). These findings together with the data published by Deubel (2000) are rather exciting because they contradict the hypothesis that benthic diversity decreases with increasing latitude. A comparable result was found by Brey et al. (1994) for the Antarctic Weddell Sea benthos where diversity was also higher than previously thought.

There are interesting structural differences in the meiofaunal and microbial communities in the Arctic. Results published by Soltwedel and Schewe (1998) demonstrate for stations sampled in summer 1996 north of 85°N in permanent ice-covered regions of the central Arctic Ocean, a slight northward decrease in sediment-bound phytodetrital organic matter (as indicated by chloroplastic pigments) along the Lomonosov Ridge crest. Both the enzymatic activity and biomass of bacteria from the same study do not follow the weak trend of northward decreasing values, indicating that the central part of the Arctic Ocean is largely decoupled from the primary production at the ice-edge and on the Arctic shelves. However, stations visited in 1996 were more than 750 km away from the ice-edge in the Laptev Sea where the supply of riverine nutrients from the large Siberian rivers allows a high productivity in the water column. Bacterial abundance and biomass did not correlate significantly with food availability in the sediments of the Laptev Sea continental slope but decreased approximately threefold between shelf edge (30–85 mg C cm^{-3}) and the slope (7–25 mg C cm^{-3}) at all transects, corresponding to a decrease in bacterial numbers from around 4×10^9 cells cm^{-3} to 1×10^9 cells cm^{-3} (Boetius and Damm 1998). However, bacterial numbers were much lower in sediments of the permanently ice-covered central Arctic basins and ridges, indicating the strong oligotrophy of this area (Kröncke et al. 1994). Furthermore, bacterial abundance on the Eurasian Arctic slope and in the deep-sea was at the lower end of numbers from corresponding depth ranges of temperate oceanic regions (Deming and Baross 1993). Their review of data from the NW-Atlantic showed a decrease in bacterial biomass by more than an order of magnitude with increasing water depth (100–6,000 m) related to a decrease in POC fluxes by around two orders of magnitude. Based on samples taken from the Lomonosov and Alpha Ridges (Schewe 2001), the mean meiofaunal abundance ranges up to 297×10^3 ind. m^{-2} on the Lomonosov Ridge at 1,270 m depth, 190×10^3 ind. m^{-2} at 1,470 m on top of the Alpha Ridge, and 72×10^3 ind. m^{-2} in the deep Makarov basin at 3,170 m (Table 6.2). Additionally, the size spectra of fauna inhabiting sediments in the deep basins of the central Arctic Ocean are shifted towards smaller individuals when compared with those taken at a similar water depths in the North Atlantic (see Fig. 6.1). Within the microbiota, species belonging to the nanofauna (size between 2–32 μm) account for roughly two thirds of the total microbial biomass (TMB), whereas bacteria account for 32%, and metazoan meiofauna represent only 0.5–1.5% of the TMB. This tendency towards dwarfness is most likely explained by extreme limitation of food supply (Thiel 1975).

6.5 Organic Carbon Utilization by the Arctic Benthos

Benthic communities respond to their environment by consuming organic matter reaching the sediments, utilizing some of this to grow and maintain their population size while respiring the remainder. Food sources considered in this review (Chapter 6.3) are the primary production in waters directly over the benthic site, marine material produced in adjacent areas and laterally transported into the site, and terrestrial material carried out from the

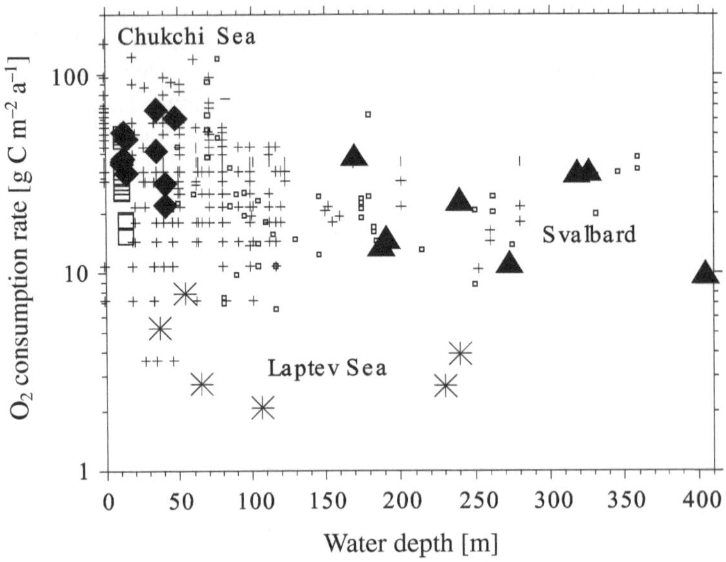

Fig. 6.5. Comparison of benthic oxygen consumption rates in Arctic continental shelf regions and in mid-latitude shelves versus water depth. Average rates in the Svalbard area and the northern Chukchi Sea are similar to those in mid-latitude shelves even though average primary productivity in overlying waters of the different regions differ by a factor of 20. The lower range in benthic rates found in the Laptev Sea suggests that overlying productivity and organic matter supply to the benthos are much less than in the other regions. Data for Arctic shelves: filled triangles – Svalbard region (Hulth et al. 1994); Stars – Laptev Sea (Boetius and Damm 1998); Diamonds – northern Chukchi Sea (summer, Devol et al. 1996); Squares – northern Chukchi Sea (winter, Devol et al. 1996). Data for mid-latitude shelves: small crosses – northeastern U.S.A. (Phoel 1982; Christensen and Altabet submitted); small squares – Washington State shelf (Pamatmat and Banse 1969; Pamatmat 1971, 1973; Devol and Christensen 1993)

coastline in the fluvial transports. This latter type of material is considered to be relatively poor in nutritional quality, much of it being relatively refractory, and thus its capability to support benthic populations is uncertain. Nevertheless, in areas where lateral inputs of material are negligible, the size and the total metabolism of the community should vary in proportion to the local primary productivity and should be inversely correlated with increasing water depth.

6.5.1 Arctic Continental Shelves

Because the depth range of most shelves is similar, comparison of the benthic community metabolism between the various Arctic shelves may perhaps indicate differences in regional productivity. However, a surprising fact recognized independently by several investigators is that benthic carbon remineralization rates from the Arctic shelves appear to be similar to coastal sediments in temperate and tropical regions, thus suggesting that the prevailing low temperature is not a factor (Glud et al. 1998, Kostka et al. 1999; Rysgaard et al. 1998).

Figure 6.5 presents rates of total oxygen consumption by shelf sediments versus water column depths from three Arctic shelves and two mid-latitude shelves (oxygen consumption has been converted to carbon release based on Redfield ratios). Data from the northern Chukchi Sea range between 15 and 70 g C m^{-2} y^{-1} (Devol et al. 1996) and from near Svalbard range between 10 and 40 g C m^{-2} y^{-1} (Hulth et al. 1994). Based on many data from the shelf areas off of the northeastern United States (small crosses on Figure 6.5), consumption rates range between 3.5–130 g C m^{-2} y^{-1} and show only a moderate decrease in mean value with increasing depth. Sediment results from the Washington State continental shelf (small squares, Fig. 6.5) have a similar range, and decrease approximately the same amount with increasing water depth. Within each region, variability is large since the data may have been collected in different years, in different seasons, and from randomly selected sites which may differ in their depositional and sedimentary characteristics. Since these sources of variability are common to all shelf environments, it would be expected that collection of a sufficiently large number of samples from a shelf region would yield a reliable overall average, and that averages among shelves should reflect differences in the ecosystem richness. The dichotomy is that the four shelves shown here appear to differ little in their total sediment oxygen demand, and yet primary production rates from these four regions differ by an order of magnitude. Productivities in the northern Chukchi Sea may be about 40 g C m^{-2} y^{-1} (Cota, unpublished data), while those in the seasonally ice-free waters near Svalbard have been estimated to be 80-100 g C m^{-2} y^{-1}. In the northeastern U.S. shelf, primary production rates vary between 200 and 400 g C m^{-2} y^{-1} (O'Reilly and Busch 1984), while off Washington State, overlying productivities average 740 g m^{-2} y^{-1} (Perry et al. 1989). Christensen (1989) discussed this dichotomy and suggested that because the total benthic carbon oxidation is the sum of all oxic consumption occurring near the sediment surface and the deeper anaerobic consumption, it was the anaerobic respiration occurring deeper in the sediments that showed much greater variation between rich and poorer sites. Models of oxygen consumption within sediments (Christensen 2000) indicate that relatively small fluxes of usable organic matter reaching the sediments would support average oxygen consumption rates at the lower shelf levels; these models are based on the condition that oxygen is supplied from the bottom waters by molecular diffusion. Bioturbation and bioirrigation may enhance the oxygen supply from the overlying bottom waters, but the respiration rate of the infauna generally is limited to a several-fold increase in the oxygen supply above the molecular diffusive level (Christensen et al. 1984; Glud et al. 1998). The consequence of this limitation by the oxygen supply is that most shelves with primary productivities above some lower limit should have relatively similar rates of benthic oxygen consumption, as demonstrated in Figure 6.5. The final set of Arctic data in Figure 6.5 is from the outer Laptev Sea shelf and slope (Boetius and Damm 1998). These data appear to be consistently lower than the other four data sets by about an order of magnitude. Our interpretation is that the sediments in this remote Arctic region receive much less usable organic matter than do sediments in either the northern Chukchi Sea or the shelf areas near Svalbard. The paucity in organic matter supply to sediments in the Laptev Sea may be due to extremely low productivities in the overlying waters or due to lateral export of the productivity without it being deposited on the underlying sediments. Boetius and Damm (1998) concluded that lateral advection was not significant.

The above discussion suggests that total oxygen consumption rates by shelf sediments may not always reflect overlying productivity because the supply of oxygen from the overlying water may limit this rate. Macrobenthos and epibenthic megafauna may not be subject to the same restriction since motility may allow them direct exposure to bottom water oxygen, so it is possible that macrobenthic estimators may more closely relate to overlying productivity than does total sediment oxygen consumption largely carried out by microbial organisms. Macrobenthos are typically included within the chambers or cores used for oxygen consumption rates, thus their metabolic rates represent a portion of the total rate. The organic carbon demand of macrozoobenthos from the Kara Sea was estimated using values from the literature on annual production, annual respiration and carbon remineralization from other regions of the Arctic (Klages et al. 2003). The estimated rates were 3.5–43.2 mg C m^{-2} d^{-1} for 54 stations located in the nearshore adjacent to the Ob and Yenisei river mouths to the outer shelf of the Kara Sea from depths ranging between 10–68 m. Similar calculations for the northern shelf of the Laptev Sea indicate rates of approximately 2.53 mg C m^{-2} d^{-1} (Deubel 2000; Seiler 1999). These rates represent up to 25% of the total sediment oxygen demand.

In contrast, megafauna are typically excluded from laboratory incubations for total oxygen demand, so that their rates would add to the total demand by the benthic community. Most common in the Arctic are the brittle stars (Ophiuroidae). In the Barents Sea, the carbon remineralization and demand of megabenthic brittle stars were compared with bulk carbon requirements of the infaunal sediment community (including bacteria, meiofauna and small macrofauna), which was estimated from oxygen uptake rates of whole sediment cores (Piepenburg et al. 1995). Brittle star carbon uptake varied over a very broad range (one to three orders of magnitude) whereas sediment community carbon demand fluctuated by only a factor of 2. This variability corresponded to the pronounced small- and meso-scale patchiness in ophiuroid spatial distribution. Wherever brittle stars occurred in dense beds, they dominated the benthic carbon flux and mineralized an amount of carbon that was equivalent to the highest total sediment carbon uptake recorded in the study area. Maximum remineralization rates estimated for brittle star assemblages in the Barents Sea were about 12 mg C m^{-2} d^{-1} which is equivalent to the total carbon demand (both growth and respiration) of about 22 mg C m^{-2} d^{-1} (Piepenburg 2000). During summer (June to September) these assemblages consume up to 9% of the average net daily primary production and up to 31% of the vertical flux of POC (Piepenburg 2000). However, in shelf trenches of the Barents Sea, where brittle stars are comparatively rare, their contribution to the total benthic carbon mineralization is negligible. On the eastern Greenland shelf, brittle star assemblages remineralize up to 5 mg C m^{-2} d^{-1}, equivalent to a carbon demand of 10 mg C m^{-2} d^{-1} and consume up to 4% of the average net daily summertime primary production and up to 14% of the vertical flux of POC (Piepenburg 2000). Again, these organisms are rare in the Greenland shelf trenches. In the Laptev Sea, maximum remineralization rates of brittle star assemblages were about 6 mg C m^{-2} d^{-1} (carbon demand of about 11 mg C m^{-2} d^{-1}; Piepenburg 2000), representing up to 7% of the autochthonous primary production and up to 20% of the sedimented POC during an average summer day. It should be noted, however, that allochthonous carbon sources were not included in these calculations although they might be particularly important in this region. Similarly, Ambrose et al. (2001) estimated carbon remineralization rates of the assemblages of up to 13.66 mg C m^{-2} d^{-1} in the northern Chukchi Sea.

On the Arctic shelves where water depth shows only minor correlation with the benthic rates (Fig. 6.5), seasonal variations in organic matter flux may perhaps be significant. In Young Sound in northeast Greenland, a significant increase in sedimentation of organic material occurred after sea ice retreat in mid-July and was followed by rapid remineralization of the easily degradable fraction of the settling organic material (Rysgaard et al. 1998). The respiration rates returned to their basic lower level within 1 month and remained low for the rest of the season. In contrast, no seasonal differences in total sediment oxygen demand were observed between winter and later summer near Point Barrow on the Chukchi Sea (Fig. 6.5, Devol et al. 1996). Larger organisms likely show highly seasonal distributions but few data are available to examine this hypothesis.

6.5.2
Central Arctic Ocean

On the continental slopes and in the deep-sea, both depth and surface productivity may strongly influence fluxes of sinking organic matter, and this influence has been quantified with many correlations based on sediment trap results (Pace et al. 1987; Berger et al. 1988; Lampitt and Antia 1997; and many others). However, Christensen (2000) noted that measurements of oxygen consumption rates of aerobic deep-sea sediments nearly always indicated greater carbon inputs to the deep-sea than predicted by most of the commonly-used sediment trap regressions, and that these trap relationships may underestimate the carbon consumption by deep-sea sediments by as much as an order of magnitude (Fig. 6.6). A new relationship between carbon flux and productivity, independent of sediment trap data, was estimated based on deep-sea sediment oxygen consumption rates (Christensen 2000) and this relationship would be directly comparable to the Arctic benthic data. We have plotted the few deep-sea sediment oxygen consumption rates from the Svalbard area (Hulth et al. 1994), the Laptev Sea (Boetius and Damm 1998), and the central Arctic (Christensen 2003) at various depth ranges (500–1500 m, 1500–2500 m, and 2500–3500 m) against the estimated productivities above these sites (Fig. 6.6).

In the Svalbard region at about 2,000 m, total oxygen consumption rates, when converted to carbon release based on Redfield ratios, indicate carbon oxidation rates of about 8 g C m^{-2} y^{-1}, while in the central Arctic and near the Laptev Sea, rates at the same depth range approximate 1 g C m^{-2} y^{-1}. However, Macdonald et al. (1993) required a flux of 0.5–0.6 g C m^{-2} y^{-1} across the 1,500 m boundary in the Canada Basin to satisfy the basin O_2 drawdown – a number that seemed awfully high at the time. Most of these Arctic data exceed the sediment trap predictions, but they approximate the predictions of the sediment-based relationship of Christensen (2000). These few data indicate that the deep Arctic sediments have very low benthic consumption rates. Nevertheless, the observed rates appear consistent with those found in other oceans when the decrease in sinking export with lower productivity is considered. Furthermore, the differences in sediment oxygen demand between the Svalbard area and the central Arctic can be explained by the regional differences in the overlying productivity.

We compared the role of macrobenthos in these deep-sea sediments with the total sediment consumption. The average macrobenthic demand in deep Arctic ranged between 0.85 mg C m^{-2} d^{-1} (0.31 g C m^{-2} y^{-1}) in the deep continental slope off of the Laptev Sea, 0.46 mg C m^{-2} d^{-1} (0.17 g C m^{-2} y^{-1}) in the deep sea close to the North Pole, 0.047 mg C m^{-2} d^{-1} (0.017 g C m^{-2} y^{-1}) for the Makarov Basin, and 0.013 mg C m^{-2} d^{-1} (0.005 g C m^{-2} y^{-1}) for the central deep Amundsen Basin (Fig. 6.6). Biomass data from the Amerasian part of the Arctic published by Paul and Menzies (1974) indicate that the median macrobenthic carbon demand at different depths ranged between 0.1 and 1.41 mg C m^{-2} d^{-1} (0.04–0.51 g C m^{-2} y^{-1}). Thus, the macrobenthos could represent a significant proportion of the total sediment consumption.

In this review we also attempt to estimate the organic carbon demand necessary to maintain benthic communities in the central Arctic Ocean

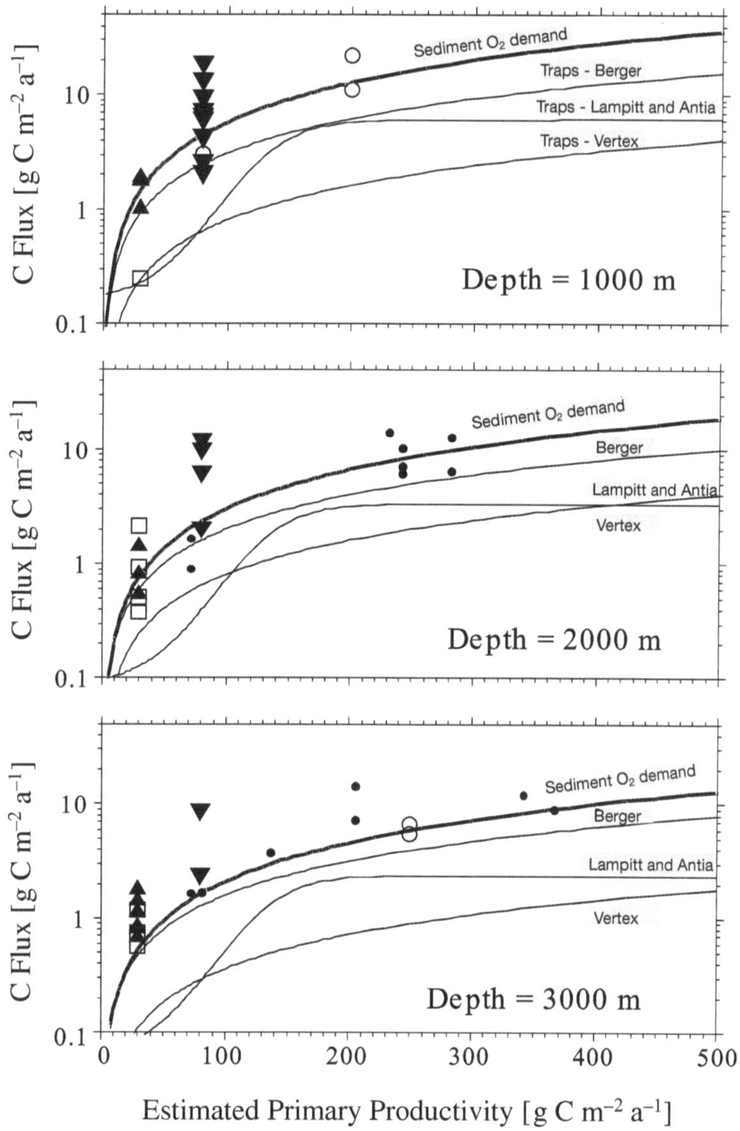

Fig. 6.6. Benthic oxygen consumption rates at different water depths in deep-sea Arctic slope and basins versus estimated primary productivity. Benthic rates are from stations within 500 m of the listed water depth. Lines represent several regressions of sinking organic carbon fluxes at the listed water depth versus overlying productivity. Arctic benthic data greatly exceed predictions based on sediment trap regressions but follow the relationship from other oceans between benthic rates and productivities. Benthic rates: open inverted triangles – Svalbard region (Hulth et al. 1994); open normal triangles – Laptev Sea (Boetius and Damm 1998); open squares – central Arctic (Christensen 2003); open circles – deep Norwegian, Greenland Seas (Graf et al. 1995); small darkened circles – ice free oceans used in regression by Christensen (2000). Regressions: thick line – aerobic deep-sea sediments (Christensen 2000); thin lines – several sediment trap evaluations (Berger et al. 1988; Lampitt and Antia 1997; Vertex given by Pace et al. 1987)

in the same way as Seiler (1999) did for the East Greenland benthos, Deubel (2000) for a part of the Eurasian Arctic and Klages et al. (2003) for the Kara Sea shelf. The methodology which is described in some detail in Figure 6.7 follows a "bottom up" approach trying to interpret processes from static faunal biomass data, e.g. how much organic carbon is necessary to maintain the somatic production of a sampled population (Fig. 6.8). Based on measured biomass (total wet weight [ww] per area) from quantitative box corer samples we applied empirical relationships between biomass, annual production, annual respiration, and carbon remineralization. Although this is a rather rough methodology which, of course, contains several uncertainties (no data for large areas because they are still un-

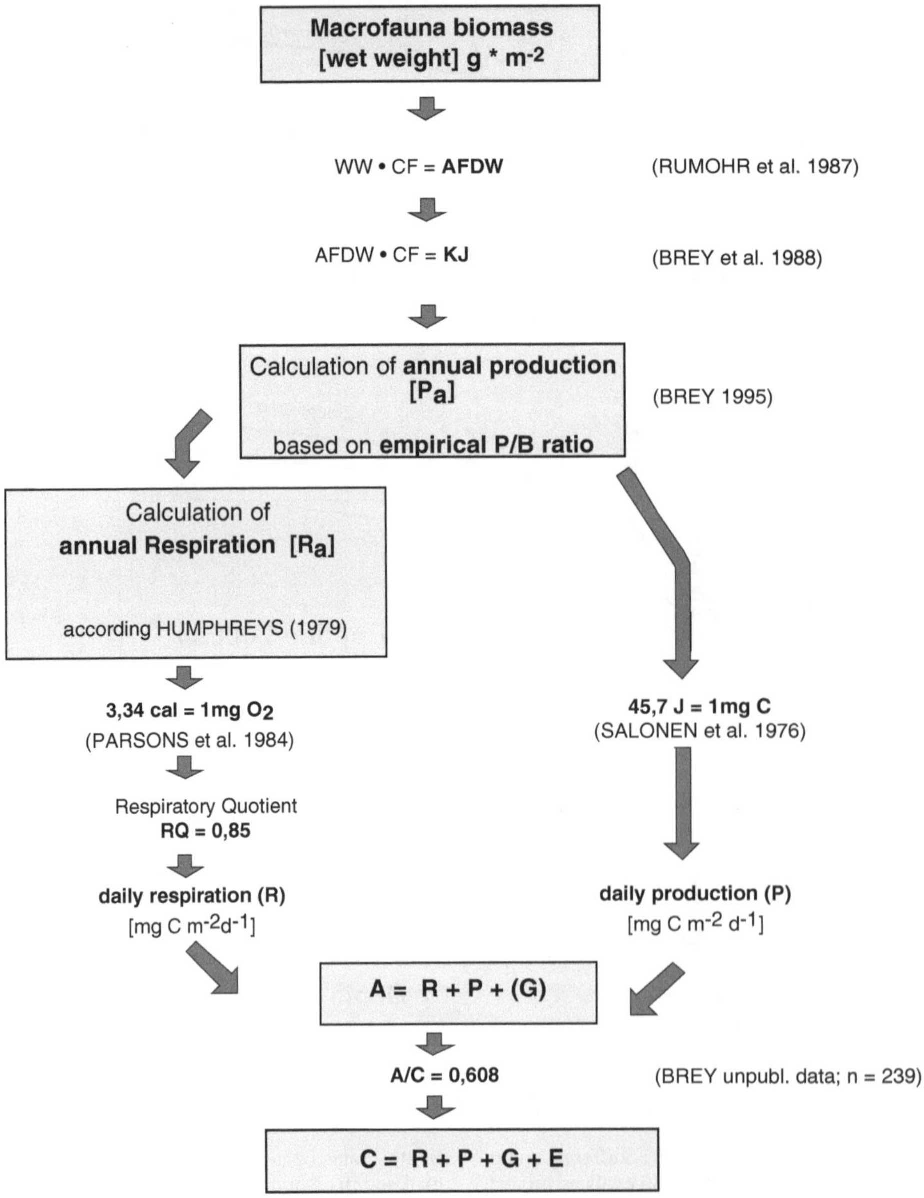

Fig. 6.7. To calculate annual production of major taxonomic groups sampled in the central Arctic Ocean, published empirical conversion factors of wet weight biomass into ash free dry weight (Rumohr et al., 1987) and kilojoule (Brey et al., 1988) were applied. The P/B ratio was calculated for each taxonomic group using an empirical relationship published by Brey (1995) which considered bottom water temperature, water depth, mode of life, feeding type and taxa. The P/B ratio served to calculate the annual somatic production (P_a) of the population at a given sampling station. Based on P_a, the annual respiration (R_a) was calculated according to regression equations given by Schwinghamer et al. (1986) and Humphreys (1979). The equation of Schwinghamer et al. (1986) basically considers P/R ratios of benthic invertebrates as published by Banse (1979) and Banse and Mosher (1980) and is expressed as $\log_{10}R_a = 0.367 + 0.993 \log_{10}P_a$ [kcal m^{-2} a^{-1}]. We also calculated annual respiration specifically for molluscs, crustaceans and all other invertebrates (except insects, molluscs and crustaceans) according to equations given by Humphreys (1979) which are expressed by as $\log_{10}R_a = 1.775 + 0.716 \log_{10}P_a$ [cal m^{-2} a^{-1}], $\log_{10}R_a = 0.607 + 0.969 \log_{10}P_a$- [cal m^{-2} a^{-1}], $\log_{10}R_a = 1.041 + 0.812 \log_{10}P_a$ [cal m^{-2} a^{-1}], respectively. Respiration was converted into oxygen equivalents (Parsons et al., 1984) and subsequently expressed as caloric units. Applying a generally accepted respiratory quotient (RQ) of 0.85 for benthic invertebrates we obtained the daily respiration and carbon demand (mg C m^{-2} d^{-1}). The daily production was calculated by the equation given by Salonen et al. (1976). Unpublished data for 239 species (Brey, pers. comm.) imply an assimilation (A) to carbon demand (C) ratio of 0.608. This ratio allowed the estimation of carbon demand of the macrozoobenthos for different regions of the central Arctic Ocean. Abbreviations used: WW – wet weight, CF – conversion factor, AFDW – ash free dry weight, KJ – Kilo Joule, Pa – annual production, Ra – annual respiration, A – assimilation, R – respiration, P – somatic production, G – gonadal production, E – egestion, C – carbon demand (from Klages et al. 2003)

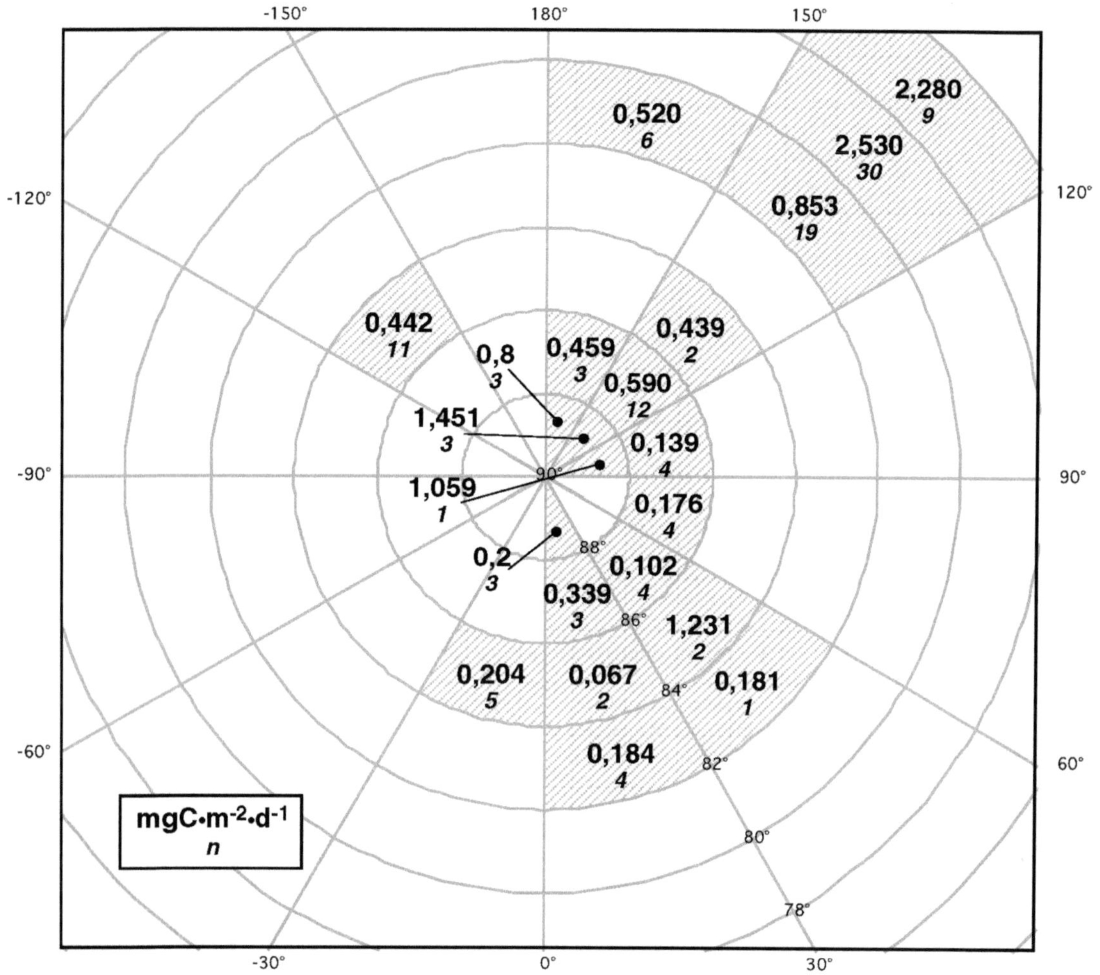

Fig. 6.8. Calculated daily macrobenthic carbon demand [mg C m^{-2} d^{-1}] in the central Arctic Ocean. Based on the total area of about 4,489 × 10^6 km^2 the macrofauna of the central Arctic Ocean consumes 1,083,000 tonnes of C_{org} (median 0.661 mg C m^{-2} d^{-1}) annually

sampled, data of empirical relationships do not necessarily fit to polar deep-sea organisms, etc.) the result (an estimate by rule of thumb) of about 1 million tons of organic carbon which is necessary for the somatic production of the macrozoobenthos is in good agreement with the carbon budget given by Sakshaug (Chapter 3) for the central Arctic Ocean. On the other hand, the estimate for the Kara Sea macrozoobenthos alone is between 5–7 million tons (Klages et al. 2003). If the Kara Sea, which contributes <20% of the Arctic shelves, is representative, then benthic carbon utilization on the Arctic shelves dwarfs that of the central Arctic Ocean basins.

Biological communities operating under food limitation are thought to evolve toward high utilization efficiencies with low loss rates, thus, because of the low productivity of the Arctic, it might be expected that carbon burial rates in Arctic deep-sea sediments would be very low. In similar fashion, Canfield (1994) and Jahnke (1996) described the relationship between the proportion of carbon buried in sediments versus sediment accumulation rates (Fig. 6.9; see also Müller and Suess 1979; Stein 1986, 1990). Conceptually, lower sediment deposition rates allow incoming organic material to reside within the active biotic zone of the sediments for longer periods of time, and this results in a greater fraction of the material becoming remineralized rather than buried. Based on those authors' formulations, the low deposition rates found in the Amundsen Basin (0.5–2.0 g of dry sediment m^{-2} y^{-1}, Stein et al. 1994 a,b; Jones et al. 1995; Darby et al. 1997; see Chapter 7.9 for details) would result in only 1–2% of the incoming organic matter being preserved by burial. For comparison, we directly calculated the proportion of organic carbon that is permanently buried based on the typical range

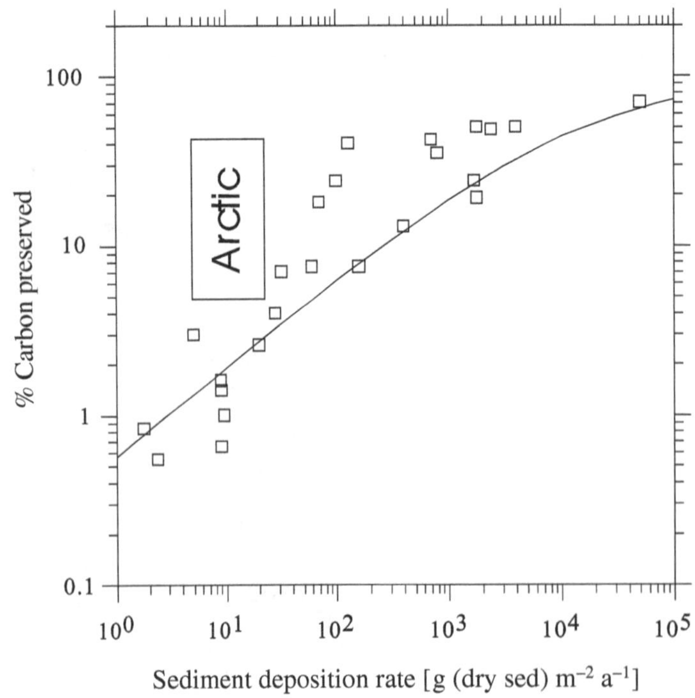

Fig. 6.9. The relationship between percentage of carbon preserved by burial versus sediment deposition rate. Open squares are for world-wide aerobic sediments (Canfield 1994) and the line is the relationship of Jahnke (1996). Carbon burial in Arctic (Amundsen Basin) sediments (box) appears anomalously high based on sediment accumulation rates of Stein et al. (1994a, b), Jones et al. (1995) and Darby et al. (1997), and benthic oxygen consumption rates of Boetius and Damm (1998) and Christensen (2003) and may reflect the high proportion of terrigenous material in Arctic deep-sea sediments

of organic carbon content of sediments below 5 cm depth and the typical range of total oxygen demand (centered around 0.8 g C m^{-2} y^{-1}). The percentage of carbon preserved by burial ranged between 6 and 28% (Fig. 6.9), and is much greater than that expected from the Canfield relationship. The burial percentage would be increased if the material arriving at the sediment surface were recalcitrant unusable organics. Thus it is likely that the high proportions of terrestrial material indicated in these deep Arctic sediments (Stein et al. 1994a, b; Fahl and Stein 1997, 1999; Stein and Fahl 2000) could be directly influencing the capacity of these sediments to preserve organic carbon (for details see Chapters 7.9 and 8). It is not known if the terrestrial material received by the Arctic deep-sea sediments is a usable food source for deep-sea benthos.

6.6 Summary and Concluding Remarks

- The central Arctic Ocean is an oligotrophic area with relatively low benthic biomasses. However, even below the permanent ice cover, primary productivity is high enough to fuel benthic life at an activity comparable to other oligotrophic oceanic regions.
- Food availability is the most important factor in determining structure and function of outer shelf, margin and Arctic deep sea benthic communities over the whole size spectra (bacteria to megafauna). Accordingly, a persistent change of the primary production regime caused by, for example, climate-change induced alteration of the sea-ice cover would most likely lead to changes in species composition and standing stocks of the Arctic benthos.
- The Arctic Ocean is characterized by its large riverine inflow and the concurrent transport of terrestrial minerals and organic matter. Deposition of this material is visible in the sedimentary record but does not seem to be an important food resource to the benthic communities.
- Temperatures close to the freezing point do not to affect the abundance and activity of benthic organisms, since respective data from temperate oceanic regions were in the same range for similar water depths.
- Echinoderms, mainly brittle stars, play a major role in benthic carbon cycling on the shelves and upper slope.
- Like elsewhere in the ocean a decrease in average organism size with increasing water depth is observed for the Arctic Ocean benthos. Even more strongly pronounced is the, on average, smaller size spectra of meiofauna organisms of the central Arctic Ocean compared to other deep sea regions at comparable water depths.
- Quantitative data on benthic biomass, activity and carbon utilization are almost missing for the central Arctic Ocean especially with regard to annual and interannual variability. Our calculation of the annual carbon demand of the macrobenthos must be considered as a very preliminary estimate.
- The currently available oxygen consumption rates for Arctic deep-sea communities suggest that the organic matter supply to the deep Arctic Ocean is more than an order of magnitude greater than expected from commonly used sediment trap regressions of sinking organic carbon fluxes. However, the Arctic data are consistent with benthic rates from other oceans after compensating for the low productivities found under the ice-pack.

Chapter 7

Organic Carbon in Arctic Ocean Sediments: Sources, Variability, Burial, and Paleoenvironmental Significance

7.1 Organic Carbon in Arctic Ocean Sediments: A General Introduction

R. Stein, R.W. Macdonald

In the preceding chapters, organic carbon sources, pathways and transformation processes at the sea floor have been presented and discussed. In the following chapter, particulate organic carbon in Arctic Ocean sediments, its composition, burial and paleoenvironmental significance will be discussed. With respect to the Arctic Ocean's organic carbon sedimentary records, we concentrate on surface sediment data and sediment cores representing the late Quaternary time interval, because information about organic carbon accumulation during pre-Quaternary (i.e., Jurassic to Tertiary) and its paleoenvironmental implications is very sparse (e.g., Thiede et al. 1990). For completeness, however, a short overview on pre-Quaternary records will be presented here.

7.1.1 Pre-Quaternary (Jurassic-Cretaceous) Organic Carbon Records

The present-day Arctic Ocean is fringed by several large sedimentary basins (e.g., North Alaska Slope, Arctic Canada, the Barents Shelf and western Siberia) that are highly productive in terms of gas and oil (Fig. 7.1.1). The hydrocarbons known from these circum Arctic basins probably derived from organic-carbon rich mudrocks ("black shales") which were deposited under marine conditions during Mesozoic times (Leith et al. 1992; see Larson 1991 for a global view). In the Barents Sea, for example, data from petroleum exploration drill holes representing Jurassic/Cretaceous time intervals indicate that these black shales are characterized by very high organic carbon (OC) contents (typically 15–25% with peak values of >30%) of a mixed marine/terrigenous origin (Fig. 7.1.1; Mørk and Bjorøy 1984; Arhus 1991; Arhus et al. 1990; Leith et al. 1992; Langrock et al. 2003a, 2003b; Mutterlose et al. 2003). It is currently still under discussion whether the high accumulation of organic carbon reflects increased primary production or increased preservation due to restricted bottom water circulation and anoxia (e.g., Mørk and Bjorøy 1984; Leith et al. 1992; Langrock et al. 2003a, 2003b; Mutterlose et al. 2003).

For the central Arctic Ocean, only in four short cores from the Alpha Ridge have older pre-Neogene sediments been recovered (Jackson et al. 1985; Clark et al. 1986; Clark 1988; Firth and Clark 1998): Core FL533 (Early Maastrichtian), Core FL437 (Late Maastrichtian), Core FL422 (middle Eocene), and Core CESAR-6 (Late Maastrichtian) (Fig. 7.1.2). In addition, sediments recovered in three cores from the southeastern flank of the Northwind Ridge are believed to be of Cretaceous age (Phillips et al. 1990).

The upper Cretaceous sediments from the Alpha Ridge consist of organic-carbon-rich black mud containing a diverse assemblage of acritarchs, prasinophytes and dinoflagellate cysts (Core FL533), and almost pure laminated siliceous oozes with excellently preserved diatoms and silicoflagellates (Cores FL-437 and CESAR-6) (Dell'Agnese and Clark 1994; Firth and Clark 1998). In general, these data suggest a warmer (ice-free) Arctic Ocean with strong seasonality and high paleoproductivity, most likely associated with upwelling conditions (Clark et al. 1986). Differences in sediment composition between the cores may have been caused by lateral and temporal variations in nutrient conditions, oceanic currents, bottom-water oxygen levels, and basin topography (Firth and Clark 1998). Without doubt, the very different paleogeographic boundary conditions during Cretaceous times have to be considered when interpreting these data. At that time, the Eurasian Basin did not exist, the Lomonosov Ridge fragment belonged to the Eurasian continental block, and the Alpha Ridge was probably located close to this block (Fig. 7.1.2; Lawver and Scotese 1990; see Chapter 1.3).

The black shales recovered in the lowermost (Early Maastrichtian) 67 cm of the 348 cm long

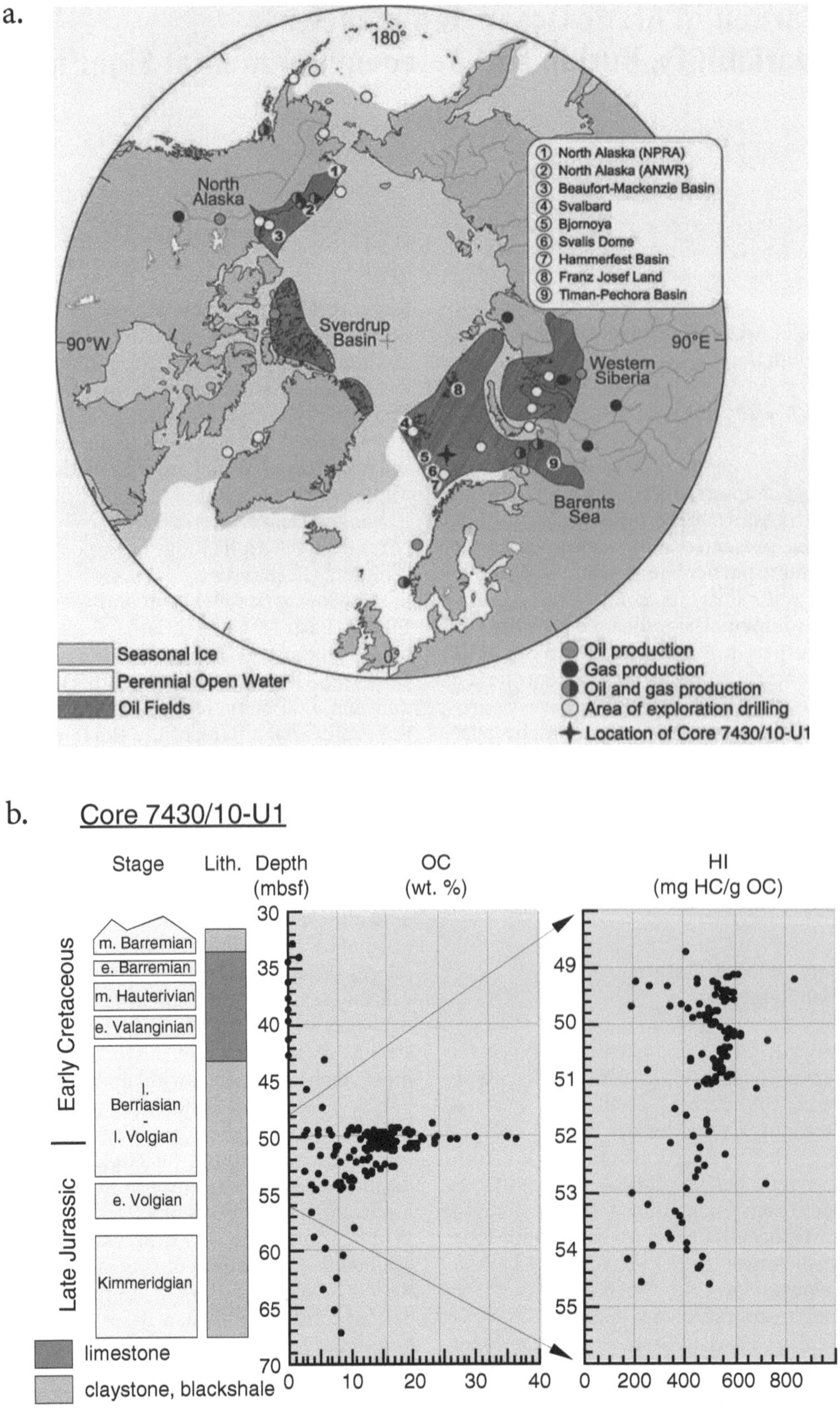

Fig. 7.1.1. (A) Distribution of major circum-Arctic oil and gas fields (modified from Bakke et al. 1998). (B) Organic carbon content and Rock-Eval data from upper Jurassic/lower Cretaceous sedimentary rocks of Barents See Core 7430/10 (74°12.47′N, 30°14.44′E, 335 m of water depth) (data from Langrock et al. 2003a, 2003b; Mutterlose et al. 2003)

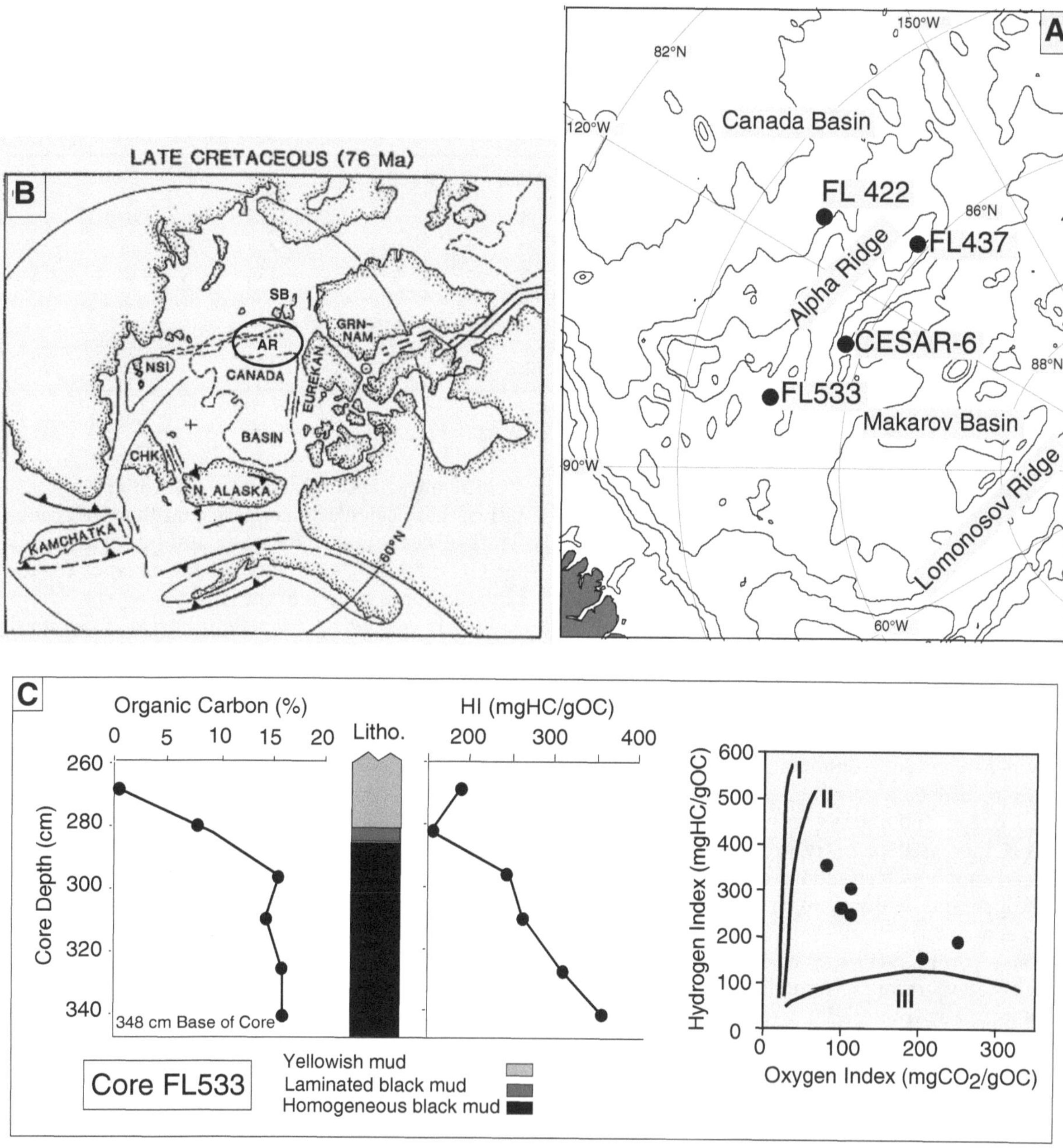

Fig. 7.1.2. (A) Map with location of sediment cores FL422, FL437, FL533, and CESAR-6 on the Alpha Ridge (for overview map see Fig. 7.9.1); (B) Reconstruction of the Late Cretaceous Arctic Ocean (from Smith 1987; Lawver and Scotese 1990). AR Alpha Ridge, SB Svalbard, NSI New Siberian Islands, GRN-NAM pole of opening for Greenland-North America; (C) Organic carbon, hydrogen index and oxygen index data and lithologies of the lowermost Early Maastrichtian section of Core FL533 (data from Clark et al. 1986; Firth and Clark 1998).

Core FL533 are characterized by very high OC contents of up to almost 16% and hydrogen index values of about 250 to 350 mgHC/gOC, and Tmax values of about 420 °C (Fig. 7.1.2; Clark et al. 1986), indicating an immature, mixed terrigenous-marine type of organic matter. Whether these black shales recovered in core FL533 resulted from anoxic conditions in an isolated local basin, a depositional environment under an oceanic water mass exhibiting an oxygen minimum, rapid burial and/or high terrigenous OC input remains an open question. Based on the mixed type of organic matter and the morphologically well-preserved nature of the organic matter in Core FL533, Clark et al. (1986) suggest a limited transport of organic detritus from the nearby Greenland-Svalbard-Scandinavia con-

Table 7.1.1. Calculation of total sediment (TS) and organic carbon (OC) budgets using different linear sedimentation rates (LSR) and OC contents. Typical mean values for Cretaceous North Atlantic black shales have been used for wet bulk density (WBD), porosity and dry density (DD) (see Stein et al. 1986 and further references therein). For the size of the Cretaceous Arctic Ocean we used 50 % of the modern size following Clark (1977)

	OC = 5%	OC = 10%	OC = 15%	OC = 5%	OC = 10%	OC = 15%
Size (10^3 km^2)	2250	2250	2250	2250	2250	2250
LSR (cm ky^{-1})	2	2	2	5	5	5
WBD (g cm^{-3})	2	2	2	2	2	2
Porosity (%)	40	40	40	40	40	40
DD (g cm^{-3})	1.6	1.6	1.6	1.6	1.6	1.6
TS (g cm^{-2} ky^{-1})	3.2	3.2	3.2	7.9	7.9	7.9
OC (g cm^{-2} ky^{-1})	0.16	0.32	0.5	0.4	0.8	1.2
TS Budget (10^6 t y^{-1})	71	71	71	178	178	178
OC Budget (10^6 t y^{-1})	**3.6**	**7.1**	**10.7**	**8.9**	**17.8**	**26.7**

tinental block (see Fig. 7.1.2) and rapid burial being the most probable cause of black shale formation rather than anoxia.

Upper Cretaceous black shales were also recovered in petroleum exploration drill holes from the Sverdrup Basin and the Beaufort-Mackenzie Basin (Leith et al. 1992), the origin of which is also still under discussion (i.e., high paleoproductivity vs. deep-water anoxia). The Arctic Ocean black shales are significantly younger than the wide-spread Aptian-Albian and Cenomanian/Turonian black shales recorded in large parts of the Cretaceous world ocean and related to regional (to global?) anoxia, increased primary production and/or increased supply of terrigenous organic matter (e.g., Schlanger and Jenkyns 1976; Arthur et al. 1984; Stein et al. 1986).

Based on the very limited number of available Mesozoic OC records from the Arctic Ocean, we are a long way from giving a robust estimate of an organic carbon budget. To provide at least an estimate of OC burial rates of the Cretaceous Arctic Ocean (excluding the shelf areas), we have constructed Table 7.1.1. The Late Cretaceous configuration of the Arctic Ocean is not known exactly but very probably was about 50 % smaller than at present (Clark 1977). A modern size of 4500×10^3 km^2 (see Chapter 1.1) implies an area of about 2250×10^3 km^2 for the Late Cretaceous Arctic Ocean. Using typical density, porosity and sedimentation rate values determined for Cretaceous black shales from the North Atlantic (see Stein et al. 1986 and further references therein) and an OC content of 10 %, an OC burial rate of 7.1×10^6 t y^{-1} is implied, a value which is about three times higher than that calculated for the modern central Arctic Ocean (see Chapter 7.9, Table 7.9.3). If higher sedimentation rates are used, which are not unrealistic (see Clark et al. 1986), the OC burial rates may have been significantly higher (Table 7.1.1). Although very tentative, these calculations imply that the Cretaceous Arctic Ocean may have been an important sink for OC (and CO_2) and, thus, of importance for the global climate system of that time.

Finally we wish to re-emphasize that these estimates are still very speculative due to the weak data base. To perform a more detailed reconstruction of the paleoceanographic and paleoenvironmental evolution of the Arctic Ocean through Mesozoic-Cenozoic times and the related organic carbon cyle, long well-dated drill cores from different key areas are urgently needed. These cores hopefully will be drilled within the new era of the Integrated Ocean Drilling Program (IODP) in the near future (Hovland 2001).

7.1.2 Modern and Late Quaternary Organic Carbon Records

The amount and composition of particulate organic carbon preserved in the sedimentary records are controlled by different factors such as terrigenous input, primary production, transformation processes in the water column and at the seafloor as well as bulk sedimentation rates. All these factors are highly variable in the different areas/environments of the Arctic Ocean. Thus, the sedimentary organic carbon records of the various marginal seas and the central Arctic Ocean (Fig. 7.1.3) are discussed separately in the following subchapters.

- The Beaufort Sea and Alaskan Shelf (Fig. 7.1.4): narrow shelf sea, enormous sediment input by the Mackenzie River;
- The North Bering-Chukchi Sea (Fig. 7.1.5): low terrigenous input, high productivity supported by nutrient import from the Pacific Ocean;

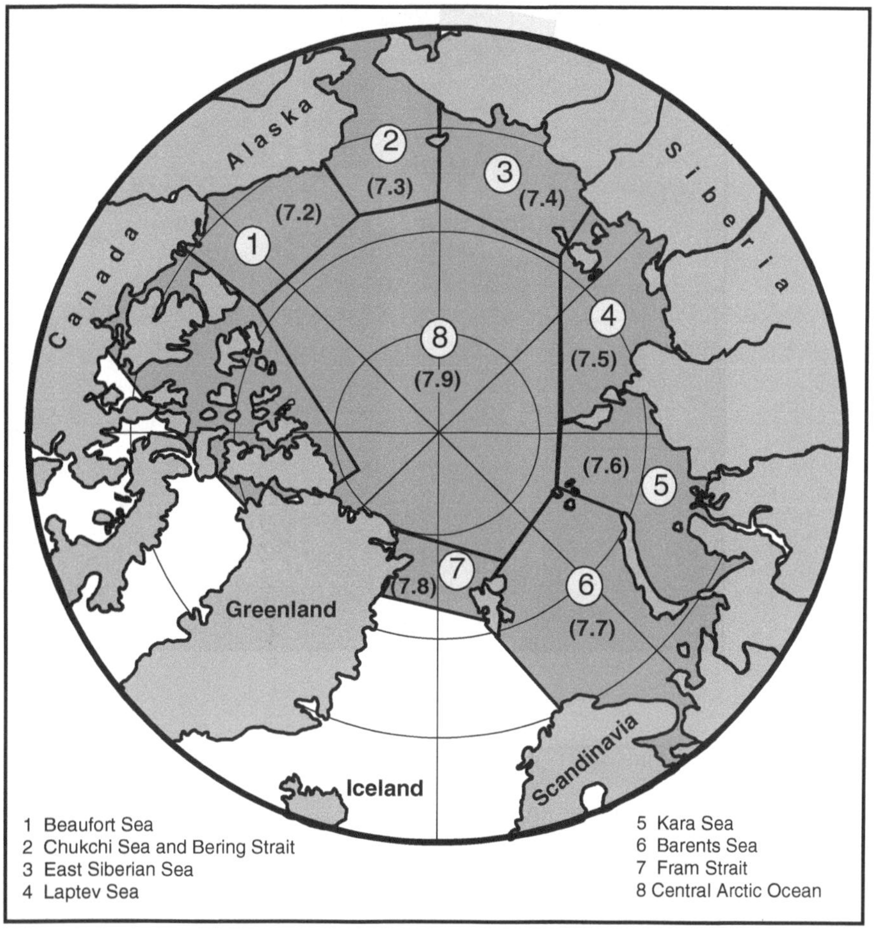

Fig. 7.1.3. Overview map of the Arctic Ocean with areas (1–8) discussed in Chapters 7.2 to 7.9

- The East Siberian Sea (Fig. 7.1.6): a very broad shallow shelf sea, Pacific-water influence, low riverine input, but high coastal erosion input;
- The Laptev Sea (Fig. 7.1.7): broad shallow shelf sea, high riverine and coastal erosion input;
- The Kara Sea (Fig. 7.1.8): broad shelf sea with deep major troughs, high riverine and coastal erosion input, some Atlantic-water inflow from the Barents Sea;
- The Barents Sea (Fig. 7.1.9): very broad shelf sea with major depressions/sinks for sediments, low terrigenous input, higher primary production due to import of nutrients from the Atlantic Ocean;
- Fram Strait and Yermak Plateau (Fig. 7.1.9): deep-water connection between the Atlantic and Arctic oceans, strong influence of Atlantic water, partly increased primary production;
- The central Arctic Ocean (Fig. 1.1.1): highly structured in basins and ridges, varying terrigenous input, influence of turbidity currents, most of the year permanently covered by sea ice, low primary production.

There is an uneven data base for the various marginal seas and the central Arctic Ocean. Much more data, for example, are available from the Beaufort, Laptev, and Kara seas than for the East Siberian Sea and the central Arctic Ocean. Furthermore, a larger number of well-dated (AMS^{14}C) sediment cores are available only from the Laptev and Kara Seas. Thus, for the latter seas, the modern budget can be set properly in the context of sedimentation changes during the Holocene. For the shelf area of the Canadian Archipelago as well as large parts of the central Arctic, no data at all are available. Accordingly, due to the lack of data, our budget calculations remain preliminary for some areas.

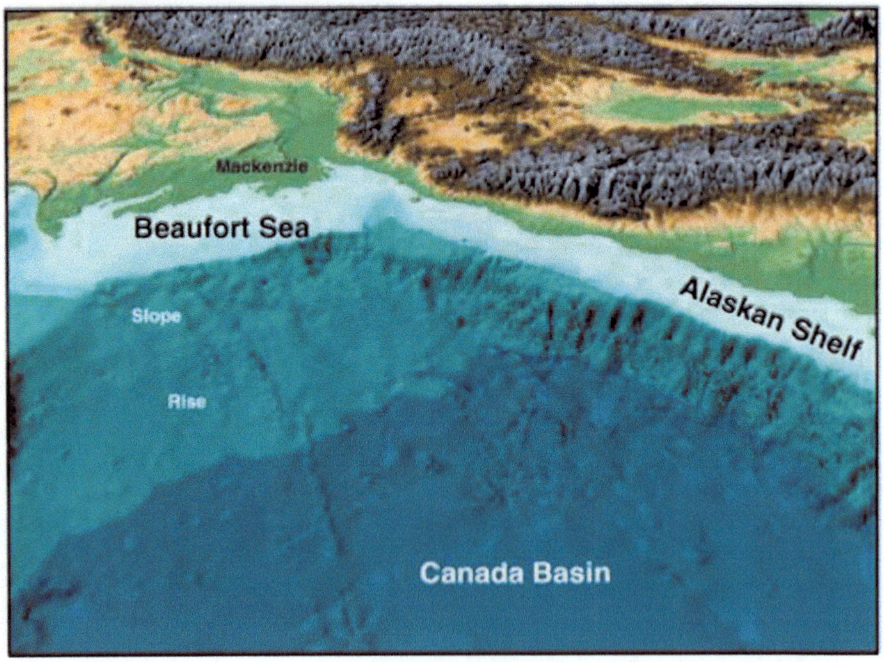

Fig. 7.1.4. The Beaufort Sea and Alaskan Shelf (Map based on IBCAO data and compiled by J. Hatzky, using the IVS Fledermaus software)

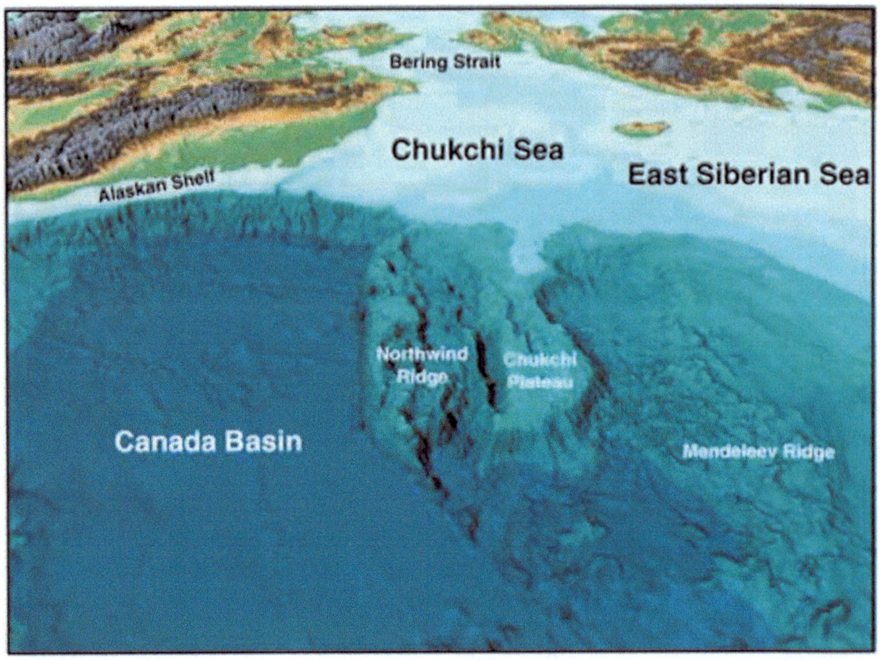

Fig. 7.1.5. The North Bering and Chukchi Sea (Map based on IBCAO data and compiled by J. Hatzky, using the IVS Fledermaus software)

Fig. 7.1.6. The East Siberian Sea (Map based on IBCAO data and compiled by J. Hatzky, using the IVS Fledermaus software)

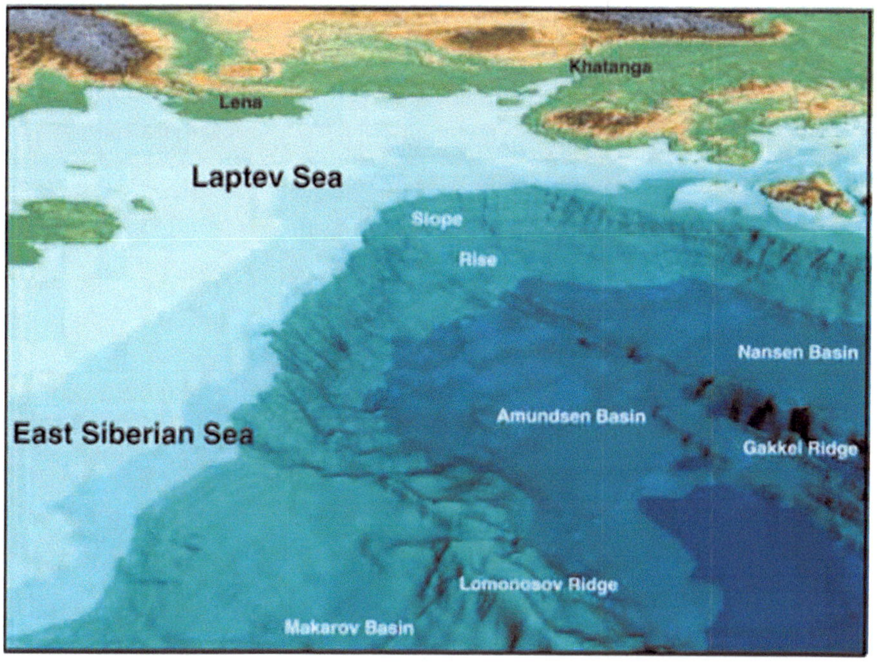

Fig. 7.1.7. The Laptev Sea (Map based on IBCAO data and compiled by J. Hatzky, using the IVS Fledermaus software)

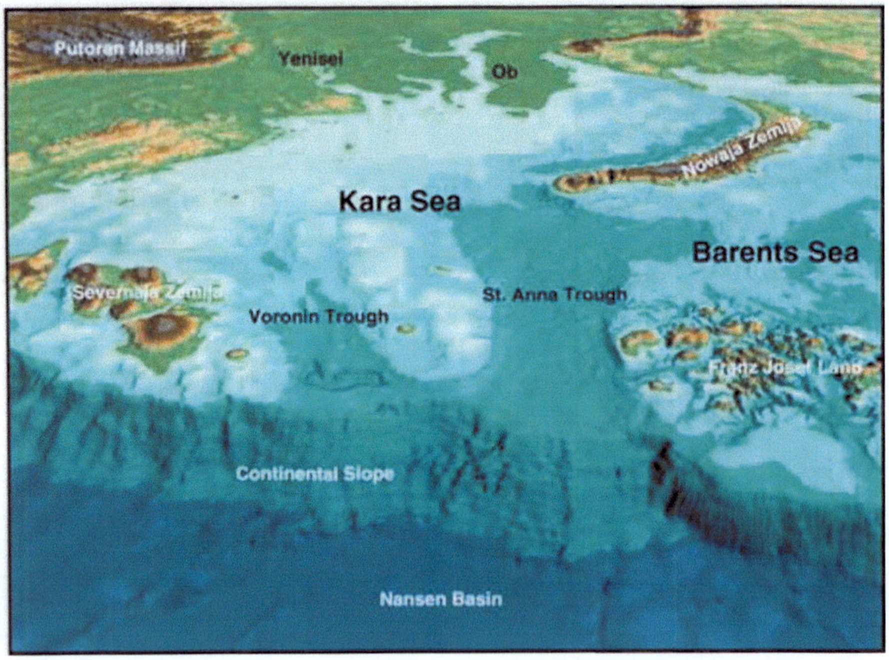

Fig. 7.1.8. The Kara Sea (Map based on IBCAO data and compiled by J. Hatzky, using the IVS Fledermaus software)

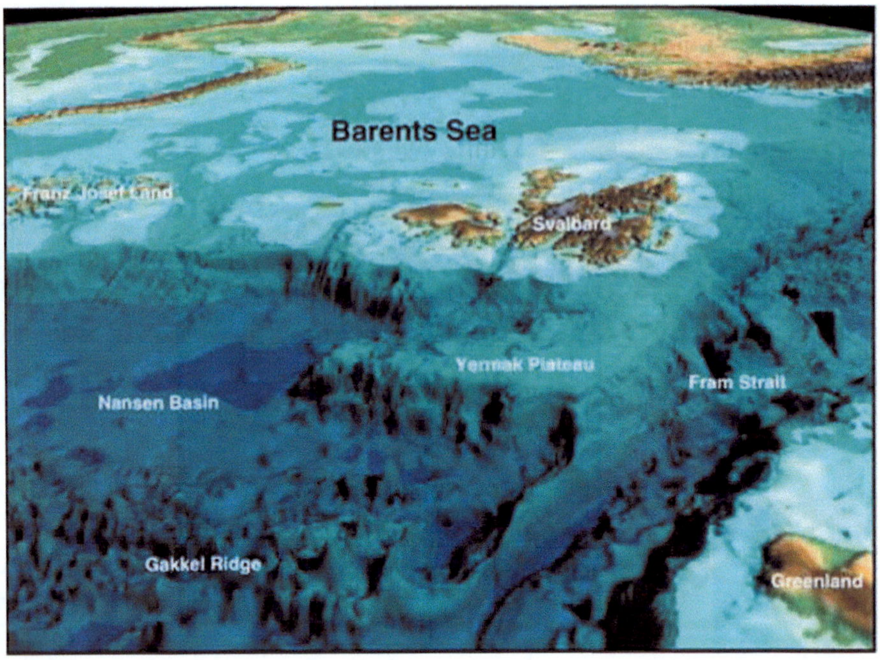

Fig. 7.1.9. The Barents Sea, Fram Strait and Yermak Plateau
(Map based on IBCAO data and compiled by J. Hatzky, using the IVS Fledermaus software)

7.2 The Beaufort Sea: Distribution, Sources, Fluxes and Burial of Organic Carbon

R.W. Macdonald, A.S. Naidu, M.B. Yunker, C. Gobeil

7.2.1 Introduction

The margin of the Beaufort Sea (Fig. 7.2.1) comprises the Alaskan shelf (44,000 km^2), the Mackenzie shelf (65,000 km^2) and the Banks Island shelf (23,000 km^2). These shelves share common large-scale oceanographic features such as the general anticyclonic ice drift in the Beaufort Gyre (McLaren et al. 1987), a flaw-lead system over the middle shelf (Stirling and Cleator 1981; Reimnitz et al. 1994; Macdonald et al. 1995) and a cyclonic current along the shelf slope/break (the Beaufort Undercurrent; see Aagaard 1984). Furthermore, surface marine water for these shelves derives primarily from the Pacific Ocean via Bering Strait, this water having undergone various degrees of processing within the Arctic en route. However, due to very different local settings, each shelf must be considered a distinct biogeochemical domain. The large troughs formed by the Barrow Canyon, the Mackenzie Trough, the Amundsen Gulf and the trench at M'cLure Strait conveniently demarcate these shelves (Fig. 7.2.1).

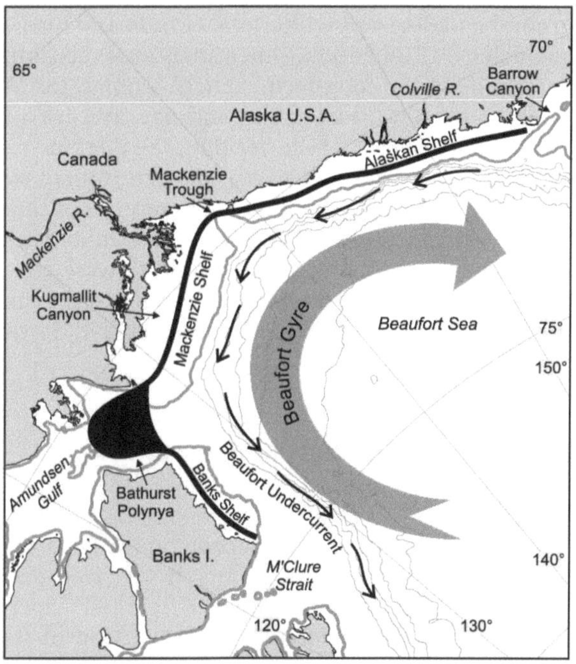

Fig. 7.2.1. A map showing the disposition of the Beaufort Shelves and features discussed in the text. The thick, dark line over the central shelves demarcates the flaw lead system which often includes a large opening (Bathurst polynya) at the entrance to Amundsen Gulf (Stirling and Cleator 1981; Carmack and Macdonald 2002)

The Alaskan shelf (see Fig. 7.1.4), in the western Beaufort Sea, is relatively narrow (40–100 km), and has many small rivers which discharge intense, short-duration, relatively small inputs of water and sediments. For example, the largest river, the Colville, discharges most of its annual inflow ($\sim 16 \times 10^9$ m^3 runoff and 5.8×10^6 t sediment) during a three-week period at spring (Arnborg et al. 1967; Walker 1974). The sediments supplied from these Alaskan rivers, which may be locally important, are augmented by coastal erosion of ice-bonded, poorly-consolidated material estimated at 2.3×10^6 t y^{-1} (Rachold et al. 2000). On the other hand, substantial quantities of sediment are probably exported from the shelf by ice rafting. For example, Reimnitz et al. (1993) estimated that during three winter months, sediment transport by ice across a 1 km N-S shelf transect amounted to 67×10^3 t. Eastward currents along the coast together with upwelling in summer (Hufford 1974; Mountain 1974) supply the nutrients to support primary production of which ice algae provide an important component especially early in spring (Horner and Schrader 1982).

The Mackenzie Shelf is broad (~150 km) and marked by a large submarine valley at its eastern end and smaller submarine valleys (e.g., Kugmallit and Ikit Troughs – see Blasco et al. 1990) over the central shelf. These troughs are likely important locations for import and export of water and associated properties (see, for example Carmack and Macdonald 2002) and have also been important locations for sediment deposition during the Holocene (Blasco et al. 1990). The shelf is overwhelmed by the Mackenzie River which annually supplies ~330 km^3 of runoff plus $\sim 124 \times 10^6$ t of sediment (Macdonald et al. 1998; see Chapter 2) making this the largest single source of sediments to the Arctic Ocean. As in many other Arctic Ocean locations, the coastline of this shelf experiences rapid erosion (Harper 1990; Solomon and Forbes 1994) partly due to thermal degradation of ice-bonded sediment and partly due to long-term response to gradual inundation from sea-level rise (Héquette et al. 1995). The large inflow of the Mackenzie River together with the relatively small area of the shelf imply a short flushing time (0.5–1 year, Macdonald et al. 1987; Omstedt et al. 1994). The annual inflow spread uniformly over the shelf would amount to about 5 m of fresh water, making this shelf by far the most estuarine in the Arctic

(Macdonald 2000). Primary production over the shelf, estimated from nitrate budgets at about 20 g C m^{-2} y^{-1} (Macdonald et al. 1987), is supported partly by nutrients from the Mackenzie River and partly by upwelling over the northwest shelf. Extensive, multi-disciplinary investigations in the Mackenzie River, delta and over the shelf, conducted during the 1970s (e.g., the Beaufort Sea Project) and the 1980s (the Northern Oil and Gas Action Plan), have produced data and literature relevant to regional sediment and organic carbon cycling (e.g., Pelletier 1975; Thomas et al. 1986; Macdonald et al. 1987; Macdonald and Thomas 1991; Telang et al. 1991; Yunker et al. 1993; Yunker et al. 1995; Hill et al. 2001).

The Banks shelf receives little direct input from land, but probably is overridden occasionally by the Mackenzie plume (Milne and Herlinveaux 1975; Giovando and Herlinveaux 1981). One of the more important features of this region is the large flaw lead that often forms the northerly extension of the Bathurst polynya at the entrance to Amundsen Gulf (Fig. 7.2.1, Stirling and Cleator 1981). This large polynya appears regionally important for biology (Stirling et al. 1993; Carmack and Macdonald 2002; Stirling 2002), and may very well sustain relatively large marine organic carbon fluxes due to high primary production. The Bathurst polynya will be the locus of a large interdisciplinary study (Canadian Arctic Shelf Exchange Study (CASES)) which should develop a proper understanding of this region's biogeochemical cycles.

7.2.2 Data Base

Sediment compositional data presented here are extracted from published and unpublished literature sources. For the Alaskan Beaufort shelf, sediment data for organic carbon and nitrogen and $\delta^{13}C$ have been assembled mainly from two reports (Naidu 1985; Naidu et al. 2000, ~80 samples). For this work, sediment samples were taken from the upper 3 cm of the sediments collected by grab samplers or gravity cores. Details on sample preparation and methods of analyses for $\delta^{13}C$ and organic carbon and nitrogen are given in Naidu et al. (1993, 2000). Briefly, samples were stored frozen in polyethylene bags. Before analysis, the subsample was treated with 10% HCl to remove carbonate, was then washed to remove excess acids and salt, oven dried and pulverized. Individual splits were analyzed using three functionally similar mass spectrometers with $\delta^{13}C$ being referenced to V-PDB with standard error assessed at ± 0.2‰. Organic carbon and nitrogen were reported on a weight to weight basis and were analyzed on a P.E. Model 240B CHN analyzer or a Europa 20/20 isotope ratio mass spectrometer with precision reported as better than 10%.

For the Mackenzie Shelf, the primary sources for sediment data include Pelletier (1975; 198 samples analyzed for organic carbon, grain size); Naidu et al. (2000; 2001; OC, N, $\delta^{13}C$, *n*-alkanes), Gobeil et al. (1991; ~60 samples from six sediment cores for OC, N) and Yunker et al. (1990, 1991, 1993, 1995, 2002; OC, *n*-alkanes) with methods described in those publications. The data base in Naidu et al. (2000) and subsequent updates can be accessed at www.sfos.uaf.edu/pubs/naidu/. For the Pelletier (1975) dataset, plots have been produced here from the original data (courtesy, S. Solomon). Samples were collected at about 200 sites (20 km intervals) between 1972 and 1975 using Van veen or Dietz-LaFond grab samplers from ships or helicopter. These samples were subsequently analyzed for texture (sieve/pipette), organic carbon and other parameters. Organic carbon was analyzed using LECO instrumentation (combustion). For the Naidu et al. (2000, 26 samples) data, grab samples or sediment core surface sections have been used with methods described as above. For the Gobeil et al. data (1991) organic carbon profiles were produced for a limited number of sediment cores collected from the Mackenzie Shelf and slope regions. For these, freeze-dried sediments from box-cores were ground and determined for total carbon and nitrogen using a CHN Perkin Elmer analyzer. Precision, determined from long-term control samples (SRM MESS-1; n = 146), was 2.8% and the average of these samples was 2.7% lower than the documented SRM value. Organic carbon was determined by correcting the total carbon for the presence of carbonate as determined through Ca determinations (precision ~5%). Comparative samples were analyzed for organic carbon and nitrogen following the more sophisticated protocol of Hedges and Stern (1984).

7.2.3 Distribution and Sources of Organic Carbon in Surface Sediments

The distribution of organic carbon on the Beaufort Shelves is influenced by inputs from river discharge, especially for the Mackenzie Shelf which receives a disproportionately large portion of its inorganic sediments from the Mackenzie River. Sediment grain size provides a useful predictor of organic carbon (OC) content with higher OC percentages generally associated with fine sediments (Fig. 7.2.2). However, this control does not likely operate in the same way for the Alaskan and

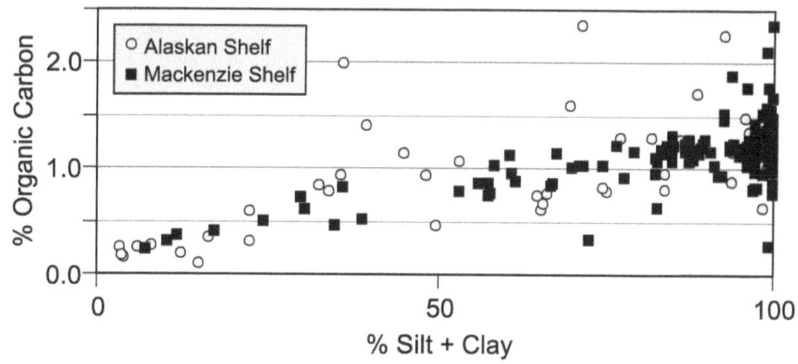

Fig. 7.2.2. A plot showing the distribution of organic carbon as a function of grain size distribution (% silt plus clay) for the Alaskan Beaufort shelf (open circles) and the Mackenzie shelf (closed squares). Data taken from Pelletier (1975) and Naidu (2000) (courtesy of S. Solomon)

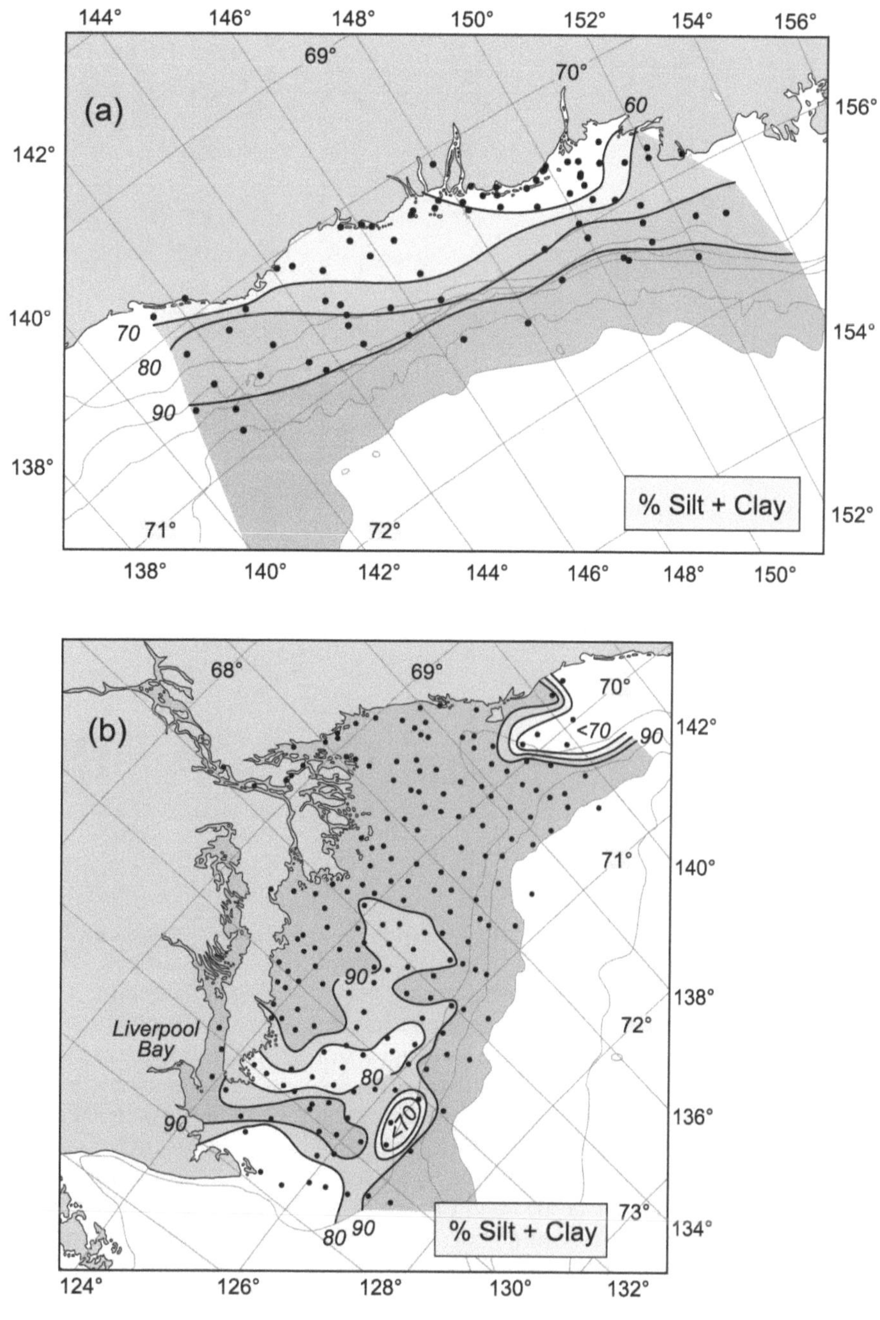

Fig. 7.2.3. Contour plots showing the distribution of fine sediments as represented by % fine particulate < 63 μm (silt plus clay) for a) the Alaskan shelf and b) the Mackenzie shelf. Data taken from Naidu (2000) and Pelletier (1975)

Fig. 7.2.4. The distribution of organic carbon in surface sediments over a) The Alaskan Shelf (Naidu et al. 2000) and b) The Mackenzie Shelf (Pelletier 1975) and c) (c, s. S. 181)

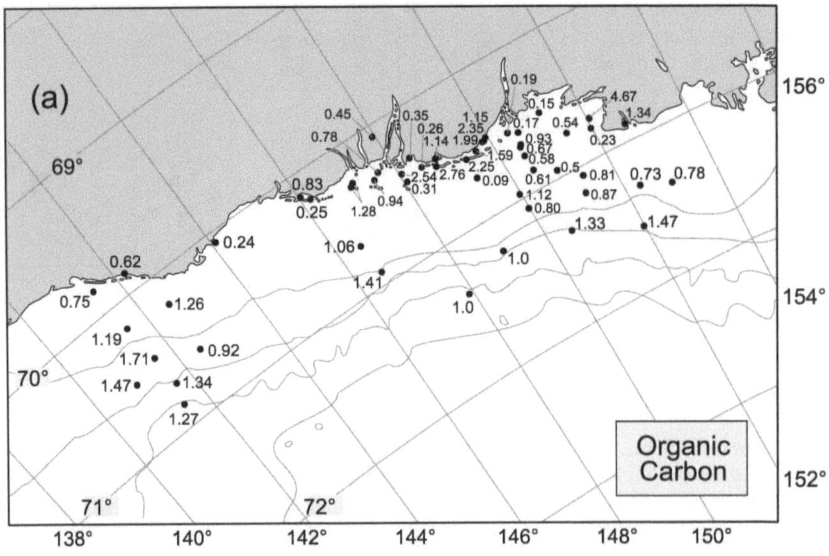

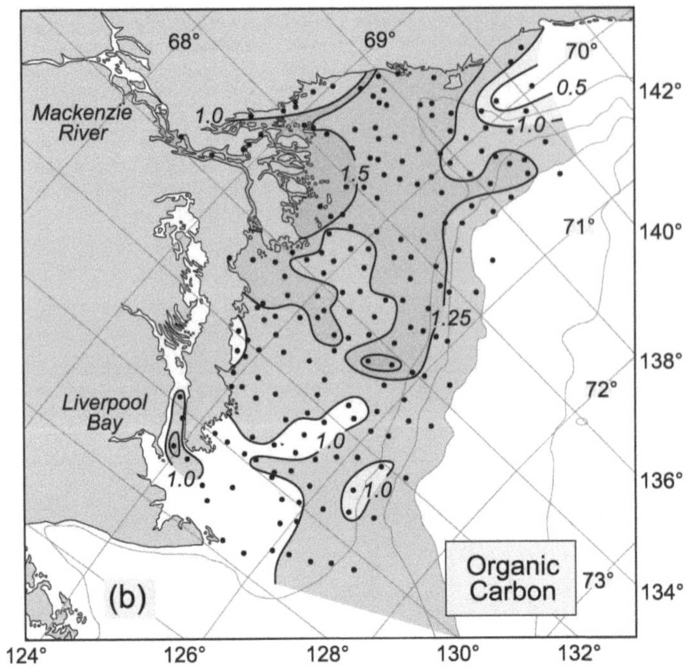

Mackenzie shelves. By comparison, we infer that relatively less fine-grained sediments are found over much of the Alaskan shelf (Fig. 7.2.3a) probably due to a combination of factors including relatively low modern supply of fine material from rivers and coasts, winnowing out of the fines consequent to reworking of shelf sediments by currents, waves and ice, gouging by ice and dilution by deposition of coarse material transported by ice (e.g., see Pelletier 1975; Barnes et al. 1987; Barnes et al. 1988; Rearic et al. 1990; Reimnitz et al. 1993). Over the adjacent Alaskan slope, sediments tend to be finer, comprising over 90% silt plus clay and containing 1–1.5% OC. The coarser shelf sediments usually contain <1% OC, often <0.5% OC, but exhibit occasional high "spot" values (almost 5% in one extreme case).

The Mackenzie Shelf provides a strong contrast to the Alaskan shelf. Although the Mackenzie slope has uniformly fine sediments similar to those of the Alaskan slope (Fig. 7.2.3b), fine sediments are also found distributed over substantial portions of the inner shelf including shallow regions off the Mackenzie's Delta. The continuing large supply of

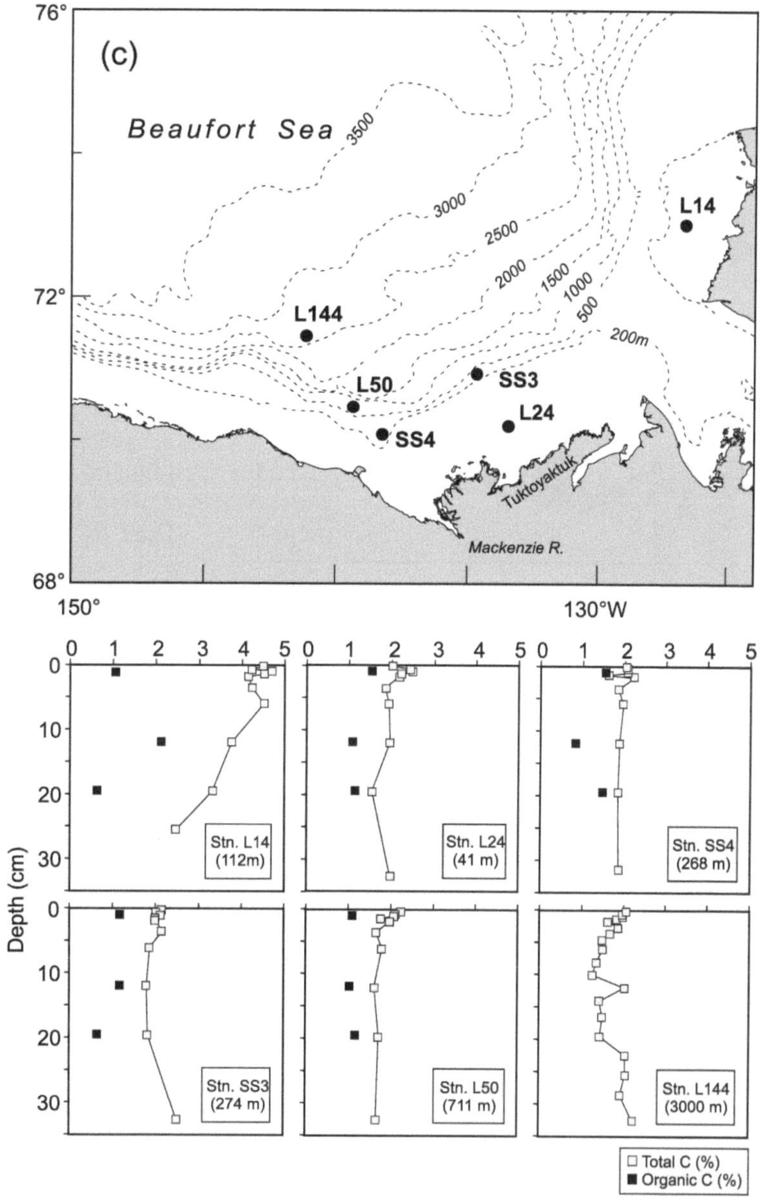

Fig. 7.2.4. c) Vertical profiles for organic carbon from sediment box cores collected at the indicated stations (data from Gobeil et al., unpublished). The Alaskan Shelf data in panel a) have not been contoured due to the thinness of the data for many areas of the shelf

fine fraction (silt and clay) from the Mackenzie River apparently overwhelms any effect of winnowing by ice scour, storm waves or currents, all of which are known to be agents of sediment reworking on the inner shelf (Hill and Nadeau 1989; Blasco et al. 1998; Hill et al. 2001; Carmack and Macdonald 2002). It is noteworthy that away from the Mackenzie River mouths, toward the northeast and northwest corners of the shelf, coarser sediments are found. A band of fine sediments is also observed extending from the entrance of Liverpool Bay (Fig. 7.2.3b).

Too few surface sediment samples from the Alaskan Shelf have been analyzed for organic carbon to allow meaningful contouring of the data. Based on the available surface sediment samples (Fig. 7.2.4a), the organic carbon content for Alaskan slope sediments usually exceed 1% with the highest value being 1.7% OC. Although there is much scatter in the shelf samples, they appear generally to contain lower OC concentrations, many of the data points falling between 0.25 and 0.9% OC. Near shore, there are a few locations with OC concentrations exceeding 2%.

For the Mackenzie Shelf, the effect of grain-size distribution on OC content can be seen due to a much denser data set which allows meaningful contouring (Fig. 7.2.2 and compare Fig. 7.2.4b with Fig. 7.2.3b). Sediments with the highest organic carbon content are associated obviously with fine-

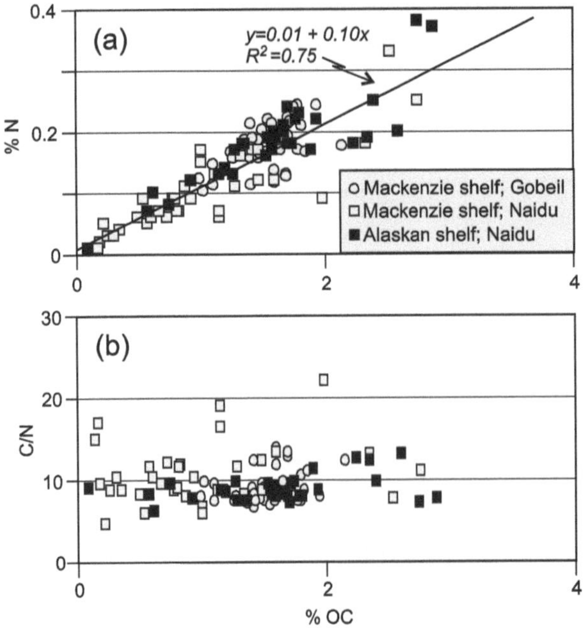

Fig. 7.2.5. a) The relationship between organic carbon and total nitrogen in surface sediments from the Beaufort shelves. The intercept (0.01) is not significantly different from zero implying little or no inorganic nitrogen is included. Data are from Gobeil (Mackenzie shelf, unpublished), open circles; Naidu et al. (2000) (open squares – Mackenzie Shelf; closed squares – Alaskan Shelf). b) C/N ratio plotted against % OC (symbols and data sources as in Figure 5a)

grain sediments of the shelf. However, there is an important difference between the distribution of organic carbon and the distribution of fine sediments. In Figure 7.2.3b, the region contoured as >90% fine particulates contains for the most part sediments with >99% fine particulates (<63 µm). Despite the uniform, fine grain size, the organic carbon within this region shows a clear spatial pattern with the highest OC values observed nearer the Mackenzie River mouths (1.25–1.5%) and lesser OC in sediments from the slope (1–1.25%). This implies the Mackenzie River to be an important source of the carbon distributed over much of the southwest shelf.

There are very few data for the vertical distribution of organic carbon distribution in sediment cores from the Beaufort shelves. Gobeil et al (1991, and unpublished data) have analyzed five shelf cores and one basin core (Fig. 7.2.4c). Although most of the cores show evidence of slight organic matter burndown in the top few centimeters of the core, the general trend with depth is one of relatively uniform concentration with the organic carbon representing about two-thirds of the total carbon (1.5–2%). The core collected at L14 on the Banks Shelf is an exception; total carbon ranges between 2.5–4.7% in this core of which only one third is organic carbon.

The sources of carbon can often be more specifically – or more incisively – determined from the composition of the organic matter (e.g., organic carbon to nitrogen ratio (C/N), *n*-alkanes and other biomarkers) or from the isotopic ratios ($\delta^{13}C$) of the organic carbon itself. For the application of these tracers to understanding of the biogeochemistry of arctic sediments we refer to Chapter 1.4. Biomarkers for the Beaufort shelves speak eloquently of a system strongly influenced by terrigenous carbon inputs (river input or coastal erosion) augmented by marine organic carbon (Venkatesan et al. 1983; Yunker et al. 1991, 1993; Steinhauer and Boehm 1992; Yunker and Macdonald 1995; Goñi et al. 2000; Naidu et al. 2000).

The C/N ratio of sediments provides an indication of the relative contributions of marine organic to terrestrial organic matter (see Chapter 1.4). Marine organic matter (OM) exhibits a C/N ratio of about 5–8 whereas terrigenous OM may have values >15 (Minagawa et al. 1991; Meyers 1997). However, inorganic nitrogen, bound especially to the clay mineral illite, sometimes imparts a confounding factor that needs to be evaluated (Schubert and Calvert 2001; see Chapter 1.4). In the case of the Beaufort shelf sediments there is a strong correlation between C and N, and based on the intercept (0.01 on the N axis, Fig. 7.2.5a) we infer that inorganic N is a very minor fraction for sediments from this region. The C/N ratios (Fig. 7.2.5b) lie predominantly in the range 6–14 implying varied mixtures of terrigenous and marine OM with no obvious relationship between % OC and the C/N ratio. The distribution of C/N ratio shows a broad spatial trend for the Alaskan margin, with values decreasing from 14 in the inshore to 9 on the shelf and 8 in the slope region (Naidu 1985).

As shown by Naidu et al. (2000), the organic carbon stable isotope composition ($\delta^{13}C$) of shelf sediments exhibits a clear spatial pattern (Fig. 7.2.6). Traditional assignments of the $\delta^{13}C$ composition of marine and terrestrial OM for this region vary. For the terrigenous OM, a number of determinations for the Mackenzie and Colville Rivers and for peat suggest a reliable value for $\delta^{13}C$ to be −26.5 to −27‰ (Goñi et al. 2000; Naidu et al. 2000; Ruttenberg and Goñi 1996; Schell 1983). A single, well-constrained value for the marine end-member $\delta^{13}C$ is more difficult to determine. Naidu et al., (2000), after examining their data and reviewing published literature values for the region (Schell 1983; Parsons et al. 1989; Minagawa et al. 1991; Ruttenberg and Goñi 1996; Cooper et al. 1998; Schell et al. 1998; Goñi et al. 2000) concluded that that Beaufort Sea marine

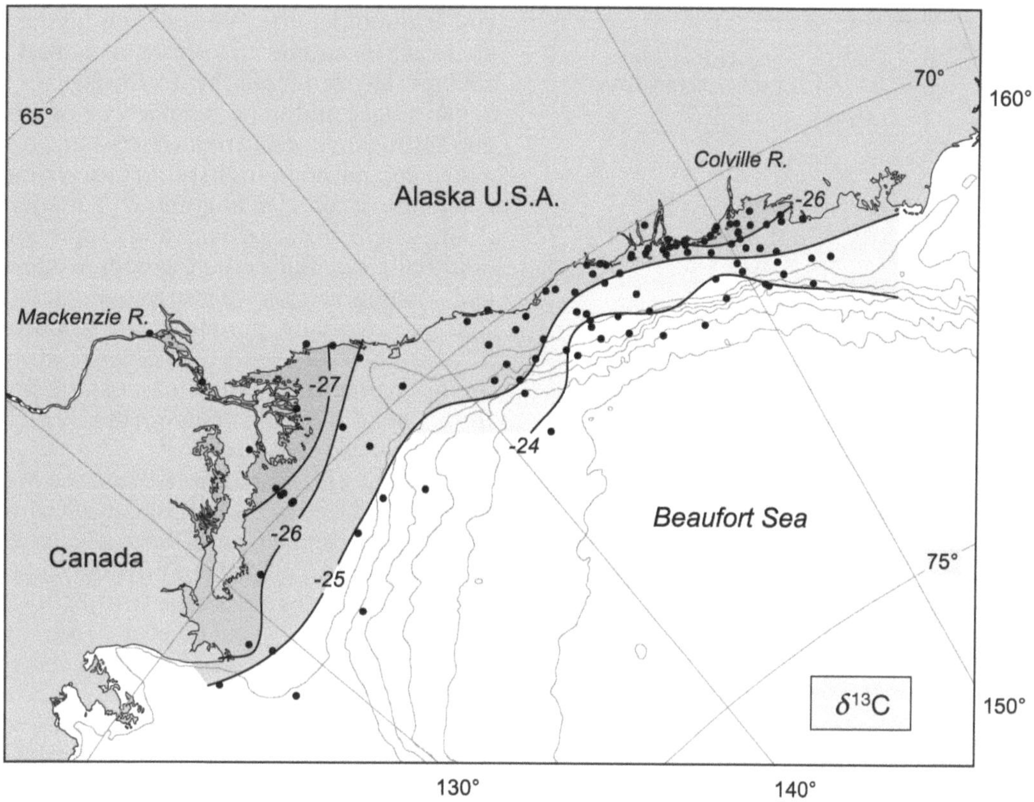

Fig. 7.2.6. A contour plot showing the distribution of $\delta^{13}C$ values for surface sediments collected on the Beaufort shelves and slope (data are from Naidu et al. 2000)

end member was about −24‰. This value agrees well with the lower range of estimates (−23.5 to −24.5) given by Goñi et al. (2000, and see their Table 5) although these authors ultimately favored a very heavy assignment for marine carbon at about −20.2‰.

The first-order interpretation of the spatial distribution of $\delta^{13}C$ (Fig. 7.2.6) suggests strong terrestrial inputs of POC near shore with progressively greater contribution from marine primary production toward the outer shelf and interior ocean. Furthermore, there is an east to west broad trend in sediments accumulated along the outer Beaufort shelf manifested as a greater ^{13}C enrichment (i.e., less negative values) toward the west. Interestingly, this same east to west trend has been observed in zooplankton (Saupe et al. 1989) and the bowhead whales that feed on them as they migrate through the region (Schell 1989). On the Mackenzie shelf, Parsons et al. (1989) provide compelling evidence based on $\delta^{13}C$ composition that terrigenous carbon supports a significant component of the inner shelf's aquatic foodweb.

Together, the C/N ratio and $\delta^{13}C$ data (Fig. 7.2.7a) suggest the shelf sediments predominantly to be admixtures of terrigenous carbon (C/N ~10 to 20, $\delta^{13}C$ ~ −27‰) and marine carbon (C/N ~6–7, $\delta^{13}C$ ~ −24‰ to −23.4‰). Rough estimates of percentage composition plotted as tie-lines on the figure show that almost all of the sediments contain over 30% terrigenous carbon and the majority over 50%. Several samples plotted on Fig. 7.2.7a (circled group at lower left of figure) suggest that there must be a source of organic carbon with depleted, terrigenous-like $\delta^{13}C$ composition (< −29‰ in one case) but low, marine-like C/N ratio. This is likely terrigenous material because many of these samples are associated with the Mackenzie River. The wide range in C/N ratios for terrigenous OM explains why the C/N spatial distribution map provides little insight on organic carbon sources whereas the $\delta^{13}C$ map (Fig. 7.2.6) presents a remarkably clear pattern. A covariance between $\delta^{15}N$ and $\delta^{13}C$ in sediments for the entire study area (Fig. 7.2.7b) further secures the above interpretation. Samples depleted in ^{13}C ($\delta^{13}C$ ~ −27‰) tend also to have low $\delta^{15}N$ (1–3‰) indicative of terrigenous material whereas samples with significant marine organic carbon ($\delta^{13}C$ > −24) have higher $\delta^{15}N$ values typical of marine particulate organic carbon (Naidu et al. 2000).

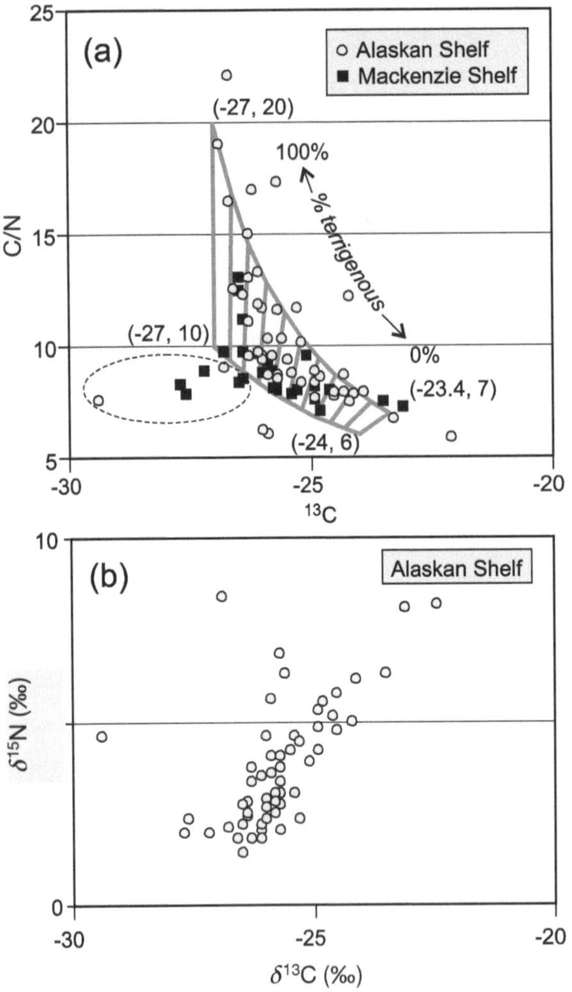

Fig. 7.2.7. a) A plot displaying the relationship between $\delta^{13}C$ and C/N for samples from the Alaskan Shelf (closed squares) and Mackenzie Shelf (open circles). Approximate organic carbon composition (terrigenous vs marine) implied by these data is shown by assuming a marine end-member of $\delta^{13}C$ = -23.4 to -24‰, C/N = 6 to 7 and a terrigenous organic carbon end-member of $\delta^{13}C$ = -27‰ and C/N = 10 to 20. Tie lines placed between the two resulting envelopes refer to percent organic carbon by source. The circled area to the bottom left outlines points from the Mackenzie Delta that exhibit terrestrial $\delta^{13}C$ values but low C/N ratios b) The relationship between $\delta^{13}C$ and $\delta^{15}N$ in samples from the Beaufort Shelf (for samples collected at the sites shown in Fig. 7.2.6). Data are from Naidu et al. (2000)

The composition of *n*-alkanes in sediments is often used to infer the source of organic matter in sediments. Freshwater and terrestrial biological systems typically produce two groups of odd-carbon alkanes: the algal alkanes (n-C_{17} with usually smaller amounts of n-C_{19} or n-C_{15}) and the more refractory plant wax alkanes (odd carbon alkanes from n-C_{23} to n-C_{33}, with a maximum typically at n-C_{27} or n-C_{29} (Eglinton and Hamilton 1967; Meyers and Ishiwatari 1993). Whereas the higher alkanes are relatively specific to terrestrial plants, the lower alkanes are produced by freshwater or marine aquatic algae and may, therefore, be of mixed origin. Petroleum hydrocarbon, if present, provides an added complication in that it has an alkane profile containing a complex mixture of primarily shorter chain length hydrocarbons and isoprenoids with essentially no odd carbon predominance (Peters and Moldowan 1993). Because biological production preferentially produces odd- numbered *n*-alkanes, the odd-even predominance at n-C_{27} can provide a practical method to assess the source and freshness of biogenic components in arctic sediments (Belicka et al. 2002).

Alkane profiles for samples from the Mackenzie River, from lakes in the Mackenzie Delta and from the Mackenzie Shelf have lower alkane fractions (n-C_{13}-C_{20} plus pristane and phytane) that are uncorrelated with higher alkane (n-C_{21}-C_{34}) fractions ($r^2 = 0.01$, $\nu = 84$) (Yunker et al. 2002). In most Mackenzie River and shelf samples the resolved, lower alkanes appear as a broad envelope of peaks that maximises at n-C_{17}, pristane or n-C_{18} reflecting a petroleum rather than an algal origin (Yunker et al. 1993). Algal-produced n-C_{17} appears as a prominent, additional peak in a number of Mackenzie River suspended particulate samples (particularly those collected in July or September). For the higher alkanes, most suspended particulate and sediment samples have a well-defined odd-even predominance consistent with terrestrial, vascular plant material. Several Mackenzie River suspended particulate samples collected in 1994 exhibited particularly elevated higher alkane concentrations (Fig. 7.2.8a) with little or no odd-even predominance leading Yunker et al. (2002) to propose recent inputs of oil.

When *n*-alkane concentrations are expressed on a per-litre basis, hydrocarbon concentrations in the Mackenzie River are linearly correlated with concentrations of suspended particulate matter (SPM) ($p < 0.001$, $\nu = 34$; Fig. 7.2.9). Within a much lower SPM range, Fernandes and Sicre (2000) have recently described a correlation between *n*-alkane concentrations and OC and SPM values in two large Eurasian arctic rivers, the Ob' and the Yenisei. In the Mackenzie, the higher *n*-alkanes exhibit a much better correlation with SPM than do the lower *n*-alkanes (Fig. 7.2.9). This suggests either that vascular plant debris is strongly associated with suspended particulate material, or that the lower *n*-alkanes are more easily lost or degraded or have multiple sources. Correlations between these two *n*-alkane parameters and OM are very poor (r^2 = 0.085–0.086), however, suggesting that the simple

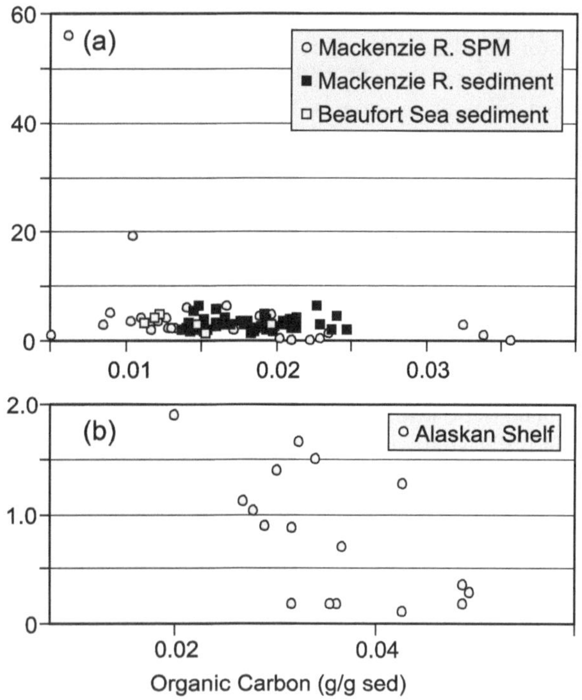

Fig. 7.2.8. A plot of (C_{27} + C_{29} + C_{31}) *n*-alkanes versus organic carbon for a) Mackenzie River suspended and bottom sediment, and bottom sediment from the shelf and b) the inner Alaskan Shelf (data from Naidu 1985)

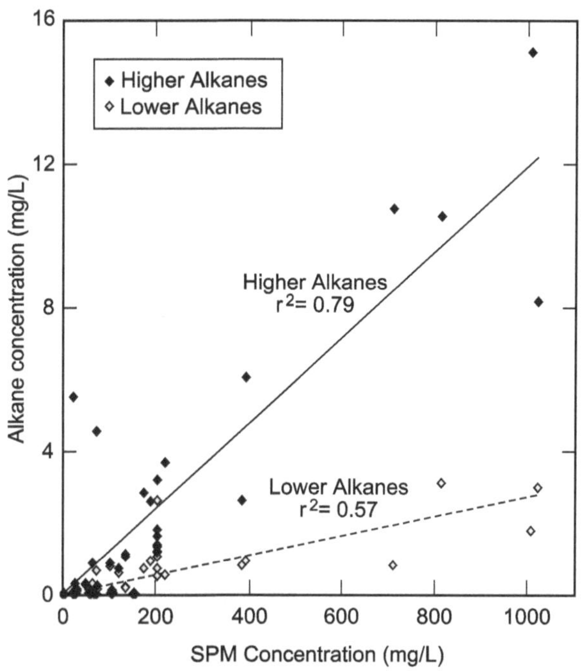

Fig. 7.2.9. Concentration of the lower and higher *n*-alkanes as a function of suspended particulate concentration for the Mackenzie River (adapted from Yunker et al. 2002)

relationship between soil erosion and riverine hydrodynamics inferred for the Ob' and Yenisei Rivers (Fernandes and Sicre 2000; see Chapter 7.6) does not apply to the Mackenzie River.

In the Mackenzie River and Beaufort Sea the most specific indicators of vascular plant inputs are the two sterols, β-sitosterol and stigmasterol (24-ethylcholest-5-en-3β-ol and 24-ethylcholest-5,22-en-3β-ol, respectively), and the triterpenoids α-amyrin and β-amyrin (Yunker and Macdonald 1995). The correlation between organic carbon and the principal vascular plant *n*-alkanes (*n*-C_{27}-C_{31} odd carbon) ($r^2 = 0.095, v = 82$) is insignificant as is the correlation between these odd *n*-alkanes and the two vascular plant sterols ($r^2 = 0.00005$ ($v = 14$), Yunker et al. 1995). In contrast, the correlation between these two sterols and α-amyrin plus β-amyrin is highly significant ($r^2 = 0.76$, $v = 14$; $p < 0.001$). The good correlations between the sterols and triterpenoids, but the poor correlation with the *n*-alkanes is consistent with mixed petroleum and vascular-plant sources for the *n*-alkanes.

The similarity between higher alkane composition of the Mackenzie River and Mackenzie Shelf sediments led Yunker et al. (1991; 1995) to conclude that allochthonous inputs from the river (suspended and dissolved) are the major influence on shelf organic geochemistry. In most Mackenzie River suspended particulate samples, nC_{17} and nC_{19} exhibit little enhancement over the adjacent, even-carbon *n*-alkanes suggesting either that algal production in the river is relatively low or that organic carbon from this source is relatively short lived due to rapid bacterial turnover (Yunker et al. 1993). Surprisingly, there is little evidence of enrichment by lower alkanes in the marine sediments of the shelf, even close to the shelf edge, suggesting that marine production of these light alkanes is quite transient, leaving little or no record in the sediments.

Peat, which tends to be dominated by higher plant alkanes with a maximum normally at nC_{31}, but occasionally also at nC_{27} (Yunker et al. 1991), is another potential source of organic carbon to the Mackenzie Shelf. Mackenzie River samples have a distinct predominance of *n*-C_{33} alkane (Yunker et al. 1993, 2002) suggesting that peat probably constitutes an important component of the organic material transported by the river. A shift of the maximum from nC_{27} in river samples toward nC_{29} in shelf samples suggests that additional inputs of peat from coastal erosion are likely to be minor, at least regionally. Using the hydrocarbon composition data together with river flow and suspended load data, Yunker et al (1991) estimated that the Mackenzie River alone supplied over 90 t y^{-1} of $\sum(nC_{27}+nC_{29}+nC_{31})$ alkane, carried mostly by the particulate phase. The composition of coastal soils and peats and their present rate of erosion imply an additional flux of about 21 t y^{-1} of these alkanes to the shelf.

A recent study of the Alaskan shelf (Naidu et al. 2001) found that high molecular weight *n*-alkanes greater than C_{21} were the major components in the alkane fraction of all sediment samples collected along the inner shelf from about 144°W to 152°W, an observation in keeping with previous published work (Steinhauer and Boehm 1992). The dominant maximum at C_{27} (occasionally C_{29}) together with odd/even ratios (summed from n-C_{15} to n-C_{34}) of 2 to 4.8 clearly signaled the importance of terrestrial plant waxes. Likely sources of this material include local rivers (e.g., the Colville) and widely distributed erosion of coastal sediments containing peat and other plant material (e.g., see Yunker et al. 1991). However, the concentrations of the higher *n*-alkanes in the Alaskan sediments were much lower than those for the Mackenzie River and shelf (compare Fig. 7.2.8a with Fig. 7.2.8b) and, like these latter samples, exhibited no significant correlation with organic carbon content.

7.2.4 The Effect of Sea-level Rise During the Holocene

Stratigraphic investigations of cores from the Mackenzie Shelf (Hill et al. 1985, 1993, 2001; Campeau et al. 2000) have developed a regional eustatic sea-level curve that shows local sea level has risen by 140 m over the past 27 ky. The rate of sea-level rise (SLR) was relatively rapid during the mid Holocene (0.7–1.4 cm y^{-1}) but slowed to a maximum probable rate of about 0.25 cm y^{-1} over the past 3 ky. Sea level has been shown to dramatically alter the transport of terrestrial particulates to the shelves and basins of the Arctic Ocean (see discussion of the Laptev and Kara Sea shelves in Chapters 7.5 and 7.6, and Stein et al. 2003a).

On the Mackenzie shelf, the continuous Holocene SLR had a number of important consequences on the depositional processes, especially in the context of Mackenzie River sediment discharge and the location of sedimentation sites. The Mackenzie Trough and Kugmallit Canyon have acted as channels for the Mackenzie River when the sea was at its lowest level, with discharge of sediment and freshwater occurring at the shelf break 9 ky BP (Fig. 7.2.10). With inundation, the discharge locations withdrew southward. Present coastal erosion, which occurs for most of the coast at rates from 5 ~20 m y^{-1} (Fig. 7.2.11, Harper et al. 1985), must be seen as the continuation of the Holocene marine transgression.

Locations for pervasive ice scour, presently at water depths from 10–20 m (Fig. 7.2.12), must also have changed with SLR (Blasco et al. 1998). Disturbance of shelf sediments by scour is an important process unique to ice-covered shelves that aerates, mixes and transports sediment, influencing organic carbon preservation (Hulthe et al. 1998). Scouring reworks surface sediments completely in two to five decades (Barnes et al. 1987; Rearic et al. 1990; Blasco et al. 1998) and the resultant rough topography provides widely distributed, small-scale sediment traps and altered benthic habitats (Blasco et al. 1998; Carmack and Macdonald 2002).

The deposition of Mackenzie River sediments at the outer shelf margin during the period 9 to 5 ky BP (Fig. 7.2.10) probably destabilized the shelf slope region leading to enhanced turbidity flows. For example, Grantz et al. (1999) suggest that Holocene sedimentation in the Amerasian Basin is made up of a slow 'pelagic' accumulation interrupted with more voluminous episodic turbidites. For both processes, the Mackenzie River has probably been the largest regional contributor of sediments with its imprint evident as far away as the foot of the Northwind Ridge (~1000 km from the Mackenzie delta). For the past 8 ky, the Canada Abyssal Plain has witnessed a continual rain of ice-rafted clastic sediment (= 1 cm ky^{-1}) with 1–5 m of episodically placed turbidites at intervals from 830 to 3450 years (Grantz et al. 1996). The source of the turbidites has been assigned to the Mackenzie Delta (or its clinoform) based on physiography and organic carbon $\delta^{13}C$ (–27 to –25‰) values. Over long periods, shelf/slope failures triggering turbidity flows may account for a significant component of the Mackenzie Shelf sediment and organic carbon budgets.

7.2.5 Burial Rates of Organic Carbon and Budget

Organic carbon budget for the Mackenzie Shelf

We base the organic carbon budget (Fig. 7.2.13) on the one given by Macdonald et al., (1998). Since those authors provided a detailed description of the data sources and methods of estimation, these will be presented only briefly here. Because the Mackenzie River provides so much sedimentary material, an accurate assessment of the sediment composition is crucial to establishing a regional carbon budget for the shelf. Fortunately, sufficient data have been collected for over two decades to constrain the river's discharge (330 km^3 y^{-1}) and sediment load (124 × 10^6 t y^{-1}). However, far fewer data are available for organic carbon (e.g., Iseki et al. 1987; Telang et al. 1991; Whitehouse et al. 1989)

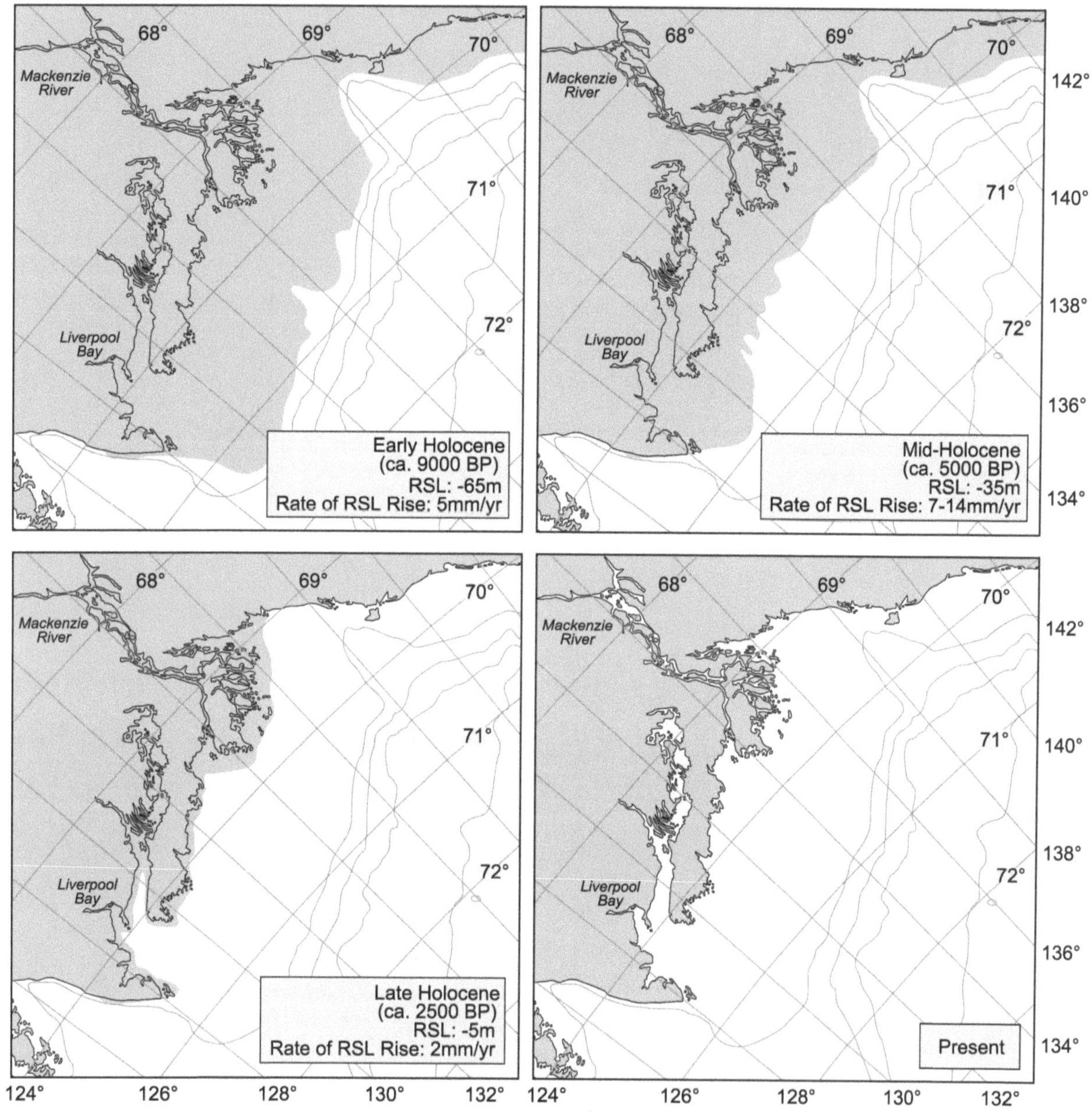

Fig. 7.2.10. A schematic diagram showing the inundation of the continental shelf of the Mackenzie shelf based on sea-level curves developed by Hill et al. (1985; 1986). Figure modified from Héquette et al. (1995)

or biomarkers (Yunker and Cretney 1995; Yunker et al. 1993; Yunker et al. 1991) making the estimate for organic carbon less robust. Sediment loadings of the Mackenzie River are based on long-term hydrological data analyzed by Carson et al. (1998) who provide an estimate for sediment entering the top of delta (Fig. 7.2.13). A difficulty with the large, multi-distributary delta, however, is estimating what portion of the sediment (and organic carbon) accumulates within the upper delta, and how much passes into the estuary (see for example Hill et al. 2001). Based on rating curves and other data (Carson et al. 1998), it appears that about half of the Mackenzie's sediment load presently does not make it to the sea. Nevertheless, for the open shelf, the Mackenzie provides the largest single source of sediment (59×10^6 t y^{-1}) followed by coastal erosion (7.1×10^6 t y^{-1}). Given the small length of Mackenzie Shelf coastline (600–700 km) compared to the larger Eurasian shelves, the supply from coastal erosion is respectable, made small only by comparison with sediment from the Mackenzie. Macdonald et al. (1998) estimated the total shelf sediment accumulation using seismic data by assuming the widely observed near-surface

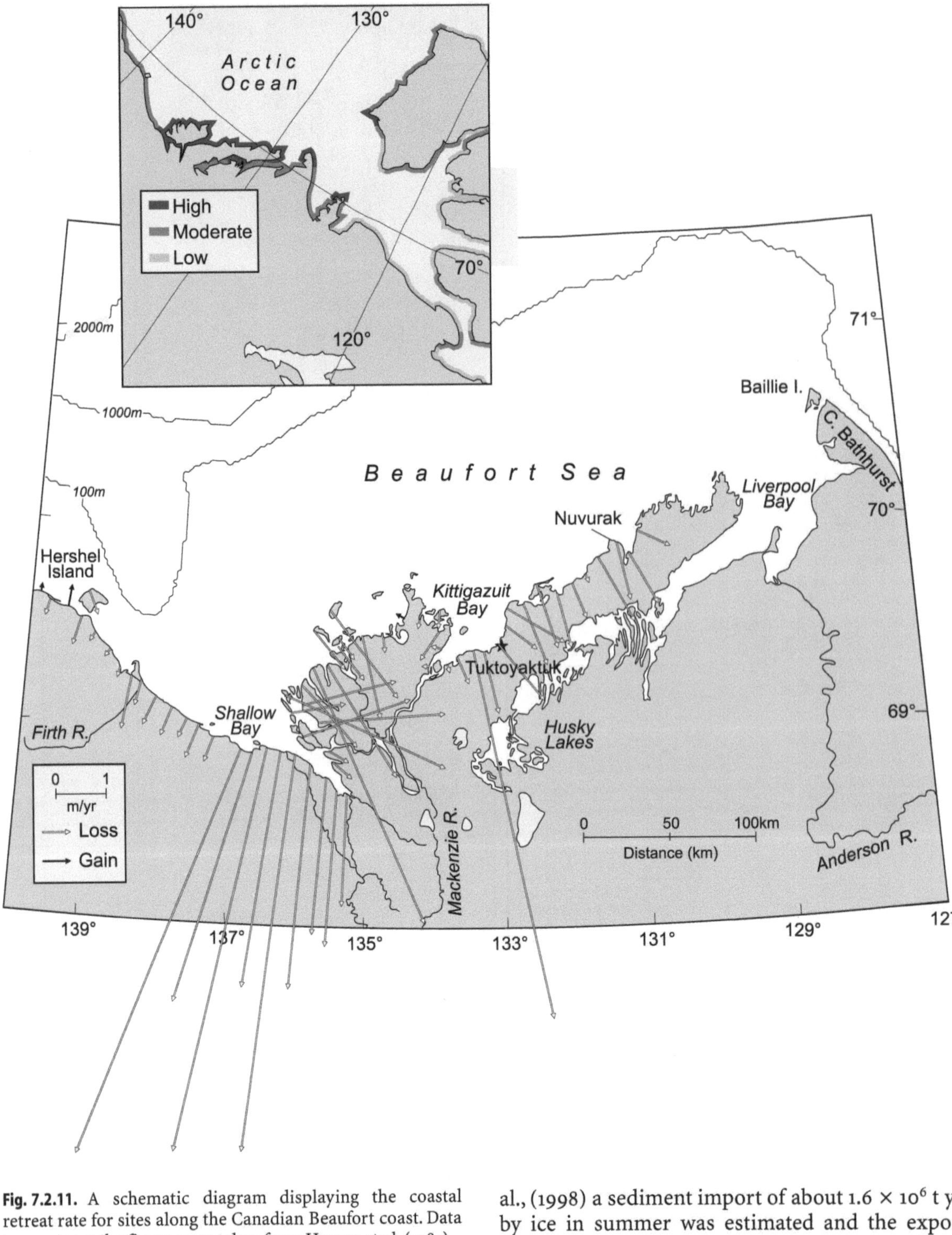

Fig. 7.2.11. A schematic diagram displaying the coastal retreat rate for sites along the Canadian Beaufort coast. Data to construct the figure were taken from Harper et al. (1985)

unconformity represents the Holocene transgressive surface. Sediment isopach maps and RSL curves were considered and a number of caveats were given in providing the long-term shelf sedimentation estimate of 53×10^6 t y^{-1}. In Macdonald et al., (1998) a sediment import of about 1.6×10^6 t y^{-1} by ice in summer was estimated and the export of sediment by ice, turbidity currents and exchange at the margin (18×10^6 t y^{-1}) was calculated by balancing the shelf sediment budget. In Figure 7.2.13 (top), we have simply presented a net export of sediment (13.1×10^6 t y^{-1}) which would include turbidity flows, surface water exchange and net ice export. According to Eicken (see Chapter 2.4,

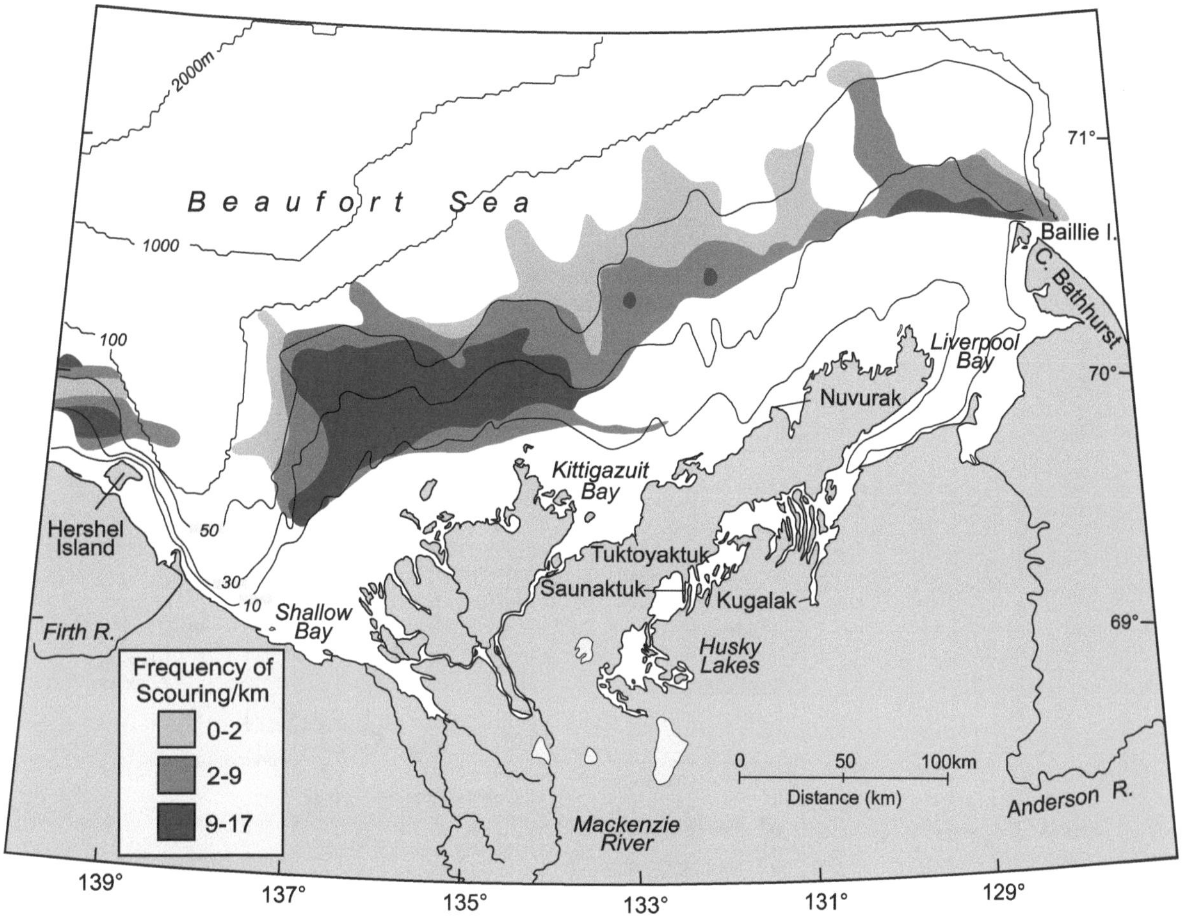

Fig. 7.2.12. A contour plot showing that the most intense frequency of ice scours on the sediment bottom are found at depth ranges from 8 to 35 m. (Modified from Milne and Herlinveaux (1975))

Table 2.12), the export of sediments and organic carbon by ice is relatively small for the Beaufort Sea (0.075×10^6 t y^{-1} SPM and 0.0005×10^6 t y^{-1} POC), leaving the other two processes to account for the majority of the export. Although there are no quantitative estimates of the export of suspended sediment in from the shelf, either at the surface or near the bottom, there is ample evidence that such export occurs (Giovando and Herlinveaux 1981; Hill and Nadeau 1989; Carmack and Macdonald 2002)

The organic carbon budget, described in detail in Macdonald et al. (1998), is presented separately for marine carbon (Fig. 7.2.13 middle) and terrestrial carbon (Fig. 7.2.13 bottom). Of significance is the fact that although marine primary production provides the largest source of organic carbon for the shelf (3×10^6 t y^{-1}), much of this carbon must be remineralized or exported such that only a small portion is preserved in sediments compared to terrestrial carbon. The budget (Fig. 7.2.13) implies that buried carbon may constitute as little as 10% from marine sources. Macdonald et al. (1998) concluded that the shelf was net-heterotrophic because the estimated loss (metabolism) of terrestrial carbon exceeded the estimated burial of marine carbon – despite the fact that this shelf has a large inorganic sediment supply and, therefore, the means to bury carbon. These authors noted the large dissolved organic carbon (DOC) supply from the Mackenzie River (1.3×10^6 t y^{-1}), but assumed that most of it simply passed unchanged across the shelf due to the short residence time of surface waters. However, further work is needed to determine the fate of terrestrial DOC over the shelf as it comprises almost half the organic carbon supply of the Mackenzie River (Telang et al. 1991). Terrestrial DOC metabolized or oxidized during transit across the shelf would increase the net heterotrophy of the shelf, perhaps significantly (see Johannessen et al. 2003).

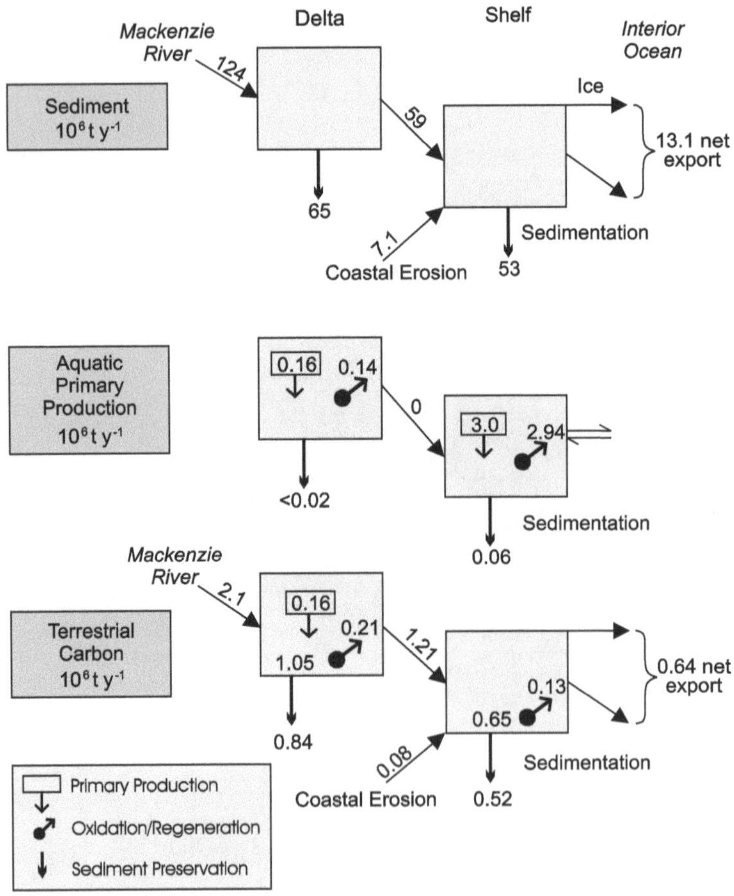

Fig. 7.2.13. A box model of the Mackenzie Delta and shelf showing in the top panel the budget for total sediments, in the middle panel the budget for autochthonous primary produced carbon and in the bottom panel the budget for allochthonous terrigenous organic carbon (modified from the budget given in Macdonald et al. (1998))

Organic carbon budget for the Colville Delta

We have calculated a tentative budget (Fig. 7.2.14) for particulate organic carbon (POC) for the 450 km^2 area of the marine facies of the Colville Delta, Alaskan Beaufort Sea coast (Fig. 7.2.1). This budget is based on data drawn from several publications (Schell and Zieman 1983, and references therein; Naidu et al. 1984, 1999, 2000; Parrish 1987). The deltaic marine facies is the offshore region that derives its sediment predominantly from the Colville River, the largest river in the North Slope of Alaska (Naidu and Mowatt 1975; Naidu et al. 1984). In this region, marine production comes from microalgae (benthic, epontic and pelagic) and local macroalgae (kelp beds). The marine primary production rate in the nearshore of the Alaskan Beaufort Sea is ~25 g C m^{-2} y^{-1} (Parrish 1987). Additional sources of the POC input are the Colville River runoff and erosion of the peat-enriched shoreline. The average annual output of POC from the Colville River is 0.17×10^6 t C y^{-1} (Schell and Zieman 1983). We calculate a net annual mass output of POC from erosion of the 72 km coast of the Colville Delta to be about 0.089 t C y^{-1}, based on a mean erosion rate of 1.4 m y^{-1} (Naidu et al. 1984), an average height of 3 m of the permafrost-dominated (75% ice) coastal bluff, an average thickness of 1.82 m of peat (Schell and Zieman 1983), and 1.18 m non-peaty deposits in the bluffs, and an assumed 60% OC in peat and 1.5% OC in non-peat sediments. If the POC derived annually from terrigenous sources were to be deposited evenly over the entire area of the Colville Delta marine facies, the amount of POC thus accumulated would be 378 g C m^{-2} y^{-1} from the Colville River and 202 g C m^{-2} y^{-1} from shoreline erosion. The POC derived from marine primary production (25 g C m^{-2} y^{-1}) would increase the supply of carbon to 605 g C m^{-2} y^{-1} for potential deposition within the above area. However, the average mass of POC currently accumulating within the delta is 160 g C m^{-2} y^{-1}, as calculated from an average sediment accumulation rate of 1 g cm^{-2} y^{-1} (Naidu et al. 1999) and a mean POC content of 1.6% for Colville deltaic sediments (Naidu et al. 2000, www.sfos.uaf.edu/pubs/naidu/). Thus it is suggested that about 72% of the pool of POC potentially available for deposition must be lost by advection, remineralization or biological con-

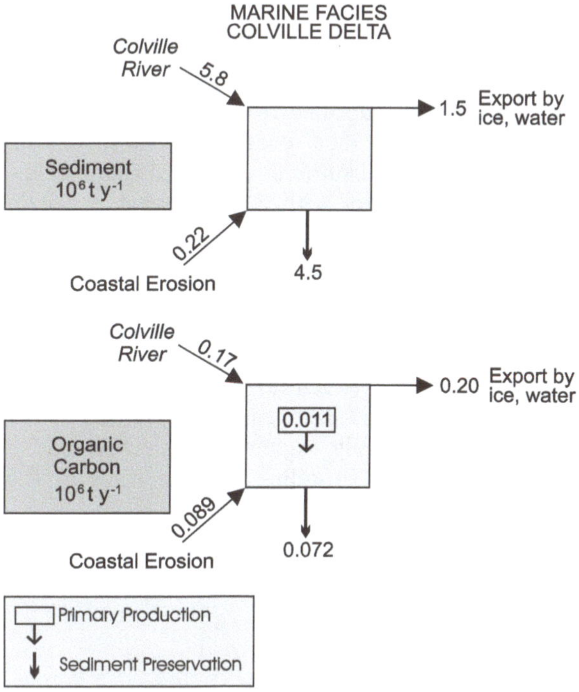

Fig. 7.2.14. A box model for the Colville delta (marine facies) showing in the top panel the budget for total sediments and in the bottom panel the budget for organic carbon

sumption. As discussed below, we contend that much of the loss likely occurs through advection by ice.

Arnborg et al. (1967) noted that 43% of the annual discharge and 73% of the total inorganic suspended load (5.8×10^6 t) are discharged from the Colville River during a three-week period at spring breakup (late May-mid June). This discharge extends as a turbid overflow wedge over sea ice to about 10 km offshore, resulting eventually in thick deposition on the ice of the entrained suspended sediments and POC (Walker 1974). With the subsequent breakup of the sea ice most of the fluvial sediment and POC deposit is transported to the shelf by ice rafting. Consequently, deposition of a large proportion of the Colville-derived sediment and associated POC is bypassed at the delta-front platform located off the river mouth and, thus, lost to the shelf (Naidu and Mowatt 1975). Advection provides another mechanism to transport riverine suspended matter and marine production toward the interior ocean, both for the narrow Alaskan Shelf (Fig. 7.2.15) and for the broader Mackenzie Shelf (Macdonald et al. 1999, and see satellite image of the Mackenzie Plume on the cover of the journal issue). In Figure 7.2.15, regions of

Fig. 7.2.15. The distribution of Chl a (mg m^{-3}) over the central Alaskan Shelf for August 25, 2002 based on SeaWiFS satellite imagery and algorithms. Bathymetry out to the shelf edge (100m) is contoured in red and to the left (70.5°N, 151°W) is the Colville River and at center (70.25°N 148°W) the Sagavanirktok River. Image provided by SeaWiFS Project, NASA/Goddard Space Flight Center and ORBIMAGE

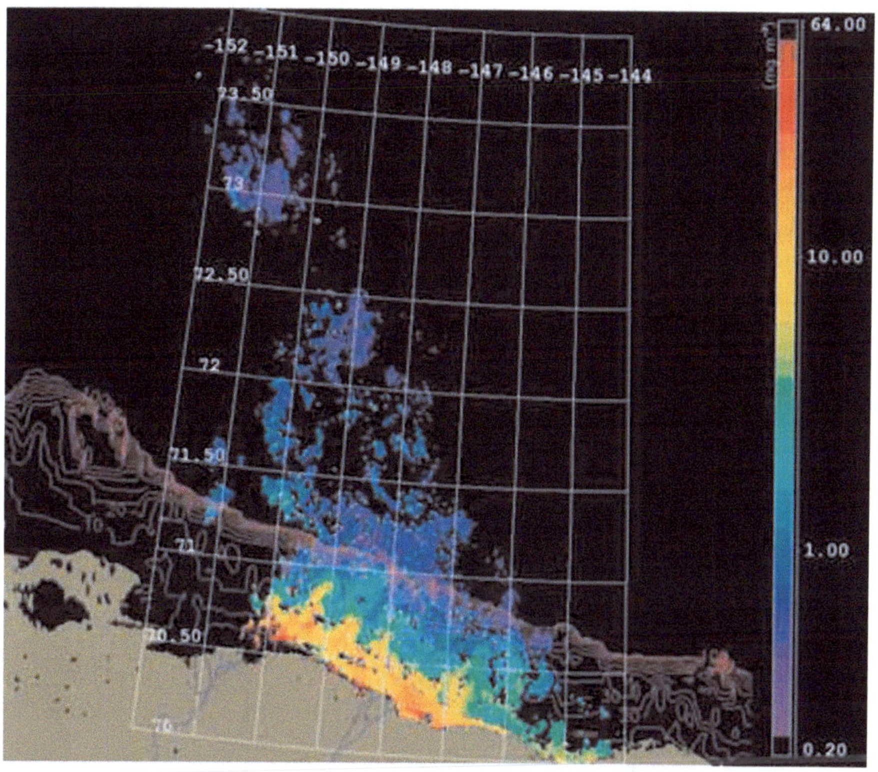

higher Chla spaced at 25–30 km intervals along the coast to the east of the Colville River probably mark baroclinic instabilities which would provide a mechanism to transport nearshore suspended sediments and terrestrial and marine organic carbon to the outer shelf. These export processes together with the narrowness of the Alaskan Shelf and the limited supply of inorganic material to support burial suggest that this shelf generally captures only small amounts of organic matter, either terrigenous or marine. Considering the mean $\delta^{13}C$ value of −26.2‰ sediment POC of the marine facies of the Colville Delta, the $\delta^{13}C$ values of the terrestrial and marine end-members, −27‰ and −23.5‰ respectively (Naidu et al. 2000), and using a mixing equation (Schulz and Calder 1976, and see Chapter 1.4) we estimate that 77% of the POC deposited in the Colville Delta marine facies is terrestrial and the rest is marine-derived.

Our estimation of the relative importance of the mass of POC supplied from the two terrestrial sources, fluvial versus coastal erosion, to the marine deltaic environment of the Alaskan Beaufort Sea is consistent with that for the Mackenzie Delta, Canadian Beaufort Sea (Macdonald et al. 1998). Both the investigations indicate that the mass of POC derived from fluvial discharge significantly exceeds that from coastal erosion, which is in contrast to the estimate for the Lena Delta of Laptev Sea, Russia (Rachold et al. 2000). Presumably, the regional disparity can be explained by the presence of relatively higher coastal cliffs and a magnitude higher shoreline erosion rates in the Lena Delta (Rachold et al. 2000).

7.2.6 Summary and Concluding Remarks

- The shelves of the Beaufort Sea are disconnected from one another by large trenches and, because they span both American and Canadian waters, have seen no large scale, coordinated research with the possible exception of a recent effort to construct regional surface maps using historical data supplemented with analyses of archived sediments for organic carbon content, C/N and stable isotope ($\delta^{13}C$, $\delta^{15}N$) composition (Naidu et al. 2000).
- Of the Alaskan, Mackenzie and Banks Shelves, only the former two have been studied sufficiently to discuss organic carbon cycling and sources.
- The shelf sediments have organic carbon contents, 0.5–2%, that correlate with sediment grain size. The continuing large supply of fine material from the Mackenzie shelf together with relatively less ice-rafting of gravel results in finer sediment accumulations on that shelf and, accordingly, higher organic carbon concentrations in surface sediments.
- Terrestrial inputs of organic matter pervade both shelves. However, the C/N ratio proves a poor tool for assessing organic matter composition and $\delta^{13}C$ appears to provide a more reliable way to distinguish carbon sources in sediments. In general, over 50% of the sediment organic carbon appears to derive from terrigenous sources.
- Higher *n*-alkanes (terrestrial plant waxes), seen in virtually all sediment samples, corroborate the conclusion based on $\delta^{13}C$ values that terrestrial organic carbon pervades shelf sediments.
- The shelves share many oceanographic features, but their sedimentary setting is very different because the Alaskan shelf is relatively starved of terrigenous inputs from rivers whereas the Mackenzie shelf receives large input from the Mackenzie River which within the Arctic is the fourth largest by flow and absolutely the largest by sediment discharge.
- A secondary source of sediment for both regions is coastal erosion especially as it often contains carbon rich peat deposits.
- Taken together, the budgets displayed in Figures 7.2.13 and 7.2.14 suggest that sediment burial in the Mackenzie Delta/Shelf sediments and in the Colville Delta amounts to 123×10^6 t y^{-1} accounting for about 1.5×10^6 t C y^{-1}. The remaining Alaskan Shelf likely accounts for a small additional burial of sediments and carbon supplied by several small rivers and coastal erosion.

- To improve these budgets, better data are required for 1) Holocene record of sedimentation rate and location (AMS^{14}C dated sediment cores), 2) Organic content and composition of the Mackenzie River, 3) the export from the shelf of sediment and organic carbon in ice or water. For the Alaskan shelf, a budget for the Colville nearshore has been constructed but should be expanded to include the entire shelf, setting in context the inputs from rivers, coastal erosion and marine primary production.
- Dissolved organic carbon appears to be potentially an important terrigenous component of the shelf carbon budget and yet little is known about its rate of supply and whether it behaves conservatively and transits the shelf, or whether it gets trapped or metabolized over the shelf.
- Ice scour is an important sedimentary process for both the Alaskan and Mackenzie shelves and transport of sediment by ice is undoubtedly important over the Alaskan shelf. However, virtually no quantitative measurements have been made of the export of carbon and sediments by this process.
- Little is known about sediment and organic carbon accumulation on the slopes of the Beaufort Sea and yet these regions appear to contain fine sediments with relatively high organic carbon content (1–1.5%). Episodic sediment gravity flows at these slopes appear to be the single most important long-term sediment supply for the Canada Basin.

7.3 The Continental Margin of the North Bering-Chukchi Sea: Concentrations, Sources, Fluxes, Accumulation and Burial Rates of Organic Carbon

A. S. Naidu, L. W. Cooper, J. M. Grebmeier, T. E. Whitledge, M. J. Hameedi

7.3.1 Introduction

The seasonally ice-covered contiguous continental margins of Bering and Chukchi Seas are some of the largest margins in the world; the U.S. portion comprises more than 50% of the total Exclusive Economic Zone of the United States. These margins have several unique environmental features that set them apart from other high Arctic margins. The region is bounded in the east and west by continents, as opposed to the rest of the arctic margins, which have land to the south and open ocean to the north. Additionally, there are several sharp cross and along shelf gradients in environmental parameters. For example, nutrient-rich basin waters upwelling along the continental slope of the Bering Sea and advecting onto the adjacent northwest Bering shelf successively give way along their migration northward to nutrient-poor waters. Across the North Bering-Chukchi margin there are also broad east to west gradients in nutrient concentration, temperature, current and salinity. A factor that contributes to the latter gradients is the presence of relatively larger, nutrient-poor fluvial flux in the east and the advection of nutrient-rich basin waters in the west. As is to be expected, the primary production rate parallels the nutrient concentrations in overlying waters and has, within the nutrient-enriched plume the highest rate (up to 720–840 g C m^{-2} y^{-1}; Springer and McRoy 1993) among all the adjacent arctic margins. The gradients have combined to develop a distinct north-south aligned hydrographic front that bisects the study area into two broad water masses (Walsh et al., 1989, and references therein).

There are additional characteristics of the North Bering-Chukchi Sea that are different from rest of the arctic margins. The recent paleogeographic history of the area has been quite distinct, inasmuch as most of the region during the late Pleistocene was not glaciated and, instead, was a fluvial-dominated coastal plain. This is reflected in the present sea bottom topography, which determines the present pathways of current flows (Fig. 7.3.1), and also extensively in the nature of surficial sediments, which influence the distribution of benthic community there. Further, the Bering-Chukchi Sea region has low degree of seasonal sea ice cover relative to other Arctic shelves and the longest migration of the ice edge between seasons. All these lateral and latitudinal environmental gradients no doubt influence the regional budget of organic carbon. It is likely that the budget of organic carbon in the very productive Northern Bering-Southern Chukchi Sea differs from the other arctic shelves which are relatively less productive. The major goal of this chapter is to synthesize published and unpublished data on the concentrations, sources, net fluxes, and rates of accumulation, burial, remineralization of organic carbon (OC), and rates of sulfate reduction and oxygen uptake in sediments of the continental shelf of north Bering-Chukchi Sea. An additional goal is to develop a 'working budget' for components of the OC cycle for high and low productive regions of the Chukchi shelf. These data, in conjunction with additional in-put on fluxes and accumulation and burial rates of OC will provide a basis to establish a more comprehensive budget for comparison with other arctic margins and the global carbon budget. Furthermore, these data will allow us to evaluate the organic carbon budget estimate published by Romankevich and Vetrov (2001). Based on OC distribution patterns of surface sediments and Holocene sediment thicknesses, these authors give an annual OC burial rate of 0.23×10^6 t for the entire Chukchi Sea.

Study Area and Environmental Setting

The study area comprises the relatively shallow (10–60 m deep) North Bering-Chukchi Sea, which is bounded by Russia in the west and Alaska in the east (Figs. 7.1.5 and 7.3.1). Portions of the Bering Sea are biologically among the most productive regions of the world margins, supporting large bodies of commercial (above 50% of the U.S. Fishery) production and subsistence fisheries harvests. Many of the environmental features of the study area have been described (Sharma 1979; Hood and Calder 1981; Smith and Grebmeier 1995; Walsh et al. 1989; Coachman and Hansell 1993; Feder et al. 1991a, b; Loughlin and Ohtani 1999, among others). Therefore, only a brief description of the area follows.

The hydrodynamics of the Northern Bering-Chukchi Sea region are complex. Water masses from various sources entrained in distinct currents maintain their identity along lengthy pathways. Mean current flow of Pacific-derived water is northward into the Arctic during most of the year. In the Bering and Chukchi Seas two Pacific-origin water masses, the Anadyr (AW) and Bering Shelf (BS) waters, are combined to form the modified Bering Shelf-Anadyr water (BSAW), which is confined to the northwest Bering Sea (Fig. 7.3.1). Flowing eastward along the Alaskan coast is Alaska Coastal Water (ACW), which is heavily influenced by the Yukon River and numerous smaller drainages. In the summer the AW is relatively colder (<1.5°C), more saline (32.5) and carries higher concentrations of inorganic nutrients (NO_3 = >20 µM) than the ACW (T= >2°C, S = <31.8, negligible NO_3) or the BS (NO_3 = <1.0 µM) (Whitledge et al. 1988, 1992; Walsh et al., 1989; Feder et al., 1994; Cooper et al. 1997). The nutrient-rich upwelled water from the Bering Basin is advected onto the adjacent shelf and entrained in the AW as the Anadyr Stream (Fig. 7.3.1). The BSAW and ACW move northward (net flow of 0.8 Sv in summer and 0.2 Sv in winter) into the Chukchi Sea, via the narrow (85 km) and shallow (54 m) Bering Strait (Fig. 7.3.1). These two water masses are delineated by a sharp vertical hydrographic front (HF), which extends N-S from Chirikov Basin to southcentral Chukchi Sea (Fig. 7.3.1). The BSAW plume, located west of the front and within Gulf of Anadyr, Chirikov Basin and Southern Chukchi Sea (Fig. 7.3.1), sustains high annual primary productivity (mean 470 g C m^{-2} y^{-1} and as great as 720–840 g C m^{-2} y^{-1}, Springer and McRoy 1993). Farther north in the Chukchi Sea the productivity to the west of the front declines to ~250–300 g C m^{-2} y^{-1}, whereas east of the front it is ~50–80 g C m^{-2} y^{-1} (Sambrotto et al.1984; Walsh et al.1989; Grebmeier et al.1995). However, there are large seasonal oscillations of nutrient concentrations (Whitledge and Luchin 1992) and productivity, with >70% of the annual productivity compressed within the episodic spring bloom, a phenomenon typical of the arctic seas (Dunbar 1968).There are several observations of East Siberian Current which promote speculation about transport of particulate organic carbon (POC) into central Chukchi Sea and possibly into the Arctic Basin (Weingartner et al. 1999). Additional exchanges across the Chukchi shelf break may influence some water mass properties (Weingartner et al. 1998) but there is little evidence for offshore sources of dissolved organic carbon (DOC) on the shelf.

Relatively large concentrations of ammonium have been observed on both the northern Bering and Chukchi shelves (Whitledge et al. 1988, 1992) which indicates relatively high rates of decomposition of organic matter (Walsh et al. 1981). The locations of the large ammonium concentrations south of St. Lawrence Island, the central Chirikov Basin, and north of Bering Strait are thought to be located over a series of sites with high OC in the sediments (Whitledge et al. 1992). Benthic production is closely coupled to pelagic production. A predominant portion of the phytodetritus, which is ungrazed in the water column, settles on the sea floor, sustaining a rich macrobenthic community (Highsmith and Coyle 1992; Grebmeier 1993; Feder et al. 1994; Grebmeier et al. 1995; Grebmeier and Cooper 1995). As a result, large populations of benthic-feeding marine mammals and birds serve as apex predators in the food chain (Grebmeier and Harrison 1992).

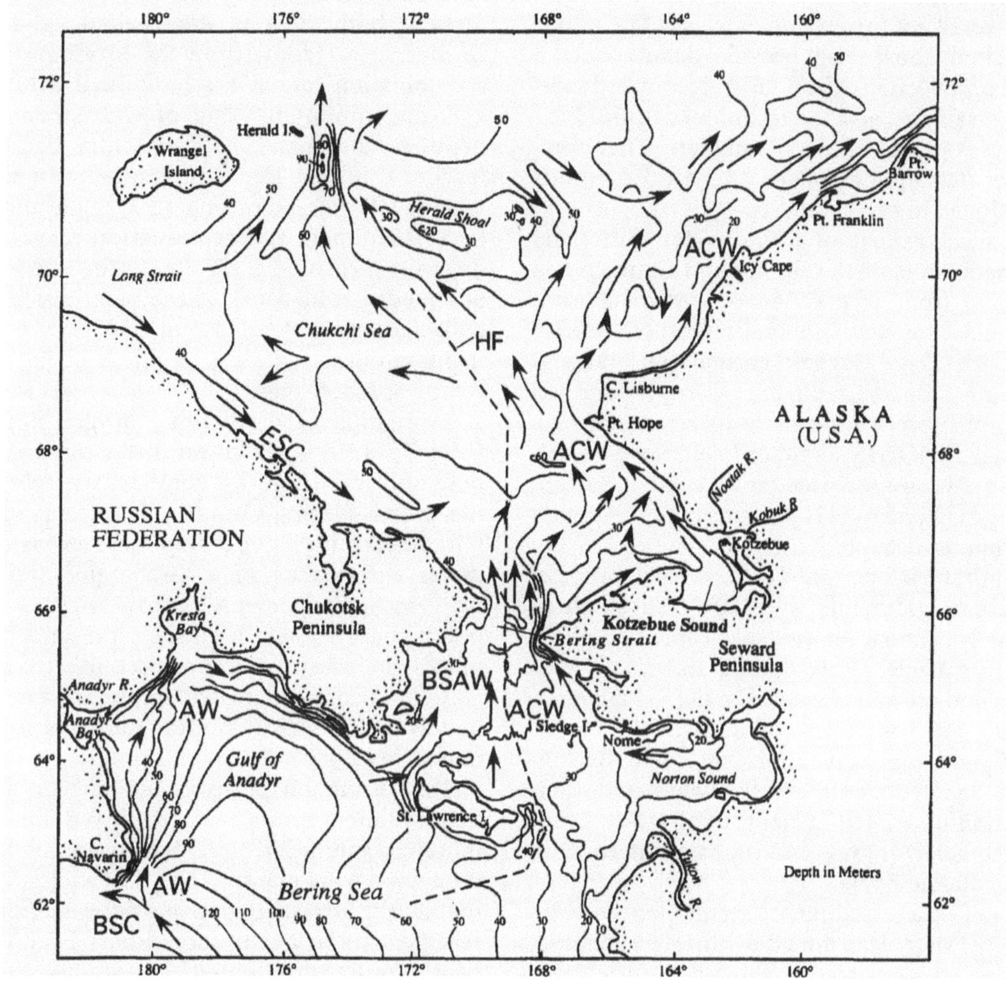

Fig. 7.3.1. Bathymetry, Hydrographic Front (HF), and major current systems in the north Bering-Chukchi Sea (after Coachman et al. 1975; Nihoul et al. 1993). ACW: Alaska Coastal water; BSC: Bering Slope Current; AW: Anadyr Water; BSAW: Bering Sea- Anadyr Water; ESC: East Siberian Current

The eastern coastal region of the Northern Bering-Chukchi Sea is dominated by deltas of the Yukon, Kuskokwim, Noatak and Kobuk rivers; the hydrochemistry of these rivers are described in Telang et al. (1991). In the west the Anadyr is a minor river, which has minimal effects on the distribution of DOC in Anadyr Bay but may be a significant source of POC (Agatova et al. 1999). The nearshores east and south of the Seward Peninsula are characterized by numerous polynyas. Sea ice covers the north Bering-Chukchi Sea for 7–9 months, and its margins can sustain substantial production (Niebauer and Alexander 1985). The sea floor sediments of the Russian-Alaskan margin consist of a mosaic of poorly-sorted gravelly to muddy contemporary to relict sediments, displaying no spatial trend, and which are frequently reworked by ice gouging, marine mammals and occasional storm action (Naidu 1988).

7.3.2 Database, Material and Methods

For this synthesis the original data for the concentrations of sediment organic carbon (OC) and total nitrogen (N), C/N ratios (wt/wt), carbon isotope ratios ($\delta^{13}C$ ‰), flux and mass accumulation rates of total sediment and OC, percents organic carbon remineralization in sediment, sulfate reduction rates, and estimates of the benthic oxygen uptake rates are tabulated from several published papers (Grebmeier 1987, 1993; Grebmeier et al. 1988, 1989; Feder et al. 1994; Grebmeier and Cooper 1994, 1995; Baskaran and Naidu 1995; Naidu et al. 2000; Cooper et al. 2002) and unpublished reports (Feder et al. 1989, 1991). Our synthesis is based on analyses of about 400 surface grab sediment samples and selected gravity cores from the study area. The cores, sectioned onboard the ship, and the grab samples were kept frozen until analysis. The publications cited above also provide details on the methods of collection of bottom grab, core and sediment trap samples and the techniques utilized for the analyses of the above parameters. Therefore, only brief descriptions of the analytical methods are mentioned in this synthesis.

The concentrations of C and N for individual surface sediments of the Bering and Chukchi Sea are derived from several sources (Sharma 1979; Feder et al. 1989, 1991; Grebmeier 1993; Grebmeier and Cooper 1995). Organic carbon (OC) and N contents on CO_3-free sediments were determined either by an isotope ratio mass spectrometer or a P.E. Model 240B CHN analyzer (Naidu et al. 1991). The C/N ratios are reported on a weight-to-weight basis of the OC and total N percents in the sediment organic matter (Chapter 1.4).

The carbon isotope ratios ($\delta^{13}C$; see Chapter 1.4) of the above sediments were analyzed by three functionally similar mass spectrometers (V.G. Micro-mass Model 602E, Europa 20.20, and V.G. SIRA II), and are all referenced to the V-PDB standard, with a standard error of analysis of ±0.2‰. The $\delta^{13}C$ and C/N values for the North Bering-Chukchi Sea sediments are tabulated and their distribution illustrated in Naidu et al. (1993, 2000) and Grebmeier (1993), and updated here with supplemental new data.

The net mass sediment accumulation rates (Table 7.3.1) were determined by the ^{210}Pb method, using alpha spectrometry (Baskaran and Naidu 1995). The sediment (total) and OC advective flux to the sea floor were estimated for selected 9 sites in the Chukchi Sea (Table 7.3.1) by collecting particles settling into a trap (designed after Hong 1986) deployed 5 m above the sea floor, in August when the sites were free of sea ice.

The mass accumulation rate of OC (Table 7.3.1) was ascertained from the product of the net accumulation (g cm^{-2} y^{-1}) at a site and the OC concentration, mg/g, in the top 1-cm of sediment at that site (Baskaran and Naidu 1995). It was possible to estimate the percent of total OC remineralized in sediment only for two out of six locations investigated in the Chukchi Sea (Feder et al. 1991). This study on two cores (SU5 and SU10) was over a core depth interval of 17cm, where net linear decreases in OC with core depths were observed (r: −0.88 and −0.90 for SU5 and SU10 respectively). This estimation was based simply on the difference between the total C concentration (on a dry weight basis) in the top 1-cm of a sediment core sample and the concentration of OC measured at depth in the core, following the 'G' model approach (Berner 1980, 1982; Martens and Klump, 1984). The OC burial rates at both stations were calculated following Gordon et al. (2001) from the product of the net accumulation rate at the individual core location and the minimum value of OC encountered at depth in the core (i.e., "refractory background" level of OC). The depth-integrated mineralization rate of OC was according to Marten and Klump (1984). Sediment sulfate reduction rates (mM m^{-2} d^{-1}) were measured at six sites in the south Chukchi Sea (Feder et al. 1991). These measurements were collected on short (16–20 cm) HAPS core samples, using radiotracer ($^{35}SO_4^{2-}$) methodologies (Jorgensen 1978; Albert 1985).

Estimation of the benthic oxygen uptake rate (mM O_2 m^{-2} d^{-1}), which estimates the organic carbon utilization rate by benthos, was obtained in the North Bering-Chukchi Sea from 1984–1999 (Grebmeier 1987, 1993, unpublished data; Grebmeier and McRoy 1989; Grebmeier and Cooper 1994, 1995; Cooper et al. 2002). Each of the data points reported here for a site represents the mean of two replicate measurements conducted on HAPS (133 cm^{-2}) cores incubated shipboard for approximately 12 hr at in-situ bottom temperatures (Grebmeier 1988, 1989).

For estimation of sediment particle and POC flux sediment traps were deployed for about a month in summer at fairly widely spaced sites, and for a few days at one site (KW3) in winter. To collect the yearly fluxes of total particles and POC, the sediment traps ideally should have been deployed continuously throughout an annual cycle, but this was not possible due to logistical constraints. Another limitation in the estimation is that we assume that the particles intercepted in the traps are primary settling detritus advected vertically downward from ambient overlying waters containing insignificant amounts of resuspended particles from the sea bottom.

The data on the concentrations of mud, OC, and N, C/N, $\delta^{13}C$, and benthic oxygen uptake rates corresponding to sediments of the individual locations of the study area can be accessed at www.sfos.uaf.edu/pubs/naidu/.

7.3.3 Distribution and Sources of Organic Carbon in Surface Sediments

The concentrations of OC and N vary widely, ranging from 0.1 to 2.8% (Fig. 7.3.2) and 0.01 to 0.28%, respectively. Generally, the concentrations of organic carbon in sediments of the North Bering-Chukchi are similar to those (0.5 to 1.5%) of most world shelves (Premuzic et al. 1982; Romankevich 1984). The exceptions are northwest and

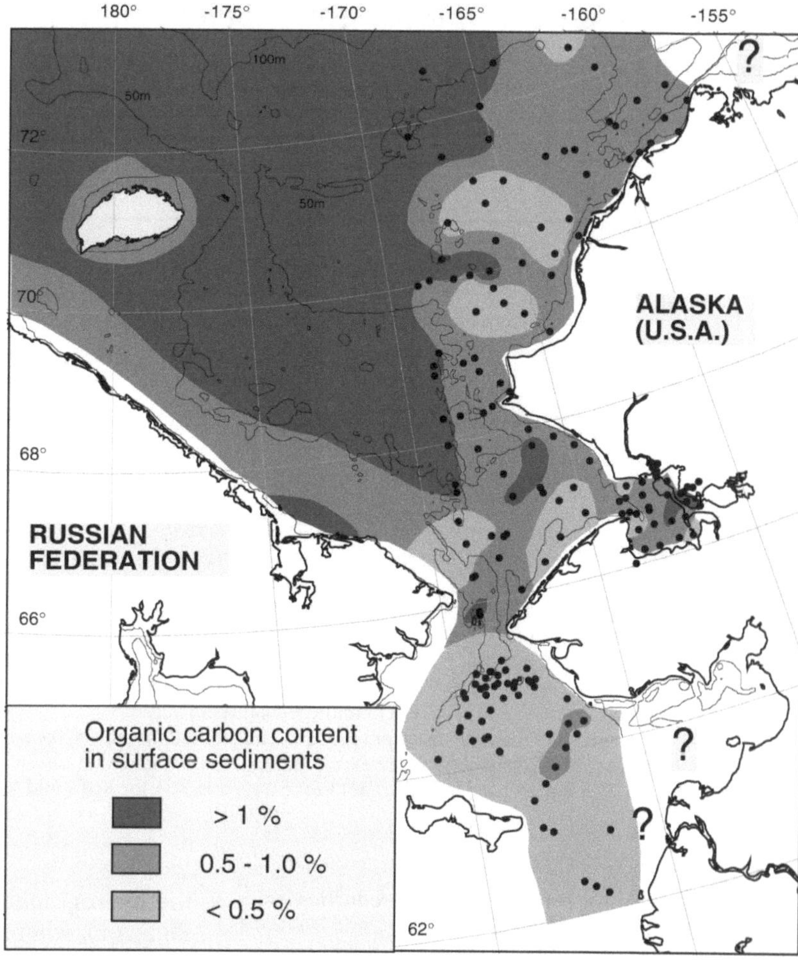

Fig. 7.3.2. Distribution of organic carbon (%) in surface sediments of North Bering-Chukchi Sea. For the area east of 170°W, OC contents >1% predominently represent values between 1 and 1.5%. For the area west of 170°W where no data points are shown, the OC distribution pattern has been taken from Romankevich and Vetrov (2001)

southeast Gulf of Anydyr and northwest Chukchi Sea where locally higher (1.5 to 2.8%) OC occurs (Grebmeier 1993). Despite the large scatter in the OC vs. N plots there is a net significant covariance between the two (Fig. 7.3.3). There are no significant regional distributional trends for OC and N within the study area (Feder et al. 1989, 1991). In the Chukchi Sea, where extensive set of samples have been analyzed for grain size and OC, we find a net significant correlation ($p < 0.05$, $n = 80$) between OC and mud contents in sediments despite a large scatter in the plots.

The factors that control the organic carbon concentrations in marine sediments have been a topic of intense debate (e.g. Mayer 1994; Hedges and Keil 1995 among many others). In North Bering-Chukchi Sea we found no consistent across shelf distributional pattern in OC. It might have been expected that relatively higher concentrations of OC would occur in sediments underlying more productive waters, such as in the region west of

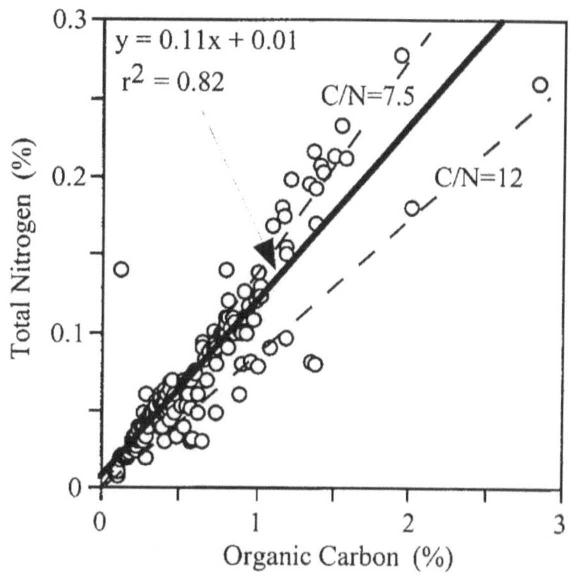

Fig. 7.3.3. The organic carbon (%) versus total nitrogen (%) plot of sediments from North Bering-Chukchi Sea

Table 7.3.1. Fluxes of total sediment (TSF) and particulate organic carbon (OCF), mass accumulation rates of sediments (MARS) and of OC (MAROC), percents OC remineralized (ROC), burial rates of OC (BROC) and sulfate reduction rates (SRR) at selected stations in Chukchi Sea (after Baskaran and Naidu 1995; Feder et al. 1989, 1991). All values, except ROC and SRR are in mg cm^{-2} y^{-1}. The ROC and SRR are in percents and mM m^{-2} d^{-1} respectively

Stat.	Lat. N.	Long. W.	Water Depth (m)	TSF	OCF	MARS	MAROC	ROC	BROC	SRR
KS1	66.62°	163.00°	13	660	14.52	88	1.4	–	–	1.0
KW2	66.58°	163.00°	14	737	–	–	–	–	–	1.3
SU1	68.17°	168.00°	51	596	13.49	–	–	–	–	–
SU4	67.73°	166.00°	43	1517	24.73	–	–	–	–	–
SU8	66.52°	168.00°	34	4335	76.73	–	–	–	–	7.8
SU11	67.04°	165.75°	28	1036	25.07	31	0.1	–	–	2.9
KS14	66.83°	163.90°	24	686	–	88[a]	0.8	–	–	1.3
KS16	67.40°	164.45°	26	874	16.87	–	–	–	–	–
CH13	72.52°	164.13°	48	510	1.85	167	2.3	–	–	–
CH21	71.20°	164.13°	42	–	115	–	1.2	–	–	–
CH25	72.63°	167.08°	51	–	83	–	1.3	–	–	–
CH38	70.70°	167.38°	52	–	165	–	0.4	–	–	–
CH39	71.87°	168.26°	48	–	90	–	0.1	–	–	–
CH40	70.28°	167.90°	45	–	119	–	1.2	–	–	–
SU5	67.03°	169.00°	50	4335[b]	76.73[b]	159	2.0	52	1.0	23.6
SU10	70.15°	166.57°	42	510[c]	1.85[c]	119[d]	1.2	44	0.67	1.3
KW3	66.18°	162.00°	14	660	14.52	88[a]	–	–	–	–

a, b, c, and d extrapolated from KS1, SU8, CH13 and CH40, respectively.
For sites SU5, SU10 and KW3 the total sediment and POC fluxes, and for sites KS14 and KW3 the mass accumulation rates were unavailable. The values for these locations are extrapolated from adjacent stations, which have similar productive and depositional regimes (e.g. SU8 data is used for SU5, CH13 for SU10, and KS1 for KS14 and KW3).

Table 7.3.2. Comparison of the Primary production rates (PPR), TSF, OCF, MARS, MAROC, ROC, BROC and DIOCMR for sites SU5 and SU10, which are representatives of relatively very productive and less productive regions, respectively, for Chukchi Sea. All values are in mg cm^{-2} y^{-1}, except ROC and DIOCMR, which are in percents and mol m^{-2} y^{-1} respectively

PARAMETER	SU5	SU10
Primary Production (PPR)	84[a]	7
Total sediment flux (TSF)	4335	510
(Particulate) organic carbon flux (OCF)	76	2
Mass accumulation rate of sediment (MARS)	159	119
Mass accumulation rate of OC (MAROC)	2.03	1.2
POC remineralized (ROC)	52	44
Burial rate of POC (BROC)	1.0	0.67
Sulfate reduction rate	23.6	1.3
Depth-integrated OC remineralization rate, (DIOCMR)	0.89	0.44

a: Based on the highest value reported (Springer and McRoy 1993).

the hydrographic front (marked HF on Fig. 7.3.1; Fig. 7.3.2), where relatively higher fluxes of OC to the sea floor have been recorded (Tables 7.3.1 and 7.3.2). Furthermore, sediment grain size will also be an important factor controlling OC concentrations, based upon the covariance between sediment mud and OC contents generally observed here and elsewhere (Blackburn 1987; Keil et al. 1994; Mayer 1994). On the North Bering-Chukchi Sea shelf a mosaic of sediment types (gravelly to muddy; Naidu 1984) occur. The large scatter in the OC-mud plots suggest that granulometry is not the sole factor influencing the sediment OC concentrations in the study area. It is possible that the effect of higher fluxes and deposition of OC is compensated by coarser sediments (probably low strand relict lag deposit), or that organic carbon remineralization rates and preservation efficiency vary regionally among sediments with the same grain size distribution.

The sediment $\delta^{13}C$ and C/N ratios have cross-shelf distribution trends. Generally, in the Bering Sea $\delta^{13}C$ progressively increases from −24‰ in the coastal region in the east, especially off the Yukon River and Norton Sound, to −20‰ in the west within the Gulf of Anadyr (Fig. 7.3.4). Farther inshore along the Gulf margin and the Anadyr

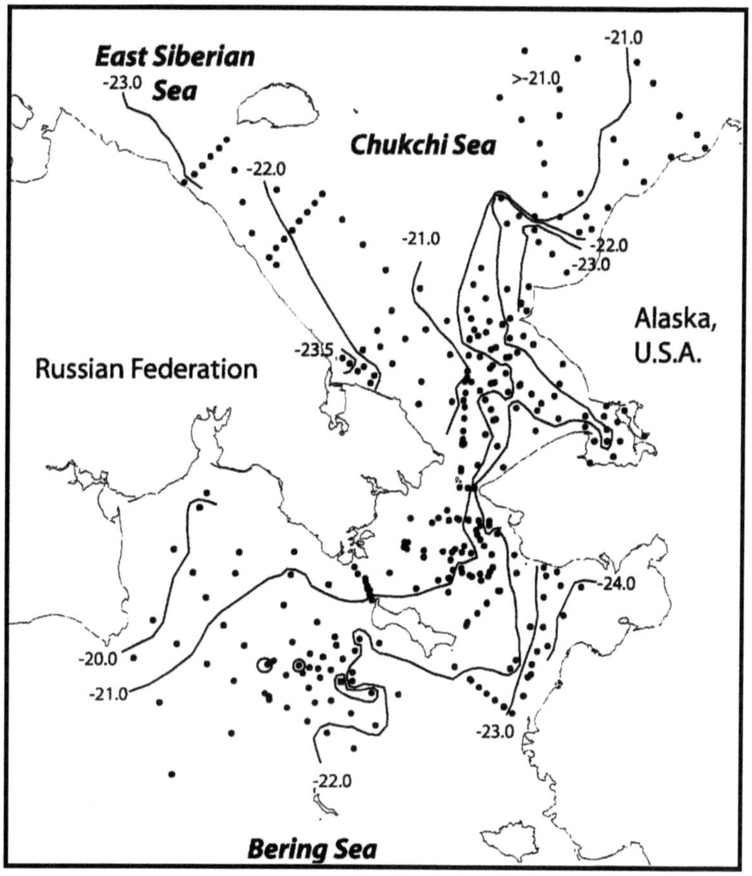

Fig. 7.3.4. Distribution pattern of $\delta^{13}C$ (‰) in sediments of North Bering-Chukchi Sea

Bay and Anadyr Estuary the sediments are distinguished by the lowest $\delta^{13}C$ values, (mean of 8 samples: −26.6‰) (Naidu et al. 1993, 2000). Likewise, in the Chukchi Sea a progressive increase in $\delta^{13}C$ across the shelf is observed from the east to about the central shelf, and then farther west the values decrease to −22‰ or less near the East Siberian coast (Fig. 7.3.4). Within the entire study area a progressive cross-shelf decrease from the east to west in sediment C/N is clearly observed, with relatively higher values concentrated off major fluvial systems (Naidu et al. 1993; Fig. 7.3.5).

The sources of organic carbon and organic matter in sediments of north Bering-Chukchi Sea are discussed in detail by Blackburn (1987) and Naidu et al. (1993, 2000), using the distribution patterns in sediment $\delta^{13}C$ and C/N as proxies (see Chapter 1.4).These sets of data on $\delta^{13}C$ and C/N have been supplemented since publication of the Naidu et al. (2000) work and updated versions are shown in Figures 7.3.4 and 7.3.5. Generally, in the North Bering Sea sediments there is an east to west, across shelf progressive increase in $\delta^{13}C$ and a parallel decrease in C/N. In the Chukchi Sea a net decrease in $\delta^{13}C$ and an increase in C/N occur from about mid shelf to the nearshore in east and west. Plots between $\delta^{13}C$ versus C/N values for Chukchi Sea sediments indicate a significant correlation between the two ($p < 0.05$, $n = 93$). The lateral increase in $\delta^{13}C$ and corresponding decrease in C/N are explained by a net increase in marine- versus terrigenous-dominated particulate organic carbon and organic matter respectively in sediments (see Chapter 1.4). This conclusion is based on the presence in North Bering-Chukchi Sea of two end members, marine and terrigenous, with distinctly different $\delta^{13}C$ signatures, −21.2‰ and −27‰ respectively (Naidu et al. 1993, 2000). Therefore, a simple mixing equation (Calder and Parker 1968; Fry and Sherr 1984; see Chapter 1.4), considering the $\delta^{13}C$ values of the end members and of sediment C for a location, can account for the relative abundances of marine and terrigenous C at that location. Likewise, the sediment C/N is used to infer the sources of organic matter and by implication OC as well, assuming that the C/N end member values of marine and terrigenous organic matter are <5–7 and >15 respectively (Naidu et al. 1993, 2000; see Chapter 1.4). It is further suggested by

Fig. 7.3.5. Distribution pattern of C/N in sediments of North Bering-Chukchi Sea

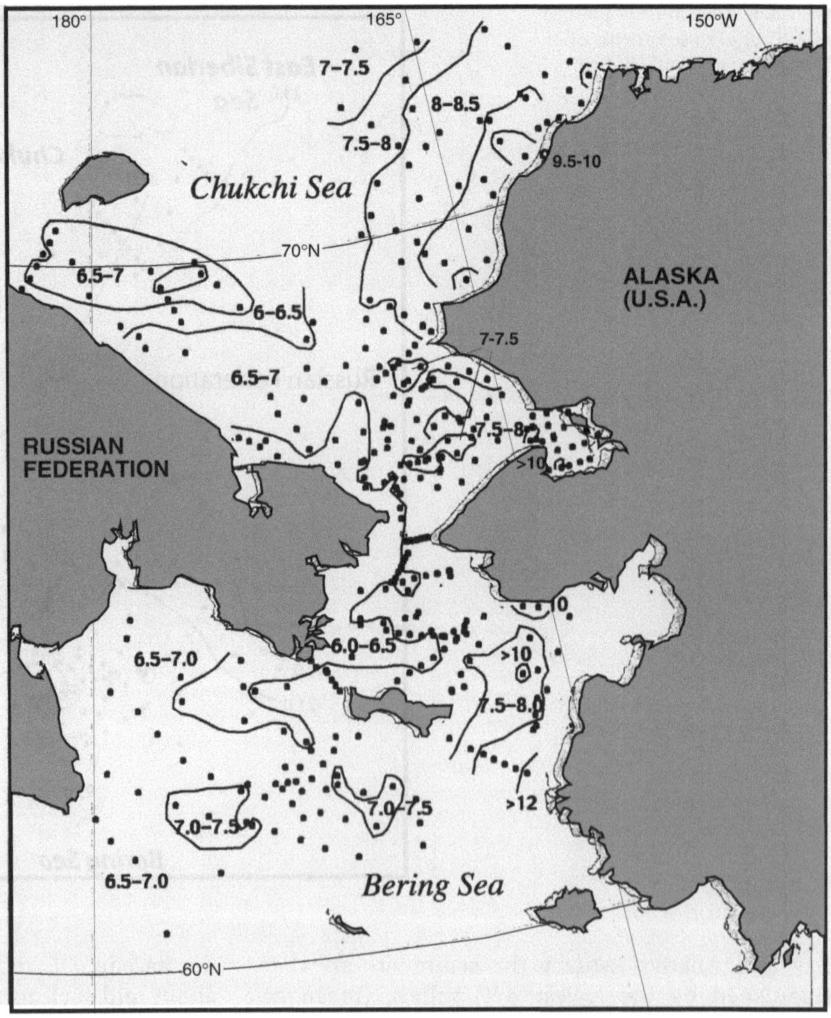

the above authors, based on the sediment $\delta^{13}C$ and C/N, that much larger rates of deposition of marine organic matter occur under relatively more productive waters, such as within the region west of the hydrographic front (compare Figs. 7.3.1, 7.3.4, and 7.3.5).

Recently, Schubert and Calvert (2001) found that significant quantities of bound inorganic nitrogen (N_{bou}, see Chapter 1.4) occur in Arctic Ocean deep-sea illitic sediments. Likewise, Müller (1977) reported large concentrations of ammonium ion adsorbed by clays in Pacific deep-sea sediments. These studies imply that the use of C/N_{total} instead of C/N_{org} (where N_{org} is $N_{total} - N_{bou}$, Chapter 1.4) may provide misleading assignments of the sources of marine sediment organic matter. While these studies were conducted on deep-sea sediments, it is possible that the high illite content of Bering-Chukchi Sea (Naidu et al. 1981) may preferentially adsorb ammonium ions (N_{bou}) and, thus, affect the C/N value (which is based on N_{total}). It is, therefore, possible that the lower C/N ratios in sediments generally in the region west of the hydrographic front in North Bering-Chukchi Sea (Fig. 7.3.5) are actually due to larger N_{bou} and not due to N_{org} of total N as has been implied relating to the organic matter sources (marine versus terrigenous). A plot of OC versus N values of the North Bering-Chukchi Sea sediments shows a significant covariance between the two, with the intercept at 0% of C corresponding to 0.01% N (Fig. 7.3.3). This suggests that only a very minor portion of the total N in the sediment could be N_{bou}. It should be noted that in the Chukchi Sea sediments the concentrations of illite (and presumably N_{bou}) decrease from the nearshore to the central shelf (Naidu et al. 1981), a factor that would go against the possibility discussed above. We suggest, therefore, that the C/N_{total}-based inferences on the provenances of organic matter in the north Bering-Chukchi Sea

(Naidu et al. 1993, 2000) might be clarified by more detailed N partitioning investigations (Chapter 1.4) and additional biomarker studies (Hayes et al. 1989; Jasper and Gagosian 1989; Prahl and Muehlhausen 1989; Fahl and Stein 1999; Stein et al. 1999; see Chapter 1.4) could provide a more robust discrimination of organic carbon sources to the sediments in the Bering-Chukchi shelf.

7.3.4 Fluxes, Accumulation and Burial Rates and Remineralization of OC, and Benthic Oxygen Uptake Rates

It is assumed that the sediment mass and OC accumulation rates at the nine core sites (Table 7.3.1) are measures of the depositions of sediments and admixed OC on a constant depositional rate basis. Considering the shallow water depths of the shelf we suggest that in the Chukchi Sea the relative differences in the OC flux between locations reflect the primary productivity rates of overlying waters and the resulting mass of phytodetritus advecting down (Table 7.3.2). It should be noted that the yearly flux data (Tables 7.3.1 and 7.3.2) are computed by us from various one-time, short-term real field data that were reported on a daily rate basis by Baskaran and Naidu (1995). The computation assumed that the daily-based data is a representative mean only for approximately four months of the productive open water season, and that only 10% of the yearly fluxes of sediment and particulate OC occur during the rest of the eight months when primary production is negligible under sea ice cover (Baskaran and Naidu 1995).

Clearly, there is an across shelf east-west increase in the sulfate reduction rate (Table 7.3.1), with the lowest rate of sulfate reduction in Kotzebue Sound and the adjacent southeast Chukchi Sea region (e.g., Stations SU11, KS14, KS1 and KW3) and greatest sulfate reduction far offshore region (e.g., SU5 and SU8) in areas that also have high benthic biomass (Grebmeier et al.1988). These lateral variations in sulfate reductions in sediments are most likely due to a progressive east-west increased remineralization resulting from greater deposition of labile, marine-derived organic carbon in the west. Our findings are, therefore, consistent with Blackburn's (1987) suggestion that there is a close correlation between microbial remineralization rates and the quality of organic matter and OC in sediments of Bering-Chukchi Sea. The benthic sediment oxygen uptake rates in the Chukchi shelf also show an across-the-shelf pattern, with values progressively increasing from the eastern nearshore to the mid shelf and then decreasing towards the East Siberian coast (Fig. 7.3.6). In the north Bering Sea the highest values are southwest of St. Lawrence Island and south of Bering Strait in Chirikov Basin, with values subsequently decreasing in a concentric pattern about the maxima (Fig. 7.3.6). The regional variations in the oxygen uptake rates are attributed primarily to the relative abundance of the benthic biomass, which in turn is determined by the lateral variations in marine vs. terrigenous OC and the total flux of OC depositing at the sea floor (Grebmeier et al. 1988).

As discussed earlier, it was possible to estimate the percentages of organic carbon mineralized and the burial rate of OC at two sites (SU5, a high productive site and SU10 a less productive site) in the Chukchi Sea (Tables 7.3.1 and 7.3.2). At four other sites investigated the OC was well mixed within the sediment cores precluding an estimate of the remineralization. Although the burial rate of OC at SU5 is about a factor of two higher than at SU10 site, the percent OC remineralized at these contrasting sites is similar (Tables 7.3.1 and 7.3.2). The regional difference in burial rate of OC is most likely due to the relatively higher accumulation rate of sediment and OC at the SU5 location. The burial efficiencies of OC (after Gordon et al. 2001) for SU10 and SU5 are 46% and 64% respectively. This between station difference is assigned mainly to the differences in gross sediment and OC accumulation rates, which is consistent with the widely held view for marine deposits (Müller and Suess 1979; Henrichs and Reeburgh 1987; Rullkoetter 2000).

The regional variations in the rates of remineralization and burial of carbon and the sediment sulfate reduction, as discussed above, are consistent with the hypothesis formulated at the beginning of our investigations. Prior to the estimations of the rates, we had hypothesized that in Chukchi Sea the rates at productive sites in midshelf (e.g., SU5) would be significantly higher than in sediments at less productive sites in the nearshore (e.g., SU10). The underlying premise was that in highly productive waters relatively more marine-derived labile (more easily metabolizable) organic carbon will be deposited and that in less productive waters more land-derived refractory carbon will be accumulated (Blackburn 1987; Grebmeier 1987; Feder et al 1991; Meyers and Eadie 1993). The regional sediment $\delta^{13}C$ and C/N values in sediment trap samples (Baskaran and Naidu 1995) and bottom sediments (Naidu et al. 1993, 2000) indicate that the OC deposited at SU5 is indeed, as hypothesized, predominantly marine and, presumably therefore, relatively more labile than OC sedimenting at SU10.

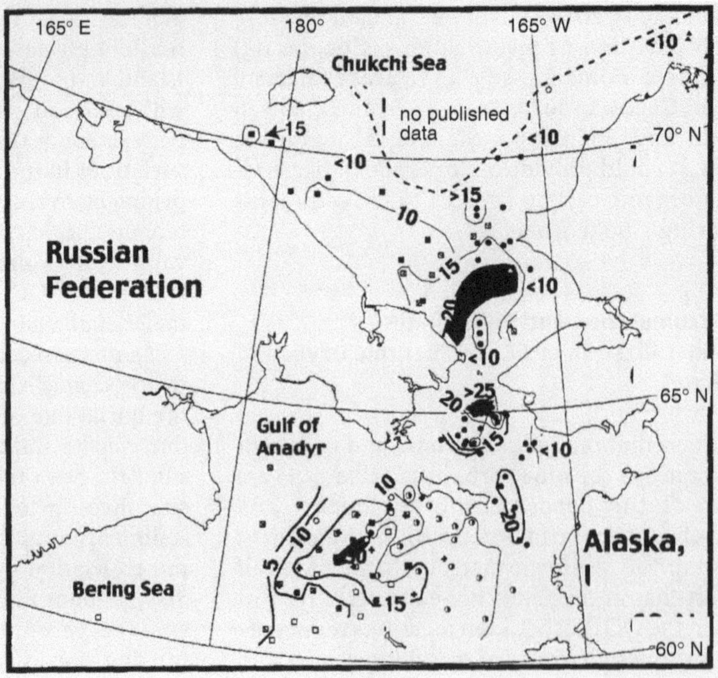

Fig. 7.3.6. The benthic oxygen uptake rates (mM O_2 m^{-2} d^{-1}) across the shelf of North Bering-Chukchi Sea

Tentative Mass Balance of OC

Assuming SU5 and SU10, represent offshore high productive, and inshore, less productive portions of the Chukchi Sea respectively, our database in conjunction with the primary production rate provide a tentative basis to calculate aspects of the (particulate) organic carbon mass balance for these regions of contrasting productivity (Fig. 7.3.7). A basic assumption in the ensuing calculation of the 'working budget' is that almost the entire initial source of particulate OC is marine primary production (phytodetritus) from ambient waters overlying the above two sites. That this assumption is reasonable is supported by the C/N ratio (6–7) observed for suspended particles of surface and near bottom waters throughout the Chukchi Sea shelf (Feder et al. 1991). However, it should be noted that the nearshore North Bering-Chukchi Sea off the major rivers, receive significant inputs of land-derived particulate OC from coastal erosion and/or fluvial outflow (Naidu et al.1993, 2000) as demonstrated by sediment $\delta^{13}C$ and C/N values (Figs. 7.3.4 and 7.3.5). A potential complication to the budget calculation is that significant amounts of OC in the Chukchi Sea waters could be allochthonous marine OC, imported from the north Bering Sea (Walsh et al. 1989; Feder et al. 1994). However, no quantitative estimates are available on the import fluxes of terrigenous and allochthonous organic carbon and their distribution in Chukchi Sea. Furthermore, little is known about the fate of dissolved organic carbon and its importance for the shelf OC budget.

It appears that about 90% of the OC derived from primary production from the productive region and about 29% from the relatively less productive waters settles (flux) to the sea floor. Further, about 3% of the total organic carbon flux at SU5 and 60% at SU10 finally accumulate in the bottom surficial sediments. The rest of the OC reaching the sea floor must, therefore, be lost through resuspensionand advection, remineralization and/or uptake by benthic organisms. The significantly higher benthic biomass and benthic respiration rates at the more productive sites (Grebmeier 1987, Grebmeier et al. 1989; Fig. 7.3.6) imply that most of the settling particulate OC escapes grazing by pelagic organisms, gets to the bottom and is eventually consumed by benthos. The subsequent losses due to remineralization of the sedimented OC at the SU5 and SU10 sites, are estimated to be 52% and 44% respectively. The depth-integrated remineralization rate at SU5 is higher by a factor of 2 than at SU10 (Table 7.3.2). Eventually, OC is buried at SU5 and SU10 sites at the rates of 1.0 mg cm^{-2} y^{-1} and 0.67 mg cm^{-2} y^{-1} respectively (Table 7.3.2, Fig. 7.3.7).

A more comprehensive organic carbon budget for the Chukchi Sea incorporating allochthonous sources of marine and terrestrial OC, autochthonous production, export from the shelf and burial in shelf sediments must be established before valid

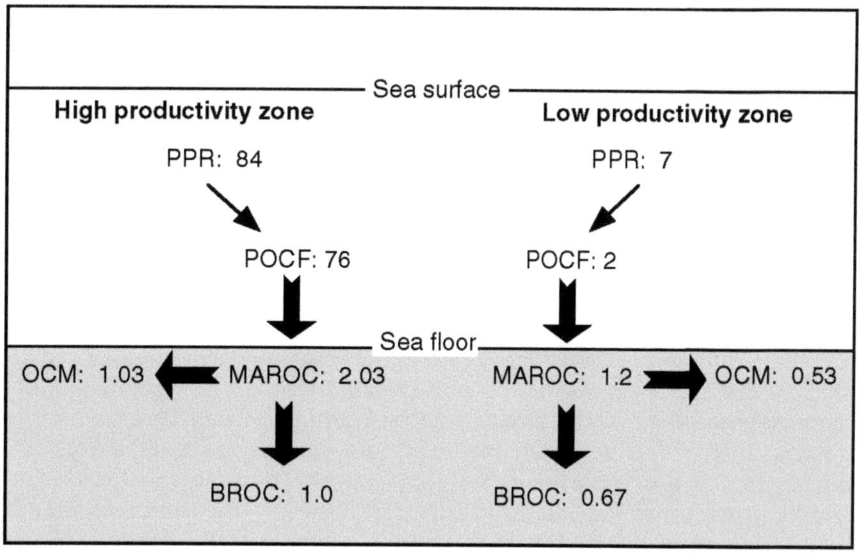

Fig. 7.3.7. A schematic illustration of the mass balance of particulate organic carbon (POC) in sediments for the high (Site SU5) and low productive (SU10) zones of the Chukchi Sea shelf. PPR: primary productivity rate of POC; POCF: POC flux; MAOC: Mass accumulation rate of OC; OCM: OC lost by remineralization; BROC: OC burial rate. All values are in mg cm^{-2} y^{-1}

comparisons can be made with other margins of the Arctic (for example, Walsh and McRoy 1986; Grebmeier 1987; Macdonald et al. 1998; Stein and Fahl, 2000; see Chapters 7.2, 7.5 and 7.6). Further data such as presented for sites SU5 and SU10, are needed in order to confirm the first rough organic carbon budget estimates published by Romankevich and Vetrov (2001) who report a burial rate for the Chukchi Sea to be 0.23 × 10^6 t OC y^{-1}. Progress towards this objective will also contribute to wider environmental research goals, such as assessing the potential role of sediments of the shelves and ocean basins as sinks for OC and in sequestration of natural and anthropogenic CO_2 (Walsh et al. 1981; Berner 1992; Hedges et al. 1997; Naidu et al. 2000; de Haas et al. 2002).

7.3.5
Summary and Concluding Remarks

- In general, little is known about the concentrations, sources, fluxes, accumulation and burial rates, and remineralization of OC of sediments of the North Bering-Chukchi Sea shelf. In the study area the OC concentrations range from 0.1 to 2.8 %, which are close to the values generally reported for shelf sediments (0.5 to 1.5 %). The OC distribution shows no across shelf pattern. Although there is a significant correlation in OC and mud contents, wide scatter in the binary plots suggests that grain size is not the sole factor determining sediment OC.
- Generally, an across shelf, from east to west, increase in δ^{13}C and decrease in C/N in sediments are found. On the Chukchi Shelf on going from the low productive nearshore to the highly productive midshelf, there are also increases in the rates of OC remineralization, sulfate reduction and benthic oxygen uptake. The above trends reflect larger flux and deposition of marine OC relative to terrigenous OC in the midshelf.
- A tentative budget for OC is formulated based on two sites, one representing the high productive (84 mg C cm^{-2} y^{-1}) zone and the other representing low productive (7 mg C cm^{-2} y^{-1}) zone (Fig. 7.3.7). About 90 % and 74 % of the primary production of carbon from the productive and less productive waters, respectively settles to the bottom of which about only 3 % and 60 % respectively finally accumulate in the bottom sediments. The rest of the OC flux at the two sites is lost by advection, remineralization or consumption by benthic organisms. Subsequent remineralization of the accumulated OC is 52 % and 44 % respectively. Further measurements are necessary to calculate a more comprehensive OC budget for the Alaskan arctic shelf.

7.4 The East Siberian Sea: Distribution, Sources, and Burial of Organic Carbon

V. I. PETROVA, G. I. BATOVA, A. G. ZINCHENKO, A. V. KURSHEVA, E. V. NARKEVSKIY

7.4.1 Introduction

The East Siberian Sea (see Fig. 7.1.6) is the most shallow, most ice-bound and most poorly-explored marginal sea of the Arctic Ocean. The first visual descriptions of this sea were made in 1910–1911 during the course of ice-breaker cruises by "Taimyr" and "Vaigach", and the first grain-size and lithological data were obtained from sediment samples collected during the ice-breaker "Severny Polyus" cruise in 1946 (Geology of the USSR, 1970). Systematic studies of bottom sediments began in 1952–64 and were performed by the Arctic and Antarctic Scientific Research Institute (AANII) and the Research Institute for Geology of the Arctic (NIIGA). As a product, the first map series was compiled by Yu. P. Semyonov in 1961 (Semyonov, 1965). Between 1976 and 1989, organic geochemical properties of bottom sediments were characterized during integrated geological and geophysical investigations (Danyushevskaya et al. 1980; Romankevich et al., 1982; Danyushevskaya et al., 1990). More recently, from 1991 to 1995, data have been obtained using more sophisticated analytical techniques which permit an evaluation of the components of organic matter in bottom sediment at the molecular level (Petrova et al. 2001). Here, we present these and other data obtained during field investigations conducted over a number of years.

Topography

The seabed of the East Siberian Sea is smooth, has a low gradient, and is clearly defined only at the shelf break where it deepens into the Amerasian Basin of the Arctic Ocean. Correspondingly, the low gradients directed toward the interior ocean from the continent, islands and submarine rises provide little energy to support lithodynamic transport. Barriers associated with island uplifts, the largest of which is situated to the west of Wrangel Island (Fig. 7.4.1) or, less commonly, with submarine rises, are few. There are no large basins that could serve as terminal traps for sediment in the East Siberian Sea shelf. A few over-deepened sections of the Kolyma and the Indigirka submarine valleys may be sediment traps. However, considering that they are not fully closed, these channels probably provide only intermediate depocenters. A portion of the Kolyma submarine valley from the present

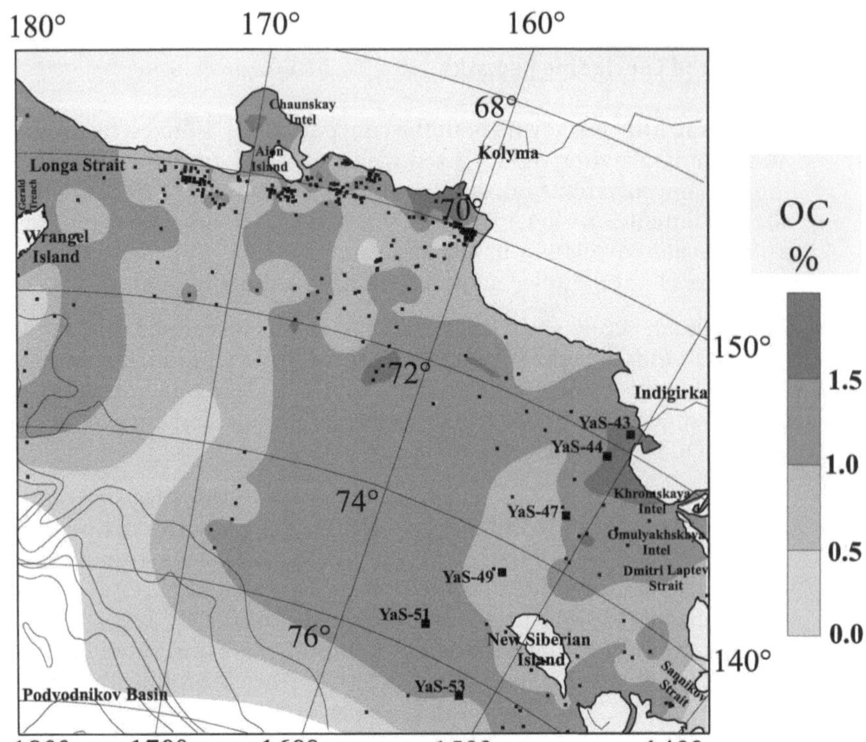

Fig. 7.4.1. The distribution of total organic carbon (OC) content in surface sediments from the East Siberian Sea

river mouth to the northern extremity of Aion Island more closely resembles a trap, and Long Strait contains a semi-closed basin bounded by the protrusion of the southern submarine slope of the Wrangel Island uplift (Zinchenko 2000).

Hydrology, oceanic circulation, tidal currents, sea ice

The general water circulation patterns suggest sediment supply over the outer shelf from currents originating in the Laptev and Chukchi Seas. In addition, the Siberian Coastal Current transports water eastward through Sannikov and Dmitri Laptev Straits continuing along the East Siberian coast and through Long Strait toward the Chukchi Sea (Proshutinsky et al., 1995; Weingartner et al., 1999). The largest velocities are typically associated with currents along the continental slope where water circulates around the New Siberian Islands and Wrangel Island. However, the Siberian Coastal Current transports material mainly southeastward – i.e., in a direction favorable for depositing sediments in the submarine valleys oriented across the currents – and in the semi-enclosed basin in Long Strait (Anonymous 1980).

The oceanographic conditions of the southwestern East Siberian Sea are strongly impacted by 213–250 $km^3\ y^{-1}$ of river inflow (see Gordeev 2000; Chapter 2). The influence of runoff can be seen out to 72–74°N in areas adjacent to the Chukchi Sea, out to 76°N in the eastern part of the East Siberian Sea and as far as 79°N to the north of the New Siberian Islands due to the supply of freshened water from the Laptev Sea (Belov 1976). Toward the northeast and east, Arctic waters and Pacific waters from the Chukchi Sea respectively dominate the oceanography (Romankevich et al., 1982).

Sedimentation processes

Sedimentation in the East Siberian Sea is controlled by several predominant processes including: coastal erosion of terrigenous material, supplied through thermal abrasion of ice-bonded sediments; river discharge; sea bottom washout and; to a minor extent, biogenesis. Thus, the composition of bottom sediments primarily reflects the geological composition of source areas. Low rates of chemical and biological processes in aquatic environments and bottom sediments under polar climate conditions limit any contribution from non-terrigenous sources to sediments (Romankevich et al., 1982).

Bottom topography and hydrological and hydrodynamic conditions of the East Siberian Sea shelf also exert appreciable control on sedimentation. Ice cover, which persists during most of the year over the shelf and all year in the interior ocean, reduces currents and wave action on the bottom. This produces stagnant zones leading to the deposition of fine sediment at shallow depths – a feature typical of polar sedimentogenesis (Semyonov et al. 1965; Romankevich et al., 1982; Kosheleva and Yashin 1999).

Lithology of bottom sediments

The lithology of bottom sediments of the East Siberian Sea reflects the geological composition of sediment source regions and, to a large degree, the bottom topography and hydrological conditions of the shelf (Semyonov 1965; Romankevich et al., 1982; Danilov et al. 1994; Nikiforov 1996; Kosheleva and Yashin 1999).

The western part of the shelf, whose catchment and coasts are made up of loose, frozen Quaternary deposits, is distinguished by low rates of accumulation of sediments with uniform grain-size and mineralogical composition. For example, in regions near the Indigirka and Kolyma River mouths, the thickness of recent deposits, evaluated by coring, amounts only to tens of centimeters. The light fraction in sediments comprises feldspar, plagioclases, orthoclase, mica and rock fragments. Hydromica is the dominant clay mineral admixed with lesser amounts of kaolinite. The heavy fraction in sediments is dominated by amphibole and epidote, accompanied by dark ore minerals and pyroxene with lesser amounts of zircon, sphene, garnet and apatite.

The coast of the eastern part of the sea is composed essentially of metamorphosed sedimentary and igneous rocks that supply a wide range of clastic material (from boulders to clays). The thickness of Holocene sediments is low because of low wave energy and the rocky character of the coasts. The mineralogical composition of the sediment light fraction and the distribution of clay minerals are similar to those in the western part of the shelf. The heavy fraction of sediments is dominated by dark ore minerals, pyroxene, epidote and less common amphibole.

Possible sources of organic carbon

The transport of organic matter (OM) by water can occur in dissolved, colloidal and suspended phases. Although there are very few studies of the elemental and molecular composition of dissolved and suspended OM in the eastern Arctic (Gauwet and Sidorov 1996; Peulvé et al. 1996; Lara et al. 1998; Saliot et al. 1996; Kiselyov et al. 2000), several important conclusions can be drawn from their findings. The dissolved riverine OM probably derives from terrigenous and not autochthonous sources. This is suggested by the isotopic composition of the suspended material in rivers and by a positive correlation between silicate concentration and dissolved OC and nitrogen. Furthermore, suspended particulate organic matter (POM) in rivers contains almost exclusively compounds produced by terrestrial vegetation (polysaccharides, phenolics, long-chained *n*-alkanes and *n*-alkenes having a high odd-number index) thus unambiguously suggesting the POM to be terrigenous with little if any contribution from aquatic production. The OM composition of suspended sediment in surface waters of the shelf shows a decrease in influence of terrigenous sources as revealed by a substantial decrease in the content of polysaccharides, phenol compounds and products of protein destruction, and an increase in autochthonous sources. The bimodal distribution of *n*-alkanes, *n*-alkenes and alkadienes points to the contribution from both aquatic and terrigenous sources of OM. Long-chain *n*-alkanes and *n*-alkenes, present in suspended sediments at intermediate water depths, suggest the influence of terrestrial vegetation. The predominant characteristic of suspended material in intermediate depths corresponding to Arctic water masses was the presence of *n*-alkanes, alkenes, alkadienes in the C_{14-18} range, probably of microbial origin. Organic matter, derived from the abrasion of loose Quaternary deposits along the coastal margin, is essentially terrigenous as suggested by isotope composition ($\delta^{13}C_{org}$ from −27.16 to −27.54‰).

In summary, aquatic transport supplies predominantly terrigenous OM to the bottom sediments of the East Siberian shelf, and most of the marine organic production, evident in surface waters, is metabolized during sedimentation.

Quantitative estimates of the production of autochthonous organic carbon for the East Siberian Sea shelf remain speculative due to inadequate study of biological structure and primary production of the shelf waters (Romankevich and Vetrov, 2000). However, the presence of compounds associated with phytoplankton production suggests that individual fractions of OM exhibit sufficient stability to contribute to OC distribution.

Atmospheric aerosols contain organic matter which may account for as much as 50% of the weight of insoluble particles of dust (Simoneit 1977; Gagosian et al. 1981; Lipiatou and Albaiges 1994; Simoneit and Elias 2000; Shevchenko 2000; see Chapter 2.5). For arctic aerosols, the content of Corg and OM has been found to be as high as 30% and 60%, respectively (Shevchenko 2000). Matthiessen (1999) showed that pollen and spores, found in suspended particulate material and in bottom sediments of the Arctic seas, were supplied mainly through eolian transport. Continental aerosols contain a vegetation component as suggested by the presence of phenolic compounds of lignin (vanillin and lilac) typical of woody material (Simoneit and Elias 2000; Peresypkin 2000). Comparison of the mineral composition of soils and aerosols with suspended and bottom sediments near river mouths and over the adjacent shelf imply that aerosol transport provides a significant part of the terrigenous OM in the eastern Arctic (Serova and Gorbunova 1997).

7.4.2
Data Base, Material, and Methods

The samples include bottom sediments collected at 535 stations in the course of research cruises conducted by VNIIOkeangeologia in the period from 1964 to 1995. The carbonate and organic carbon contents of sediments were determined by Knopp's method (Korchagina and Chetverikova 1976).

A more detailed analysis of the OM composition of sediments followed the scheme outlined in Danyushevskaya et al. (1990). Sediments were Sohxlet extracted; chemical and group composition was studied for bituminoids; chromatographic fractionation was used for the separation of total hydrocarbons and gas-liquid chromatography was used for *n*-alkanes. Individual composition of alkanes was determined on a gas-chromatograph with a flame-ionization detector under the following conditions: steel-packed columns of 3.5 m × 3 mm in size, SKTFT-50X stationary phase (5% on Colorchrom), helium as the carrier gas (40 ml/min), and linear programming of the column temperature from 80 to 320 °C.

7.4.3
Distribution and Sources of Organic Carbon in Surface Sediments

The content and distribution of total organic carbon (OC) in bottom sediments are controlled by sediment supply, grain-size composition of the sediments and the depositional environment (Fig. 7.4.1). Over the East Siberian Sea shelf, Holocene sedimentation is characterized by uniform terrigenous deposition. Sediment grain size is monotonous: the most common pelite-aleuritic varieties of sediment are accompanied by mictite and silt shale; sandy sediments are localized in the western part of the water basin (region around New Siberian Islands) and in the coastal zone (Kosheleva and Yashin 1999).

Sediment organic carbon is distributed almost normally with most sediments containing 0.5–1.5%

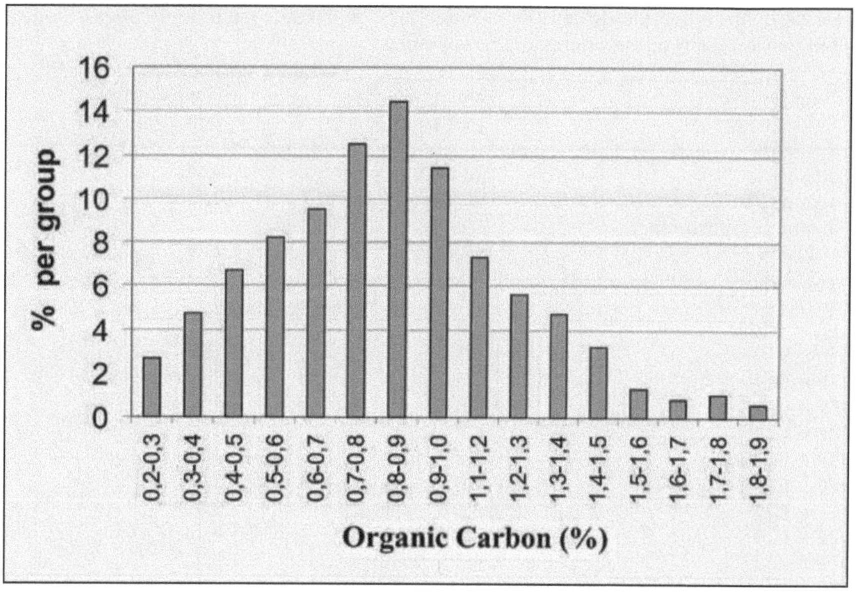

Fig. 7.4.2. A distribution histogram of organic carbon (OC) in the surface sediments from East-Siberian Sea (total number of samples = 535)

OC (Fig. 7.4.2). The higher values (1.5–2%) are associated with sediments near the Indigirka and Kolyma River mouths whereas the lowest values (<0.5%) are associated with sandy sediments adjacent to islands of the western shelf, in local wash and re-deposition zones, and in sediments of the continental shelf (Fig. 7.4.1).

Statistical parameters, summarizing OC content in surface (0–20 cm) and deep (>20 cm) sediments of the East Siberian Sea (Table 7.4.1), concur with the well-known geochemical relationship between grain size and organic carbon content (increase in average OC content from sand to silt to pelite is 0.57 to 0.87 to 1.00%), and illustrate the effect of organic metabolism (0.93% for surface sediments, 0.88% for deep sediments).

However, the statistical parameters characterizing OC variation are not consistent with the lithological monotony of East Siberian shelf sediments, being comparable to those for the polygenetic sediments of the Barents Sea ($V_x = 41.0\%$; (Romankevich and Vetrov 2001). This suggests that hydrodynamic and geomorphological parameters play an important role in the formation of a heterogeneous OC distribution within the East Siberian shelf's lithologically homogeneous sediment field.

Composition of Organic Carbon in Surface Sediments

The composition of *n*-alkanes in estuarine/deltaic sediments of the East Siberian Sea (Table 7.4.2) shows a strong predominance of high-molecular odd-number compounds (n-C_{22-37} to 88%, OEP to 6.7) indicative of poorly altered terrigenous organic material. This has been further confirmed by comparing alkane hydrocarbon composition in typical sediments containing dispersed OM with the composition of a sediment band visibly rich in vegetable debris.

Table 7.4.1. Organic carbon (OC) content (%) in surface sediments of the East Siberian Sea

	statistical parameters					
sediments	n	$X_{ar.m}$	s^2	s	(V*100%)	s_r
sands	11	0.57	0.04	0.20	35.09	0.34
silts	149	0.87	0.12	0.34	39.08	0.39
pelites	183	1.00	0.09	0.30	30.00	0.30
mictites	118	0.82	0.08	0.29	35.37	0.35
surface	461	0.93	0.10	0.32	34.41	0.34
submerged	74	0.88	0.04	0.20	22.73	0.23

n number of samples
$X_{ar.m}$ arithmetic mean
s^2 dispersion
s standard deviation
s_r variation
V*100% coefficient of variation

Table 7.4.2. Mean organic carbon (OC) content and *n*-alkane content and composition in surface sediments from the East-Siberian Sea. OEP = odd-even predominance index

Parameter (mean)	Estuary/Deltaic (n = 8)	Shelf/Coastal (n = 25)	continental margin (n = 6)
OC, %	1.96	0.64	0.53
n-alkanes, %	0.0006	0.0003	0.0003
n-alkane composition:			
<*n*-C21, %	12.0	21.7	59.0
>*n*-C22,%	87.7	75.1	31.0
<*n*-C21/>*n*-C22	0.1	0.3	2.1
n-C17/*n*-C29	0.0	0.5	2.4
Pr/Ph	0.3	1.3	2.1
OEP *n*-C17	1.3	1.4	1.2
OEP *n*-C19	1.6	1.4	1.0
OEP *n*-C21	1.8	1.5	1.1
OEP *n*-C23	2.2	1.8	1.3
OEP *n*-C25	2.7	2.1	1.6
OEP *n*-C27	3.6	2.9	2.3
OEP *n*-C29	4.6	4.1	3.3
OEP *n*-C31	6.7	5.7	4.3
OEP <*n*-C21	1.2	1.4	1.1
OEP >*n*-C23	3.7	3.3	2.6
i-C19/*n*-C17	0.1	0.6	0.6
i-C20/*n*-C18	0.2	0.5	0.4
i-C19+20/*n*-C17+18	0.1	0.5	0.5

Table 7.4.3. Station location, water depths, organic carbon (OC) content, *n*-alkane content and composition in surface sediments from the East-Siberian Sea. OEP = odd-even predominance index. For locations see Figure 7.4.1.

Parameter	YaS-43	YaS-44	YaS-47	YaS-49	YaS-51	YaS-53	PS-60	PS-57
Latitude, N	71 63	72 08	73 50	74 50	75 50	76 13	80 28	81 19
Longitude, E	151 18	151 17	152 47	153 43	154 48	153 04	150 45	150 46
Water depth, m	5.5	12	20	17	31	38	1685	2640
OC, %	1.58	1.04	0.77	0.61	0.72	0.43	0.55	0.48
n-alkanes, %	0.0005	0.0005	0.0003	0.0001	0.0002	0.0001	0.0003	0.0003
n-alkanes composition:								
<*n*-C21, %	13.8	11.6	8.8	9.7	14.3	19.7	43.9	67.6
>*n*-C22, %	86.1	88.3	90.8	90.2	84.4	77.4	48.3	23.3
<*n*-C21/ >*n*-C22	0.2	0.1	0.1	0.1	0.2	0.3	0.9	2.9
n-C17/ *n*-C29	0.0	0.0	0.0	0.0	0.1	0.2	1.8	2.7
Pr/Ph	0.0	0.0	1.5	0.0	0.9	1.3	4.1	1.1
OEP *n*-C17	1.7	1.5	1.3	0.8	2.5	1.9	1.6	0.9
OEP *n*-C19	1.5	1.6	1.6	1.5	1.6	1.6	0.9	1.1
OEP *n*-C21	1.6	1.7	1.8	1.7	1.6	1.4	1.2	1.1
OEP *n*-C23	2.1	1.9	2.0	1.9	1.8	1.7	1.5	1.2
OEP *n*-C25	2.6	2.3	2.4	2.3	2.1	2.1	1.9	1.5
OEP *n*-C27	3.0	3.1	3.4	3.1	2.7	3.0	3.3	1.8
OEP *n*-C29	3.9	4.3	4.7	4.5	3.8	3.9	5.5	2.5
OEP *n*-C31	5.5	5.9	7.0	6.2	5.6	5.2	7.5	3.0
OEP <*n*-C21	1.6	1.6	1.6	1.3	1.9	1.7	1.2	1.0
OEP >*n*-C23	3.4	3.5	3.9	3.6	3.2	3.2	3.9	2.0
i-C19/*n*-C17	0.0	0.0	0.4	0.0	0.7	0.8	0.4	0.8
i-C20/*n*-C18	0.0	0.0	0.2	0.0	1.2	1.0	0.2	0.6
i-C19+20/*n*-C17+18	0.0	0.0	0.3	0.0	0.9	0.8	0.4	0.7

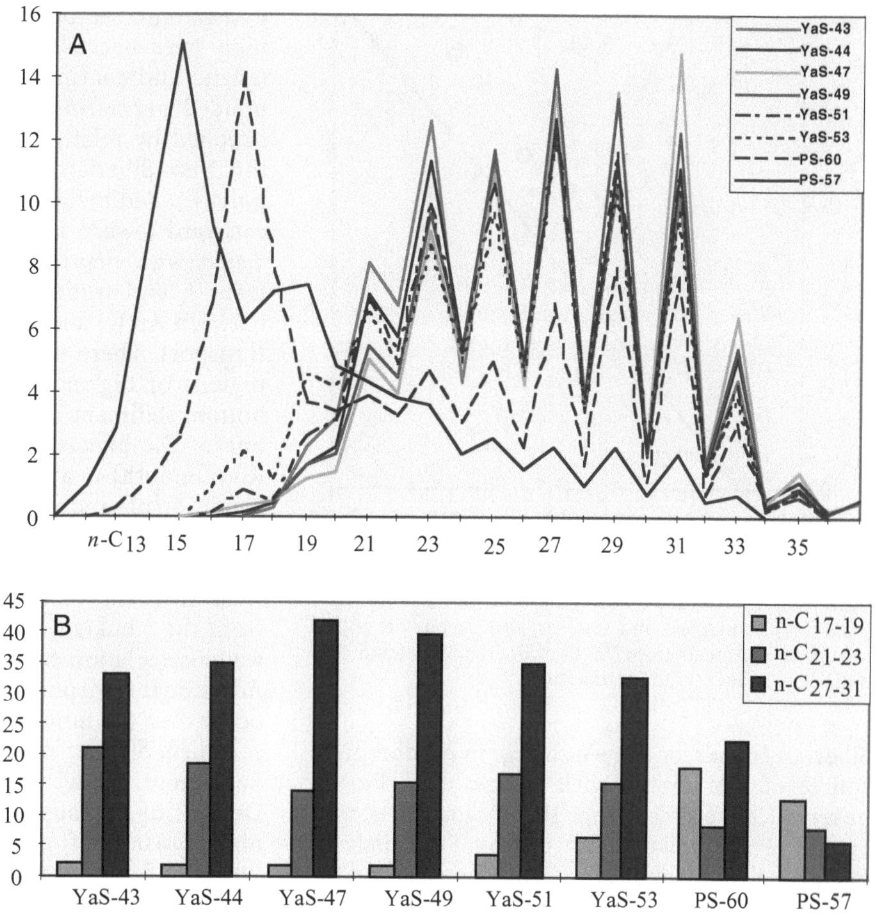

Fig. 7.4.3. *n*-alkane (C_{11}–C_{37}) distribution (A) and composition (B) in surface sediments along a transect from Indigirka River to Podvodnikov Basin. For location see Figure 7.4.1

In shelf/coastal sediments from the East Siberian Sea, the *n*-alkane distribution pattern corresponds to terrigenous OM. However, to the north-west in the region of the New Siberian polynya behind the fast ice (Popov et al. 2000), the influence of aquatic production of organic material becomes apparent ($<n$-C_{21} = 36.6–51.7%). These lower carbon number *n*-alkanes in sediments are likely supplied by enhanced biological productivity in the adjacent northeastern part of the Laptev Sea (Fahl and Stein, 1997).

In continental margin sediments from the shelf edge and slope, the proportion of alkanes from aquatic production is consistently large ($<n$-C_{21} = 65.4–67.7%). However, the odd/even indices (Table 7.4.2) suggest better preservation of terrigenous material (OEP $<n$-C_{21} = 1.0–1.2; OEP $>n$-C_{23} = 1.8–3.9) probably due to its association with mineral surfaces which protect it from metabolism during long-distance transport (Hedges and Oades 1997; Petrova et al. 1999).

The change in the primary source of organic material accompanying change in lithofacies condition of sediments is clearly evident along a section from the Indigirka River estuary to the Podvodnikov Basin (Table 7.4.3; Fig. 7.4.3). The *n*-alkane spectral distribution in sediments (Fig. 7.4.3) shows a consistent, successive decrease in proportion of the terrigenous components and an increase in proportion of marine production components.

The correlation between absolute concentrations of the long-chain *n*-alkanes in µg/gSed and OC (g/gSed) (Fig. 7.4.4) indicates that the contribution of *n*-alkanes is higher in the East Siberian Sea in comparison to the Kara and Laptev seas (see Chapters 7.5 and 7.6). Off the Indigirka River, maximum *n*-alkane concentrations of 16 to 23 µg/gSed (or 1000 to 1600 µg/gOC) were determined. Furthermore, in the absence of long-chain *n*-alkanes the OC content is still > zero (Fig. 7.4.4) which suggests the presence of (marine) organic carbon that does not contain long-chain *n*-alkanes (see Chapter 1.4).

In similar sections across the Laptev Sea shelf, the molecular patterns of alkane hydrocarbons show either persistence of OM composition, as seen on the profile from the Chaunskaya Inlet to the New

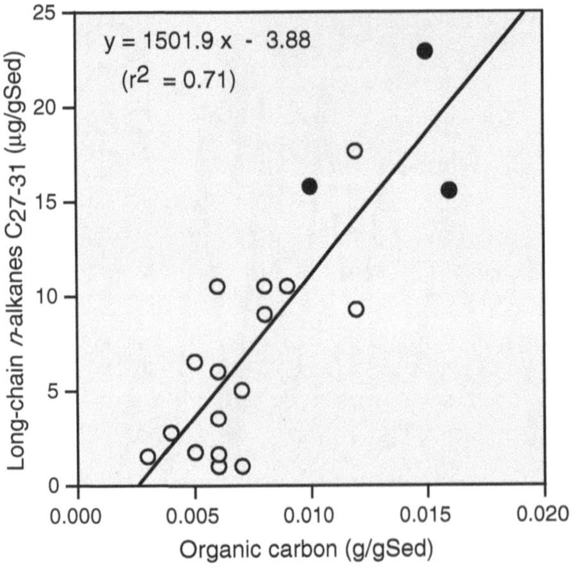

Fig. 7.4.4. Correlation between long-chain *n*-alkanes (C_{27} + C_{29} + C_{31}) in µg/gSed and total organic carbon in g/gSed in surface sediments from the East Siberian Sea. Solid dots indicate samples off the Indirka River

Siberian Islands, or show a change in predominant source of OM in the shelf-pelagic direction, as observed on profiles from the Blagoveshchensky Strait to Podvodnikov Basin, or from Smirnitskogo Bay to Podvodnikov Basin. In the former case, the sediments contain homogeneous terrigenous OM (n-C_{23-31}); in the latter case a change from predominantly terrigenous components to predominantly marine productivity components is suggested by the alkane patterns.

Thus, transport of particulates by water or atmosphere essentially supplies terrigenous OM adsorbed on mineral particles to the sedimentary basin of the East Siberian Sea. This OM, which is deposited on the sea floor without substantial change during transport, is augmented by OM from marine production. However, most of the marine OM is rapidly metabolized such that surface sediments ultimately contain terrigenous OM enriched by the most stable compounds from marine production.

Factors controlling organic carbon supply and sedimentation in the East Siberian Sea

Based on shelf topography and currents (Fig. 7.4.5), three regions differing in sediment supply, transport and accumulation of OC, can be distinguished: the western (New Siberian–Kolyma), the eastern (Wrangel), and the marginal-shelf (divided into the western and the eastern domains).

In the western region, the OC is supplied by river runoff rich in products of permafrost degradation. Coastal and sea-bottom thermo-abrasion is common here because extensive parts of the coast (island and continental) are composed of poorly-bonded permafrost soils. Organic carbon is also supplied by relatively high marine production in the New Siberian branch of the Great Siberian polynya. Sedimentary material transports southeastward toward Long Strait. Frontal zones, associated with depth gradients ("geomorphological traps"), and fronts in the zones of convergence between Arctic and Pacific water act as barriers to transport. There is no clear relationship between regions of higher OC concentration and type of bottom sediment in the western region. Moreover, one of the highest OC concentrations (Indigirka River mouth) is associated with the unfavourable condition of sandy sediment (Kosheleva and Yashin 1999).

In the eastern region, OC is supplied partly by water from the western part of the shelf and partly from the Chukchi Sea (Fig. 7.4.5) due to cyclonic water circulation to the north of Wrangel Island. Additional transport of sedimentary material may occur over the inner shelf toward Long Strait with a return of part of it in the southern Chukchi Sea where it may deposit in the South Chukchi Depression. OC supply through river runoff, thermal abrasion of permafrost and other sources typical of the western region is drastically reduced here. Correlation of higher OC concentrations with sediments that contain a substantial proportion of amorphous silica suggests that marine primary production is an important OC source (Romankevich et al., 1982; Kosheleva and Yashin 1999).

The marginal-shelf region includes part of the shelf conterminal to its edge on the south; it is distinguished by an evident bottom gradient towards the Podvodnikov Basin. Organic matter appears to be supplied here from adjacent regions (Laptev and Chukchi seas). Furthermore, sediment can be supplied by descending lithodynamic flows along the submarine valleys and hollows referred to as the Pra-Indigirka system.

Thus, OC distribution in the East Siberian Sea is very heterogeneous. Regions of higher concentrations tend to be located in geomorphological traps of the first kind (e.g., depressions, submarine valleys and separate small sink holes). Geomorphological traps of the second kind may be found at fronts, especially where they are oriented normal to the direction of transportation. Examples include sediment traps north of Blagoveshchensk Strait and in Long Strait. Commonly, the presence of a geochemical barrier will work together with a geomorphological trap, for example in the Kolyma

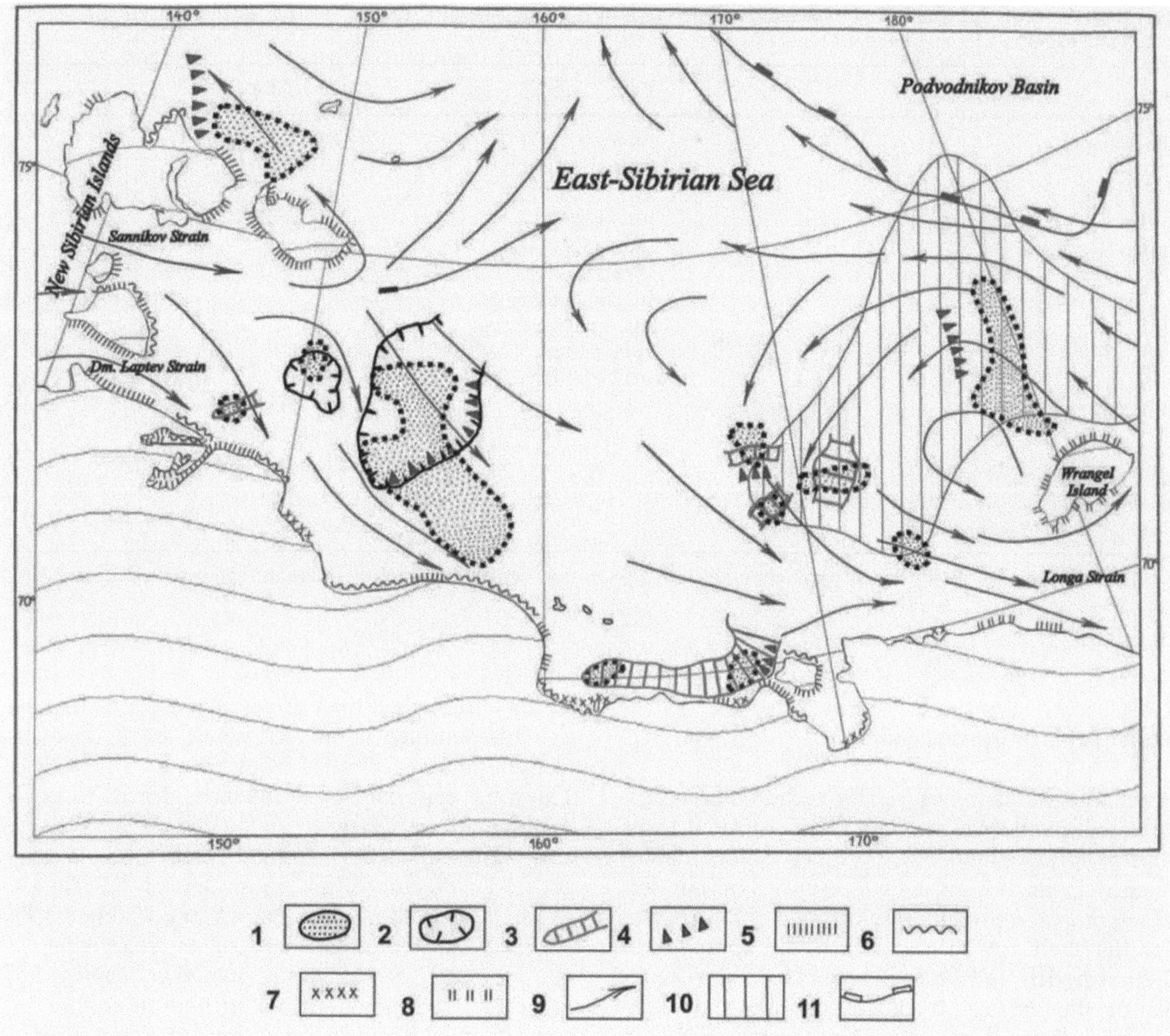

1 Areas of relatively high OC concentrations in the bottom sediments
2 Depressions and sinks in the sea bottom
3 Fragments of submarine valleys
4 Screening slopes (slopes preventing/barriering sediment matter transportation)
5 Termoabrasive coasts
6 Coasts formed by tidal currents
7 Coasts formed by estuarine processes (i. e., accumulation of riverine and marine sediments)
8 Lagoonal coasts
9 Major current directions (from Atlas of the oceans, Arctic ocean, MO USSR, 1980)
10 Bottom sediments with increased soluble silica content (after Kosheleva and Yashin, 2000)
11 Shelf edge

Fig. 7.4.5. A schematic diagram showing the major currents and shelf topography that provide the geomorphological and hydrological controls for organic carbon (OC) distribution in bottom sediments of the East Siberian Sea (based on the Geomorphological map of USSR scale 1:2 500 000, 1987)

and the Indigirka submarine valleys and parts of the submarine valley opposite the mouth of the Omulyakhskaya and the Khromskaya inlets. Most of the recognized depocenters appear to contain temporary deposits which can be resuspended and transported because the topography does not form a closed basin. The most important terminal depocenter, the Podvodnikov Basin, may have its sediment supply hampered by low relief energy and the presence of lateral flows in the marginal-shelf region. An important depocenter for East Siberian sediments is the almost isolated South Chukotka Depression, connected with the outer shelf via the narrow Gerald Trench.

Table 7.4.4. Sources, flux and burial of organic carbon (OC) in the East-Siberian Sea. Data source: Romankevich and Vetrov (2001) and Chapter 2 for allochtonous input

Source	Flux	OC 10^6 t y^{-1}
allochtonous	riverine	**1.90**
	coastal erosion	**2.20**
	aeolian	**0.18**
	groundwater	0.43
authochtonous	phytoplankton	7.00*
	ice algae	4.70
	macrophytobenthos	–
	microphytobenthos	3.50
	zooplankton	0.50
	zoobenthos	1.60
Total flux to seafloor		**13.2 (75.3% from suspended OC)**
Mineralization in the water column		**4,5 (34% from total flux)**
Mineralization on the surface of the seafloor		**7.7 (89%)**
Burial in bottom sediments		**0.96**

* This value implies a total primary production of about 7 g C m^{-2} y^{-1} which is much lower than the range given by Sakshaug of 25–40 g C m^{-2} y^{-1} (Chapter 3, Table 3.7).

7.4.4 Burial Rates of Organic Carbon

A reliable evaluation of organic carbon flux to bottom sediments requires data from sediment traps and/or dated sediment core profiles. Unfortunately, such data do not exist for the East-Siberian sea. However, an approximation of the carbon balance for the shelf can be made using the approach of Romankevich and Vetrov (2001) (see Table 7.4.4). Given the average thickness of Holocene sediments, estimated at 1.75 m (Kosheleva and Yashin 1999), the bulk rate of organic carbon mass accumulation in sediments is 110 mg m^{-2} ky^{-1}. Prorated for the entire shelf area of 987×10^3 km^2 (Chapter 1.1) this sedimentation rate would be equivalent to the annual accumulation of 0.96×10^6 t of OC. This very approximate estimate conforms to estimates made by integration for the Barents, Kara and Chukchi Seas (Romankevich and Vetrov 2001). Furthermore, the summary of OC fluxes estimated for the East Siberian Sea (Table 7.4.4) suggests that total flux of OC to sediments is approximately 13.2×10^6 t C per year. The burial in bottom sediments estimated here, therefore, is consistent with remineralization efficiencies estimated at 81% for the Kara Sea to 99% for the Chukchi Sea.

7.4.5 Summary and Concluding Remarks

About 1×10^6 t y^{-1} of organic carbon have been buried in the East-Siberian Sea during the Holocene. This rate of burial is similar to estimates for the Laptev Sea (Stein and Fahl, 2000; see Chapter 7.5) and other Arctic Seas (Romankevich and Vetrov 2001). The lack of reliable information on the bioproductivity of the East Siberian Sea, modern sedimentation rates, the intensity of OM mineralization in the water column and at the sea floor and, hence, the efficiency of OM preservation in sediments hinder the construction of an accurate OC budget for the region. Further investigation on all of these topics and, especially, on (well-dated) sediment cores is urgently required.

7.5 The Laptev Sea: Distribution, Sources, Variability and Burial of Organic Carbon

R. Stein, K. Fahl

7.5.1 Introduction

The Laptev Sea (Figs. 7.1.7 and 7.5.1) is one of the broad and shallow Eurasian Arctic marginal seas and covers almost 500,000 km^2. The average water depth is about 50 m and decreases to 15–20 m in the shallow southern part (Holmes and Creager 1974; Jakobsson 2002). Several submarine valleys, developed in the last glacial during times of lowered sea-level by erosion of the rivers, cross the shelf from the south to the north (Holmes and Creager 1974; Kleiber and Niessen 1999). Holocene sedimentation mainly takes place in these submarine valleys (e.g., Kuptsov and Lisitzin 1996; Stein and Fahl 2000; Bauch et al. 2001a,b). One of the most important characteristics of the Laptev Sea is its high freshwater discharge, reaching about 745 km^3 y^{-1}, of which about 525 km^3 y^{-1} is contributed by the Lena River alone; the corresponding values for total suspended matter supply reach values of about 28.6 × 10^6 t y^{-1} and 20.7 × 10^6 t y^{-1} (Holmes et al. 2002; Chapter 2.2). This river discharge plays a key role in controlling the temperature and salinity structure of surface water masses, the formation of sea ice, as well as biological processes (e.g., Aagaard and Carmack 1989; Gordeev et al. 1996; Stein 1998; Kassens at al. 1999). Most of the fine-grained sediment load supplied by the Lena River is not deposited in the Lena Delta but transported farther offshore. The delta acts as a filter for the sand and gravel fraction but does not store a considerable amount of river suspension (Rachold et al. 2000, 2002).

With respect to sea ice, the Laptev Sea shows the highest net ice production rates in the Arctic Ocean (Kassens et al. 1999). Large amounts of sediment (including organic matter) may become incorporated into the sea ice in the Laptev Sea shelf area and transported onto the open ocean (e.g., Pfirman et al. 1989; Nürnberg et al. 1994; Eicken et al. 1997; see Chapter 2.4).

The dominant factors controlling the organic carbon input into the Laptev Sea are coastal erosion and river discharge (Stein 1998; 2000 and further references therein). Sediment input by coastal erosion, which may reach values twice as high as the sediment supply by rivers (58.4 × 10^6 t y^{-1} versus 28.6 × 10^6 t y^{-1}), results in total organic carbon supply rates of 1.8 × 10^6 t y^{-1} (see Chapter 2.2 and 2.3). Concerning the organic-carbon cycle of the Laptev Sea, river discharge is important for two reasons. On one hand, there is the direct supply of terrigenous (dissolved and particulate) organic matter, reaching in total 6.8 × 10^6 t y^{-1}. The annual supply of dissolved and particulate organic carbon of the Lena River, as the main contributor, is about 3.6 and 1.2 × 10^6 tons, respectively (Rachold and Hubberten 1999; see Chapter 2.2). On the other hand, the rivers transport large amounts of nutrients (e.g., Martin et al. 1993; Gordeev et al. 1996; Holmes et al. 2002) controlling primary production. For the inner Laptev Sea, an average primary production of about 25–40 g C m^{-2} y^{-1} is estimated (Chapter 3, Table 3.7). Relatively high contents of chlorophycean algae, phaeopigments and specific marine biomarkers in the surface sediments near the river mouths and on the shelf, indicating increased primary productivity, document this fluvial influence (Heiskanen and Keck 1996; Fahl and Stein 1997; Nadina and Bauch 1999; Kunz-Pirrung 1999; Boucsein and Stein 2000). Eolian organic carbon input into the Laptev Sea is of minor importance (< 0.1 × 10^6 t y^{-1}; see Chapter 2.5). Processes important for the export of organic matter out of the Laptev Sea include sea-ice export (Chapter 2.4), ocean currents, and turbidity currents (Stein and Korolev 1994; Stein et al. 1999a; Stein and Fahl 2000).

Within a German-Russian research project initiated in the Laptev Sea area to study the land-ocean interaction and its paleoclimatic interrelationships (Kassens et al. 1999), detailed investigations on the characterization and quantification of the organic matter and its variability in space and time were performed. The results of these studies were published in a large number of papers: e.g., Stein and Nürnberg (1995), Stein (1996, 1998), Fahl and Stein (1997 1999), Bauch et al. (1999, 2000, 2001a, b), Stein et al. (1999a, 2001), Boucsein and Stein (2000), Knies et al. (2000), Stein and Fahl (2000), Boucsein et al. (2001, 2002), Mueller-Lupp et al. (2000), and Fahl et al. (2001). Supplementary to the investigations on marine sediments, the organic matter within the rivers themselves was studied (e.g., Lara et al. 1998; Rachold and Hubberten 1999). Furthermore, it should be mentioned that additional investigations were carried out by other groups, e.g., within the framework of the international program SPASIBA (e.g., Heiskanen and Keck 1996; Peulvé et al. 1996; Saliot et. al. 1996).

In this synthesis paper we (1) summarize results of these studies related to the distribution, composition, variability, and accumulation of organic carbon and (2) present a total sediment and organic carbon budget.

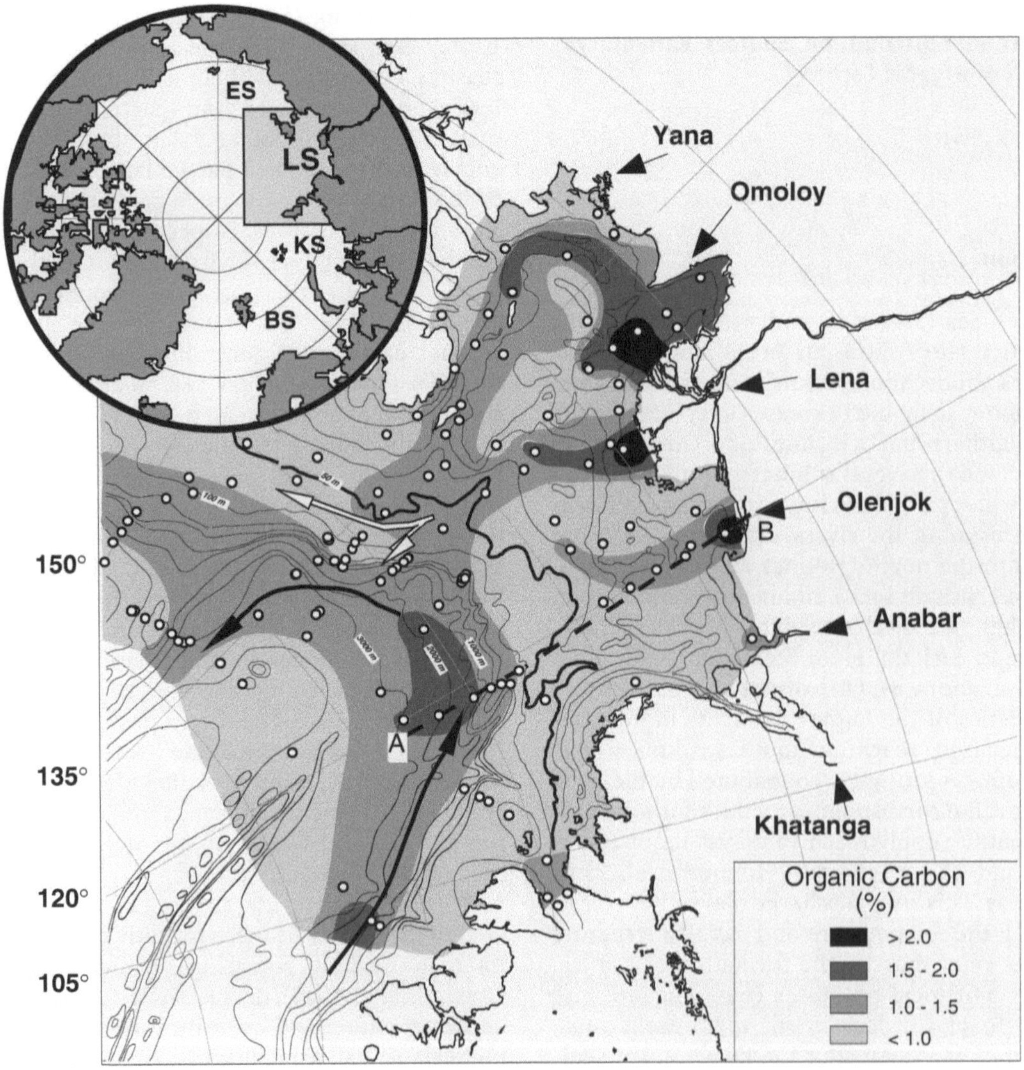

Fig. 7.5.1. Distribution map of total organic carbon in surface sediments from the Laptev Sea continental margin (open circles indicate sampling sites). Main rivers (Yana, Omoloy, Lena, Olenjok, Anabar and Khatanga) are shown. Black arrows: flow of Atlantic water mass along the Eurasian continental margin. Bold depth contour (50 m): approximate position of coastline during times of lowered sea level at about 11 Cal. kyrs. BP. White arrows: major probable pathways of riverine material transport between 9–11 Cal kyrs. BP. A-B (stippled line) marks transect of maceral composition shown in Chapter 1.4, Fig. 1.4.4. Upper left: regional overview of the Arctic Ocean showing the Eurasian continental shelf seas (BS Barents Sea; KS Kara Sea; LS Laptev Sea; ES East Siberian Sea)

7.5.2 Data Base, Material and Methods

Total organic carbon (OC), total nitrogen (TN), hydrogen (HI) and oxygen index (OI) values, stable carbon isotopes of the organic matter, maceral composition as well as biomarker data were used to determine the amount and composition of the organic carbon fraction in surface sediments as well as sediment cores. A short description of the methods and use of the different proxies as organic-carbon-source indicator are given in Chapter 1.4.

To interpret the organic-geochemical records in terms of paleoenvironmental changes with time, and to calculate sedimentary and organic carbon budgets, a chronological framework is necessary. For the Holocene, this framework is based mainly on AMS^{14}C datings (Bauch et al. 1999, 2001b; Stein and Fahl 2000). In addition, magnetic susceptibility records were used. In most of the cores, the base of the Holocene is characterized by a prominent decrease in magnetic susceptibility that can be used to correlate the cores from the Laptev continental margin (e.g., Stein et al. 1999a; Kleiber and Niessen 2000; Stein and Fahl 2000). For the sediment cores from the Laptev Sea continental slope representing older, pre-Holocene intervals, i.e., oxygen isotope stages (OIS) 2 to 6, the stratigraphy is based on oxygen isotope stratigraphy, paleomagnetostratigraphy, biostratigraphy (especially dinoflagellates), lithostratigraphy, and magnetic susceptibility records (Müller 1999; Stein et al. 1999a, 2001; Matthiessen et al. 2001).

Based on the chronological framework, linear sedimentation rates (LSR) were interpolated between age-fixed points (see Bauch et al. 2001b). Then, total sediment (TS) and total organic carbon (OC) accumulation rates were calculated using LSR, wet bulk density (WBD) values, and a grain density (GD) of 2.65 g cm^{-3} (see van Andel et al. 1975; Weber et al. 1997):

$$\text{Porosity PO (\%)} = \frac{(GD - WBD)}{(GD - 1.026)} \cdot 100$$

$$\text{TS-AR (g cm}^{-2}\text{ ky}^{-1}) = LSR * (WBD - 1.026\ PO/100)$$

$$\text{OC-AR (g C cm}^{-2}\text{ky}^{-1}) = \text{TS-AR} \cdot OC/100$$

7.5.3
Distribution and Sources of Organic Carbon in Surface Sediments

The distribution of organic carbon in the Laptev Sea surface sediments is strongly influenced by the huge river discharge. Loads of total suspended matter and particulate organic carbon reach values of about 28.6×10^6 t y^{-1} and 1.3×10^6 t y^{-1}, respectively (Gordeev et al. 1996; Rachold and Hubberten 1999; Holmes et al. 2002; see Chapter 2.2). Most of this material accumulates within the inner Laptev Sea. This supply of terrigenous organic carbon by rivers is reflected in the relatively high total organic carbon contents of the surface sediments of the Laptev Sea with values ranging between about 0.5 and 2.3% (Fig. 7.5.1). Maximum OC values of up to 2% and more occur in the vicinity of the eastern Lena Delta, off the Olenjok river mouth and southwest of the New Siberian Islands, and the central part of the lower Laptev Sea continental slope (Fig. 7.5.1). In the eastern and middle Laptev Sea, these OC maxima follow submarine channels which can be traced back to the river mouths. These submarine channels are the main modern (late Holocene) depocenters in the inner Laptev Sea, where fine-grained riverine material accumulates (e.g., Kuptsov and Lisitzin 1996), as also reflected in the distribution of clay and silt content of the surface sediments (Fig. 7.5.2a).

The predominantly terrigenous origin of the organic matter of the Laptev Sea is shown in the organic geochemical bulk parameters, maceral composition as well as biomarker data (Fig. 7.5.2).

Organic Geochemical Bulk Parameters

Areas of high OC concentration commonly correspond to low HI values (<100 mgHC/gOC) and light stable carbon isotope composition ($\delta^{13}C_{org}$ = –26.5‰; Mueller-Lupp et al. 2000; Fahl et al. 2001) pointing to a terrigenous origin (Figs. 7.5.2b and 7.5.2d). The lightest $\delta^{13}C_{org}$ values recorded near the mouth of the Lena River are in good agreement with the composition of the particulate organic matter of the Lena itself (–27.1‰; Rachold and Hubberten 1999). High C/N ratios generally also correlate with the high OC values. These high C/N ratios, however, only reach values of 7 to 10 (Fig. 7.5.2c). Assuming that typical values for marine and terrigenous organic matter are about 6 and >15, respectively (Bordowskiy 1965; Scheffer and Schachtschabel 1984; Hedges et al. 1986; Meyers 1997), would indicate a more mixed origin of the organic matter. It has to be considered, however, that the C/N ratios are calculated as "OC/N_{tot}" ratios and not as "OC/N_{org}" ratios. If the amount of inorganic nitrogen bound as ammonium in clay minerals (N_{bou}), especially in the clay mineral illite (Stevenson and Cheng 1972; Müller 1977), is high, the measured C/N ratios may be biased too low (see Chapter 1.4.2). In a plot of N_{tot} versus OC (Fig. 7.5.3), there is an intercept of 0.05 and 0.10% N_{tot} at 0% OC, indicating that there is a significant fraction of inorganic nitrogen in the samples. The data points with the higher N_{bou} value of about 0.10% mainly represent the central and eastern Laptev Sea which is characterized by high illite contents (on average about 50%; Rossak et al. 1999). After a correction of the N_{bou} content (see Chapter 1.4.2), the majority of the corrected C/N ratios vary between 10 and 25, indicating a dominantly terrigenous organic carbon origin.

Based on relatively heavy $\delta^{13}C_{org}$ values of about –23‰ measured in the upper Laptev Sea slope sediments, one could assume a higher proportion of marine organic matter which is, however, not supported by other proxies such as the macerals (see below). The heavy isotope values, however, could also be explained by contribution from sea-ice algae, although the dominant organic carbon source is still terrigenous (see Chapter 1.4.2).

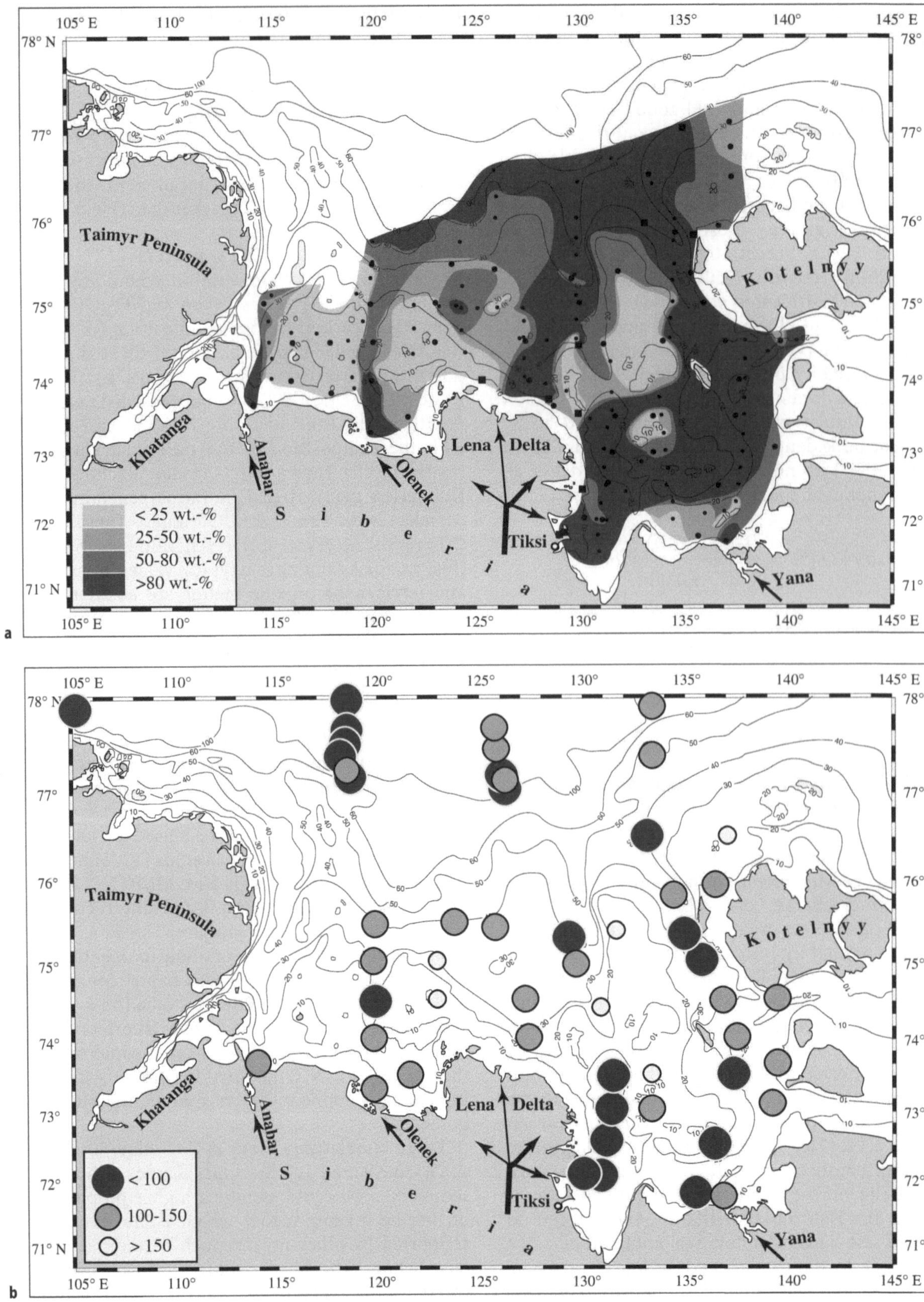

Fig. 7.5.2. Grain-size data and proxies for organic carbon composition in surface sediments from the Laptev Sea; 2a, 2b, 2e (supplemented by new unpublished data), and 2f from Fahl et al. (2001); (a) Distribution of silt and clay (< 63 µm); (b) Hydrogen index values in mg HC/g OC

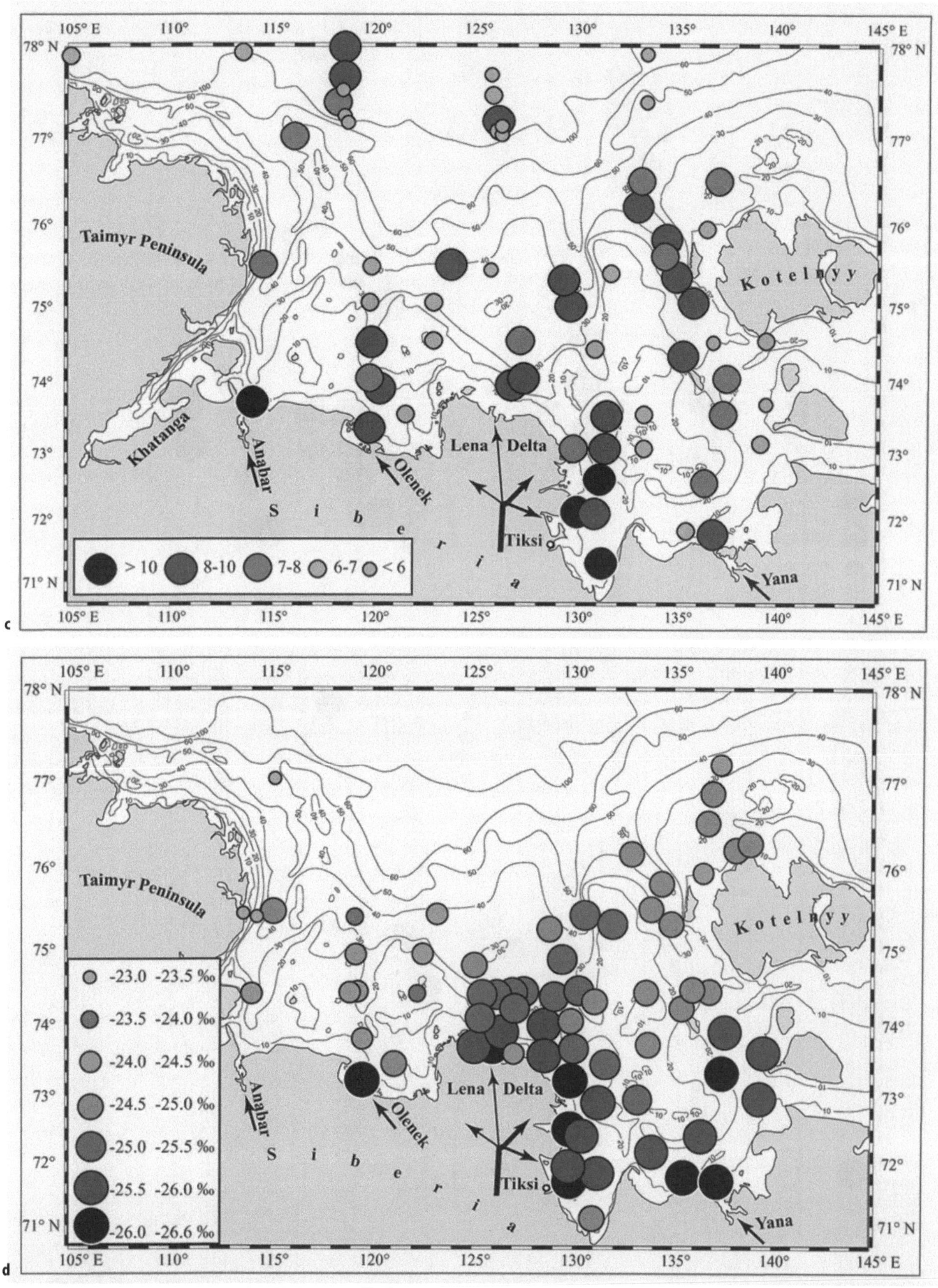

Fig. 7.5.2 (continued)
(c) C/N ratios; (d) Distribution of $\delta^{13}C_{org}$

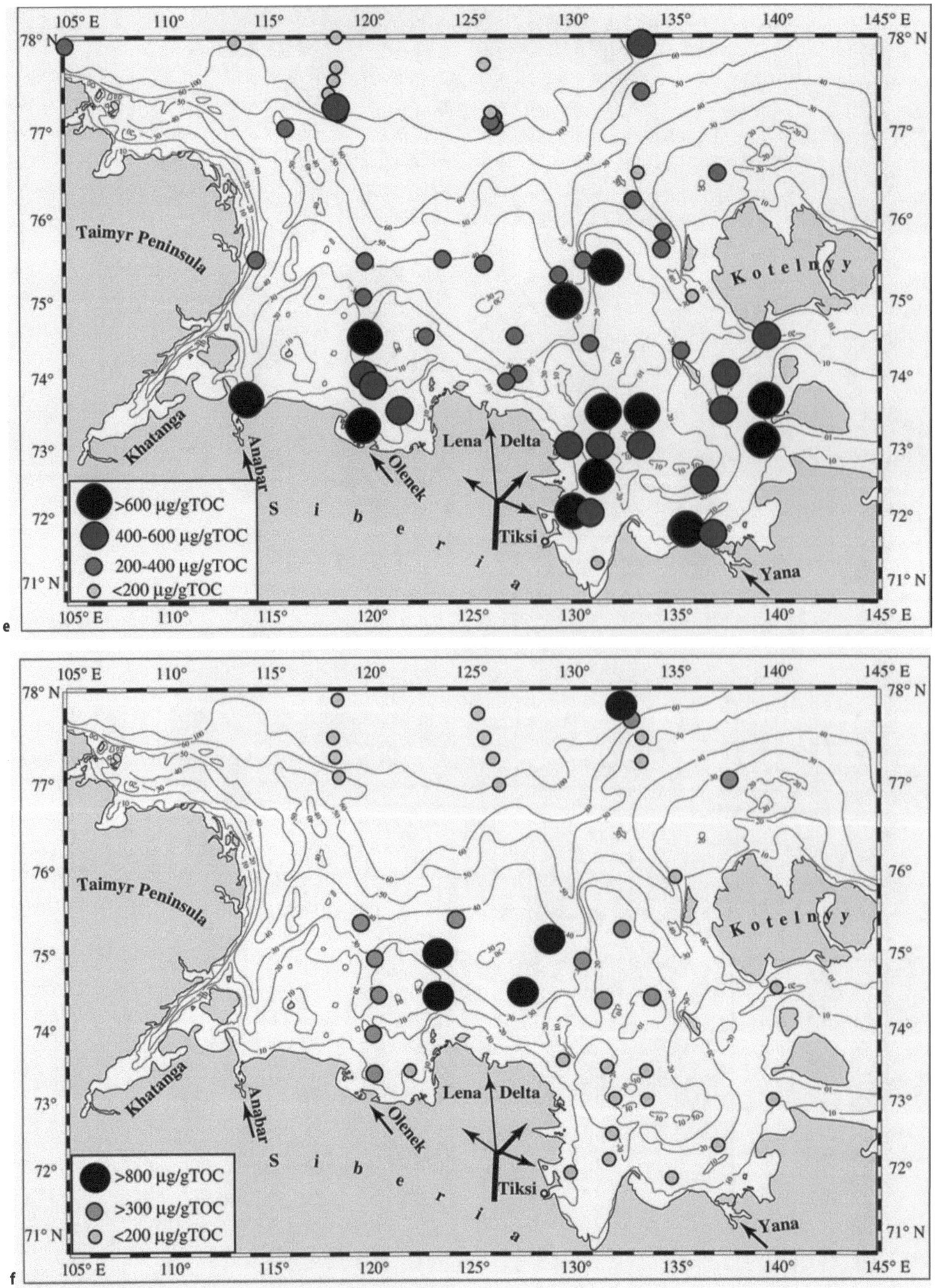

Fig. 7.5.2 (continued)
(e) Distribution of long-chain *n*-alkane (C_{27} + C_{29} + C_{31}) concentrations (µg/gOC)
(f) Distribution of the sum of 16:1(n-7) and 20:5(n-3) fatty acids (in µg/gOC)

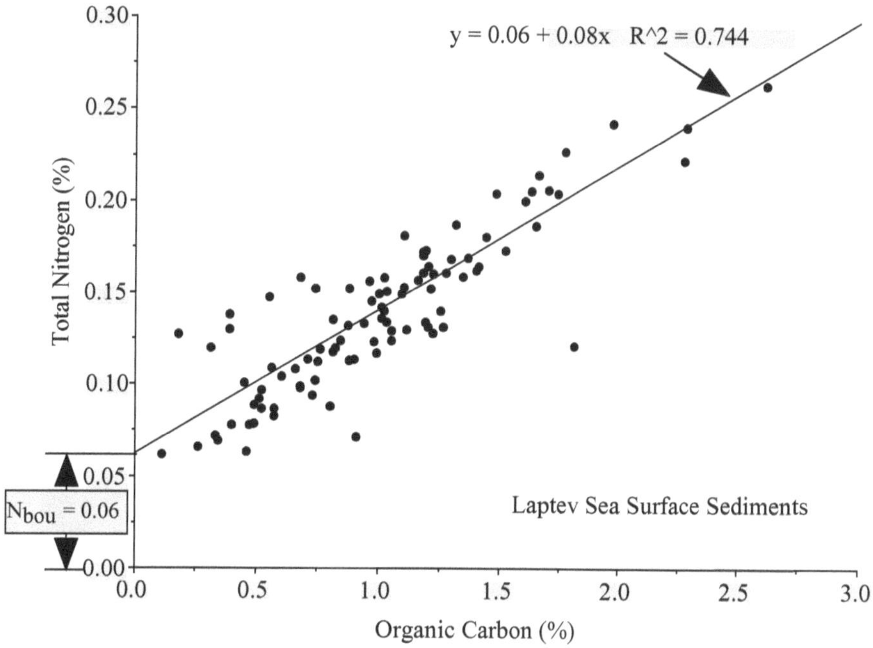

Fig. 7.5.3. Correlation between total organic carbon (OC) and total nitrogen (N_{tot}) in surface sediments from the Laptev Sea. The intercept of 0.05 to 0.10 % N_{tot} at 0 % OC implies the fraction of inorganic nitrogen contained by the samples

The OC maximum along the lower continental slope characterized by low HI values and high C/N ratios (Figs. 7.5.1, 7.5.2b, and 7.5.2c), indicating a terrigenous source, may be related to the inflow of Atlantic water masses laterally transporting (organic-carbon-enriched) suspended matter (see, e.g., Knies et al. 2000; Stein et al. 2001) and/or input from the Laptev Sea shelf (see 7.5.5).

Maceral composition

Based on the abundances of terrigenous macerals (i.e., vitrinite/huminite, inertinite, terrigenous liptinites, detritus, freshwater alginites) and marine macerals (lamalginite, dinoflagellate cysts, marine litptinito-detrinite), the amount of marine versus terrigenous organic matter in the Laptev Sea surface sediments was estimated quantitatively (see Boucsein and Stein 2000). In the inner Laptev Sea, the sediments contain organic matter that is about 95 % of terrigenous origin and, directly off the river mouths, values as high as 97 to 99 % are reached (Fig. 7.5.4). Furthermore, close to the river mouths minor but significant amounts of fresh-water algae are found (3 to almost 10 %), indicating aquatic, land-based primary production (see Chapter 1.4.3, Fig. 1.4.5; Boucsein and Stein 2000). Towards the north, with increasing distance from the land, the amount of terrigenous macerals decreases whereas marine macerals occur in significant amounts, and for the outer Laptev Sea shelf, an average proportion of 85 % of terrigenous organic matter was determined (Fig. 7.5.4). At the upper continental slope, marine organic matter may contribute as much as 20 to 30 % of the total organic matter. The same trend is also obvious in the grain size of the terrigenous organic particles. Towards the north, the size of the huminite/vitrinite and inertinite particles decreases while the amount of detritus <10 µm increases (see Chapter 1.4.3, Fig. 1.4.5; Boucsein and Stein 2000), which is caused by a stronger fragmentation and sorting due to long-distance transport.

Distribution of long-chain *n*-alkanes

The dominantly terrigenous source is verified by the distribution of odd long-chain *n*-alkanes C_{27}–C_{31} (Fig. 7.5.2e; Fahl and Stein 1997). The highest concentrations of these biomarkers, which are generally accepted as an indicator for a terrigenous higher plant source of the organic matter (see Chapter 1.4.4), were found in the vicinity of the eastern Lena Delta as well as off the rivers Anabar, Olenek and Yana. There, the *n*-alkane concentrations reach values of about 600 to >1000 µg/gOC. Towards the continental slope, the concentration of long-chain *n*-alkanes decrease to values of <250 µg/gOC (Fig. 7.5.2e), indicating a decreasing terrigenous influence.

To estimate the relative proportion of terrigenous and marine organic carbon more quantitatively, the correlation between absolute concentrations of the long-chain *n*-alkanes in µg/gSed and OC (g/gSed) can be used (Fig. 7.5.5; see Chapter 1.4.4; Prahl and Muehlhausen 1989; Fernandes and Sicre 2000). For the Kara Sea, for example, Fernandes and Sicre (2000) found that the Ob and Yenisei

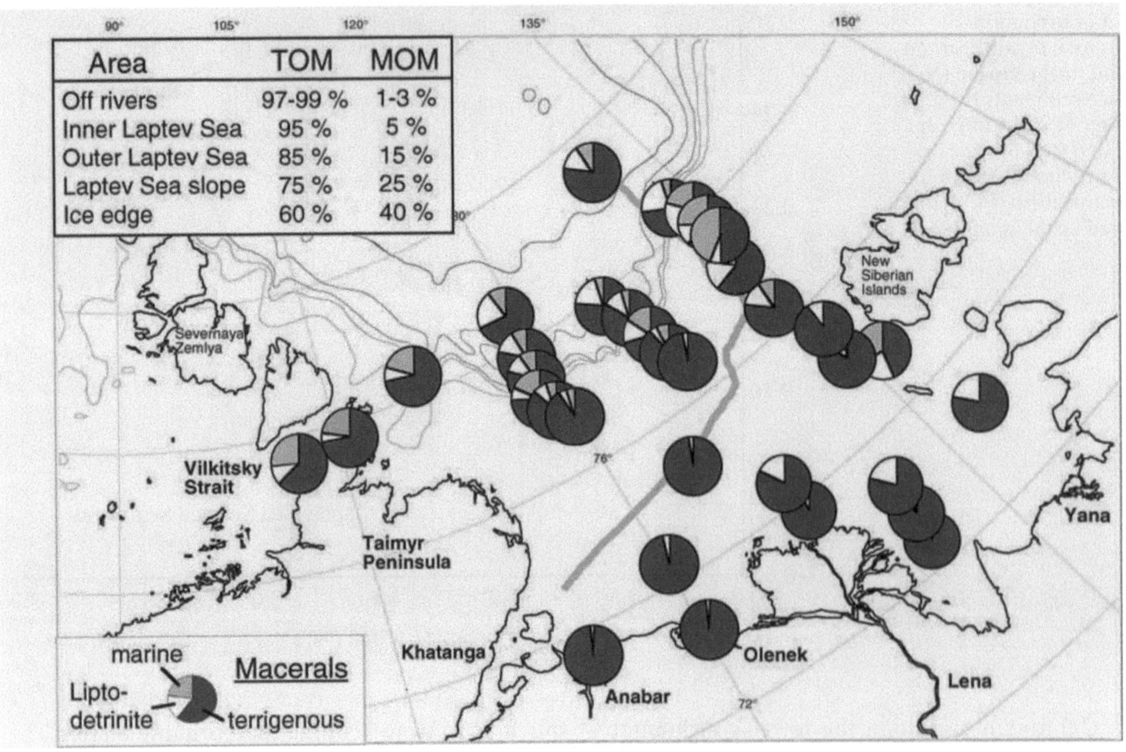

Fig. 7.5.4. Maceral composition in Laptev Sea surface sediments (from Boucsein and Stein 2000). Based on the maceral data, average values of terrigenous and marine organic matter for sediments from the inner and outer Laptev Sea and the adjacent slope were calculated and listed. Dark gray line indicates location of ice edge in September 1995

rivers supply long-chain *n*-alkanes to the Kara Sea in a relatively constant ratio to the organic carbon content (ALKOC ratio). The Kara Sea ALKOC ratio is 451 based on the sum of odd *n*-alkanes C_{25}–C_{31}, and 369 based on the sum of odd *n*-alkanes C_{27}–C_{31} (see Chapter 7.6). Assuming that a decrease in the ALKOC ratio in the sediments north of the estuaries is caused by dilution with marine organic carbon that does not contain long-chain *n*-alkanes (see Prahl and Muehlhausen 1989), the TerrOC equation from Chapter 1.4.4 can be used to calculate the proportion of terrigenous organic carbon.

For the samples close to the river mouths in the Laptev Sea where maximum long-chain *n*-alkane values >600 µg/gOC were determined and the proportion of terrigenous organic carbon is almost 100% based on maceral data, a positive correlation between long-chain *n*-alkanes and OC with a slope (mean ALKOC ratio) of 750 is obvious (Fig. 7.5.5):

$$\text{Sum } C_{27+29+31} = 750 \times OC + 0.17 \; (r^2 = 0.914)$$

Using this approach, the proportions of terrigenous organic carbon vary between 80 and 40% in the central and outer Laptev Sea, with decreasing values towards the north. These values are lower than those based on maceral data and should be used with caution. It has to be considered that the mean ALKOC ratio of 750 used for the calculation is quite variable when looking at samples from the different river mouths in detail (off Anabar: 630; off Olenek: 880; off Lena: 830; off Yana: 1075). Furthermore, terrigenous organic carbon input by coastal erosion with an unknown ALKOC ratio has to be taken into account.

Despite the clear dominance of the terrigenous organic matter, increased hydrogen indices, heavy $\delta^{13}C_{org}$ values, and biomarker data suggest the local presence of significant concentrations of marine organic matter, e.g., in the central Laptev Sea and some areas of the upper continental slope. Highest concentrations of the fatty acids 16:1(n-7) and 20:5(n-3) which are mainly synthezised by diatoms (e.g., Kates and Volcani 1968; Kattner et al. 1983; Fahl and Kattner 1993), were measured close to the ice edge (Fig. 7.5.2f). These values correlate with high concentrations of chlorophyll *a* and phaeopigment in the surface-water mass (Boetius et al. 1996) as well as biogenic opal in surface sediments (Stein and Nürnberg 1995), and are interpreted as a sign of increased productivity resulting in increased preservation of marine organic matter in the sediments. The highest value of marine macerals (44% of the organic matter) was also observed

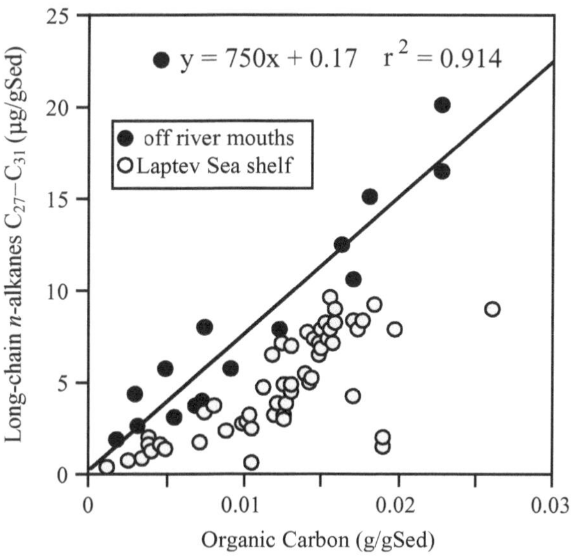

Fig. 7.5.5. Correlation between absolute concentrations of long-chain *n*-alkanes (C_{27} + C_{29} + C_{31}) (µg/gSed) and total organic carbon content (g/gSed) in Laptev Sea surface sediments. The solid points, for which the regression curve has been calculated, represent samples mainly from areas close the rivers mouths. The open circles represent samples from the open Laptev Sea shelf (see Chapter 1.4.4 for further background information)

near the ice-edge (Fig. 7.5.4), supporting the organic geochemical bulk parameters as well as the biomarker data.

The overall predominance of terrigenous organic matter and the very low proportion (1–10%) of marine organic matter preserved in the inner Laptev Sea surface sediments is in agreement with data on oxygen uptake in Laptev Sea near-surface sediments that point to a very low content of labile marine organic compounds in these sediments (Boetius and Damm 1998). This predominance of terrigenous organic carbon in sediments from the Laptev Sea continental margin, i.e., about 90% for the shelf and 80% for the continental slope sediments, is different from other world ocean (non-Arctic) shelf areas where the proportion of marine organic matter preserved in the sediments reaches much higher values. In sediments from the Washington coastal margin (Prahl and Muehlhausen 1989) and the shelf off East Liberia (Westerhausen et al. 1993), for example, about 60% of the organic carbon is of terrigenous origin. For the Mackenzie and Kara shelves, on the other hand, very similar values to those obtained for the Laptev Sea shelf were determined (see Chapters 7.2 and 7.6).

7.5.4 Late Quaternary Organic Carbon Records and Paleoenvironment

The late Quaternary paleoenvironmental evolution of the Laptev Sea continental margin area and the related sediment as well as organic carbon inputs are strongly controlled by changes in river discharge, coastal erosion, sea-ice cover, and oceanic circulation patterns (e.g., Stein 1998, 2000; Bauch et al. 1999, 2001a, b; Stein et al. 1999a; Stein and Fahl 2000, and further references therein). Furthermore, the sediment input along the Eurasian continental margin has been influenced by the waxing and waning of continental ice sheets (e.g., Velitchko et al. 1997; Svendsen et al. 1999). At least during the Saalian (OIS 6) glaciation, when the Laptev Sea was also covered by a huge ice sheet, related processes may have influenced the organic carbon input at that time. During glacial sea-level lowstand, the shallow Laptev Sea was subaerially exposed (Fig. 7.5.6) and terrigenous (fluvial) sediments were transported farther north to the shelf edge and, following down-slope transport by turbidity currents and slumps, may have become another important factor controlling the sedimentation at the continental slope and adjacent deep sea (e.g., Stein et al. 1999a; Kleiber and Niessen 2000).

The (Deglacial to) Holocene Record

For the Holocene time interval, a number of AMS^{14}C dated sediments cores are available (see Bauch et al. 2001b for summary) which allows a detailed reconstruction of the changes in sediment and organic carbon fluxes and their relationship to environmental changes (Bauch et al. 1999, 2001a, b; Fahl and Stein 1999; Stein et al. 1999a, 2001; Mueller-Lupp et al. 2000; Stein and Fahl 2000; Naidina and Bauch 2001).

In most of the investigated cores, the Holocene interval is characterized by relatively high OC contents (Fig. 7.5.7). Holocene OC contents around 1% are most typical. Organic-geochemical bulk parameters (hydrogen indices, C/N ratios, $\delta^{13}C_{org}$ values), biomarker data, and maceral composition all indicate the dominance of terrigenous organic matter, reflecting the terrigenous supply being the most important factor controlling organic carbon deposition at the Laptev Sea continental margin throughout the Holocene.

In the shelf cores KD9502-14 and PS2725-5, HI values of are < 100 mgHC/gOC throughout the entire record. In a HI vs. OI ("van-Krevelen-type") diagram, all data points plot into the "Krevelen-type III" field (Fig. 7.5.8), reflecting the predominance of immature/fresh terrigenous organic carbon. In both cores, C/N ratios do not show high-

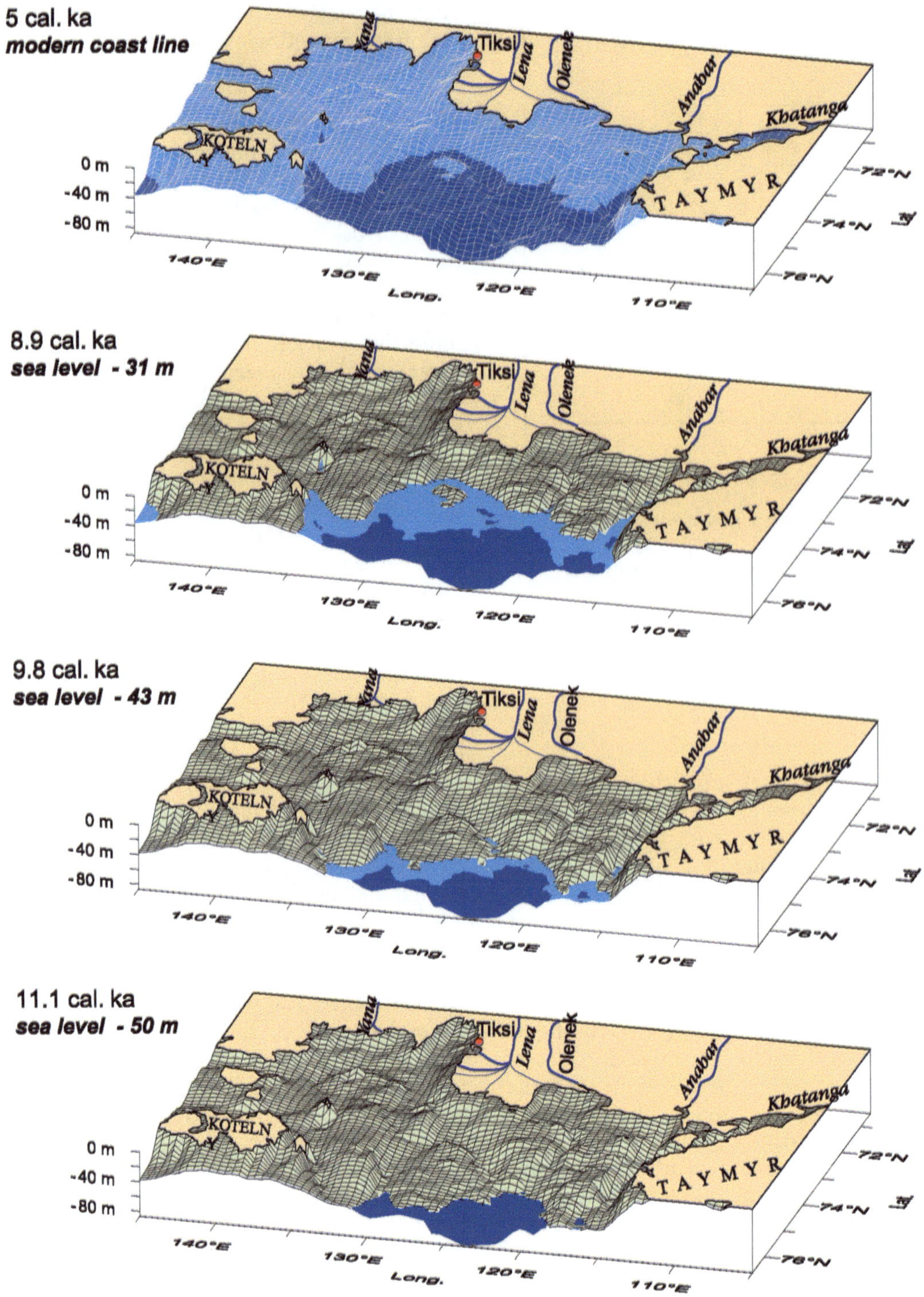

Fig. 7.5.6. Reconstruction of the Laptev Sea transgression showing the difference in areal flooding at various time intervals. The topograpic map is based on Russian navigation charts and the bathymetric data obtained during several marine German-Russian expeditions. Note that the modern shelf topography does not reflect the actual paleosurface prior to inundation (from Bauch et al. 2001b)

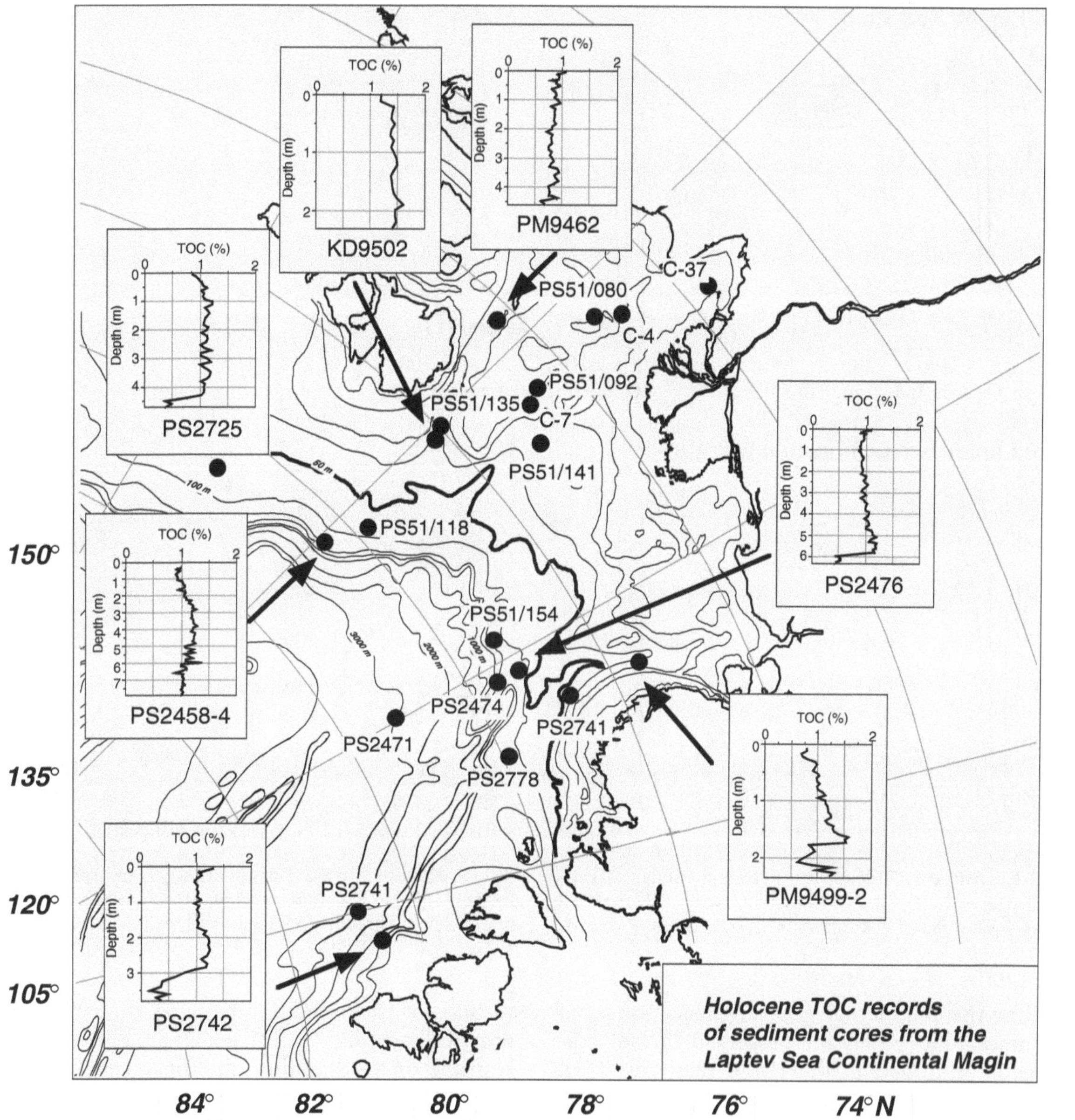

Fig. 7.5.7. OC distribution in selected sediment cores from the Laptev Sea, representing the Holocene. Records of cores PM9499-2, PM9462-2, and KD9502 based on data from Bauch et al. (1999, 2001a). The locations of the cores considered for calculation of sediment and organic carbon budgets (see Table 7.5.1) are shown

amplitude variations and the lowest values were recorded the the uppermost two samples (Fig. 7.5.9). Average C/N ratios of 10 and 7 were recorded in cores KD9502-14 and PS2725-5, respectively. Correcting for inorganic nitrogen, recalculated C/N ratios increase to about 15 and 11, respectively, and point to the dominance of terrigenous organic matter. In the uppermost part of both cores, some increased amount of marine (/aquatic) organic matter might be present. A high proportion of terrigenous matter is also indicated by relatively light $\delta^{13}C_{org}$ values ranging between −25.5 and −26.5‰ (Mueller-Lupp et al. 2000). The biomarker composition supports the predominantly terrigenous character of the organic carbon in both cores (Fig. 7.5.9). In general, the concentrations of long-chain terrestrial plant *n*-alkanes ($C_{27} + C_{29} + C_{31}$) are relatively high (300–400 µg/gOC), comprising more than 50% of the total *n*-alkanes. The concentrations of the corresponding short-chain aquatic phytoplankton *n*-alkanes ($C_{17} + C_{19}$), are rather low (10 µg/gOC). Thus, the ratio of long- to short-chain *n*-alkanes ranges between 25 and 40.

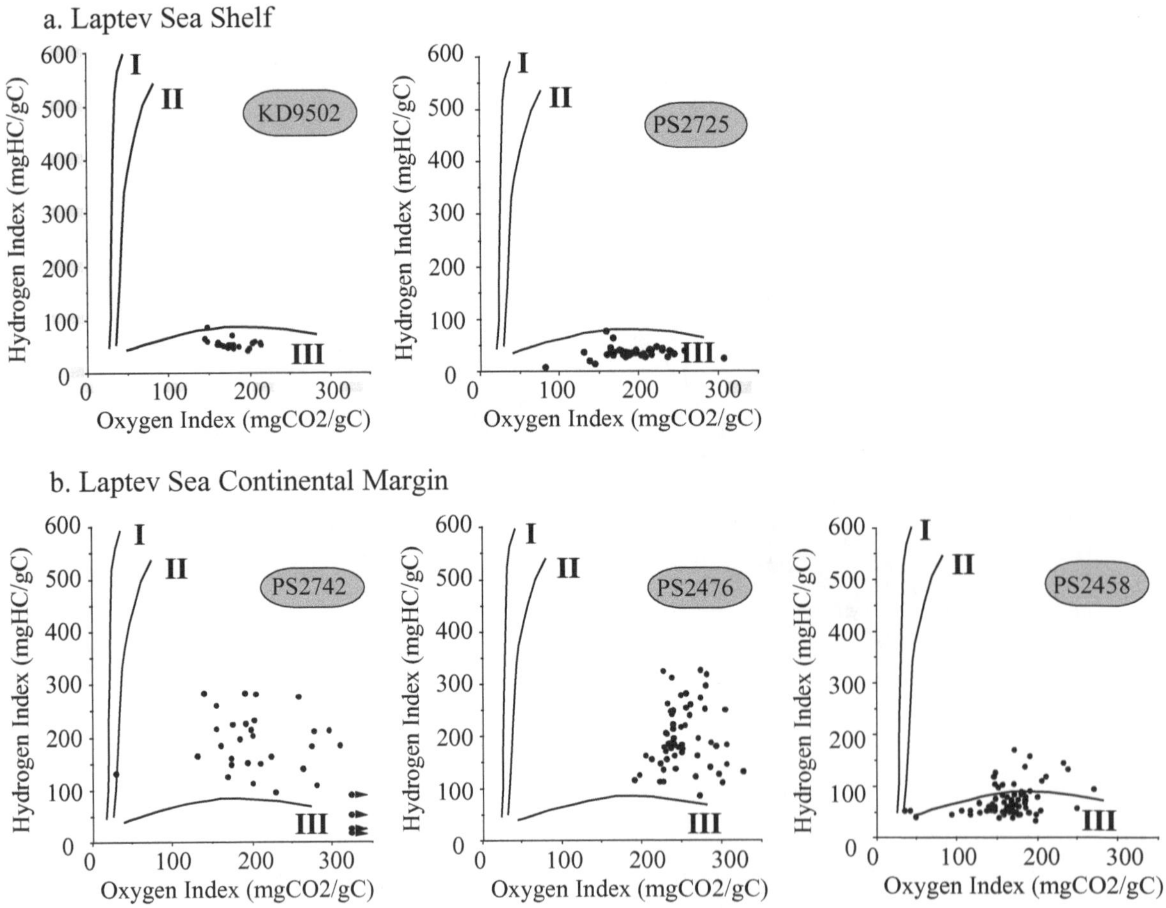

Fig. 7.5.8. Hydrogen index versus oxygen index ("van-Krevelen-type") diagrams for (a) the shelf cores KD-9502 and PS2725-5 and (b) the continental slope cores PS2742-5, PS2476-4, and PS2458-4. Roman numerals indicate different kerogen types: I and II marine, III terrigenous organic matter (classification after Tissot and Welte 1984)

Along the continental slope, higher amounts of marine organic matter are preserved in the sediments, as indicated in the HI/OI diagrams (Fig. 7.5.8, Cores PS2742-5, PS2476-4, PS2458-4) as well as the biomarker distributions in Core PS2458-4 (Fig. 7.5.10). Hydrogen indices are increased up to 300 mgHC/gOC, and the ratio of long- to short-chain *n*-alkanes reach values of 2 to 6. Furthermore, a major change in organic carbon composition is obvious during the Holocene time interval, shown in the record of Core PS2458-4. In the lower core section (>10 Cal. kyrs. BP) low HI values and relatively high concentrations of long-chain *n*-alkanes ($C_{27}+C_{29}+C_{31}$) (Fig. 7.5.10) indicate predominantly terrigenous organic matter. The upper core section (<10 Cal. kyrs. BP), on the other hand, is characterized by relatively high HI values (up to 200 mgHC/gOC) suggesting a more marine character of the organic material (although the terrigenous origin remains dominant). Increased amounts of marine organic matter is also indicated by higher concentrations of short-chain fatty acids, dinosterol, and brassicasterol (Fig. 7.5.10), which are derived mainly from diatoms and/or dinoflagellates (e.g., de Leeuw et al. 1983; Nichols et al. 1984; Volkman et al. 1993). Furthermore, a relatively high proportion of marine organic carbon is confirmed by micropaleontological and maceral data (Fig. 7.5.10). Marine particulate organic matter (lamalginites and dinflagellate cysts) occurs in significant amounts of about 10% in this interval of Core PS2458-4 (Fig. 7.5.10; Boucsein et al. 2000). Approximately a further 10% of the macerals belongs to the group of liptodetrinite which cannot be identified as either marine or terrigenous particles because of their small size. Thus, in total some higher value of marine organic matter (about 20%) may be preserved in the Holocene continental margin sediments. The maceral records of Laptev Sea continental margin cores PS2476-4 and PS2742-5 show very similar trends to that of Core PS2458-4; terrigenous organic carbon accounts for about 90%

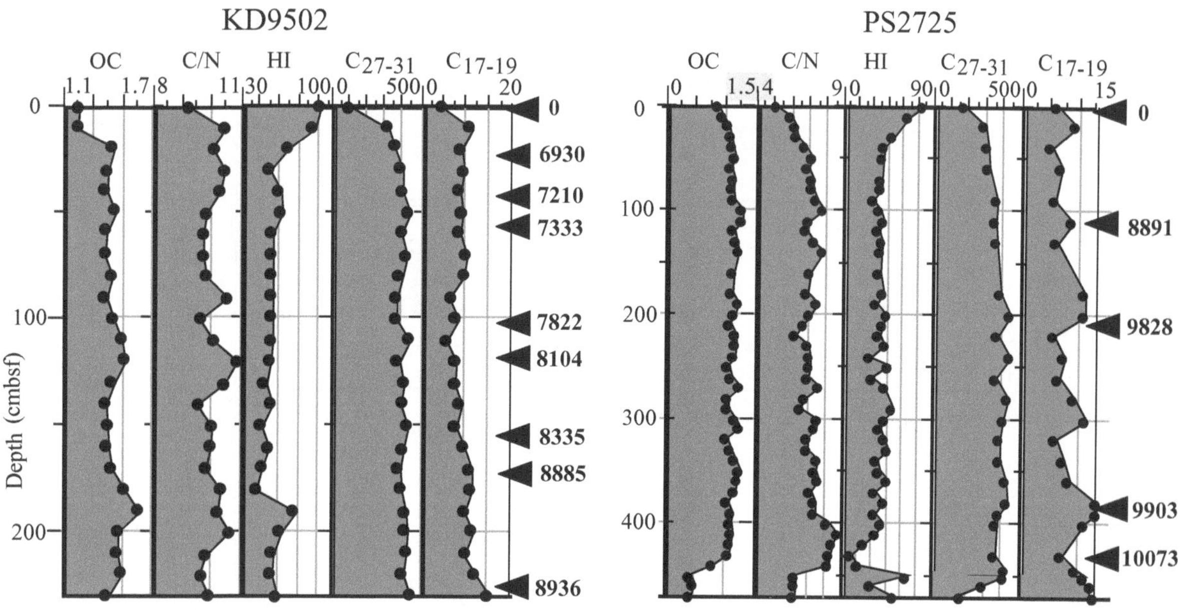

Fig. 7.5.9. Distribution of OC, HI values, long-chain *n*-alkanes ($C_{27}+C_{29}+C_{31}$), and short-chain *n*-alkanes ($C_{17}+C_{19}$) in sediment cores KD9502 and PS2725, both from the Laptev Sea shelf

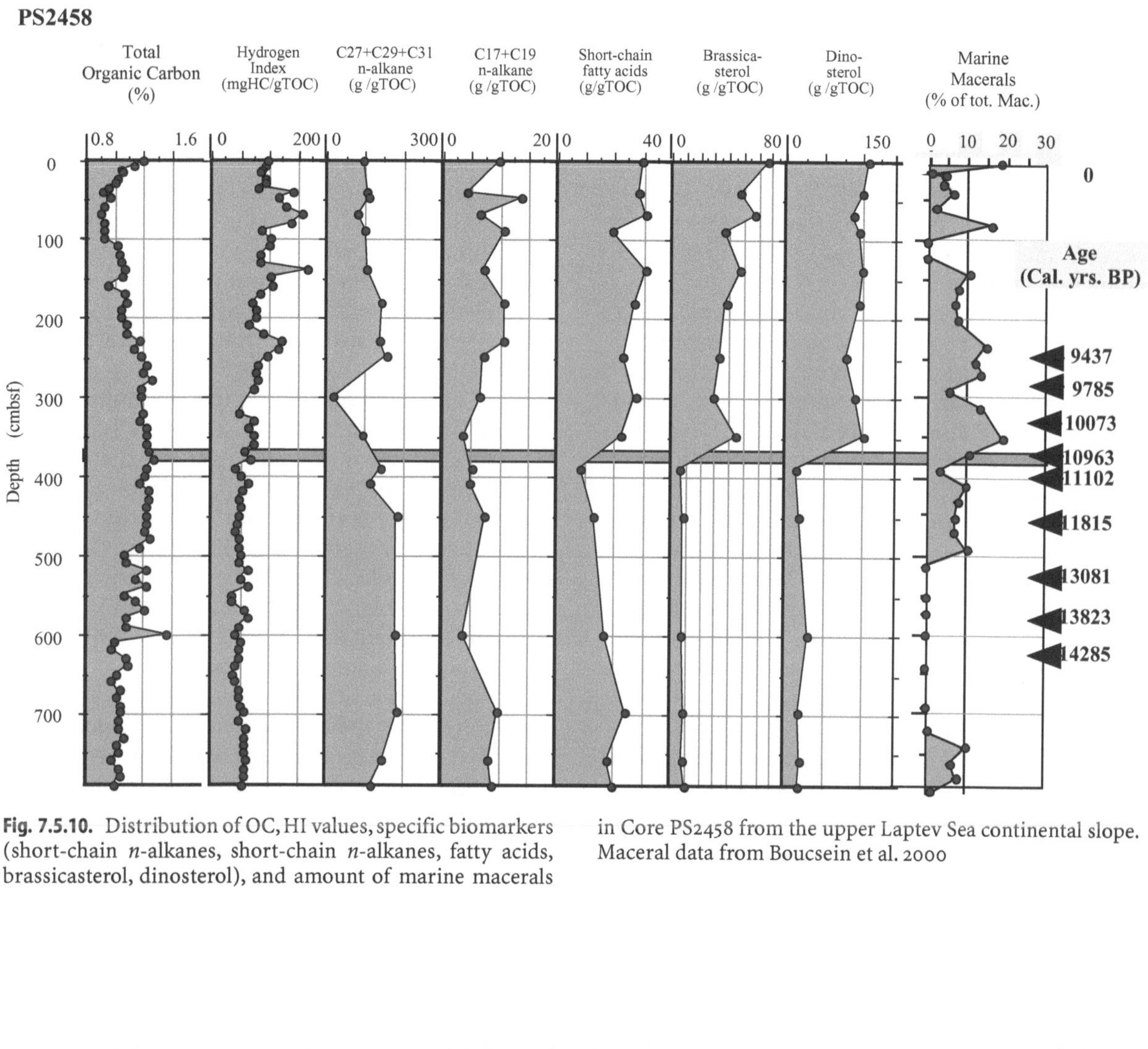

Fig. 7.5.10. Distribution of OC, HI values, specific biomarkers (short-chain *n*-alkanes, short-chain *n*-alkanes, fatty acids, brassicasterol, dinosterol), and amount of marine macerals in Core PS2458 from the upper Laptev Sea continental slope. Maceral data from Boucsein et al. 2000

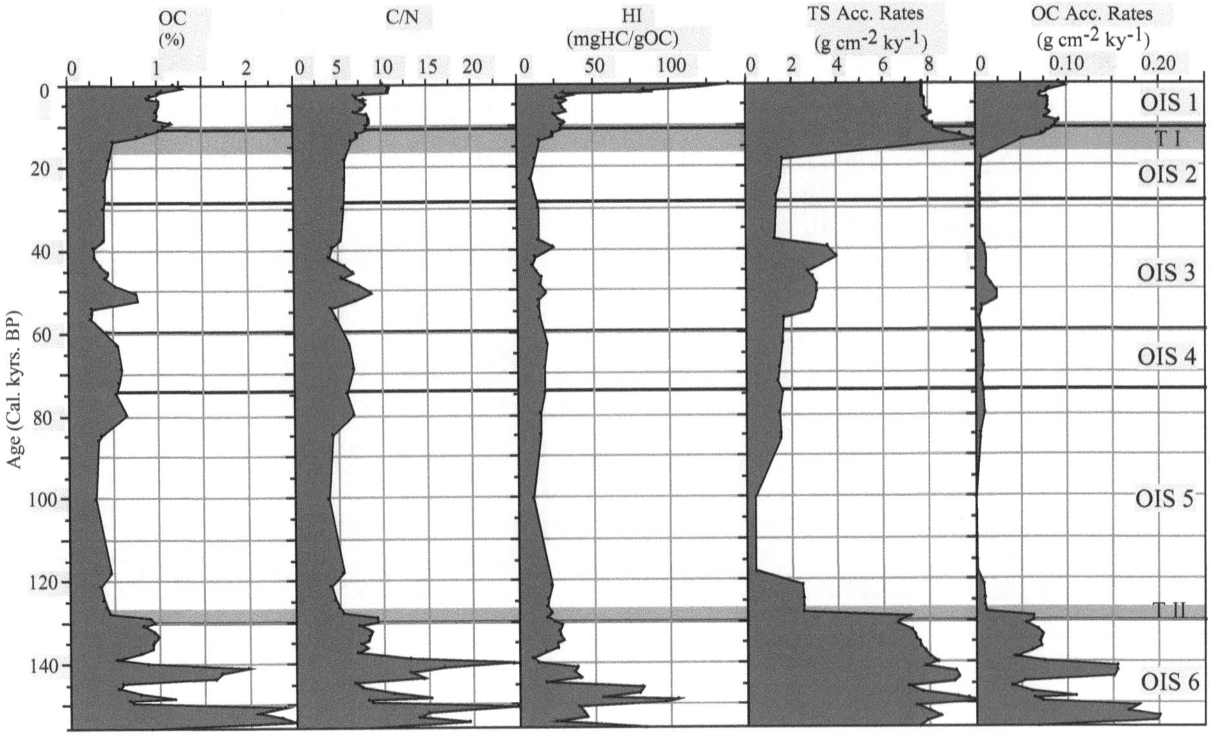

Fig. 7.5.11. Distribution of OC, C/N ratios, hydrogen indices, and accumulation rates of total sediments and total organic carbon in Core PS2741 from the western mid Laptev Sea continental slope, representing the last about 150 Cal kyrs. BP (data from Knies et al. 2000)

(80%) of the total organic matter, whereas marine organic matter reaches values of only 10% (20%) (Boucsein et al. 2000, 2002).

The higher proportion of marine organic matter in the Laptev Sea continental slope sediments during the last about 10 Cal kyrs. BP, is probably a reflection of the decreased supply of terrigenous organic matter (i.e., reduced dilution) (see discussion below). In addition, warm Atlantic water masses flowing as a boundary current along the entire Eurasian continental margin and causing increased surface-water productivity, appears to have been intensified near 9–10 Cal kyrs. BP (Knies and Stein 1998; Knies et al. 1999, 2000; Matthiessen et al. 2001; Boucsein et al. 2002). This paleoceanographic event may also explain increased amounts of marine organic matter being preserved at the Laptev Sea continental margin.

In order to get information about the spatial and temporal variability of organic carbon flux and to relate this variability to paleoenvironmental changes during Holocene times, consideration of organic carbon accumulation rates is necessary (e.g., Mueller-Lupp et al. 2000; Stein and Fahl 2000; Bauch et al. 2001a). Here, these data will be discussed in relationship with budget calculations in Chapter 7.5.5.

The pre-Holocene (OIS 6 to OIS 2)

For the pre-Holocene, i.e., the last glacial and older, the data base is much sparser because only a few sediment cores representing oxygen isotope stages (OIS) 2 to 6 are available from the Laptev Sea continental slope (Stein et al. 2001). Furthermore, the age control of these cores is much less precise than for the Holocene cores (Matthiessen et al. 2001; Stein et al. 2001). As an example, organic carbon records from two selected cores (PS2741 and PS2471) are shown here to illustrate that during pre-Holocene times additional factors have influenced the organic carbon deposition at the Laptev Sea continental margin. For further data and a more detailed discussion of the pre-Holocene evolution in organic-carbon accumulation at the Eurasian continental margin we refer to Knies et al. (2000, 2001) and Stein et al. (2001).

At the Laptev Sea continental slope, OIS 6 is characterized by the occurrence of prominent dark gray intervals with increased amounts of terrigenous organic carbon, indicated by maximum OC values (and accumulation rates) and minimum hydrogen index values (Fig. 7.5.11). During this time interval, extended continental ice sheets are

Fig. 7.5.12. Organic-geochemical bulk parameters in the sedimentary sequence of Core 2471: Total organic carbon contents (%), hydrogen index values (mgHC/gOC), and total organic carbon/total nitrogen (C/N) ratios. Gray intervals indicate turbidites (T1, T2, and T3)

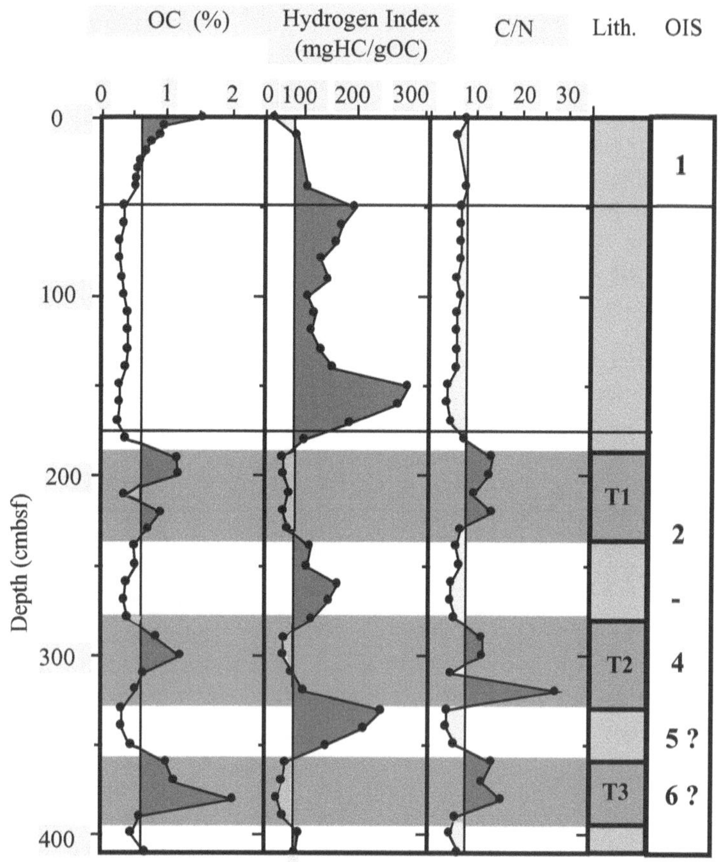

proposed for western (Barents-Kara-Sea) as well as eastern Eurasia (e.g., Velitchko et al. 1997; Svendsen et al. 1999; Knies et al. 2001). Probable source areas of the terrigenous organic matter deposited within cores PS2741-1 are the Mesozoic rocks on Franz-Josef-Land (Elverhoi et al. 1989) and the coal deposits in the area between northern Taimyr Peninsula and the Lena River district (Anon 1978). During ice advance the organic-carbon-rich bedrocks were eroded, and material was transported towards the adjacent continental margin. During OIS 5, OIS 4 and upper OIS 3, OC values were significantly reduced probably due to reduced terrigenous sediment input.

Whereas the Barents Sea area including Svalbard and Franz-Josef-Land was covered by a huge continental ice sheet during the last glacial OIS 2 (e.g., Mangerud and Svendsen 1992; Elverhoi et al. 1995; Mangerud et al. 1998; see Fig. 1.2.5), an extended continental ice sheet was absent in the Laptev Sea area; only smaller local glaciations in mountain areas and on Severnaya Zemlya occurred (e.g., Arkhipov et al. 1986; Velitchko et al. 1997; Svendsen et al. 1999). This different climatic evolution in the western and eastern Eurasian continental margin areas is documented in the organic carbon records. In the Laptev Sea area, the supply of terrigenous (organic) matter was significantly reduced due to the lack of ice sheets and decreased river discharge (Sidorchuk and Panin 1996), resulting in low OC values and low organic carbon accumulation rates (Fig. 7.5.11). Along the Barents Sea continental margin, on the other hand, the flux of (terrigenous) organic carbon increased markedly during OIS 2 (Knies et al. 1999). In Core PS2138-1 from this area, absolute maxima in accumulation rates of bulk sediment and organic carbon of 38 g cm^{-2} ky^{-1} and 0.2 g cm^{-2} ky^{-1}, respectively, were reached (Chapter 7.7).

Another mechanism influencing organic carbon fluxes at the Laptev Sea continental slope/rise and adjacent deep sea, is turbidity currents, as shown in the organic carbon record of Core PS2471 (Fig. 7.5.12). The turbidites have significantly increased organic carbon contents (1 to 2%) in comparison to the normal hemipelagic sediments. The organic matter in the turbidites displays a strong terrigenous signal indicated by low hydrogen index values (<100 mgHC/gOC) and high C/N ratios (>10) (Fig. 7.5.12). The terrigenous origin is also indicated by lignin phenol maxima in the three turbidite intervals of Core PS2471 (J. Lobbes, unpubl. data 1997).

Although the exact age of these turbidites is still under discussion (Stein et al. 1999a; Matthiessen et al. 2001), it is, however, likely that the turbidites were deposited during times of lowered sea level (i.e., OIS 2 to 4 or even OIS 6). At those times, high amounts of terrigenous material could have been transported close to the shelf edge resulting in an instability of the upper slope and gravitational down-slope transport of material. By this mechanism, huge amounts of organic matter may have been deposited in the deep sea environment. Turbidites characterised by high organic carbon contents are also described from the Canada Basin (Grantz et al. 1996 1999). Furthermore, the deep Arctic basins are characterised in general by the common occurrence of turbidites (e.g., Fütterer 1992). Thus, organic-carbon transport by turbidity currents is probably a major source for organic carbon deposited in the central Arctic Ocean (see Chapter 7.9).

7.5.5
Accumulation Rates and Budget of Total Sediment and Organic Carbon

Total sediment (TS) and total organic carbon (OC) accumulation rates of 20 sediment cores from the Laptev Sea and adjacent continental slope (Fig. 7.5.7; Table 7.5.1) were used to evaluate temporal and spatial changes in total sediment and organic carbon flux during Holocene times and its controlling mechanisms as well as to calculate a budget of total sediment and organic carbon. Most of these cores have an excellent AMS^{14}C-based chronology (see Bauch et al. 2001b and further references therein). The data listed in Table 7.5.1, provide the basis for the estimates of average accumulation rates of the different areas and the further budget calculations presented in Tables 7.5.2 to 7.5.5.

For the budget calculation, the Laptev Sea and the adjacent continental slope were divided into several different domains: the area off the Lena Delta, the southeastern, eastern, central, and western part of the inner Laptev Sea (water depths of about 20–50 m), the very shallow areas with water depths < 20 m, the outer shelf and shelf break area (water depths of 50–150 m), the upper slope (water depths of 150–1000 m) and the lower slope (water depths of 1000–2000 m), and the continental rise (water depths of 2000–3000 m) (Table 7.5.1). Areas with water depths >2000 m (Laptev Sea Rise) were were counted as "central Arctic areas" (see Chapter 7.9). For the shallow inner Laptev Sea < 20 m water depth, which is strongly influenced by reworking, redeposition and erosion due to wave-induced bottom currents and sediment entrainment into sea ice (Are 1994; Eicken et al. 1997; Lindemann 1998), a zero sediment balance is assumed. Furthermore, two approaches were followed. First, average Holocene TS and OC budgets were calculated using mean Holocene TS and OC accumulation rates. This is a common approach, when budget calculations are based on thicknesses of Holocene deposits obtained from seismic profiles (e.g., Macdonald et al. 1998; Romankevich et al. 2000), or based on sediment cores which do not have a time control precise enough to resolve different time intervals within the Holocene (e.g., Stein and Fahl 2000). From the Laptev Sea, however, a large number of well-dated sediment cores is available (Bauch et al. 2001b). Thus, in a second approach, TS and OC budgets were calculated for different Holocene time intervals characterized by different rates of sediment input and accumulation rates.

To distinguish between terrigenous and marine components in the budget calculation, the proportion of terrigenous and marine organic carbon was estimated using maceral data (Figs. 7.5.4 and 7.5.10; see discussion above). For the inner and outer Laptev Sea, 95% (5%) and 85% (15%), respectively, are estimated to be of terrigenous (marine) origin. At the continental slope, average amount of marine organic carbon preserved in the sediment cores is lower than in the surface sediments (Figs. 7.5.4 and 7.5.10; Boucsein and Stein 2000; Boucsein et al. 2002), probably due to early diagenetic remineralisation of labile marine organic compounds. Thus, average values of 80% terrigenous and 20% marine organic carbon were used for the continental slope sediment cores. During the early Holocene (9–11 Cal. kyrs. BP), i.e., at times of lowered sea level (see Fig. 7.5.6) and increased terrigenous sediment supply towards the outer shelf and upper continental slope, the proportion of terrigenous organic matter was higher (see discussion below). This is also indicated in the maceral data. At Core PS2458-4, >90% (<10%) of organic matter is of terrigenous (marine) origin at dates older than 10 Cal kyrs. BP (Fig. 7.5.10). For the slope cores PS2742-5 and PS2476-4, similar values were determined (Boucsein et al. 2002). Thus, 95% (5%) and 90% (10%) are assumed to be of terrigenous (marine) origin at the outer shelf and continental slope, respectively, between 9 and 11 Cal. kyrs. BP.

Average Holocene total sediment and organic carbon budgets

Average Holocene accumulation rates of TS and OC in the eastern and central inner Laptev Sea are about 30 to 50 g cm^{-2} ky^{-1} and 0.5 and 1 g cm^{-2} ky^{-1}, respectively, with maximum values occurring in

Table 7.5.1. Linear sedimentation rates (LSR) and accumulation rates of total sediment (TS) and total organic carbon (OC) for Holocene time intervals (for location of cores see Fig. 7.5.7). LSR of cores C-4, C-7, C-8, C-11, and C-37 are based on radiocarbon datings of bulk organic matter (Kuptsov and Lisitzin 1996) and, thus, should be taken with great caution (see further discussion in the text). For all other cores, except 3, LSR are based on AMS^{14}C datings of marine bivalves (Bauch et al. 1999, 2001b; Stein and Fahl 2000). For cores PS2474, PS2476 and PS2471, where no AMS^{14}C dates are available, chronology is based on correlation with AMS^{14}C-dated sediment cores (Stein and Fahl 2000; Stein et al. 1999). TS accumulation rates (LSR x DD) were calculated using a mean dry density (DD) value of 0.9 g cm^{-3} (see Stein and Fahl 2000), for Core PM9462, mean DD values were taken from Bauch et al. 2001a. OC accumulation rates were calculated using TS accumulation and mean OC values (Bauch et al. 1999 2001a; Mueller-Lupp et al. 2000; Stein and Fahl 2000)

Core	Latitude (°N)	Longitude (°E)	Water depth (m)	Linear Sedimentation Rates (cm ky^{-1})				TS Accumulation Rates (g cm^{-2} ky^{-1})				OC Accumulation Rates (g cm^{-2} ky^{-1})				Ref.
				Holocene	0–7 Calky	7–9 Calky	9–11 Calky	Holocene	0–7 Calky	7–9 Calky	9–11 Calky	Holocene	0–7 Calky	7–9 Calky	9–11 Calky	
Inner Laptev Sea (Modern WD 20-50 m)																
Off Lena Delta																
C-37	71,617	130,367	10	< 24	24	n.d.	n.d.	< 22	22	n.d.	n.d.	n.d.	0,4	n.d.	n.d.	4
PS51/080-13	73,459	131,638	21	< 32	32	n.d.	n.d.	29	29	n.d.	n.d.	0,72	0,72	n.d.	n.d.	1
C-4	73,167	131,000	26	< 170	170	n.d.	n.d.	< 153	153	n.d.	n.d.	< 2.7	2,7	n.d.	n.d.	4
Southeastern Laptev Sea																
PM9462-4	74,503	136,005	27	50	50	50	n.d.	50	45	65	n.d.	0,45	0,41	0,52	n.d.	1
PS51/092-12	74,593	130,139	32	59	41	176	n.d.	53	37	158	n.d.	0,7	0,74	2,07	n.d.	1
PS51/092-13	74,594	130,137	32	50	50	n.d.	n.d.	45	45	n.d.	n.d.	0,9	0,9	n.d.	n.d.	1
C-7	74,883	129,983	37	< 159	159	n.d.	n.d.	< 143	143	n.d.	n.d.	< 2.2	2,2	n.d.	n.d.	4
Eastern and central Laptev Sea																
PS51/141-2	75,227	128,641	42	35	11	11	194	32	10	10	175	0,41	0,13	0,13	2,27	1
KD9502-14	76,192	133,117	46	25	11	82	n.d.	23	10	74	n.d.	0,33	0,14	1,11	n.d.	2
C-8	75,400	130,050	48	n.d.	15	n.d.	n.d.	n.d.	14	n.d.	n.d.	n.d.	0,18	n.d.	n.d.	4
PS51/135-4	76,165	133,243	51	41	11	93	124	37	10	84	112	0,54	0,15	1,21	1,62	1
Western Laptev Sea																
PM9499-2	75,501	115,545	48	14	4	31	31	14	4	28	28	0,15	0,04	0,34	0,34	2
Outer Laptev Sea (Modern WD 50–90 m)																
Western Laptev Sea																
PS51/159-10	76,767	116,032	60	31	13	17	113	28	11	15	102	0,28	0,12	0,15	1,02	1
Eastern Laptev Sea																
C-11	76,867	130,083	66	n.d.	11	11	n.d.	n.d.	10	10	n.d.	n.d.	0,14	0,14	n.d.	4
PS2725-5	78,657	144,135	77	43	13	13	266	39	12	12	239	0,43	0,13	0,13	2,63	3

Table 7.5.1 (continued)

Core	Latitude (°N)	Longitude (°E)	Water depth (m)	Linear Sedimentation Rates ($cm\ ky^{-1}$)				TS Accumulation Rates ($g\ cm^{-2}\ ky^{-1}$)				OC Accumulation Rates ($g\ cm^{-2}\ ky^{-1}$)				Ref.
				Holo-cene	0–7 Calky	7–9 Calky	9–11 Calky	Holo-cene	0–7 Calky	7–9 Calky	9–11 Calky	Holo-cene	0–7 Calky	7–9 Calky	9–11 Calky	
Laptev Sea Shelf Break (Modern WD 90–150 m)																
PS51/118-2	77,892	132,237	114	52	2	2	388	47	2	2	349	0,47	0,02	0,02	3,49	1
PS51/118-3	77,892	132,199	122	37	8	8	402	33	7	7	362	0,33	0,07	0,07	3,62	1
Upper Continental Slope (Modern WD 150–1000 m)																
PS51/154-11	77,276	120,610	270	18	12	18	90	14	10	14	72	0,14	0,1	0,14	0,72	1
PS2778-2	77,978	113,065	341	38	29	29	55	30	23	23	44	0,3	0,23	0,23	0,44	3
PS2476-4	77,390	118,193	521	50	n.d.	n.d.	n.d.	40	n.d.	n.d.	n.d.	0,4	n.d.	n.d.	n.d.	3
PS2458-4	78,167	133,398	983	34	27	27	89	27	22	22	71	0,3	0,22	0,25	0,82	1
Lower Continental Slope (WD 1000–2000 m)																
PS2474-3	77,670	118,575	1494	20	n.d.	n.d.	n.d.	16	n.d.	n.d.	n.d.	0,16	n.d.	n.d.	n.d.	3
PS2742-5	80,788	103,815	1890	22	n.d.	n.d.	n.d.	18	n.d.	n.d.	n.d.	0,18	n.d.	n.d.	n.d.	3
Continental Rise (WD 2000–3000 m)																
PS2741-1	81,105	105,395	2400	12	n.d.	n.d.	n.d.	10	n.d.	n.d.	n.d.	0,1	n.d.	n.d.	n.d.	3
PS2471-4	79,152	119,793	3047	2	n.d.	n.d.	n.d.	2	n.d.	n.d.	n.d.	0,02	n.d.	n.d.	n.d.	3

References for chronology and linear sedimentation rates:
1: Bauch et al., 2001b; 2: Bauch et al., 1999; 3: Stein and Fahl, 2000; 4: Kuptsov and Lisitzin, 1996).

Table 7.5.2. Accumulation rates and budget of total sediment and OC (Holocene average 0–11 Cal. kyrs. BP). Values are based on data shown in Table 7.5.1.

Area (Modern WD)	Size (km^2) (Modern)	Accumulation Rates ($g\ cm^{-2}\ ky^{-1}$)		Budget ($10^6\ t\ y^{-1}$)		Total Accumulation (10^9 t)	
		Total sediment	OC	Total sediment	OC	Total sediment	OC
Inner Laptev Sea (WD < 20 m)		0	0	0	0	0	0
Inner Laptev Sea (WD 20–50 m)							
Off Lena Delta	10000	100	2,5	(10) 6.4[1]	(0.25) 0.16[1]	70,4	1,76
Southeastern part	20000	50	1	(10) 8.2[2]	(0.20) 0.16[2]	90,2	1,76
Eastern and central part	60000	32	0,48	19,2	0,29	211,2	3,19
Western part	30000	14	0,15	4,2	0,05	46,2	0,55
Outer shelf (WD 50–90 m)							
Eastern and central part	50000	39	0,43	19,5	0,22	214,5	2,42
Western part	20000	28	0,28	5,6	0,06	61,6	0,66
Shelf Break (WD 90–150 m)	10000	40	0,4	4	0,04	44	0,44
Total Laptev Sea				**67,1**	**0,98**	**738,1**	**10,78**
Upper Slope (150–1000 m)	50000	32	0,32	16	0,16	176	1,76
Lower slope (1000–2000 m)	50000	17	0,17	8,5	0,09	93,5	0,99
Total Laptev Sea slope				**24,5**	**0,25**	**269,5**	**2,75**

[1] assuming accumulation 0–7 Cal kyrs. BP.
[2] assuming accumulation 0–9 Cal yrs. BP.

Table 7.5.3. Average accumulation rates of total sediment and OC for different time intervals (0–11 Cal kyrs BP). Values are based on data shown in Table 7.5.1.

Area (Modern WD)	Total sediment accumulation rates ($g\ cm^{-2}\ ky^{-1}$)			OC Accumulation Rates ($g\ cm^{-2}\ ky^{-1}$)		
	0–7 Calky	7–9 Calky	9–11 Calky	0–7 Calky	7–9 Calky	9–11 Calky
Inner Laptev Sea (WD < 50 m)						
Off Lena	100	0	0	2,5	0	0
Southeastern part	45	100	0	1,1	1,5	0
Eastern and central part	11	70	140	0,15	1,1	1,9
Western part	4	28	28	0,04	0,34	0,34
Outer shelf (WD 50–90 m)						
Eastern and central part	11	11	240	0,13	0,13	2,63
Western part	11	15	102	0,12	0,15	1,02
Shelf Break (WD 90–150 m)	5	5	350	0,05	0,05	3,5
Upper Slope (150–1000 m)	22	22	65	0,22	0,23	0,65
Lower slope (1000–2000 m)	(12)	(12)	(34)	(0.12)	(0.13)	(0.36)

the southeastern part (Table 7.5.2). Directly off the Lena Delta, i.e., in the area with maximum OC contents determined in the surface sediments (Fig. 7.5.1), even higher values of about 100 g cm^{-2} ky^{-1} and 2.5 g cm^{-2} ky^{-1} are probably reached. In the western inner Laptev Sea, the TS and OC accumulation rates are significantly lower (14 g cm^{-2} ky^{-1} and 0.15 g cm^{-2} ky^{-1}, respectively). For the outer Laptev Sea including the shelf break area, TS and OC accumulation rates of 28 to 39 g cm^{-2} ky^{-1} and 0.28 to 0.43 g cm^{-2} ky^{-1}, respectively, were determined. Down-slope, average TS and OC accumulation rates decrease from 32 to 17 g cm^{-2} ky^{-1} and 0.32 to 0.17 g cm^{-2} ky^{-1}, respectively (Table 7.5.2).

Taking these average accumulation rates, an average Holocene TS and OC budget was determined. The calculated budget is an update of the earlier calculation by Stein and Fahl (2000), in-

Table 7.5.4. Budget of total sediment and OC for different time intervals (0–11 Cal kyrs BP). Values are based on data shown in Table 7.5.1.

Area (Modern WD)	Size (km²) (Modern)	Total sediment (10^6 t y^{-1})			OC (10^6 t y^{-1})		
		0–7 Calky	7–9 Calky	9–11 Calky	0–7 Calky	7–9 Calky	9–11 Calky
Inner Laptev Sea (WD < 50 m)							
Off Lena	10000	10	0	0	0,25	0	0
Southeastern part	20000	9	16[1]	0	0,22	0.29[1]	0
Eastern and central part	60000	7	34[1]	21[2]	0,09	0.53[1]	0.29[2]
Western part	30000	1	7[1]	2[2]	0,01	0.08[1]	0.03[2]
Outer shelf (WD 50–90 m)							
Eastern and central part	50000	6	6	120	0,07	0,07	1,32
Western part	20000	2	3	20	0,02	0,03	0,2
Shelf Break (WD 90–150 m)	10000	0,5	0,5	35	0,01	0,01	0,35
Total Laptev Sea		**35,5**	**66,5**	**198**	**0,67**	**1,01**	**2,19**
Upper Slope (150–1000 m)	50000	11	11	33	0,11	0,12	0,33
Lower slope (1000–2000 m)	50000	(6)	(6)	(17)	(0.06)	(0.07)	(0.18)
Total Laptev Sea slope		**17**	**17**	**50**	**0.17**	**0.19**	**0.51**

[1] assuming 80% of area due to lowered sea level.
[2] assuming 25% of area due to lowered sea level.

Table 7.5.5. Total accumulation of total sediment and OC for different time intervals (0–11 Cal kyrs BP). Values are based on data shown in Table 7.5.1.

Area (Modern WD)	Size (km²) (Modern)	Total sediment (10^9 t)			OC (10^9 t)			Total Accumulation (10^9 t)	
		0–7 Calky	7–9 Calky	9–11 Calky	0–7 Calky	7–9 Calky	9–11 Calky	Total sediment	OC
Inner Laptev Sea (WD 20–50 m)									
Off Lena	10000	70	0	0	1,75	0	0	70	1,75
Southeastern part	20000	63	32[1]	0	1,54	0.58[1]	0	95	2,12
Eastern and central part	60000	49	68[1]	42[2]	0,63	1.06[1]	0.58[2]	159	2,27
Western part	30000	7	14[1]	4[2]	0,07	0.16[1]	0.06[2]	25	0,29
Outer shelf (WD 50–90 m)									
Eastern and central part	50000	42	12	240	0,49	0,14	2,64	294	3,27
Western part	20000	14	6	40	0,14	0,06	0,4	57	0,6
Shelf Break (WD 90-150 m)	10000	3,5	1	70	0,07	0,02	0,7	75	0,79
Total Laptev Sea		**248,5**	**133**	**396**	**4,69**	**2,02**	**4,38**	**775**	**11,09**
Upper Slope (150–1000 m)	50000	77	22	66	0,77	0,24	0,66	165	1,6
Lower slope (1000–2000 m)	50000	(42)	(12)	(35)	(0.42)	(0.14)	(0.36)	(89)	(0.85)
Total Laptev Sea slope		**119**	**34**	**101**	**1.23**	**0.38**	**1.02**	**297**	**2.83**

[1] assuming 80% of area due to lowered sea level.
[2] assuming 25 % of area due to lowered sea level.

cluding the new AMS^{14}C-chronology from a large number of sediment cores (Bauch et al. 2001b). According to the Stein and Fahl (2000) calculation, the values are higher (20×10^6 t y^{-1} for TS and 0.4×10^6 t y^{-1} for OC) directly off the Lena Delta and lower (13×10^6 t y^{-1} for TS and 0.2×10^6 t y^{-1} for OC) in the main part of the inner Laptev Sea in comparison to those calculated in this paper (Table 7.5.2). In total, the new Laptev Sea budget for TS and OC accumulation, however, is similar to the earlier calculations, giving values of 67.1×10^6 t y^{-1} and 0.98×10^6 t y^{-1} for TS and OC, respectively

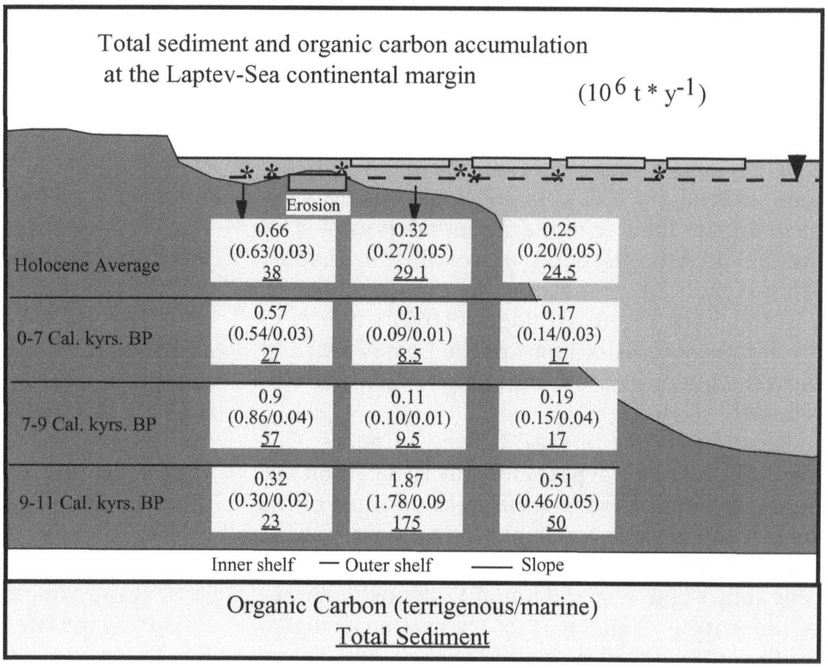

Fig. 7.5.13. Summary of total sediment and total organic carbon accumulation and burial, respectively, in the Laptev Sea and adjacent continental slope in $10^6 \times t\ y^{-1}$. Estimates of marine and terrigenous organic carbon proportion is based on maceral data (see Fig. 7.5.4 and text)

(Fig. 7.5.11, Table 7.5.2). The OC burial rates can be further separated into about 0.9×10^6 t y^{-1} terrigenous organic carbon and about 0.08×10^6 t y^{-1} marine organic carbon (Fig. 7.5.13). For the inner Laptev Sea, our earlier calculation was based only on ^{14}C datings of bulk organic matter (Kuptsov and Lisitzin 1996) which may give less reliable results due to the presence of older as well as younger organic material. Thus, the new calculation based on AMS^{14}C datings of marine bivalves and considering data from additional sediment cores, is more convincing for that part of the Laptev Sea. At the continental slope, another 24.5×10^6 t y^{-1} and 0.25×10^6 t y^{-1} of TS and OC is accumulating on average. Separating the OC burial rate into its terrigenous and marine proportion, this gives values of 0.2×10^6 t y^{-1} and 0.05×10^6 t y^{-1} of terrigenous and marine organic carbon, respectively.

During the entire Holocene (0–11 Cal. kyrs. BP), 738×10^9 t of TS and 10.8×10^9 t of OC (9.9×10^9 t terrigenous and 0.9×10^9 t marine organic carbon) have been accumulated in the Laptev Sea, and about 270×10^9 t of TS and 2.75×10^9 t of OC (2.2×10^9 t terrigenous and 0.55×10^9 t marine organic carbon) at the Laptev Sea continental slope (Table 7.5.2).

Total sediment and organic carbon budgets for three Holocene time intervals

The Holocene sedimentation on the Laptev Sea continental shelf and adjacent slope was strongly influenced by the post-glacial sea-level rise and the related flooding of the shelf area (e.g., Stein and Fahl 2000; Bauch et al. 2001, 2001b). In a synthesis paper on the chronology of the Holocene transgression in the Laptev Sea, Bauch et al. (2001b) estimated that the flooding of the 50 m, 43 m, and 31 m isobaths was completed by approximately 11.1, 9.8, and 8.9 Cal. kyrs. BP (Fig. 7.5.6). During the early Holocene (i.e., >9 Cal. kyrs. BP), large parts of the southern Laptev Sea were subaerial, and main sedimentation took place farther to the north on the outer shelf and upper continental slope. Bearing this in mind, the Holocene was subdivided into three intervals (0–7, 7–9, and 9–11 Cal. kyrs. BP). At these age boundaries, major breaks in sedimentation rates from high to significantly lower values are recognizable (Bauch et al. 2001b). Average accumulation rates and, finally, budgets were calculated for these three intervals.

For the calculation of budgets of the 7–9 and 9–11 Cal. kyrs. BP time intervals, the reduced size of the deposition areas in the inner Laptev Sea due to the lowered sea level, must be considered. Since isostatic sea-level changes are of minor importance in the Laptev Sea area due to the absence of a major ice sheet in this region during the Last Glacial Maximum (Dunayev and Pavlidis 1988; Svendsen et al. 1999; Mangerud et al. 2002), the global sea-level curve can be adapted to our study area and a first-order reconstruction of the change in land/sea distribution due to sea-level rise is possible based on topographic maps (Fig. 7.5.6; Bauch et al. 2001b). The modern shelf topography, however, does not reflect the actual paleosurface because the major submarine paleo-channels were filled with sediments during the early Holocene (e.g., Kleiber and Niessen 1999). Thus, a precise estimate of the depositional areas in the inner Laptev Sea is difficult. As a first approximation, we assume that during the time interval 7–9 Cal. kyrs. BP, there was no sedimentation in the area of the modern depocenter off the Lena, and the rest of the area of the inner Laptev Sea was reduced by 20%. During the time interval 9–11 Cal. kyrs. BP, we assume no sedimentation in the entire southeastern part of the Laptev Sea, and the remaining part of the inner Laptev Sea area was reduced by 75%.

During the time interval between 9 and 11 Cal. kyrs. BP, the main depocenter was situated on the outer shelf and near the shelf break. In most of the cores from this area, accumulation rates were about one order of magnitude higher than the late Holocene values (Table 7.5.3). Annual TS deposition rates of 140×10^6 t y^{-1} and 35×10^6 t y^{-1} were calculated for the outer Laptev shelf and the shelf-break area, respectively (Table 7.5.4). Only about 23×10^6 t y^{-1} were accumulating in the paleochannels of the inner Laptev Sea. For the whole Laptev Sea, a value of almost 200×10^6 t y^{-1} was calculated (Fig. 7.5.10, Table 7.5.4), a value which is 5-6 times as high as the late Holocene value (see below). At the continental slope, about 50×10^6 t y^{-1} were accumulating at that time, i.e., three times more than during the Holocene. Increased coastal (and seafloor) erosion and increased supply of riverine matter farther to the north account for the maximum TS accumulation in the outer shelf/shelf break area.

The OC burial rates show similar trends (Fig. 7.5.13, Table 7.5.4). A maximum was determined for the outer shelf/shelf break area, where 1.87×10^6 t y^{-1} of organic carbon were buried (1.78×10^6 t y^{-1} terrigenous and 0.09×10^6 t y^{-1} marine organic carbon). In the inner Laptev Sea, the OC burial rate reaches a value of about 0.32×10^6 t y^{-1} (i.e., 0.3×10^6 t y^{-1} terrigenous and 0.02×10^6 t y^{-1} marine organic carbon). At the continental slope, 0.51×10^6 t y^{-1} (0.46×10^6 t y^{-1} terrigenous and 0.05×10^6 t y^{-1} marine organic carbon) were buried. In total, 396×10^9 t of total sediment and 4.38×10^9 t of OC (4.16×10^9 t terrigenous and 0.22×10^9 t marine organic carbon) were deposited in the Laptev Sea within 2000 years (Table 7.5.5). The corresponding values for the adjacent continental slope are 101×10^9 t of total sediment and about 1.02×10^9 t of OC (0.92×10^9 t terrigenous and 0.22×10^9 t marine organic carbon).

As sea level rose, the coastline and the main depocenters shifted to the south, resulting in diminished sedimentation on the outer shelf (Bauch et al. 1999, 2001b; Stein and Fahl 2000; Stein et al. 2001). This evolution is clearly reflected in the budget calculations of the 7–9 Cal kyrs. BP time interval (Fig. 7.5.13, Table 7.5.4). At the outer shelf, the sediment accumulation decreased to a value of $<10 \times 10^6$ t y^{-1}, the OC burial rate to a value of 0.11×10^6 t y^{-1} (of which 10% is of marine origin). The main depocenter is now much farther to the south in the inner Laptev Sea, where the TS accumulation reaches almost 60×10^6 t y^{-1}, and 0.9×10^6 t y^{-1} of OC (i.e., 0.86×10^6 t y^{-1} terrigenous and 0.04×10^6 t y^{-1} marine organic carbon) were buried. In total 133×10^9 t of TS and about 2.02×10^9 t of OC (1.92×10^9 t terrigenous and 0.1×10^9 t marine organic carbon) were deposited in the Laptev Sea during this time interval. At the continental slope 34×10^9 t of total sediment and 0.38×10^9 t of OC (0.3×10^9 t terrigenous and 0.08×10^9 t marine organic carbon) were accumulated at the same time (Table 7.5.5).

During the late Holocene, i.e., the last 7 Cal. kyrs. BP, TS and OC accumulation rates were significantly lower than during the early Holocene, an observation which is characteristic for the entire Laptev Sea (Tables 7.5.1 and 7.5.2). At about 7 Cal. kyrs. BP, the sea level was probably close to the modern level, although the modern depositional environment in the Laptev Sea was probably fully established no earlier than 5 Cal. kyrs. BP (Bauch et al. 2001b). In our budget calculation we regard the 0–7 Cal. kyrs. BP time interval as representing approximately modern conditions.

For this time interval (0–7 Cal. kyrs. BP), our calculation gives a TS deposition rate of 32×10^6 t y^{-1} and 8.5×10^6 t y^{-1} for the inner and outer Laptev Sea, respectively (Fig. 7.5.13, Table 7.5.4), indicating that the main modern depocenter is in the inner Laptev Sea. In the same areas, 0.57×10^6 t y^{-1} (0.54×10^6 t y^{-1} terrigenous and 0.03×10^6 t y^{-1} marine organic carbon) and 0.1×10^6 t y^{-1} (0.09×10^6 t y^{-1} terrigenous and 0.01×10^6 t y^{-1} marine organic carbon) of OC, respectively, are buried. At the adjacent continental slope 17×10^6 t y^{-1} of TS are accumulating, and 0.17×10^6 t y^{-1} of OC (0.14×10^6 t y^{-1} terrigenous and 0.03×10^6 t y^{-1} marine organic carbon) are buried.

In total, about 250×10^9 t of TS and 4.7×10^9 t of OC (4.4×10^9 t terrigenous organic carbon and 0.3×10^9 t marine organic carbon) were buried in the Laptev Sea during the last 7 Cal. kyrs. BP (Table 7.5.5). The corresponding values for the continental slope are 119×10^9 t of TS and 1.19×10^9 t of OC (0.98×10^9 t terrigenous and 0.21×10^9 t marine organic carbon).

Very recently, Rachold et al. (2002) published a total sediment budget of the Laptev Sea for the last

5 Cal. kyrs. BP. The budget is based on the assumption that three selected sediment cores (PM9499, PS51/092-12, and PM9462), which were also included in our calculation (see Table 7.5.1), are representative for the sediment accumulation in the western, central, and eastern Laptev Sea, respectively. According to their budget, a total sedimentation in the Latev Sea amounts to 304×10^9 t for the last 5000 years. This number is significantly higher than that determined here (250×10^9 t in 7000 years, i.e., 179×10^9 t in 5000 years). It is likely that the three selected sediment cores, all of which are from the inner Laptev Sea, are not representative for the sedimentation in the whole Laptev Sea during late Holocene times because sedimentation rates are significantly higher there than in the outer Laptev Sea (Table 7.5.1). Thus, the value obtained by Rachold et al. (2002) seems to be too high.

Summing-up the total TS and OC accumulation values for the three time intervals to get the Holocene total, 775×10^9 t of sediments and 11.1×10^9 t of organic carbon were accumulated in the Laptev Sea during the last 11 Cal. kyrs. BP (Table 7.5.5). These numbers are quite similar to those obtained from Holocene mean values (Table 7.5.2). This indicates, that mean Holocene sedimentation rate data (e.g., Stein and Fahl 2000) may give reliable average budget values, but it also indicates that, in the Laptev Sea, it is inappropriate to compare mean Holocene sediment output values with modern sediment input values. Rather, the sedimentary and organic carbon budget of the last 7 Cal. kyrs. BP representing approximately modern environmental conditions, should be used.

Mass Balance Calculation of Input and Output (Late Holocene)

In the following, the budget of the 0–7 Cal. kyrs. BP time interval is related to the modern TS and OC input values (see Chapter 2), and a simple input-output box model is developed (Fig. 7.5.14) similar to that obtained by Macdonald et al. (1998) for the Beaufort Sea.

(i) Total sediment

From the 87.3×10^6 t y^{-1} of sediments supplied annually by rivers and coastal erosion, and – as a minor contributor – by winds (see Chapter 2), 35.5×10^6 t y^{-1} (41%) are accumulating on the Laptev Sea shelf. Taking the sediment export by sea ice to be 10×10^6 t y^{-1} (11% of the input) (Chapter 2.4), 41.8×10^6 t y^{-1} (48%) are transported towards the continental slope where another 17×10^6 t y^{-1} (20% of the initial input) are accumulating. This leaves about 24.8×10^6 t y^{-1} (28% of the initial input) which can be further transported by bottom and turbidity currents towards the continental rise

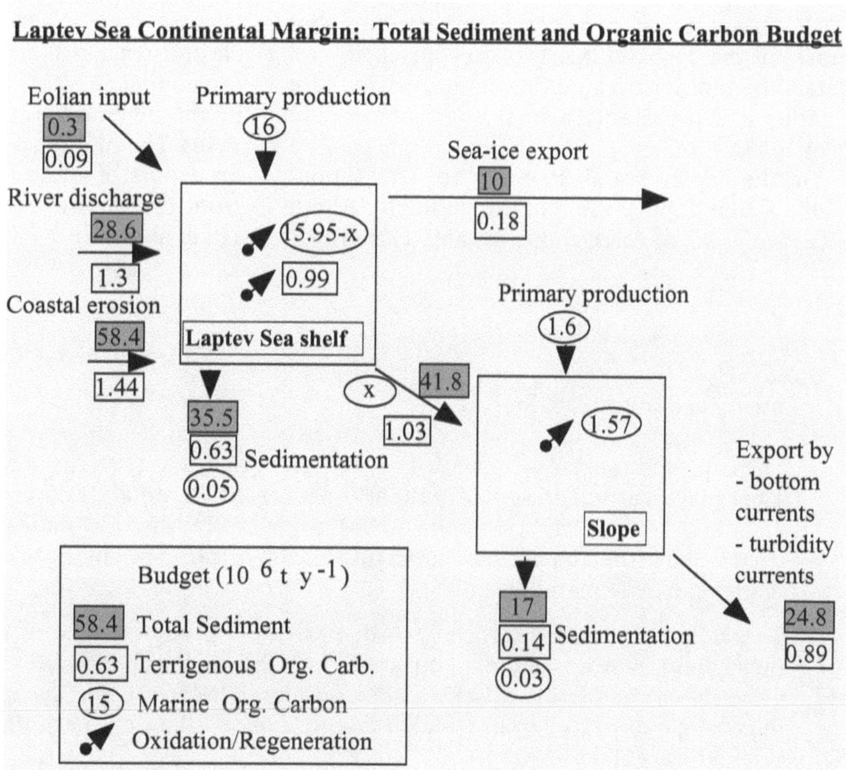

Fig. 7.5.14. Box model of the total sediment and particulate organic carbon budget at the Laptev Sea continental margin for the late Holocene time interval (0–7 Cal. kyrs. BP). All numbers are in 10^6 t y^{-1}. "x" indicates that part of the marine POC might be exported from the shelf towards the slope (see Liu et al. 2000). For data source see Figure 7.5.13 and Tables 7.5.1 and 7.5.4

and deep ocean basins. Bottom transport of sediments by the Novosibirsk Current from the Laptev Sea continental margin into the Amundsen Basin is estimated to reach about 6.2×10^6 t y^{-1} (Eicken et al. 1997). The remaining 18.6×10^6 t y^{-1} of sediment are available to be transported towards the Nansen Basin (see Chapter 7.9).

(ii) Terrigenous Organic Carbon

In the Laptev Sea, the annual supply of terrigenous particulate organic carbon (POC) by rivers is about 1.3×10^6 t y^{-1} (see Chapter 2). The total organic carbon input by coastal erosion is estimated to reach about 1.8×10^6 t y^{-1}. A clear distinction between POC and DOC seems to be difficult (see discussion Chapter 2.2). Most of the OC supplied by coastal erosion, however, is probably POC and can be related to organic carbon values of the marine sediments, because the coastal erosion values are also based on organic carbon measurements in coastal sediments. For our budget calculation, we assume that 80% of the OC equals POC (i.e., about 1.44×10^6 t y^{-1}). Eolian organic carbon input is of minor importance, reaching values of $< 0.1 \times 10^6$ t y^{-1} (see Chapter 2). Thus, we estimate a total input of terrigenous POC of about 2.83×10^6 t y^{-1}. Following Ittekkot (1988), we assume that 35% of this terrigenous POC (i.e., 0.99×10^6 t y^{-1}) belongs to the labile (metabolizable) fraction and becomes rapidly oxidized. In the budget calculation, the terrigenous dissolved organic carbon (DOC) has not been considered. It is assumed that most of the DOC either transits the shelf and enters the interior ocean or becomes oxidized (see Chapter 4; Opsahl and Benner 1997; Macdonald et al. 1998).

On the Laptev Sea shelf, 0.63×10^6 t y^{-1} (about 22% of the input) are buried annually. About 0.18×10^6 t y^{-1} of terrigenous organic carbon (or 6% of the initial input) is exported by sea ice (Chapter 2). The remaining 1.03×10^6 t y^{-1} of terrigenous organic carbon can escape to the slope, where 0.14×10^6 t y^{-1} (or 5% of the initial input) are buried. About 0.9×10^6 t y^{-1} of terrigenous organic carbon, which is about 30% of the initial input into the Laptev Sea, are supplied towards the interior ocean via ocean and turbidity currents.

(iii) Marine Organic Carbon

Data on average annual primary production in the seasonally ice-covered Laptev Sea are difficult to obtain (see Chapter 3). They are based on short-term measurements (production rates/day) and estimates of the productivity period (months/year). Using an average primary production value of 30 gC m^{-2} y^{-1} for the Laptev Sea, an annual input of about 16×10^6 t y^{-1} of marine organic carbon can be calculated (see Chapter 3). In the sediments of the Laptev Sea shelf, only < 0.5% of the primary production, i.e., about 0.05×10^6 t y^{-1}, are buried. That means, > 99% of marine organic carbon is remineralized or exported.

At the continental slope, another 0.03×10^6 t y^{-1} of marine organic carbon are buried. Assuming a primary production input of about 1.6×10^6 t y^{-1} (based on an average primary production value of 50% of the Laptev Sea, i.e., 16 gC m^{-2} y^{-1}; see Chapter 3, Table 3.7), this would result in a burial of 2% of the primary produced organic carbon, a value which might be somewhat too high. In general, only about 5% of the primary produced carbon reaches the sea floor at a depth of 1000 m (Suess 1980; Berger et al. 1989), and about 1% or less might be buried in the slope sediments (Walsh 1989). The difference, however, can be explained by an export of some marine organic carbon debris from the shelf towards the slope (see Walsh 1989; Liu et al. 2000).

7.5.6
Summary and Concluding Remarks

- The Laptev Sea surface sediments are characterized by OC contents between 0.5 and more than 2%. The organic carbon preserved in the sediments is predominantly of terrigenous origin. In the inner and outer Laptev Sea, 95% (5%) and 85% (15%) of the organic matter, respectively, is terrigenous (marine). At the continental slope, on the other hand, about 25% of the sedimentary organic matter can be of marine origin.
- The Holocene organic carbon records indicate a predominance of terrigenous organic matter throughout. Between about 7 and 10 Cal kyrs. BP, major changes in supply of total sediment and terrigenous and marine organic carbon occur, related to the post-gacial sea-level rise and changes in coastal erosion, Siberian river discharge, and/or Atlantic water inflow along the Eurasian continental margin.

- During pre-Holocene times, glacial erosion and turbidity currents have influenced the sediment and organic carbon fluxes and accumulation at the Laptev Sea continental slope/rise.
- As an average for the Holocene, burial rates of about 67×10^6 t y^{-1} of total sediments, 0.9×10^6 t y^{-1} of terrigenous organic carbon, and 0.08×10^6 t y^{-1} of marine organic carbon were calculated for the Laptev Sea shelf. In total, 738, 9.9, and 0.9×10^9 t of total sediment, terrigenous and marine organic carbon, respectively, have been accumulated on the Laptev Sea shelf during the last 11 Cal. kyrs. BP. Presently at the continental slope, 24.5×10^6 t y^{-1} of total sediment, 0.2×10^6 t y^{-1} of terrigenous organic carbon, and 0.05×10^6 t y^{-1} of marine organic carbon are buried, based on Holocene average values. For the total Holocene accumulation, the corresponding values are 270, 2.2, and $0.55 \times o^9$ t.
- The budgets for the three different Holocene time intervals indicate distinct changes in sediment accumulation which are related mainly to changes in sediment input by coastal erosion and river discharge. Between 9 and 11 Cal. kyrs. BP, the main depocenter for total sediment as well as organic carbon was situated near the outer shelf/shelf break. Sediment input by coastal (and seafloor) erosion and river discharge must have been significantly higher than during the late Holocene.
- For the late Holocene time interval (0–7 Cal. kyrs. BP), a mass balance for total sediments implies that about 35.5×10^6 t y^{-1} (41% of the initial input) is accumulating on the Laptev Sea shelf and 17×10^6 t y^{-1} (20%) on the slope, 10×10^6 t y^{-1} (11%) is exported by sea ice, and 24.8×10^6 t y^{-1} (28%) is exported by currents towards the interior ocean. For terrigenous organic carbon, 0.63×10^6 t y^{-1} (22% of the input) is buried on the Laptev Sea shelf, 0.14×10^6 t y^{-1} (5%) is buried on the slope, 0.18×10^6 t y^{-1} (6%) is exported by sea ice and 0.9×10^6 t y^{-1} (about 30%) is exported via currents towards the deep sea. For marine organic carbon, 0.05×10^6 t y^{-1} of is buried in Laptev Sea shelf sediments (about 1% of the primary production), and 0.03×10^6 t y^{-1} in slope sediments.

7.6 The Kara Sea: Distribution, Sources, Variability and Burial of Organic Carbon

R. Stein, K. Fahl

7.6.1 Introduction

The Kara Sea (see Fig. 7.1.8), the third largest arctic marginal sea with an area of 926,000 km², has a mean water depth of 130 m (see Chapter 1; Jakobsson 2002). In the central and southern Kara Sea, water depths <100 m and <50 m are predominant (Fig. 7.6.1). A common feature in this part of the Kara Sea is the presence of small channels, 1 to 10 m in depth, probably generated during the last sea-level lowstand (Johnson and Milligan 1967; Stein 2001; Niessen and Dittmers 2002) and important as pathways for modern sediment transport and accumulation (see below). A prominent structure in the southwestern Kara Sea is the Novaya Zemlya Trough with maximum depths of >430 m. Major physiographic provinces of the northern Kara Sea are the St. Anna Trough and the Voronin Trough (Figs. 7.1.8 and 7.6.1). These troughs, representing the deepest part of the Kara Sea with maximum depths of about 640 m and 400 m, respectively (Johnson and Milligan 1967; Pavlov and Pfirman 1995), connect the Kara Sea shelf with the open Arctic Ocean. Thus, they are important for the water exchange between the shelf and the open ocean as well as for the sediment transport from the Kara Sea into the central Arctic. Both troughs are separated by the shallow Central Kara Plateau comprising islands and regions with water depths <50 m (Pavlov and Pfirman 1995).

The Kara Sea surface-water circulation shows a pronounced seasonal variability caused by the wind field, river run-off and ice formation (Harms et al. 2000). During spring and summer, characterized by strong river run-off, the general circulation pattern is anticyclonic. During autumn and winter, the circulation is almost reversed showing a very pronounced eastward-flowing coastal current (Fig. 7.6.1; Pavlov and Pfirman 1995; Harms et al. 2000).

The Kara Sea differs from the other Eurasian shelf seas due to the strong influence of river discharge. It receives about 50% (1480 km³ y^{-1}) of the total river run-off discharged to the entire Eurasian Arctic (see Chapter 2; Holmes et al. 2002). Most of this discharge is related to the major rivers Ob and Yenisei contributing 404 km³ y^{-1} and 620 km³ y^{-1},

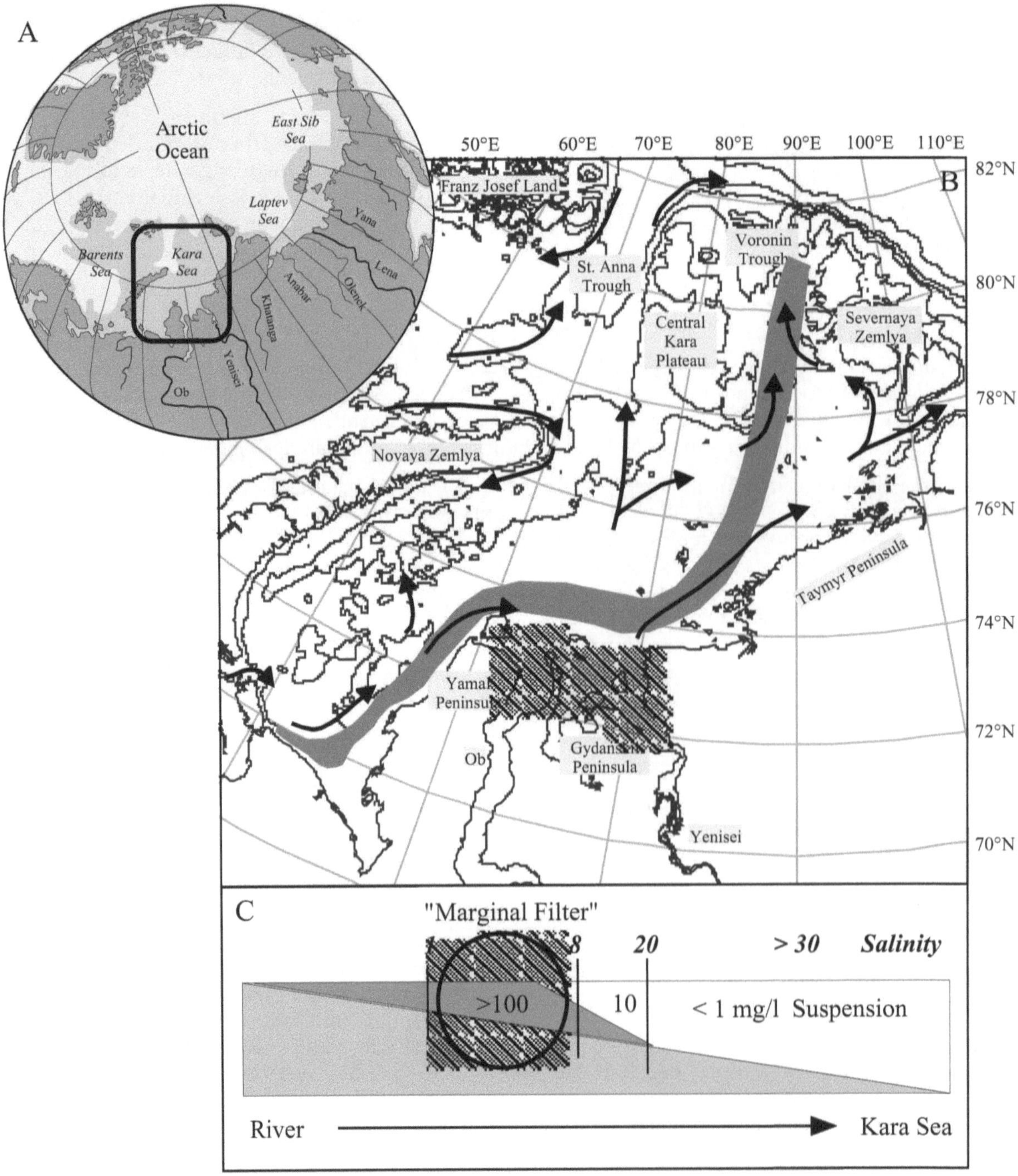

Fig. 7.6.1. (A) Overview map of northern Eurasia and the adjacent Arctic Ocean indicating the major Eurasian rivers draining into the Arctic Ocean. The study area of the Kara Sea is marked. (B) Map of the Kara Sea showing the surface-water circulation pattern during autumn/winter and the distribution of polynyas (Pavlov and Pfirman 1995; Harms et al. 2000). The area of the marginal filter is indicated as hatched field. The southwestern Kara Sea south of 76°N and west of 70°E has not been included in our studies (see Section 7.6.5). (C) Scheme of the marginal filter (according to Lisitzin 1995)

respectively. The discharge exhibits strong seasonal and interannual variations (Pavlov and Pfirman 1995; Gordeev et al. 1996; Holmes et al. 2002). The maximum discharge rate is observed in June, when about 45–65% of the annual freshwater run-off and 80% of the annual suspended matter are released (Shiklomanov and Skakalsky 1994; Gordeev et al. 1996). The modern discharge of total suspended matter by the Ob and Yenisei rivers is 15.5×10^6 t y^{-1} and 4.7×10^6 t y^{-1}, respectively. The dissolved and particulate organic carbon supply by the same rivers is estimated to reach about 3.68×10^6 t y^{-1} and 4.86×10^6 t y^{-1}, and 0.36×10^6 t y^{-1} and 0.17×10^6 t y^{-1}, respectively (Chapter 2). For the Yenisei, a total suspended matter supply of 14.4×10^6 t y^{-1}, which is about three times higher than the modern value, has been measured for times preceding the dam constructed near Krasnoyarsk in 1967 (Telang et al. 1991; Holmes et al. 2002). The input of particulate organic matter prior to the dam was probably also three times higher than today (i.e., about 0.5×10^6 t y^{-1}).

Most of the suspended matter is trapped in the estuaries where freshwater and salt water mixes (salinities of about 2 to 10) and rapid accumulation (precipitation) of fine-grained suspension occurs due to coagulation processes (Fig. 7.6.1). According to Lisitzin (1995), more than 90% of the suspended matter (including particulate organic carbon) and about 30% of the dissolved matter probably accumulate within this so-called "marginal filter". This results in thick sequences of young (Holocene), soft sediments overlying the pre-Holocene basement in the Ob and Yenisei estuaries (Stein 2001; Dittmers et al. 2003; Stein et al. 2003a). In the Yenisei area north of about 73°30′N, the thickness of the Holocene sedimentary cover approaches zero. The total loss of dissolved organic carbon in the marginal filter appears to be low (3 to 11%; Lisitzin 1995; Köhler et al. 2003; see Chapter 4).

Concerning the total sediment input into the Kara Sea, coastal erosion is probably more important than river discharge. Romankevich and Vetrov (2001) give an estimate of 109×10^6 t y^{-1} of total sediment annually supplied by coastal erosion. Based on this number, the input of total organic carbon is estimated to reach about 1×10^6 t y^{-1}. Most of this organic matter is probably particulate matter (see Chapter 2 for details).

Sea-ice distribution is another important feature for several reasons. For about eight months (from October to May), almost the entire Kara Sea is covered with ice of different types (fast ice, drift ice, stable ice massifs) (e.g., Pavlov and Pfirman 1995). Seaward of the fast ice zone, an area of flaw leads/polynyas occurs, kept ice-free by the prevailing offshore winds (Fig. 7.6.1). The sea-ice cover influences biological processes, water mass stratification and sedimentary processes. Significant amounts of sediments can be incorporated into the sea ice by different processes, such as suspension freezing, anchor ice formation, deposition of river sediments onto a flooded fast ice cover, and subsequently exported from the shelf towards the open ocean (e.g., Pfirman et al. 1990; Reimnitz et al. 1993; Dethleff et al. 1993; Nürnberg et al. 1994; Eicken et al. 2000; see Chapter 2.4). From the Kara Sea, about 2.4×10^6 t y^{-1} of total sediment and about 17×10^3 t y^{-1} of terrigenous particulate organic carbon are exported annually towards the Arctic Ocean interior (Chapter 2.4, Table 2.11). Another important aspect related to sea ice and sediment transport is the fact that freezing of sea water results in brine formation. The dense saline water may flow through submarine channels and move resuspended matter in the bottom nepheloid layer towards the outer shelf, across the shelf break, and into the Arctic basins (Lisitzin 1995). With this process, which is mainly active during the autumn to winter period, sediments as well as organic matter may escape the marginal filter.

During approximately the last 10 years, several major expeditions studying the hydro-bio-geo-system were conducted in the Kara Sea. Examples include four multidisciplinary expeditions between 1991 and 1994 carried out by VNIIOkeangeologia St. Petersburg (Kulikov et al. 1999; see Stein et al. 1999b for further data and references). A milestone in Kara Sea research, the Kara Sea Expedition of RV "Dmitry Mendeleev" in 1993 (Lisitzin and Vinogradov 1995), resulted in numerous new data and concepts on sedimentary processes and the ecosystems of the Kara Sea as well as the Ob and Yenisei estuaries (e.g., Lisitzin 1995, and numerous other papers in the special issue of Oceanology, English Translation, Vol. 34, No. 5, 1995). Recently, a Russian-German research project on "Siberian River Run-off (SIRRO)" was initiated to study in detail the freshwater discharge, its influence on biological, geochemical, and geological processes in the Kara Sea, and its variability in space and time (Stein et al. 2003b and further references therein). Within SIRRO, five expeditions with RV "Akademik Boris Petrov" were carried out in 1997, 1999, 2000, 2001, and 2002, and a huge number of sediment samples for multidisciplinary studies were obtained (Matthiessen and Stepanets 1998; Stein and Stepanets 2000, 2001, 2002; Schoster and Levitan 2003).

Concerning studies on distribution and sources of organic carbon in the Kara Sea, the comprehensive studies by E. Romankevich and colleagues from the Shirshov Institute Moscow (e.g., Romankevich et al. 1982, 2000; Romankevich and Vetrov 2001) and V.I. Gurevich from VNIIOkeangeologia St. Petersburg (Gurevich 1995) have to be mentioned. Studies on organic carbon sources, pathways and sinks are also a major objective of the Russian-German SIRRO project (Boucsein et al. 1999; Fahl et al. 2003; Kodina et al. 2003; Köhler et al. 2003; Kraus et al. 2003; Stein et al. 2003a). Using stable carbon isotopes, Kodina et al. (2003) investigated the contribution of biogenic methane to ikaite formation in the southern Kara Sea.

7.6.2
Data base, Material and Methods

In this synthesis paper, total organic carbon (OC), total organic carbon/total nitrogen (C/N) ratios, Rock-Eval parameters (hydrogen index and Tmax values), stable carbon isotopes of the organic matter, maceral composition as well as specific biomarkers (*n*-alkanes) have been used to determine the amount and composition of the organic carbon fraction in surface sediments and sediment cores. For further information about methods and use of the different proxies as organic-carbon-source indicator we refer to Chapter 1.4. Data presented in this paper are for the most part newly obtained within the SIRRO Project, augmented with published data as outlined in Table 7.6.1.

The chronological framework needed to interpret the organic-carbon records in terms of paleoenvironmental change with time and to calculate linear sedimentation rates as well as total sediment and total organic carbon accumulation rates, is based primarily on AMS^{14}C datings (Stein et al. 2003a). The AMS^{14}C dates are δ^{13}C-normalized and corrected for a reservoir age of 440 years (Mangerud and Gulliksen 1975). Calendar years were calculated according to Stuiver et al. (1998). In addition, magnetic susceptibility and wet bulk density records (Dittmers et al. 2003) were used for core correlation.

7.6.3
Distribution and Sources of Organic Carbon in Surface Sediments

In the southern Kara Sea, the highest OC values occur in the Ob and Yenisei estuaries and the submarine channels towards the north (1.5 to 2.5%; Fig. 7.6.2). The shallower areas outside the channels are characterized by significantly lower OC values (<1%). This distribution pattern is related to the

Table 7.6.1. Summary of references of data used in this chapter

Parameter	References
Surface sediments	
(1) Southern Kara Sea	
Organic carbon (OC)	this paper; Stein (1996); Romankevich and Vetrov (2001)
C/N ratios	this paper; Krishnamurthy et al. (2001)
Stable carbon isotopes of Corg	Fernandes and Sicre (2000); Krishnamurthy et al. (2001); Fahl et al. (2003)
Rock-Eval pyrolysis indices	this paper
Maceral composition	Boucsein et al. (1999)
Biomarker (*n*-alkanes)	this paper; Fernandes and Sicre (2000); Fahl et al. (2003)
(2) Northern Kara Sea	
Organic carbon (OC)	Andrew and Kravitz (1974), Stein (1996); Romankevich and Vetrov (2001)
C/N ratios	Stein (1996)
Rock-Eval pyrolysis indices	Stein (1996)
Biomarker (*n*-alkanes)	this paper
Sediment cores	
(1) Southern Kara Sea	
Organic carbon (OC)	this paper; Stein et al. (2003a)
C/N ratios	this paper
Rock-Eval pyrolysis indices	this paper; Stein et al. (2003a)
Biomarker (*n*-alkanes)	Stein et al. (2001, 2003a)
(2) Northern Kara Sea	
Organic carbon (OC)	Stein and Knies (2000); this paper
C/N ratios	Stein and Knies (2000); this paper
Rock-Eval pyrolysis indices	Stein and Knies (2000); Stein et al. (2001); this paper
Maceral composition	Boucsein et al. (2002)

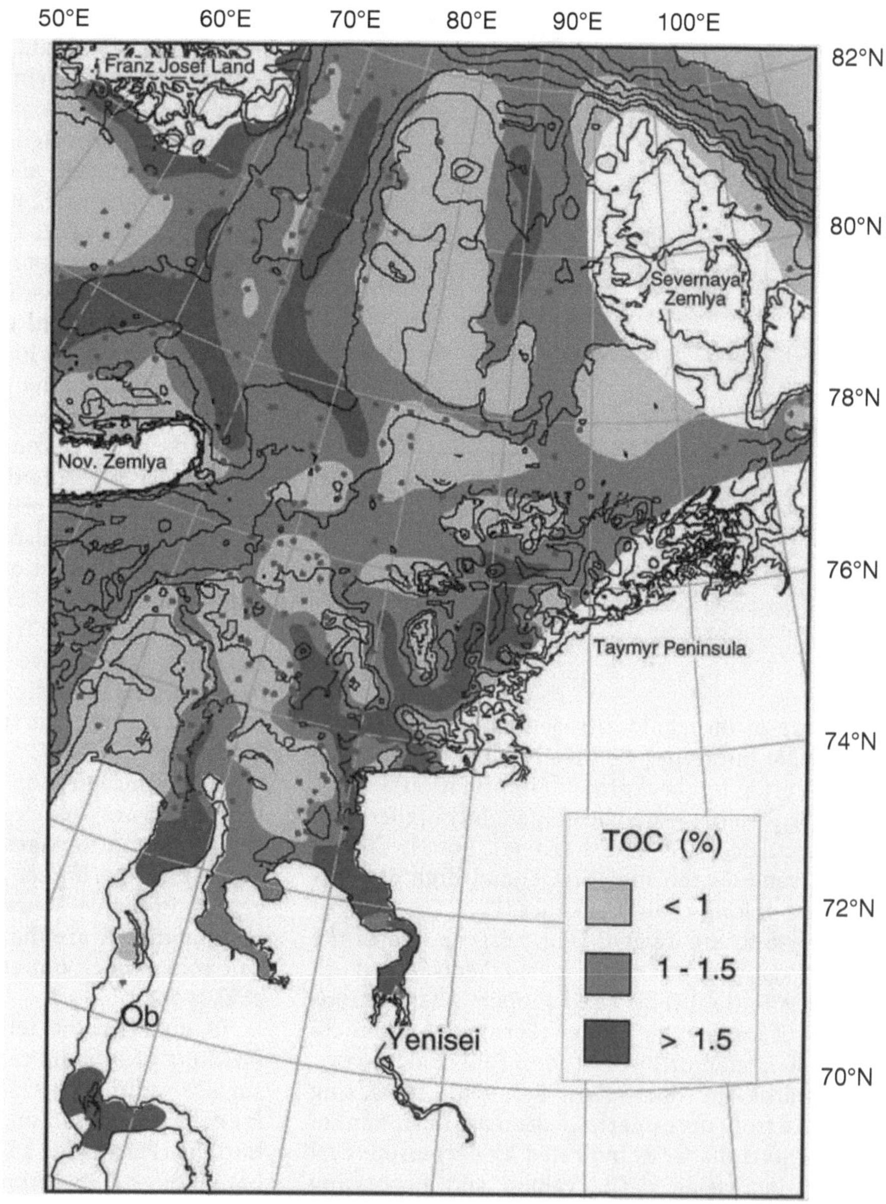

Fig. 7.6.2. Distribution map of total organic carbon concentrations in Kara Sea surface sediments. For references see Table 7.6.1.

riverine suspended matter input. The estuaries, functioning as marginal filters, capture and accumulate most of the fine-grained suspended sediment including the organic matter (Lisitzin 1995). This is reflected in absolute OC maxima > 2% measured in the central estuaries, as well as in the positive correlation between OC content and amount of fine fraction < 63 µm (Fig. 7.6.3).

In the northern Kara Sea, the OC distribution map displays several general features (Fig. 7.6.2). Elevated OC values of 1 to 2% occur in the St. Anna Trough and the Voronin Trough, separated by characteristically low OC values in the Central Kara Plateau. Maximum OC values in the St. Anna Trough are concentrated in its eastern central part. High OC values of >1.5% also occur within a narrow rim around Franz Josef Land and in the deeper part (water depth >300 m) of the area between Franz Josef Land and Novaya Zemlya. The former maximum is probably related to input of terrigenous matter from Franz Josef Land (see below), the latter related to organic carbon input from the Barents Sea via Atlantic water inflow (see Fig. 7.6.1).

The predominantly terrigenous origin of organic carbon in the southern Kara Sea and, especially in the estuaries, is clearly reflected in the high C/N ratios between 9 and 14 (Fig. 7.6.4). As shown in a OC vs. N_{tot} diagram (Fig. 7.6.5) Kara Sea surface sediments probably do not contain significant

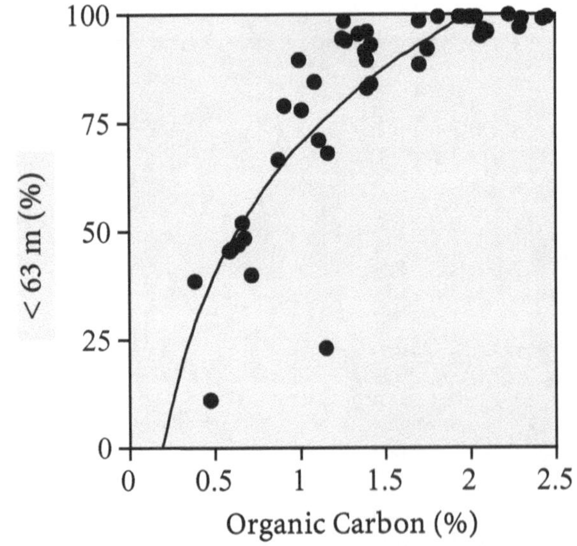

Fig. 7.6.3. Correlation between total organic carbon content (%) and percentages of fine fraction < 63 µm in surface sediments from the Kara Sea

amounts of inorganic nitrogen as implied by the negligible intercept (< 0.01% N_{bou}) at OC = 0 (see Chapter 1.4 for background information). Thus, "OC/N_{org}" values would be only slightly higher than the calculated C/N ratios (10–15). Low hydrogen index values < 100 mgHC/gOC and high concentrations of long-chain *n*-alkanes ($C_{27} + C_{29} + C_{31}$) of about 350 to 410 µg/gOC (Fig. 7.6.4) as well as the light $\delta^{13}C_{org}$ values of about –27‰ (Yenisei) and –28 to –28.7‰ (Ob) (Fig. 7.6.6) support a terrigenous origin of the organic matter (Fernandes and Sicre 2000; Krishnamurthy et al. 2001; Fahl et al. 2003).

Towards the open Kara Sea, with increasing distance from the estuaries, a decrease in amount of terrigenous matter is indicated by decreasing C/N ratios, increasing $\delta^{13}C_{org}$ values, and decreasing long-chain *n*-alkanes (Figs. 7.6.4 to 7.6.6). Based on the C/N ratios as well as the low hydrogen index values, a dominance of terrigenous organic matter is implied for the central Kara Sea up to 78°N and in the eastern Kara Sea. A transport of riverine material towards the north and northeast probably occurs via eastward-flowing (near-bottom) currents through the submarine channels and such a mechanism may also explain the terrigenous organic carbon signal. This mechanism is also reflected in the OC distribution pattern (Fig. 7.6.2). Inorganic tracers, i.e., clay minerals and magnetic susceptibility, determined on the same set of surface sediment samples imply a modern sediment transport from the estuary towards the northeast (Wahsner et al. 1999; Dittmers et al. 2003; Stein et al. 2003a).

Although terrigenous organic matter clearly seems to be dominant in the entire southern and central Kara Sea, marine organic matter may locally become enriched. A prominent OC maximum of about 2.5%, supported by three samples, occurs in the area off the Ob and Yenisei estuaries at about 74°N (Fig. 7.6.2), i.e., just in the zone of polynyas (see Fig. 7.6.1). These samples are characterized by very high absolute concentrations of polyunsaturated fatty acids, indicative for fresh marine organic matter (Fahl et al. 2003). A possible explanation for this local enrichment of marine organic carbon being preserved in the sediments might be an increased primary production related to ice-free polynya conditions and nutrient release from the ice edge. In polynyas and near the ice edge distinctly elevated primary production has been determined (e.g. Nelson et al. 1989; see Chapter 3). A very similar situation has been observed in the Laptev Sea (Fahl and Stein 1997; see Chapter 7.5.3).

In the St. Anna Trough area of the northern Kara Sea, the relative amount of marine organic matter seems to be significantly higher than in the southern Kara Sea, as suggested from the HI values > 100 mgHC/gOC, C/N ratios < 7, and long chain *n*-alkane concentrations of < 150 µg/gOC (Fig. 7.6.4). Exceptions are sites close to Franz Josef Land where local inputs of terrigenous organic matter are indicated by the higher C/N ratios and the low HI values. Probable source areas of the terrigenous organic matter are the organic-carbon-rich Mesozoic rocks which out-crop on the islands (Elverhoi et al. 1995).

To quantify the terrigenous and marine proportions of organic carbon in the inner Kara Sea surface sediments, several approaches were used: (1) estimates based on long-chain *n*-alkane distributions (Fernandes and Sicre 2000); (2) estimates based on stable carbon isotopes of the organic matter (Fernandes and Sicre 2000; Krishnamurthy et al. 2001); and (3) estimates based on maceral composition (Boucsein et al. 1999).

The *n*-alkane aproach

As outlined in Chapter 1.4.4, Fernandes and Sicre (2000) found that the Ob and Yenisei rivers supply long-chain *n*-alkanes into the Kara Sea in a relatively constant ratio to the organic carbon content (ALKOC ratio = 451). This relationship is represented by the following equation:

$$C_{25+27+29+31} = 451 \times OC + 0.117 \quad (r^2 = 0.95)$$

where the sum of odd *n*-alkanes C_{25}–C_{31} is given in µg/gSed. and OC in g/gSed. (see Fig. 1.4.6). Assuming that a decrease in the ALKOC ratio in the

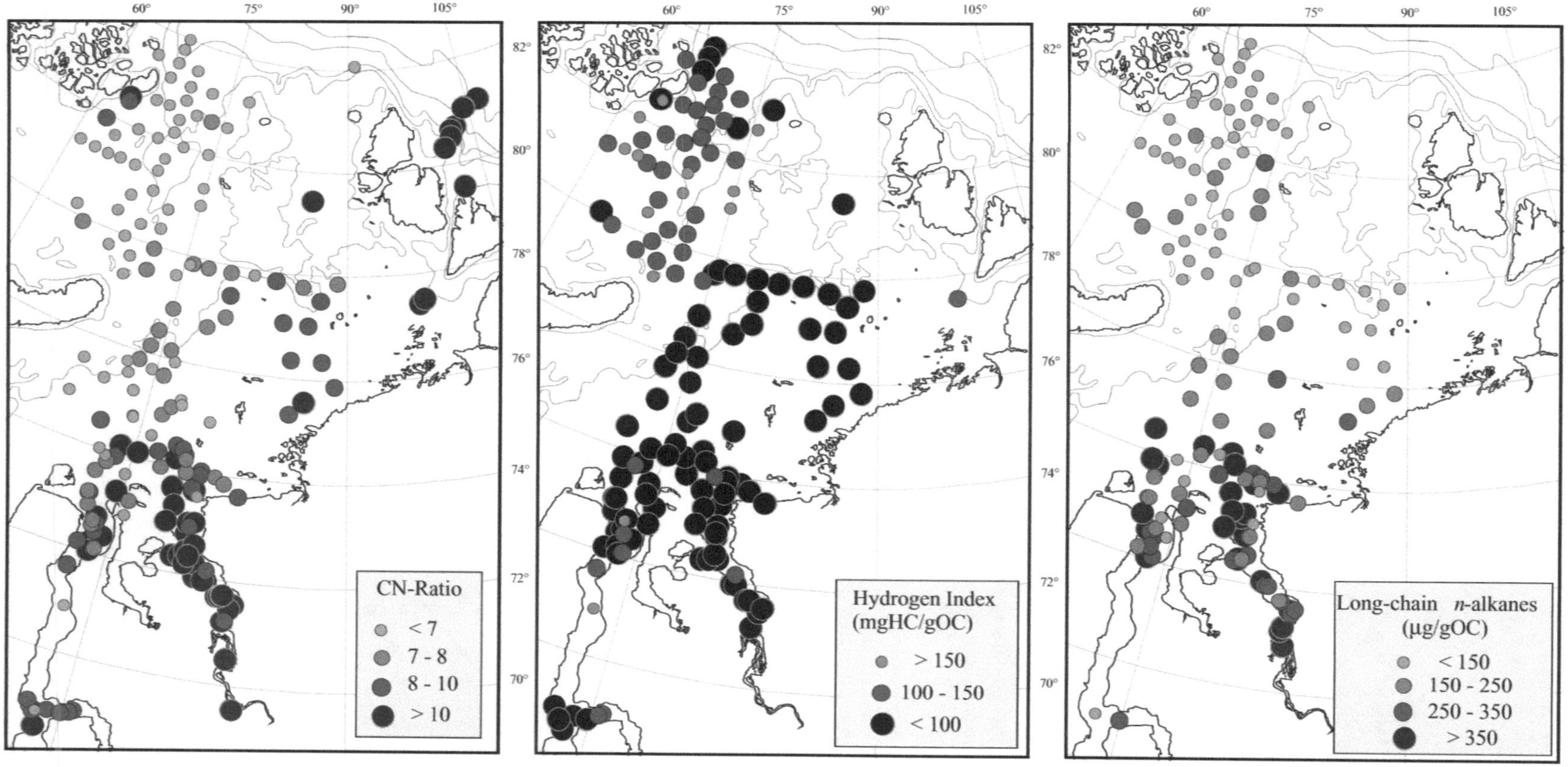

Fig. 7.6.4. (a) Distribution of C/N ratios in Kara Sea surface sediments. (b) Distribution of hydrogen index values (in mgHC/gOC) in Kara Sea surface sediments. (c) Distribution of long-chain *n*-alkanes ($C_{27} + C_{29} + C_{31}$ in µg/gOC) in Kara Sea surface sediments. In the distribution map of C/N ratios and long-chain *n*-alkanes, data from Krishnamurthy et al. (2001) and Fernandes and Sicre (2000), respectively, were included

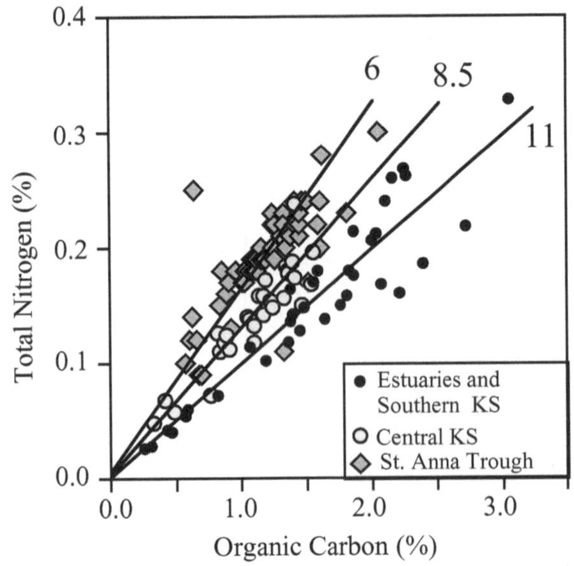

Fig. 7.6.5. Correlation between total organic carbon (%) and total nitrogen (%) in Kara Sea surface sediments. Lines with numbers indicate C/N ratios. Samples from the southern and central Kara Sea ("Akademik Boris Petrov" expeditions 1999, 2000 and 2001) and northern Kara Sea ("Professor Logachev" Expedition 1994)

sediments north of the estuaries is caused by dilution with marine organic carbon that does not contain long-chain *n*-alkanes (see Prahl and Muehlhausen 1989), the TerrOC equation from Chapter 1.4.4 can be used to calculate the proportion of terrigenous organic carbon. In the rivers and the estuaries, the organic matter is almost entirely of terrigenous origin (see Yenisei Transect, Fig. 7.6.7). North of the mixing zone (marginal filter) of Ob and Yenisei up to about 76°N, where unfortunately only six samples were measured by Fernandes and Sicre (2000), the proportion of terrigenous organic carbon ranges between 66 and 89% of the total organic carbon (Fernandes and Sicre 2000).

To compare the Fernandes and Sicre (2000) data with our data from the Kara and Laptev seas as well as the St. Anna Trough where we used the sum of odd *n*-alkanes C_{27} to C_{31} as a proxy for terrigenous input, we re-plotted the Fernandes and Sicre data set, using only the sum of C_{27} to C_{31}. Accordingly:

$$C_{27+29+31} = 369 \times OC + 0.104 \quad (r^2 = 0.95)$$

implying an ALKOC ratio of 369. The correlation coefficient and calculated proportions of terrigenous organic carbon remain the same (see Chapter 1.4, Fig. 1.4.6).

As shown in Figure 7.6.8, our data from the Ob Estuary fit very well with the data set of Fernandes and Sicre (2000) whereas the Yenisei values plot slightly below the regression line. This suggests that the ALKOC ratio might be slightly lower (326) for the Yenisei. Additional data points, however, are necessary to prove whether this difference is significant. Using our data set, the proportions of terrigenous organic carbon estimated from *n*-alkane data vary between 100 and 84% in the Ob Estuary, decreasing to 60% at 74°N. For the Yenisei, they range from about 95 to 73% in the estuary. The

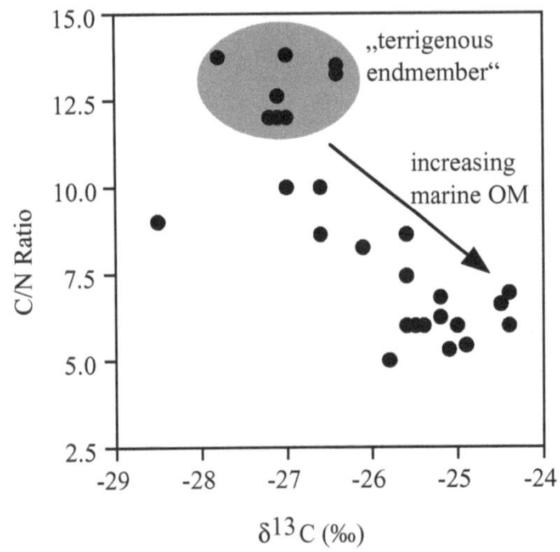

Fig. 7.6.6. Correlation between $\delta^{13}C_{org}$ values and C/N ratios in surface sediments from the Ob and Yenisei estuaries and the adjacent southern Kara Sea (data from Krishnamurthy et al. 2001). Grey-shadded area marks samples from the Ob and Yenisei estuaries ("terrigenous endmember")

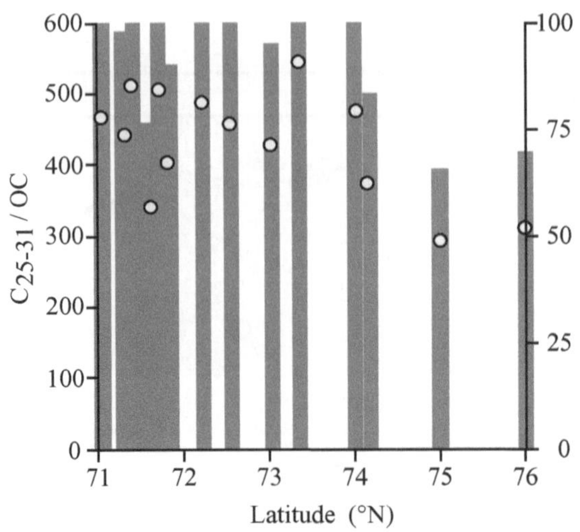

Fig. 7.6.7. Estimates of relative proportions of terrigenous organic matter based on long-chain *n*-alkane data ($C_{25} + C_{27} + C_{29} + C_{31}$) in the Yenisei transect (data from Fernandes and Sicre 2000). Long-chain *n*-alkane data are given in μg/gOC. For further explanation see text

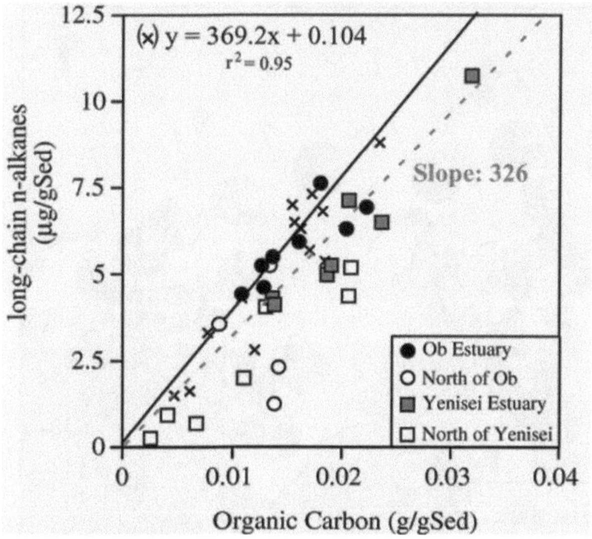

Fig. 7.6.8. Correlation between long-chain n-alkanes (C_{27} + C_{29} + C_{31}) in µg/gSed and total organic carbon in g/gSed. The regression line and equation are related to the Fernandes and Sicre (2000) data (crosses)

Yenisei values would be 85 to 100%, if the lower ALKOC ratio of 326 is used.

In the St. Anna Trough area, the proportion of terrigenous organic carbon reaches values of about 20 to 50% when using the *n*-alkane approach. These values, however, are certainly too low. Here, organic carbon input from other sources (e.g., Franz Josef Land, Barents Sea, Arctic Ocean) may be important, for which the ALKOC ratio is not known.

The isotope approach

To estimate the marine and terrigenous organic carbon proportions in the sediment from the Ob and Yenisei estuaries and the adjacent southern Kara Sea, the stable isotope approach has been used by Fernandes and Sicre (2000) and Krishnamurthy et al. (2001) (see Chapter 1.4.2 for background and formula). For the calculation, Fernandes and Sicre (2000) chose an average terrestrial endmember of −27.8‰, and two different marine endmembers were used, −19.1‰ and −16.0‰. The latter, very heavy value, was derived from the correlation between relative concentrations of long-chain *n*-alkanes (as a proxy for terrigenous input) and $\delta^{13}C_{org}$ values (for details and discussion see Fernandes and Sicre 2000). Using this heavy value of −16.0‰, the proportions of terrigenous organic carbon range from about 70 to 80% in the southern Kara Sea north of the mixing zone (marginal filter) (Fig. 7.6.9), which are similar to values obtained using the *n*-alkane approach. Based on a marine endmember of −19.1‰, the corresponding values

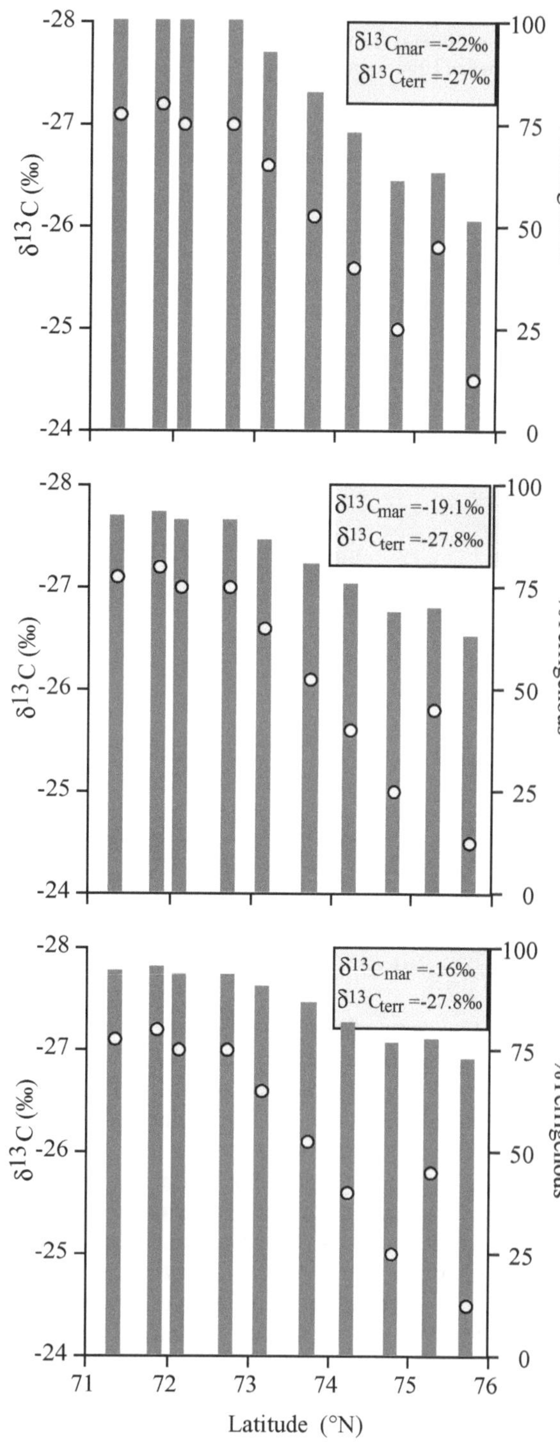

Fig. 7.6.9. Estimates of relative proportions of terrigenous organic matter based on $\delta^{13}C_{org}$ values (data from Krishnamurthy et al. 2001), using various $\delta^{13}C_{org}$ values for the terrigenous and marine endmembers. For further explanation see text

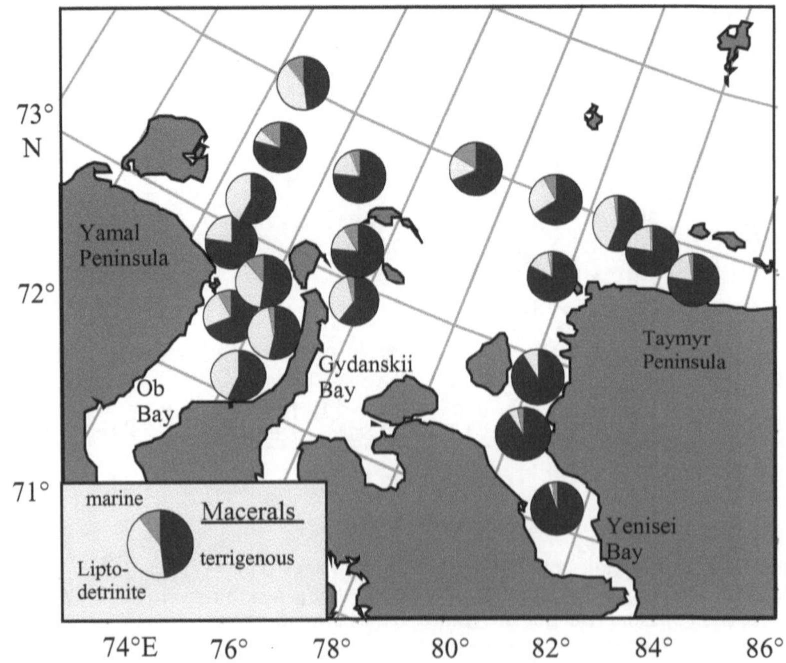

Fig. 7.6.10. Distribution of macerals in surface sediments from the inner Kara Sea (Boucsein et al. 1999)

are lower, reaching 60 to 73% (Fernandes and Sicre 2000). Krishnamurthy et al. (2001) have used the isotope approach in almost exactly the same study area. They used −27‰ and −22‰ as terrestrial and marine endmembers, respectively. Consequently, their estimates of the proportion of terrigenous organic carbon are lower (50–70%) in the southern Kara Sea (Fig. 7.6.9).

The maceral composition approach

Another approach to estimate relative proportions of marine and terrigenous organic matter in marine sediments is the use of the maceral composition (Boucsein et al. 1999, 2000, 2002; Boucsein and Stein 2000; for method and details see also Chapter 1.4.3). Based on maceral data, 90 to 95% of the organic matter is of terrigenous origin in the Yenisei Estuary (Fig. 7.6.10). Towards the north, there is a relative decrease in terrigenous organic matter. At 74°N, 20–25% of the organic matter is probably of marine origin. For the Ob Estuary, proportions of terrigenous organic carbon seem to be lower reaching about 65 to 80%, and no clear decrease towards the north is obvious. Especially in the Ob Estuary, however, the amount of liptodetrinites, which does not allow a clear organic carbon source identification, is very high (Fig. 7.6.10). Major proportions of these liptodetrinites are certainly of terrigenous origin (Boucsein et al. 1999). Thus, the given estimates of terrigenous organic carbon have to be regarded as minimum values.

From these results we can conclude that the estimates of the marine and terrigenous proportions of the organic matter using the three different approaches all indicate the predominance of terrigenous organic carbon in the southern Kara Sea with a decreasing trend towards the north. Terrigenous organic carbon proportions of about 80 to 100% were characteristic in the Yenisei Estuary, whereas north of the marginal filter, the values distinctly decrease. Differences between the approaches, however, are obvious when looking at details. Based on the experience in the Laptev Sea (Boucsein and Stein 2000; Boucsein et al. 2000, 2002; see Chapter 7.5), maceral analysis where visible marine and terrigenous organic particles can be distinguished and counted, seems to give very reliable results. Unfortunately, however, the database on macerals is very limited for the Kara Sea and restricted to the area south of 74°N. Furthermore, high amounts of liptodetrinites which can be of either terrigenous or marine origin, make a clear identification of the organic carbon source more difficult. The difficulty of the $\delta^{13}C$ approach, i.e., the less well constrained marine endmember, has already been mentioned above. In general, the *n*-alkane approach seems to give reliable results in and close to the estuaries where a single (riverine) source can be assumed. In the central and northern Kara Sea, where organic carbon input from other areas (e.g., Franz Josef Land, Barents Sea and Arctic Ocean) is probably more important and the ALKOC ratio (see above) of these different organic carbon

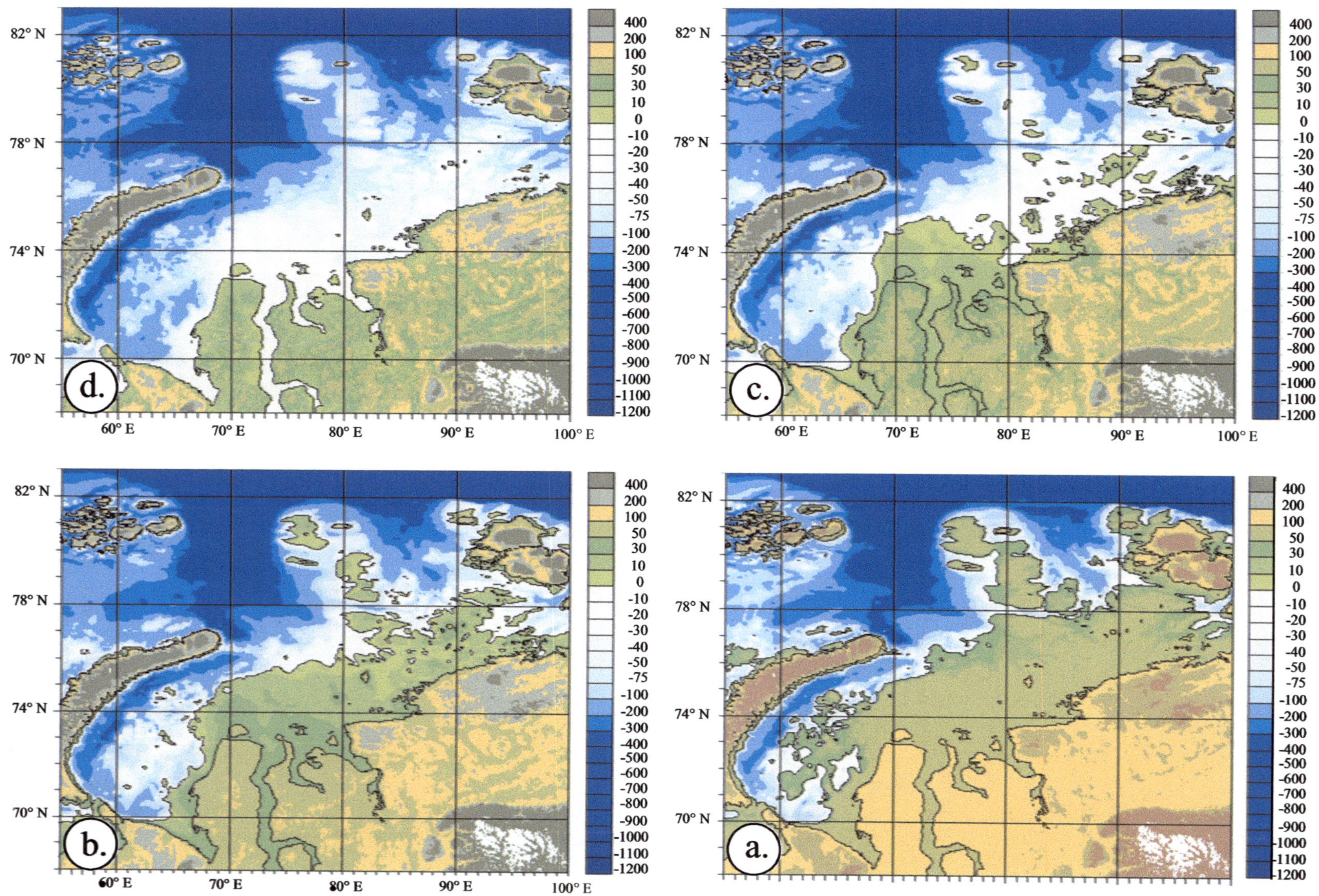

Fig. 7.6.11. Land/sea distribution in the Kara Sea at times of lowered global sea level (according to Fairbanks 1989). The maps have been constructed simply by using the 30 m, 50 m and 100 m isolines of the modern bathymetry (assuming no or negligible isostatic rebound); thus, they should be used as first approximation. (a) about 18 Cal. Kyrs. BP (−100 m), (b) 11 Cal kyrs. BP (−50 m), (c) 9 Cal kyrs. BP (−30 m), and (d) modern situation. Maps were drawn based on the IBCAO data set (I. Harms, Institute of Oceanography, Hamburg University)

sources may be different, these estimates have to be intepreted with great caution. When interpreting the terrigenous and marine organic carbon data derived from the various methods, one should have these problems in mind.

7.6.4 Late Quaternary organic Carbon Records and Paleoenvironment

Similar to the Laptev Sea, the late Quaternary paleoenvironmental evolution of the Kara Sea and the related sediment as well as organic carbon input have been strongly influenced by changes in river discharge, coastal erosion, sea-ice cover, and oceanic circulation patterns as well as the waxing and waning of the Barents Kara Ice Sheet (e.g., Polyak et al. 1997; 2000, 2002; Velitchko et al. 1997; Svendsen et al. 1999; Lubinski et al. 2001; Mangerud et al. 2002; Stein et al. 2002, 2003a). During the Last Glacial Maximum (LGM), major parts of the Kara Sea were covered by a huge ice sheet, the northeastern margin of which is still poorly known (e.g., Svendsen et al. 1999; Polyak et al. 2000, 2002; Stein et al. 2002). Thus, glacigenic processes likely influenced the organic carbon input. Furthermore, at that time the shallow Kara Sea was subaerially exposed due to the lowered glacial sea level (Fig. 7.6.11; e.g., Fairbanks 1989), and terrigenous (fluvial) sediments were transported much farther to the north. With the postglacial sea-level rise, the shallow Kara Sea became widely flooded, and intensified seafloor/coastal erosion and increased river discharge resulted in very high sediment accumulation on the shelf (e.g., Polyak et al. 1997; Stein et al. 2003a and further references therein).

Studies of the late Quaternary paleoenvironmental history based on sediment cores from the Kara Sea are still very rare. Levitan et al. (1995), who performed sedimentological and micropaleontological investigations of sediment cores obtained from the "Dmitry Mendeleev" in 1993 (Lisitzin and Vinogradov 1995), related their data to changes in depositional environment during the last Glacial to postglacial (Holocene). Due to the absence of any absolute AMS^{14}C age control, precise climatic reconstructions and correlations with similar data sets from the Barents and Laptev seas have not been possible. Based on a study of four AMS^{14}C dated sediment cores from the St. Anna Trough, more detailed information about the glacial/deglacial history and related Atlantic water and freshwater inflows in the northern Kara Sea became available (Polyak et al. 1997; Lubinski et al. 2001). Based on eleven AMS^{14}C dated sediment cores from the southern Kara Sea obtained within the SIRRO Project, a very precise picture can be constructed of the changes in terrigenous siliciclastic and organic carbon fluxes in the southern Kara Sea during (deglacial to) Holocene times and their relationship to climate change and sea-level rise (Stein et al. 2003a). We are aware of no studies on sediment cores representing the older pre-LGM time intervals. In this paper, we concentrate on the (deglacial to) Holocene time interval. For changes in organic carbon fluxes and composition and its relationship to paleoenvironmental changes during the pre-Holocene (last glacial) we refer to Stein et al. (2001) and Boucsein et al. (2002).

(Postglacial to) Holocene Changes in Amount and Compostion of Organic Carbon

For the Holocene, a large number of OC records obtained within the SIRRO Project are available for the southern to central Kara Sea and the Ob and Yenisei estuaries (Fig. 7.6.12). Based on these records, OC contents between about 1 and 2% are most typical in the estuaries. In the southern and central Kara Sea, these values vary between 0.5 and 1.2%. In the lower part of some cores, representing the early Holocene (see below), the OC contents are somewhat higher, typically around 1 to 1.5%. In the northern Kara Sea, the OC records show similar trends (Fig. 7.6.12).

Concerning the composition of the organic carbon, low hydrogen indices <100 mgHC/gOC are predominant throughout the Holocene time interval in all our southern Kara Sea records (Stein et al. 2003a), suggesting a dominantly terrigenous origin throughout. In a "van-Krevelen-type" diagram, the data points fall mainly into the "kerogen type III" field (Fig. 7.6.13). In most of the diagrams, higher Tmax values around 430 °C were measured in sediments of early Holocene age whereas the late Holocene sediments have Tmax values of <400 °C (Fig. 7.6.13). The higher early Holocene values probably indicate the presence of more refractive reworked (fossil) organic matter (see below).

Looking at C/N ratios on a transect of sediment cores from the estuaries towards the northern Kara Sea, some general trends are obvious (Fig. 7.6.14). At cores BP99-04/7 and BP00-38/2, located in the northern Yenisei and Ob estuaries, respectively, relatively high C/N ratios between 9 and 12 are typical throughout the records with the maximum values more typical in the lower (early Holocene) part of the records, indicating the dominance of terrigenous organic matter. In the cores from the central Kara Sea, the relatively low C/N ratios in the late Holocene time interval, varying between 6 and 9, point to a decreased amount of terrigenous organic carbon in the sediments. Significantly

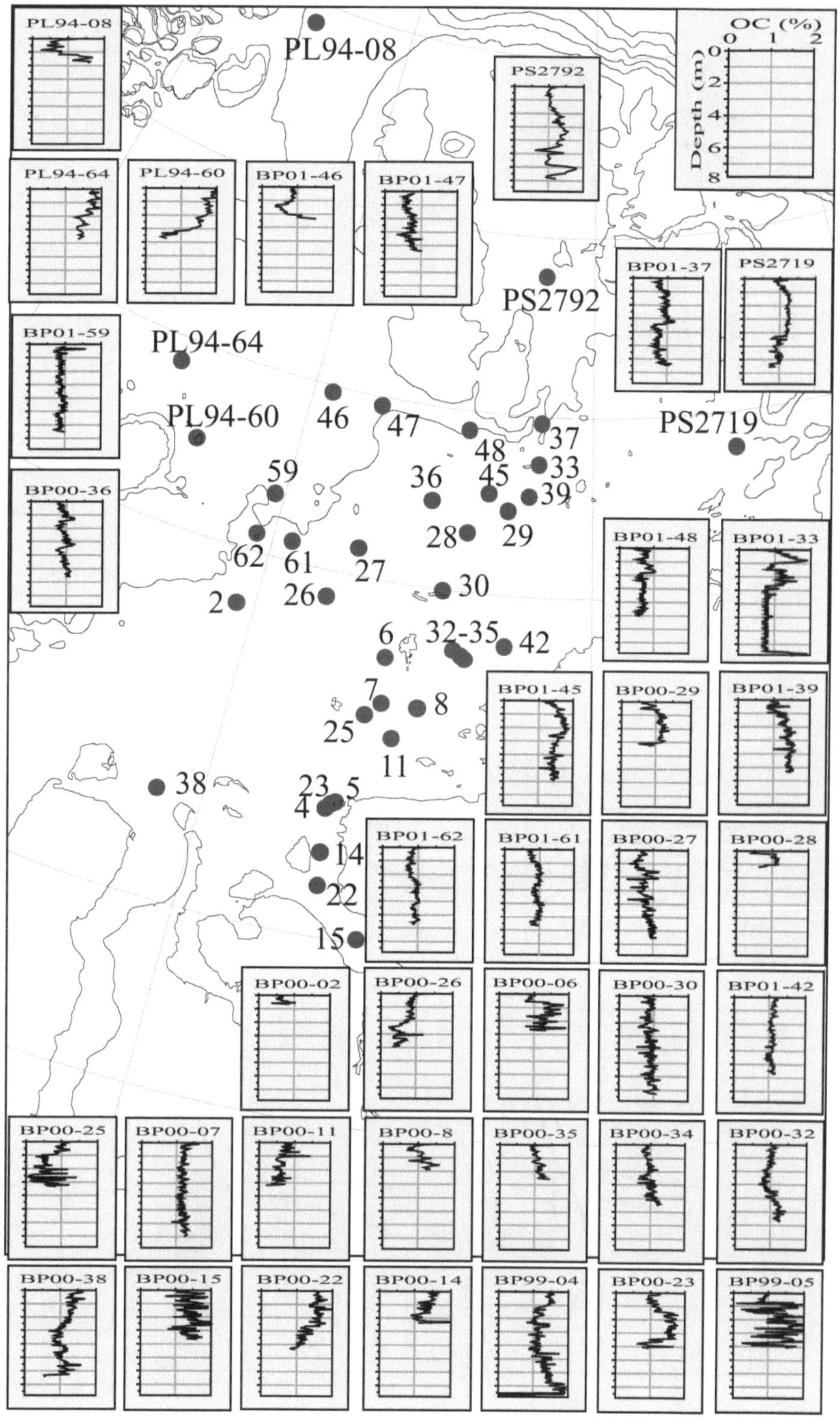

Fig. 7.6.12. Total organic carbon (%) records of sediment cores from the Kara Sea (depth scale in 1m steps). Vertical line in the middle of the records indicates 1% organic carbon concentration. On the map, the location of the cores is shown

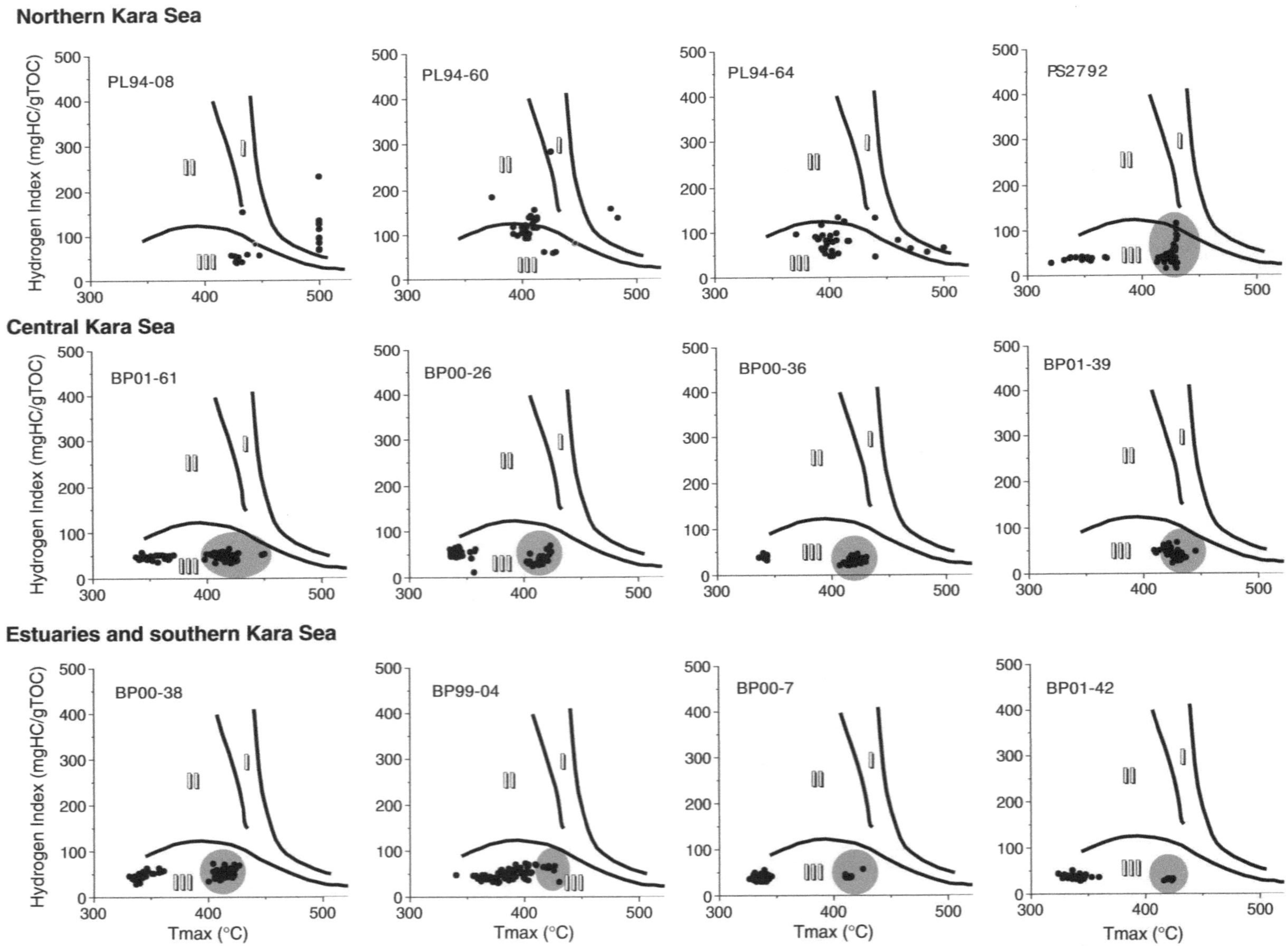

Fig. 7.6.13. Diagrams of hydrogen indices vs. Tmax values in sediment cores from the Kara Sea. Gary-shaded area indicates samples representing an early Holocene age

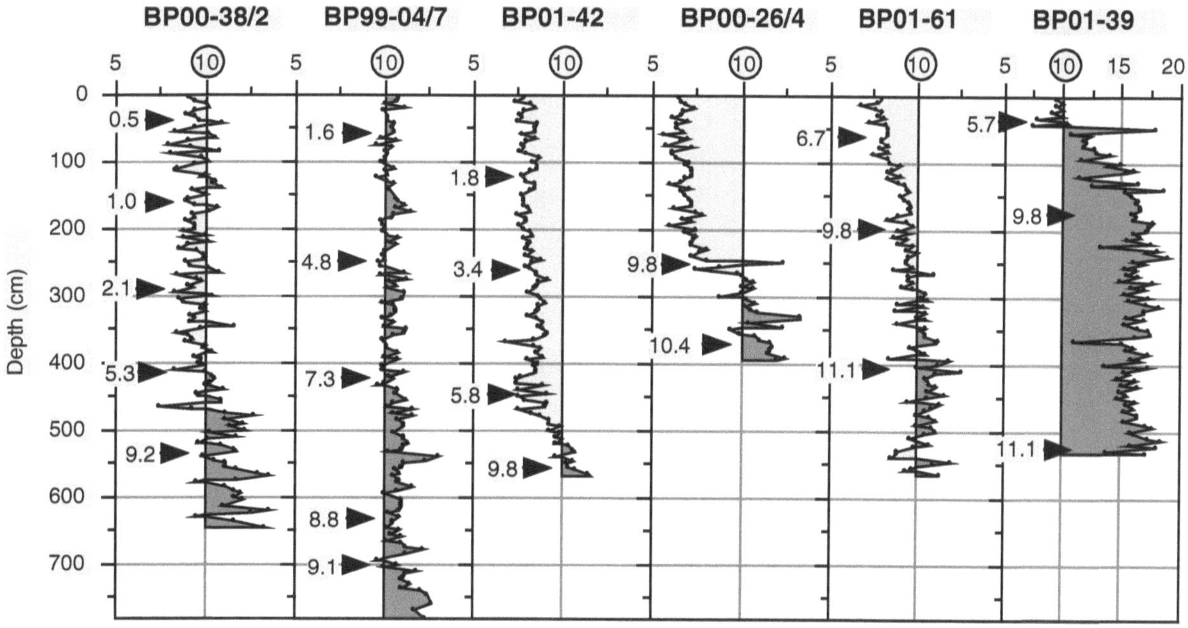

▲ **Fig. 7.6.14.** C/N ratios of selected sediment cores from the Kara Sea (for location of cores see Fig. 7.6.12). For reference, a C/N ratio of 10 is indicated and intervals with C/N ratios >10 are shaded in gray. Numbers at black triangles are ages (Cal. kyrs. BP)

Fig. 7.6.15. Correlation between total organic carbon (%) and total nitrogen (%) in (A) southern Kara Sea cores BP99-04/7 (Yenisei Estuary) and BP00-38/2 (Ob Estuary) and (B) central Kara Sea cores BP00-26/4 and BP01-39/2 (for location of cores see Fig. 7.6.12) ▼

higher C/N ratios of 10 to 18 were measured in the early Holocene time interval (Fig. 7.6.14). At cores BP00-26 and BP01-61, these higher C/N ratios of 10 to 13 are similar to those recorded in the estuary cores BP99-04 and BP00-38 (Figs. 7.6.14 and 7.6.15), suggesting higher terrigenous (riverine) organic matter supply at these more northern stations. At Core BP01-39, the Early Holocene C/N ratios are significantly higher than even the riverine values. Here, an additional supply from other terrigenous sources (e.g., by coastal erosion) is assumed (see discussion below). Coincident with the higher C/N ratios, Tmax values increased to values of >410 to 430 °C in the Early Holocene (Figs. 7.6.13 and 7.6.16), supporting a change in the organic carbon source.

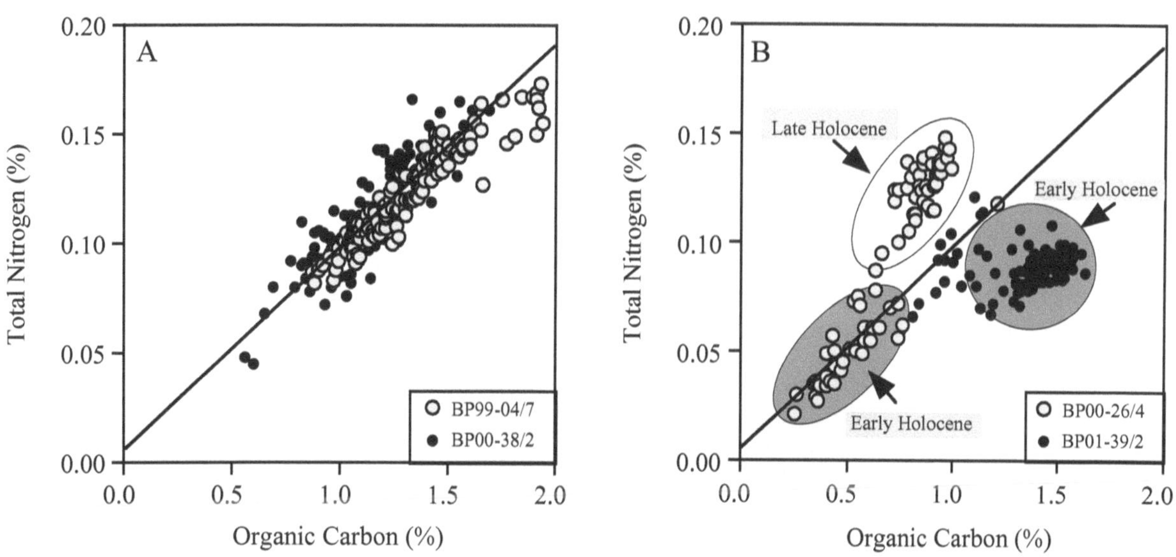

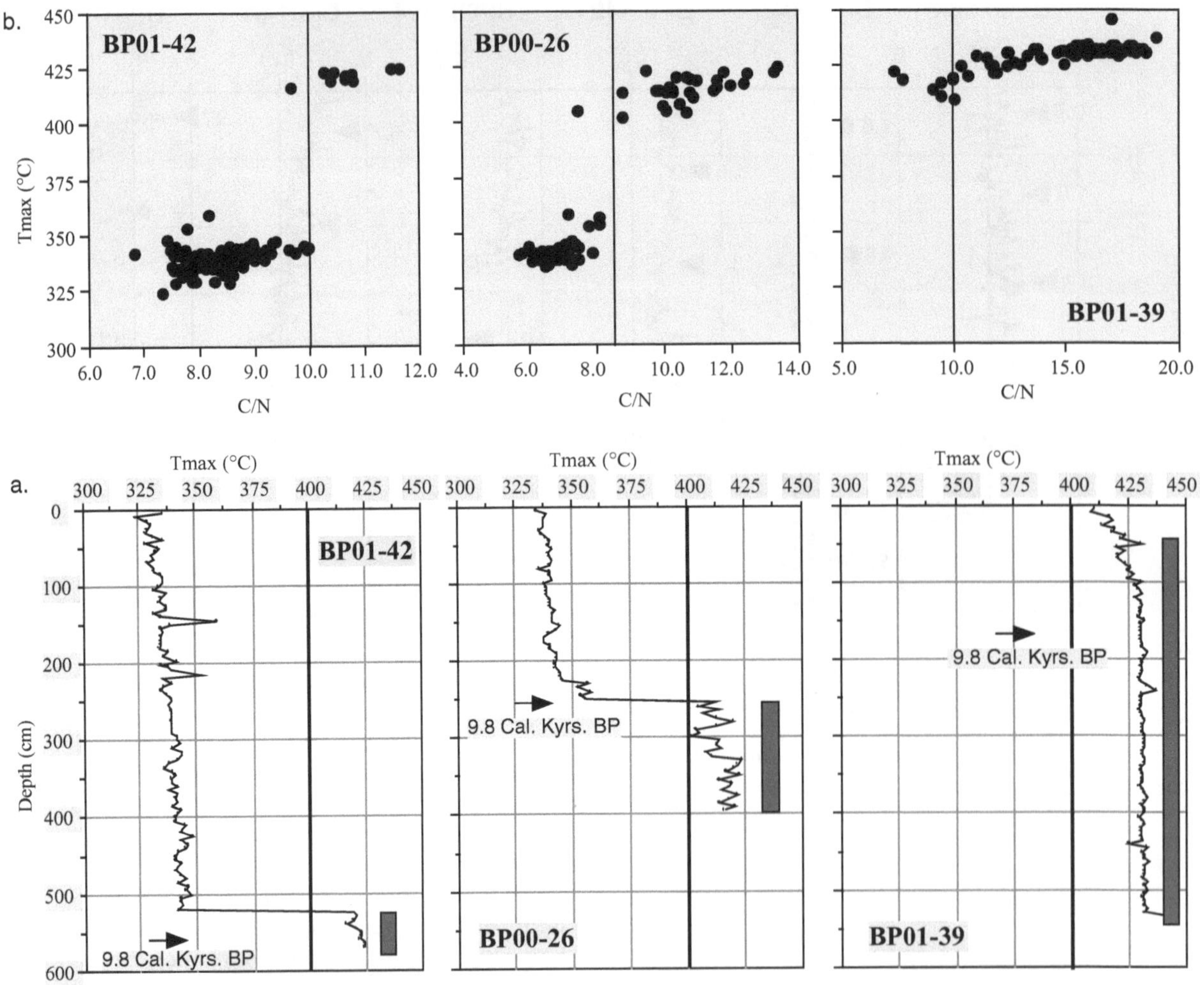

Fig. 7.6.16. Tmax values vs. depth plots (a) and correlation between C/N ratios and Tmax values (b) for three selected sediment cores from the southern and central Kara Sea (for location of cores see Fig. 7.6.12). The gray bars indicate the early Holocene interval characterized by the presence of more refractive (fossil) organic matter. AMS^{14}C age was determined on bivalves (Stein et al. 2003a)

The higher Tmax values may suggest the presence of more fossil organic matter (see Wagner and Henrich 1994).

In the southern Kara Sea, a long-chain *n*-alkane record is available only from the AMS^{14}C dated sediment Core BP99-04/7 from the northern Yenisei Estuary. Highest long-chain *n*-alkane concentrations occur in the lowermost (> 9 Cal kys BP) and uppermost (< 2 Cal kyrs. BP) parts of the record (Fig. 7.6.17). Using the *n*-alkane data (see 7.6.3), the proportion of terrigenous organic carbon varies between 70 and 100% indicating the importance of riverine organic matter at this site.

For the northern Kara Sea, two sediment cores (PL9408 and PL9460) representing the last Glacial to Holocene time interval are available from the St. Anna Trough (Fig. 7.6.18). The glacigenic diamicton recovered near the base of of these cores provides evidence for grounded glacier ice, which probably reached the shelf edge during the LGM (Polyak et al. 1997). Based on AMS^{14}C ages, the minimum age for the retreat of the grounded ice from the central deep St. Anna Trough is about 13,300 yrs B.P. (Polyak et al. 1997). The diamicton at the northern Core PL94-08, located close to Franz Josef Land, is characterized by maximum OC values, very low hydrogen indices, high C/N ratios and almost 100% proportions of terrigenous macerals (Fig. 7.6.18; Stein and Knies 1999; Boucsein et al. 2002). Terrigenous organic matter with a Franz Josef Land origin was probably supplied to this core site by glaciers.

During the postglacial to Holocene time interval, OC values significantly increased in the southern St. Anna Trough (Cores PL94-60 and PL94-64; Figs. 7.6.12 and 7.6.18), but decreased in the northern

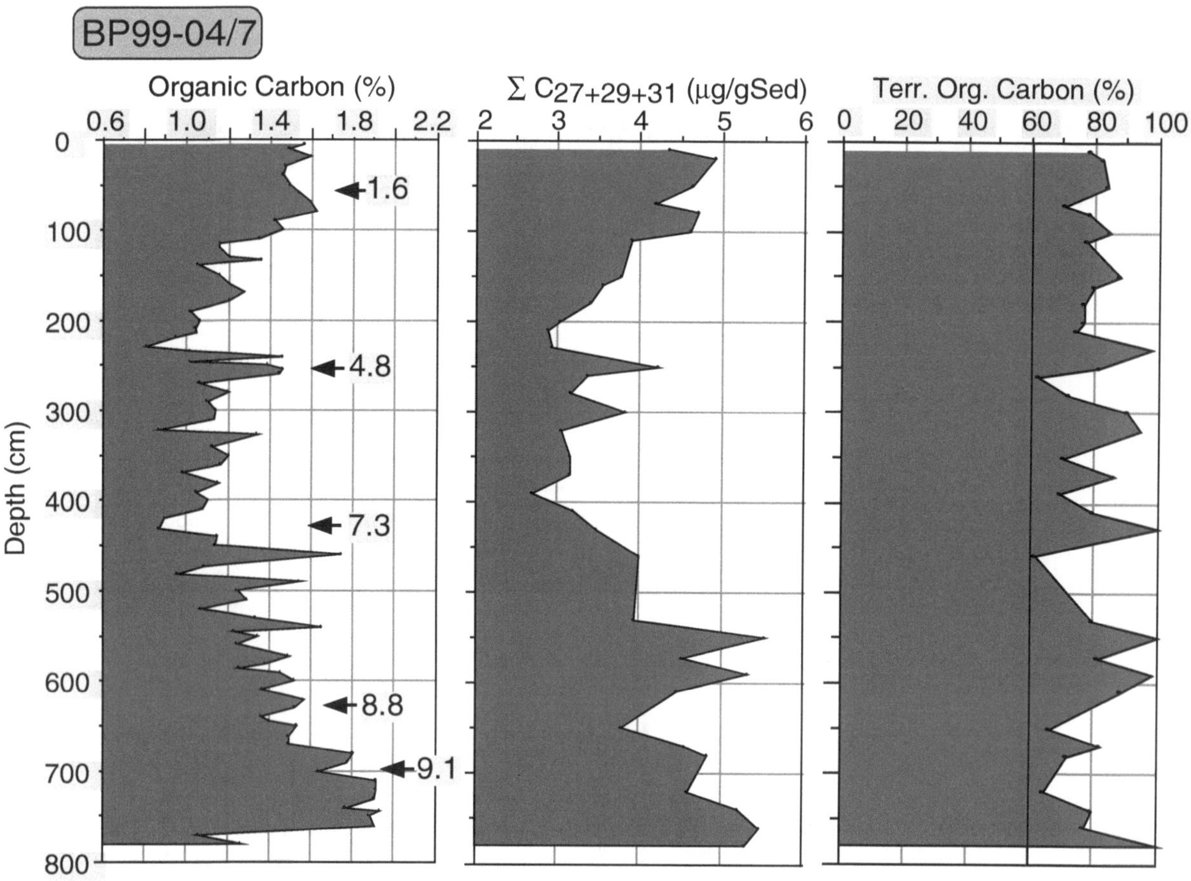

Fig. 7.6.17. Distribution of total organic carbon, long-chain n-alkanes (C_{27} + C_{29} + C_{31} in µg/gSed), and estimates of the relative proportion of terrigenous organic carbon (based on long-chain *n*-alkanes) at Core BP99-04/7 (Northern Yenisei Estuary). Numbers with arrows indicate ages (Cal. kyrs. BP)

Core PL94-08. The relative amount of marine organic carbon seems to have increased at that time, as indicated by an increase in HI values and a decrease in C/N ratio (Fig. 7.6.18). The predominant proportion of organic carbon, however, remains of terrigenous origin, as deduced from maceral data. Terrigenous organic carbon accounts for about 70 to 80% of the total organic matter, whereas marine organic matter reaches values of up to 15% (Fig. 7.6.18; Boucsein et al. 2002). About 5 to 15% of the macerals belong to the group of liptodetrinite which cannot be identified as either marine or terrigenous particles because of their small size. Thus, up to a maximum of 30% of the organic carbon might be of marine origin. The higher proportion of marine organic matter at the northern St. Anna Trough during the Holocene is explained by the influence of Atlantic water masses flowing as a boundary current along the Eurasian continental margin (see Knies and Stein 1998; Stein et al. 2001; Boucsein et al. 2002). This paleoceanographic influence appears to have been intensified near 9 Cal kys BP.

Holocene Changes in Accumulation Rates of Total Sediment and Organic Carbon

Accumulation rates of total sediment and organic carbon give information about the spatial and temporal variability of total sediment and organic carbon flux. All data from the eleven AMS^{14}C dated sediment cores from the Kara Sea and a detailed discussion in terms of paleoenvironmental changes are presented in Stein et al. (2003a). Here, we summarize general aspects of relevance to the total sediment and organic carbon budget. The changes in accumulation rates in the inner and central Kara Sea are strongly related to the post-glacial sea-level rise. In the earliest Holocene (10–11 Cal. krs. BP), maximum accumulation rates were recorded in the central Kara Sea. Accumulation rates of total sediment and total organic carbon reach very high values of about 150 to 700 g cm^{-2} ky^{-1} and about 1.5 to 10 g cm^{-2} ky^{-1} respectively (Figs. 7.6.19 and

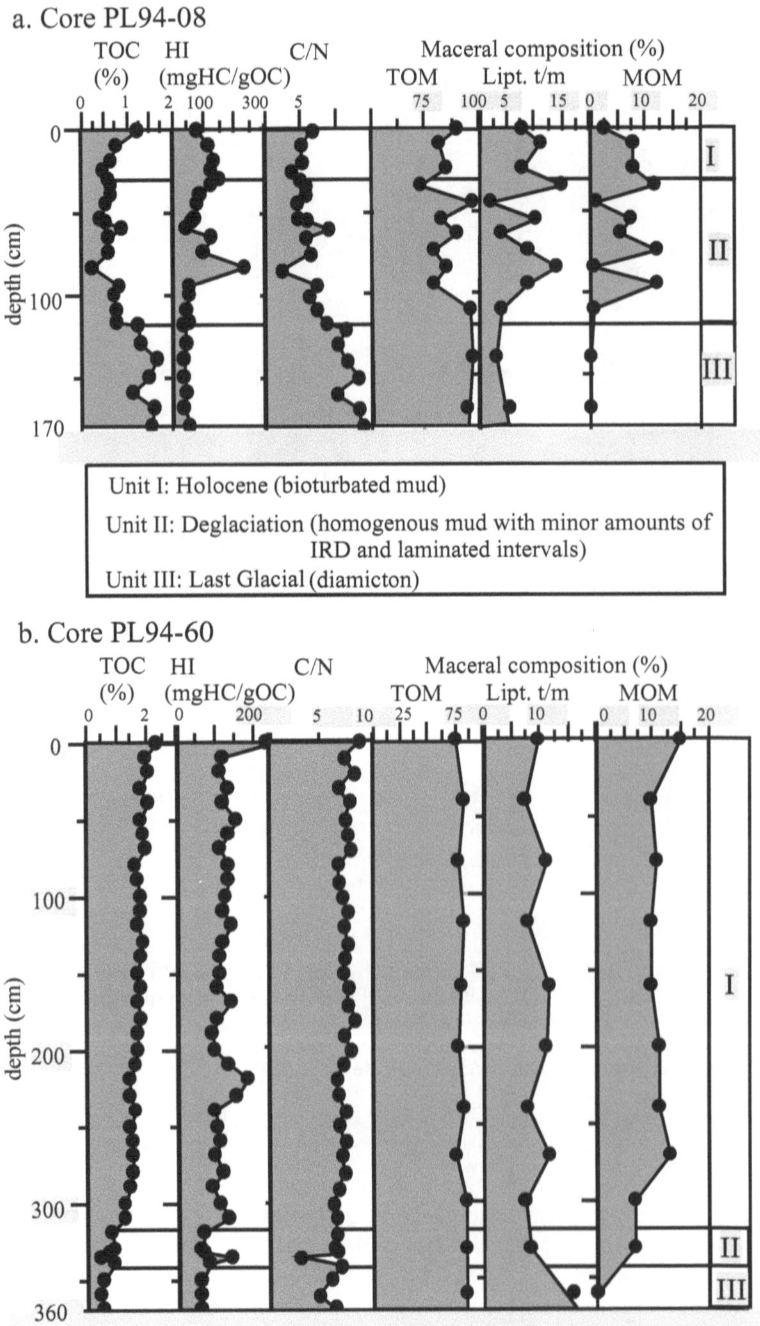

Fig. 7.6.18. Distribution of total organic carbon, hydrogen index values, C/N ratios and maceral composition at cores PL94-08 and PL94-60 (for location of cores see Fig. 7.6.12). TOM = terrigenous organic matter. MOM = marine organic matter. Lipt.t/m = terrigenous and marine lipto-detrinite. Data from Stein and Knies (1999) and Boucsein et al. 2002

7.6.20). At that time, the sea level was about 40 to 50 m lower than today and large parts of the inner Kara Sea were subaerial. Terrigenous sediments including organic matter were transported much more to the north, and the main depocenter was situated more toward the central Kara Sea (Fig. 7.6.21). At the eastern Kara Sea Core PS2719/2, a distinct maximum in total sediment and (terrigenous) organic carbon supply of about 90 g cm^{-2} ky^{-1} and 1.1–1.3 g cm^{-2} ky^{-1} respectively, also appears between about 9 and 11 Cal kys.BP (Stein et al. 2001). Farther to the north, in the interior St. Anna Trough, significantly increased sedimentation rates were also recorded at that time (Herlihy 1996; Polyak et al. 1997).

Maximum accumulation rates were contemporaneously determined in a large number of sediment cores from the Laptev Sea (e.g., Bauch et al. 2001; Chapter 7.5). The correspondence between these cores means that between 9 and 11 Cal. kyrs. BP the supply of terrigenous (organic) matter sharply increased along the entire Kara Sea – Laptev Sea

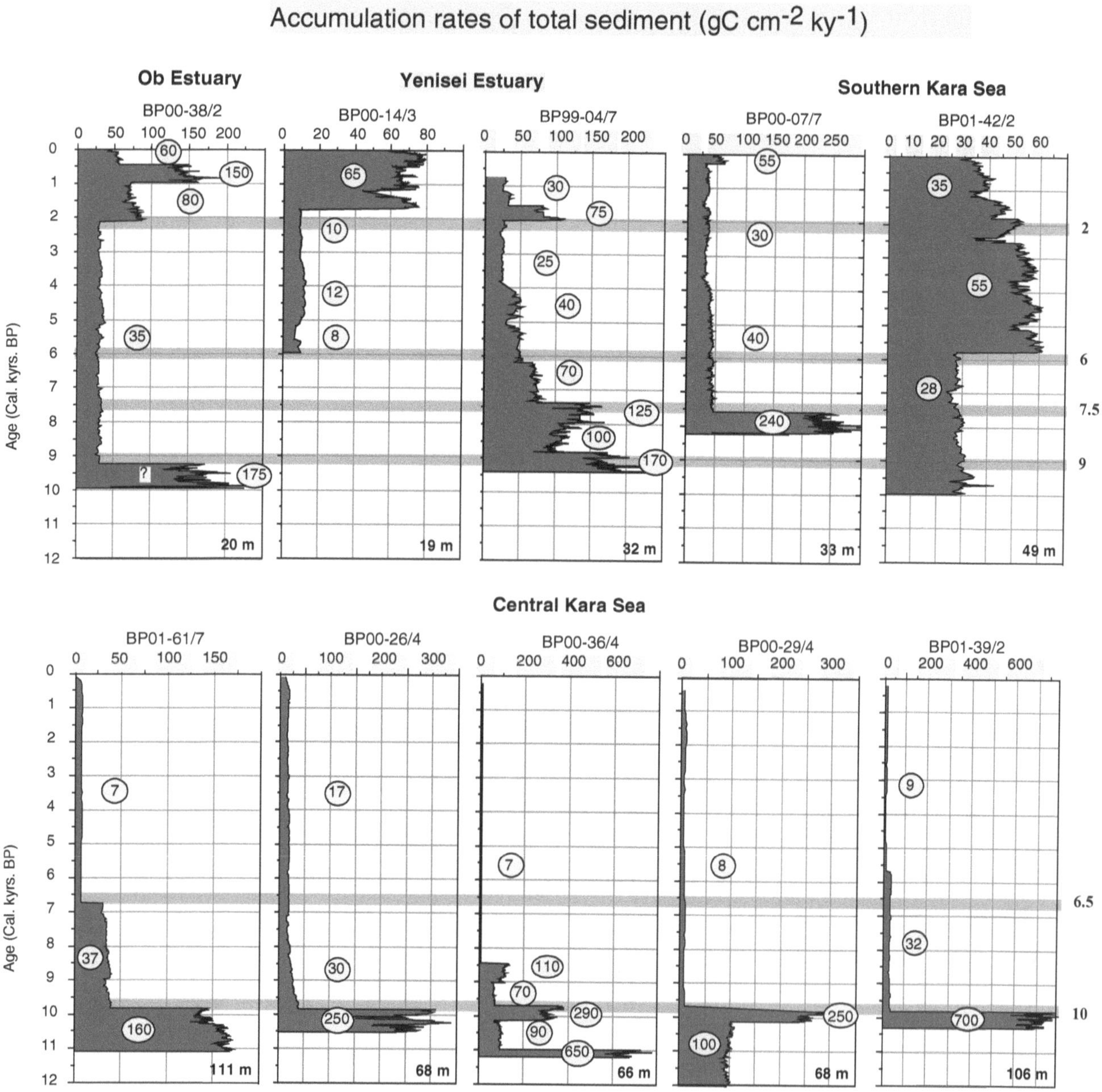

Fig. 7.6.19. Accumulation rate records of total sediments from the AMS-^{14}C dated sediment cores (for location of cores see Fig. 7.6.21 and Table 7.6.2). Circled numbers indicate mean values for specific time intervals. The horizontal gray bars with numbers (Cal. kyrs. BP) at the right side highlight changes in accumulation rates. In the lower right corner of each record, the modern water depth is given

continental margin. This distinct maximum in terrigenous organic carbon input occurred at a time of a major postglacial sea-level rise following the "Younger Dryas Cooling Event" (Fairbanks 1989, 1990), when the shallow inner Kara Sea and Laptev Sea shelf became widely flooded for the first time after the LGM, resulting in large-scale sea floor/coastal erosion and, thus, a distinctly increased supply of terrigenous (organic) matter (e.g., Polyak et al. 1997; Stein and Fahl 2000; Bauch et al. 2001a, b). The river discharge probably increased at the same time, transporting large amounts of riverine material towards the core locations. Near 9 Cal ka, accumulation rates distinctly decreased on the central/outer shelf (Figs. 7.6.19 and 7.6.20).

With increasing sea level, sedimentation in the central Kara Sea became drastically reduced, and the main depocenter shifted towards the south (Fig. 7.6.21), reaching approximately "modern" condition during the last 6 Cal. kyrs. BP. Near 2 Cal.

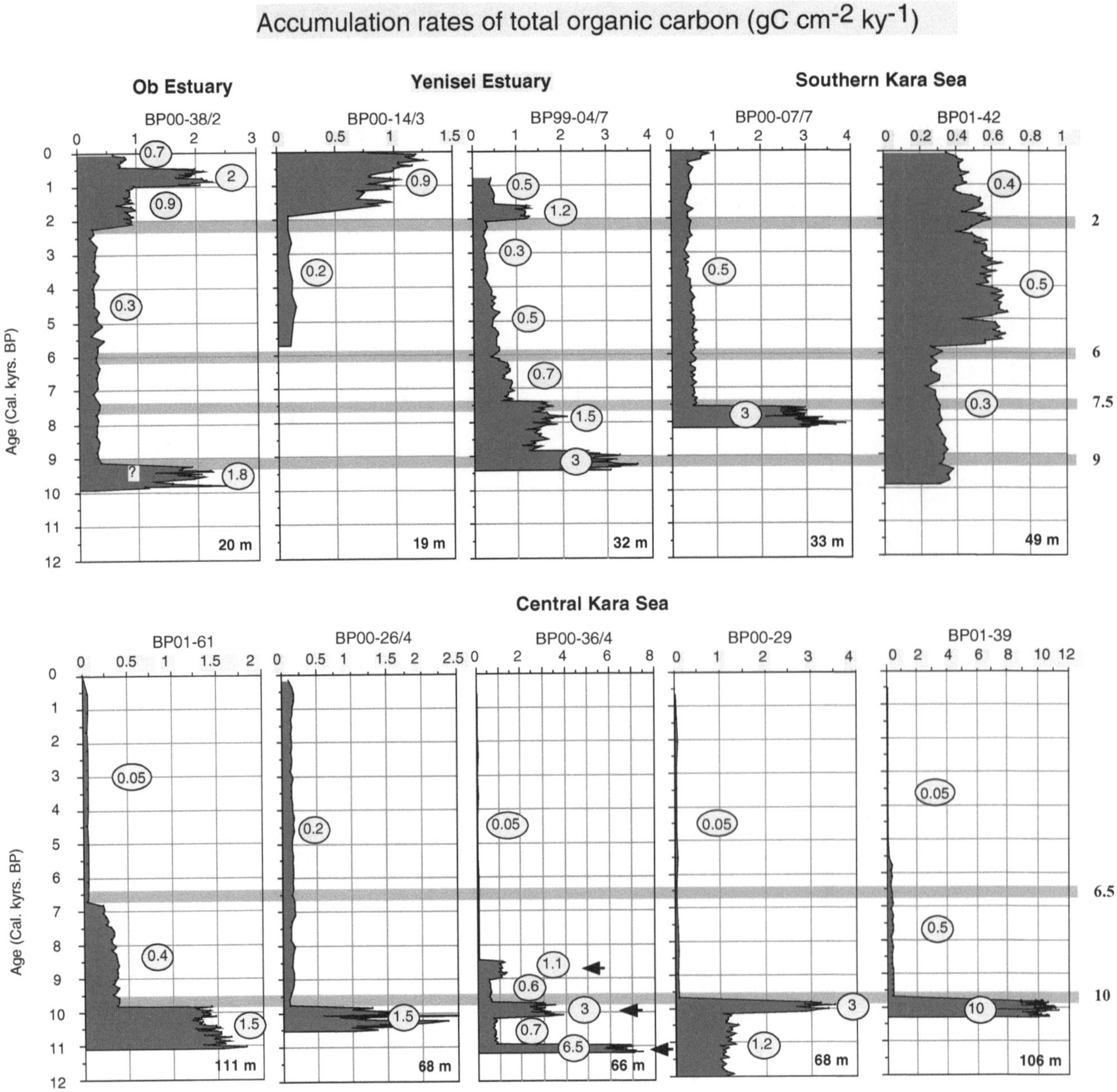

Fig. 7.6.20. Accumulation rate records of total organic carbon from the AMS-^{14}C dated sediment cores (for location of cores see Fig. 7.6.21 and Table 7.6.2). Circled numbers indicate mean values for specific time intervals. The horizontal gray bars with numbers (Cal. kyrs. BP) at the right side highlight changes in accumulation rates. In the lower right corner of each record, the modern water depth is given

kyrs. BP, accumulation rates at Core BP99-04/7 increased to values of up to about 75 g cm^{-2} ky^{-1}. This change coincided with a distinct decrease in grain size, probably indicating reduced current velocity (Stein et al. 2003a). Under these conditions, increased amounts of fine-grained sediment accumulated in the estuary as indicated in the elevated accumulation rates (Fig. 7.6.19), and organic matter became enriched in the finer-grained sediments, indicated by the elevated OC values (Figs. 7.6.12 and 7.6.17) as well as increased accumulation rates of organic carbon (Fig. 7.6.20). Similar contemporaneous trends, i.e., the increase in total sediment and organic carbon accumulation rates, are also observed in the Yenisei Core BP00-14/3 and the Ob Core BP00-38/2. At the same time, accumulation rates decreased in the southern Kara Sea Core BP01-42/2 (Figs. 7.6.19 and 7.6.20), suggesting an increase in effectiveness of the marginal filter as sediment trap.

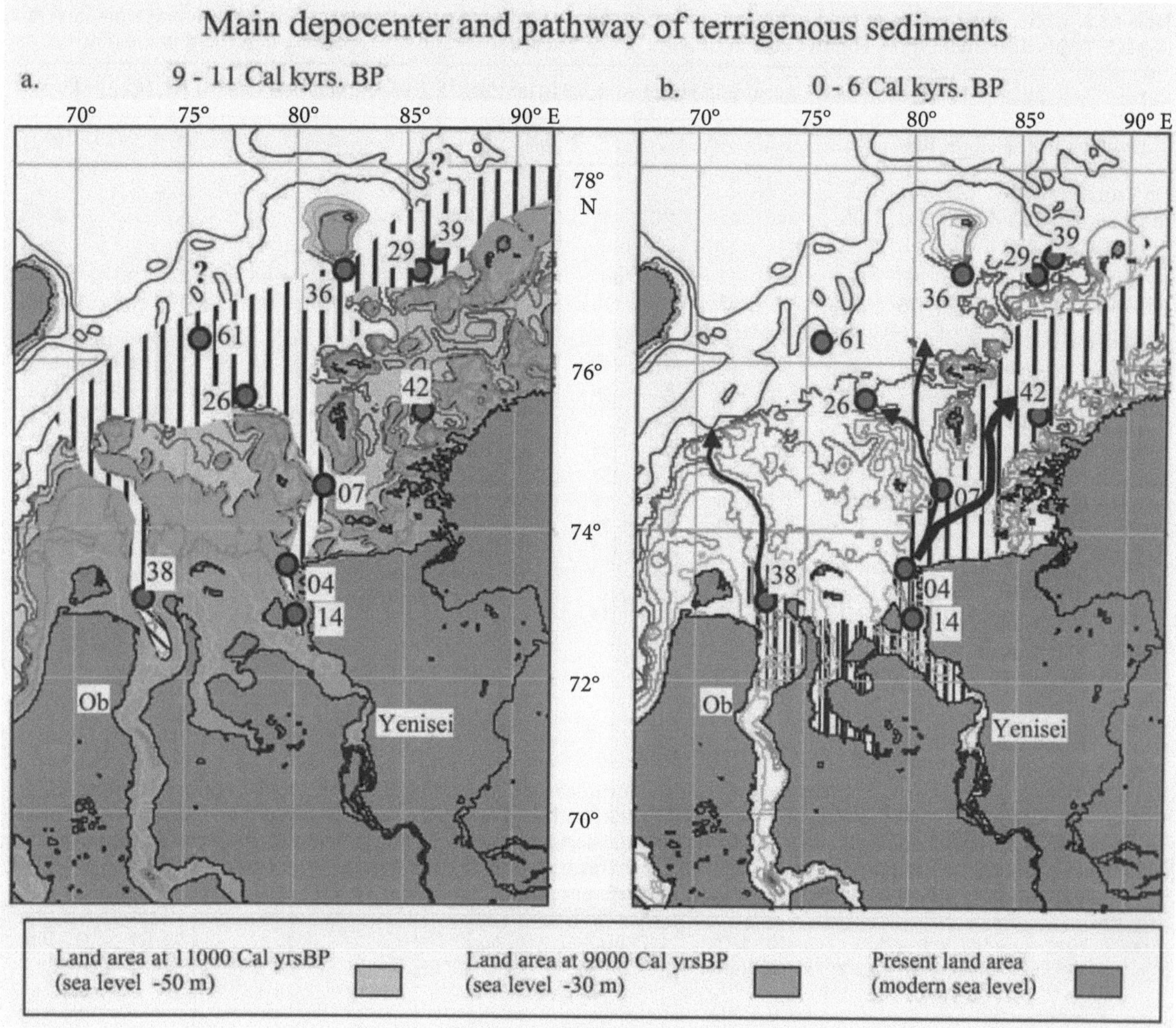

Fig. 7.6.21. Main depocenter and pathways of terrigenous matter during (a) the early Holocene (9–11 Cal. kyrs. BP) and (b) the late Holocene (0–6 Cal. kyrs. BP). The southern Kara Sea with land-ocean distribution at 9000 and 11000 Cal. Yrs. BP when the sea level was approximately 30 m and 50 m, respectively, lower than today (Fairbanks 1989), is shown. The maps have been constructed simply by using the 30 m and 50 m isolines of the modern bathymetry (and assuming no or negligible isostatic rebound); thus should be used as first approximation. Gray circles with numbers indicate location of cores for which records are presented in Figures 7.6.19 and 7.6.20 (04: BP99-04/7; 38: BP00-38/2, etc.). Black arrows and triangles indicate pathways of terrigenous input by rivers and coastal erosion, respectively. The main depocenter is marked as hatched area

7.6.5 Budget of Total Sediment and Organic Carbon

As for the Laptev Sea (see Chapter 7.5), a budget of total sediment and total organic carbon was calculated for the entire Holocene (0–11 Cal. kyrs. BP) as well as for specific Holocene time intervals (Tables 7.6.2 to 7.6.6). The latter was possible due to the large number of AMS^{14}C sediment cores available for the southern Kara Sea (Table 7.6.2). For the calculation of terrigenous and marine organic carbon burial rates we used average relative concentration values as summarized in Table 7.6.4. Based on the discussion in Chapter 7.6.3, one should bear in mind the difficulty in providing accurate estimates of the marine and terrigenous organic carbon proportions.

Table 7.6.2. Accumulation rates of total sediment (TS) and total organic carbon (OC) determined for different time intervals in AMS^{14}C dated sediment cores. For area code see Figure 7.6.22. The database and references are described in the text

Core	Lat. N	Long. E	Area	Accumulation rates of TS (g cm^{-2} ky^{-1})						Accumulation rates of OC (g cm^{-2} ky^{-1})					
Time intervals (Cal. Kyrs. BP)				0–2	2–6	6–7.5	7.5–9	9–10	10–11	0-2	2-6	6-7.5	7.5-9	9-10	10-11
Ob Estuary															
BP00-38	73,197	73,239	Ob	100	35	35	35	35	175	1,2	0,3	0,3	0,3	0,3	1,8
Yenisei Estuary															
BP00-14	72,931	79,790	Yen	65	12	nd	nd	nd	nd	0,9	0,2	nd	nd	nd	nd
BP99-04	73,415	79,675	Yen	50	35	70	110	170	nd	0,8	0,4	0,7	1,5	3	nd
Southern Kara Sea															
BP00-07	74,658	81,141	S4	35	30	40	240	nd	nd	0,5	0,5	0,5	3	nd	nd
BP01-42	75,437	86,062	S5	35	55	28	30	30	nd	0,45	0,6	0,27	0,32	0,35	nd
Central Kara Sea															
BP00-26	75,709	77,960	C4	17	17	17	25	30	200	0,2	0,2	0,2	0,2	0,2	1,5
BP01-61	76,215	75,886	C4	7	7	30	37	37	160	0,05	0,05	0,22	0,35	0,4	1,5
BP00-29	76,936	85,763	C7	8	8	8	8	35	150	0,06	0,06	0,06	0,08	0,35	1,8
BP00-36	76,962	81,963	C7	7	7	7	70	70	500	0,05	0,05	0,05	0,4	1,9	4
BP01-39	77,112	86,748	C7	9	9	32	32	32	700	0,08	0,08	0,35	0,41	0,38	9,8
St. Anna Trough															
PL94-67	78,300	70,000	N1	5	9	42	15	71	23	0,09	0,15	0,71	0,2	0,92	0,28
Easternmost Kara Sea															
PS2719	77,608	97,538	C7	8	8	10	40	95	95	0,1	0,1	0,1	0,5	1,2	1,2

Table 7.6.3. Average Holocene (0–11 Cal. kyrs. BP) values for linear sedimentation rate (LSR), total organic carbon (OC), dry density (DD), accumulation rates of total sediment (TS AccR) and total organic carbon (OC AccR), budget of total sediment (TS Budget) and total organic carbon (OC Budget), and total accumulation of total sediment (TS Tot. Acc.) and total organic carbon (OC Tot. Acc.) during the last 11 Cal. kyrs. BP. For area code and size see Figure 7.6.22 (for areas Ob-1 to Ob-4 and Yen-1 to Yen-4 see Dittmers et al. 2003). The database and references are described in the text

Area	Size (10^3 km^2)	Hol. Sed. thickness (m)	LSR (cm ky^{-1})	OC (%)	DD (g cm^{-3})	TS AccR (g cm^2 ky^{-1})	OC AccR (g cm^2 ky^{-1})	TS Budget (10^6 t y^{-1})	OC Budget (10^6 t y^{-1})	TS Tot. Acc. (10^9 t)	OC Tot. Acc. (10^9 t)
Ob Estuary											
Ob-1	2,5	17,5	159	2	0,9	143,1	2,862	3,578	0,072	39,353	0,787
Ob-2	3,8	12,5	114	1,5	0,9	102,6	1,539	3,899	0,058	42,887	0,643
Ob-3	5,9	7,5	68	1,5	0,9	61,2	0,918	3,611	0,054	39,719	0,596
Ob-4	8,7	3	27	1	0,9	24,3	0,243	2,114	0,021	23,255	0,233
Total	20,9							**13,2**	**0,205**	**145,2**	**2,26**
Yenisei Estuary											
Yen-1	1,8	17,5	159	2	0,9	143,1	2,862	2,576	0,052	28,334	0,567
Yen-2	2	12,5	114	1,5	0,9	102,6	1,539	2,052	0,031	22,572	0,339
Yen-3	3,8	7,5	68	1,5	0,9	61,2	0,918	2,326	0,035	25,582	0,384
Yen-4	6,5	3	27	1	0,9	24,3	0,243	1,580	0,016	17,375	0,174
Total	14,1							**8,5**	**0,13**	**93,9**	**1,46**
Gydanskii Bay											
GB	15,1	5	45	1	0,9	40,5	0,405	**6,1**	**0,06**	**67,3**	**0,67**
Southern Kara Sea											
S1	28,1	0,3	2	0,5	0,9	1,8	0,009	0,506	0,003	5,564	0,028
S2	3,3	3,5	32	1	0,9	28,8	0,288	0,950	0,010	10,454	0,105
S3	24,4	0,3	2	0,8	0,9	1,8	0,014	0,439	0,004	4,831	0,039
S4	6,2	5	45	1,1	0,9	40,5	0,446	2,511	0,028	27,621	0,304
S5	41,9	5	45	1,1	0,7	31,5	0,347	13,199	0,145	145,184	1,597
S6	37,4	0,3	2	0,8	0,9	1,8	0,014	0,673	0,005	7,405	0,059
Total	141,3							**18,3**	**0,19**	**201,1**	**2,13**

Table 7.6.3 (continued)

Area	Size (10^3 km²)	Hol. Sed. thickness (m)	LSR (cm ky⁻¹)	OC (%)	DD (g cm⁻³)	TS AccR (g cm² ky⁻¹)	OC AccR (g cm² ky⁻¹)	TS Budget (10^6 t y⁻¹)	OC Budget (10^6 t y⁻¹)	TS Tot. Acc. (10^9 t)	OC Tot. Acc. (10^9 t)
Central Kara Sea											
C1	24,1	3,5	32	1,2	0,9	28,8	0,346	6,941	0,083	76,349	0,916
C2	4,3	1,3	12	1,2	0,9	10,8	0,130	0,464	0,006	5,108	0,061
C3	28,5	0,3	2	0,8	0,9	1,8	0,014	0,513	0,004	5,643	0,045
C4	18,1	5	45	0,8	0,9	40,5	0,324	7,331	0,059	80,636	0,645
C5	32,9	1,3	12	1	0,9	10,8	0,108	3,553	0,036	39,085	0,391
C6	13,1	0,3	2	0,8	0,9	1,8	0,014	0,236	0,002	2,594	0,021
C7	37,7	5	45	1,2	1	45	0,540	16,965	0,204	186,615	2,239
C8	10,3	0,3	2	0,5	0,9	1,8	0,009	0,185	0,001	2,039	0,010
Total	169							**36,2**	**0,39**	**398,1**	**4,33**
Northern Kara Sea											
N1	33,9	3,5	32	1,2	0,9	28,8	0,346	9,763	0,117	107,395	1,289
N2	59,6	1,3	12	1,1	0,9	10,8	0,119	6,437	0,071	70,805	0,779
N3	7,7	0,3	2	0,5	0,9	1,8	0,009	0,139	0,001	1,525	0,008
N4	81,8	0,3	2	0,5	0,9	1,8	0,009	1,472	0,007	16,196	0,081
N5	23,1	3,5	32	1,1	0,9	28,8	0,317	6,653	0,073	73,181	0,805
N6	29,3	1,3	12	1,1	0,9	10,8	0,119	3,164	0,035	34,808	0,383
N7	40,6	0,3	2	0,5	0,9	1,8	0,009	0,731	0,004	8,039	0,040
Total	276							**28,4**	**0,31**	**311,9**	**3,38**
Total (estuaries and southern to eastern, and northern Kara Sea)											
	636							**111**	**1,3**	**1221**	**14,2**
Southwestern Kara Sea											
	235							**83**	**0,83**	**913**	**9,13**
Total Kara Sea (Holocene 0–11 Cal.Kyrs. BP)								**194**	**2,1**	**2134**	**23,3**

Table 7.6.4. Average Holocene (0–11 Cal. kyrs. BP) values of relative concentrations, burial rate, and total Holocene accumulation of terrigenous (TerrOC) and marine (MarOC) organic carbon. In brackets, the values for the late Holocene (0–6 cal. kyrs. BP) are given. The database and references are described in the text

Area	Rel. Conc. (%)		Burial rate (10^6 t y⁻¹)			Total accumulation (10^9 t)		
	TerrOC	MarOC	OC	TerrOC	MarOC	OC	TerrOC	MarOC
Ob Estuary	90	10	0,21	0,19	0,02	2,26	2,03	0,23
Yenisei Estuary	95	5	0,13	0,12	0,01	1,46	1,39	0,07
Gydanskii Bay	95	5	0,06	0,05	0,01	0,67	0,64	0,03
Southern Kara Sea	85	15	0.19 (0.23)	0.16 (0.20)	0.03 (0.03)	2,13	1,81	0,32
Central Kara Sea	80	20	0.39 (0.10)	0.31 (0.08)	0.08 (0.02)	4,33	3,46	0,87
Northern Kara Sea	75	25	0.31 (0.07)	0.23 (0.05)	0.08 (0.02)	3,38	2,54	0,84
Total (estuaries and southern to eastern, and northern Kara Sea)								
			1.29 (0.80)	1.06 (0.69)	0.23 (0.11)	14,2	11,9	2,3
Southwestern Kara Sea	80	20	0.83 (0.58)	0.66 (0.46)	0.17 (0.12)	9,13	7,3	1,83
Total Kara Sea			**2.1 (1.38)**	**1.72 (1.15)**	**0.4 (0.23)**	**23,3**	**19,2**	**4,1**

Table 7.6.5. Average early Holocene (9–11 Cal. kyrs. BP) values for linear sedimentation rate (LSR), total organic carbon (OC), dry density (DD), accumulation rates of total sediment (TS AccR) and total organic carbon (OC AccR), and budget of total sediment (TS Budget) and total organic carbon (OC Budget). The total areas became reduced because part of the early Holocene Kara Sea was subaerially exposed due to the lowered sealevel. For area code see Figure 7.6.22. The database and references are described in the text

Area	Size (10^3 km^2)	LSR (cm ky^{-1})	OC (%)	DD (g cm^{-3})	TS AccR (g cm^2 ky^{-1})	OC AccR (g cm^2 ky^{-1})	TS Budget (10^6 t y^{-1})	OC Budget (10^6 t y^{-1})
Ob Estuary								
Ob	2,5	122	1,3	0,9	110	1,430	**2,8**	**0,04**
Yenisei Estuary								
Yen	1,8	190	2	0,9	170	3,400	**3,1**	**0,06**
Southern Kara Sea								
S2	1	55	1	0,9	50	0,500	0,500	0,005
S4	1,5	111	1	0,9	100	1,000	1,500	0,015
S5	10	55	1	0,9	50	0,500	5,000	0,050
Total	12,5						**7,0**	**0,07**
Central Kara Sea								
C1	24,1	111	1,2	0,9	100	1,200	24,100	0,289
C4	18,1	133	1	0,9	120	1,200	21,720	0,217
C5	32,9	55	1	0,9	50	0,500	16,450	0,165
C7a	25	390	1,3	1	350	4,550	87,500	1,138
C7b	12,5	105	1	0,9	95	0,950	11,875	0,119
Total	112,6						**161,6**	**1,93**
Northern Kara Sea								
N1	33,9	55	1,2	0,9	50	0,600	16,950	0,203
N2	59,6	22	1,1	0,9	20	0,220	11,920	0,131
N5	23,1	55	1,3	0,9	50	0,650	11,550	0,150
N6	29,3	22	1,1	0,9	20	0,220	5,860	0,064
Total	145,9						**46,3**	**0,55**
Total (estuaries and southern to eastern, and northern Kara Sea)	275						**221**	**2,7**
Southwestern Kara Sea	??						**??**	**??**

Average Holocene Budget

The calculation of the average Holocene budget is based on the distribution map of average sedimentation rates shown in Figure 7.6.22. This map was produced from Holocene sediment thickness charts (Bjelvin and Holm 1993; Gurevich 1995), sediment echograph profiling results (Stein et al. 2000; Stein 2001; Niessen and Dittmers 2002), penetration values of the gravity corer into soft (Holocene) sediments (Matthiessen et al. 1999; Stein and Stepanets 2000, 2001, 2002), AMS^{14}C-dated sediment cores (Stein et al. 2003a), and the correlation of additional sediment cores to the AMS^{14}C-dated sediment cores using magnetic susceptibility records (Dittmers et al. 2003). The study area was divided into four main domains: The Ob and Yenisei estuaries and Gydanskii Bay, the southern Kara Sea, the central Kara Sea, and the northern Kara Sea (Fig. 7.6.22). The southwestern part of the Kara Sea (see Fig. 7.6.1) has not been included in our studies. The data needed for the budget were taken from Vetrov and Romankevich (Chapter 7.7). Each of the main domains was further subdivided into subareas. For all domains, average sedimentation rates, OC values and dry density values and, as the final result, accumulation rates and total sediment and organic carbon budgets were calculated (Table 7.6.3; see Chapter 7.5.2 for formulas). For the OC accumulation rates, OC contents of sediment cores (Fig. 7.6.12) as well as the OC distribution in surface sediments (Fig. 7.6.2) were considered. The dry density values were calculated from wet bulk density values (see Chapter 7.5) obtained by multi-sensor core logging (Dittmers et al. 2003). The budget presented in this study is related to a total area of about 870,000 km^2.

Table 7.6.6. Average late Holocene (0–6 Cal. kyrs. BP) values for linear sedimentation rate (LSR), total organic carbon (OC), dry density (DD), accumulation rates of total sediment (TS AccR) and total organic carbon (OC AccR), and budget of total sediment (TS Budget) and total organic carbon (OC Budget). For area code see Figure 7.6.22 (for areas Ob-1 to Ob-4 and Yen-1 to Yen-4 see Dittmers et al. 2003). The database and references are described in the text

Area	Size (10^3 km^2)	LS (cm ky^{-1})	OC (%)	DD (g cm^{-3})	TS AccR (g cm^2 ky^{-1})	OC AccR (g cm^2 ky^{-1})	TS Budget (10^6 t y^{-1})	OC Budget (10^6 t y^{-1})	TS Tot. Acc. (10^9 t)	OC Tot. Acc. (10^9 t)
Ob Estuary										
Ob-1	2,5	159	2	0,9	143,1	2,862	3,578	0,072	21,465	0,429
Ob-2	3,8	114	1,5	0,9	102,6	1,539	3,899	0,058	23,393	0,351
Ob-3	5,9	68	1,5	0,9	61,2	0,918	3,611	0,054	21,665	0,325
Ob-4	8,7	27	1	0,9	24,3	0,243	2,114	0,021	12,685	0,127
Total	20,9						**13,2**	**0,21**	**79,2**	**1,23**
Yenisei Estuary										
Yen-1	1,8	159	2	0,9	143,1	2,862	2,576	0,052	15,455	0,309
Yen-2	2	114	1,5	0,9	102,6	1,539	2,052	0,031	12,312	0,185
Yen-3	3,8	68	1,5	0,9	61,2	0,918	2,326	0,035	13,954	0,209
Yen-4	6,5	27	1	0,9	24,3	0,243	1,580	0,016	9,477	0,095
Total	14,1						**8,5**	**0,13**	**51,2**	**0,80**
Gydanskii Bay										
GB	15,1	45	1	0,9	40	0,400	**6,0**	**0,06**	**36,2**	**0,36**
Southern Kara Sea										
S1	28,1	2	0,5	0,9	1,8	0,009	0,506	0,003	3,035	0,015
S2	3,3	32	1	0,9	28	0,280	0,924	0,009	5,544	0,055
S3	24,4	2	0,8	0,9	1,8	0,014	0,439	0,004	2,635	0,021
S4	6,2	33	1,1	0,9	30	0,330	1,860	0,020	11,160	0,123
S5	41,9	44	1,1	0,7	40	0,440	16,760	0,184	100,560	1,106
S6	37,4	2	0,8	0,9	1,8	0,014	0,673	0,005	4,039	0,032
Total	141,3						**21,2**	**0,23**	**127,0**	**1,35**
Central Kara Sea										
C1	24,1	8	1,2	0,9	7	0,084	1,687	0,020	10,122	0,121
C2	4,3	3	1,2	0,9	3	0,036	0,129	0,002	0,774	0,009
C3	28,5	2	0,8	0,9	1,8	0,014	0,513	0,004	3,078	0,025
C4	18,1	11	0,8	0,9	10	0,080	1,810	0,014	10,860	0,087
C5	32,9	3	1	0,9	3	0,030	0,987	0,010	5,922	0,059
C6	13,1	2	0,8	0,9	1,8	0,014	0,236	0,002	1,415	0,011
C7	37,7	11	1,2	1	10	0,120	3,770	0,045	22,620	0,271
C8	10,3	2	0,5	0,9	1,8	0,009	0,185	0,001	1,112	0,006
Total	169						**9,3**	**0,10**	**55,9**	**0,59**
Northern Kara Sea										
N1	33,9	6	1,6	0,9	5	0,080	1,695	0,027	10,170	0,163
N2	59,6	2	1,1	0,9	2	0,022	1,192	0,013	7,152	0,079
N3	7,7	2	0,6	0,9	1,8	0,011	0,139	0,001	0,832	0,005
N4	81,8	2	0,5	0,9	1,8	0,009	1,472	0,007	8,834	0,044
N5	23,1	6	1,1	0,9	5	0,055	1,155	0,013	6,930	0,076
N6	29,3	2	1,1	0,9	2	0,022	0,586	0,006	3,516	0,039
N7	40,6	2	0,5	0,9	1,8	0,009	0,731	0,004	4,385	0,022
Total	276						**7,0**	**0,07**	**41,8**	**0,43**
Total (estuaries and southern to eastern, and northern Kara Sea)										
	636						**65**	**0,8**	**391**	**4,8**
Southwestern Kara Sea (70 % of Holocene average)										
	235						**58**	**0,58**	**348**	**3,48**
Total Kara Sea (late Holocene 0–6 Cal. kyrs. BP)							**123**	**1,4**	**739**	**8,2**

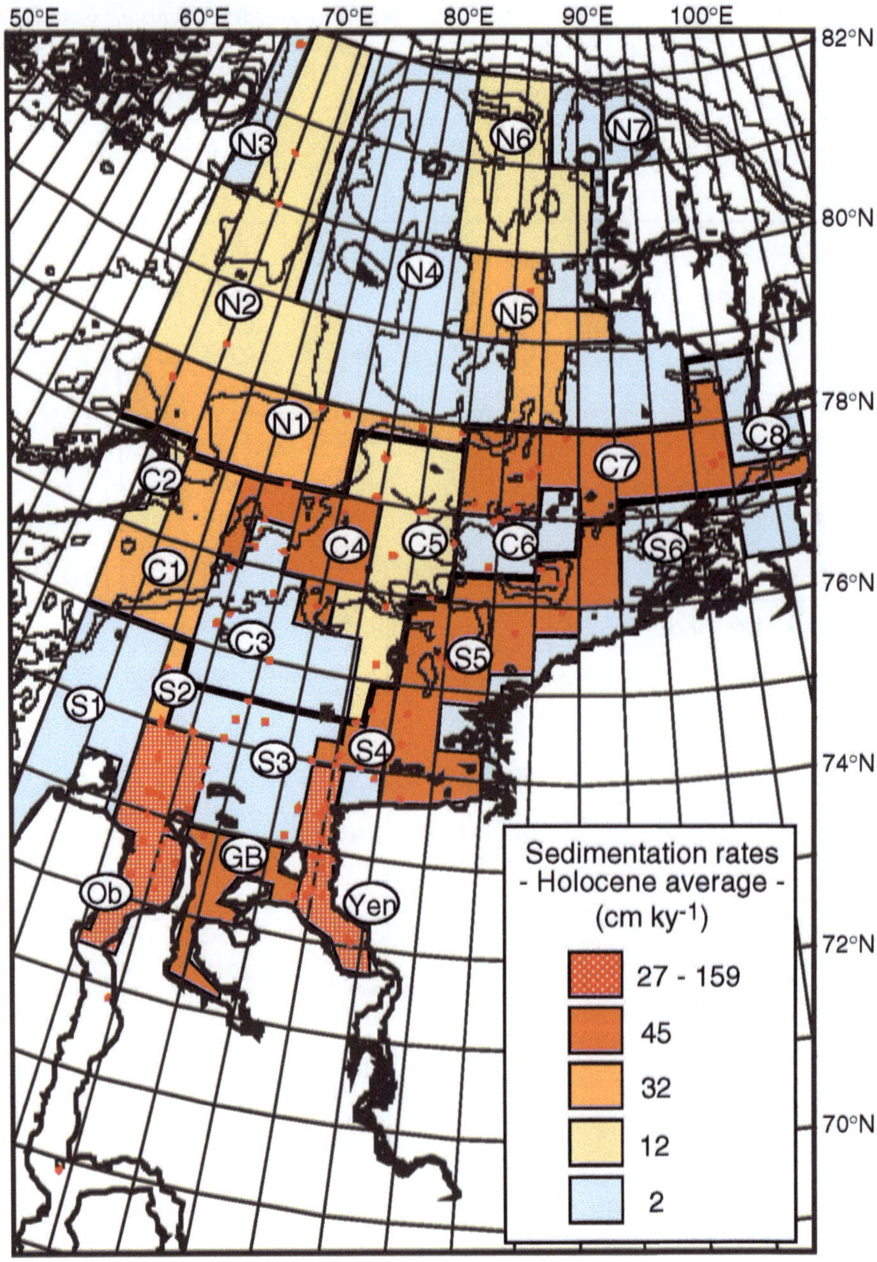

Fig. 7.6.22. Distribution map of average Holocene linear sedimentation rates in the study area. The whole area has been divided into four main domains: (1) the marginal filter zone with the Ob and Yenisei estuaries and Gydanskii Bay; (2) the southern Kara Sea with subareas S1 to S6; (3) the central Kara Sea with subareas C1 to C8; and (4) the Northern Kara Sea with subareas N1 to N7. The data base and references are described in the text (see also Table 7.6.3)

In the Ob and Yenisei estuaries, average annual total sediment accumulation is about 13.2×10^6 t y^{-1} and 8.5×10^6 t y^{-1}, respectively (Dittmers et al. 2003). In Gydanskii Bay, a part of the marginal filter, probably another 6.1×10^6 t y^{-1} of total sediment is accumulating. That means, approximately 27.8×10^6 t y^{-1} of total sediment are trapped in the marginal filter (Fig. 7.6.23), which implies a total sediment accumulation of about 306×10^9 t during the last 11 Cal. kyrs BP (Table 7.6.3). The annual burial rates for total organic carbon in the Ob and Yenisei estuaries reach values of 0.21×10^6 t y^{-1} (0.19×10^6 t y^{-1} terrigenous, 0.02×10^6 t y^{-1} marine organic carbon) and 0.13×10^6 t y^{-1} (0.12×10^6 t y^{-1} terrigenous and 0.01×10^6 t y^{-1} marine organic carbon), respectively. In the entire marginal filter system (including Gydanskii Bay) about 0.4×10^6 t y^{-1} of total organic carbon (0.36 terrigenous, 0.04 marine) are buried (Fig. 7.6.23), i.e., in total 4.39×10^9 t of OC were accumulated within the last 11 Cal. kyrs. BP (Tables 7.6.3 and 7.6.4).

In the southern, central, and northern Kara Sea, 18.3×10^6 t y^{-1}, 36.2×10^6 t y^{-1} and 28.4×10^6 t y^{-1}, respectively, of total sediment is accumulating,

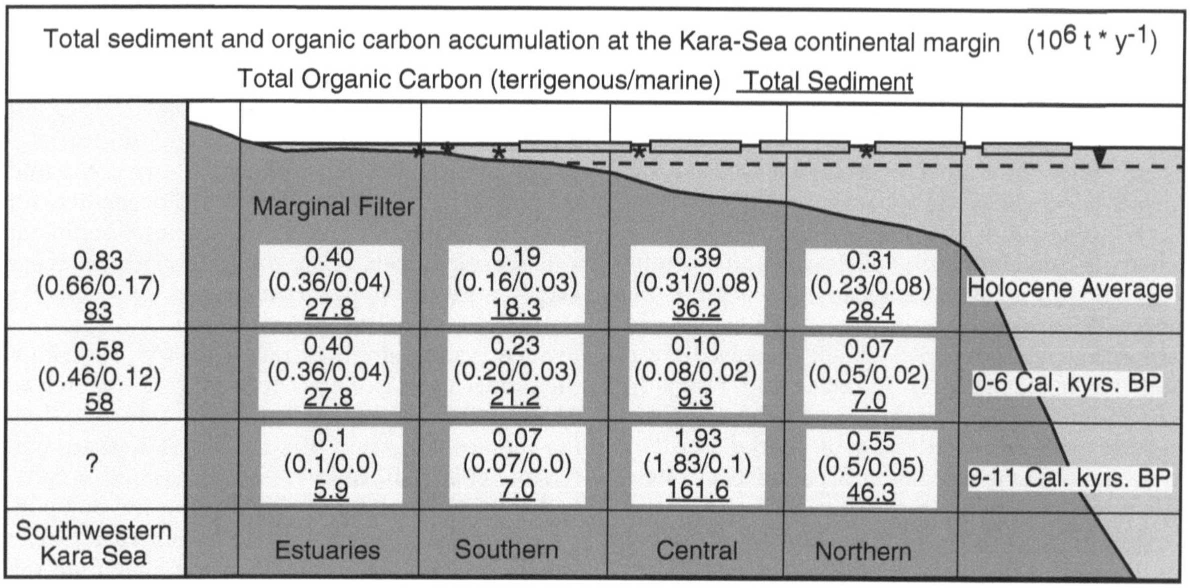

Fig. 7.6.23. Summary of total sediment and total organic carbon accumulation and burial, respectively, in the Kara Sea in units of 10^6 t y^{-1}. The data for the southwestern Kara Sea were taken from Vetrov and Romankevich (Chapter 7.7). For database and further details see Tables 7.6.2 to 7.6.6.

the corresponding burial rates for total organic carbon are 0.19×10^6 t y^{-1} (0.16×10^6 t y^{-1} terrigenous, 0.03×10^6 t y^{-1} marine organic carbon), 0.39×10^6 t y^{-1} (0.31×10^6 t y^{-1} terrigenous, 0.08×10^6 t y^{-1} marine organic carbon), and 0.31×10^6 t y^{-1} (0.23×10^6 t y^{-1} terrigenous, 0.08×10^6 t y^{-1} marine organic carbon) (Fig. 7.6.23, Tables 7.6.3 and 7.6.4).

For the estuaries and the southern, eastern and northern Kara Sea, which have been investigated in detail (see Fig. 7.6.22), a total of 111×10^6 t y^{-1} total sediment and 1.29×10^6 t y^{-1} total organic carbon are accumulating (Table 7.6.3). Splitting the latter number into its terrigenous and marine proportions, 1.06×10^6 t y^{-1} and 0.23×10^6 t y^{-1}, respectively, are obtained (Table 7.6.4). For the southwestern part of the Kara Sea not included in our study but investigated by Vetrov and Romankevich (see Chapter 7.7), a burial of total organic carbon of 0.83×10^6 t y^{-1} is given. Assuming an average OC content of 1% (see Chapter 7.7), 83×10^6 t y^{-1} of total sediment are accumulating.

For the entire Kara Sea, we estimate an average Holocene accumulation of 194×10^6 t y^{-1} of total sediment and 2.1×10^6 t y^{-1} of total organic carbon. The organic carbon burial rate can be divided into 1.7×10^6 t y^{-1} of terrigenous and 0.4×10^6 t y^{-1} marine organic carbon. Based on these numbers, a total of 2134×10^9 t of total sediment and 23.3×10^9 t of total organic carbon (19.2×10^9 t of terrigenous, 4.1×10^9 t of marine origin) were accumulated in the whole Kara Sea during the last 11 Cal. kyrs.BP (Tables 7.6.3 and 7.6.4).

Total sediment and organic carbon budgets for selected Holocene time intervals

Early Holocene (9–11 Cal. kyrs. BP)

During the early Holocene (9–11 Cal. kyrs. BP), when the sea level was 40–50 m lower than today, the main depocenter was situated much more to the north in the central Kara Sea (Fig. 7.6.21b), as clearly reflected in the accumulation rate records (Figs. 7.6.19 and 7.6.20; see above). At that time, most parts of the shallow southern Kara Sea as well as Central Kara Plateau (see Fig. 7.6.1) were subaerial, and sedimentation was restricted to the deeper areas (Fig. 7.6.22; Table 7.6.5). In the Yenisei and Ob estuaries, sedimentation probably continued in the deep central (fluvial?) channels.

Based mainly on the total sediment and total organic carbon records of the AMS^{14}C dated sediment cores (Figs. 7.6.19 and 7.6.20), a budget calculation was performed (Table 7.6.2). Whereas the southern Kara Sea was of very minor importance as an accumulation area, huge amounts of sediments and organic carbon were deposited in the north (Fig. 7.6.21b). Maximum rates were calculated for the central Kara Sea, where about 162×10^6 t y^{-1} of total sediment and 1.9×10^6 t y^{-1} of total organic carbon were accumulating (Fig. 7.6.23). These values are more than four times higher than the Holocene average values for the central Kara Sea. In the northern Kara Sea, the burial rates reached values of about 46×10^6 t y^{-1} and 0.55×10^6 t y^{-1}, respectively, which is still two times higher than the Holocene average for this area. In total, 221×10^6 t y^{-1} of total sediment and 2.7×10^6 t y^{-1} of total organic carbon were accumulating in the southern, central and northern Kara Sea (Table 7.6.5). This is almost exactly twice the Holocene average. Based on the organic-geochemical bulk parameters (C/N ratios and HI values) as well as maceral data, the organic matter is predominantly of terrigenous origin (Figs. 7.6.13 to 7.6.18). That means, at that time the supply of terrigenous sediment as well as organic carbon must have been doubled. Increased river discharge into the central Kara Sea and, probably even more important, proximity to the coastline together with increased rates of coastal erosion, have caused these maximum accumulation rates. The same evolution has been described for the Laptev Sea (e.g., Stein et al. 2001; Bauch et al. 2001b; see Chapter 7.5). Parts of the organic material were probably more refractive/reworked as suggested from higher Tmax values determined in the early Holocene samples (Figs. 7.6.13 and 7.6.16).

Late Holocene (0–6 Cal. kyrs. BP)

During the last 6 Cal. kyrs. BP, approximately modern conditions can be assumed. The main depocenter was situated within the marginal filter and in the southeastern part of the Kara Sea (Fig. 7.6.21a). Within the marginal filter zone, i.e., the Ob and Yenisei estuaries and Gydanskii Bay, we assume the same values of accumulation of total sediment and total organic carbon as given as Holocene average (Table 7.6.3). The main difference between the early and late Holocene situations occurs in the southern, central, and northern Kara Sea (Fig. 7.6.23, Table 7.6.6). Whereas the accumulation of total sediment and total organic carbon increased in the southern Kara Sea reaching values of 21.2×10^6 t y^{-1} and 0.23×10^6 t y^{-1}, respectively, the accumulation drastically decreased in the central and northern parts of the Kara Sea. For the central Kara Sea, the burial rate of total sediment and total organic carbon were no more than 9.3×10^6 t y^{-1} and 0.1×10^6 t y^{-1}, respectively. The corresponding values for the northern Kara Sea are 7×10^6 t y^{-1} and 0.07×10^6 t y^{-1} (Fig. 7.6.23, Table 7.6.6).

In total, 65×10^6 t y^{-1} of total sediment and 0.8×10^6 t y^{-1} of total organic carbon (i.e., 0.69×10^6 t y^{-1} of terrigenous, 0.11×10^6 t y^{-1} of marine origin) accumulate annually in the southern, central and northern Kara Sea. For the southwestern Kara Sea, values of about 70% of the Holocene average have been assumed (Table 7.6.6; see Chapter 7.7). That means, for the whole Kara Sea a total accumulation of 123×10^6 t y^{-1} of total sediment and 1.38×10^6 t y^{-1} of total organic carbon were calculated. The latter value can be separated into 1.15×10^6 t y^{-1} of terrigenous and 0.23×10^6 t y^{-1} of marine organic carbon.

Mass Balance Calculation of Input and Output

As for the Laptev Sea (Chapter 7.5), we compared the budget calculated for the late Holocene interval with the modern input values of total sediment and organic carbon (Fig. 7.6.24). Terrigenous input values were taken from Chapter 2, with one exception. For the Yenisei, we have used a total suspended matter supply of 14.4×10^6 t y^{-1} instead of the modern value of 4.7×10^6 t y^{-1} (see Chapter 2). The higher value was measured at a time preceding the 1967 construction of a dam near Krasnoyarsk (Telang et al. 1991; Holmes et al. 2002) and, accordingly, is the more appropriate value for our purpose. For the input of particulate organic matter we also used a value three times higher than the modern (post-1967) one (i.e., about 0.5×10^6 t y^{-1}). For coastal erosion input, we have used the published value of 109×10^6 t y^{-1} (Romankevich and Vetrov 2001). According to first unpublished results from ongoing studies on coastal erosion in the Kara Sea, this value (as well as the terrigenous OC input) might have to be reduced by a factor of 2 to 3 (A. Vasiliev, pers. comm. 2003). We also lack on quantitative data of sediment and OC input from the Barents Sea via Atlantic water inflow, which is assumed from the organic carbon distribution in surface sediments (see Fig. 7.6.2). Future research is needed to fill these gaps. Furthermore, it has to be mentioned that the input of dissolved organic and inorganic matter is not included in the model.

Total Sediment

The input of total sediment via winds (0.55×10^6 t y^{-1}), rivers (40.6×10^6 t y^{-1}), and coastal erosion (109×10^6 t y^{-1}) can be summed up to about 150×10^6 t y^{-1}. In the marginal filter zone, about 27.8×10^6 t y^{-1} are accumulating. Assuming that coastal erosion is of minor significance in the protected estuaries, and that riverine input is predominant (see Dittmers et al. 2003), suggests that during the last 6 Cal.kyrs. BP about 70% of the input of riverine suspended matter has been trapped in the marginal filter. This value is lower than that given by Lisitzin (1995) who estimated – based on recent measurements of suspended matter concentration – that 90% of the suspended matter is accumulating in the estuaries. This value would become even lower if the dissolved inorganic matter is considered in the budget calculation (Dittmers et al. 2003). The accumulation rate records from the estuaries, however, indicate a further increase during the last 2 Cal. kyrs. BP (Figs. 7.6.19 and 7.6.20)

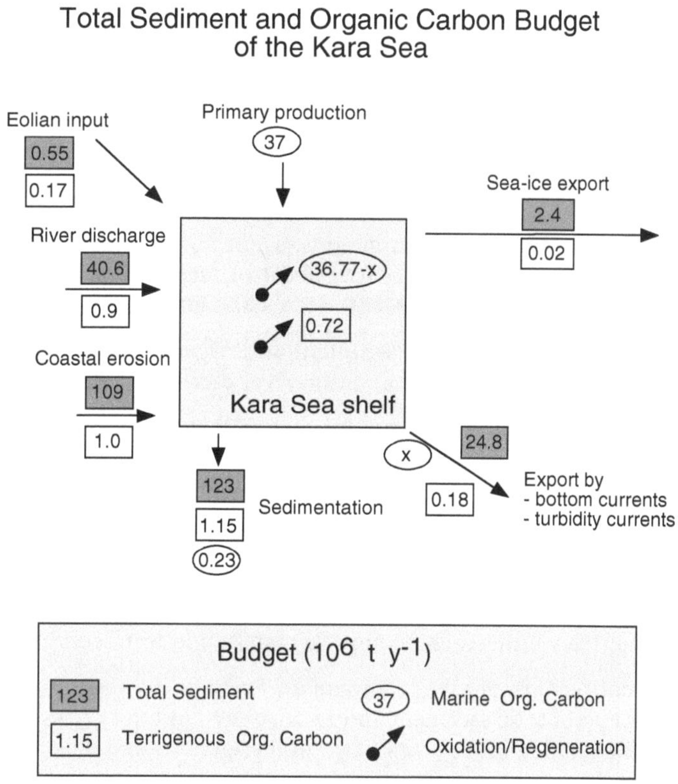

Fig. 7.6.24. Box model of the total sediment and particulate organic carbon budget for the Kara Sea. For data source see Figure 7.6.23 and Tables 7.6.2 and 7.6.6. For comments on coastal erosion values see text.

which may suggest a higher accumulation within the marginal filter. On the open Kara Sea shelf, about 95.2×10^6 t y^{-1} of total sediment are accumulating. This results in a total of 123×10^6 t y^{-1} for the entire Kara Sea including the estuaries, which is 82% of the total terrigenous sediment input. 24.8×10^6 t y^{-1} (or 17% of the total input) of sediments are exported via currents towards the slope and deep sea, and another 2.4×10^6 t y^{-1} (or <2% of the total input) are exported via sea ice (Fig. 7.6.24).

Terrigenous organic carbon

In the Kara Sea, the annual supply of terrigenous particulate organic carbon (POC) by rivers is about 0.9×10^6 t y^{-1} using the "pre-dam" value of 0.5×10^6 t y^{-1} for the Yenisei River (see Chapter 2; Telang et al. 1991; Holmes et al. 2002). The organic carbon input by coastal erosion is estimated to reach about 1×10^6 t y^{-1}, most of it probably related to particulate organic carbon (Romankevich et al. 2000). This value, however, may overestimate the coastal erosion input, as suggested from new ongoing studies (see above). The eolian organic carbon input is about 0.17×10^6 t y^{-1} (see Chapter 2). Based on published data, we estimate a total input of terrigenous POC of about 2.07×10^6 t y^{-1} for the budget calculation. As for the Laptev Sea budget, we assume that 35% of this terrigenous POC (i.e., 0.72×10^6 t y^{-1}) belongs to the labile (metabolizable) fraction and becomes rapidly oxidized (see Ittekkot 1988). We have not considered the terrigenous dissolved organic carbon (DOC), most of which is assumed to transit the shelf and enter the interior ocean (see discussion in Chapter 4).

For the whole Kara Sea shelf, 1.15×10^6 t y^{-1} (about 56% of the input) are buried. About one third (0.36×10^6 t y^{-1}) is trapped within the marginal filter zone. Export by sea ice seems to be of very minor significance (0.02×10^6 t y^{-1} or about 1% of the input). The remaining 0.18×10^6 t y^{-1} of terrigenous organic carbon (about 9% of the initial input) can be exported towards the interior ocean via ocean and turbidity currents.

Marine organic carbon

The database on primary production for the Arctic shelves, especially the Kara Sea, is sparse. Estimates of annual primary production ranging between 30 and 50 gC m^{-2} y^{-1} are given in Chapter 3 (Table 3.7). Such levels of primary production would imply a total marine organic carbon input of 28 to 65×10^6 t y^{-1} (average of about 37×10^6 t y^{-1}). In the Kara Sea sediments, only 0.23×10^6 t y^{-1} (or 0.6% of the primary production) are buried which implies that >99% of the marine organic carbon is remineralized or exported.

7.6.6 Summary and Concluding Remarks

- The Kara Sea surface sediments display organic carbon (OC) contents between 0.2% and up to 3%. The higher values occur in the Ob and Yenisei estuaries, submarine channels, the St. Anna and Voronin troughs, and the areas close to Franz Josef Land. The organic carbon preserved in the sediments is predominantly of terrigenous origin. Towards the open Kara Sea, with increasing distance from the estuaries, a decrease in amount of terrigenous matter is indicated by decreasing C/N ratios, increasing $\delta^{13}C_{org}$ values, and decreasing long-chain *n*-alkanes.
- The post-glacial to Holocene sediment and organic carbon input have been strongly influenced by sea-level rise and related changes in river discharge and coastal erosion.
- In the earliest Holocene (10–11 Cal. krs. BP), maximum accumulation rates were recorded in the central Kara Sea. Accumulation rates of total sediment and total organic carbon reach very high values of about 150 to 700 g cm^{-2} ky^{-1} and about 1.5 to 10 g cm^{-2} ky^{-1}, respectively. At that time, the sea level was about 40 to 50 m lower than today and large parts of the inner Kara Sea were subaerial. Terrigenous sediments including organic matter were transported much more to the north, and the main depocenter was situated more toward the central Kara Sea. With increasing sea level, sedimentation in the central Kara Sea became drastically reduced, and the main depocenter shifted towards the south, reaching approximately "modern" condition during the last 6 Cal. kyrs. BP.
- For the entire Kara Sea, we estimate an average Holocene (0–11 Cal. kyrs. BP) accumulation of 194×10^6 t y^{-1} of total sediment and 2.1×10^6 t y^{-1} of total organic carbon. The organic carbon burial rate can be divided into 1.7×10^6 t y^{-1} of terrigenous and 0.4×10^6 t y^{-1} marine organic carbon.
- For the late Holocene time interval (0–6 Cal. kyrs. BP), a mass balance for total sediments implies that about 123×10^6 t y^{-1} (82% of the initial input) is accumulating on the Kara Sea shelf and 24.8×10^6 t y^{-1} (or 17% of the input) and 2.4×10^6 t y^{-1} (or <2% of the total input) are exported by currents and sea ice, respectively, towards the interior ocean. For terrigenous organic carbon, 1.15×10^6 t y^{-1} (about 56% of the initial input) is buried on the Kara Sea shelf, and 0.18×10^6 t y^{-1} (about 9% of the input) is exported towards the interior ocean via currents. The export via sea ice is <1% of the input. During the late Holocene, only 0.23×10^6 t y^{-1} (or 0.6% of the primary production) are buried in the sediments which implies that >99% of the marine organic carbon is remineralized or exported.
- We still lack crucial measurements of total sediment and organic carbon inputs by coastal erosion, and sediment core data from the northern Kara Sea and adjacent continental margin. Future research focussed on these aspects should help to fill these gaps.

7.7 The Barents Sea: Distribution, Sources, Variability and Burial of Organic Carbon

A. VETROV, E. ROMANKEVICH

7.7.1 Introduction

The Barents Sea (see Fig. 7.1.9), 1424×10^3 km^2 in size, wholly occupies the westernmost shelf of the Eurasian Arctic. Its average depth is 222 m but the bottom relief is characterized by a complex alternation of numerous elevations and troughs, the deepest of which is the Bear Island Trough (600 m).

The climate of the Barents Sea reflects its location within the Arctic and the warming effect of the inflowing water from the North Atlantic. Interaction of Atlantic and Arctic air masses leads to frequent storms which, together with large areas of open water, produce wave heights up to 8 m in the coastal zone. On average, 5–6 storm days per month occur in the Barents Sea during winter, 1–2 during summer, and 12-15 during autumn.

The exchange of water with the Atlantic Ocean strongly influences the hydrology and hydrochemistry of the Barents Sea. Warm, salty water (4–12 °C, 35) from the North Atlantic flows into the Barents Sea with the Nordkapp current which splits into two branches, one proceeding along the Murmansk coast into the White Sea and the

Pechora Sea, and the other turning northeastwards where it submerges under the cooler and fresher Arctic water. Warm Atlantic water and cold Arctic water form a complex current system with a characteristic cyclonic circulation.

Despite its location north of the Arctic Circle, the southwestern part of the Barents Sea remains free from ice throughout the year due to heat transported with Atlantic water masses. During winter only a few bays along the Murman coast freeze, and the drift-ice edge stops at Cap Kanin Nose. A long growth period (in comparison with the other Arctic shelf seas), together with strong winds and currents to aerate water and supply nutrients from deep water into the euphotic zone promote active photosynthesis in the Barents Sea during the polar day. With a mean annual productivity of 110 mg C m^{-2} per day, the Barents Sea can be classified as mesotrophic. In the Atlantic water masses, productive regions are located in the southwestern Barents Sea (250–330 mg C m^{-2} per day, Vedernikov and Demidov 1997; Slagstad and Stole-Hensen 1991), near the Murman Peninsula (500–600 mg C m^{-2} per day, Bobrov 1985) and in a small region south of the Franz Josef Archipelago. For Arctic water masses, typical primary production is estimated to be 50–140 mg C m^{-2} per day (Slagstad and Stole-Hensen 1991). Photosynthesis occurs in the water column (phytoplankton), in the ice (Melnikov 1989; Druzhkov et al. 2001; Gosselin et al. 1997), in macrophytes (Blinova 1969; Makarov and Shoshina 1986; Makarov 1998) and in microphytobenthos at the littoral zone and on the seafloor (Kuznetsov 1991; Kucheruk et al. 1999).

The Barents Sea receives relatively small riverine inflow (163 km^3 y^{-1}) compared to the Siberian Arctic seas (see Chapter 2.2) with the result that over the Barents shelf salinity differs little from the average oceanic salinities of source water. The Pechora River (130 km^3 y^{-1} on average) supplies most of the runoff with many of the remaining rivers completely freezing in winter. Somewhat offsetting the small discharges, rivers of the Barents and White seas contain greater average concentrations of DOC + POC (13.8 mg l^{-1} with 90% in dissolved form) than other Eurasian Arctic rivers. The high DOC content comprises fulvic and humic substances supplied by inland swamps. Most of the organic matter supplied by the Pechora River sediments is settled in Pechora Bay. Highest DOC is observed in waters of Onega and North Dvina rivers discharging into the White Sea (see Chapter 2.2).

Coastal erosion is another important terrigenous source of organic carbon (OC) in the Barents Sea (see Chapter 2.3). Low-relief shores, widespread permafrost, large tidal ranges (1–3 m at most locations and up to 5 m in narrow bays) and a general rise in sea level (1–2.5 mm y^{-1}) during the past century all promote coastal erosion.

Eolian input of organic carbon into the Barents Sea amounts to about one quarter to almost one half of the river input of POC (Shevchenko et al. 1999; Romankevich et al. 2000, 2000a; Shevchenko 2000; see Chapter 2.5). The outlet glaciers of Novaya Zemlya, Svalbard and Franz Josef Land are only minor sources of organic carbon.

7.7.2
Data Base, Materials and Methods

The distribution of OC in surface sediments of the Barents Sea was first reviewed in the 1960s (Klenova 1960, 1961; Lapina and Belov 1961). Landmark biogeochemical studies of surface sediments in the seas of the former Soviet Union (Gorshkova 1970; 1974) were followed by other more general studies and reviews (see, for example, Romankevich et al. 1982; Danyushevskaya et al. 1978; 1985; 1990). During the 1990s there has been an increase in the number and complexity of programs to determine composition, fluxes, distribution and accumulation of organic matter in the Barents Sea region (see Gurevich 1995; Kosheleva and Yashin 1999; Petrova 2000; Petrova et al. 2001; Romankevich and Vetrov 2001).

Over many years, data from the Shirsov and other institutes on various forms of carbon in Holocene bottom sediments have been collated into the "Carbon" databank at Ocean Chemistry Laboratory of the P.P. Shirsov Institute of Oceanology. Presently, the data include percent concentrations of OC, $CaCO_3$ and opal in the bottom sediments, sedimentation rates, physical properties and lithological descriptions of sediments. The databank PANGAEA also has been used to obtain organic carbon data from the Barents Sea.

OC in bottom sediments from the Barents Sea has been measured by various analytical methods, namely Knopp's method during the 1950s–1970s (Uspensky et al. 1966) and later by a method using high temperature combustion in oxygen flow with a LECO or Soviet analog instruments (AN 7529 or the more sensitive AN 7560). Reasonable agreement between the two methods (8–10% difference) is observed.

7.7.3
Distribution and Sources of Organic Carbon in Surface Sediments

OC content and its accumulation rate in the sediments depend on a number of factors including; i) photosynthetic organic matter flux to the sea-

Table 7.7.1. Organic carbon contents of surface sediments and sediment types of the Barents, Pechora and White seas (% per dry sediment)

Sea	Type of sediment	Organic carbon (%)				
		N	Min	Max	Average	St.dev.
Barents	Sand	121	0.02	1.75	0.36	0.32
	Silt	127	0.19	2.96	1.17	0.61
	Pelite	250	0.09	3.28	1.38	0.67
	All sediments	498	0.02	3.28	1.08	0.72
Pechora	Sand	72	0.02	0.70	0.22	0.16
	Silt	14	0.19	1.18	0.68	0.25
	Pelite	30	0.09	2.01	1.04	0.58
	All sediments	116	0.02	2.01	0.49	0.48
White	Sand	3	0.14	2.22	1.14	1.04
	Silt	8	0.80	2.01	2.37	0.48
	Pelite	48	0.29	4.37	2.02	0.95
	All sediments	59	0.14	4.37	1.89	0.94

floor, ii) terrigenous organic carbon fluxes, iii) granulometric composition of the bottom sediments and iv) efficiency of organic matter burial and preservation. The cumulative effect of these factors on OC accumulation in sediments of the Barents Sea varies greatly producing a patchy, mosaic distribution of OC.

OC content in the upper layer of the surface sediments of the Barents Sea (Table 7.7.1) ranges from 0.02% (coarse sand) to 3.3% (pelite) with an average of 1.08%. This value falls between averages calculated for the Pechora Sea (0.49%) and the White Sea (1.89%). In the Barents Sea minimal, maximal and average OC concentrations increase from sand (average 0.26%) to silt (average 1.17%) to pelite (average 1.38%) suggesting that a major part of the organic matter (OM) is sorbed to clay mineral + oxide surfaces which are sorted during particle transport.

The Barents, White and Chukchi seas are the most productive regions among the Eurasian Arctic seas, which is reflected in the higher gradient of OC from sand to silt to pelite. In contrast, pelites from low production regions contain less OM. Apparently the enormous amounts of DOC supplied by Siberian rivers do not support significant carbon fluxes into sediments possibly due to slow DOC turn-over rates leading to wider transport.

An OC distribution map of surface sediments (0–5–10 cm) in the Barents, northern White and southwestern Kara seas (Fig. 7.7.1) has been compiled based on earlier published maps (Romankevich et al. 1982; Danyushevskaya et al. 1990; Gurevich 1995; more than 1000 sites) and the "Carbon" database (~600 sites including data obtained during Russian expeditions into the Barents and Kara seas between 1993 and 1998). The map has been constructed by taking into account bottom topography, grain-size distribution, and hydrological conditions. Mapping was accomplished by computer programs originally developed to forecast expected OC for regions where measurements were scarce or absent (see Romankevich and Vetrov 2001).

The OC distribution in the Barents Sea sediments depends strongly on sediment grain-size and bottom topography with greatest OC concentrations (~3%) found in clayey sediments of the trenches. Mean OC content in the clayey (1.2–1.4%) and sandy (0.2–0.4%) sediments are close to those observed for sediments of the adjacent continental margin (Danyushevskaya et al. 1990). In the Barents Sea, enhanced wave and current activity favor mechanical differentiation of the sedimentary matter which then defines the distribution of grain size and OC.

Tides reach 3 m near the Scandinavian coast, 4–5 m in the narrow bays of the Barents Sea, 7 m in the Throat (Gorlo) of the White Sea, and 10 m in the mouth of the Mezene River. Tides do not exceed 1 m close to the western coast of the Novaya Zemlya and 0.5–1 m in the Kara Sea. Associated strong tidal currents lead to the deposition of mainly coarse-grained sediments in the coastal zone, whereas fine-grained sediments are transported to deep-water. As a result, in the southern open part of the Barents Sea adjacent to the continent (where intensive terrigenous supply is accompanied by biologically favorable conditions), sediments are represented mostly by sands containing little OM (OC content < 0.5 and 0.5–1.0% predominate). The muds of the troughs and depressions are the richest in OM (Fig. 7.7.1).

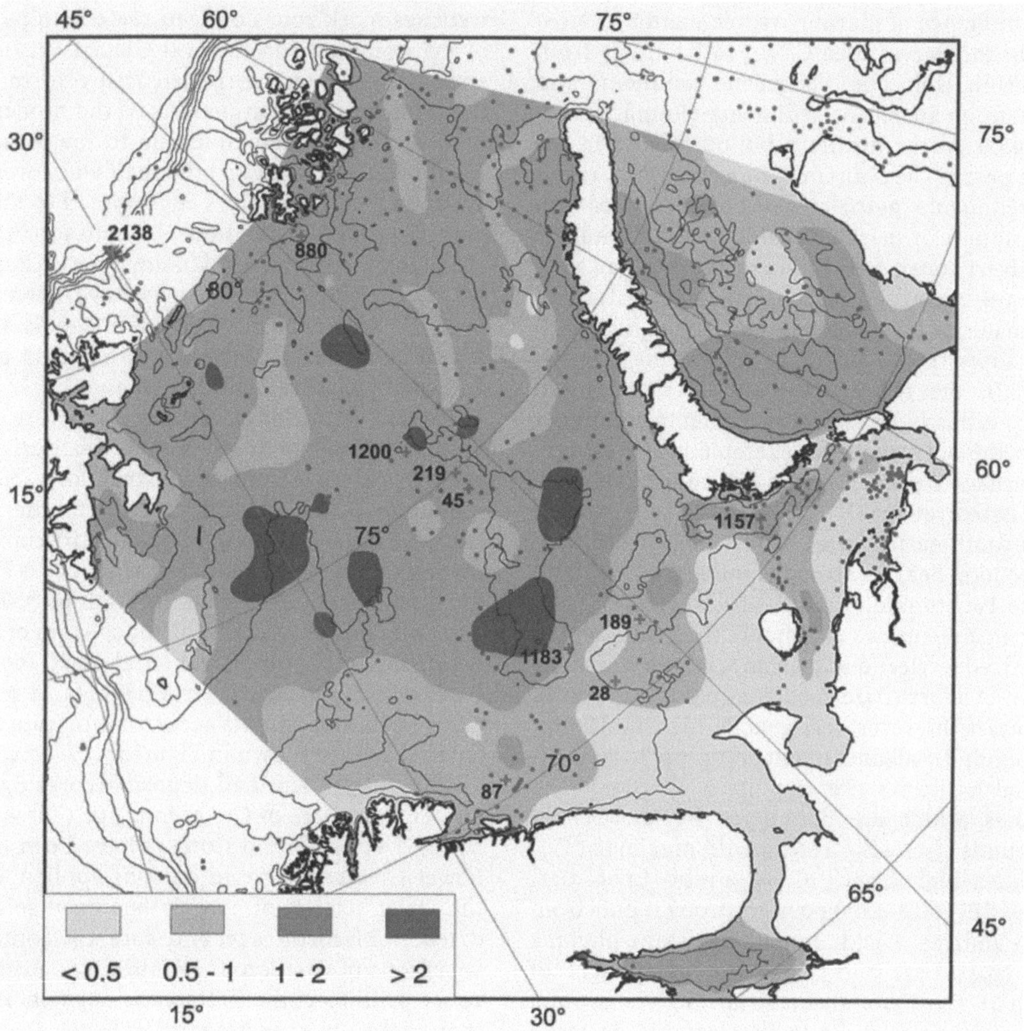

Fig. 7.7.1. The distribution of total organic carbon (OC %) in surface sediments of the Barents Sea and location of sediment cores shown in Figures 7.7.2 to 7.7.4.

OC-rich sediments (> 2%) are also located near the foot of submarine rises (for example nearby Central Bank, Great Bank) which tend to exhibit a rich development of zoobenthos. High bioproductivity is supported in these locations by contiguous cold and warm water masses (polar front) and gravitational settling of animal debris down the slope to produce OC-rich zones. High OC concentration is also observed in the back-eddies of islands (for example to the east of Kolguev Island, 1–2%).

In the south-western Kara Sea, intensive differentiation of the sedimentary matter results in the sedimentation of coarse-grained matter in the zones of high hydrodynamic activity usually at depths shallower than 100 m, while fine-grained matter accumulates in more quiescent troughs (Novaya Zemlya Trough), depressions, and topographic lows (sedimentary traps). The tidal amplitude is less in the Kara Sea than in the Barents Sea, and frigid climate promotes a longer period of ice cover which leads to much calmer water promoting sinking of fine particulate matter. The OC distribution in bottom sediments correlates with the pelitic fraction with about one half of the south-western Kara Sea area exhibiting silty clay with OC content of 1.0–1.5%. Nutrient-rich water of the northern current along Novaya Zemlya promotes bioproductivity in this region.

Sources of Organic Carbon

The group and elementary composition of bitumens and humic acids, the molecular composition of hydrocarbons, alkanes, and arenes, and the C/N ratios all indicate a mixed allochtonous-autochtonous (terrigenous/marine) origin of organic matter in sediments of the Barents Sea with a predominance of marine organic matter in most areas. In the Barents Sea C/N_{org} ratios range from 10 to 40 in Holocene-Pleistocene sediments and from 8 to 20 in recent sediments (Romankevich et al. 1982). At the northern boundary and in the middle part of the sea (Great Bank) OC/N_{org} ratios in the sediments (9–10) suggest mixed sources with high amounts of marine organic matter (Schubert and Calvert 2001). $\delta^{13}C$ values of −23.8 to −24.2‰, which are isotopically heavier than terrestrial C_3 plant material (C_4 plant are lacking in high latitude), support the presence of marine organic matter. In the regions affected by river runoff OC/N_{org} ratios in the surface sediments are higher than in the open shelf; for example in the estuary of Severnaya Dvina River the C/N ratio may reach 35±15 (Artemyev 1996).

The south-eastern part of the Barents Sea (i.e., the Pechora Sea) is strongly influenced by land and the Pechora's discharge. Nevertheless organic matter in the surface sediments from this region has a mixed terrigenous/marine origin in which the fraction of short-chain *n*-alkanes predominates. In some cases, however, terrigenous high-molecular compounds (*n*-alkanes) contribute up to 52% of the total *n*-alkanes (Peresypkin et al. 2002). The *n*-alkanes, which are dominated by short-chain compounds ($\sum C_{10}$–$C_{22} > 66\%$) with maxima at C_{16}, C_{18}, C_{20}, a high C_{10}–C_{22}/C_{23}–C_{40} ratio (2.06–3.21) and low CPI (0.26–1.01) point to strong inputs from marine sources. A wide range of pristane/phytane ratios (0.44–1.22) and significant component of mid-chain C_{18}–C_{22} *n*-alkanes (up to 45% of total *n*-alkane content) indicate that aerobic bacterial transformation of organic matter is an important source of alkanes. Usually, organic matter in the surface sediments of the Pechora Sea undergoes little or no diagenesis and is supplemented by fresh detrital matter. Most of the terrestrial matter supplied by Pechora River is trapped in Pechora Bay which is separated from the open sea by the Gulyaevskii Koshki barrier islands.

7.7.4 Late Quaternary Organic Carbon Records and Paleoenvironment

The Barents Sea shelf comprises for the most part stratum plains or, rarely, base plains covered by an uneven layer of unconsolidated or poorly-consolidated Cenozoic sediments overlapping stratas of Mesozoic and Upper Paleozoic deposits. The plains are intersected by shelf and intrashelf trenches manifested as depressions along the continuation of linear margin morphostructures reflecting structures of folded basement. The shelf trenches mark zones of deep-seated faults. Retreat of the ocean during the Last Glacial Period had a strong impact on morphostructure formation of the modern shelf. Large parts of the modern shelf became land at that time due to lower sea level. It has been proposed that the shelf was covered by a grounded ice sheet who's edge spread to the upper part of the continental slope down to 500 m (Polyak et al. 1995; Polyak and Mikhailov 1996). Alternatively, several glacial shields may have been located over the archipelagoes and some shoals, between which small sea basins were covered by pack ice and shelf glaciers (Pavlidis et al. 1998).

Present knowledge of the structure of the bottom sediments in the Barents Sea derives from geophysical data and reconstructions based on sediments of the adjacent continental margin and from sub-aquatic drilling. Mass-spectroscopic radiocarbon dating of the sediment sequence is available for a few locations (Gataullin et al. 1993; Elverhoi and Solheim 1983; Hald et al. 1999; Knies et al. 1999; Lubinski et al. 1996; Polyak et al. 1995, 1997, 2000; Polyak and Solheim 1994; Murdmaa et al. 2003). OC distribution and its accumulation rate in some dated cores are shown in Figures 7.7.2–7.7.4.

The oldest identified deposits (cores 87, 45 and 219; Fig. 7.7.2) are of Cretaceous age (Tarasov et al. 2000). Deposits from Core 45 have been dated as Lower Cretaceous based on spore/pollen analysis, the lower section of core 87 has been seismically dated to Mesozoic age, and core 219 contains associations of Cretaceous foraminifera. Average OC concentration in the Cretaceous deposits is higher than in the upper sediments.

A long-duration hiatus in sedimentation occurs during glacial times with the entire Tertiary missing in the record. The most ancient Pleistocene deposits containing a rich association of benthic foraminifera (Mikulino Interglacial period, oxygen isotope stage 5e, main stage of boreal transgression) are also characterized by higher OC (1.5%, core 45) than sediments from later periods. Identified zoogene remains indicate that the water depth during the Mikulino Interglacial period was greater than at present (Tarasov et al. 2000). The Mikulino deposits contain large amounts of pollen from woody plants (Pavlidis et al. 1992; Samoilovich et al. 1993) suggesting that forests may have spread as far north as the Barents shore (Pavlidis et al. 1998). These deposits, identified in core 87, have been dated to the lower Weichselian (OIS 5 a–d). Global cooling and a sea level drop during this period are accompanied by reduced bioproduction and OC content in the sediments.

The Weichselian glacial deposits (OIS 4–2) comprise (Fig. 7.7.2) mainly dense, dark-gray clay with

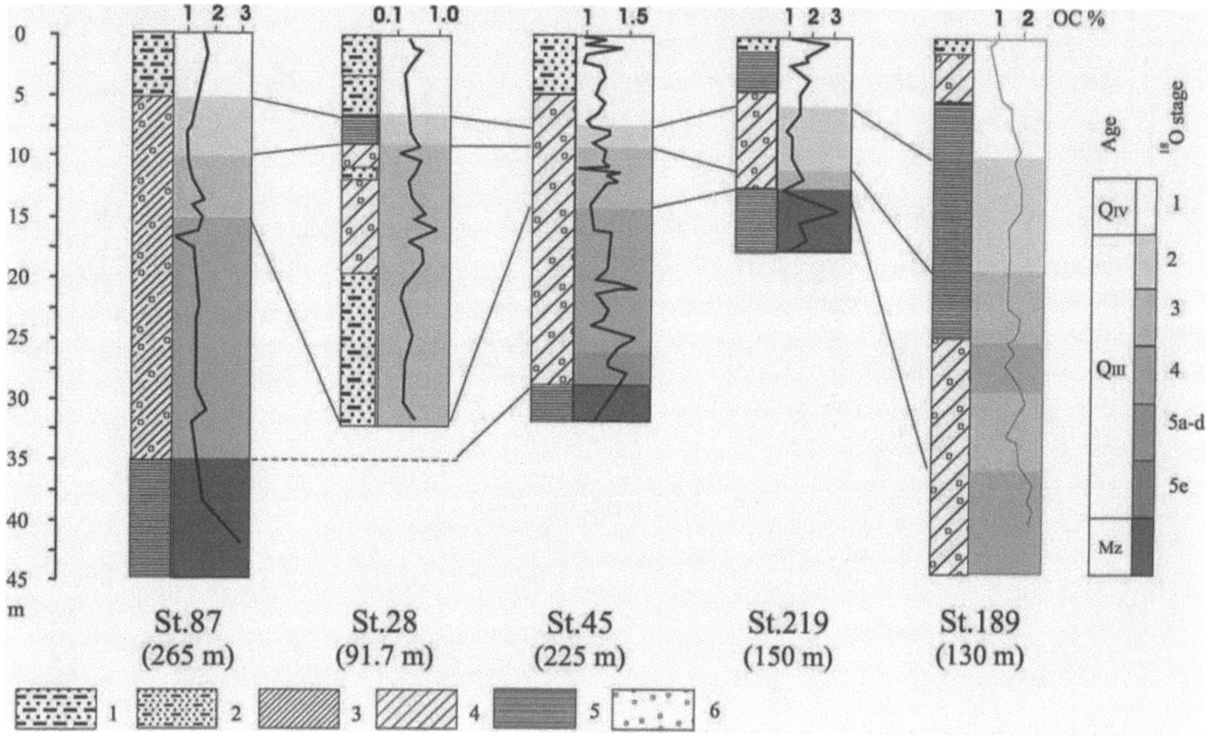

Fig. 7.7.2. Total organic carbon (OC) in selected cores from the Barents Sea (Tarasov et al. 2000). 1 – fine silty mud; 2 – silty pelite mud; 3 – moraine clay (glacial marine clay; diamicton); 4 – clay; 5 – gravel, pebble. For location of cores see Figure 7.7.1.

gravel and pebble and extremely low shell content. In the coastal regions foraminifera are poorly preserved and have been redeposited from older interglacial layers. In the central parts of the sea, however, the shell composition does not suggest redeposition. *Retroelphidium clavatum*, found in the core 189, are able to survive extreme conditions such as below-zero temperatures, low productivity, low oxygen and significant freshening. *Cassidulina reniform* and *C. teretis*, which require higher salinity, are found in the Central basin (core 45) along with *Islandiella helenae*. Increased concentration of these species is presently observed in regions characterized by high seasonal productivity and seasonal ice-cover, implying the continuous influence of Atlantic waters in the Central basin during the Weichselian (Tarasov et al. 2000). OC content varies during this period but does not differ significantly from the Holocene record. On the whole these core sections support the hypothesis that there was not a huge, solid ice-sheet over the entire Barents Sea during the Last Glacial Period.

The end of OIS 2 is associated with deglaciation. According to lithological, micropaleontological and geochemical evidence two stages of deglaciation are identified in the Barents Sea (Pavlidis et al. 1998; Polyak and Solheim 1994; Murdmaa and Ivanova 1999) – an early stage (III) and a late stage (II) (Fig. 7.7.3). The transition between these two stages is estimated to have occurred at about 13 Cal. kyrs. BP based on AMS^{14}C dates (Polyak and Solheim, 1994; Polyak et al. 1997; Hald et al. 1999). Unit III represents a proximal glaciomarine deposit that was formed during the deglaciation at a time of intensive iceberg discharge and large supply of fine-grained suspended material (glacial flour) (Murdmaa and Ivanova 1999; Duplessy et al. 2001). The dark-gray color of the mud usually indicates anoxic conditions of early diagenesis caused by a weak oxygen penetration into nearbottom water during a long ice cover period. The presence of well-preserved shells of modern species of foraminifera and pteropods indicates marine sedimentary conditions during that period. OC content was low (core 1200). At the late stage of deglaciation (unit II) the deposits are formed mainly from nepheloid fluxes containing particulate matter supplied by melting glaciers. Freshening and cold surface water did not promote bioproductivity and OC content remains at the level of the early stage of deglaciation.

In the Barents Sea, the Holocene is characterized by an absence of gravitational fluxes, a prevalence of hemipelagic and nepheloid sedimentation

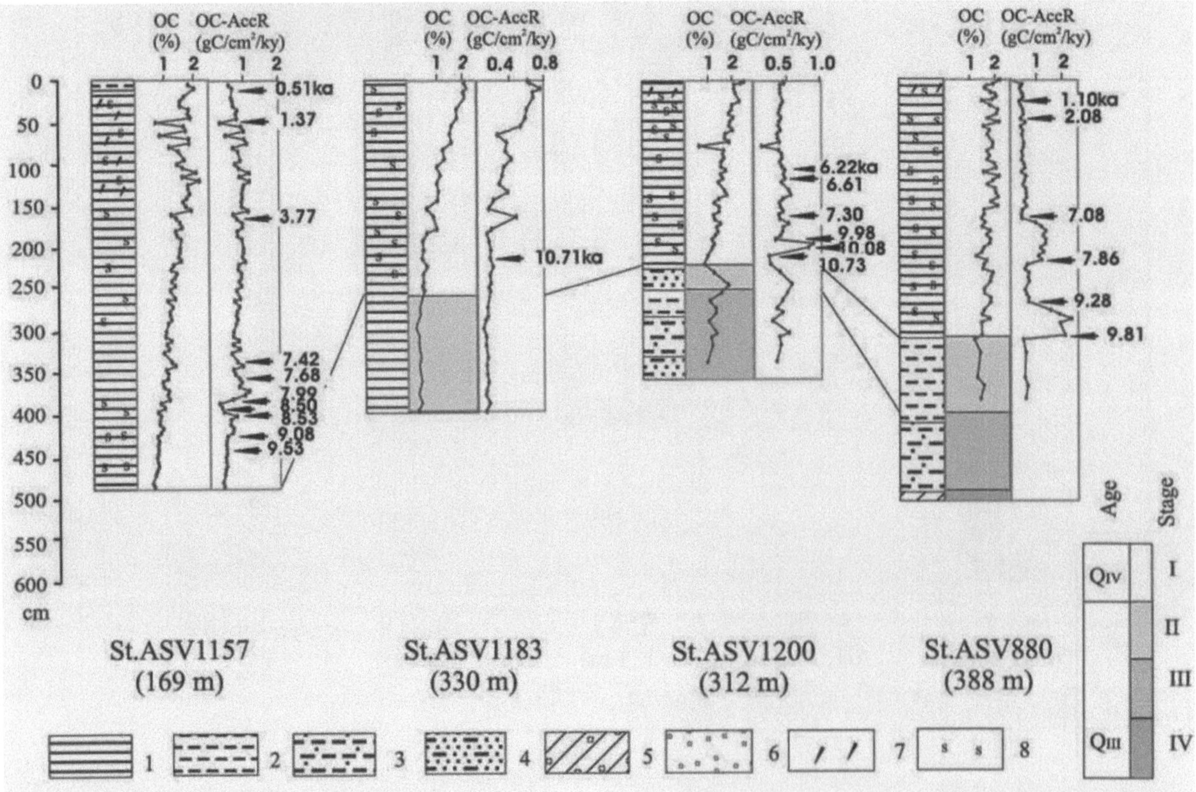

Fig. 7.7.3. Total organic carbon (OC) and its accumulation rate in selected cores from the Barents Sea (Murdmaa and Ivanova 1999; Murdmaa et al. 2003). 1 – pelite mud; 2 – silty pelite mud; 3 – sandy silty pelite mud; 4 – pebble loam (diamicton); 5 – gravel, pebble; 6 – polychaeta tubes; 7 – hydrotroilite. For location of cores see Figure 7.7.1.

(Murdmaa and Ivanova 1999), and an increase in bioproductivity and "biofilter sedimentation" of fine suspended material from surface water. Sediments have been redeposited by bottom currents from the submarine elevations into the depressions. In the early Holocene the glacial source of organic matter was still significant along with modern terrigenous sources. The Holocene is characterized by an increased inflow of warm Atlantic water masses into the Barents Sea which affected hydrological and ice conditions, stratification of water, and hence bioproductivity in various parts of the region. During the Holocene, sedimentary OC generally increases (up to more than 2%; Fig. 7.7.3) and the OC accumulation rate ranges from 0.043 to 0.5 g C cm^{-2} ky^{-1} for shallow-water sands of the southeastern part of the Barents Sea (Levitan et al. 2000), and at least 0.3 g C cm^{-2} ky^{-1} for deep-water sediments (Fig. 7.7.3). OC accumulation rate exhibits a maximum (1–1.5 g C cm^{-2} ky^{-1}) in some sequences (for example, cores 1157 and 880) during the period associated with the Holocene climatic optimum (6.9–7.85 Cal. kyrs. BP). The fact that the accumulation rate maximum is not accompanied by a maximum in OC content, implies that the climatic optimum experienced an increase in both bioproduction and supply of terrigenous material (with terrigenous OM) caused by glacier melting and thermoabrasion. Accordingly, the enhanced terrestrial flux diluted the authochtonous flux to the seafloor. A maximum of OC accumulation rate in the Early Holocene (cores 880 and 1200) is possibly associated with iceberg discharge and increase in bioproductivity supported by enhanced water exchange with the Atlantic Ocean.

Indicators of terrigenous/authigenous composition of OM in sediments (C/N, Hydrogen index, $\delta^{13}C_{org}$, $C_{17,19}/C_{27,29,31}$ n-alkane ratio) allow the determination of OM sources and conditions of its accumulation in sediments along the northern Barents Sea margin (PS2138-1, Fig. 7.7.4; Knies and Stein 1998; Knies et al. 1999). The Atlantic water has been shown to enter the Arctic over the last 150 ka (Knies et al. 1999). A peak in accumulation rate of OC between 25 and 23 Cal kyrs. BP (Fig. 7.7.4) can be explained by enhanced Atlantic water advection resulting in increased bioproductivity, and more favorable burial and preservation conditions of authochtonous organic matter due to the large flux of fine-grained, organic-rich terrigenous material supplied by retreating glaciers.

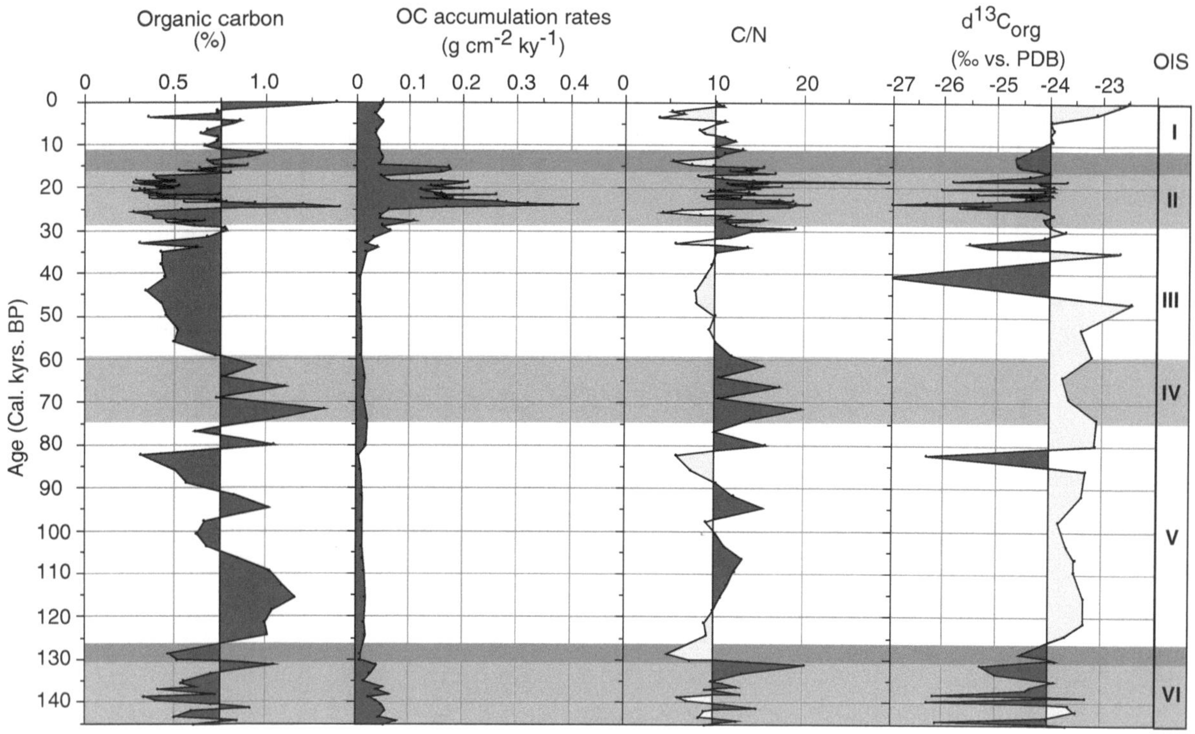

Fig. 7.7.4. Total organic carbon (OC), OC accumulation rates, C/N ratios, and stable carbon isotopes of the organic matter in Core PS2138-1 from the Barents Sea continental slope (data from Knies et al. 1999). For location of core see Figure 7.7.1.

Higher gas and bituminous content of Barents Sea sediments shows the most active zone for biochemical organic matter transformation to be within the upper layer (0–0.6 m; more pronounced in the socle shoals) where CO_2 and methane concentrations are much higher than background. Bitumens of these deposits are augmented by oxygenated compounds, resins and asphaltenes (Romankevich et al. 1982; Danyushevskaya et al. 1990). Microbiological investigations of the sediments and observed intensive gas generation in incubated bottles reveal the potential for biochemical transformations of the organic matter.

In the upper 0.7–7.0 m of the sediments, two stages of diagenetic OM transformations are observed within the socle plains and shelf plateaus. For the first stage, occupying the 0.7–1.5 (2.0) m layer, there are observed increases in humic-acid content of the OM, in the number of oxygen-containing resin components in bitumens, in CO_2 concentration and, to a lesser degree, in gaseous methane. The physical appearance of sediments, their color and lithological-mineral composition, and the prevalence of ferric iron show that transformation of OM has taken place in an oxygenous environment with saprophyte microflora activity and local sulphate reduction. Only about 6% of OC is diagenetically lost during transformation of Fe_2O_3 into FeO.

For the second stage (>2 m), the content of bitumens, oils and hydrocarbons in the OM increases, but humic acid content and gas-phase CO_2 decrease. Diagenetic loss of OC during sulphate reduction increases to 11%. On the whole, diagenetic loss of OC during ferric iron reduction to ferrous form under anoxic conditions amounts to 15% with some of the shelf cores exhibiting OC loss as high as 24%. Organic matter diagenesis is accompanied by an increase in hydrocarbon gases in the sediments (methane and sometimes ethane) through anaerobic microbiological generation.

7.7.5 Burial rates and Organic Carbon Budget

The OC accumulation rate map of the Barents and southwestern Kara Seas (Fig. 7.7.5) has been compiled using the OC distribution (Fig. 7.7.1) and thickness of Holocene deposits obtained by seismic profiling (Gurevich 1995) assuming the Holocene to comprise the past 11 Cal. kyrs. BP. The sedimentation rate estimated from seismic data ranges from 1 to 100 cm ky^{-1} and may be overestimated in some cases due to the difficulty of identifying the

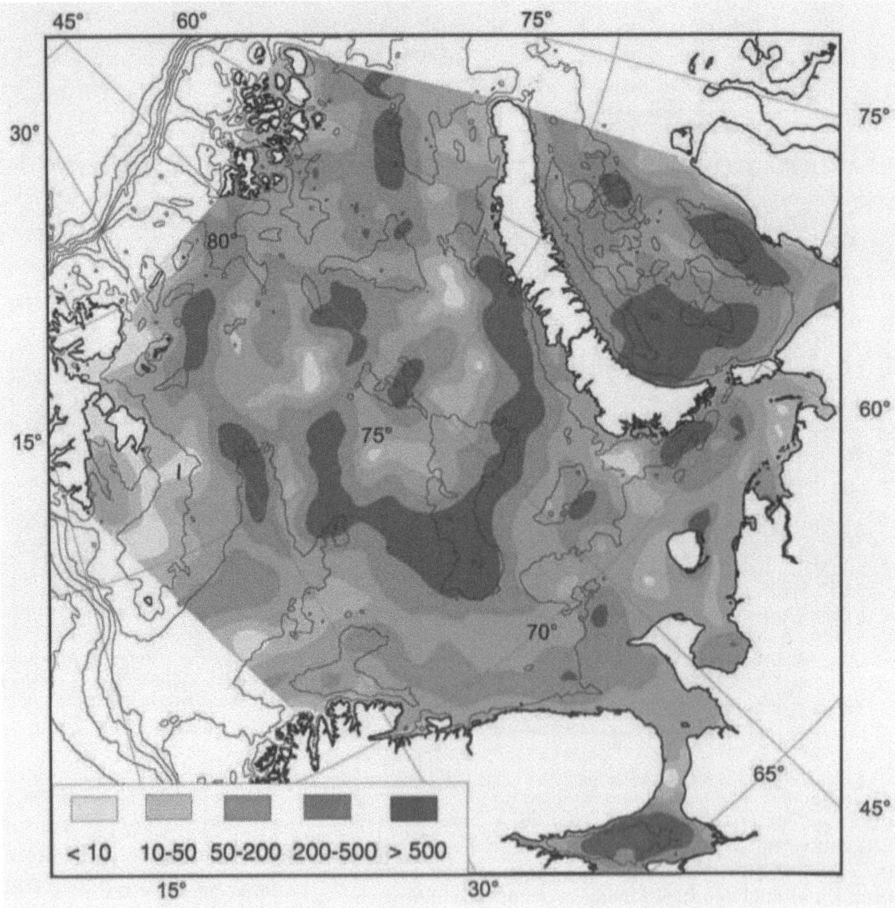

Fig. 7.7.5. Accumulation rate of total organic carbon in the Barents Sea (mg C cm^{-2} ky^{-1}) based on OC distribution (Figure 7.7.1) and sediment accumulation during the Holocene

Holocene-Pleistocene boundary in the complex sediment sequences of the Barents Sea. There is, however, fairly good agreement between sedimentation rate estimated by radiocarbon (Solheim and Elverhoi 1996; Levitan et al. 2000, 2000a; Andreeva et al. 2000) and by seismic methods. Nevertheless, radiocarbon dating gives higher sedimentation rates in the Russkaya Gavane (up to 800 cm ky^{-1}, Murdmaa et al. 2000) and in the Kola Bay (256 cm ky^{-1}, Andreeva et al. 2000) than does seismic profiling (<50 cm ky^{-1}). High sedimentation rates are observed not only in the redeposited sediments of depressions characterized by anoxic conditions but also in the hydrodynamically active southern coastal zones where sand- and silt-grained terrigenous material is accumulated.

In the Barents Sea, OC accumulation rate varies from zero (absence of sediments or erosion) to 500 mg C cm^{-2} ky^{-1} with most values lying within a range of two orders of magnitude. In the Barents and Kara seas, the largest amounts of OC (>500 mg C cm^{-2} ky^{-1}) are deposited on the accumulation plains: Central, Medvejinskaya, to the west of the Central elevation and to the north of the Persey bank; in the depressions of the North-Eastern Plateau; and on the accumulative plains St Anna and Novozemelskaya.

Altogether, the total mass of OC presently accumulating in Barents Sea sediments is estimated at 3.4×10^6 t C y^{-1}. This value, based on surface sediment OC concentrations and average Holocene sedimentation rates, may be an overestimate. The few available cores (Fig. 7.6.3) show that average sedimentation rate in the Holocene is about 30% higher than recent sedimentation. Thus, the amount of organic carbon presently being buried is estimated as $2.5–3.4 \times 10^6$ t C y^{-1}. Furthermore, the cores show that the average OC content of Holocene sediments is about 70% of surface sediment values. Thus, the average amount of organic carbon buried during the Holocene is estimated approximately at 2.4×10^6 t C y^{-1} which implies a total Holocene organic carbon accumulation of 26.4×10^9 t C. In the White Sea about 0.37×10^6 t C y^{-1} is buried or 0.26×10^6 t C y^{-1} taking into account the reduced sedimentation rates during the late Holocene. Total organic carbon burial in the White Sea during the Holocene is estimated to be about 2.9×10^9 t C and, similarly, about 0.83×10^6 t C y^{-1}

Table 7.7.2. Fluxes and masses of TOC in the Barents and White seas. Note that the phytoplankton values differ from those given in Chapter 3, Table 3.7.

Fluxes and masses	Barents Sea	White Sea
Fluxes of TOC from land, 10^6 t C y^{-1}:		
River runoff, DOC	5.72	
River runoff, POC	0.63	
Wave abrasion, POC	0.5	0.3
Eolian transfer, POC	0.25	0.02
Underground runoff, DOC	0.33	
Total fluxes of DOC (F_{DOC})	6.05	
Total fluxes of POC (F_{POC})	1.7	
Total fluxes of TOC from land (F_{TER})	7.75	
Bioproduction, 10^6 t C y^{-1} *1		
Phytoplankton	55 (0.66)	1.5 (0.02)
Ice algae	3	
Macrophytobenthos	0.066 (0.063)	0.064 (0.056)
Microphytobenthos	5.3 (0.35)	0.009 (0.0009)
Zooplankton	8.3 (3.3)	0.22 (0.1)
Zoobenthos	3.2 (7.9)	0.1 (0.3)
Photosynthetic production (F_{PHSYN}), 10^6 t C y^{-1}	63.4	1.6
$F_{PHSYN} + F_{POC}$, 10^6 t C yr^{-1}	66.7	
$F_{PHSYN} : F_{TER}$	100:12	
$F_{PHSYN} : F_{POC}$	100:2.6	
Mass of DOC, 10^6 t	550	45
Mass of POC, 10^6 t	20	1.4
Flux onto water-bottom interface, 10^6 t C y^{-1} *2:		
Production of phytoplankton (PP_{PLANK})	15.3 (28%)	0.82 (55%)
PP_{PLANK} + production of ice algae (PP_{ice})	16.1 (28%)	
$PP_{PLANK} + PP_{ICE}$ + production of phytobenthos (F_{PHSYN})	21.8 (34.4%)	0.89 (56.7%)
Total flux (marine + terrigenous) (F_{BOTTOM}) *3	24.4 (36.6%)	
Mineralization of $PP_{PLANK} + PP_{ICE}$ in water column, 10^6 t y^{-1} (% of $PP_{PLANK} + PP_{ICE}$)	41.9 (72%)	0.6 (40%)
Mineralization on seafloor, 10^6 t C y^{-1} (% of F_{BOTTOM})	21.7 (85%)	
Burial in the bottom sediments, 10^6 t C y^{-1}	3.4 (2.5) *4	0.37 (0.26) *4
Coefficient of fossilization, % of F_{PHSYN}	5.4	23
Coefficient of fossilization, % of $F_{PHSYN} + F_{POC}$	5.6 (4.0) *4	

*1 Maximal biomass (10^6 t) in summer is presented in brackets; *2 Percent of initial flux of TOC is given in brackets; *3 Sources include total production of photosynthesis, POC of river runoff, wave abrasion and eolian transfer; *4 with correction on modern sedimentation rate

is buried in the south-western part of the Kara Sea (using average sedimentation rate for the Holocene) (Fig. 7.7.5).

In Table 7.7.2 and Figure 7.7.6, an organic carbon budget including the main OC sources of the Barents and White sea sediments is shown (for data base see also Romankevich and Vetrov, 2001 and references therein). The schematic diagram does not include carbon exchange between adjacent ocean areas or ocean-atmosphere CO_2 exchange for which data are still too scarce to provide a robust assessment. The main sources of OC in the Barents and White seas are photosynthetic production, river runoff and coastal erosion. Groundwater inflow bypassing the rivers and aeolian transport are less important in the area (see Chapter 2).

Photosynthetic production of phytoplankton was estimated using chlorophyll concentrations obtained by CZCS satellite radiometer during 1978–1986 and both chlorophyll concentration and net primary production measured on board research vessels (Vinogradov et al. 2000). Ice algae bioproduction has been estimated using an average net primary production of 30 mg C m^{-2} per day (Gosselin et al. 1997) taking into account the seasonal ice conditions and duration of the polar day

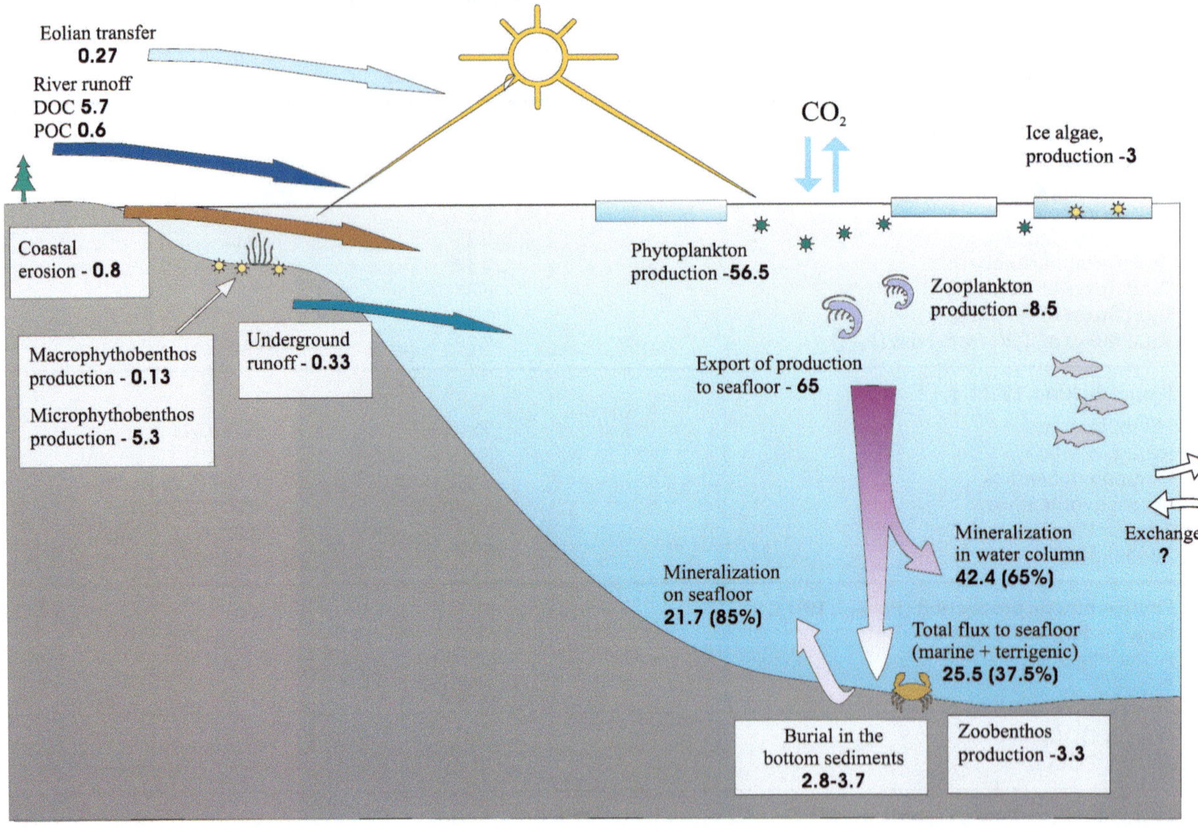

Fig. 7.7.6. Fluxes (10^6 t C y^{-1}) and balance of organic carbon in the Barents and White Seas. Note that the phytoplankton values differ from those given in Chapter 3, Table 3.7.

(Romankevich and Vetrov 2001). Assessments of annual bioproduction of phytobenthos, zooplankton and zoobenthos were made using available literature data on biomasses, bioproductivity and P/B coefficients for specific regions of the sea with extrapolation to areas of different biological communities (Romankevich and Vetrov 2001). The share of the macrophytobenthos production in the Barents Sea is small (~0.1%). In the White Sea the production of macrophytes contributes slightly more (4.3% of phytoplankton production). Microphytobenthos production is approximately two orders of magnitude higher than that of macrophytobenthos (11.2% of phytoplankton production) (Romankevich and Vetrov 2001). The value has been estimated separately for the Pechora Sea using mean biomass of microphytobenthos of 22 g C m^{-2} at depths of 5–25 m (Kucheruk et al. 1999) and a P/B coefficient of 0.05 per day (Kuznetsov et al. 1998), and for other parts of the Barents Sea using mean bioproduction rate 10 mg C m^{-2} h^{-1} (Kuznetsov 1991).

OC supplied by river runoff is higher than previously estimated (4.65 10^6 t C y^{-1}, Romankevich and Vetrov 2001) due to a more complete inclusion of discharge from small rivers (see Chapter 2.2). The ratio of POC to DOC in the shallow Barents Sea is 1:27, similar to the World Ocean (1:29) (Romankevich and Vetrov 1997), but different from the influx ratio (1:4.3). High rate of coastal erosion is supported by the active destruction of permafrost which is widely distributed (see Chapter 2.3). For estimates of coastal erosion published data from Romankevich and Vetrov (2001) have been used. Recently, estimates of coastal erosion in the Arctic seas have been revised on the basis of new research methods including airplane photography and satellite observations. To estimate OC flux from groundwater, data for river runoff in winter have been used because groundwater contributes most of the recharge at that time. Average DOC concentration in underground runoff at this season is estimated to be 6.5 mg l^{-1}. Recently improved estimates of aeolian input have been taken from Chapter 2.5. Snow and ice are significant only as agents of transfer and redistribution of carbon supplied by the various sources (see Chapter 2.4).

The total OC flux to the seafloor includes both marine and terrigenous components which are often estimated from sediment-trap data. Unfortunately, sediment trap data for this region are scarce and trap results in shallow water are often biased by sediment resuspension (see Chapter 5). Here, we have estimated marine organic carbon flux to the seafloor using Tseitlin's (1993) relationship between OC flux (F_c), depth (z) and primary production (PP), setting z = 50 m for all depths <50 m (Romankevich and Vetrov 2001). Because this relationship was developed for the mid-latitudes (and refer to Chapter 8 for comments on the applicability of these formulae to the Arctic), we have checked its applicability to Arctic seas by comparing formula projections with calculations of OC fluxes in the Greenland and Norwegian seas based on benthic-lander experiments (Schlueter and Pfannkuche 2000; Schlueter et al. 2000). Estimated annual OC fluxes to the seafloor in the North Atlantic for 500 m and 1000 m horizons (10.3 and 4.5×10^6 t C yr^{-1}, primary production is 150–160 $\times 10^6$ t C yr^{-1}; Schlueter and Pfannkuche 2000; Schlueter et al. 2000) agree well with values derived from the formula $F_c = 33$ PP/z (10.2 and 5.1×10^6 t C yr^{-1}). Accordingly, this formula yields a total flux of authochtonous OM to the seafloor of the Barents and White seas with 1° resolution of 16.1×10^6 and 0.82×10^6 t C y^{-1}, respectively, and 21.8×10^6 t and 0.89×10^6 t C y^{-1} together with phytobenthos production.

The estimate of allochtonous OM flux to seafloor assumes that all terrigenous POC reaches the seafloor. In standard filtration methods, all colloids are included in the DOM fraction which, for the deltaic environment of the Lena River, amounts to 20 to 55% of the DOM (Cauwet and Sidorov 1996). Our results from ultrafiltration using Amicon technology and fractionation on molecular weights show that the behavior and chemical composition of various colloid fractions strongly differs from that of real dissolved (ultrafiltered) OM. Some terrigenous DOM is mineralized to CO_2 and nutrients, some is transported across the shelf, and some is transformed to colloids and POM which may then be buried in the bottom sediments. The main transformation of DOM to POM occurs at the land-sea boundary and at other biogeochemically active sites (estuaries, water-sediment interface, etc). Here, we assume that approximately 20% of the DOC (other estimations give a value of about 5%; see Chapter 4) is transformed into POC as result of flocculation, sorption and bioassimilation, and the freeze-melt cycle in this area (Romankevich and Vetrov, 2001). In spite of progress in recent investigations of DOC in the Arctic (Wheeler et al. 1997; Opsahl et al. 1999; Kattner et al. 1999) the fate of the rest of DOC from river runoff in the Barents Sea is still unknown.

Total OC flux reaching the seafloor in the Barents and White seas ($F_{bottom:}$; Table 7.7.2, Fig. 7.7.6) is estimated at 25.5×10^6 t C y^{-1}, with components including total photosynthetic production ($F_{phsyn} = 65 \times 10^6$ t C y^{-1}), POC from river runoff ($F_{poc} = 0.63 \times 10^6$ t C y^{-1} + 20% $F_{doc} = 0.92 \times 10^6$ t C y^{-1}), OC from coastal erosion (0.8×10^6 t C y^{-1}) and eolian input (0.27×10^6 t C y^{-1}). Average total OC flux to the seafloor is estimated to be 37.5% of the initial flux.

One of the major locations for OM transformation and decomposition (second only to the photic zone) is the seafloor, and especially the uppermost layer of the sediments where a large part of OM flux reaching the seafloor is mineralised (~20.6×10^6 t C y^{-1} or 85%). Here, we have estimated remineralization as the difference between OC flux reaching the seafloor and OC preserved in deep sediments, which assumes negligible export of carbon towards the interior ocean.

The exchange of organic carbon between the Barents Sea and the adjacent areas (i.e., the continental slope in the north and the Kara Sea in the east) cannot be calculated with any confidence due to a lack of data on the spatial distribution of POC and DOC at the boundaries of the Barents Sea (Romankevich and Vetrov 2001). We believe that only a small part of the OM (and inorganic matter) is exported from the Barents Sea to the continental slope and interior ocean. In this respect, the Barents Sea strongly differs from the Beaufort and Laptev Seas, where a significant export of total sediments and organic matter occurs (see Chapter 7.2 and 7.5). The width of the Barents Sea shelf is large (~1500 km) compared to those of the Laptev and Beaufort shelves (~500 km and 150 km, respectively) possibly explaining why most particulate organic matter cannot reach the continental slope. Furthermore, there are many local depocenters of sediments with high organic carbon content (i.e., natural sediment and organic carbon traps), generated by the irregular relief of the Barents shelf. These depocenters play an important role in the mass balance.

The coefficient of fossilisation (ratio of buried OC to primary production) is estimated for the Barents and White seas to be 5.4% and 23% correspondingly. The latter very high value implies that OC in the White Sea's sediments is represented mainly by terrigenic organic matter, but this needs to be evaluated directly using *n*-alkane, lignin and C/N data. The coefficient of fossilisation as a ratio of buried OC to total photosynthetic and POC fluxes from land is estimated at 5.6%, which is far higher than for the pelagic part of the World Ocean (0.02%) or for continental margins (0.8–1.3%) (Romankevich and Vetrov 1997). Thus, the degree of closure of carbon cycle in the Barents Sea is about 94.5%. Such high imbalance of the cycle (biosynthesis + input from land – sediment burial) is one of the remarkable features of the carbon cycle in the Arctic seas. High coefficients of fossilisation characteristic for the Arctic seas are likely supported by the shallowness of the seas, the considerable terrigenous inputs, low temperature and higher resistance of much of the organic matter to bacterial attack. It is noted that arctic marine organic matter has a higher content of lipids which tend hardly to be assimilated by microflora.

A comparison of the amount of organic carbon buried in the White Sea (0.37 10^6 t C y^{-1}) with the POC flux into this area (about 1.0 10^6 t C y^{-1}) suggests that about 70–80% of this POC and about 95% of the DOC are transferred into the Barents Sea. It should be noted, however, that several White Sea bays of the White Sea having high sedimentation rates of organic-rich material have been neglected in the present calculations. These areas, if included, might significantly change the estimates of sedimentation, OC accumulation, and exchange of organic carbon between the Barents and White seas.

The ratio of OC preserved in bottom sediments (2.5 to 3.4 × 10^6 t C y^{-1}) to POC flux from land is estimated to be about 1.8 to 2.5 for the Barents Sea, implying that a large fraction of organic matter in these sediments is supplied from marine production. In recent sediments of the Barents Sea, marine organic matter accounts for about 70% of the OC. This sharply contrasts OM in bottom sediments from the Kara, Laptev and East-Siberian seas, where the terrestrial component comprises ≥50%, undoubtedly due to the huge Siberian river inputs (see Chapters 7.4, 7.5, and 7.6).

7.7.6. Summary and Concluding Remarks

- The most important features that distinguish the Barents Sea from other Arctic seas and that strongly influence the distribution and accumulation of organic matter, are (i) strong influence of the Atlantic Ocean on the supply and genesis of organic matter, (ii) relatively small river inflow and coastal erosion, (iii) moderate ice formation, and (iv) prolonged growing period and higher bioproductivity.
- Organic carbon distribution in the surface sediment is complex and depends mainly on sediment grain size and bottom topography. OC content ranges from 0.02% (coarse sand) to 3.3% (pelite) with an average of 1.08%. In sands, which are distributed widely in shallow, nearshore zones, differentiation of sediments is most active due to tides and waves, and average OC content is as low as 0.36%. In the muddy sediments of depressions and troughs OC amounts to 1.4%. Organic matter in surface sediments has a marine/terrigenous mixed character (averaging about 70% marine) with a predominance of marine components even in the south-eastern part of the Barents Sea, which is more influenced by terrigenous input.
- Holocene OC accumulation rates range from 0 to >500 mg C cm^{-2} ky^{-1} with a tendency toward an increasing proportion of marine organic components with time. A maximum OC accumulation rate (1.5 g C cm^{-2} ky^{-1}) is associated with the Holocene climatic optimum.
- The amount of organic carbon presently accumulating is estimated at 2.5–3.4 × 10^6 t C y^{-1}. The average amount of organic carbon buried during the Holocene is about 2.4 × 10^6 t C y^{-1} implying a total organic carbon accumulation during the Holocene of 26.4 × 10^9 t C. In the White Sea about 0.26–0.37 × 10^6 t C y^{-1} is buried and the total amount of organic carbon buried during the Holocene is estimated to be about 2.9 × 10^9 t C. The coefficient of fossilization is estimated at 5.8% which is higher than estimates for the pelagic part of the World Ocean (0.02%) or for the continental margins (0.8–1.3%). The accumulation of significant amounts of organic carbon in this region, which has relatively low annual primary production, is due to the shallow water depth, significant terrigenous organic carbon supply and low temperature reducing rate of organic matter destruction.

7.8 Northern Fram Strait and Yermak Plateau: Distribution, Variability and Burial of Organic Carbon and Paleoenvironmental Implications

D. Birgel, R. Stein

7.8.1 Introduction

The environment of the Fram Strait, the only deep-water connection of the Arctic Ocean to the world oceans via the North Atlantic (Fig. 7.8.1; see Fig. 7.1.9), is influenced by the distribution of sea-ice and two opposing current systems. The northward flowing West Spitsbergen Current (WSC) transports warm, near-surface water (Manley 1995; Rudels et al. 2000) to the Northern Fram Strait. About 22% of the northward flowing Atlantic waters are re-circulated within the RAC (Return Atlantic Current) between 78 and 80°N, west of Svalbard. At 80°N the WSC splits into the Svalbard Branch (ca. 33% of the WSC waters) and the Yermak Branch (ca. 45% of the WSC waters). On the western side of the Fram Strait, the East Greenland Current (EGC) transports cold and low-salinity water southwards along the eastern continental margin of Greenland. (Fig. 7.8.1).

Primary production in ice-covered areas of western Fram Strait is limited by sea-ice cover, and influenced by the predominant water mass. Productivity in the interior Arctic Ocean is generally low (0.09 g C m^{-2} d^{-1}) (Wheeler et al. 1996; see Chapter 3). At marginal ice zones and oceanic fronts in the Fram Strait, however, primary productivity exhibits strong fluctuations and may exceed 1 gC/m^2/day (Hirche et al. 1991). The accumulation of organic carbon in sediments depends not only on the supply from primary productivity, but also on selective degradation in sediments. Efficient vertical transport through the water column by formation of aggregations ("ballast effect") (Ittekkot et al. 1992; Knies and Stein 1998) and increased lateral transport by strong currents enable a higher preservation of organic carbon in the sediments. In this region, the WSC is capable of transporting large amounts of suspended organic matter to the ice-covered regions in northern Fram Strait (Rutgers van der Loeff et al. 2002).

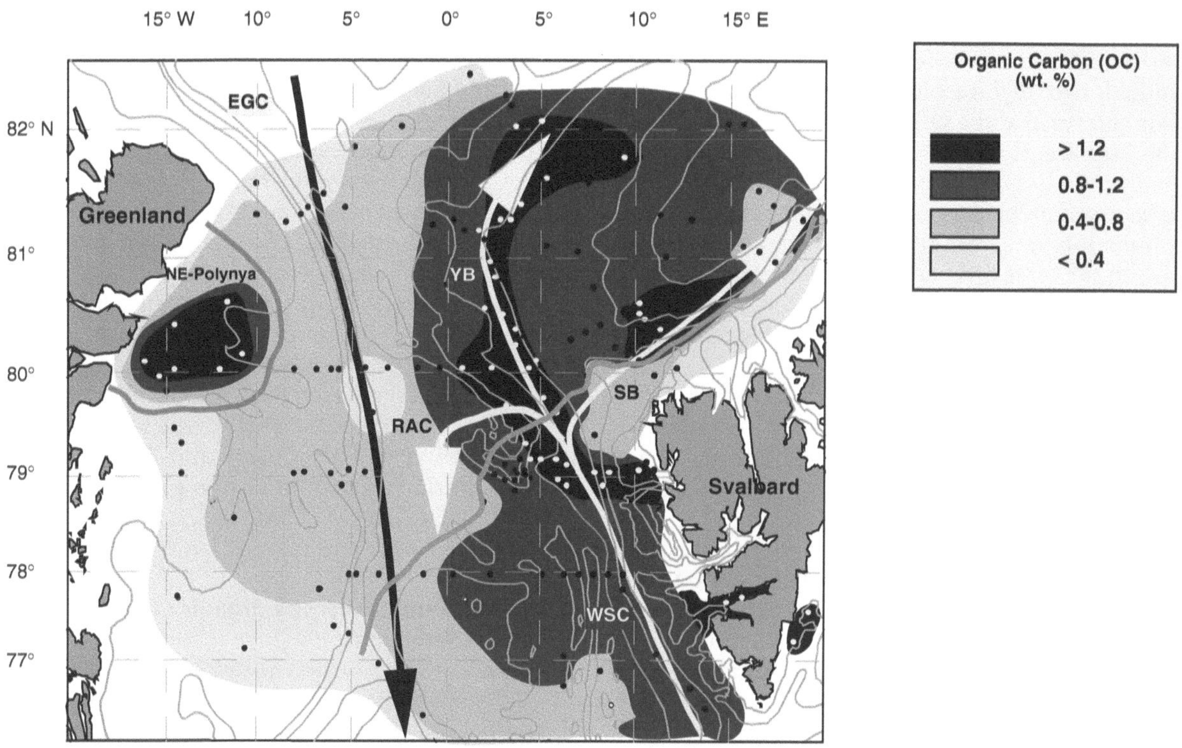

Fig. 7.8.1. Distribution map of total organic carbon in surface sediments from the Fram Strait. Black/White dots indicate OC surface sample points (Hebbeln and Berner 1993; Stein et al. 1994a; Kierdorf 2001; this paper). White arrows: West Spitsbergen Current (WSC) and its extensions: the Yermak Branch (YB), the Svalbard Branch (SB), and the Return Atlantic Current (RAC). Black arrow: East Greenland Current (EGC). Light grey line: average summer sea-ice distribution. Dark grey line: NE-Greenland polynya

Numerous studies have dealt with paleoceanography and the associated organic carbon accumulation in the sediments of Fram Strait and adjacent regions during the last glacial/interglacial cycle (e.g., Hebbeln 1992; Hebbeln et al. 1994; Elverhøi et al. 1995; Andersen et al. 1996; Hebbeln and Wefer 1997; Hebbeln et al. 1998; Notholt, 1998; Vogt et al. 2001, Taylor et al. 2002). However, in most of the sedimentary records a low temporal resolution prevents the identification of short-term climatic fluctuations, like those reconstructed from high-resolution terrestrial ice-core records, (e.g. GISP 2/GRIP; Grootes et al. 1993). Occasionally, short-term events recorded as enhanced organic matter accumulation have been found in cores from the northern Fram Strait/Yermak Plateau region (Knies and Stein 1998; Vogt et al. 2001). These events are caused by a rapid incorporation of organic matter in fine-grained material followed by rapid transfer to the seafloor.

Rapidly changing climatic and oceanographic conditions can be recorded exceptionally well by undisturbed deep-sea sediments, particularly in the distribution and variability of organic carbon in sediments. Yet, there exists little information about the regional response during the last deglaciation and the potential influence of terrigenous material on marine sedimentation of organic carbon in northern Fram Strait. To address this problem, we studied two high-resolution cores spanning the time intervals of the last glacial, the last deglaciation and the Holocene. Here, we present data on the distribution and sources of organic carbon in surface sediments and in long sediment cores. Accumulation rates of total sediment and organic carbon for three different time intervals are calculated and an organic carbon budget for Fram Strait/ Yermak Plateau is presented for the Holocene.

7.8.2 Data Base, Material and Methods

The surface sediment samples (0–1 cm) and long sediment cores were recovered during the RV *Polarstern* expeditions ARK XIII/2 (Stein and Fahl 1997), ARK XV/2 (Jokat et al. 1999) and ARK XVI/1 and 2 (Krause and Schauer 2000). Sampling was carried out with a giant box corer and a multicorer, and long sediment cores were recovered with a gravity corer and a Kastenlot. All samples were stored at −30 °C until further treatment. Total organic carbon (OC), hydrogen indices (HI) and stable carbon isotopes ($\delta^{13}C_{org}$) were determined on homogenised bulk samples. Unhomogenised subsamples were extracted and purified for biomarker determination according to a method modified from Folch et al. (1957) and Bligh and Dyer (1959). Here, we produce a synthesis from previously published data (Hebbeln 1992; Hebbeln and Berner 1993; Hebbeln et al. 1994; Stein et al. 1994a, 1994b, 1994c; Hebbeln and Wefer 1997; Notholt 1998; Vogt et al. 2001) augmented with new results.

OC was determined using a LECO CS analyser. The determination of hydrogen indices (HI) is a useful proxy to estimate the proportions of terrigenous and marine organic matter in sediments (Tissot and Welte 1984). Analysis was performed by a Rock-Eval 6 (Espitalié et al. 1977). To examine the composition of the organic matter in more detail, we determined $\delta^{13}C_{org}$ values as well as specific biomarkers, i. e., long-chain *n*-alkanes (C_{27}, C_{29}, C_{31}) as a proxy for land-derived vascular plant material (e.g. Prahl and Muelhausen 1989; Yunker et al. 1995), and brassicasterol (24-Methyl-cholesta-5,22-dien-3β-ol) as a typical marker of diatoms and coccolithophores (Nichols et al. 1984; Meyers, 1997). For further organic-geochemical parameters as proxies for the identification of organic carbon sources we refer to Chapter 1.4. The stable carbon isotopes were measured using a mass spectrometer (ANCA-SL 20-20 Europa Scientific). Biomarkers were determined by gas chromatography on a Hewlett-Packard 6890 and gas chromatography/ mass spectrometry on a Hewlett-Packard GC/MS 5890 (details of the analytical procedure are given in Fahl and Stein 1997; 1999).

To interpret the organic-carbon records and to calculate accumulation rates and an organic carbon budget it is necessary to obtain a precise stratigraphical framework using AMS ^{14}C dates and stable oxygen isotopes of the planktonic foraminifera *Neogloboquadrina pachyderma sin.* Age models for the investigated cores have been established (Hebbeln 1992; Hebbeln et al. 1994; Elverhøi et al. 1995; Notholt 1998; Vogelsang et al. 2001; Vogt et al. 2001; Nørgaard-Pedersen et al. in press) and radiocarbon ages have been transformed to calibrated ages according to Bard (1998) and Stuiver et al. (1998). Additional age information was obtained by correlating OC-curves of selected cores.

7.8.3 Distribution and Sources of Organic Carbon in Surface Sediments

The OC surface map (Fig. 7.8.1) implies two obvious domains. The first domain, western Fram Strait, shows OC values of 0.2 to 0.8 %, suggesting low organic carbon input and/or diminished preservation in surface sediments. The surface sediments of the Arctic Ocean/ Fram Strait peripheral zone have a predominantly terrigenous source (Stein et al.

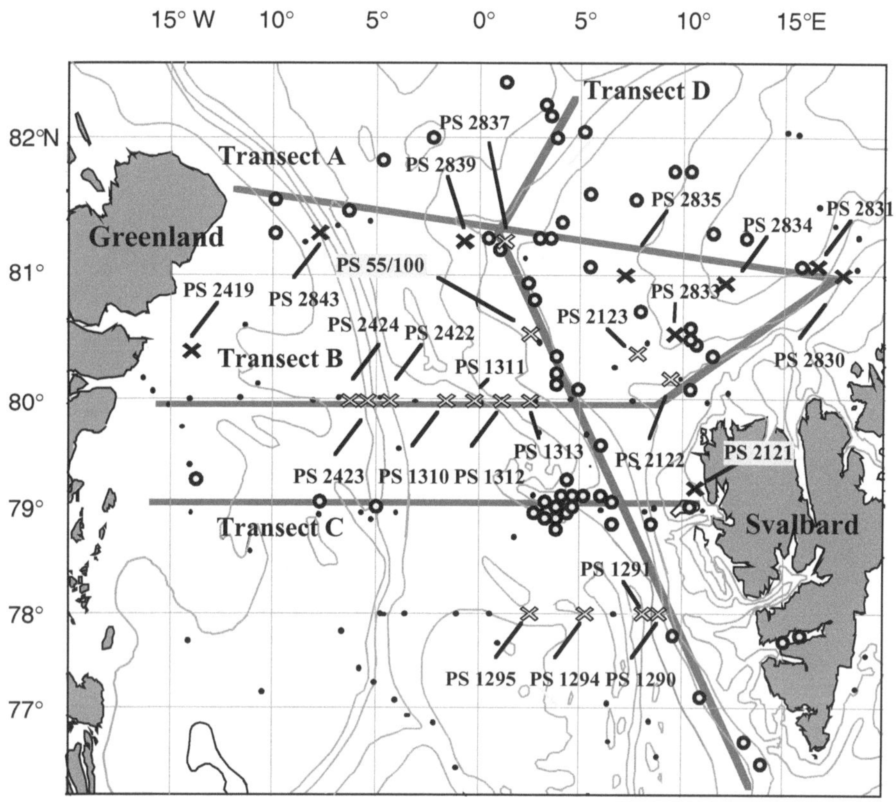

Fig. 7.8.2. Map of the study area. White crosses: Locations of cores for which accumulation rates were calculated for all three intervals (see Table 7.8.1) (Hebbeln 1992; Notholt 1998; Vogt et al. 2001; this paper). Black crosses: Core locations for which only Holocene accumulation rates were calculated (Stein and Fahl 1997). White large dots: Surface samples in which hydrogen indices and $\delta^{13}C_{org}$ values were determined (this paper). Black small dots: Surface samples with hydrogen indices or $\delta^{13}C_{org}$ values (Hebbeln and Berner 1993; Stein et al. 1994a; Kierdorf 2001). Dark grey lines: Transects A–D with surface sediment data (see Fig. 7.8.3)

1994a; Hebbeln and Berner 1993) and are mainly composed of inorganic material transported by glaciers. In the three transects A, B, and C (Fig. 7.8.2), the lowest OC values were found in sediments of the Greenland shelf and slope.

Low hydrogen indices imply a dominantly terrigenous source for organic carbon (Fig. 7.8.3). The $\delta^{13}C_{org}$ values are equivocal on first sight. In transect B light values toward the Greenland margin indicate a dominantly terrigenous source (Fig. 7.8.3). In contrast, slightly heavier values are observed in transects A and, especially under the NE-polynya (Transect A), where values up to −20‰ were observed. These heavier isotopic compositions might be produced by the incorporation into sediments of higher proportions of ice-algae, which are reported to be isotopically heavy. Gibson et al. (1999) found a carbon isotopic composition of sea-ice algae up to −8‰.

For transects A and D, we obtained additional information from long-chain *n*-alkane and brassicasterol data (Fig. 7.8.4). In transect A, long-chain *n*-alkane concentrations >70 µg/gOC are typical whereas brassicasterol values are low (40 µg /gOC), with two exceptions at stations PS 2837 and PS 2836 at the western Yermak Plateau slope and at stations PS 2830 and PS 2831 in the Svalbard Branch of the WSC. There, a stronger contribution of marine phytoplankton (brassicasterol: 40–110 µg/gOC) and a minor contribution of terrigenous organic matter (long-chain n-alkanes <70 µg/gOC) were observed. In transect D, we found a lower terrigenous organic matter flux to the surface sediment, indicated by lower long-chain *n*-alkane values (<70 µg/gOC). On the other hand a stronger contribution of marine organic matter is indicated by significantly increased brassicasterol values (>50 µg/gOC) in large parts of the Yermak Branch of the West Spitsbergen Current (Fig. 7.8.4). Detailed investigations were performed on similar samples by Birgel et al. (subm.) on fatty acids. These compounds represent more than 70% of the analysed biomarkers and enable an improved insight in organic carbon distribution in the northern Fram Strait.

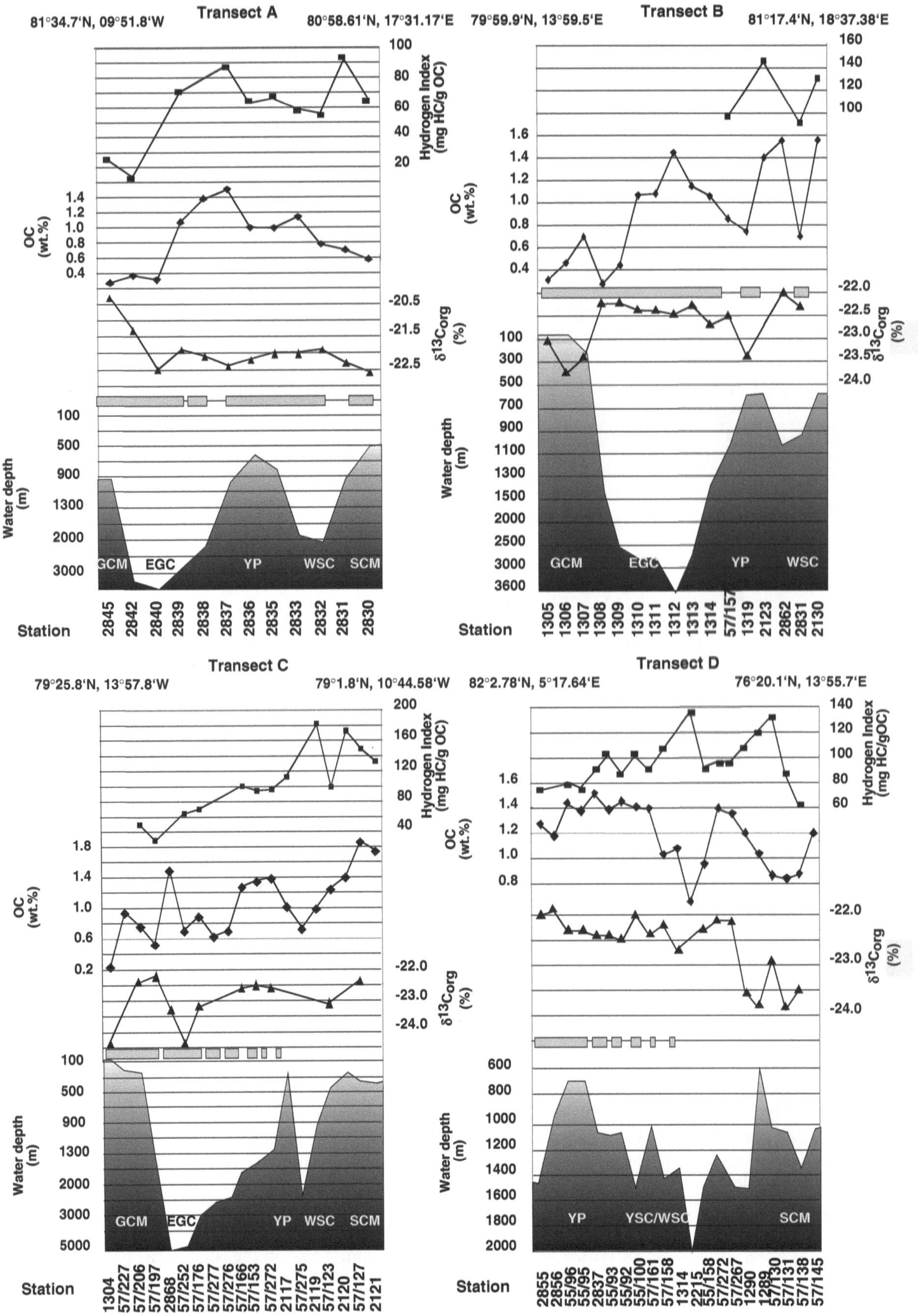
Transect A
81°34.7'N, 09°51.8'W
80°58.61'N, 17°31.17'E
Hydrogen Index (mg HC/g OC)
OC (wt.%)
δ13Corg (%)
Water depth (m)
GCM EGC YP WSC SCM
Station
2845 2842 2840 2839 2838 2837 2836 2835 2833 2832 2831 2830
Transect B
79°59.9'N, 13°59.5'E
81°17.4'N, 18°37.38'E
OC (wt.%)
δ13Corg (%)
Water depth (m)
GCM EGC YP WSC
Station
1305 1306 1307 1308 1309 1310 1311 1312 1313 1314 57/157 1319 2123 2862 2831 2130
Transect C
79°25.8'N, 13°57.8'W
79°1.8'N, 10°44.58'W
Hydrogen Index (mg HC/g OC)
OC (wt.%)
δ13Corg (%)
Water depth (m)
GCM EGC YP WSC SCM
Station
1304 57/227 57/206 57/197 2868 57/252 57/176 57/277 57/276 57/166 57/153 57/272 2117 57/275 2119 57/123 2120 57/127 2121
Transect D
82°2.78'N, 5°17.64'E
76°20.1'N, 13°55.7'E
Hydrogen Index (mg HC/gOC)
OC (wt.%)
δ13Corg (%)
Water depth (m)
YP YSC/WSC SCM
Station
2855 2856 55/96 55/95 2837 55/93 55/92 55/100 57/161 57/158 1314 2215 55/158 57/272 57/267 1290 1289 57/130 57/131 57/138 57/145

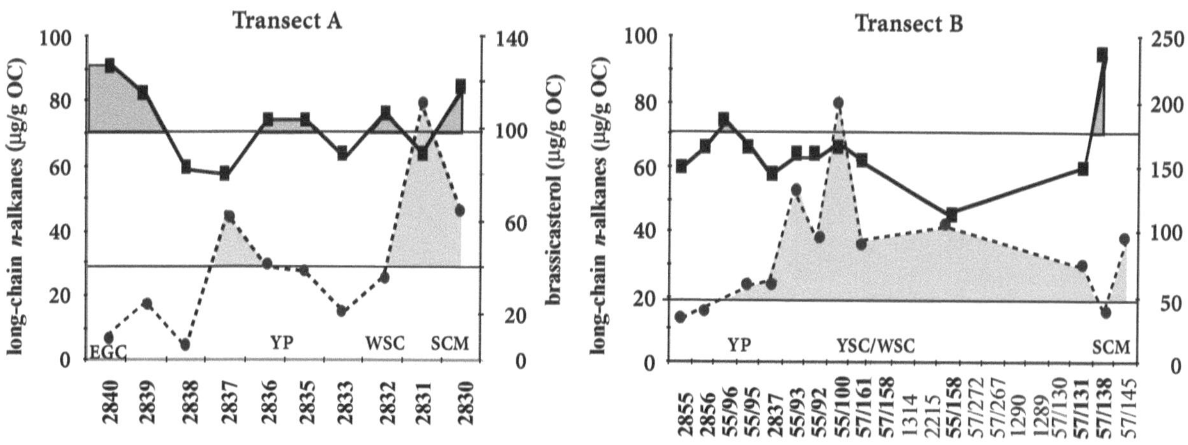

Fig. 7.8.4. Long chain *n*-alkane and brassicasterol distributions of transects A and D in the northern Fram Strait. Bold numbers on the x-axis indicate stations, for which biomarker samples are available. Y-axis: black squares represent long-chain *n*-alkane (terrigenous organic matter) concentrations (µg/gOC), black dots are representing brassicasterol (marine organic matter) concentrations (µg/gOC). Dark grey areas: predominant terrigenous organic material deposition, indicated by high long-chain *n*-alkane concentrations. Grey areas: predominant marine organic matter deposition, indicated by high brassicasterol concentrations. For abbreviations see Figure 7.8.3.

Enhanced vertical transport, especially in the autumn, has been observed in the NE-polynya (Bauerfeind et al. 1997; Ramseier et al. 1997). This transport is possibly supported by the formation of new sea-ice and the associated formation of superdense water. Such a mechanism would lead to accelerated transport to the seafloor of aggregates of sea-ice algae mixed with terrigenous detritus (Bauerfeind et al. 1997). Most of the marine organic matter is degraded, 40 to 70% of the organic matter is re-circulated in the euphotic zone (Deming et al. 1995), and less than 7% of the particulate organic matter reaches the seafloor. Pesant et al. (2002) showed that storms accelerate the vertical transport of primary produced organic matter. They noted that 21 to 60% of the primary produced material, especially diatoms, could be exported very quickly in this way from the upper water column to the seafloor, especially in summer months. It is possible that our stations in the NE polynya were affected by such storm events which would promote enhanced preservation of marine organic matter in the surface sediments. The East Greenland slope and the central part of the Fram Strait (Fig. 7.8.3) are dominated by OC values of 0.4 to 0.8%. Hydrogen indices increase slightly from west to east, remain below 100 mg HC/gOC, typical of predominantly terrigenous organic matter. The $\delta^{13}C_{org}$ distribution with values of −22 to −22.5‰ suggests a different composition of organic matter in the investigated sediments than from hydrogen indices. Normally, stable carbon isotopes of planktonic organic matter show $\delta^{13}C_{org}$ values of −19 to −20‰. In sediments of high latitudes, however, a depletion of ^{13}C has formerly been observed (Rau et al. 1982, 1991; Schubert and Calvert 2001). Terrestrial organic matter shows values between −26 to −28‰, normally. Therefore, $\delta^{13}C_{org}$ values, which lie between −22 and −22.5‰, indicate a mixture of marine and terrigenous organic matter in our investigated areas. From biomarker records, we record a significant deposition of long-chain *n*-alkanes, indicating a distinct terrigenous organic matter component in the organic flux to the surface sediments (Fig. 7.8.4).

The second domain, eastern Fram Strait, is characterised by OC values >0.8%, indicating a higher flux and/or better preservation of organic matter in the sediments (Fig. 7.8.1). In the deeper parts of the western Svalbard and Yermak slope (Svalbard Branch/Yermak Branch) higher OC values were measured (Fig. 7.8.1). The hydrogen indices suggest a stronger contribution of marine organic matter in the surface sediments. Although the northernmost transect A remains dominated by terrigenous organic matter, transects B and C show values

◀ **Fig. 7.8.3.** Transects A–D in the Northern Fram Strait. For locations see Figure 7.8.2. OC-values (see Fig. 7.8.1), hydrogen indices and $\delta^{13}C_{org}$ values (Hebbeln and Berner 1993; Stein et al. 1994a; this paper) on the x-axis, water depth on the y-axis. Transects A–C are West-East transects from the cold-water dominated western part of the Fram Strait to the warm-water dominated eastern part. Transect D is a North-South profile following the WSC and the Yermak Branch extensions (YB/SB). GCM = Greenland Continental Margin; EGC = East Greenland Current; YP = Yermak Plateau; WSC = West Svalbard Current; SCM = Svalbard Continental Margin; YSC = Yermak Slope Current

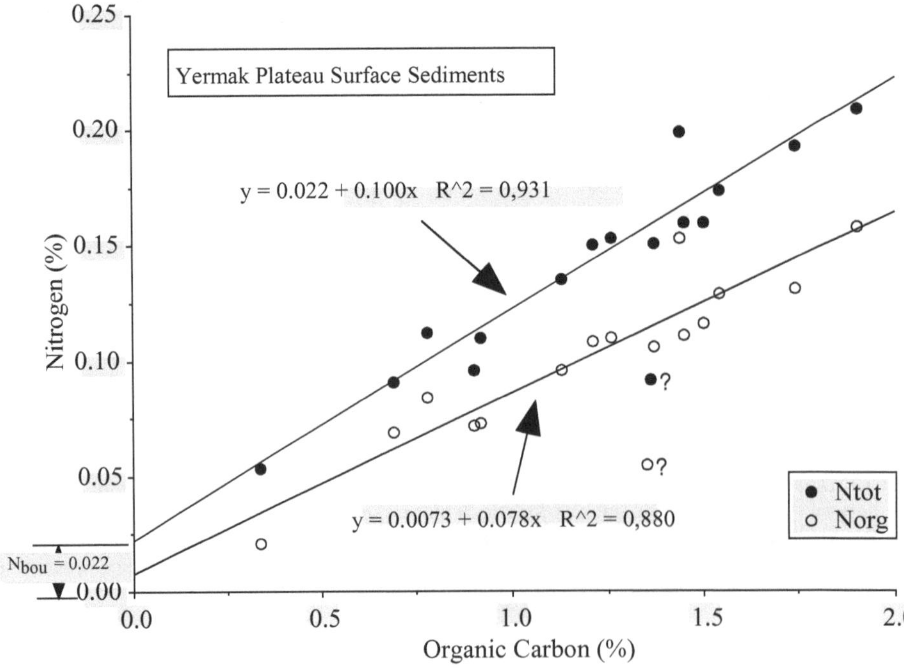

Fig. 7.8.5. Correlation between total nitrogen (N_{tot}) and total organic carbon (OC (wt.%)) and between organic nitrogen (N_{org}) and total organic carbon for the Yermak Plateau (ARK XIII/2) surface samples (Schubert and Calvert 2001)

>100 mg HC/gOC (Fig. 7.8.3), indicating a stronger influence of the WSC and partial open-water conditions. The $\delta^{13}C_{org}$ values obtained for eastern Fram Strait are similar to those for central Fram Strait, varying between −22 to −22.5‰, which are slightly heavier than the (−23 to −23.5‰) range determined for similar samples collected from the Yermak Plateau by Schubert and Calvert (2001).

The OC/N_{tot} ratios in marine organic matter typically vary between 8 and 10 (Redfield et al. 1963), whereas terrestrial organic matter has ratios of >20 (Scheffer and Schachtschabel 1984). Schubert and Calvert (2001) measured OC/N_{tot} ratios on the Yermak Plateau between 7 and 10, thus suggesting a predominant marine composition of organic matter. However, the plot of OC versus total nitrogen intersects the nitrogen axis at 0.022% (Fig. 7.8.5; Schubert and Calvert 2001), implying a significant amount of inorganic nitrogen (N_{bou}) in these sediments (see Chapter 1.4 for background information). Correcting the nitrogen data for inorganic nitrogen, accordingly, yields OC/N_{org} ratios between 9 and 16, indicating a more mixed marine/terrigenous source, comparable with our hydrogen indices and carbon isotope values.

Maximum OC values were obtained in the easternmost area of the northern Fram Strait (0.8–1.9%) (Fig. 7.8.3). In the northernmost section of transect D, $\delta^{13}C_{org}$ values are heavy to intermediate, and hydrogen indices vary between 80–140 mg HC/gOC, indicating a distinct increase of marine organic matter in this part of transect D (Fig. 7.8.3). The supply of terrigenous material, which is transported from the Arctic Ocean by sea-ice (Dethleff et al. 2000) and released during melting when ice meets warm Atlantic water, is suggested from increased long-chain *n*-alkane values (60–90 µg/gOC) in this domain (Fig. 7.8.4). The two branches of the WSC, the Yermak Branch flowing to the north and the Svalbard Branch flowing in a north-east direction to the north of Svalbard (Fig.1), transport WSC water to the perennially ice-covered Yermak Plateau (Manley 1995).

Suspended particulate organic matter is supplied by the WSC from the southern Svalbard slope areas and the Svalbard shelf and fjord zones to the northernmost Fram Strait (Fig. 7.8.3). Soltwedel et al. (2000) detected higher bottom water temperatures reaching an average 0.8°C in 1000 m water depth (Yermak Branch) and 2.0°C in 500 m water depth (Svalbard Branch) in areas influenced by the Yermak Branch/Svalbard Branch. High concentrations of chlorophyll *a* in the sediments from this region indicate fresh marine organic matter transported laterally by Atlantic water. This suggestion is supported by increased values of marine-derived brassicasterol (50–200 µg/gOC) in transects A and D (Fig. 7.8.4). The strong influx of nutrient-rich, warm Atlantic water supports exportable production at the marginal ice-zones northwest of Svalbard and at the frontal zones of the southern Fram Strait (Heimdal 1983; Andersen 1989). Aggregation of the dominantly terrigenous organic matter from the Svalbard archipelago and the Barents Sea

(Berner and Wefer 1993), with minor amounts of primary produced marine organic matter in the Fram Strait, enables a quick vertical transport to the seafloor ("ballast effect"; see above). Primary production in the northern Fram Strait is supported additionally by the release of nutrients from melting sea-ice (see above). The organic matter incorporated in the sea-ice may originate from as far away as Siberia (Dethleff et al. 2000).

In the northern part of transect D, HI-values (80–100 mg HC/gOC) and $\delta^{13}C_{org}$ values (–22 to –22.5‰) in the Yermak Branch area (Fig. 7.8.3) indicate mixed sources of organic matter. The HI-values, however, suggest predominantly terrigenous organic matter (see above) whereas the $\delta^{13}C_{org}$ values indicate a mixed terrigenous/marine organic matter source, transported via the WSC to the northernmost positions. Increased concentrations of long-chain *n*-alkanes and a decreased flux of brassicasterol to the seafloor suggest a relative increase in terrigenous organic carbon (Fig. 7.8.4). In transect B, the HI and $\delta^{13}C_{org}$ values are comparable to those measured along transect D. Only for the Yermak Plateau transects B/D were high long-chain *n*-alkane, low HI and light $\delta^{13}C_{org}$ values observed. It seems likely that terrigenous organic matter is better preserved here than marine organic matter, similar to the shelf deposits off East Greenland (brassicasterol <50 µg/gOC, long-chain n-alkanes >70 µg/gOC). In the Kongsfjorden outlet (transect C), a higher marine organic content is implied (HI >100 mg HC/gOC), and intermediate $\delta^{13}C_{org}$ values (–22.5 to –23‰) were obtained. We propose that comparable processes operate in the vicinity of the Svalbard fjords as those observed over the shelf and upper slope regions of the Laptev Sea where primary production is increased by riverine nutrient input (Boetius and Damm 1998). Nutrient-rich suspensions are released from melting glaciers in the proximal fjords supplying nutrients to these regions and to the adjacent shelf where primary production is already well-supported by the warm, Atlantic waters flowing in from south. The consequent increased sedimentation rates and the lateral and vertical influx of organic matter lead to enhanced preservation in surface sediments.

7.8.4 Late Quaternary Organic Carbon Records and Paleoenvironment

Five cores from eastern Fram Strait (Fig. 7.8.2; PS2837, PS55/100, PS2122, PS1294, PS1290) were examined in detail for OC, $\delta^{13}C_{org}$-values, and hydrogen indices (our data together with data from Hebbeln et al. 1994; Knies 1994; Elverhøi et al. 1995; Vogt et al. 2001). Based on AMS ^{14}C-ages and the OC records three time slices were chosen to show the development during the last 30,000 calendar years BP, the last glacial (Oxygen Isotope Stage – OIS – 2), the deglaciation, and the Holocene interglacial.

The last Glacial (30–18 Cal. kyrs. BP) (OIS 2)
In the eastern Fram Strait, the last glacial maximum is characterised by heavy $\delta^{18}O$-values (Hebbeln et al., 1994; Elverhøi et al., 1995), and a maximum extent of the Svalbard Barents Sea Ice Sheet (SBIS) (Mangerud et al. 1992; Mangerud and Svendsen 1992; Mangerud et al. 1998; Mangerud et al. 2002) which was supported by increased moisture supply. This latter was a result of two Atlantic water advances, reaching the Fram Strait (Nordway Events = NE) between 34 and 28 (NE 2) and 24 and 18 (NE-1) Cal. kyrs. BP (Hebbeln et al. 1994; Dokken and Hald 1996; Hald et al. 2001). In our sediment cores, the NE-2 and NE-1 events are characterised by low OC values, mixed $\delta^{13}C_{org}$-values (–23 to –24‰) and terrigenous HI- values (10–30 mg HC/gOC) (Figs. 7.8.6 and 7.8.7) in eastern Fram Strait. During the NE-1 and NE-2 events, the entire Svalbard archipelago was covered by the SBIS with the result that no terrigenous or freshwater material could be exported from the fjords. In the NE-1/2 horizons of core PS2837, at least 8% of the benthic foraminifers came from the Atlantic (Wollenburg et al. subm.). Palaeoproductivity values calculated from benthic foraminifers, however, remained low due to a decreased supply of nutrients in the surface waters. In addition, low organic carbon values might derive from slightly increased numbers of benthic foraminifers (Wollenburg et al. subm.), which would have fed on organic carbon in the sediments. Except for ice-rafted detritus with low OC values, only small amounts of organic matter were transported to the seafloor.

From 26.9 to 23.8 Cal. kyrs. BP, an enormous OC peak was detected in all cores from the eastern Fram Strait (Fig. 7.8.6). This organic carbon event (Event I) (Knies and Stein 1998) can be assigned to the maximum ice-advance of the SBIS, an advance that started ca. 28 Cal. kyrs. BP and reached its maximum extension 23.6 Cal. kyrs. BP (Andersen et al. 1996; Knies et al. 2000). During this period, the Svalbard glaciers reached the shelf edge (Mangerud and Svendsen 1992; Mangerud et al. 2002), and fine-grained organic material was eroded from the shelves and exported over the shelf edge by the prograding ice (Fig. 7.8.6).

In contrast, Dokken and Hald (1996) postulated a short-term melting of the SBIS during the time of

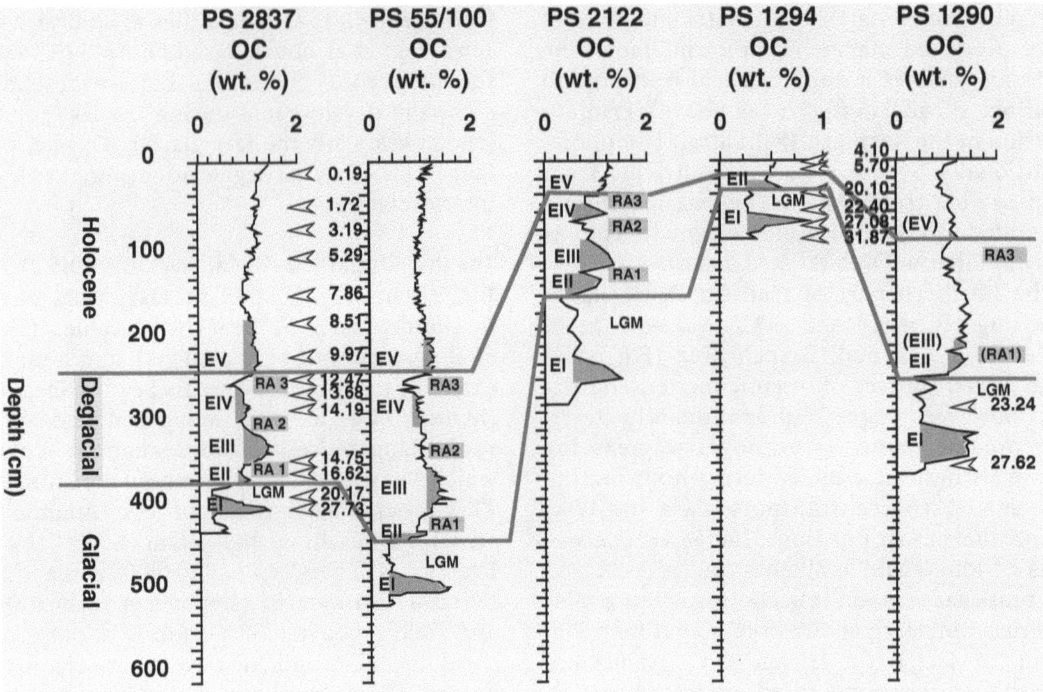

Fig. 7.8.6. OC records of 5 sediment cores from different locations in the Fram Strait (see Table 7.8.1, Fig. 7.8.1) used for core correlation (Hebbeln et al. 1994; Notholt 1998; Vogt et al. 2001). Glacial, deglacial and Holocene intervals are indicated. Arrows mark calibrated ages Cal. Kyrs. BP, calculated from ^{14}C ages using the equations of Bard (1998) and Stuiver et al. (1998). Original ^{14}C ages and age models for this paper are from Hebbeln (1992); Hebbeln et al. (1994); Notholt (1998); Vogelsang et al. (2001); Vogt et al. (2001); and Nørgaard-Pedersen et al. (in press). E I-V: high preservation events with an increased influx of organic carbon (Knies and Stein 1998; this paper). RA 1–3: re-advances of the SBIS (Svendsen et al. 1996; Mangerud et al. 1998) with decreased preservation and influx of organic carbon to investigated core positions

Event I. A quick retreat of glaciers and a massive outflow of suspended, fine-grained, organic-rich material to the areas of the northern seas were described (cited below). An additional source of OC-rich sediments (with a dominantly terrigenous origin as indicated by light $\delta^{13}C_{org}$ values) may have been the Spitsbergenbanken and the Barents Sea shelves (Bjørlykke et al. 1978; Elverhøi et al. 1995). The terrigenous organic matter accumulating in sediments at this time, however, was accompanied by a significant amount of marine organic matter, indicated by an increased hydrogen index and a maximum in brassicasterol (50 µg/gOC) (Fig. 7.8.7). Primary production probably did not increase during this period (Wollenburg et al. subm.) and the inflow of Atlantic water may even have decreased or temporarily stopped. The increased supply of suspended, fine-grained material led to aggregation of terrigenous debris with marine organic matter and, consequently accelerated transport through the water column (Knies and Stein 1998). Diminished feeding on organic matter by benthic foraminifers brought on by a deterioration in their living conditions due to short-term high accumulation rates probably also improved preservation of organic carbon in the sediment. Unfortunately, there are too few dated intervals to assess with confidence how long Event I lasted but it may have been only a few hundred years, comparable to the following events II to IV during the deglacial phase (see below). Event I was detected in all investigated cores of the eastern Fram Strait (Figs. 7.8.6 and 7.8.7).

Event I was identified in several cores from the Eastern Greenland margin (Nam et al. 1995), the Vøring Plateau, Northern Iceland Plateau (Wagner and Henrich, 1994), as well as in the southern Fram Strait (Hebbeln et al. 1994; Elverhøi et al. 1995), the northern Barents Sea (PS 2138; Knies and Stein 1998), and the northern Yermak Plateau (PS 2212; Vogt et al. 2001). The widespread effect of this event may indicate an intense and short-term retreat of large ice-shields all over the North Atlantic and the Northern Seas linked to an enormous outflow of suspended material. After the influx of fine-grained OC had ceased, the sedimentation of coarse-grained material and ice rafted detritus (IRD) increased from 24 Cal. kyrs. BP on (Hebbeln et al.

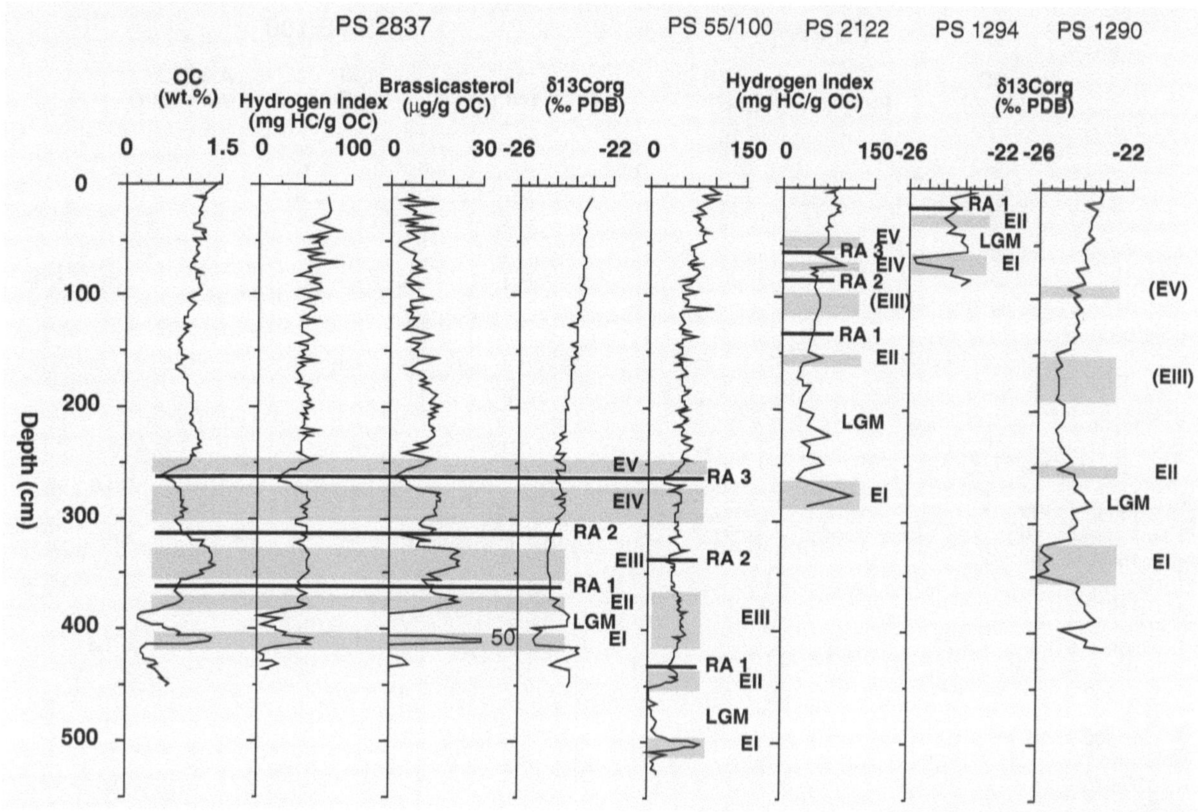

Fig. 7.8.7. Bulk organic-geochemical parameters $\delta^{13}C_{org}$ for cores PS 1290, PS1294 and PS 2837 and hydrogen indices for cores PS2122, PS2837 and PS55/100. For core PS2837 OC-values and Brassicasterol vs. depth are also shown. Data from cores PS1294 and PS1290 are from Hebbeln (1992) and Hebbeln and Wefer (1997), data from core PS2122 are taken from Vogt et al. (2001). E I–V: high preservation events with an increased influx of organic carbon to the sediments. RA 1–3: re-advances of the SBIS (after Svendsen et al., 1996 and Mangerud et al. 1998)

1994; Elverhøi et al. 1995; Andersen et al. 1996; Birgel 1999). A renewed influx of Atlantic water in southern Fram Strait (NE-1; Hebbeln et al. 1994; Dokken and Hald 1996) triggered the final build-up of the SBIS to the shelf edge. At the Last Glacial Maximum (LGM) the glaciers around Spitsbergenbanken and Svalbard reached the shelves and produced conditions comparable with the NE-2-Event (Andersen et al. 1996). Erosion of organic-carbon-rich siltstones from the Svalbard shelf regions and lateral transport in the WSC to the northern Yermak Branch and Svalbard Branch areas immediately ceased.

The Deglaciation (18–11 Cal. kyrs BP)

With the onset of the deglaciation (Termination I) a rapid shift to lighter oxygen isotope values was recorded in the Arctic Ocean and the southern Fram Strait (Jones and Keigwin 1988; Hebbeln et al. 1994; Stein et al. 1994b, 1994c; Knies and Stein 1998). These values were interpreted as a meltwater anomaly reflecting the onset of the SBIS decay (Elverhøi et al. 1995). A shift to lighter $\delta^{18}O$-values at the beginning of the deglaciation can be correlated with increasing OC values and HI indices in all investigated cores (Figs. 7.8.6 and 7.8.7). Due to insufficient age points, the entire deglaciation of the SBIS (15.7–11.9 Cal. kyrs. BP) was described by Knies and Stein (1998) as one single OC-event (Event II). However, there is evidence for a short-term OC-Event II (16.0 to 15.4 Cal. kyrs. BP) in our cores and three following additional short-term OC-Events (see below). These deglacial pulses mirror the successive decay of the SBIS with three re-advances toward glacial-like conditions in the Yermak (PS2837, PS55/100) and Svalbard branches (PS2122) of the WSC (Figs. 7.8.6 and 7.8.7).

From a comparison of all cores in the investigated area (Fig. 7.8.6, Table 1), we infer the SBIS to have commenced disintegration from ca. 18 Cal. kyrs. BP on (Fig. 7.8.6). The first signal was accompanied by an increased input of IRD and coarse-grained inorganic material (c.f., Elverhøi et al. 1995; Knies and Stein 1998). Accumulation rates remained low at first (Table 7.8.1; Fig. 7.8.8), plausibly because the ice-sheet of the Svalbard shelf was marine-based

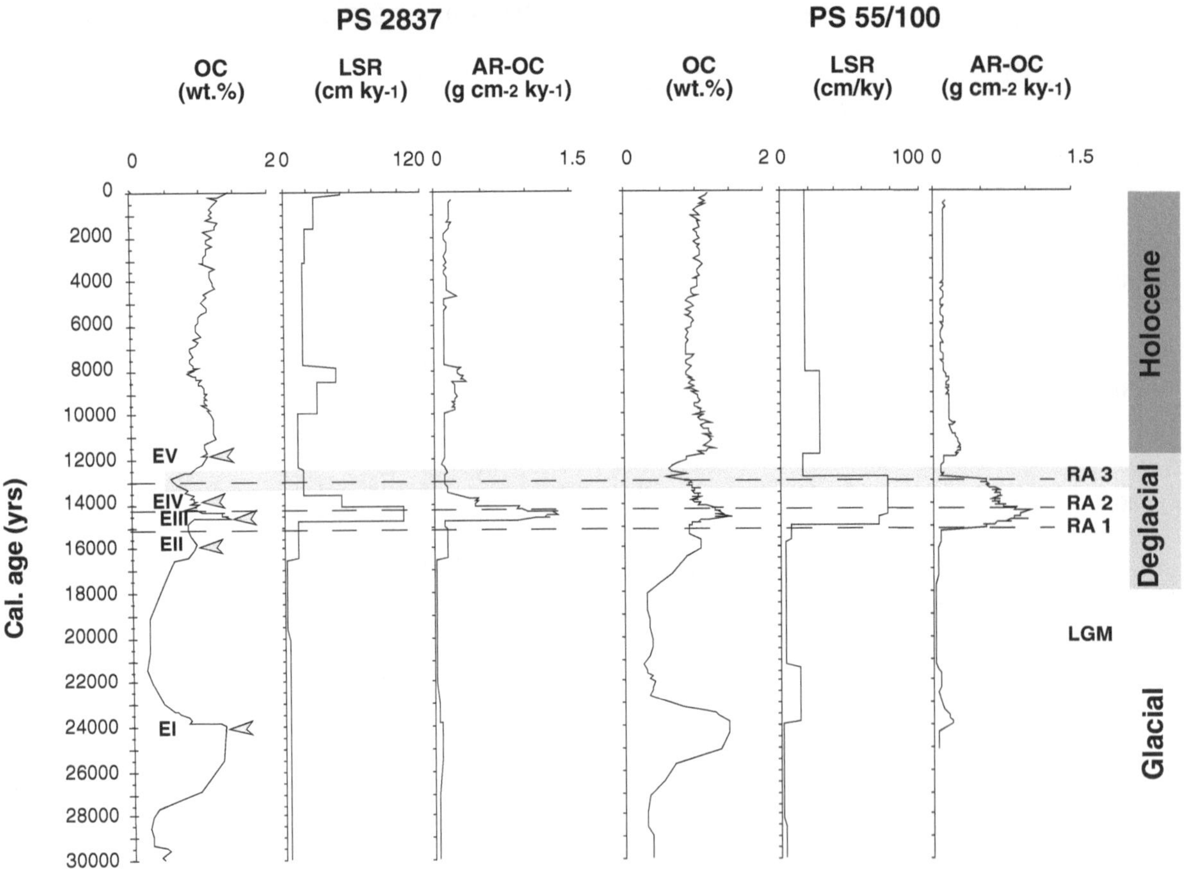

Fig. 7.8.8. Total organic carbon contents, linear sedimentation rates and accumulation rates of organic carbon of the cores PS2837 and PS55/100 versus calendar years. E I-V: high preservation events with an increased influx of organic carbon. RA 1-3: re-advances of the SBIS

incorporating only minor amounts of sediment into the sea-ice (Andersen et al. 1996). Beginning from 16 Cal. kyrs. BP on, more organic matter was deposited in eastern Fram Strait (Event II, 16 to 15.4 Cal. kyrs. BP) (Figs. 7.8.6 and 7.8.7). The deglacial material was probably composed of a mixture of terrigenous organic matter from a Svalbard source (HI: ca. 50 mg HC/gOC, $\delta^{13}C_{org}$ values: -23 to -24‰), re-suspended organic matter from a southern source (Hebbeln et al. 1994), and primary produced marine organic matter (Brassicasterol) (Fig. 7.8.7). An increased preservation of marine organic matter was caused by the "ballast effect" (Ittekkot et al. 1992; Knies and Stein 1998). Mangerud et al. (1992) presented evidence that the Svalbard glaciers did not disappear from the shelf areas until 15.1 Cal. kyrs. BP and therefore Svalbard cannot be the main and only source area of the OC peaks of Event II detected in our cores. After Event II, a short-term re-advance (RA-1; Figs. 7.8.5 and 7.8.6) of the Svalbard glaciers is suggested (Svendsen et al. 1996), accompanied by low OC values and low accumulation rates (Figs. 7.8.6 and 7.8.8). The following OC-event (Event III, Figs. 7.8.5 and 7.8.6) appears in the Yermak Branch cores (PS2837, PS55/100 and PS2122), with major changes in the flux of organic matter. Event III is another temporary OC-event (14.7 to 14.4 Cal. kyrs. BP), accompanied by the rapid Bølling short-term warm event. Linear sedimentation rates increased from values of 10 cm ky^{-1} up to 120 cm ky^{-1} in the Yermak Branch (Fig. 7.8.8). A massive deposition of clay at the position of PS 2837 accompanied by a major decrease of IRD to almost zero, was detected in the Yermak Branch area (Hass pers. comm.; Birgel 1999). OC values >1.1% were recorded in all cores from the eastern Fram Strait with light $\delta^{13}C_{org}$ and low HI, indicating a dominantly terrigenous source (Figs. 7.8.6 and 7.8.7), but a distinct portion of marine organic matter (brassicasterol) was preserved as well (see Event II).

This second, short-term breakdown, described as major melting of the glaciers on Svalbard, was paralleled by an enormous input of suspended material (Elverhøi et al. 1995; Andersen et al. 1996). Coarse particles were sedimented only on the shelf and the nearby slope areas of Svalbard. Enormous amounts of suspended organic material (Fig. 7.8.8) were transported laterally from the fjords, northwards by the WSC and downslope by turbidity currents. A smaller re-advance of glacial conditions (RA-2) was evident in cores PS2837, PS55/100 and PS2122, with a further decreased input of organic matter. This second cooling during the deglacial phase was immediately followed by a third, short-term melting event (Event IV, 14.2 to 13.7 Cal. kyrs. BP), comparable with the major deglacial Event III, but less pronounced (Figs. 7.8.5 and 7.8.6). Accumulation rates of organic matter were less during Event IV than during Event III (Fig. 7.8.8). The major melting of the SBIS and the maximum input and preservation, however, appeared during Event III. The last re-advance of the Svalbard glaciers (RA-3) occurred from 13.7 to 11.9 Cal. kyrs. BP (PS2837, Fig. 7.8.6), related to the Younger Dryas cold period (Hebbeln et al., 1998). Conditions during this phase were probably comparable with the LGM (Fig. 7.8.6). All organic parameters show minimum values and low accumulation rates of organic matter (Fig. 7.8.8) because of rapid prograding glaciers and the re-establishing of "LGM-like" conditions (Figs. 7.8.6 and 7.8.7) with ice-covered fjords at least in eastern Svalbard (Mangerud et al., 2002) and a reduced influx of fine-grained organic material from the shelves. Cold conditions lasted for ca. 1500 years (Fig. 7.8.8). Near 12 Cal. kyrs. BP the last deglaciation event (Event V) with a comparable input of organic matter as Events II-IV (Figs. 7.8.6 and 7.8.7) introduced stable climatic conditions. The stable inflow of Atlantic water established at this time increased the northward advection of heat leading to increased surface water temperatures indicated by low $\delta^{18}O$ values of planktonic foraminifera (Duplessy et al. 2001). The advected heat might also have initiated an earlier retreat of the glaciers and the sea-ice to modern distributions (Mangerud et al. 1992). OC values were comparable to late Holocene values and remained stable (0.9–1.1 wt.%) until 8 Cal. kyrs. BP.

The Holocene (11–0 Cal. kyrs. BP)

The Holocene is characterised by stable and warmer conditions in eastern Fram Strait (Hald et al. 1996) until ca. 9 Cal. kyrs. BP (PS 2837; Figs. 7.8.6 and 7.8.8). The glaciers retreated to their modern positions with only minor re-advances. The cessation of the supply of suspended material from glacial melting material produced a relatively low flux of organic matter to the seafloor (Figs. 7.8.6 to 7.8.8). Only between 8.40 to 8.13 Cal. kyrs. BP (Fig. 7.8.8, PS2837) is a short-term build-up of the Svalbard glaciers evident as indicated by low OC values. This short-term cooling event, observed synchronously in several locations on the northern hemisphere (e.g., Alley et al. 1997), may have been due to the collapse of the Laurentide ice-sheet releasing large amounts of meltwater. The main reason for this short-time cooling event, however, remains unclear. After this last cooling event, warmer conditions were established again and the modern marine and terrestrial environment remained stable until the present.

7.8.5 Accumulation Rates and Organic Carbon Budget

To quantify the organic carbon burial and its spatial and temporal variability in the study area, accumulation rates of organic carbon have been calculated for the three time intervals discussed above. A total sediment (TS) and total organic carbon (OC) budget have been estimated for the Holocene. During this time interval relatively stable environmental conditions can be assumed and there is a better database available than for the glacial and deglacial periods.

The last Glacial and Deglaciation

During the glacial period the environmental conditions in the eastern Fram Strait changed drastically (Fig. 7.8.9a). The accumulation rates of organic carbon were influenced by massive ice-sheets on the continents and by sea-ice, which covered almost the entire Fram Strait. Only a narrow gateway in eastern Fram Strait enabled Atlantic water to flow to the northern areas (Fig. 7.8.9a). Particulate organic and inorganic glacial debris were released from melting icebergs and sea-ice and transported laterally to northern areas (Fig. 7.8.9a).

At that time, TS and OC accumulation rates of 10.1 to 12.7 g cm^{-2} ky^{-1} and 0.07 to 0.1 g cm^{-2} ky^{-1}, respectively, were reached in the Svalbard Branch (Table 7.8.1; Fig. 7.8.9a). Compared with deglacial and Holocene accumulation rates, these values are significantly lower. At the Yermak Branch core positions, 4.6 to 6.0 g cm^{-2} ky^{-1} of TS and 0.03 to 0.04 g cm^{-2} ky^{-1} of OC were accumulated. These values are five times less (TS), and ten times less (OC) than sedimentation rates during the deglacial phase. In this phase, an increased deposition of TS and OC occurred (Fig. 7.8.9b), especially under the Yermak Branch area (on average 25.9 and 0.28 g cm^{-2} ky^{-1}, respectively) and on the West Svalbard slope (on average 15.7 and 0.18 g cm^{-2} ky^{-1}, respectively). These increased TS and OC accumulation rates are responses to the decay of the SBIS (see above).

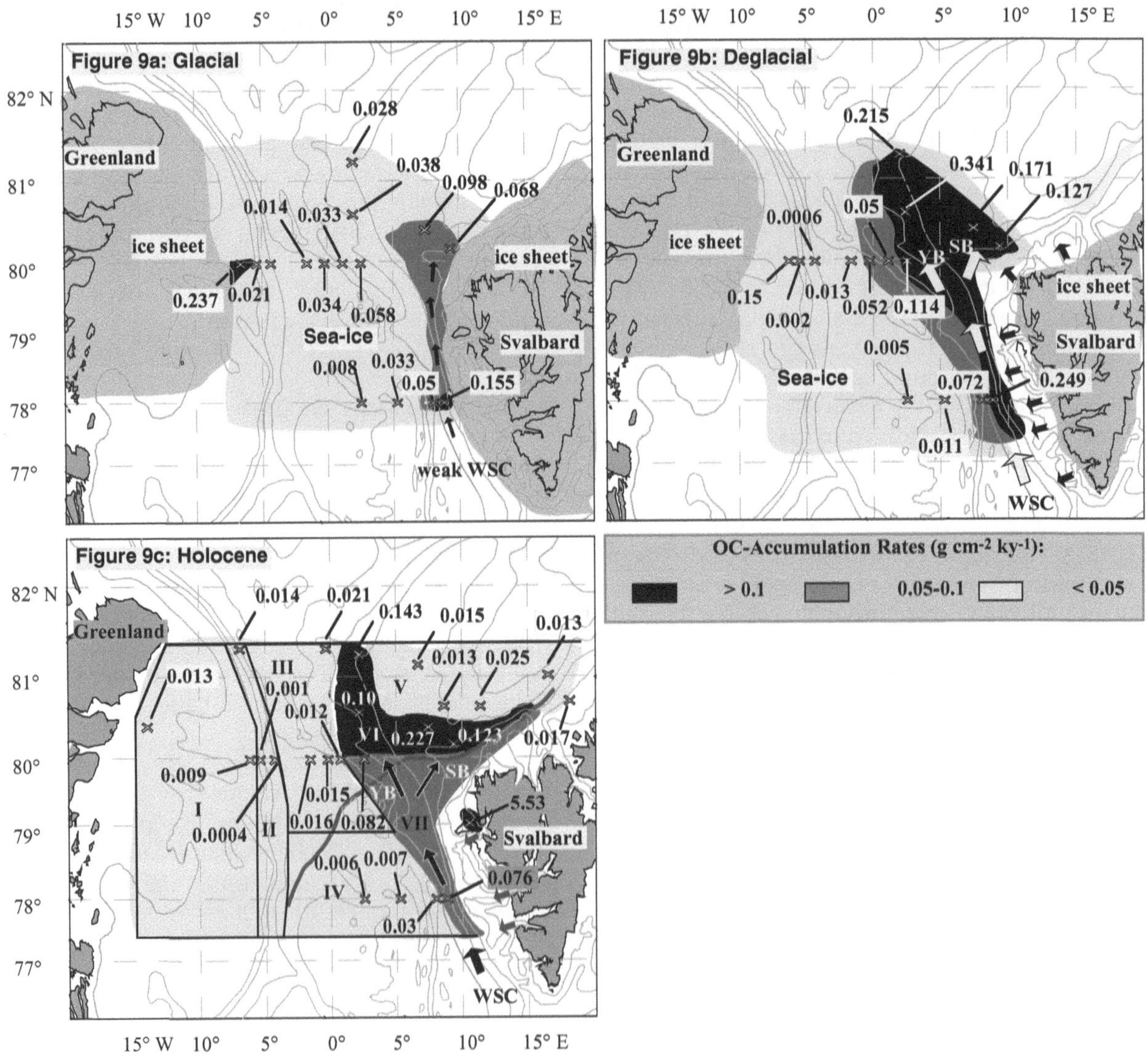

Fig. 7.8.9. The last Glacial (30–18 Cal. kyrs. BP), the Deglaciation (18–11 Cal. kyrs. BP) and the Holocene (11–0 Cal. kyrs. BP). (a–c) phases: OC accumulation rates (g cm^{-2} ky^{-1}; Table 7.8.1) and average ice distribution for the different time slices (after Andersen et al., 1996; Svendsen et al., 1996, 1999; Mangerud et al. 2002). Crosses mark positions, for which linear sedimentation rates and accumulation rates were calculated (see Table 7.8.1).
9a): Black arrows: Inflow of Atlantic water and open-water conditions (NE 1/2) (after Hebbeln et al. 1994).
9b): White arrows: increased inflow of warm, Atlantic water masses (WSC), in the Northern Fram Strait (divided into the YB = Yermak Branch, and SB = Svalbard Branch) asynchronous with events I–V. Black arrows: Influx of organic matter from the collapsing Svalbard-Barents Sea Ice Sheet (SBIS) synchronous with events I–V.
9c): Black arrows: increased inflow of warm, Atlantic water masses (WSC), in the Northern Fram Strait divided into the YB = Yermak Branch, and SB = Svalbard Branch. Grey arrows: Influx of organic matter from the collapsing Svalbard-Barents Sea Ice Sheet (SBIS). Roman numbers I–VII: Areas, for which an organic carbon budget was calculated (see Tables 7.8.1 and 7.8.2). Grey, thick line: Average summer sea-ice edge in the investigated area

Table 7.8.1. Average linear sedimentation rates (LSR), bulk and organic carbon accumulation rates (AR-TS, AR-OC) for the glacial, deglacial and Holocene time periods. The calculation of accumulation rates is based on mean values of total organic carbon (OC), LSR and dry density (DD). WD = water depth. See Figure 7.8.2 for core positions

	Core location	WD	LSR	OC	DD	AR-TS	AR-OC
		m	cm ky^{-1}	wt.%	g cm^{-3}	g Sed. cm^{-2} ky^{-1}	g C cm^{-2} ky^{-1}
East Greenland Shelf (Area I)							
PS 2419[a]	80°27′N, 13°40′W	412					
Holocene			6	0,15	1.44*	8,64	0,01
PS 2424[a]	80°02′N, 05°44′W	445					
Holocene			2,9	0,21	1.44*	4,15	0,009
Deglacial			26,9	0,42	1.32*	35,53	0,15
Glacial			41,2	0,38	1.5*	61,8	0,24
East Greenland Slope (Area II)							
PS 2843[b]	81°34′N, 07°17′W	2527					
Holocene			4,2	0.23′*	1,45	6,09	0,01
PS 2423[a]	80°02′N, 05°26′W	829					
Holocene			0,8	0,31	0.6*	0,45	0,001
Deglacial			1,1	0,2	0.86*	0,94	0,002
Glacial			7,2	0,3	1.02*	7,31	0,02
PS 2422[a]	80°01′N, 04°13′W	1938					
Holocene			0,3	0,21	0.6*	0,2	0,0004
Deglacial			0,3	0,24	0.86*	0,28	0,0006
Northern Central Fram Strait (Area III)							
PS 2839[b]	81°24,3′N, 01°01′W	2933					
Holocene			3,3	0.81#	0,8	2,64	0,02
PS 1310[c]	80°00′N, 00°82′W	2628					
Holocene			2,4	0,74	0,94	2,22	0,02
Deglacial			2,4	0,45	1,21	2,85	0,01
Glacial			3,4	0,38	1,09	3,71	0,01
PS 1311[d]	79°97′N, 00°20′E	2628					
Holocene			2,5	0,66	0,79	1,98	0,01
Deglacial			6,9	0,78	0,97	6,71	0,05
Glacial			7,8	0,43	1,02	7,93	0,03
PS 1312[d]	79°99′N, 01°17′E	3593					
Holocene			2,5	0,56	0,86	2,15	0,01
Deglacial			3,1	1,12	1,44	4,42	0,05
Glacial			4,8	0,59	1,15	5,53	0,03
Southern Central Fram Strait (Area IV)							
PS 1295[e]	77°99′N, 02°46′E	3112					
Holocene			1,4	0,53	0,74	1,06	0,006
Deglacial			1,4	0,42	0,84	1,18	0,005
Glacial			2,3	0,39	0,92	2,15	0,008
PS 1294[e]	78°00′N, 05°34′E	2677					
Holocene			1,8	0,5	0,84	1,47	0,007
Deglacial			2,8	0,45	0,91	2,5	0,01
Glacial			3,9	0,4	0,93	3,61	0,03
PS 1291[d]	78°00′N, 08°07′E	2408					
Holocene			3,2	0,71	1,3	4,16	0,03
Deglacial			8,4	1,12	0,76	6,43	0,07
Glacial			3,9	0,81	1,56	6,13	0,05

Table 7.8.1 (continued)

	Core location	WD	LSR	OC	DD	AR-TS	AR-OC
		m	cm ky^{-1}	wt.%	g cm^{-3}	g Sed. cm^{-2} ky^{-1}	g C cm^{-2} ky^{-1}
Yermak Plateau (Area V)							
PS 2835[b]	81°06,1′N, 07°04′E	855					
Holocene			2,5	0.75#	0,84	2,1	0,02
PS 2834[b]	80°54,8′N, 09°49′E	1024					
Holocene			2,5	0.78#	0,71	1,78	0,01
PS 2833[b]	80°58,4′N, 11°49′E	1969					
Holocene			3,8	0.86#	0,78	2,96	0,03
PS 2831[b]	81°06′N, 16°52′E	1024					
Holocene			2,1	0.53#	1,19	2,5	0,01
PS 2830[b]	80°58,4′N, 17°28′E	513					
Holocene			2,5	0.44#	1,57	3,92	0,02
(Yermak/Svalbard Branch) (Area VI)							
PS 2837	81°13′N, 02°22′E	1042					
Holocene			22,6	1,09	0,58	13,08	0,14
Deglacial			24	1,04	0,86	20,64	0,22
Glacial			4,1	0,61	1,12	4,56	0,03
PS 55/100	80°29′N, 02°56′E	1538					
Holocene			20	1,01	0,53	10,59	0,11
Deglacial			38	1,1	0,82	31,13	0,34
Glacial			6,2	0,62	0,98	6,03	0,04
PS 2123[f]	80°16′N, 09°85′E	574					
Holocne			11,4	1,38	1,44	16,46	0,23
Deglacial			10,9	1,15	1,36	14,85	0,17
Glacial			7	0,77	1,83	12,74	0,1
PS 2122[f]	80°39′N, 07°54′E	702					
Holocene			9,3	1,08	1,22	11,35	0,12
Deglacial			9	1,06	1,33	11,98	0,13
Glacial			7,8	0,67	1,3	10,11	0,07
West Svalbard Slope (Area VII)							
PS 1313[d]	79°98′N, 02°81′E	2623					
Holocene			11,5	0,9	0,79	9,09	0,08
Deglacial			12,3	1,01	0,92	11,33	0,11
Glacial			7,8	0,58	1,29	10,02	0,06
PS 1290[g]	78°00′N, 08°72′E	1522					
Holocene			12,8	1,01	0,59	7,54	0,08
Deglacial			12,9	1,24	1,56	20,12	0,25
Glacial			15,7	0,86	1,15	18	0,16
West Svalbard Shelf PS 2121[h]	79°01′N, 10°43′E	337					
Holocene			97,9	2,16	2,62	256,4	5,54

[a] Notholt (1998),
[b] Stein and Fahl (1997), preliminary age model and LSR (see cruise report)
[c] Hebbeln (1991)
[d] Hebbeln (1992)
[e] Hebbeln and Wefer (1997)
[f] Vogt et al. (2001)
[g] Elverhoi et al. (1995)
[h] Müller (1995)

* = estimated DD, no DD was measured for this cores.
*′ = estimated OC from nearby surface samples (this paper)
\# = averaged OC values from surface samples (this paper)
YB/SB = Yermak Branch/Svalbard Branch of the WSC (see Fig. 8 and 9).

Taylor et al. (2002) calculated average linear sedimentation rates for the North Atlantic up to the southern tip of the West Svalbard slope. They subdivided the last 27,000 years in two time slices, the Late Weichselian (27–12 Cal. kyrs. BP) and the Holocene (12–0 Cal kyrs. BP). Sedimentation rates of 10–15 cm ky^{-1} and OC accumulation rates of 0.08–0.12 g cm^{-2} ky^{-1} were calculated for the Late Weichselian in the West Svalbard slope zone. Our linear sedimentation rates are comparable with their values for both time slices. The OC accumulation rates of the West Svalbard slope zone from Taylor et al. (2002) are between 0.08–0.12 g cm^{-2} ky^{-1} for the Late Weichselian. We found comparable glacial values of 0.06–0.16 g cm^{-2} ky^{-1}, but values of 0.11 to 0.25 g cm^{-2} ky^{-1} in the deglacial. Accumulation rates in our investigation areas change significantly from the glacial to the deglacial. The disintegration history of the SBIS must be taken into account more precisely, and therefore the glacial and deglacial time slices should be discussed separately.

The western part of the Northern Fram Strait remained under a heavy sea-ice sheet in the deglacial period and Greenland and the adjacent shelf areas were covered by massive ice-sheets (Fig. 7.8.9b). However, the average accumulation rates (34.6 for TS and 0.13 g cm^{-2} ky^{-1} for OC) are twice as high as during the deglacial phase. The increased supply of glacial material from Greenland is obvious. The deep-sea environment of western Fram Strait was not affected by environmental changes and OC accumulation rates, therefore, did not change noticeably in this area over the last 30,000 years (Table 7.8.1).

The Holocene

Mean Holocene OC accumulation rates in the Northern Fram Strait/ Yermak-Plateau vary between 0.0004 g cm^{-2} ky^{-1} (OC) on the eastern Greenland slope and 0.227 g cm^{-2} ky^{-1} in the Svalbard Branch of the WSC (Fig. 7.8.9c). The corresponding values for TS accumulation are 0.2 and 16.6 g cm^{-2} ky^{-1}. The maximum accumulation rates, however, were obtained on the western Svalbard shelf where 256 (TS) and 5.5 g cm^{-2} ky^{-1} (OC) were attained (Table 7.8.1). For the western part of the Fram Strait, i.e., in the EGC-dominated areas I and II (Fig. 7.8.9c), only 0.2 to 6.1 g cm^{-2} ky^{-1} (TS) and 0.0004 to 0.01 g cm^{-2} ky^{-1} (OC) were estimated. Even in the highly productive open-water zone of the NE-polynya (Fig. 7.8.1) the accumulation rates of organic matter were low for the last 11 Cal. kyrs. BP (Fig. 7.8.9c, Table 7.8.1). In central Fram Strait (areas III and IV; Fig. 7.8.9c), slightly increased values compared with the NE-Greenland areas I and II were obtained with 1.1 to 4.2 g cm^{-2} ky^{-1} (TS) and 0.006 to 0.03 g cm^{-2} ky^{-1} (OC) (Table 7.8.1).

The northernmost Area (V) is characterised by low accumulation rates of TS (1.78 to 3.9 g cm^{-2} ky^{-1}) and OC (0.01 to 0.03 g cm^{-2} ky^{-1}; Table 7.8.1, Fig. 7.8.9c). A perennial sea-ice cover, low primary productivity, less light and nutrients result in decreased production of organic carbon on the Yermak Plateau. Furthermore, the Yermak Plateau is not influenced strongly by Atlantic water masses and lateral and vertical transport of nutrients and suspended organic matter by the WSC are weakened or interrupted.

Unlike the low accumulation rates of Area I-V, areas VI and VII showed the highest Holocene accumulation rates of organic material (7.5 to 16.5 g cm^{-2} ky^{-1} for TS and 0.08 to 0.23 g cm^{-2} ky^{-1} for OC) in the Fram Strait. The maximum values in Area VI are comparable with those documented in other continental margins of the world oceans (see e.g., Stein 1991). The accumulation rates of organic matter increase monotonically from the southernmost tip of Area VII following the WSC up to the Svalbard and Yermak branches at average water depths between 600 and 1600 m (Fig. 7.8.9c).

Holocene Total Sediment and Organic Carbon Budgets

The investigated region with a total size of 280,000 km^2 was subdivided into eight areas. Budgets of total sediment (TS) and total organic carbon (OC) were calculated from the accumulation rates in Table 7.8.1 (see also Table 7.8.2, Fig. 7.8.9c).

During the entire Holocene (0-11 Cal. kyrs. BP), 165 × 10^9 t of TS and 0.95 × 10^9 t of OC were deposited in the Northern Fram Strait/Southern Yermak Plateau area (Table 7.8.2). Based on the organic-carbon source indicators (hydrogen indices; $\delta^{13}C_{org}$-values) determined in a selected set of sediment cores (see above), about 45% of the organic carbon (i.e., 0.43 × 10^9 t) is estimated to be of marine origin and 55% (i.e., 0.52 × 10^9 t) of terrigenous origin. The main depocenter was determined to be in the vicinity of the WSC (areas VI and VII), where 65% (or 0.63 × 10^9 t) of the OC and 36% of the TS (or 59.9 × 10^9 t C) were buried in 20% of the total area. The deep-sea environments and the Yermak Plateau (areas II–V) together received 21% of the TS and 23% of the OC of the total budget. On the East Greenland shelf (Area I) 71 × 10^9 t of TS (43% of total budget) and only 0.11 × 10^9 t of OC (11% of total budget) were accumulated.

Table 7.8.2. Flux, budget and total accumulation of total sediment (TS) and OC for the last 11 Cal. kyrs. BP (Holocene)

Deposition of organic carbon	Approx. area (km²)	Flux (g cm⁻² ky⁻¹)		Budget 10⁶ t C/year		Total Accumulation (10⁹ t)	
		Total sediment	OC	Total sediment	OC	Total sediment	OC
(East Greenland Shelf (Area I)	101.000	6,40	0,01	6,46	0,01	71,10	0,11
East Greenland Slope (Area II)	17.000	2,25	0,005	0,38	0,0009	4,21	0,01
Northern Central Fram Strait (Area III)	35.000	2,25	0,02	0,79	0,007	8,66	0,08
Southern Central Fram Strait (Area IV)	41.000	2,23	0,01	0,91	0,004	10,06	0,05
Yermak Plateau (Area V)	31.000	3,32	0,02	1,03	0,006	11,31	0,07
SB/YB (Area VI)	19.000	12,87	0,15	2,45	0,03	26,90	0,31
West Svalbard Slope (Area VII)	36.000	8,32	0,08	3,00	0,03	32,95	0,32
(West Svalbard Shelf)	no cores	256,42	5,54	not calculated	not calculated	not calculated	not calculated
Total Holocene	**280.000**			**15**	**0,09**	**165,2**	**0,95**

7.8.6 Summary and Concluding Remarks

- In general, the preservation and burial of organic matter in the Fram Strait/Yermak Plateau area are controlled by: (1) strong lateral transport of suspended matter by Atlantic water during the Holocene and by meltwater plumes during the deglacial phase and probably during the glacial phase; (2) enhanced vertical transport facilitated by water masses containing increased amounts of suspended organic matter and therefore better capturing and preserving of marine organic carbon through aggregation with terrigenous organic matter; (3) increased primary production due to upwelling of nutrients near the sea-ice edges, on the shelves and in the fjords; and (4) preferential deposition and preservation of organic carbon due to high bulk accumulation rates.
- During approximately the last 30,000 cal. yrs., increased primary production has not been the major reason for increased preservation of organic carbon in the sediments. Rather, it is the lateral input of suspended terrigenous organic matter from the continents and the associated "ballast effect" which have resulted in a quick vertical transport and increased preservation of marine organic matter in the sediments. High accumulation rates of inorganic material, however, are needed to support aggregation and deposition of the organic carbon. Furthermore, allochthonous organic matter (both marine and terrigenous) is transported to the northernmost Fram Strait by strong currents.
- In total, 0.95×10^9 t C of total organic carbon have been deposited in the Northern Fram Strait/Southern Yermak Plateau area during the Holocene, i.e., the last 11 Cal. kyrs. BP. Of this, about half is terrigenous, half marine. Most of the organic carbon (65%) has been buried in sediments under Yermak/Svalbard branch, which accounts for only 20% of the total area.

7.9 The Central Arctic Ocean: Distribution, Sources, Variability and Burial of Organic Carbon

R. Stein, C. Schubert, R.W. Macdonald, K. Fahl, H.R. Harvey, D. Weiel

7.9.1 Introduction

Although it is generally accepted that the Arctic Ocean is sensitive to climate change and an important component of the global climate system, this region remains the last major physiographic province of the earth whose short- and long-term geological history is not very well known. Since the first recovery and description of deep-sea sediments during the famous 1893–1896 FRAM-Expedition of Fridtjof Nansen (Nansen 1897; Böggild 1906), the progress of gaining a better understanding of the Arctic Ocean system and its relationship to global change has been slow in comparison to other ocean regions. This lack of knowledge is caused mainly by the major technological/logistic difficulties in reaching this permanently ice-covered region with normal research vessels and in retrieving long, undisturbed sediment cores. On one hand, the available sediment samples and data from the central Arctic Basins are derived from drifting ice islands such as T-3 (e.g., Clark et al. 1980) and CESAR (Jackson et al. 1985). On the other hand, a restricted number of ship expeditions have been carried out (see Chapter 1.3, Table 1.3.2) which have greatly enhanced our knowledge on Arctic Ocean (paleo-) environments.

Modern Environmental Setting

The modern environmental setting, i.e., the modern hydrography as well as sea-ice characteristics, are described in more detail in Chapter 1.2. Here, we summarize only a few aspects important for the input, distribution, and accumulation of organic carbon.

Today, two major current systems dominate the surface-water circulation in the Arctic Ocean: the anticyclonic Beaufort Gyre in the Amerasian Basin and the Transpolar Drift in the Eurasian Basin (Fig. 7.9.1). The latter system transports sea ice from the Siberian shelves through the central Arctic Ocean to the Fram Strait. Through Fram Strait, two major currents exchange water between the Arctic and the world ocean (see Chapter 1.2). The cold, ice-transporting East Greenland Current is the main current out of the Arctic Ocean. In contrast, the Westspitsbergen Current, an extension of the North Atlantic-Norwegian Current, carries warm, relatively saline water into the Arctic Ocean where it cools and extends at intermediate water depths into the eastern Arctic Basins (Aagaard et al. 1985; Carmack 1990; Rudels et al. 1994; Fahrbach et al. 2001). Atlantic water dominates the Eurasian Basin whereas surface waters (0–400 m) of the Amerasian Basin are strongly influenced by Pacific water entering the Arctic Ocean through Bering Strait. Of importance for the thermohaline circulation in the Arctic Ocean is the formation of brines on polar shelves, i.e., cold, saline, and well-oxygenated water masses which sink over the continental margin into the deep basins (Aagaard et al. 1985). The export of sea ice through Fram Strait and its melt in the Greenland-Norwegian Sea together with the export of copious amounts of freshwater from rivers play an important role in controlling the deep-water formation in the northern North Atlantic and, thus, is of global significance (Aagaard and Carmack 1989; Untersteiner and Carmack 1990).

Another important phenomenon of the Arctic Ocean is the permanent sea-ice cover with its strong seasonal variation in the marginal (shelf) areas (see Chapter 1.2). This sea-ice cover has a distinct influence on marine biota, oceanic circulation, and surface albedo, which all are major controls on climatic change. A more closed sea-ice cover restricts primary production due to low light influx in the surface waters, e.g. only 0.1% of incident light penetrates snow and ice (assuming 8 cm of snow on 3.5 m of sea-ice) to reach the surface water (Andersen 1989). For the deep Arctic Ocean, low annual primary production values of about 15 g C m^{-2} are given, assuming a 120-day growth season (Gosselin et al. 1997; see Chapter 3). Ice algal production is particulately important in the central Arctic where it can contribute over half of the total primary production (Gosselin et al. 1997). However, productivity may be very high at the ice edge due to supply of nutrients and stratification of surface waters during melting phases and ice-edge upwelling, resulting in phytoplankton blooms (see Chapter 3).

A large proportion of the sea ice is "dirty ice" containing high amounts of sediment as well as terrigenous organic carbon (Pfirman et al. 1989, 1997; Wollenburg 1993; Nürnberg et al. 1994; see Chapter 2.4). Most of the sediment is incorporated into the sea ice in the Siberian shelf areas and then transported as ice-rafted debris (IRD) through the central Arctic Ocean via the Transpolar Drift. In areas of extensive melting, sediment

Fig. 7.9.1. Map of the Arctic Ocean showing average organic carbon (OC) contents (%) for different areas. Question marks highlight areas where no data are available. Black dots indicate OC data points (data points from the Barents Sea and East Siberian Sea are not included; see Chapters 7.4 and 7.7). The location of data transects A, B, and C shown in Figures 7.9.2 to 7.9.4, are marked as black lines. Gray arrows mark the two major surface water current systems (Beaufort Gyre and Transpolar Drift)

particles are released and deposited at the sea floor. In these areas, IRD may dominate the supply and accumulation of terrigenous material. Sediment input from sea ice may also be relatively more important on Arctic Ocean ridges and elevations protected from sediment supply by turbidity currents (e.g., Stein and Korolev 1994; Behrends et al. 1999; Wahsner et al. 1999). In the deep Arctic basins, on the other hand, turbidity currents may become the dominant source for sediment and organic carbon input (e.g., Fütterer 1992; Stein et al. 1994a; Grantz et al. 1999; Svindland and Vorren 2002).

7.9.2
Data Base, Material and Methods

In this synthesis paper, organic carbon content (OC), organic carbon/total nitrogen (C/N) ratios, Rock-Eval parameters (hydrogen index and Tmax values), carbon stable isotopes of the organic matter, as well as specific biomarkers (*n*-alkanes) have been used to determine the amount and composition of the OC fraction in surface sediments and sediment cores. Most of the sediment data presented in this paper have been extracted from the published literature. Data on OC content as well as geochemical bulk parameters (i.e., C/N ratios, Rock-Eval pyrolysis data, and $\delta^{13}C_{org}$ values) in the Eurasian Arctic Ocean were obtained from Stein et al. (1994a, b, c), Schubert and Stein (1996, 1997), Knies et al. (2000), Schubert et al. (2001), Schubert and Calvert (2001), and Stein et al. (2001). In addition, nitrogen stable isotope composition has been used as a proxy for nutrient utilization and primary production (Schubert et al. 2000; Schubert and Calvert 2001). From the Eurasian Arctic Ocean area, *n*-alkane data are also available (Schubert and Stein 1996; Knies et al. 2000; Fahl, unpubl. data). For the Amerasian Arctic Ocean, the data base, especially for sediment cores, is more limited. Data on OC content, C/N ratios and *n*-alkane distribution have been obtained from Darby et al. (1989), Clough et al. (1997), Cranston (1997), Grantz et al. (1999), and Belicka et al. (2002).

Further details about the use of proxies as organic-carbon-source indicators are described in Chapter 1.4. For the description of the various methods applied to produce the data, we refer to the cited literature.

For the interpretation of the organic-geochemical records in terms of paleoenvironmental changes versus time, and for the calculation of sedimentary and OC budgets, a chronological framework is necessary. This framework is based mainly on AMS^{14}C dates (Zahn et al. 1985; Stein et al. 1994b, c; Darby et al. 1997; Nørgaard-Pedersen et al. 1998, 2002; Grantz et al. 1999; Knies et al. 2000, 2001). Furthermore, oxygen isotope stratigraphy, paleomagneto-stratigraphy, biostratigraphy (e.g., dinoflagellates and coccoliths), lithostratigraphy, and magnetic susceptibility records have been considered (Gard 1986; Baumann 1990; Gard and Backman 1990; Nowaczyk and Baumann 1992; Gard 1993; Nowaczyk et al. 2000; Stein et al. 2001; Matthiessen et al. 2001; Nørgaard-Pedersen et al. 2002).

7.9.3
Distribution and Sources of Organic Carbon in Surface Sediments

In comparison to the Arctic marginal seas (see Chapters 7.2 to 7.7), the data base on OC distribution in the central Arctic Ocean is very sparse (Fig. 7.9.1). Thus, it is not possible to produce an OC distribution map for the entire Arctic Ocean. Information on the OC content and composition is available only from restricted areas and transects. Although the ice-covered central Arctic Ocean is of low productivity, the OC content in the surface sediments is generally high compared to other world open ocean areas and varies between 0.4 and about 2% (Stein et al. 1994a). Based on the existing data, the continental slopes are characterized by OC contents of about 1% (Fig. 7.9.1). In the major basins such as the Canada Basin, the Amundsen Basin and the Makarov Basin, OC values of 0.7 to 1% are typical. On the central Gakkel Ridge as well as the Lomonosov Ridge, the OC contents seem to be lower (0.3 to 0.6%) (Figs. 7.9.2 and 7.9.3). Toward the Eurasian continental margin, OC contents increase to 0.8 to 1.0% in surface sediments over the Lomonosov Ridge, with higher values more typical for the Amundsen Basin side (Fig. 7.9.4).

Although there is no clear correlation between OC content and grain size, OC contents > 0.5% occur mainly in fine-grained sediments with a silt plus clay content of > 90% (Fig. 7.9.5). These fine-grained sediments are concentrated in the basins whereas on the oceanic ridges and plateaus higher fractions of sand tend to be observed (Stein et al. 1994a).

A likely cause for the high OC content of Arctic surface sediments is the relatively high contribution of terrigenous organic matter derived from the surrounding marginal seas. Major transport processes involved are ocean currents and input via sea ice as well as turbidity currents (Stein et al. 1994a and further references therein). The latter process is especially important for the deep ocean basins (Stein et al. 1994a; Grantz et al. 1999). A remaining question, however, is what are the quantitative contributions the marine and terrigenous proportions of the organic matter preserved in the central Arctic Ocean sediments. Studies based on different OC source proxies give very different results (e.g., Stein et al. 1994a; Schubert and Stein 1996, 1997; Schubert and Calvert 2001).

Organic-geochemical Bulk Parameters as Source Indicator

The generally low hydrogen index values of < 100 mgHC/gOC measured in surface sediments from the Eurasian Basin (Figs. 7.9.2 and 7.9.6)

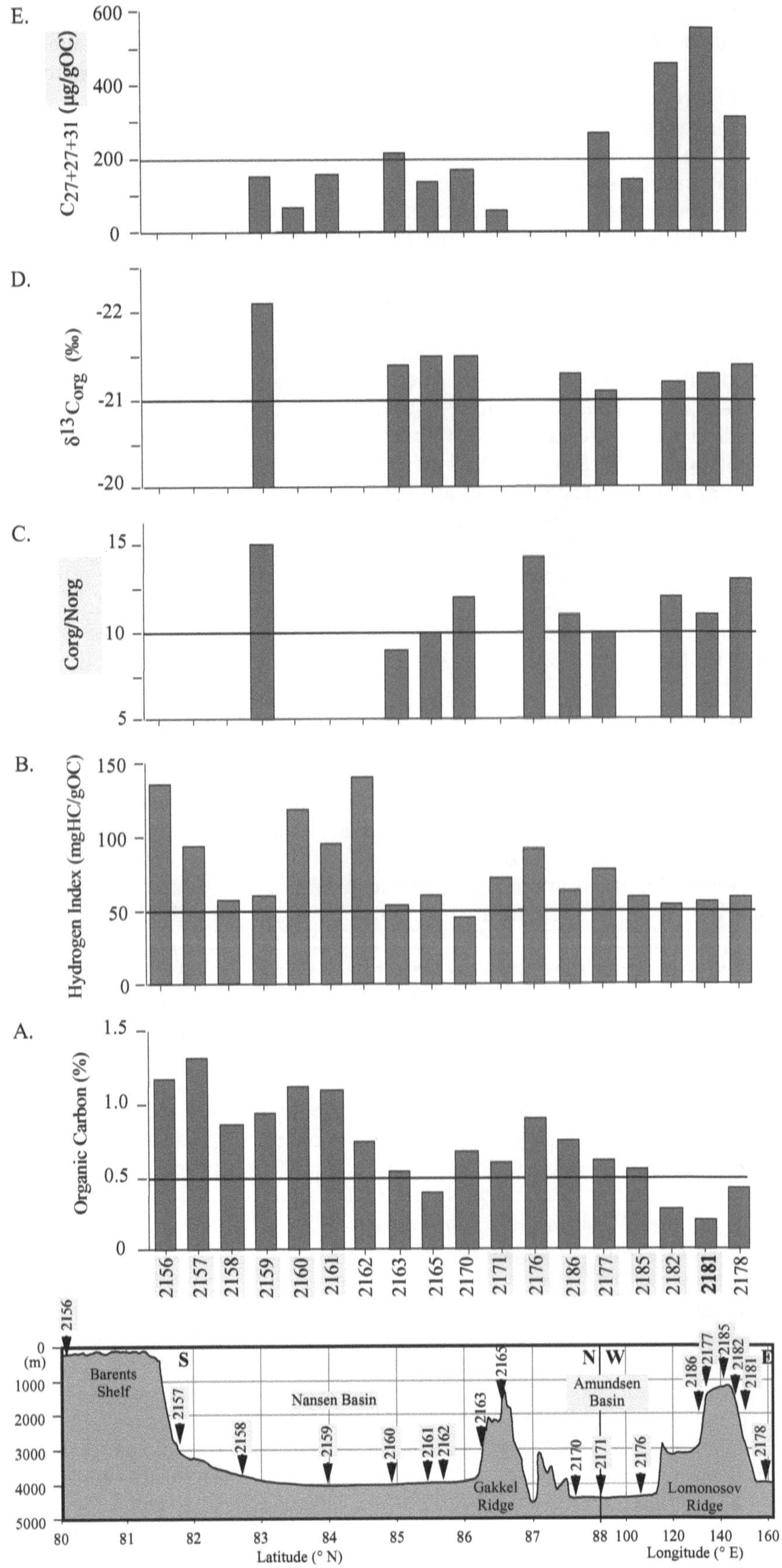

Fig. 7.9.2. Transect A from the Barents Sea shelf across the Nansen Basin, Gakkel Ridge Amundsen Basin and Lomonosov Ridge towards the Makarov Basin (for location see Fig. 7.9.1). (A) Organic carbon contents; (B) hydrogen index values; (C) C_{org}/N_{org} ratios; (D) carbon stable isotopes of the organic matter; and (E) concentrations of long-chain *n*-alkanes (sum of $C_{27} + C_{29} + C_{31}$)

imply a clear predominance of terrigenous organic matter. Only close to the Svalbard and Barents Sea continental margin, where elevated hydrogen indices of 100 to 250 mgHC/gOC were determined, is increased preservation of marine organic matter indicated. These increased marine organic carbon values are probably caused by increased (ice-edge) surface-water productivity due to reduced sea-ice cover and increased nutrient supply, both triggered by the inflow of warm Atlantic water via the Westspitsbergen Current. The somewhat higher hydrogen index values determined in surface sediments from the central Nansen Basin are related to turbidity currents causing rapid burial and increased preservation of marine organic matter (Fig. 7.9.2; Stein et al. 1994a). Hydrogen index values allow only a very general classification of the origin of the organic matter and do not provide a reliable estimate of the percentage composition of terrigenous and marine organic matter.

Low C/N ratios of 4 to 8 (Fig. 7.9.2) determined in the sediments are more typical for marine organic matter. The sediments from the central Arctic, however, contain a high proportion of inorganic nitrogen (N_{bou}) bound as ammonium in clay minerals (Fig. 7.9.7; Schubert and Calvert 2001; see Chapter 1.4). After correction for N_{bou}, the "C_{org}/N_{org}" ratios lie between 8 and 15 (Fig. 7.9.2) indicating a more mixed marine/terrigenous source for the organic matter. Across the Lomonosov Ridge, close to the Laptev Sea continental margin, C/N ratios are generally higher on the Amundsen Basin side and lower on the Makarov Basin side (Fig. 7.9.4). This may reflect a higher terrigenous influence in the Amundsen Basin and a higher marine influence in the Makarov Basin. The latter may have been caused by higher primary productivity supported by the influence of nutrient-rich Pacific waters and supply from the East Siberian Sea. This explanation is supported by biomarker data from sediment trap samples collected across the Lomonosov Ridge, indicating significantly higher amounts of marine organic matter in the water column at locations east of the Lomonosov Ridge (Fahl et al. 1997). In the Amerasian Basin, C/N ratios vary between 8 and 15 (Fig. 7.9.3; see Cranston 1997), again suggesting a mixed marine/terrigenous OC source for this part of the Arctic.

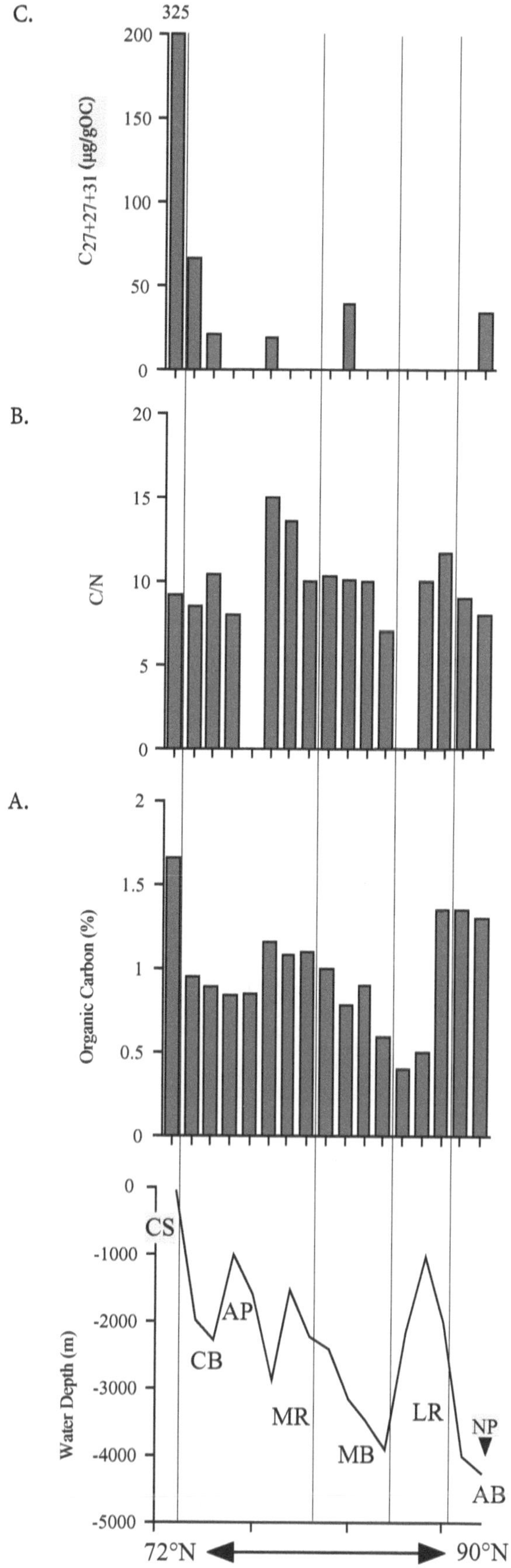

Fig. 7.9.3. Transect B from the Chukchi Sea shelf (CS) across the Canada Basin (CB), Arlis Plateau (AP), Mendeleev Ridge (MR), Makarov Basin (MB), and Lomonosov Ridge (LR) towards the Amundsen Basin (AB) (for location see Fig. 7.9.1). NP indicates North Pole. (A) Organic carbon contents; (B) C/N ratios; and (C) concentrations of long-chain *n*-alkanes (sum of C_{27} + C_{29} + C_{31}) (data from Clough et al. 1997; Cranston 1997; Belicka et al. 2002)

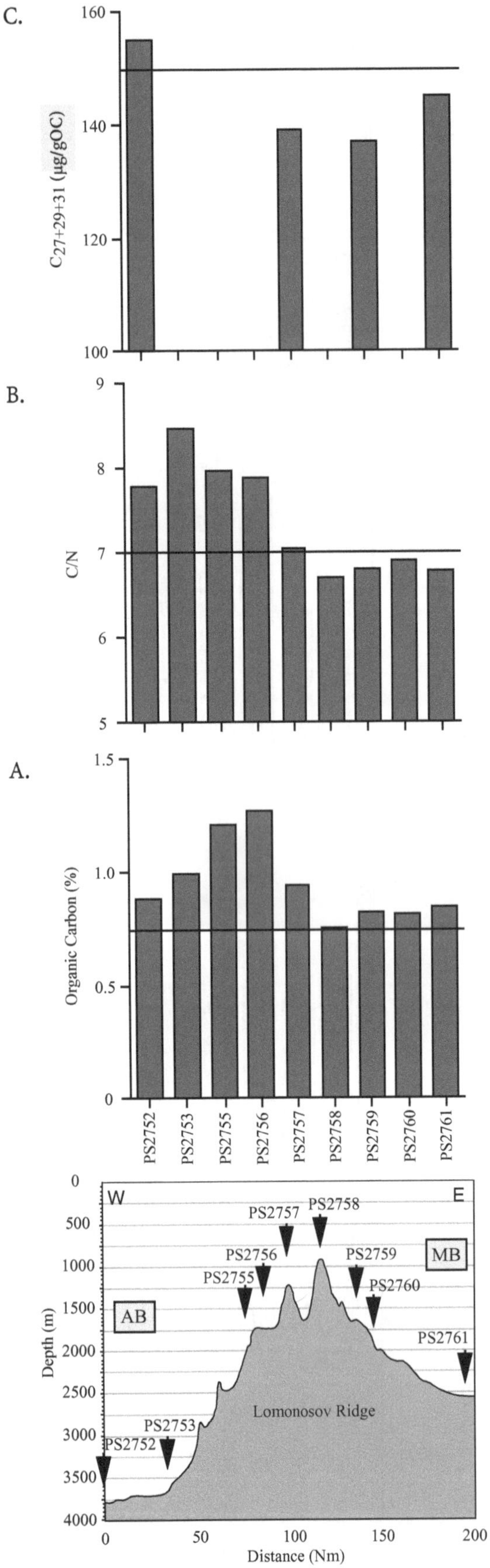

The very heavy $\delta^{13}C_{org}$ values ranging between −21.1 and −22.1‰ (Fig. 7.9.2) across the Eurasian Basin are somewhat surprising in view of the suggestion that there is a high terrigenous organic matter component in the sediments and that $\delta^{13}C_{org}$ values of high latitude phytoplankton can approach terrestrial values (see Chapter 1.4 for further discussion and references). Based on the $\delta^{13}C_{org}$ values, Schubert and Calvert (2001) estimated the maximum terrigenous contribution to be only about 30% of the organic matter in central Arctic Ocean sediments. This interpretation of the isotope data contradicts earlier conclusions about the relative abundance of terrigenous organic matter based on hydrogen indices, C/N ratios as well as biomarker data (Stein et al. 1994a; Schubert and Stein 1997). It has to be considered, however, that carbon isotope composition of phytoplankton in the Arctic is highly variable, which may limit its use as a marine/terrigenous source indicator (see Chapter 1.4; Schubert and Calvert 2001).

Long-chain *n*-Alkanes

The long-chain *n*-alkane data (see Chapter 1.4) have successfully been used to estimate the relative proportion of terrigenous organic matter in the Kara and Laptev seas (see Chapters 7.5 and 7.6). For the central Arctic, however, this approach has to be applied with caution. The abundance of long-chain *n*-alkanes may not represent the bulk terrigenous organic matter in general, as suggested from the low correlation with hydrogen index values and C/N ratios (Schubert and Stein 1997). This is probably caused by the very different proportions of long-chain *n*-alkanes found in organic matter derived from different source areas around the Arctic (see Chapters 7.2, 7.5 and 7.6). Whereas the Kara and Laptev Sea riverine organic matter input is characterized by high long-chain *n*-alkane contributions, for the Mackenzie River and, especially, the Alaskan Shelf the *n*-alkane contributions are much lower (Fig. 7.9.8). Since the organic matter accumulating in the central Arctic may derive from one or more of the different areas, and since the relative importance of the different source areas is often not known, it is difficult to use the *n*-alkane data to infer the relative proportion of terrigenous organic matter. Nevertheless, some information about the OC source can be obtained from *n*-alkane data.

◀ **Fig. 7.9.4.** Transect C from the Lomonosov Ridge (for location see Fig. 7.9.1). AB and MB indicate Amundsen Basin and Makarov Basin, respectively. (A) Organic carbon contents; (B) C/N ratios; and (C) concentrations of long-chain *n*-alkanes (sum of C_{27} + C_{29} + C_{31})

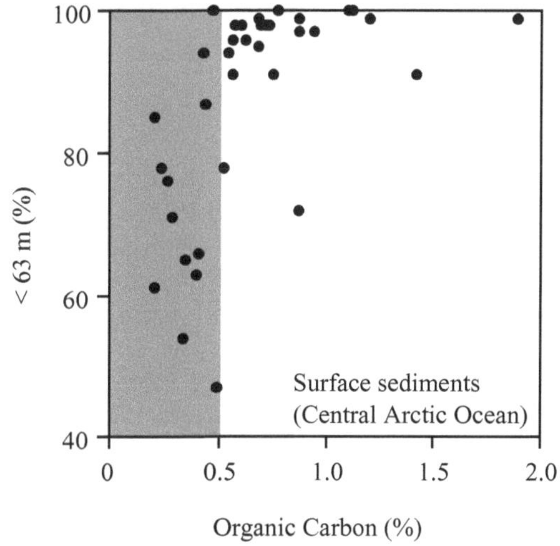

Fig. 7.9.5. Correlation between organic carbon content (%) and percentages of fine fraction < 63 μm in surface sediments from the Eurasian Basin

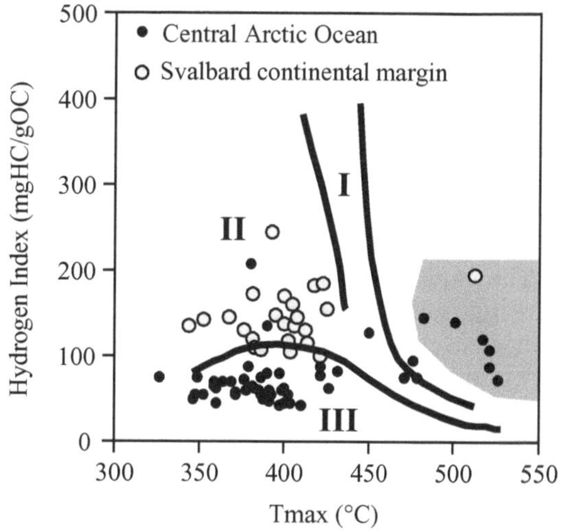

Fig. 7.9.6. Diagram of hydrogen indices vs. Tmax values in surface sediments from the central (Eurasian) Arctic Ocean and the western and northern Svalbard continental margin. Grey area indicates samples representing reworked organic matter

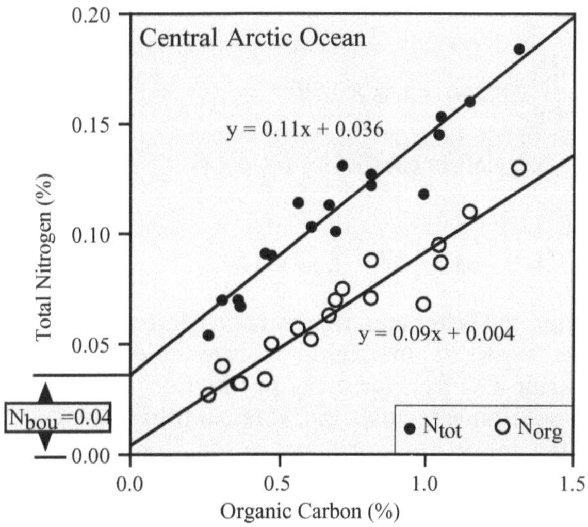

Fig. 7.9.7. Organic carbon (OC) versus total nitrogen (N_{tot}) correlation in central Arctic surface sediments. The estimated amount of inorganic nitrogen (N_{bou}) is indicated. In addition, the correlation between OC and organic nitrogen (N_{org}) is shown. See text for further explanation (data from Schubert and Calvert 2001)

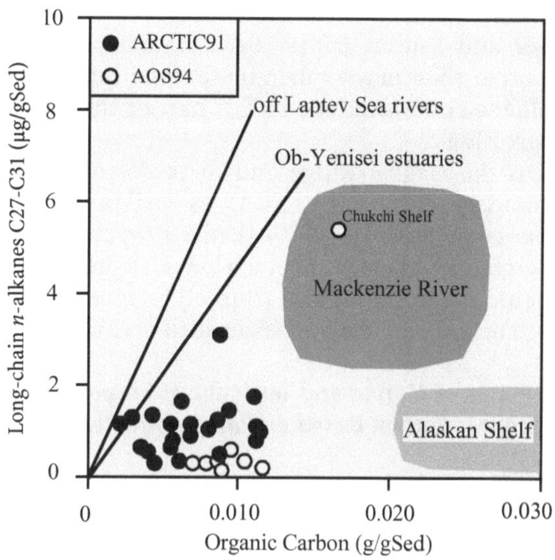

Fig. 7.9.8. Correlation between absolute concentrations of long-chain *n*-alkanes (sum of C_{27} + C_{29} + C_{31} in μg/gSed) and organic carbon content (g/gSed) in central Arctic Ocean surface sediments. The solid points represent samples from the ARCTIC 91 Expedition (Transect A, Fig. 7.9.2), the open circles represent samples from the AOS94 Expedition (Transect B, Fig. 7.9.3). In addition, the correlation lines of data points from the Laptev Sea (see Chapter 7.5) and the Kara Sea (see Chapter 7.6) are shown. The dark grey and grey areas mark the fields of data points from the Mackenzie River and Alaskan Shelf area, respectively (see Chapter 7.2)

The surface sediments from the Lomonosov Ridge are characterized by high long-chain *n*-alkane concentrations reaching values of 250 to 550 μg/gOC (Fig. 7.9.2). These values are similar to those determined in surface sediments from the inner Laptev Sea (see Chapter 7.5). Thus, it seems to be plausible that the main source area of the organic matter in the ridge sediments is the Laptev Sea. The Laptev Sea source is further corroborated by the high abundance of the clay mineral smectite (Stein et al. 1994a; Wahsner et al. 1999) and the heavy mineral clinopyroxene (Behrends et al. 1999). A transport of sediment via sea ice within the Transpolar Drift system (see Fig. 7.9.1) and its release in the central Arctic on the Lomonosov Ridge is most probable.

For the Lomonosov transect close to the Laptev Sea continental margin (Transect C, Fig. 7.9.4), i.e., the potential source area of organic matter enriched by long-chain *n*-alkanes, the long-chain *n*-alkane concentrations are surprisingly low. Since the long-chain *n*-alkane concentrations in the Laptev Sea are highly variable and generally significantly lower in the outer part of the shelf (see Chapter 7.5), direct advection by water currents (instead of sea-ice transport from the inner and central Laptev Sea) of material from the outer shelf may explain these lower long-chain n-alkane concentrations in this part of the Lomonosov Ridge.

On the Gakkel Ridge and in the Nansen and Amundsen Basin (Fig. 7.9.2) as well as in the Amerasian Basin (Fig. 7.9.3), the proportions of long-chain *n*-alkanes are much lower, suggesting an organic matter input also from other source areas (such as the Canadian-Alaskan shelf area).

Estimates of Marine and Terrigenous Proportions of Organic Carbon Based on Paleoproductivity Equations

Another method to determine the relative proportion of terrigenous organic matter is the use of equations deduced for estimating (paleo-) productivity from sediment data. This approach has been successfully applied in open-ocean environments where the organic matter in the sediments is predominantly of marine origin (Müller and Suess 1979; Stein 1986; Sarnthein et al. 1987, 1992). The equations for estimating productivity are based on the relationship between measured surface-water productivity, estimates of decomposition rates of organic matter during its transfer from the surface water to the seafloor based on sediment trap data (Suess 1980; Betzer et al. 1984), and the OC accumulation in surficial sediments. According to sediment trap data (Betzer et al. 1984), the flux of (marine) organic carbon to the seafloor can be estimated as

$$CF = 0.409 \times PR^{1.41} / DEP^{0.63}$$

where CF (in gC m^{-2} y^{-1}) equals the flux of organic carbon at depth DEP (in m) and PR (in gC m^{-2} y^{-1}) equals the primary production of organic carbon in the surface water. This equation, which is based on sediment trap data derived from non-polar regions, appears applicable to the Arctic Ocean because new sediment trap data from the Arctic (see Chapter 5) fit this correlation. Using DEP = water depth of the seafloor, it is possible to estimate the amount of primary produced organic carbon reaching the seafloor (CF_{sf}). This amount of organic carbon is the maximum that is available to be buried in the sediment. Combining the work of Müller and Suess (1979) and Betzer et al. (1984), the OC accumulation rate (CA in gC m^{-2} y^{-1}) expressed as percent of the flux of organic carbon at the seafloor (CF_{sf}) (i.e., the OC burial efficiency; see Stein 1991 and further references therein) is positively correlated with the linear sedimentation rate LSR (in cm ky^{-1}) (see Stein 1986 for deduction of equation):

$$\frac{CA}{CF_{sf}} \times 100 = 2.33 \times LSR^{0.9}$$

(correlation coefficient r = 0.94)

or

$$CA = 2.33 \times LSR^{0.9} \times CF_{sf}/100$$

Using the latter equation, it is possible to estimate the expected maximum accumulation rate of marine OC from primary production values (see also Schubert 1995). In Table 7.9.1, examples are listed for various primary production and sedimentation rates.

The comparison of these accumulation rates with the OC accumulation rates calculated from Arctic Ocean sediment data (see Table 7.9.2) indicates that the accumulation rates based on sediment data are significantly higher than the values calculated from marine OC flux data. This difference can best be explained by the high proportion of terrigenous organic matter in the Arctic Ocean sediments. For the central Arctic Ocean areas such as, for example, the central Lomonosov Ridge, the Gakkel Ridge and the Amundsen Basin, OC accumulation rates of about 2.5 to 4.5 mgC cm^{-2} ky^{-1} have been calculated (see Table 7.9.2). Using a realistic primary production value of 10 to 20 gC m^{-2} y^{-1} (see Chapter 3) and low sedimentation

Table 7.9.1. Examples for calculated marine organic flux at the seafloor (CF_{sf} in gC m^{-2} y^{-1}) and accumulation rates (CA-1 to CA-6 in mgC cm^{-2} ky^{-1}) for different primary production (PP in gC m^{-2} y^{-1}), water depth (DEP_{sf} in m) and linear sedimentation rate values (LSR-1 to LSR-6 in cm ky^{-1}), using equations explained in the text. CA-1 to CA-6 are calculated using the corresponding LSR-1 to LSR-6 values

PP	DEP_{sf}	CF_{sf}	LSR-1	LSR-2	LSR-3	LSR-4	LSR-5	LSR-6	CA-1	CA-2	CA-3	CA-4	CA-5	CA-6
5	1000	0.051	0.5	1	3	5	10	25	0.06	0.12	0.32	0.51	0.94	2.15
5	2000	0.033	0.5	1	3	5	10	25	0.04	0.08	0.21	0.33	0.61	1.39
5	3000	0.026	0.5	1	3	5	10	25	0.03	0.06	0.16	0.25	0.47	1.08
5	4000	0.021	0.5	1	3	5	10	25	0.03	0.05	0.13	0.21	0.39	0.90
10	1000	0.135	0.5	1	3	5	10	25	0.17	0.32	0.85	1.34	2.51	5.72
10	2000	0.088	0.5	1	3	5	10	25	0.11	0.20	0.55	0.87	1.62	3.70
10	3000	0.068	0.5	1	3	5	10	25	0.09	0.16	0.43	0.67	1.26	2.86
10	4000	0.057	0.5	1	3	5	10	25	0.07	0.13	0.35	0.56	1.05	2.39
20	1000	0.360	0.5	1	3	5	10	25	0.45	0.84	2.25	3.57	6.66	15.19
20	2000	0.233	0.5	1	3	5	10	25	0.29	0.54	1.46	2.31	4.30	9.82
20	3000	0.180	0.5	1	3	5	10	25	0.23	0.42	1.13	1.79	3.33	7.61
20	4000	0.150	0.5	1	3	5	10	25	0.19	0.35	0.94	1.49	2.78	6.34
30	1000	0.637	0.5	1	3	5	10	25	0.80	1.49	3.99	6.32	11.80	26.91
30	2000	0.412	0.5	1	3	5	10	25	0.51	0.96	2.58	4.09	7.62	17.39
30	3000	0.319	0.5	1	3	5	10	25	0.40	0.74	2.00	3.17	5.91	13.47
30	4000	0.266	0.5	1	3	5	10	25	0.33	0.62	1.67	2.64	v4.93	11.24
50	1000	1.310	0.5	1	3	5	10	25	1.64	3.05	8.20	12.99	24.25	55.31
50	2000	0.846	0.5	1	3	5	10	25	1.06	1.97	5.30	8.40	15.67	35.74
50	3000	0.656	0.5	1	3	5	10	25	0.82	1.53	4.11	6.50	12.14	27.68
50	4000	0.547	0.5	1	3	5	10	25	0.68	1.27	3.43	5.43	10.12	23.09

rate values of 0.5 to 1 cm ky^{-1}, the expected maximum accumulation of marine OC would be between about 0.1 and 0.5 mgC cm^{-2} ky^{-1} for these areas (Table 7.9.1). These calculations imply that about only 5 to 20% of the organic matter is of marine origin, i.e., 80 to 95% of the organic matter in central Arctic Ocean sediments is of terrigenous origin. The individual numbers for the proportions of terrigenous organic matter derived from this approach should certainly not be overinterpreted but they add to the balance of evidence favoring a predominance of terrigenous organic matter in central Arctic Ocean sediments.

7.9.4 Late Quaternary Organic Carbon Records and Paleoenvironment

During late Quaternary times, the central Arctic Ocean has been characterized by distinct environmental changes, for example in sea-ice cover, surface-water salinity and stratification, plankton pruductivity and terrigenous sediment input (e.g., Stein et al. 1994b, c; Nørgaard-Pedersen et al. 1998 and further references therein). All these processes also influence OC flux and accumulation. Here, we give a short summary of these paleoenvironmental changes and their relationships to OC accumulation, emphasizing the last glacial (Oxygen Isotope Stage-OIS-2) and the Holocene.

LGM to Holocene Organic Carbon Record

Looking at a transect of short sediment cores from the Nansen Basin across the Gakkel and Lomonosov ridges towards the Makarov Basin (see Fig. 7.9.9 for location of cores), we see distinct variations in OC content versus depth (and time) (Fig. 7.9.10). In all records, OC contents increase in the Holocene sediments. Furthermore, OC contents are higher in the basin cores (especially the Nansen Basin) compared to the cores from the ridges. In sediments dating to the last glacial, OC contents appear to be significantly lower. Low hydrogen index values suggest a dominance of terrigenous organic matter. In the lower part of some of the records probably representing OIS 3, higher amounts of marine organic matter may have been preserved as suggested from elevated hydrogen index values. In general, C/N ratios vary between 4 and 15 (Fig. 7.9.10). The ratios are significantly higher in the Nansen Basin records than in all other records, suggesting relatively higher proportions of terrigenous organic matter in the Nansen Basin. When using C/N ratios for sediments from

Table 7.9.2. Summary of mean linear sedimentation rate and accumulation rate of total sediment (TS) and organic carbon (OC) data for the Holocene and last glacial time (LGM) intervals, calculated for dated sediment cores. References are listed in the table. For location of cores see Fig. 7.9.9

Area	Latitude	Longitude	Water depth (m)	LSR (cm ky^{-1})	TS-AccR (g cm^{-2} ky^{-1})	OC-AccR (mg cm^{-2} ky^{-1})	LSR (cm ky^{-1})	TS-AccR (g cm^{-2} ky^{-1})	OC-AccR (mg cm^{-2} ky^{-1})	Ref.
					Holocene			LGM		
Barents Sea slope										
PS2138	81°32.1′N	30°35.6′E	995	6	7	50	25	30	200	1
PS2446	82°23.8′N	40°54.5′E	2022	10	8	100	>30	>30	>330	1
East Sib. Sea slope										
PS2763	80°16.9′N	150°26.1′E	1591	12,6	10,6	60	10,2	10,4	30	2
Laptev Sea slope										
PS51/154-11	77°15.6′N	120°36.6′E	270	18	14	140	n.d	n.d.	n.d.	3
PS2778-2	77°58.7′N	113°03.9′E	341	38	30	300	n.d	n.d.	n.d.	4
PS2476-4	77°23.4′N	118°11.6′E	521	50	40	400	n.d	n.d.	n.d.	4
PS2458-4	78°10.0′N	133°23.9′E	983	34	27	300	n.d	n.d.	n.d.	3
PS2474-3	77°40.2′N	118°34.5′E	1494	20	16	160	n.d	n.d.	n.d.	4
PS2742-5	80°47.3′N	103°48.9′E	1890	22	18	180	n.d	n.d.	n.d.	4
Laptev Sea rise										
PS2741	81°06.0′N	105°22.0′E	2400	11,9	9	100	2	1,7	10	4
PS2471	79°09.0′N	119.77.4′E	3047	2	2	20	n.d.	n.d.	n.d.	4
Chukchi Rise										
92BC17	76°05.2′N	164°50.2′W	402	2	n.d.	n.d.	n.d.	n.d.	n.d.	5,6
Northwind Ridge										
88BC6	74°35.9′N	158°02.2′W	954	3,2	n.d.	n.d.	n.d.	n.d.	n.d.	5
88BC22				3,5	n.d.	n.d.	n.d.	n.d.	n.d.	6
Mendeleev Ridge										
94BC16	80°20.3′N	188°42.7W	1533	1,1	n.d.	n.d.	0,5	n.d.	n.d.	5,6
94BC17	81°15.91	178°58.1′E	2217	1,5	n.d.	n.d.	0,3	n.d.	n.d.	5,6
Morris Jesup Rise										
PS2200	85°19.6′N	14°00.0′W	1074	0,2	0,2	1	0,2	0,2	< 1	7,8
Lomonosov Ridge										
PS2177	88°02.2′N	134°55.1′E	1388	0,9	0,7	3	0,3	0,3	1	7,8
PS2185	87°31.9′N	144°08.4′E	1073	0,7	0,6	2,5	0,3	0,2	1	6,7
94BC28	88°52.3′N	140°22.8′E	2015	1,3	n.d.	n.d.	0,1	n.d.	n.d.	6
PS2757	81°09.6′N	140°12.1′E	1230	5,2	4,8	20	5,8	5,4	10	2
Gakkel Ridge										
PS2163	86°14.5′N	59°12.9′E	3040	0,7	0,5	2,9	< 0,7	< 0.5	< 1.5	8
PS2166	86°51.6′N	59°45.9′E	3636	0,9	0,8	3,5	0,4	0,2	< 1	7,8
PS2206	84°16.7′N	02°30.3′W	2993	0,7	0,5	2,6	0,5	0,4	1,1	8
Nansen Basin										
OD-041-04	84°01.8′N	11°14.3′E	3344	0,7	0,6	n.d.	0,4	0,3	n.d.	7
Amundsen Basin										
PS2170	87°35.8′N	60°53.7′E	4226	0,9	0,7	4,5	0,5	0,4	1,5	8
PS2195	86°13.7′N	9°35.6′E	3873	0,6	0,4	2	0,3	0,2	1	6,8
Makarov Basin										
PS2761	81°11.5′N	150°29.8′E	2640	2	2	10	3,4	3	10	2
Wrangel Abyssal Plain										
94BC20	83°10.6′N	174°5.9′E	3145	1	n.d.	n.d.	0,2	n.d.	n.d.	6
Canada Abyssal Plain										
88PC10	74°43.5′N	156°08.5′W	3899	145	n.d.	n.d.	n.d.	n.d.	n.d.	5
93BC13	75°48.3′N	155°07.6′W	3831	1	n.d.	n.d.	n.d.	n.d.	n.d.	5

(1) Knies et al. (2000) (2) Stein et al. (2001) (3) Bauch et al (2001b) (4) Stein and Fahl (2000)
(5) Grantz et al. (1999) (6) Darby et al. (1997) (7) Nørgaard-Pedersen et al. (1998, 2000) (8) Stein et al. (1994b,c)

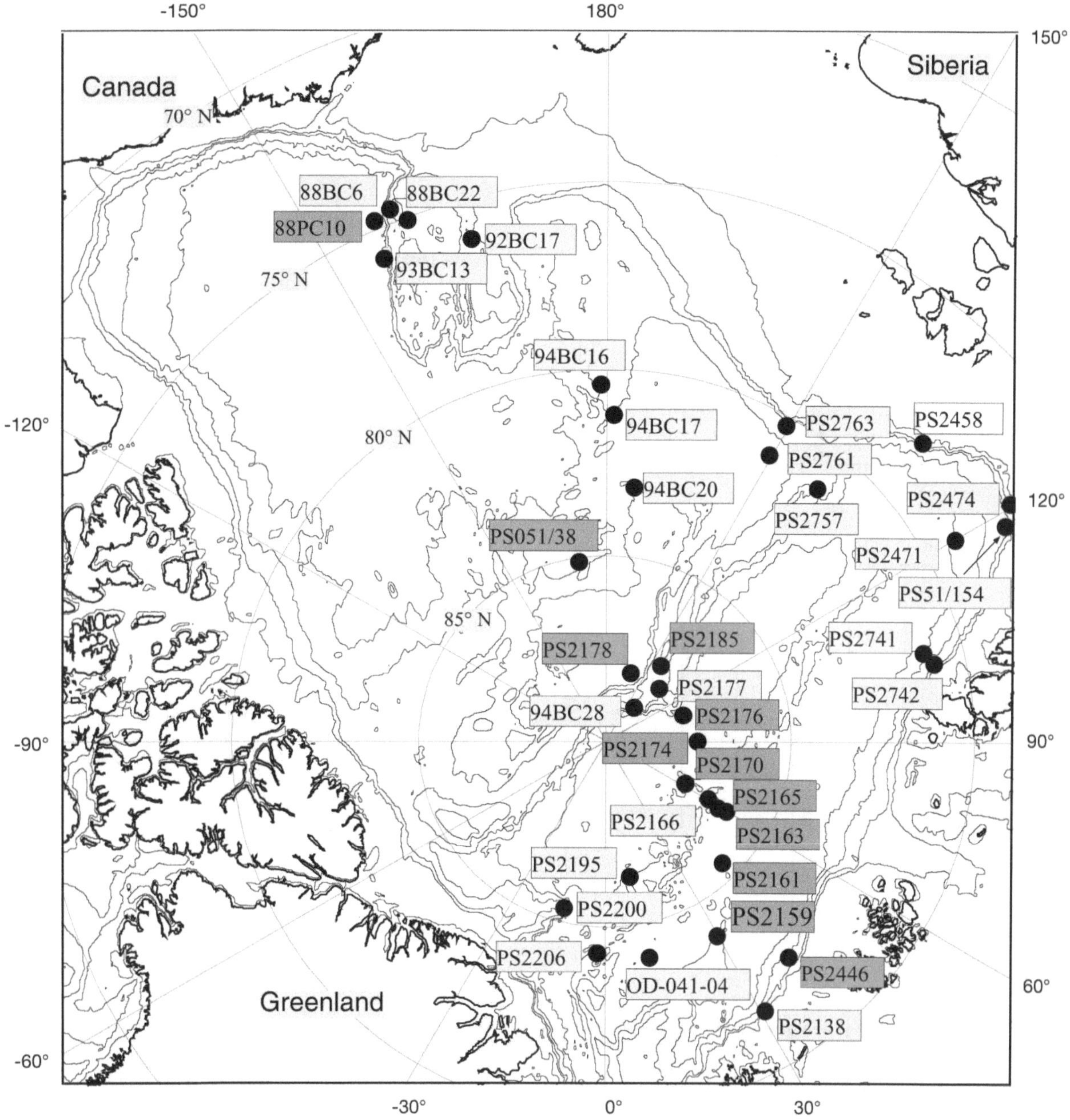

Fig. 7.9.9. Arctic Ocean map showing the location of sediment cores (numbers underlined in grey) for which records are presented in Figs. 7.9.10, 7.9.11 and 7.9.13 to 7.9.16. In addition, all sediment cores for which sedimentation and accumulation rates were calculated (see Table 7.9.2) are shown

the central Arctic Ocean sediments, the potential contribution from inorganic nitrogen needs to be weighed (see above and Chapter 1.4). With significant amounts of inorganic nitrogen, the "C_{org}/N_{org}" ratios will be higher than the C/N ratio and, accordingly, will support a more mixed terrigenous/marine origin of the organic matter in general.

To evaluate the influence of primary production on the OC accumulation in the central Arctic Ocean, Schubert et al. (2001) used nitrogen and carbon stable isotopes (Fig. 7.9.11). There are several general possibilities to explain the observed distributions (Miyaka and Wada 1967; Wada et al. 1987; Altabet and Francois 1994; Francois and Altabet 1992; Farrell et al. 1995): (1) Isotopic measurements of nitrogen and carbon represent a variable contribution of terrigenous and marine organic material to the sediment sample, (2) nitrogen isotopes increase due to higher nutrient utilization of the phytoplankton in surface waters and the related

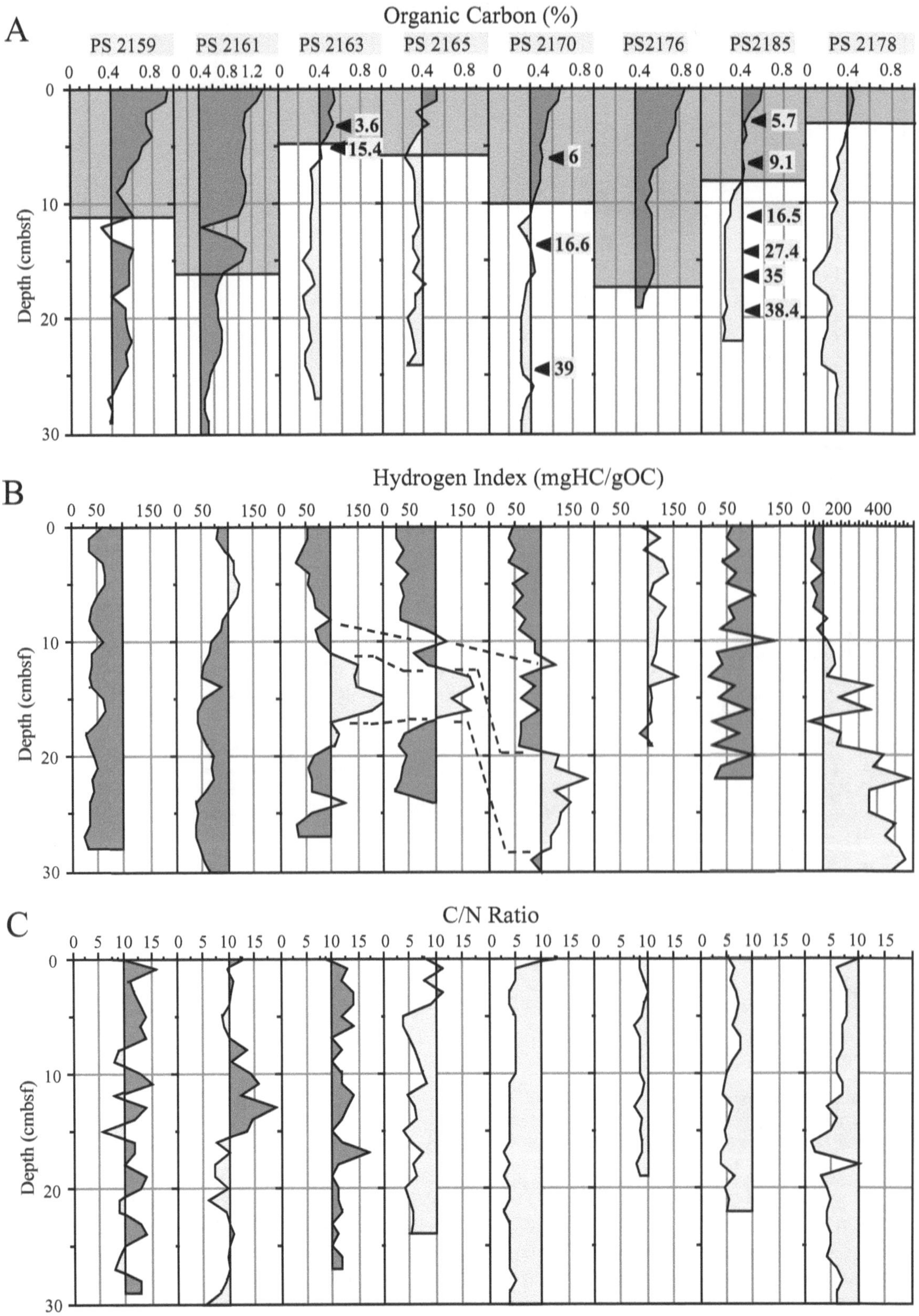

Fig. 7.9.10. Records of organic carbon (A), hydrogen index values (B), and C/N ratios (C) from selected sediment cores from the central Arctic Ocean. In the OC records, the Holocene is marked as grey-shaded interval. Numbers indicate AMS^{14}C ages in Cal. kyrs. BP. (AMS^{14}C ages from Stein et al. 1994b; Nørgaard-Pedersen et al. 1998). Correlation based on the hydrogen index values are marked as stippled lines. For location of cores see Figs. 7.9.2 and 7.9.9

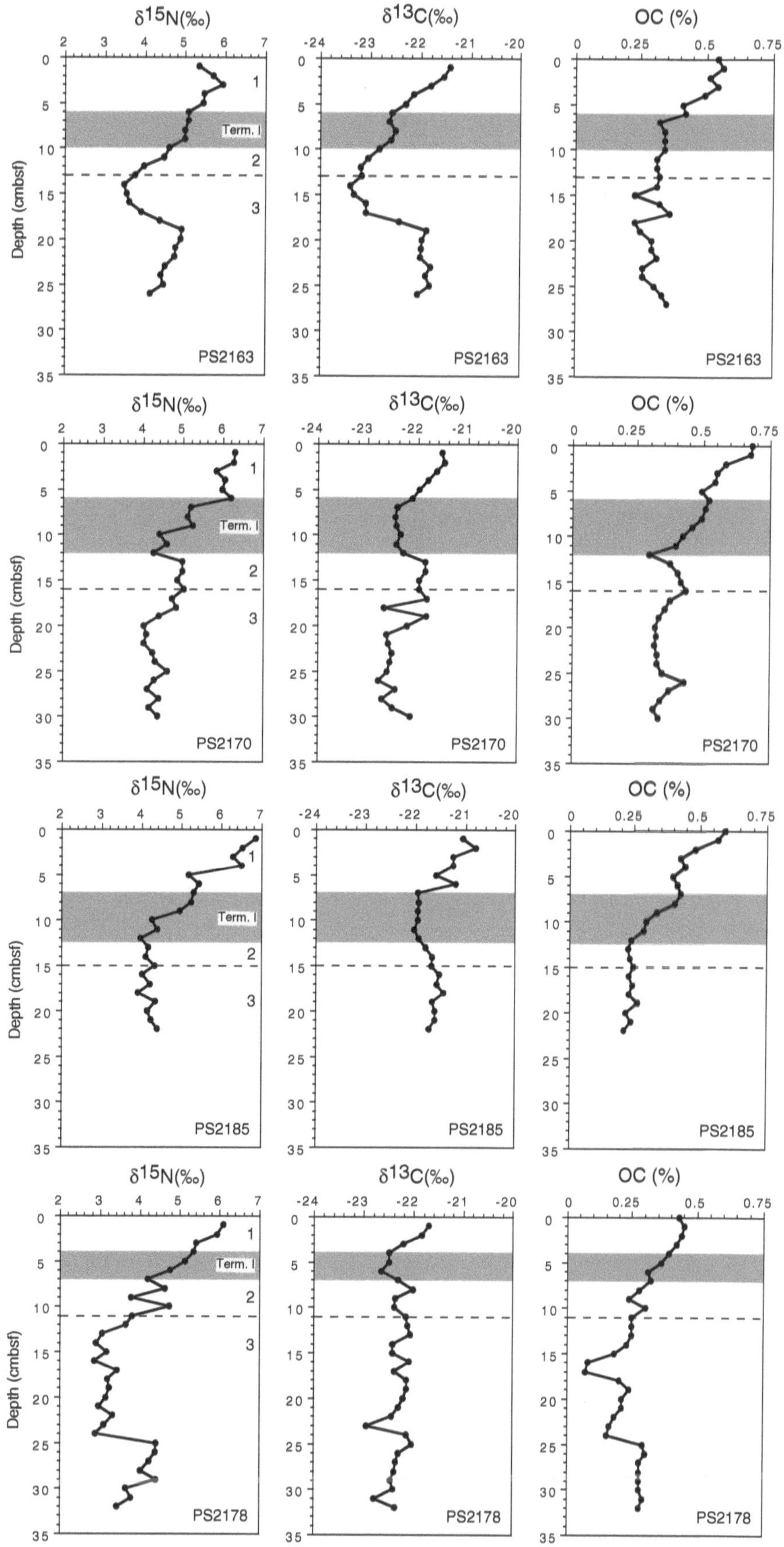

Fig. 7.9.11. Nitrogen and carbon stable isotope and organic carbon records from cores PS2163 (Gakkel Ridge), PS2170 (Amundsen Basin), PS2185 (Lomonosov Ridge), and PS2178 (Makarov Basin) (from Schubert et al. 2001). For location of cores see Fig. 7.9.9

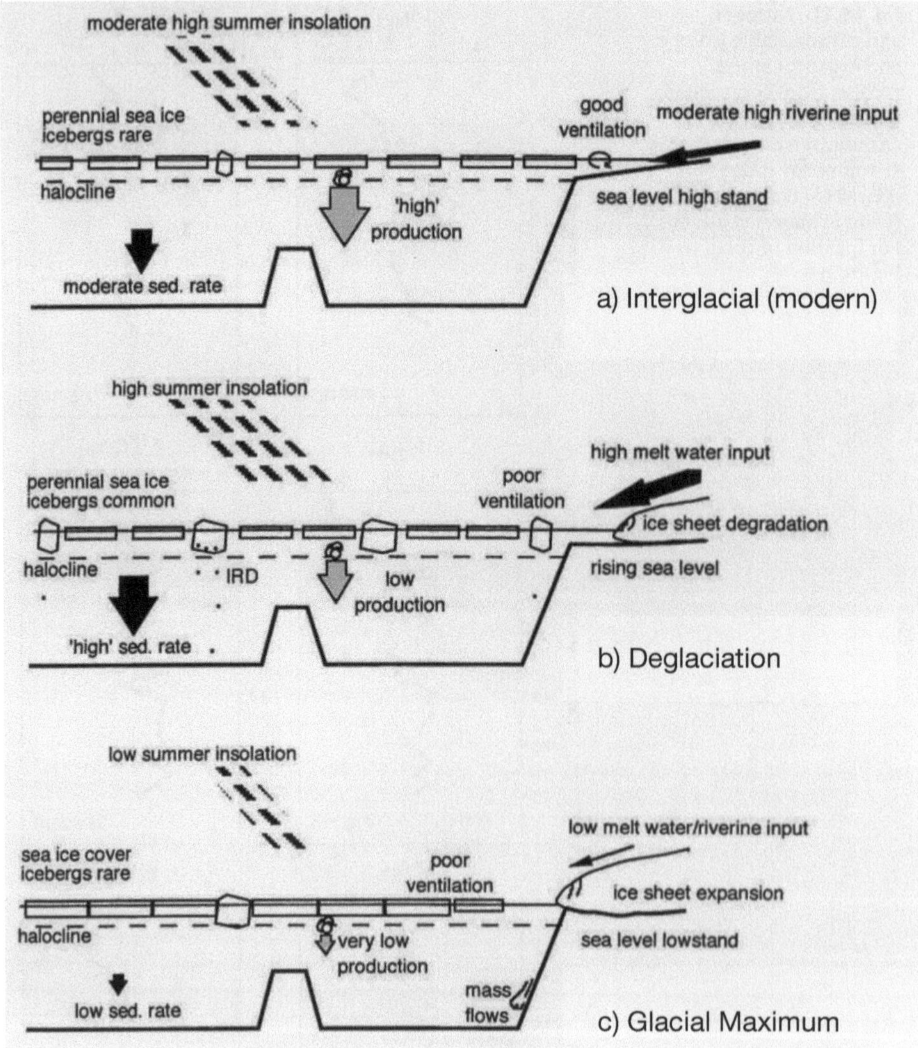

Fig. 7.9.12. Model of late Quaternary Arctic Ocean surface conditions and sedimentation patterns for three end-member climate conditions (a) interglacial or warm interstadial; (b) deglaciation period; and (c) glacial maximum (from Nørgaard-Pedersen et al. 1998; supplemented)

heavier carbon isotopic composition of a sample represents the higher contribution of marine organic matter, i.e. higher productivity of the surface waters, and (3) nitrogen isotopes in the sediment sample increase due to higher nutrient utilization of the phytoplankton in surface waters and the organic carbon isotopic composition is related to changes in the pCO_2 of the surface waters. For details of this approach we refer to Schubert et al. (2001) and Schubert and Calvert (2001).

As mentioned above and also outlined in Chapter 1.4, the use of carbon stable isotope composition as an indicator of organic-carbon source might be problematic for central Arctic Ocean environments. From biomarker investigations of Core PS2170, it has been shown that lignin and long-chain *n*-alkane concentrations, both indicators of organic matter derived from terrestrial higher plants (e.g., Leo and Barghorn 1970; Prahl and Muehlhausen 1989), do not correlate with the carbon isotope values (Schubert et al. 2001). Therefore, carbon isotope values in the central Arctic Ocean might respond more to changes in surface water pCO_2 than to the proportion of marine/terrigenous organic matter. This interpretation is further supported by nitrogen isotope values which correlate well with carbon isotopes (Schubert et al. 2001) suggesting, indeed, that the isotope distributions in sediments relate to marine nutrient utilization. The high correlation between the OC content and the nitrogen isotopes in sediment samples of the four cores can be explained in terms of productivity variations (Schubert et al. 2001). Light nitrogen isotopes are accompanied by low OC concentrations (Fig. 7.9.11) which implies low nutrient utilization is accompanied by low productivity in the surface waters and vice versa.

Low $\delta^{15}N$ values (3–4‰), observed in sections of all four cores dating to the last glacial interval (OIS 2), coincide with light $\delta^{13}C_{org}$ values (Fig.

7.9.11). As concluded earlier (Nørgaard-Pedersen et al. 1998) low sedimentation rates, low planktic productivity, and low IRD (ice-rafted debris) during OIS 2 (and uppermost OIS 3 which is very similar to stage 2 with respect to the mentioned parameters) indicate a closely packed sea-ice cover (Fig. 7.9.12). This is in good agreement with the low nitrogen isotope values during that time interval showing the very low nutrient utilization, i.e. very low productivity in the surface waters. Additionally, this interpretation is fully supported by the low OC values that represent both low concentrations of marine and terrigenous OC due to low productivity and no or very little release of terrigenous organic material from melting sea-ice (Schubert et al. 2001).

During the Holocene following the Termination I, $\delta^{13}C_{carb}$ values of the planktonic foraminifera shells of *Neogloboquadrina pachyderma* sin. determined in the same cores, distinctly increase, as do the carbonate and OC concentrations (Fig. 7.9.11; Stein et al. 1994b, c). The Holocene increase in $\delta^{13}C_{carb}$ to modern values recorded in most of the cores suggests that a well-oxygenated surface-water mass was extensive and that more open-ice conditions prevailed in the Arctic Ocean at the end of Termination I. Because the carbonate material is mainly composed of planktonic foraminifera, coccoliths, and ostracodes (Gard 1993; Cronin et al. 1995; Nørgaard-Pedersen et al. 1998), the reduced sea-ice cover and at least seasonally open waters may have increased surface-water productivity (Fig. 7.9.12). At the same time, carbon and nitrogen isotope values also increase steadily to modern times (Fig. 7.9.11). In general, the increase in $\delta^{15}N$ values during the upper Holocene shows the higher nutrient utilization in surface waters due to more open ocean water conditions enabling the phytoplankton to receive enough light for reproduction. The OC levels, however, are not likely to depend entirely on marine productivity because major proportions of the organic matter are of terrigenous origin as indicated by biomarker compositions (Schubert and Stein 1997), hydrogen index values and C/N ratios (Fig. 7.9.10; Stein et al. 1994b, c). The glacial-to-interglacial increase in OC accumulation rates by a factor of 2-3 (Table 7.9.2) is thus probably caused for the most part by increased supply of terrigenous organic matter from Eurasia. The flux of marine organic matter may also have increased during the Holocene, but oxic, well-ventilated deep-water conditions (inferred from the homogenous brown sediments (Fütterer 1992) and high $\delta^{13}C_{carb}$ values determined at *C. wuellerstorfi* tests) and the low sedimentation rates would have precluded its preservation (Stein et al. 1994b).

Pre-LGM Organic Carbon Records

OC records representing older time intervals (OIS 4 and older) are still very rare for the central Arctic Ocean and adjacent continental slope (Schubert 1995; Schubert and Stein 1996; Knies et al. 2001; Stein et al. 2001). Furthermore, the time control of these records is much weaker than for the AMS^{14}C-dated records from the last glacial and Holocene. As recorded for the Holocene/LGM time intervals (Fig. 7.9.10), the OC content of the basins appears to be higher than those from ridge cores. On the Alpha and Lomonosov ridges (cores PS2185 and PS051/38), the OC contents vary between 0.05 and 0.3% (Fig. 7.9.13) whereas in the Amundsen Basin Core PS2174 OC contents are significantly higher ranging between 0.4 and 1.5% (Fig. 7.9.14). In the latter core, maximum OC contents coincide with low hydrogen index values of <100 mgHC/gOC and high C/N ratios of 10 to >20, indicating high input of terrigenous organic matter. The clear dominance of long-chain *n*-alkanes over the short-chain *n*-alkanes may also support a terrigenous origin of

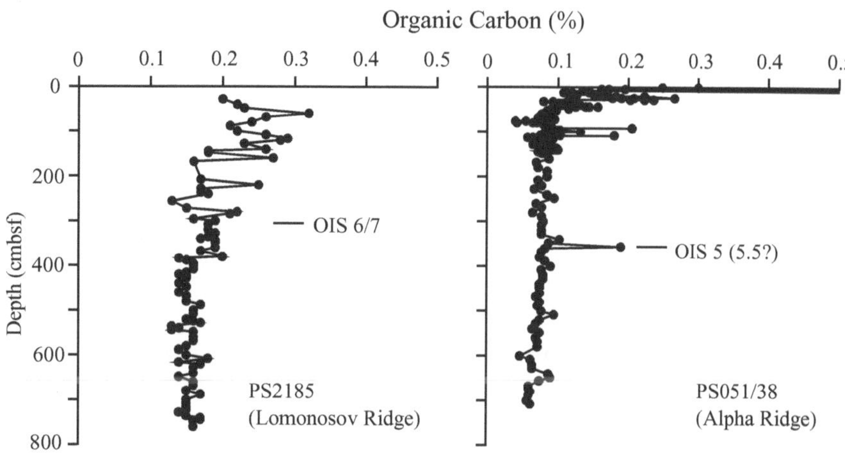

Fig. 7.9.13. Organic carbon records from Core PS2185 (Lomonosov Ridge) and Core PS051/38 (Alpha Ridge). For location of cores see Fig. 7.9.9

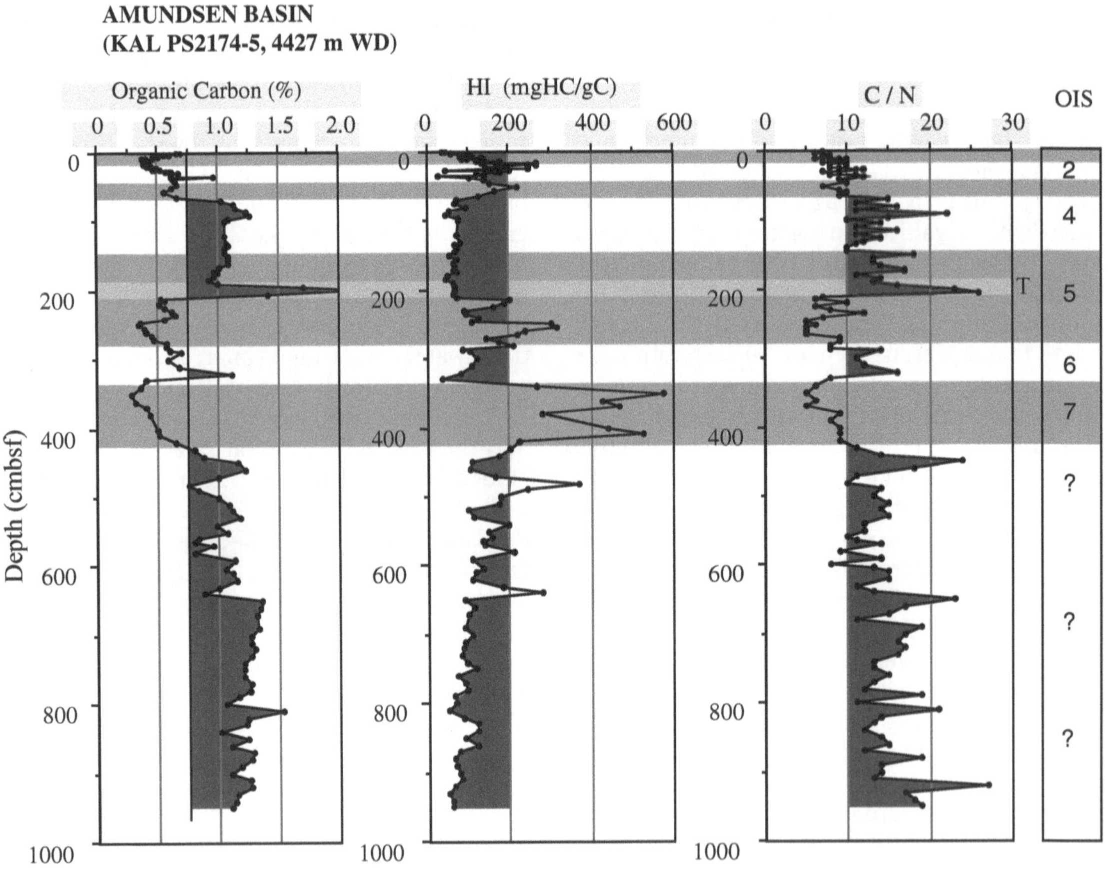

Fig. 7.9.14. Organic carbon, hydrogen index, and C/N records from Core PS2174 (Amundsen Basin). Numbers at the right side mark oxygen isotope stages. The organic carbon maximum at 200 cm below surface highlighted by a hatched bar, indicates a turbidite (T). For location of core see Fig. 7.9.9

the organic matter (Schubert and Stein 1996). The concentration of the long-chain *n*-alkanes $C_{27+29+31}$ reach high values of 200 to 600 µg/gOC which point to the Eurasian shelf seas as a potential source of the organic matter (see Fig. 7.9.8). Intervals with low OC contents most typical for the interglacials OIS 5 and 7 show higher hydrogen index values and lower C/N ratios (Fig. 7.9.14), probably due to higher surface-water production. Higher surface-water production for these intervals is also suggested from higher concentrations of short-chain *n*-alkanes and biogenic opal (Schubert and Stein 1996).

Organic Carbon Input by Turbidity Currents, Slumps and Debris Flows

Whereas on the Arctic Ocean ridges, OC accumulation derives from terrigenous organic carbon input via sea ice and ocean currents augmented by minor inputs from primary production, in the oceanic basins turbidity currents may become the prominent process for OC input and accumulation. Turbidites are very common in most (probably all) of the Arctic deep basins (e.g., Fütterer 1992; Stein et al. 1994a; Grantz et al. 1996, 1999; Svindland and Vorren 2002). In the Amundsen Basin and the Canada Abyssal Plain, for example, turbidites may be the dominant facies of the Holocene (Fig. 7.9.15) as well as pre-Holocene sedimentary sequences (Grantz et al. 1996, 1999; Svindland and Vorren 2002). Very often, these turbidites are characterized by high OC values. At Core 88PC10 obtained from the Canada Abyssal Plain (see Fig. 7.9.9 for location), the OCs of the turbidites reach values of 1 to 1.4% whereas the intercalated pelagic/hemipelagic sediments show values of 0.5 to 1.1% (Fig. 7.9.15; Grantz et al. 1999). AMS^{14}C measurements from the pelagic/hemipelagic intervals allow the turbidite sequences to be dated and sedimentation rates to be calculated. For Core 88PC10, very high sedimentation rates of about 145 cm ky^{-1} are obtained, implying very high OC accumulation rates (i.e., about 1.5 gC cm^{-2} ky^{-1}). Also for the Amundsen Basin, Svindland and Vorren (2002) determined high sedimentation rates of up to 25 cm ky^{-1} in the turbidite sequences. That means, sedimentation

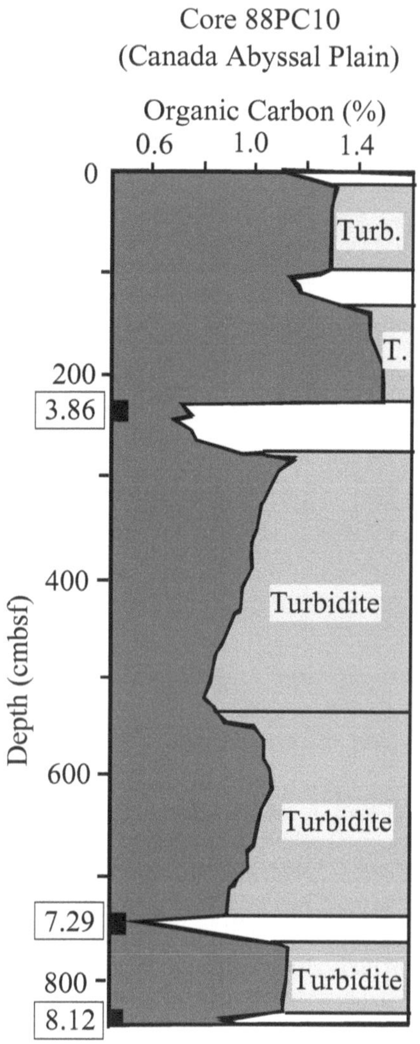

Fig. 7.9.15. Organic carbon record of Core 88PC10 (Canada Abyssal Plain) (Grantz et al. 1999). Occurrences of turbidites are marked as crossed intervals. Numbers 3.86, 7.29, and 8.12 are AMS^{14}C ages in ky (reservoir correction of 440 years) determined in the intercalated pelagic/hemipelagic intervals. The corresponding ages in calendar ky. BP are about 4.3, 8.1, and 9. For location of cores see Fig. 7.9.9

rates are 20 to more than 100 times higher than those typical for normal pelagic sediments of the central Arctic (see Table 7.9.2) indicating the importance of turbidites for the OC budget in these areas.

Along the continental slopes (see Fig. 7.5.12 in Chapter 7.5; Stein et al. 1999), slumps and debris flows also contribute to high OC accumulation. At the Barents Sea continental slope, for example, slumps and debris flows were recorded during the last glacial interval, i.e., OIS 2 (Fig. 7.9.16). These sediments are characterized by a significantly increased proportion of terrigenous OC as indicated by elevated OC values of 1 to 1.7%, high C/N ratios of 10 to 15, low hydrogen index values of < 75 mgHC/gOC (Knies et al. 2000), and light carbon stable isotopes of about −25‰ (Fig. 7.9.16). Based on AMS^{14}C dates, OC accumulation rates are > 0.33 gC cm^{-2} ky^{-1} which is more than 3 times higher than those of the „normal" Holocene not influenced by slumps and debris flows (see Table 7.9.2). That means, that along the Barents Sea continental margin, which was strongly influenced by glacigenic input of total sediment and terrigenous organic matter due to the huge LGM ice sheet (Knies et al. 2000 and further references therein), higher OC accumulation rates were typical for the last glacial interval whereas during the following Holocene interglacial, accumulation decreased. The same picture can be drawn from records of Core PS2138 (Table 7.9.2; Knies et al. 2000). This trend opposes that for the central Arctic Ocean where accumulation rates were higher during the Holocene and lower during the last glacial interval (see above; Fig. 7.9.12).

7.9.5 Accumulation Rates and Budget of Total Sediment and Organic Carbon

Based on the data summarized in Table 7.9.2, average Holocene accumulation rates of total sediment and OC vary between 7 and 40 g cm^{-2} ky^{-1} and 50 and 400 mgC cm^{-2} ky^{-1}, respectively, along the continental slopes. Towards the continental rise, the accumulation rates significantly decrease to <10 g cm^{-2} ky^{-1} and 20 to 100 mgC cm^{-2} ky^{-1}, respectively. For the abyssal plains, typical values for the non-turbidite areas are 0.5 to 2 g cm^{-2} ky^{-1} and 2 to 10 mgC cm^{-2} ky^{-1} for total sediment and OC, respectively. On the ridges, the accumulation rates are lower, reaching 0.5 to <1 g cm^{-2} ky^{-1} and 2.5 to 3.5 mgC cm^{-2} ky^{-1}. On the Lomonosov Ridge close to the Laptev Sea continental margin, higher values of about 5 g cm^{-2} ky^{-1} and 20 mgC cm^{-2} ky^{-1} were calculated for total sediment and OC (Table 7.9.2).

For the last glacial interval, the accumulation rates of total sediment and OC are significantly reduced in the central Arctic Ocean with typical values of <1 g cm^{-2} ky^{-1} and <1.5 mgC cm^{-2} ky^{-1} (Table 7.9.2). This has been caused by a reduced input of terrigenous organic matter as well as a reduced primary production due to a more closed sea-ice cover (see discussion above; Fig. 7.9.12). Only in areas which are directly influenced by glacigenic processes such as, for example, the Barents Sea continental margin (see above), are distinctly higher accumulation rates of total sediment and (terrigenous) OC calculated (Table 7.9.2; >30 g cm^{-2} ky^{-1} and >300 mgC cm^{-2} ky^{-1}, respectively).

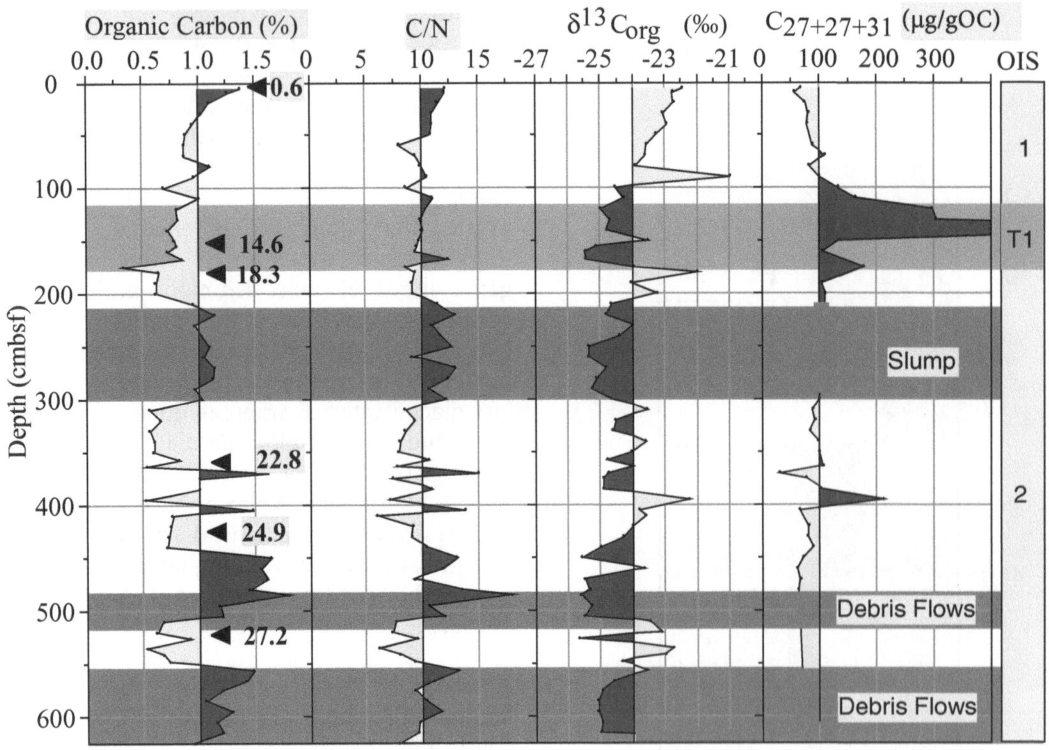

Fig. 7.9.16. Records of organic carbon, C/N ratios, carbon stable isotopes of organic matter, and concentrations of long-chain *n*-alkanes (sum of C_{27} + C_{29} + C_{31}) from Core PS2446. Slump and debris flows are marked as dark grey intervals. Oxygen Isotope Stages (OIS) 1 and 2 as well as Termination 1 (T1) are indicated. Black triangles and numbers indicate AMS^{14}C ages (in Cal. kyrs. BP) (all data from Knies et al. 2001). For location of cores see Fig. 7.9.9

Holocene Budget of Total Sediment and Organic Carbon

In comparison to the Arctic marginal seas, especially the Beaufort, Laptev and Kara seas, the budget calculation for the central Arctic has to be much more general. Due to the limited number of AMS^{14}C dated sediment cores obtained from restricted areas of the central Arctic Ocean (see Fig. 7.9.9 for core locations), only rough estimates can be given. The accumulation rate values summarized in Table 7.9.3 provide the database for the budget calculation. We have to assume that these values are typical for the different areas. This is an assumption which, of course, has to be taken with caution. Furthermore, for large parts of the central Arctic no data at all are available. Here, we use values from neighbouring areas or from similar settings. For example, for the areas of continental rises and slopes we take data mainly from the Laptev Sea continental margin where a sufficient database is available (see Chapter 7.5), assuming that such data are typical for slopes and rises. Another complication is the difficulty of estimating the areas influenced by turbidites in the basins/abyssal plains (see above). With a few exceptions, the sediment cores for which AMS^{14}C dates are available were recovered from areas not influenced by turbidites and, thus, characterized by lower sedimentation rates. For the budget calculation, we used an average sedimentation rate of 2–3 cm ky^{-1} for the abyssal plains. Sedimentation rate may be significantly higher in the turbidite areas.

Along the entire Arctic continental slopes and rises, about 107 × 10^6 t y^{-1} and 79 × 10^6 t y^{-1} of total sediment, respectively, are accumulating and about 1.3 × 10^6 t y^{-1} and 0.72 × 10^6 t y^{-1} of OC, respectively, are buried (Table 7.9.3). In the central Arctic proper, 30 and 21 × 10^6 t y^{-1} of total sediment, and 0.25 and 0.11 × 10^6 t y^{-1} of OC are accumulating in the abyssal plains and on the ridges, respectively. This gives a total burial of 237 × 10^6 t y^{-1} of total sediment and 2.4 × 10^6 t y^{-1} of OC. Most of this OC is probably of terrigenous origin, and only minor amounts derive from phytoplankton. Assuming a marine proportion of about 10 to 20%, this gives a burial rate of 0.25 to about 0.5 × 10^6 t y^{-1} of marine OC for the central Arctic including the continental slopes and rises. Then, the remaining OC, about 1.9 to 2.15 × 10^6 t y^{-1}, is of terrigenous origin.

Table 7.9.3. Summary table of total sediment and organic carbon budget, based mainly on data from Table 7.9.2. See text for further explanation

Area	Size (10^3 km^2)	LSR (cm ky^{-1})	TOC (%)	Total Sediment (g cm^{-2} ky^{-1})	Organic Carbon (g cm^{-2} ky^{-1})	Total Sediment (10^6 t y^{-1})	Organic Carbon (10^6 t y^{-1})
Continental slopes	541	22	1,2	19,8	0,24	**107**	**1,29**
Continental Rises	733					**53**	**0,46**
Mackenzie Rise	178	8	1	7,2	0,07	12,82	0,128
Canada-Greenland Rise	249	8	0,5	7,2	0,04	17,93	0,09
Barents/Kara Rise	124	8	1	7,2	0,07	8,93	0,089
Lena/Nansen Rise	67	8	1,2	7,2	0,09	4,82	0,058
Lena/Amundsen Rise	77	8	1,2	7,2	0,09	5,54	0,067
Beaufort (Alaska) Rise	23	8	1	7,2	0,07	1,66	0,017
Yermak Rise	15	8	1	7,2	0,07	1,08	0,011
Perched continental rises	362					**26**	**0,26**
Chukchi Perched Rise	43	8	1	7,2	0,07	3,1	0,031
Wrangel Perched Rise	239	8	1	7,2	0,07	17,2	0,172
Arlis Perched Rise	80	8	1	7,2	0,07	5,8	0,058
Abyssal plains	1122					**23**	**0,2**
Barents Abyssal Plain	295	3	0,8	2,7	0,022	7,965	0,064
Pole Abyssal Plain	291	2	0,8	1,8	0,014	5,238	0,042
Fletcher Abyssal Plain	42	2	0,8	1,8	0,014	0,756	0,006
Canada Abyssal Plain	494	2	0,8	1,8	0,014	8,892	0,071
Perched abyssal plains	222	3	0,8	2,7	0,022	**6**	**0,05**
Northwind Basin (isolated basin)	23	5	1	4,5	0,045	**1**	**0,01**
Ridges and plateaus	1506					**21**	**0,11**
Gakkel Ridge	295	1	0,5	0,9	0,005	2,655	0,013
Unnamed segment ridge Fram Strait	37	1	0,5	0,9	0,005	0,333	0,002
Lomonosov Ridge	307	1,5	0,4	1,4	0,005	4,145	0,017
Northwind Ridge and Chukchi Plateau	120	3	1	2,7	0,027	3,24	0,032
Yermak Plateau	35	2,5	0,8	2,3	0,018	0,788	0,006
Morris Jesup Rise	5	0,2	0,4	0,2	0,001	0,011	0
Alpha-Mendeleev Ridge	708	1,5	0,4	1,4	0,005	9,558	0,038
Total	4509					**237**	**2,4**

The calculated budget numbers have probably to be taken as minimum values. If assuming that the areas influenced by turbidites in the abyssal plains are larger, higher sedimentation rates have to be considered for the budget calculation of these areas. Furthermore, during the early Holocene, i.e., the end of the deglaciation period (see Fig. 7.9.12), sedimentation rates may have been generally higher than the modern (interglacial) values. Due to the limited number of sediment cores which would allow differentiation between late and early Holocene sedimentation, a detailed calculation of budget as has been done for the Laptev and Kara seas (see Chapters 7.5 and 7.6), can not be performed at this time for the central Arctic Ocean.

7.9.6
Summary and Concluding Remarks

- The central Arctic Ocean is characterized by relatively high organic carbon (OC) values of 0.5 to >2%. In general, OC values are higher in the basins than on the ridges and plateaus.
- It is difficult to give precise estimates of the marine and terrigenous proportions of central Arctic Ocean sediments, and very different numbers result from various published approaches. In general, however, the balance of evidence suggests that terrigenous organic carbon is predominant in most parts of the central Arctic. Significantly higher marine OC is preserved along the Svalbard continental margin, i.e., in areas influenced by the warm Atlantic water inflow, due to increased primary production.
- In comparison to the world ocean, the OC burial efficiency (i.e., the ratio between OC accumulation in near-surface sediments and flux of OC at the seafloor) in the central Arctic Ocean is significantly higher. This is caused by the high proportion of terrigenous OC buried in Arctic Ocean sediments.
- A major process controlling terrigenous OC input and its accumulation is the transport via sea ice and ocean currents. In the abyssal plain, however, turbidity currents may become the dominant process for OC preservation due to high sedimentation rate and rapid burial.
- Distinct changes in amount and composition of OC are documented in sediment cores, reflecting glacial-to-interglacial variations. During the interglacial (e.g., the Holocene), the accumulation of terrigenous as well as marine organic matter was increased, whereas during the Last Glacial significantly lower OC accumulation rates were determined. This variability has been caused by significantly higher input of terrigenous OC via sea ice and currents as well as increased primary production due to a reduced sea-ice cover during interglacial (Holocene) time, whereas during the glacial a more closed sea-ice cover resulted in a lower input of both terrigenous and marine OC.
- Only in areas directly influenced by glacigenic processes due to a major ice sheet on the neighbouring continent (such as, for example, the Barents Sea continental margin), drastically increased accumulation of organic matter occurred during the Last Glacial.
- Due to the limited number of AMS^{14}C dated sediment cores obtained from restricted areas of the central Arctic Ocean, only rough estimates of total sediment and OC budgets of the Holocene time interval can be given. In total, a burial rate of 237×10^6 t y^{-1} of total sediment and 2.4×10^6 t y^{-1} of OC were calculated.

Chapter 8

Organic Carbon Budget: Arctic Ocean vs. Global Ocean

R. Stein, R.W. Macdonald

8.1 Introduction

In this chapter we summarize the data on organic carbon (OC) sources from Chapters 2 and 3 and the data on OC burial rates from Chapters 7.2 to 7.9 to produce best estimates for total sediment and organic carbon budgets for the Arctic Ocean. To construct the organic carbon budget we have limited ourselves to particulate organic carbon (POC) and have not included the dissolved organic carbon (DOC) in our estimates. DOC certainly is an important OC source (see Chapter 4), however, we presently do not know how conservative it is having no confident estimate of which proportion becomes trapped over the shelves or metabolized during transport (see below). Based on the results presented in Chapter 4, less than 5% is buried in coastal sediments.

In order to view the Arctic Ocean OC budget in relationship to the global OC budget, a summary of global values based on literature data is also presented in this chapter.

8.2 Global Organic Carbon Fluxes: Sources and Sinks

Organic carbon (OC) in marine sediments may be either of marine or terrestrial origin. Marine organic matter is formed mainly by phytoplankton, by far the largest OC source, synthesizing about 30 to 50 Gt y^{-1} (1 Gt = 10^9 t) (Table 8.1; e.g., Berger et al., 1989; Antoine et al. 1996; Hedges et al. 1997). Typical values of primary production vary between 25–50 gC m^{-2} y^{-1} in the centers of open-ocean subtropical gyres to > 250 gC m^{-2} y^{-1} in upwelling zones (e.g., Berger et al., 1989). The marine OC produced in the high-productivity areas, however, supplies but a minor proportion of the total annual OC production of the world ocean because these areas are very small compared to the open ocean area. After leaving the euphotic zone, most of the marine OC becomes rapidly recycled during its descent through the water column and only very minor amounts reach the deep ocean seafloor. As shown from sediment trap studies, there is an exponential decrease of particulate OC flux with depth. About 10% of surficial productivity reaches a depth of 200m and perhaps 1% passes to an ocean depth of 4000 m (e.g., Suess 1980; Betzer et al. 1984). Given this distribution in vertical flux, continental shelves/upper slope areas characterized by a shallow water depth will be responsible for most of the preservation and burial of OC due to high primary production and high sedimentation rate (e.g., Berner 1989; Hedges et al. 1997; see below). Following Berger et al. (1989), the average burial rate of marine OC in the open ocean and coastal ocean is about 0.03% and 0.8% of the primary production, respectively. Taking a global marine primary production of about 30 Gt y^{-1} for the open ocean area and about 7 Gt y^{-1} for the coastal ocean (Wollast 1991; Antoine et al. 1996), this implies annual marine OC burial rates of about 0.009 Gt and 0.056 Gt, respectively (see Schlünz and Schneider 2000).

Table 8.1. Global organic carbon sources and burial rates (10^6 t y^{-1}). Primary production values from Berger et al. 1989; Siegenthaler and Sarmiento 1993; Antoine et al. 1996; River discharge values from Schlesinger and Melack 1981; Ittekkot 1988; Degens et al. 1991; Ludwig et al. 1996; Schlünz and Schneider 2001; Eolian input values from Romankevich 1984; Zafiriou et al. 1985; Organic carbon burial rates from Hedges et al. 1997

Main organic carbon sources (input)	
Primary Production	30000–50000
River input (POC)	130–200
River input (DOC)	210–230
River input (TOC)	330–430
Eolian input	100–320
Burial rate of organic carbon	
Deltaic sediments	70
Shelves and upper slopes	68
High-productivity zones	10
Shallow-water carbonates	6
Pelagic sediments	5
Anoxic basins	1
Total world ocean	160

The annual supply of terrigenous OC to the world ocean is very small in comparison to the marine OC produced by marine phytoplankton (Table 8.1). On a global scale, riverine and eolian supply are the most important processes. Data on coastal erosion, which is a major OC source in the Arctic Ocean (see Chapters 2.3 and 8.3), are not available for the global ocean.

According to Milliman and Syvitski (1992), the global total sediment discharge by rivers reaches a value of about 20 Gt y^{-1}. During the last 30 years, the global annual supply of terrigenous OC by rivers has been estimated in numerous publications giving a wide range of values (see review by Schlünz and Schneider 2001). Most of the numbers, however, fall within a relatively narrow range between about 0.3 and 0.43 Gt (e.g., Schlesinger and Melack 1981; Degens et al. 1991; Ludwig et al. 1996; Schlünz and Schneider 2001). On a global average, the ratio between dissolved (DOC) and particulate organic carbon (POC) is about 1.2–1.5 (Meybeck 1982; Ludwig et al. 1996). Thus, rivers discharge about 0.2–0.25 Gt of DOC and 0.13–0.2 Gt of POC (Ittekkot 1988; Lugwig et al. 1996; Hedges et al., 1997). According to Ittekkot (1988), about 35% of the riverine POC belongs to the labile (metabolizable) fraction and may already become oxidized in the estuaries and coastal environments, whereas the other 65% appears to be highly refractory and therefore preserved in the marginal sediments or transported farther offshore. If Ittekkot (1988) is correct, about 0.1 Gt of terrigenous (riverine) POC could be preserved and buried in marine sediments on a global scale.

For the eolian organic carbon supply the data base is very weak. The global atmospheric flux of terrigenous OC is estimated to be as large as 0.1 to 0.3 Gt (Romankevich 1984; Zafiriou et al. 1985; Hedges et al. 1997). How much of this terrigenous OC reaches the seafloor and is finally buried in marine sediments is not well known yet. In the organic-carbon-impoverished sediments of the central open-ocean regions, terrigenous OC supplied by wind may become a major proportion of the organic matter accumulated in these environments (Zafiriou et al. 1985; Prahl and Muehlhausen 1989). Gagosian and Peltzer (1987) measured high relative abundance of long-chain series of vascular plant-wax lipids (*n*-alkanes, *n*-acids, *n*-alcohols) in aerosols from the Enewetak Atoll area in the central Pacific, indicating that significant amounts of the organic matter was land-derived. In the Enewetak area, Zafiriou et al. (1985) estimated the annual eolian flux of terrigenous OC to be about 0.5 gC m^{-2} y^{-1}, which is about 1% of the marine OC produced in the central open ocean. Since the terrigenous OC that has survived considerable degradation in subaerial soils and long-distance transport to the ocean is largely resistant to further decomposition (Tissot et al., 1979; Demaison and Moore, 1980), most of this organic matter will reach the seafloor. On the other hand, from the much more labile marine OC less than 1% of the amount produced in the surface water will survive the downward transport through the water column (Suess 1980; Betzer et al. 1984). Thus, terrigenous OC supplied by wind may account for much as 50% of the total OC preserved in central open-ocean sediments (Zafiriou et al. 1985; Prahl and Muehlhausen 1989).

The total modern global burial rate of OC in near-surface sediments is estimated to be about 0.1–0.16 Gt y^{-1} (Romankevich 1984; Berner 1989; Hedges and Keil 1995). This total burial rate of OC is remarkably small, suggesting that >99% of the marine primary production as well as major proportions of the terrigenous OC introduced to the ocean must be completely remineralized (Hedges et al. 1997). According to Hedges and Keil (1995), who recalculated the data of Berner (1989) coming to a total annual burial rate of 0.16 Gt, more than 85% of the organic carbon (i.e., about 0.14 Gt) are buried in deltaic sediments and the sediments of shelves and upper slope (Table 8.1). The rest (i.e. about 0.02 Gt y^{-1}) is buried in high-productivity zones, shallow-water carbonates, pelagic (low-productivity) sediments and anoxic basins (Table 8.1).

Given the global burial rate of marine OC (0.065 Gt y^{-1}) and the amount of refractive terrigenous POC supplied by rivers (0.1 Gt y^{-1}), these alone may explain the global burial rate of total OC in marine sediments (0.16 Gt y^{-1}). However, the predominant proportion of the organic matter preserved in marine sediments would have to be of terrigenous origin. The dominance of terrigenous OC would need to be even higher if the eolian OC is added to the mass balance calculation. Data on organic carbon composition of marine sediments, on the other hand, indicate that along continental margins (exluding the Arctic; see Chapter 8.3), the main depocenter of organic carbon, the organic matter is predominantly marine-derived (e.g., Emerson et al. 1987; Emerson and Hedges 1988; Gough et al. 1993; Hedges et al. 1997). These compositional data imply that the destruction of terrigenous OC may be higher than suggested. On the other hand, the proportion of terrigenous OC being preserved in the sediments may simply be underestimated. Depending on the methods used for characterization of the OC origin, very different estimates for marine/terrigenous OC values are given. Stable carbon isotope composition is often used to estimate the proportion of terrigenous organic matter (e.g., Hedges and Mann 1979; Jasper and Gagosian 1989; Rühlemann et al. 1996; see Chapter 1.4). In the sediments off the Washington coast influenced by the Columbia River discharge, for example, sediments near the mouth of the Columbia River contain more than 90% of terrigenous OC, decreasing to less than 10% of terrigenous OC on the outer shelf and slope (Hedges and Mann 1979). Plant-wax *n*-alkane concentrations in the same area, on the other hand, indicate substantially higher percentages of terrigenous OC of 20–50% on the outer continental shelf and slope (Prahl and Muehlhausen 1989; Prahl et al. 1994). Goñi et al. (1998) re-evaluated the importance of land-derived OC in surface sediments of the Gulf of Mexico, using bulk and biomarker sedimentary compositions, and concluded that the terrigenous OC fraction was significantly underestimated in earlier studies. In future research, multiple approaches (i.e., bulk parameters, organic petrography and biomarkers) should be used to give a more robust estimate of the marine and terrigenous proportions of the OC preserved and buried in marine sediments (see Wagner and Dupont 1999).

A further concern is that the fate of DOC has not been considered in the budget estimates. There is, however, virtually no evidence that marine sediments incorporate riverine DOC beyond the small amounts that are buried in interstitial sea-water (Hedges et al. 1997). DOC is discharged directly into lighted, highly-oxygenated surface waters where, without any physically protective inorganic matrix (Hedges and Keil 1995), it is subject to photolysis and oxic biodegradation (Mopper et al. 1991). Although DOC is discharged in greater amounts than POC, it appears not to escape seawater where it is rapidly cycled by as yet poorly known mechanisms (Hedges et al. 1997; Opsahl and Benner 1997). Thus, an open question remains as to what happens to the DOC supplied by rivers.

8.3 Arctic Ocean Organic Carbon Fluxes: Sources and Sinks

In Figures 8.1 and 8.2 and Table 8.2, we have summarized the modern total sediment and organic carbon input into the Arctic Ocean marginal seas. Altogether, about 666 × 10^6 t y^{-1} of total sediment are supplied (Fig. 8.2). Terrigenous POC and DOC input are estimated at about 12.6 × 10^6 t y^{-1} and 28.1 × 10^6 t y^{-1}, respectively, with most of the terrigenous POC entering from rivers and coastal erosion (see Chapter 4). Our estimate for terrigenous DOC entering the Arctic Ocean is somewhat larger than that of Opsahl et al., (1999) (18.2 ×

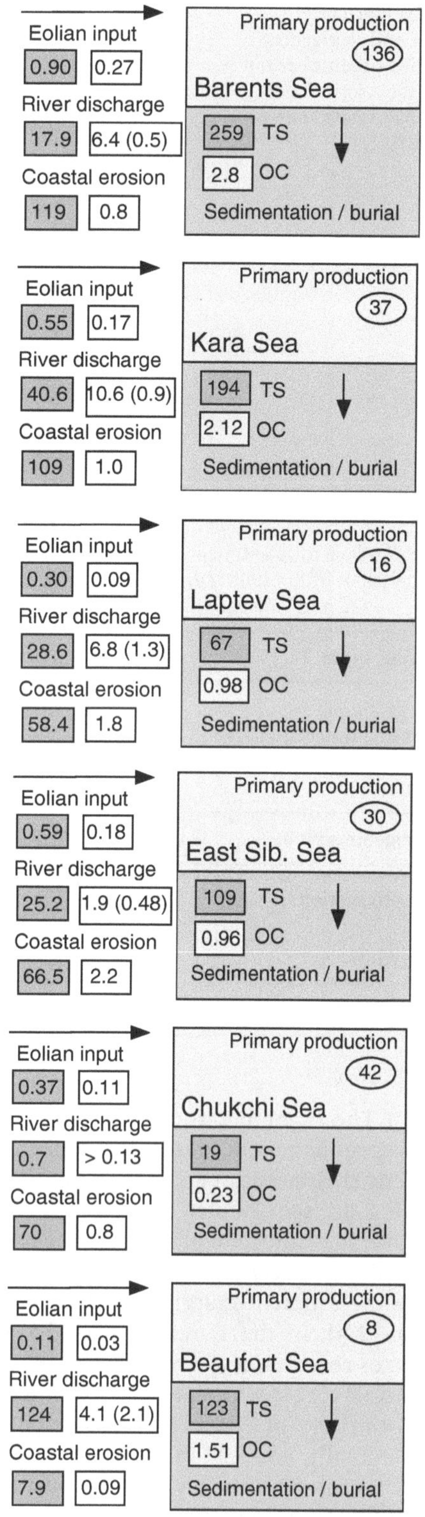

Fig. 8.1. Modern total sediment, total terrigenous OC (TOC = POC + DOC) and marine OC (primary production) input, and average Holocene burial rates of total sediment (TS) and organic carbon (OC) in the Arctic Ocean marginal seas (10^6 t y^{-1}). For riverine input, POC values are shown in brackets. For data sources see Tables 8.2 and 8.3. For discussion of Kara Sea coastal values we refer to Chapter 7.6.5

Fig. 8.2. Modern total sediment and terrigenous and marine organic carbon input in the Arctic Ocean (10^6 t y^{-1}). For data source see Table 8.2

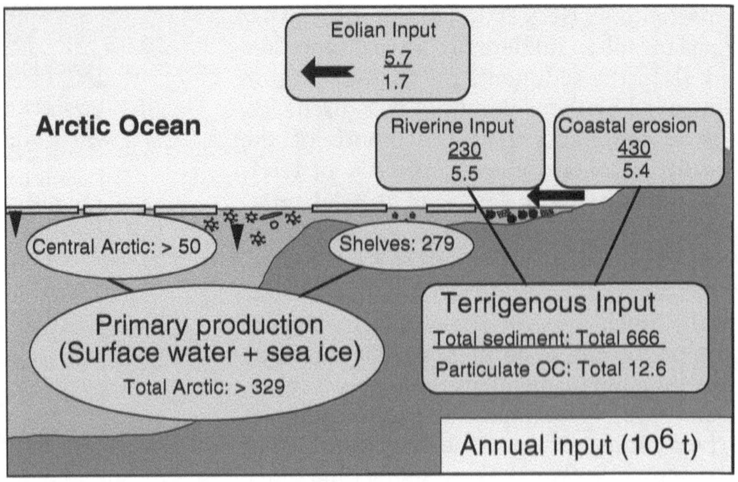

Table 8.2. Modern total sediment and terrigenous and marine organic carbon input in the Arctic Ocean (10^6 t y^{-1}). Terrigenous input data taken from Chapter 2, primary production values from Chapter 3. For the TOC input by coastal erosion we assume that 80% is POC and 20% is DOC

Terrigenous input	Total sediment	POC	DOC	TOC
Riverine input	230	5.5	24.5	30
Underground water discharge	–	–	2.3	2.3
Coastal erosion	430	(>) 5.4	(<) 1.3	6.7
Eolian input	5.7	1.7	–	1.7
Total terrigenous input	666	12.6	28.1	40.7
Primary production				
Shelves		279		
Central Arctic		> 50		
Total Arctic		> 329		

10^6 t y^{-1}) The organic carbon input from marine primary production accounts for 279×10^6 t y^{-1} over Arctic shelves and $>50 \times 10^6$ t y^{-1} in the central Arctic (Fig. 8.2; see Chapter 3).

The average Holocene burial rates of total sediment (TS) and organic carbon (OC) vary widely between the different marginal seas (Figs. 8.1 and 8.3; Table 8.3). In the Eurasian seas, TS and OC burial rates reach values of 67 to 259×10^6 t y^{-1} and about 1 to 2.8×10^6 t y^{-1}, respectively. In the Beaufort Sea, 123×10^6 t y^{-1} of TS and 1.5×10^6 t y^{-1} of OC are buried annually, whereas the lowest values for TS and OC burial (19×10^6 t y^{-1} and 0.23×10^6 t y^{-1}, respectively), were determined for the Chukchi Sea. For the central Arctic Ocean, TS and OC burial rates of 237×10^6 t y^{-1} and 2.4×10^6 t y^{-1}, respectively, were calculated (Fig. 8.3, Table 8.3). About 80% of the basin burial occurs as accumulation along the continental slopes and rises with the remainder accumulating on the central ridges and abyssal plains (Table 8.3). In total, about 1008×10^6 t y^{-1} of TS and 11×10^6 t y^{-1} of OC have been buried annually in the entire Arctic Ocean as a Holocene average (Fig. 8.3, Table 8.3). This corresponds to a total accumulation of about 11100×10^9 t (or 11.1 Gt) of TS and 121×10^9 t (or 0.12 Gt) OC during the Holocene, or, the last 11 Cal. kyrs. BP. In our calculation the shelves of the Canadian Archipelago, North Greenland and the Lincoln Sea have not been considered simply due to lack of data (Table 8.3).

Based on the origin of the OC, assigned in Chapters 7.2 to 7.9, the OC burial rates can be divided into terrigenous and marine proportions (Table 8.3). Whereas terrigenous OC is predominant in the sediments from the Beaufort, East Siberian, Laptev and Kara seas as well as the central Arctic Ocean, marine OC is much more important in the Chukchi and, especially, Barents Seas. In the Arctic marginal seas, about 5.5×10^6 t y^{-1} of terrigenous and 3.2×10^6 t y^{-1} of marine OC have been buried annually as

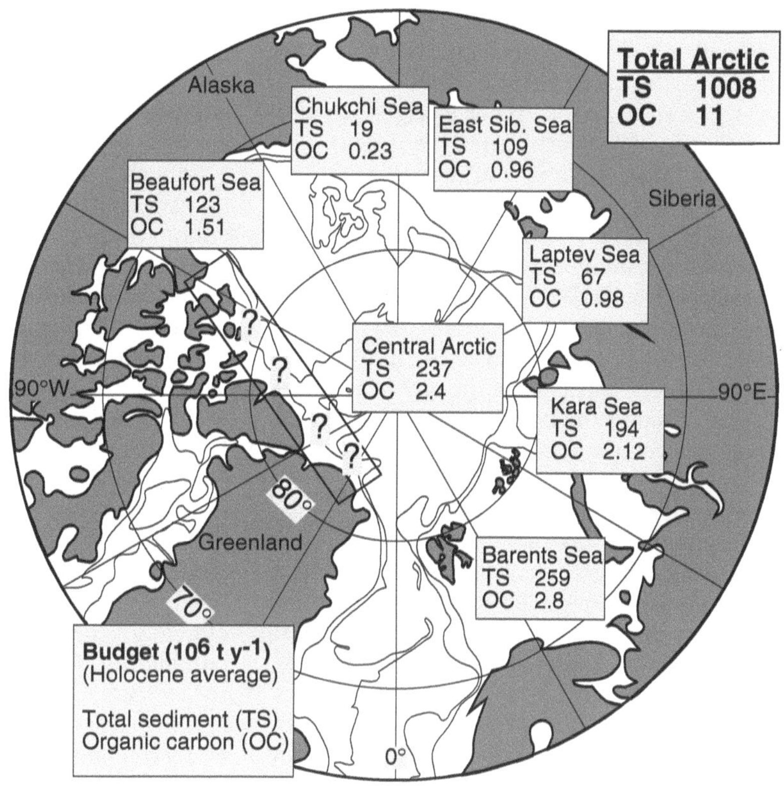

Fig. 8.3. Average Holocene burial rates (10^6 t y^{-1}) of total sediment (TS) and organic carbon (OC) over the marginal seas and basins of the Arctic Ocean. For data source and references see Table 8.3

Table 8.3. Budget (10^6 t y^{-1}) and total accumulation (0–11 Cal. kyrs. BP; 10^9 t) of total sediment and organic carbon fractions (Holocene average). The total sediment accumulation for the Barents, East Siberian and Chukchi seas were estimated from OC burial rates (2.8, 0.96, and 0.23 × 10^6 t y^{-1}, respectively) and OC percentage values (1.1, 0.88, and 1.2%). For the Chukchi Sea, the burial of total OC and OC percentage values have been taken from Romankevich and Vetrov (2001). For all other data sources see Chapters 7.2 to 7.9

Area	Size (× 10^3 km^2)	Total sediment		Organic carbon (OC)		Terrigenous OC		Marine OC	
		(10^6 t y^{-1})	(10^9 t)	(10^6 t y^{-1})	(10^9 t)	(10^6 t y^{-1})	(10^9 t)	(10^6 t y^{-1})	(10^9 t)
Continental shelves									
Beaufort Sea	178	123	1353	1.51	16.6	1.21	13.3	0.3	3.3
Chukchi Sea	620	19	209	0.23	2.5	0.11	1.2	0.12	1.3
East Siberian Sea	987	109	1199	0.96	10.6	0.67	7.4	0.29	3.2
Laptev Sea	498	67	737	0.98	10.8	0.90	9.9	0.08	0.9
Kara Sea	926	194	2134	2.12	23.3	1.72	18.9	0.4	4.4
Barents Sea (incl. White Sea)	1597	259	2849	2.8	30.8	0.84	9.2	1.96	21.6
Candian Archipelago shelf	146	?	?	?	?	?	?	?	?
North Greenland Shelf	30	?	?	?	?	?	?	?	?
Lincoln Sea	64	?	?	?	?	?	?	?	?
Continental shelves	5052	771	8481	8.68	95.4	5.45	60	3.15	34.7
Continental Slopes	541	107	1177	1.29	14.2	1.03	11.4	0.26	2.9
Continental rises	1095	79	869	0.72	7.9	0.58	6.3	0.14	1.5
Abyssal plains	1367	30	330	0.25	2.8	0.2	2.2	0.05	0.6
Ridges	1506	21	231	0.11	1.2	0.09	1.0	0.02	0.2
Total Arctic	**9555**	**1008**	**11088**	**11**	**121**	**7.4**	**80.9**	**3.6**	**39.9**

a Holocene average. For the Barents Sea, marine OC burial (about 2×10^6 t y^{-1}) accounts for >60% of the burial flux. In the central Arctic Ocean about 1.9×10^6 t y^{-1} and 0.5×10^6 t y^{-1} of terrigenous and marine OC, respectively, are buried (Table 8.3).

Taking the average global burial rate of OC of 100 to 160×10^6 t y^{-1} (or 0.1 to 0.16 Gt), the Arctic Ocean with 11×10^6 t y^{-1} accounts for about 7 to 11% of the global budget. This number is disproportionately high considering that the Arctic Ocean accounts for only 2.5% of the global ocean area. Thus, the burial efficiency of OC is significantly higher in the Arctic Ocean than the global average, which is explained by the high proportion of (more refractive) terrigenous organic matter.

Input-Output Box Model for the Modern (Late Holocene) Arctic Ocean

A distinct mismatch is evident between the annual total sediment input rate (666×10^6 t y^{-1}) and the average Holocene TS accumulation rate (771×10^6 t y^{-1} for the Arctic shelves plus 237×10^6 t y^{-1} for the Arctic interior) (Fig. 8.4). To balance this budget would require a much higher sediment input than presently occurs. This mismatch can be understood by looking at the Holocene records from the Laptev and Kara seas. For both seas – in contrast to all other areas in the Arctic – there are detailed AMS^{14}C dated sedimentary records for the Holocene time interval. These detailed records indicate a major change in accumulation of both total sediment and organic carbon through the Holocene. In the early Holocene, the accumulation rates were significantly higher than during the late Holocene, which more closely represents the modern situation, and a significantly higher sediment input via coastal erosion and river discharge is most probable (see Chapters 7.5 and 7.6). Thus, the comparison of average Holocene accumulation rates with modern input (as is often done for budget and balance calculations) is biased. It would be more valid to compare the modern input rates with late Holocene accumulation (burial) rates. For the Laptev and Kara seas, the late Holocene accumulation rates are, on average, about 60% of the mean Holocene accumulation rates. Thus, for our final input-output box model (Fig. 8.5) we have corrected the mean Holocene burial rates in Table 8.3 by multiplying them by 0.6 (e.g., 771×10^6 t y^{-1} for TS Holocene average becomes 463×10^6 t y^{-1} for late Holocene).

- **Total sediment**

From the total sediment input of 666×10^6 t y^{-1}, about 463×10^6 t y^{-1} (or about 70% of the initial input) is accumulating on the Arctic shelves (Fig. 8.5). 186×10^6 t y^{-1} are exported towards the open Arctic Ocean via currents (i.e., boundary currents, brines, turbidity currents etc.) and 14×10^6 t y^{-1} by sea ice. In the basins, another 142×10^6 t y^{-1} is accumulating. The remaining sediments, about 60×10^6 t y^{-1}, are exported from the Arctic Ocean.

- **Terrigenous particulate organic carbon**

The total terrigenous particulate organic carbon (POC) input into the Arctic shelves adds up to 11.8×10^6 t y^{-1}. Assuming that 35% of the POC belongs to the labile (metabolizable) fraction and becomes oxidized within the estuaries and coastal environments (Ittekkot 1988), a maximum of 7.7×10^6 t y^{-1} should be available for burial. On the Arctic shelves, about 3.3×10^6 t y^{-1} or 28% of the initial input are preserved and buried annually (Fig. 8.5). Therefore, 4.4×10^6 t y^{-1} are transported into the Arctic interior where another 1.1×10^6 t y^{-1} is buried. This leaves a remaining 3.8×10^6 t y^{-1} of terrigenous POC which we have accounted for in this

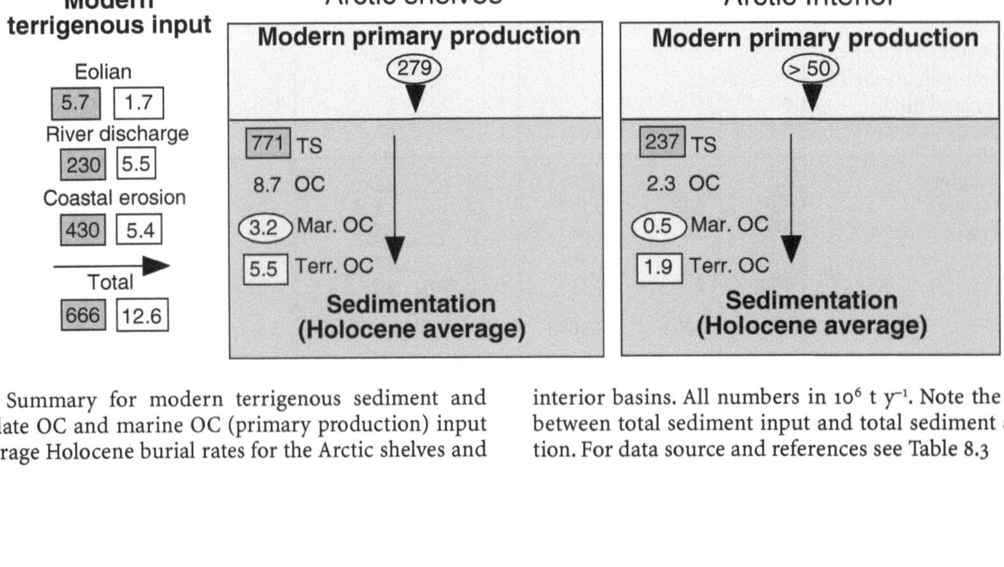

Fig. 8.4. Summary for modern terrigenous sediment and particulate OC and marine OC (primary production) input and average Holocene burial rates for the Arctic shelves and interior basins. All numbers in 10^6 t y^{-1}. Note the mismatch between total sediment input and total sediment accumulation. For data source and references see Table 8.3

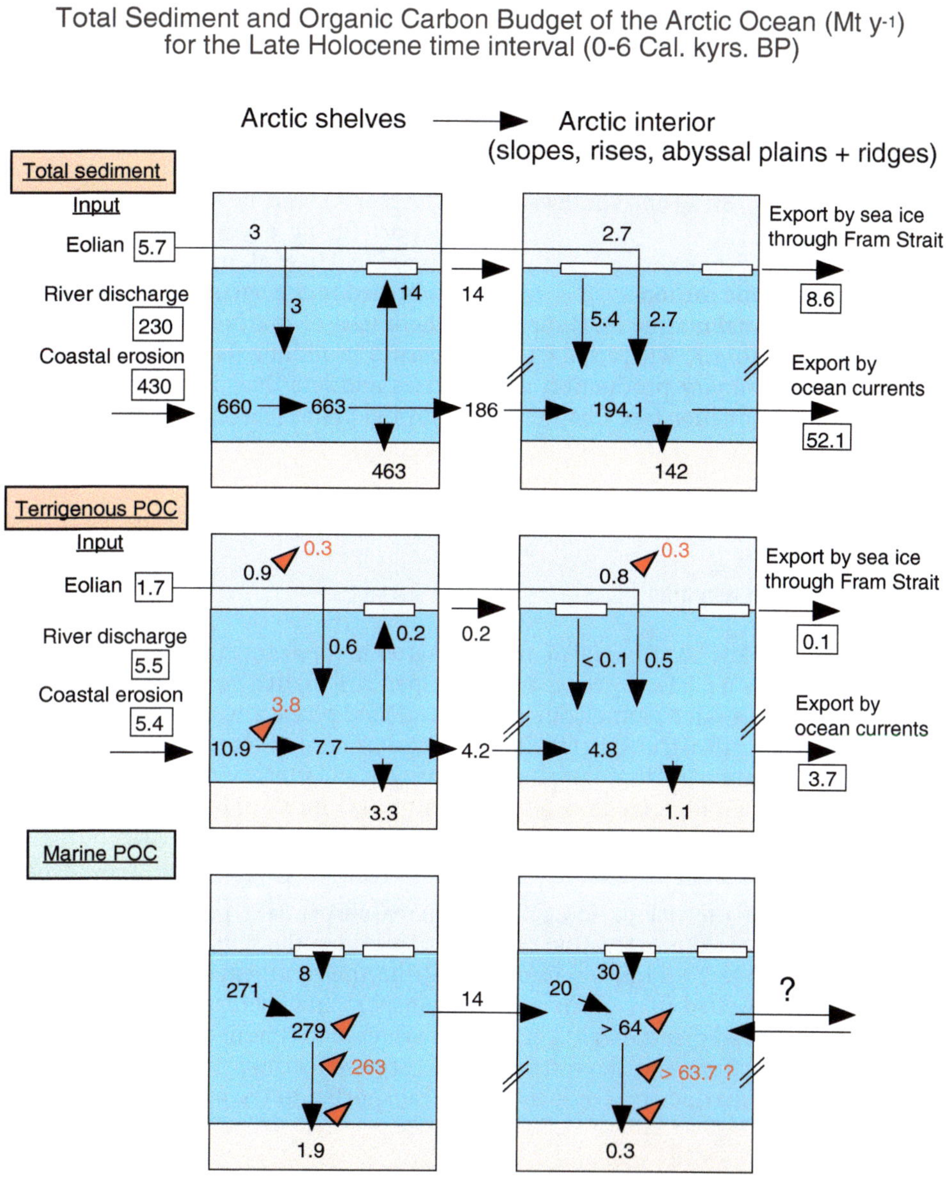

Fig. 8.5. Total sediment and particulate organic carbon input-output box model for the present-day Arctic Ocean. All numbers in 10^6 t y^{-1} (Mt y^{-1}). For data source and references see Table 8.3. Based on the results from the Laptev and Kara Seas (see Chapters 7.5 and 7.6) we assume that the late Holocene (i.e., modern Arctic Ocean) has burial rates that are only 60% of the average Holocene burial rates. For the marine OC input, sea-ice production contributes about 60% of the primary production of the whole water column in the Arctic interior compared to 3% in the Arctic shelf areas (see Chapter 3). It is assumed that 5% of the shelves' primary production (i.e., 14×10^6 t y^{-1}) is exported towards the Arctic interior (see Liu et al. 2000). For terrigenous OC, we assume that about 35% of the POC belongs to the labile (metabolizable) fraction and has already been oxidized in the estuaries and coastal environments (Ittekkot 1988). Red numbers indicate amount of organic carbon that becomes oxidized. For further explanation see text

budget by exporting it from the Arctic Ocean either in ice (0.1×10^6 t y^{-1}) or water (3.7×10^6 t y^{-1}) (Fig. 8.5). This export seems high. We cannot exclude, however, the possibility that the remineralization rate of terrigenous OC is higher than the 35% published by Ittekkot (1988) (see above) or that part of the POC is transformed into DOC. Opsahl et al. (1999) estimated the export of terrigenous DOC from the Arctic Ocean via the East Greenland Current to be 2.9–10.3 $\times 10^6$ t y^{-1} (based only on a 1 Sv flow and not including the Canadian Archipelago). If we accept the major export pathway to comprise the East Greenland Current (~3–3.5 Sv (Rudels and Friedrich 2000)) and flow through the Archipelago (~1 Sv) the export of 3.7×10^6 t y^{-1} of terrigenous POC could be accounted for by a POC concentration of about 0.03 g m^{-3} (~ 3 μM terrigenous POC). Values measured in water to the East of Greenland fall in the range of 3 to 8 μM (Lisa Miller, pers. Comm) suggesting that an important component of this POC could be terrigenous. Regrettably, we lack the required biomarker data to evaluate what part of the POC at Fram Strait and southwards is of terrigenous origin.

- **Marine particulate organic carbon**

On the Arctic shelves, the total marine POC supply is estimated at about 279×10^6 t y^{-1} where 271×10^6 t y^{-1} derives from pelagic primary production and 8×10^6 t y^{-1} from sea-ice production (see Chapter 3). Only 1.9×10^6 t y^{-1} (or about 1% of the total production) is buried in the sediments (Fig. 8.5). The input of marine POC to the Arctic interior is estimated to be about 64×10^6 t y^{-1} where 30×10^6 t y^{-1} are contributed by sea-ice production, 20×10^6 t y^{-1} by primary production in the water column (see Chapter 3), and 14×10^6 t y^{-1} is added from shelf export (using a 5% export from shelves as published by Liu et al. 2000). Of this, only 0.3×10^6 t y^{-1} is buried in the basin sediments. More than 85% of the burial of marine organic carbon in the basins occurs in sediments of the continental slopes and rises and less than 15% on the central ridges and abyssal plains (see Table 8.3).

7.8.6 Summary and Concluding Remarks

- In general, the preservation and burial of organic matter in the Fram Strait/Yermak Plateau area are controlled by: (1) strong lateral transport of suspended matter by Atlantic water during the Holocene and by meltwater plumes during the deglacial phase and probably during the glacial phase; (2) enhanced vertical transport facilitated by water masses containing increased amounts of suspended organic matter and therefore better capturing and preserving of marine organic carbon through aggregation with terrigenous organic matter; (3) increased primary production due to upwelling of nutrients near the sea-ice edges, on the shelves and in the fjords; and (4) preferential deposition and preservation of organic carbon due to high bulk accumulation rates.
- During approximately the last 30,000 cal. yrs., increased primary production has not been the major reason for increased preservation of organic carbon in the sediments. Rather, it is the lateral input of suspended terrigenous organic matter from the continents and the associated "ballast effect" which have resulted in a quick vertical transport and increased preservation of marine organic matter in the sediments. High accumulation rates of inorganic material, however, are needed to support aggregation and deposition of the organic carbon. Furthermore, allochthonous organic matter (both marine and terrigenous) is transported to the northernmost Fram Strait by strong currents.
- In total, 0.95×10^9 t C of total organic carbon have been deposited in the Northern Fram Strait/Southern Yermak Plateau area during the Holocene, i.e., the last 11 Cal. kyrs. BP. Of this, about half is terrigenous, half marine. Most of the organic carbon (65%) has been buried in sediments under Yermak/Svalbard branch, which accounts for only 20% of the total area.

Chapter 9

References

Aagaard K (1984) The Beaufort Undercurrent. In: Barnes PW, Schell DM, Reimnitz E (eds). The Alaskan Beaufort Sea; Ecosystems and Environments. Academic Press, pp 47–71

Aagaard K (1989) A synthesis of the Arctic Ocean circulation. Rapp P-v Réun Cons Int Explor Mer 188:11–22

Aagaard K, Carmack EC (1989) The role of sea ice and other fresh water in the Arctic circulation. J Geophys Res 94:14485–14498

Aagaard K, Coachman LK, Carmack EC (1981) On the halocline of the Arctic Ocean. Deep Sea Res 28:529–545

Aagaard K, Swift JH, Carmack EC (1985) Thermohaline circulation in the Arctic Mediterranean Seas. J Geophys Res 90:4833–4846

Aas E, Højerslev NK (2001) Attenuation of ultraviolet irradiance in North European coastal waters. Oceanologia 43:139–168

Aas E, Høkedal J, Højerslev NK, Sandvik R, Sakshaug E (2002) Spectral properties and UV-attenuation in arctic marine waters. In: Hessen D (ed) Radiation and Arctic ecosystems. Ecological Studies 153, Springer, Berlin Heidelberg, pp 23–72

Abramova EN (1999) Composition, abundance and population structure of spring-time zooplankton in the shelf-zone of Laptev Sea. In: Kassens H et al. (eds) Land-ocean systems in the Siberian Arctic: Dynamics and history. Springer, Berlin, pp 161–168

ACIA (2002a) http://faldo.atmos.uiuc.edu/ACIA/timeseries.html

ACIA (2002b) http://zubov.atmos.uiuc.edu/FINGERPRINT/FREQUENCY/GIF/ccc.gs

Ackman RG, Tocher CS, McLachlan J (1968) Marine phytoplankter fatty acid. J Fish Res Board Can 25:1603–1620

Agatova AI, Arzhanova NV, Torgunova, NI (1999) Organic matter in the Bering Sea. In: Loughlin TR, Ohtani K (eds) The physical oceanography of the Bering Sea. Univ Alaska Seagrant AK-SG-03, Univ Alaska Fairbanks, pp 216–283

Aibulatov NA, Matyushenko VA, Shevchenko VP, Politova NV, Potekhina EM (1999) New data about structure of lateral fluxes of suspended matter on the Barents Sea periphery. Environmental Geoscience, in press

Albert DB (1985) Sulfate reduction and iron sulfide formation in sediments of the Pamlico River Estuary, North Carolina. PH.D. Thesis Univ. N. Carolina at Chapel Hill, pp 1–181

Aldahan A, Possnert G, Scherer R, Shi N, Backman J, Bostroem K (2000) Trace-element and major-element stratigraphy in Quaternary sediments from the Arctic Ocean and implications for glacial termination. Journal of Sedimentary Research Section A: Sedimentary Petrology and Processes 70: 1095–1106

Alekin OA, Brazhnikova LV (1964) Discharge of dissolved substances from the USSR territory. Moscow, Nauka, 143 pp. (in Russian)

Alexander V (1995) The influence of the structure and function of the marine food web on the dynamics of contaminants in Arctic Ocean ecosystems. Science of the Total Environment 161:593–603

Alexander V, Niebauer HJ (1981) Oceanography of the eastern Bering Sea ice-edge zone in spring. Limnol Oceanogr 26:1111–1125

Alexandrov VYu, Martin T, Kolatschek J, Eicken H, Kreyscher M, Makshtas A (2000) Sea-ice circulation in the Laptev Sea and ice export to the Arctic Ocean: Results from satellite remote sensing and numerical modelling. J Geophys Res 105:17143–17159

Alley RB, Marotzke J, Nordhaus W, Overpeck J, Peteet D, Pielke R, et al. (2002) Abrupt Climate Change; Inevitable surprises, National Academy Press, Washington DC

Alley RB, Mayewski PA, Sowers T, Stuiver M, Taylor KC, Clark PU (1997) Holocene climatic instability: A prominent, widespread event 8200 yr ago. Geology 25: 483–486

Altabet MA, Francois R (1994) Sedimentary nitrogen isotopic ratio as a recorder for surface ocean nitrate utilization. Glob Biogeochem Cycles 8:103–116

AMAP (1998) AMAP Assessment Report: Arctic Pollution Issues. Arctic Monitoring and Assessment Program (AMAP), Oslo, Norway, 859 pp.

Ambrose W, Ahrens M, Brandt A, Dimmler W, Graf G, Gutt J, Hermann R, Jensen P, Piepenburg D, Queisser W, Renard P, Ritzrau W, Scheltz A, Thomsen L (1994) Benthos. Reps Pol Res 142: 106–112

Ambrose WG Jr, Clough LM, Tilney PR, Beer L (2001) Role of echinoderms in benthic remineralization in the Chukchi Sea. Mar Biol 139:937–949

Ambrose WG Jr, Renaud PE (1995) Benthic response to water column productivity patterns: Evidence for benthic-pelagic coupling in the Northeast Water Polynya. J Geophys Res 100: 4411–4421

Amiel D, Cochran JK, Hirschberg DJ (2002) $^{234}Th/^{238}U$ disequilibrium as an indicator of the seasonal export flux of particulate organic carbon in the North Water Polynya. Deep-Sea Res 49:5191–5209

Amon RMW, Benner R (1996) Bacterial utilization of different size classes of dissolved organic matter. Limnol Oceanogr 41:41–51

Amon RMW, Benner R (1996) Photochemical and microbial consumption of dissolved organic carbon and dissolved oxygen in the Amazon River system. Geochim Cosmochim Acta 60:1783–1792

Amon RMW, Benner R (2003) Combined neutral sugars as indicators of the diagenetic state of dissolved organic matter in the Arctic Ocean. Deep Sea Res 50/1:151–169

Amon RMW, Budeus G, Meon B (2003) Dissolved organic carbon distribution and origin in the Nordic Seas: Exchanges with the Arctic Ocean and the North Atlantic. J Geophys Res in press

Amon RMW, Fitznar H-P, Benner R (2001) Linkages among the bioreactivity, chemical composition, and diagenetic state of marine dissolved organic matter. Limnol Oceanogr 46:287–297

Amon RMW, Spitzy A (1999) Distribution of dissolved organic carbon during estuarine mixing in the southern Kara Sea. Reps Pol Res 300:102–109

Andersen ES, Dokken TM, Elverhøi A, Solheim A, Fossen I (1996) Late Quaternary sedimentation and glacial history of the western Svalbard margin. Mar Geol 133:123–156

Andersen OGN (1989) Primary production, chlorophyll, light, and nutrients beneath the Arctic sea ice. In: Herman Y (ed) The Arctic seas. Climatology, oceanography, and biology. Van Nostrand Reinhold, New York, pp 147–191

Andersen OGN (1998) Primary production, chlorophyll, light, and nutrients beneath the Arctic sea ice. In: Herman Y (ed). The Arctic Seas, Climatology, oceanography, Geology and Biology. Van Nostrand Reinhold, New York, pp 147–191

Andersen PF, Hartmann M (1992) Makrobenthos i blødbunden fra Vestgrønland. In: Andersen PF, Düwel L, Hansen OS (eds): Feltkurs i Arktisk Biologi, Godhavn 1990 Institut for Sporeplanter & Zoologisk Museum, Københavns Universitet, Denmark, pp 211–233

Anderson LG (2002) DOC in the Arctic Ocean. In: Hansell DA, Carlson CA (eds) Biogeochemistry of marine dissolved organic matter. Academic Press, San Diego, pp 665–685

Anderson LG, Björk G, Holby, O, Kattner G, Koltermann PK, Jones EP, Liljeblad B, Lindegren R, Rudels B and Swift JH (1994) Water Masses and Circulation in the Eurasian Basin: Results from the Oden 91 North Pole Expedition. J Geophys Res 99:3273–3283

Anderson LG, Chierici M, Fogelquist E (2000) Flux of anthropogenic carbon into the deep Greenland Sea. J Geophys Res 105:14339–14345

Anderson LG, Dyrssen D (1981) Chemical constituents of the Arctic Ocean in the Svalbard area. Oceanol Acta 4:305–311

Anderson LG, Dyrssen D, Jones EP (1990) An assessment of the transport of atmospheric CO_2 into the Arctic Ocean. J Geophys Res C 95:1703–1711

Anderson LG, Jones EP, Rudels B (1999) Ventilation of the Arctic Ocean estimated by a plume entrainment model constrained by CFCs. J Geophys Res 104:13423–13429

Anderson LG, Jones EP, Swift JH (2003) Export production in the central Arctic Ocean evaluated from phosphate deficits. J Geophys Res, in press

Anderson LG, Kaltin S (2001) Carbon fluxes in the Arctic Ocean – potential impact on climate change. Pol Res 20: 225–232

Anderson LG, Olsson K, Chierici M (1998) A carbon budget for the Arctic Ocean. Glob Biochem Cycles 12:455–465

Anderson LG, Olsson K, Skoog A (1994) Distribution of dissolved inorganic and organic carbon in the Eurasian Basin of the Arctic Ocean. In: Johannessen O, Muench R, Overland J (eds) The polar oceans and their role in shaping the global environment. AGU, Washington DC, pp 255–262

Andreassen I, Nöthig E-M, Wassmann P (1996) Vertical particle flux on the shelf off northern Spitsbergen, Norway. Mar Ecol Prog Ser 137: 215–228

Andreeva IA, Volkova JV, Vanshtein BG, Dzinoridze RN, Polyak LV, Cherkashev GA (2000) Variability of depositional processes in Kola Bay during Late Holocene. IMAGES Workshop "Correlation of post-glacial climatic events in the Arctic and Far East seas" IO RAS, Moscow, 1:12–13

Andrew JA, Kravitz JH (1974) Sediment distribution in deep areas of the northern Kara Sea. In: Herman Y (ed) Marine Geology and Oceanography of the Arctic Seas. Springer, Berlin, pp 231–256

Anisimova NA (1989) Distribution Patterns of Echinoderms in the Eurasian Sector of the Arctic Ocean. In: Herman Y (eds), The Arctic Seas – Climatology, Oceanography, Geology, and Biology, pp 281–302

Anning T, Nimer N, Merrett MJ, Brownlee C (1996) Costs and benefits of calcification in coccolithophorids. J Mar Syst 9:45–56

Anning, JL (1989) The development and decline of the epontic algal community in Barrow Strait, N.W.T., M Sc Thesis, University of Guelph, Ontario, Canada, 85 p

Anon, 1978. Polar Regions Atlas. Produced by the Foreign Assessment Center, Central Intelligence Agency

Anonymous (1980) Atlas of the Oceans The Arctic Ocean, MO USSR VMF

Antia AN, Bauerfeind E, Bodungen Bv, Zeller U (1993) Abundance, encystment and sedimentation of Acantharia during autumn 1990 in the East Greenland Sea. J Plankton Res 24: 99–114

Antia NJ (1976) Effects of temperature on the darkness survival of marine microplanktonic algae. Microbial Ecol 3:41–54

Antoine D, André JM, Morel A (1996) Oceanic primary production. 1. Estimation at global scale from satellite (Coastal Zone Colour Scanner) chlorophyll. Glob Biogeochem Cycles 10:57–69

Apollonio S (1959) Hydrobiological measurements on IGY Drifting Station Bravo. Trans Am Geophys Union 40: 316–319

Apollonio S (1980) Primary production in Dumbell Bay in the Arctic Ocean. Mar Biol 61:41–51

Are FE (1980) Thermal Abrasion of Sea Coasts. Nauka Publ, Moscow, 160 pp (in Russian)

Are FE (1985) Principles of shore thermoabrasion forecast. Novosibirsk Nauka Publ, 172 pp (in Russian)

Are FE (1987) Cryogenic processes of Arctic coastal zone dynamics. In: Fundamental investigations of the Earth Cryosphere. Nauka Press, Novosibirsk: 123–129 (in Russian)

Are FE (1994) Dynamics of the littoral zone of Arctic seas (State of the Art and Goals). Polarforschung 64: 123–131

Are FE (1999) The role of coastal retreat for sedimentation in the Laptev Sea. In: Kassens H, Bauch H, Dmitrenko I, Eicken H, Hubberten HW, Melles M, Thiede J and Timokhov L (eds) Land-Ocean systems in the Siberian Arctic: dynamics and history. Springer, Berlin 287–299

Arhus N (1991) The transition from deposition of condensed carbonates to dark claystones in the Lower Cretaceous succession of the southwestern Barents Sea. Norsk Geologisk Tidsskrift 71:259–263

Arhus N, Kelly SRA, Collins JSH, Sandy MR (1990) Systematic palaeontology and biostratigraphy of two Early Cretaceous condensed sections from the Barents Sea. Pol Res 8:165–194

Arkhikov EI, Brovko PF, Rybakov VF, Shuiskiy YuD (1982) Abrasive factors of sediment input into the Okhotsk Sea. In: Modern sedimentation and Quaternary morphogenesis of the Far East. Vladivostok: 165–176 (in Russian)

Arkhipov SA, Bespaly VG, Faustova MA, Glushkova O, Isaeva LL, Velichko AA (1986) Ice-sheet reconstructions. Quat Sci Rev 5:475–483

Arkhipov SA, Isayeva LL, Bespaly VG and Glushkova OY (1986) Glaciation of Siberia and northeast USSR. Quat Sci Rev 5:463–474

Arnborg L, Walker HJ, Peippo J (1967) Suspended load in the Colville River, Alaska, 1962. Geografiska Annaler 49A: 131–144
Aronson RB (1989) The ecology of *Octopus briareus* Robson in a Bahamian saltwater lake. Am Malacol Bull 7:47–56
Artemiev VE (1997) Geochemistry of organic matter in river-sea system. Kluwer, Dordrecht, 190 pp
Arthur MA, Dean WE Stow DAV (1984) Models for the deposition of Mesozoic – Cenozoic fine-grained organic-carbon-rich sediments in the deep sea. In: Stow DAV, Piper DJW (eds), Fine-Grained Sediments: Deep-Water Processes and Facies. Geol Soc London, Spec Publ 15: 527–560
Ashijan C, Smith S, Bignami F, Hopkins T, Lane P (1997) Distribution of zooplankton in the Northeast Water Polynya during summer 1992. J Mar Syst 10:279–298
Bakke T, Hameedi J, Kimstach V, Macdonald R, Melnikov S, Robertson A, et al (1998) Chapter 10: Petroleum hydrocarbons. In: Robertson A (ed). AMAP Assessment Report: Arctic Pollution Issues. AMAP, Oslo, pp 662–716
Balch WM, Fritz J, Fernandez E (1996) Decoupling of calcification and photosynthesis in the coccolithophore *Emiliania huxleyi* under steady-state light-limited growth. Mar Ecol Prog Ser 142:87–97
Balch WM, Holligan PM, Kilpatrick KA (1992) Calcification, photosynthesis and growth of the bloom-forming coccolithophore, *Emiliania huxleyi*. Cont Shelf Res 12:1353–1374
Baliño BM, Fasham MJR, Bowles MC (2000) Ocean biogeochemistry and global change. International Geosphere-Biosphere Programme IGBP Science No 2, ICSU, Stockholm, 32 pp
Balkanski Y, Monfray P, Battle M, Heimann M (1999) Ocean primary production derived from satellite data: An evaluation with atmospheric oxygen measurements. Glob Biogeochem Cycles 13:257–271
Båmstedt U, Eilertsen H-C, Tande KS, Slagstad D and Skjoldal HR (1991) Copepod grazing and its potential impact on the phytoplankton development in the Barents Sea. Pol Res 10:339–353
Båmstedt U, Karlson K (1998) Euphausiid predation on copepods in coastal waters of the Northeast Atlantic. Mar Ecol Prog Ser 172:149–168
Ban S et al (1997) The paradox of diatom-copepod interactions. Mar Ecol Prog Ser 157:287–293
Banse K (1979) On weight dependence of net growth efficiency and specific respiration rates among field populations of invertebrates. Oecologia 38:111–126
Banse K, English DC (1999) Comparing phytoplankton seasonality in the eastern and western subarctic Pacific and the western Bering Sea. Prog Oceanogr 43:235–288
Banse K, Mosher S (1980) Adult body mass and annual production/biomass relationships of field populations. Ecol Monographs 50:355–379
Barber D, Hanesiak JM, Chan W, Piwowar J (2001) Sea-ice and meteorological conditions in Northern Baffin Bay and the North Water Polynya between 1979 and 1996. Atmosphere-Ocean 39:343–359
Bard E (1998) Geochemical and geophysical implications of the radiocarbon calibration. Geochim Cosmochim Acta 6: 2025–2038
Barnard WR, Andreae MO, Iverson RI (1984) Dimethylsulfide and *Phaeocystis pouchetii* in the southeastern Bering Sea. Cont Shelf Res 3:103–113
Barnes PW, Asbury JL, Rearic DM, Ross CR (1987) Ice erosion of a sea-floor knickpoint at the inner edge of the stamukhi zone, Beaufort Sea, Alaska. Mar Geol 76:207–222
Barnes PW, Graves S, Reimnitz E (1987) Beaufort Sea coastal currents: a divergence near Barter Island, Alaska? In: Hamilton TD, Galloway JP (eds). Geologic Studies in Alaska by the U.S. Geological Survey During 1986: USGS Circular 998, pp 139–142
Barnes PW, Ravlinson SE, Reimnitz E (1988) Coastal geomorphology of Arctic Alaska, Arctic coastal processes and slope protection. ASCE
Barnes PW, Schell DM, Reimnitz E (1984) The Alaskan Beaufort Sea. Ecosystems and environments. Academic Press, New York, 466 pp
Barrie LA (1996) Occurrence and trends of pollution in the Arctic troposphere. In Wolff EW, Bales RC (eds) Chemical Exchange between the Atmosphere and Polar Snow. Springer, Berlin 93–129
Baskaran M, Naidu AS (1995) ^{210}Pb-derived chronology and the fluxes of ^{210}Pb and ^{137}CS isotopes into continental shelf sediments, East Chukchi Sea, Alaskan Arctic. Geochim Cosmochim Acta 59:4435–4448
Bauch D, Carstens J, Wefer G, Thiede J (2000) The imprint of anthropogenic CO_2 in the Arctic Ocean: evidence from planktic $\delta^{13}C$ data from water column and sediment surfaces. Deep Sea Res 47:1791–1808
Bauch HA, Cremer H, Kunz-Pirrung M (2000) Siberian shelf sediments contain clues to paleoclimatic forcing. EOS Transactions 81:233 and 238
Bauch HA, Kassens H, Erlenkeuser H, Grootes PM, Thiede J (1999) Depositional environment of the Laptev Sea (Arctic Siberia) during the Holocene. Boreas 28:194–204
Bauch HA, Kassens H, Naidina OD, Kunz-Pirring M, Thiede, J (2001a) Composition and Flux of Holocene Sediments on the Eastern Laptev Sea Shelf, Arctic Siberia. Quart Res 55:344–351
Bauch HA, Mueller-Lupp T, Taldenkova E, Spielhagen RF, Kassens H, Grootes PM, Thiede J, Heinemeier J, Petryashov VV (2001b) Chronology of the Holocene transgression at the North Siberian margin. Glob Planet Change 31: 125–141
Bauerfeind E, Bodungen Bv, Arndt K, Koeve W (1994) Particle flux, and composition of sedimenting matter, in the Greenland Sea. J Mar Syst 5:411– 423
Bauerfeind E, Garrity C, Krumbholz M, Ramseier RO, Voß M (1997) Seasonal variability of sediment trap collections in the Northeast Water Polynya. Part 2. Biochemical and microscopic composition of sedimenting matter. J Mar Syst 10:371–389
Baumann M (1990) Coccoliths in sediments of the eastern Arctic Basin. In: Bleil U, Thiede J (eds), Geological History of the Polar Oceans: Arctic versus Antarctic. NATO ASI Series C, 308, Kluwer Academic Publishers, Dordrecht, pp 437–445
Baumann MEM, Lancelot C, Brandini FP, Sakshaug E, John DM (1994) The taxonomic identity of the cosmopolitan prymnesiophyte *Phaeocystis*: a morphological and ecophysiological approach. J Mar Syst 5:5–22
Behrends M, Hoops E, Peregovich B (1999) Distribution patterns of heavy minerals in Siberian rivers, the Laptev Sea and the eastern Arctic Ocean: an approach to identify sources, transport and pathways of terrigenous matter. In: Kassens H et al. (eds), Land-Ocean Systems in the Siberian Arctic: Dynamics and History, Springer-Verlag, Berlin, 265–286
Behrenfeld MJ, Falkowski PG (1997) Photosynthetic rates derived from satellite-based chlorophyll concentration. Limnol Oceanogr 42:1–20

Bélanger S, Larouche O, Klein B, Roy S, Mei Z-P, Tremblay J-E, Vidussi F, Leblanc B (2002) Phytoplankton dynamics in the North Water polynya during 1998–2000 summers using SeaWiFS imagery. J Geophys Res, in press

Belicka LL, Macdonald RW, Harvey HR (2002) Sources and transport of organic carbon to shelf, slope, and basin surface sediments of the Arctic Ocean. Deep-Sea Research Part I 49:1463–1483

Belov NA (1976) The estimate of river drainage influence upon the formation of bottom sediments in the Arctic seas. Proceedings of AANII 323:178–187

Bender M, Jahnke R, Weiss R, Martin W, Heggie DT, Orchardo J, Sowers T (1989) Organic carbon oxidation and benthic nitrogen and silica dynamics in San Clemente Basin, a continental borderland site. Geochim Cosmochim Acta 53:685–697

Bender M, Sowers T, Labeyrie L (1994) The Dole effect and its variations during the last 130,000 years as measured in the VOSTOK ice core. Global Biogeochemical Cycles 8:363–376

Benner R (2002a) Chemical composition and reactivity. In: Hansell D, Carlson C (eds) Biogeochemistry of marine dissolved organic matter. Academic Press, New York, pp 59–90

Benner R (2002b) Molecular indicators for the bioavailability of dissolved organic matter. In: Findlay SEG, Sinsabaugh (eds) Aquatic ecosystems: Interactivity of Dissolved Organic Matter. Academic Press, pp 121–137

Benner R, Benitez-Nelson B, Kaiser K, Amon RMW (in prep.) Export of young terrigenous dissolved organic matter from rivers to the Arctic Ocean. Nature, subm

Berelson WM (2001) The flux of particulate carbon into the ocean interior: a comparison between four US JGOFS regional studies. Oceanography 14:59–67

Berge G (1962) Discolouration of the sea due to *Coccolithus huxleyi* "bloom". Sarsia 6:27–40

Berger WH, Fischer K, Lai C, Wu G (1988) Ocean carbon flux: Global maps of primary productivity and export production. In: Aregian CR (ed): Biogeochemical Cycling and Fluxes between the Deep Euphotic Zone and Other Oceanic Realms. National Undersea Research Program Research Report 88-1, NOAA, Washington DC, pp 131–176

Berger WH, Smetacek V and Wefer G (1989) Productivity of the Ocean: Past and Present. Life Sciences Research Report 44, Wiley & Sons, New York, 471 pp

Berner H, Wefer G (1990) Physiographic and biological factors controlling surface sediment distribution in the Fram Strait. In: Bleil U, Thiede J (eds). Geological History of the Polar oceans: Arctic versus Antarctic, Kluwer, Dordrecht, pp 317–335

Berner RA (1980) Early diagenesis: a theoretical approach. Princeton Univ. Press, Princeton, NJ, pp 1–241

Berner RA (1982) Burial of organic carbon and pyrite in the modern ocean: its geochemical and environmental significance. Am J Sci 282:451–475

Berner RA (1989) Biogeochemical cycles of carbon and sulfur and their effect on atmospheric oxygen over Phanerozoic time. Palaeogeogr Palaeoclim Palaeoecol 73: 97–122

Betzer PR, Showers WJ, Laws EA, Winn CD, Ditullo GR, Kroopnick PM (1984) Primary productivity and particle fluxes on a transect of the equator at 153°W in the Pacific Ocean. Deep-Sea Res 31:1–11

Bhat SG, Krishnaswamy S, Rama Lal D, Moore WS (1969) $^{234}Th/^{238}U$ ratios in the ocean. Earth Planet Sci Lett 5: 483–491

Bignami F, Hopkins TS (1993) The water mass characteristics of the Northeast Water Polynya: Polar Sea data 1992–1993, J Mar Syst 10:139–156

Billett DSM, Lampitt RS, Rice AL, Mantoura RFC (1983) Seasonal sedimentation of phytoplankton to the deep-sea benthos. Nature 302:520–522

Bilyard GR, Carey AG Jr (1979) Distribution of western Beauford Sea polychaetous annelids. Mar Bio 54:329–339

Bird RE (1984) A simple, solar spectral model for direct-normal and diffuse horizontal irradiance. Solar Energy 32:461–471

Birgel D (1999) Paläoumweltbedingungen und organischer Kohlenstoffeintrag im Spätquartär: Ergebnisse von organisch-geochemischen und sedimentologischen Untersucheungen an Sedimenten vom Yermak-Plateau und der nördlichen Fram-Strasse. Unpublished Diploma Thesis. Eberhard-Karls-Universität Tübingen/Alfred Wegener Institute for Polar and Marine Research Bremerhaven, pp 1–69

Birkenes I, Braarud T (1952) Phytoplankton in the Oslo fjord during a "Coccolithus huxleyi" summer. Avh Norske Vidensk Akad Oslo 1952(2):1–23

Biscaye PE, Anderson RF (1994) Fluxes of particulate matter on the slope of the southern Middle Atlantic Bight: SEEP-II. Deep-Sea Res II 41:459–509

Biscaye PE, Flagg, CN Falkowski, PG (1994) The shelf edge exchange processes experiment, SEEP-II: An introduction to hypotheses, results and conclusions. Deep-Sea Res II 41:231–252

Bischof JF, Darby DA (1997) Mid- to Late Pleistocene Ice Drift in the Western Arctic Ocean: Evidence for a Different Circulation in the Past. Science 277:74–84

Bischof K, Hanelt D, Tüg H, Karsten U, Brouwer PEM and Wiencke C (1998) Acclimation of brown algal photosynthesis to ultraviolet radiation in Arctic coastal waters. Pol Biol 20:388–395

Bjelvin TA, Holm JP (1993) GEOGruppen as, Tromsø, Norway, Report No. 93024.02

Björk G, Gustafsson BG, Stigebrandt A (2001) Upper layer circulation of the Nordic seas as inferred from the spatial distribution of heat and freshwater content and potential energy. Pol Res 20:161–168

Björk G, Söderkvist S, Winsor P, Nikolopoulos A, Steele M (2002) Partial return of the cold halocline layer to the Amundsen Basin of the Arctic Ocean: Implications for sea ice mass balance. Geophys Res Lett, 29:10.1029/2001GL014157, 8-1–8-4

Bjørlykke K, Bue B, Elverhøi A (1978) Quaternary sediments in the north-western part of the Barents Sea and their relation to the underlying Mesozoic bedrock. Sedimentology 25:227–246

Bjornsson H, Mysak LA, Brown RD (1995) On the interannual variability of precipitation and runoff in the Mackenzie drainage basin. Climate Dynamics 12:67–76

Blackburn TH (1987) Microbial food webs in sediments. In: Sleigh MA (ed) Microbes in The Sea. Chapter 2. Ellis Horwood/John Wiley New York, NY, pp 39–58

Blake JA, Diaz RJ (1994) Input, accumulation and cycling of materials on the continental slope off Cape Hatteras: An introduction. DEEP-SEA Res II 41:707–710

Blasco SM, Fortin G, Hill PR, O'Connor MJ, Brigham-Grette J (1990) The late Neogene and Quaternary stratigraphy of the Canadian Beaufort continental shelf. In: Grantz A, Johnson L, Sweeney JF (eds). The Geology of North America Volume L: The Arctic Ocean Region. Geol Soc Amer, pp 491–502

Blasco SM, Shearer JM, Myers R (1998) Seabed scouring by sea-ice: Scouring process and impact rates: Canadian Beaufort Shelf. In: 13th International Symposium on Okhotsk Sea & Sea Ice. Proceedings of ice scour and arctic marine pipelines workshop: Mombetsu, Hokkaido, Japan; 1998. pp 53–58

Blindheim J, Ådlandsvik B (1995) Episodic formation of intermediate water along the Greenland Sea Arctic Front. ICES CM 1995, Mini 6:1–11

Blinova EI (1969) Distribution, reserves and annual production dominating species of algae on littoral of Murman coast of the Barents Sea. Vegetal resources 5(2):163–173 (in Russian)

Bluhm BA, Piepenburg D, Juterzenka K von (1998) Distribution, standing stock, growth, mortality and production of *Strongylocentrotus pallidus* (Echinodermata Echinoidea) in the northern Barents Sea. Polar Biol 20:325–334

Blumer M Guillard RRL, Chase T (1971) Hydrocarbons of marine phytoplankton. Mar Biol 8:183–189

Bobrov YuA (1985) Production cycle of phytoplankton in the coastal zone of the Barents Sea. In: Life and conditions of its existence in the pelagic zone of the Barents Sea. Apatity, pp 166–182 (in Russian)

Bockstoce JR (1986) Whales, Ice and Men: The History of Whaling in the Western Arctic. University of Washington Press, Seattle and London, pp 1–400

Bodungen Bv, Wunsch M, Füderer H (1991) Sampling and analysis of suspended particles in the North Atlantic. In: Hurt DC, Spencer DW (eds). Marine Particles: Analysis and Characterization. Geophys Monogr Ser Am Geophys Union 63:47–53

Bodungen, Bv, Antia A, Bauerfeind E, Haupt O, Koeve W, Machado E, Peeken I, Peinert R, Reitmeier S, Thomsen C, Voß M, Wunsch M, Zeller U, Zeitzschel B (1995) Pelagic processes and vertical flux of particles: an overview of a long-term comparative study in the Norwegian Sea and Greenland Sea. Geol Rdsch 84:11–27

Boetius A, Damm E (1998) Benthic oxygen uptake, hydrolytical potentials and microbial biomass at the Arctic continental slope. Deep-Sea Res I 45:239–275

Boetius A, Nöthig E-M, Liebezeit G, Kröncke I (1996) Distribution of chlorophyll pigments as indicator for marine organic matter input in the Eurasian shelf seas and the central Arctic Ocean. Reps Pol Res 212:212–218

Böggild OB (1906) On the bottom deposits of the north polar seas. Scientific results of the Norwegian North Polar Expedition 1893–1896, 5:1–62

Böhm E, Hopkins TS, Minnet PJ (1997) Passive microwave observations of the Northeast Water Polynya interannual variability: 1978–1994, J Mar Syst 10:87–98

Bönisch G, Schlosser P (1995) Deep water formation and exchange rates in the Greenland/Norwegian Seas and the Eurasian Basin of the Arctic Ocean derived from tracer balances. Progress in Oceanography 35:29–52

Booth BC, Horner RA (1997) Microalgae on the Arctic Ocean Section, 1994: species abundance and biomass. Deep-Sea Res II 44:1607–1622

Booth BC, Smith WO jr (1997) Autotrophic flagellates and diatoms in the Northeast Water Polynya, Greenland: summer 1993. J Mar Syst 10:241–261

Booth CL, Lewin J, Postel JR (1993) Temporal variation in the structure of autotrophic and heterotrophic communities in the subarctic Pacific. Prog Oceanogr 32:57–99

Bordowskiy OK (1965) Sources of organic matter in marine basins. Mar. Geol 3:5–31

Børsheim KY (2000) Bacterial production rates and concentrations of organic carbon at the end of the growing season in the Greenland Sea. Aq Microb Ecol 21:115–123

Børsheim KY, Myklestad SM (1997) Dynamics of DOC in the Norwegian Sea inferred from monthly profiles collected during 3 years at 66 °N, 2 °E. Deep-Sea Res I 44:593–601

Børsheim KY, Myklestad SM, Sneli J-A (1999) Monthly profiles of DOC, mono- and polysaccharides in the Trondheimsfjord (Norway) during two years. Mar Chem 63: 255–272

Boucsein B and Stein R (2000) Particulate organic matter in surface sediments of the Laptev Sea (Arctic Ocean): Application of maceral analysis as organic-carbon-source indicator. Mar Geol 162:573–586

Boucsein B, Fahl K, Siebold M, Stein R (1999) Quantity and quality of organic carbon in suspended matter and surface sediments of the Ob and Yenisei estuaries and adjacent coastal areas. Reps Pol Res 300:116–126

Boucsein B, Fahl K, Stein R (2000) Late Quaternary organic matter records from the Laptev Sea continental margin: Evidence for the variability of river discharge during the last 15000 years BP. Int J Earth Sci 89:578–591

Boucsein B, Knies J, Stein R (2002) Organic matter deposition along the Kara and Laptev Sea continental margin (eastern Arctic Ocean) during last deglaciation and Holocene: Evidence from organic-geochemical and petrographical data. Mar Geol 183:67–88

Bourke RH, Garrett RP (1987) Sea ice thickness distribution in the Arctic Ocean. Cold Regions Science and Technology 13:259–280

Boyd TJ, Steele M, Muench RD, Gunn JT (2002) Partial recovery of the Arctic Ocean halocline. Geophys Res Lett 29:10.1029/2001GL014047, 2-1 – 2-4

Braarud T (1935) The Øst Expedition to the Denmark Strait in 1929. II. The phytoplankton and its conditions of growth. Hvalråd Skr 10:1–173

Braarud T, Gaarder KR, Grøntved J (1953) The phytoplankton of the North Sea and adjacent waters in May 1948. Rapp P-v Réun Cons Int Explor Mer 133:1–189

Bradley RS, Jones PD (1993) 'Little Ice Age' summer temperature variations: their nature and relevance to recent global warming trends. The Holocene 3:367–376

Brand LE (1991) Minimum iron requirements of marine phytoplankton and the implication for the biogeochemical control of new production. Limnol Oceanogr 36:1756–1771

Brandt A, Schnack K (1999) Macrofaunal abundance at 79°N off East Greenland: opposing data from epibenthic-sledge and box-corer samples. Polar Biol 22:75–81

Brassell SC, Eglinton G, Marlowe IT, Pflaumann U, Sarntheim M (1986) Molecular stratigraphy: a new tool for climate assessment. Nature 320:129–133

Brassell SC, Eglinton G, Maxwell JR, Philip RP (1978) Natural background of alkanes in the aquatic environment. – In: Aquatic Pollutants: Transformation and Biological Effects (Edited by Hutzinger O, Lelyveld IH and Zoetman BCJ). Pergamon Press, Oxford, pp 69–86

Bratbak G, Levasseur M, Michaud S, Cantin G, Fernández E, Heimdal BR, Heldal M (1995) Viral activity in relation to *Emiliania huxleyi* blooms: a mechanism of DMSP release? Mar Ecol Prog Ser 128:133–142

Brey T (1995) Empirische Untersuchungen zur Populationsdynamik makrobenthischer Evertebraten. Habilitationsschrift, Universität Bremen, 169 pp

Brey T, Klages M, Dahm C, Gorny M, Gutt J, Hain S, Stiller M, Arntz WE, Wägele JW, Zimmermann A (1994) Antarctic benthic diversity. Nature 368:297

Brey T, Rumohr H, Ankar S (1988) Energy content of macrobenthic invertebrates: general conversion factor from weight to energy. J Exp Mar Biol Ecol 117:271–278

Brinkhurst RO (1991) Benthic biology of the western Canadian continental shelf. Cont Shelf Res 8–10:737–754

Brodeur RD, Mills CE, Overland JE, Walters GE, Schumacher JD (1999) Evidence for a substantial increase in gelatinous zooplankton in the Bering Sea, with possible links to climate change. Fisheries Oceanography 8:296– 306

Brodeur RD, Ware MD (1992) Interannual and interdecadal changes in zooplankton biomass in the subarctic Pacific Ocean. Fish Oceanogr 1:2–38

Bruland KW, Bienfang PK, Bishop JKB, Eglinton G, Ittekkot VAW, Lampitt R, Sarnthein M, Thiede J, Walsh JJ, Wefer G (1989) Flux to the Seafloor. In: Berger WH et al. (eds), Productivity of the Ocean: Past and Present, Life Sci Res Rep 44, Wiley & Sons, pp 193–216

Buck KR, Nielsen TG, Hansen BW, Gastrup-Hansen D, Thomsen HA (1998) Infiltration phyto- and protozooplankton assemblages in the annual sea ice of Disko Island, West Greenland, Spring 1996. Pol Biol 20:377–381

Budéus G, Schneider W (1995) On the hydrography of the Northeast Water Polynya. J Geophys Res 100:4287–4299

Budeus G, Schneider W, Kattner G (1997) Distribution and exchange of water masses in the Northeast Water Polynya (Greenland Sea). J Mar Syst 10:123–137

Budéus G, Schneider W, Kattner G (1997) Distribution and exchange of water masses in the Northeast Water Polynya (Greenland Sea). J Mar Sys 10:123–138

Budeus G, Schneider W, Krause G (1998) Winter convective events and bottom water warming in the Greenland Sea. J Geophys Res 103:18515–18527

Budyko MI, Izrael YuA (eds) (1987) Anthropogenic changes of climate. Gidrometizdat Press, Leningrad, 406 pp (in Russian)

Buesseler KO, Bacon MP, Cochran JK, Livingston HD (1992) Carbon and nitrogen export during the JGOFS North Atlantic Bloom Experiment estimated from $^{234}Th:^{238}U$ disequilibria. Deep-Sea Res II 39:1115–1137

Bukry D (1984) Paleogene paleoceanography of the Arctic Ocean is constrained by the middle or late Eocene age of U.S.G.S. Core FL-422, Evidence from silicoflagellates. Geology 12:199–201

Bunch JN, Harland RC (1990) Bacterial production in the bottom surface of sea ice in the Canadian Subarctic. Can J Fish Aq Sci 47:1986–1995

Burd AB, Moran SB, Jackson GA (2000) A coupled adsorption-aggregation model of the POC/^{234}Th ratio of marine particles. Deep-Sea Res I 47:103–120

Burenkov VI, Gol'din YA, Gureev BA, Sud'bin AI (1995) The basic notions of distribution of optical water properties in the Kara Sea. Oceanology 35:346–357

Burkhanov V (1956) New Soviet Discoveries in the Arctic. Moscow, Foreign Languages Publishing House, pp 1–60

Buscail R, Pocklington R, Daumas R, Guidi L (1990) Fluxes and budget of organic matter in the benthic boundary layer over the northwestern Mediterranean margin. Cont Shelf Res 10:1089–1122

Bussmann I (1999) Bacterial utilisation of humic substances from the Arctic Ocean. Aquatic Microb Ecol 19:37–45

Bussmann I, Kattner G (2000) Distribution of dissolved organic carbon in the central Arctic Ocean: the influence of physical and biological properties. J Mar Syst 27:209–219

Campbell RG, Runge JA, Durbin EG (2001) Evidence of food limitation of *Calanus finmarchicus* production rates on the southern flank of Georges Bank during April 1997. Deep-Sea Res II 48:531–549

Campeau S, Héquette A, Pienitz R (2000) Late Holocene diatom biostratigraphy and sea-level changes in the southeastern Beaufort Sea. Can J Earth Sci 37:63–80

Canadian Hydrographic Service (1979) General Bathymetric Chart of the Oceans (GEBCO) Sheet 5.17. Canadian Hydrographic Service, Ottawa, scale 1:6,000,000. 1 sheet

Canfield DE (1994) Factors influencing organic matter preservation in marine sediments. Chem Geol 114:315–329

Carey AG (1991) Ecology of the North American Arctic continental shelf benthos: a review. Cont Shelf Res 11: 865–883

Carlotti F, Hirche HJ (1997) Growth and egg production of female *Calanus finmarchicus*: an individual-based physiological model and experimental validation. Mar Ecol Prog Ser 149:91–104

Carmack E (1990) Large-Scale Physical Oceanography of Polar Oceans. In: Smith WO (ed), Polar Oceanography, Part A, Physical Science, Academic Press Inc, San Diego, pp 171–222

Carmack EC Aagaard K, Swift JH, Macdonald RW, McLaughlin FA, Jones EP, Perkin RG, Smith JN, Ellis KM, Killius LR (1997) Changes in temperature and tracer distributions within the Arctic Ocean: results from the 1994 Arctic Ocean section. Deep-Sea Res II 44:1487–1502

Carmack EC, Macdonald RW (2002) Oceanography of the Canadian shelf of the Beaufort Sea: A setting for marine life. Arctic 55:29–45

Carson MA, Jasper JN, Conly FM (1998) Magnitude and sources of sediment input to the Mackenzie delta, Northwest Territories, 1974–94. Arctic 51:116–124

Carstens J, Wefer G (1992) Recent distribution od planktonic foraminifera in the Nansen Basin, Arctic ocean. Deep-Sea Res 39 (Suppl 2):507–524

Cauwet G, Sidorov IS (1996) The biogeochemistry of Lena River: organic carbon and nutrients distribution. Mar Chem 53:211–227

Cavalieri DJ, Gloersen P, Parkinson CL, Comiso JC, Zwally HJ (1997) Observed hemispheric asymmetry in global sea ice changes. Science 278:1104–1106

Charlson J, Lovelock JE, Andreae MO, Warren SG (1987) Oceanic phytoplankton, atmospheric sulphur, cloud albedo and climate. Nature 326:655–661

Chautón M (1997) Winter stocks of phytoplankton in the Trondheimsfjord and the subsequent spring bloom. Thesis, Norw Univ Sci Technol Trondheim, 78 p (in Norwegian)

Chen JH, Edwards RL, Wasserburg GJ (1986) ^{238}U, ^{234}U, ^{232}Th in seawater. Earth Planet Sci Lett 80:241–251

Christensen JP (1989) Sulfate reduction and carbon oxidation rates in continental shelf sediments, an examination of off-shelf carbon transport. Cont Shelf Res 9: 223– 246

Christensen JP (2000) A Relationship between deep-sea benthic oxygen demand and oceanic primary productivity. Oceanologica Acta 23:65–82

Christensen JP (2003) Oxygen and nutrients in deep-sea sediment pore waters of the Arctic Ocean. Deep-Sea Res, in press

Christensen JP, Devol AH, Smethie WM Jr (1984) Biological enhancement of solute exchange between sediments and bottom water on the Washington continental shelf. Cont Shelf Res 3:9–23

Churun V, Ivanov B (1998) Investigations of the hydrophysical structure in the mixing zone between fresh and saline waters in the Ob and Yenisei estuaries. Reps Pol Res 266: 11–18

Clark DL (1969) Paleoecology and sedimentation in part of the Arctic Basin. Arctic 22:233–245
Clark DL (1977) Paleontologic response to post-Jurassic crustal plate movements in the Arctic Ocean. Biol Geol 2:55–76
Clark DL (1988) Early history of the Arctic Ocean. Paleoceanography 3:539–550
Clark DL, Byers CW, Pratt LM (1986) Cretaceous black mud from the Central Arctic Ocean. Paleoceanography 1(3): 265–271
Clark DL, Whitman RR, Morgan KA, Mackay SD (1980) Stratigraphy and glacio-marine sediments of the Amerasian Basin, central Arctic Ocean. Geol Soc Amer Spec Pap 181, 57 pp
Clough LM, Ambrose WG, Cochran JK, Barnes C, Renaud PE, Aller RC (1997) Infaunal density, biomass, and bioturbation in the sediments of the Arctic Ocean. Deep-Sea Res II 44:1683–1704
Coachman LK, Aagaard K (1974) Physical oceanography of Arctic and subarctic seas. In: Herman Y (ed) Marine geology and oceanography of the Arctic Seas. Springer, New York, pp 1–72
Coachman LK, Aagaard K, Tripp RB (1975) Bering Straight: the regional physical oceanography. University of Washington Press, Seattle, pp 1–172
Coachman LK, Hansell DA (1993) ISHTAR: Inner shelf transfer and recycling in the Bering and Chukchi Seas. Cont Shelf Res 13:473–704
Coachman LK, Walsh JJ (1981) A diffusion model of cross-shelf exchange of nutrients in the Bering Sea. Deep-Sea Res 28A:819–837
Coachman LK, Whitledge TE, Goering JJ (1999) Silica in Bering Sea deep and bottom water. In: Loughlin TR, Ohtani K (eds) The physical oceanography of the Bering Sea. Univ Alaska Seagrant AK-SG-03, Univ Alaska Fairbanks, pp 285–310
Coale KH, Bruland KW (1985) ^{234}Th:^{238}U disequilibria within the California Current. Limnol Oceanogr 30:22–33
Coble P (1996) Characterization of marine and terrestrial DOM in seawater using excitation-emission matrix spectroscopy. Mar Chem 51:325–346
Cochran JK, Barnes C, Achman D, Hirschberg, DJ (1995) Thorium-234/Uranium-238 disequilibrium as an indicator of scavenging rates and particulate organic carbon fluxes in the Northeast Water Polynya, Greenland. J Geophys Res (C) 100:4399–4410
Codispoti LA (1979) Arctic Ocean processes in relation to the dissolved silicon content of the Atlantic. Mar Sci Comm 5:361–381
Coles RL, Taylor PT (1990) Magnetic anomalies. In: Grantz A, Johnson L, Sweeney JF (eds) The Arctic Ocean region (L), Boulder, Colorado, Geol Soc Amer, pp 119–132
Conover RJ, Gustavson KR (1999) Sources of urea in arcticseas: zooplankton metabolism. Mar Ecol Prog Ser 179: 41–54
Conover RJ, Herman AW, Prinsenberg SJ, Harris LR (1986) Distribution of and feeding by the copepod *Pseudocalanus* under fast ice during the Arctic spring. Science 232:1245–1247
Cooper LW, Grebmeier JM, Larsen IL, Egorov VG, Theodorakis C, Kelly HP, Lovvorn JR (2002) Seasonal variation in sedimentation of organic materials in the St. Lawrence Island polynya region, Bering Sea. Mar Ecol Prog Ser 226:13–26
Cooper LW, Larsen IL, Beasley TM, Dolvin SS, Grebmeier JM, Kelley JM, et al (1998) The distribution of radiocesium and plutonium in sea ice-entrained arctic sediments in relation to potential sources and sinks. J Envir Radioactivity 39:279–303
Cooper LW, Whitledge TE, Grebmeier JM, Weingartner T (1997) The nutrient, salinity, and stable oxygen isotope composition of Bering and Chukchi Sea waters in and near the Bering Strait. J Geophys Res (C) 102:12563–12573
Coppola L, Roy-Barman M, Wassmann P, Mulsow S, Jeandel C (2002) Calibration of sediment traps and particulate organic carbon export using ^{234}Th in the Barents Sea. Mar Chem 80:11–26
Cota GF, Legendre L, Gosselin M, Ingram RG (1991) Ecology of bottom ice algae: I. Environmental controls and variability. J Mar Syst 2: 257–277
Cota GF, Pomeroy LR, Harrison WG, Jones EP, Peters F, Sheldon WM Jr, Weingartner TR (1996) Nutrients, primary production and microbial heterotrophy in the southeastern Chukchi Sea: Arctic summer nutrient depletion and heterotrophy. Mar Ecol Prog Ser 135:247–258
Cowie GL, Hedges JI (1984) Carbohydrate sources in a coastal marine environment. Geochim Cosmochim Acta 48: 2075–2087
Cranston RE (1997) Organic carbon burial rates across the Arctic Ocean from the 1994 Arctic Ocean Section expedition. Deep-Sea Res II 44:1705–1724
Cronin TM, Holtz TR, Stein R, Spielhagen R, Fütterer DK, Wollenburg J (1995) Late Quaternary Deep Water and Mid-Depth History of the Eurasian Basin, Arctic Ocean. Paleoceanography 10:259–281
Crowley TJ, Lowery TS (2000) How warm was the Medieval Warm Period? Ambio 29:51–54
Curtis MA (1975) The marine benthos of Arctic and sub-Arctic continental shelves. A review of regional studies and their general results. Polar Res 17:595–626
Dahl E, Laubier L, Sibuet M, Stromberg JO (1976) Some quantitative results on benthic communities of the deep Norwegian Sea. Astarte 9:61–79
Dahle S, Denisenko SG, Denisenko NV, Cochrane S (1998) Benthic fauna in the Pechora Sea. Sarsia 83:183–210
Dallimore SR, Wolfe SA, Solomon SM (1996) Influence of ground ice and permafrost on coastal evolution, Richards Island, Beaufort Sea coast, N.W.T. Can J Earth Sci 33: 664–675
Dalløkken R, Sandvik R, Sakshaug E (1994) Variations in bio-optical properties in the Greenland/Iceland/Norwegian Seas. SPIE 2258:266–276
Dalpadado P, Skjoldal HR (1991) Distribution and life history of krill from the Barents Sea. Pol Res 10:443–460
Dalpadado P, Skjoldal HR (1996) Abundance, maturity and growth of the krill species *Thysanoessa inermis* and *T. longicaudata* in the Barents Sea. Mar Ecol Prog Ser 144:175–183
Daly KL (1997) Flux of particulate matter through copepods in the Northeast Water Polynya. J Mar Syst 10: 319–342
Danilov AI, Ivanov VV, Pryamikov SM, Timokhov LA (1994) The environment of the Laptev and East-Siberian seas. In: Scientific results of LAPEKS-93 expedition, Gidrometeoizdat, St. Petersburg, pp 32–45
Dansgaard W, Johnsen SJ, Clausen HB, Dahl-Jensen D, Gundesturp NS, Hammer CU, Hvidberg CS, Steffensen JP, Sveinibjournsdottir AE, Jouzel J, Bond G (1993) Evidence for general instability of past climate from a 250-kyr ice-core record. Nature 364:218–220
Danyushevskaya AI, Dibner VD, Shkatov EP et al. (1978) Complex geological-geochemical researches at the arctic shelf. Oceanology 18(1):70–78 (in Russian)

Danyushevskaya AI, Petrova VI, Yashin DS, Batova GI, Artemyev VE (1990) Organic matter in the bottom sediments of polar zones of the World Ocean. Nedra, Leningrad, pp 1–280 (in Russian)

Danyushevskaya AI, Yashin DS, Kirillov OV (1980) Geochemical peculiarity in distribution of organic carbon in bottom sediments of the Arctic Seas. Oceanology, 21(2): 281–289

Danyushevskaya AI, Yashin DS, Kirillov OV (1980) Geochemical regularities of the organic carbon distribution in bottom sediments from the Arctic seas. Oceanology 2:281–289

Danyushevskaya AI, Yashin DS, Lapina NN (1985) Distribution and diagenetic evolution of organic matter in the bottom deposits of the polar sedimentogenesis zone. In: Organic matter of the recent and old sediments. Nauka Press, Moscow, pp 54–62 (in Russian)

Darby D, Bischof J, Cutter G, de Vernal A, Hillaire-Marcel C, Dwyer G et al. (2001) New record shows pronounced changes in Arctic Ocean circulation and climate. EOS Transactions 82:601, 607

Darby DA, Bischof JF, Jones GA (1997) Radiocarbon chronology of depositional regimes in the western Arctic Ocean. Deep-Sea Res II 44:1745–1757

Darby DA, Burckle LH, Clark DL (1974) Airborne dust on the Arctic pack ice: its composition and fallout rate. Earth Planet Sci Lett 24:166–172

Darby DA, Naidu AS, Mowatt TC, Jones G (1989) Sediment composition and sedimentary processes in the Arctic Ocean. In: Herman Y (ed), The Arctic Seas – Climatology, Oceanography, Geology, and Biology, New York, pp 657–720

Dayton PK (1990) Polar benthos In: Smith WO Jr (ed) Polar Oceanography, Part B: Chemistry, Biology and Geology, pp 631–685

Dayton, PK, Mordida BJ, Bacon (1994) Polar marine community. Amer Zool 34:90–99

de Hass H, van Weering TCE, de Stigter H (2002) Organic carbon in shelf seas: sinks or sources, processes and products. Cont Shelf Res 22:691–717

de Leeuw JW, Rijpstra WIC, Schenck PA, Volkman JK (1983 Free, esterified, and residual bound sterols in Black Sea Unit I sediments. Geochim Cosmochim Acta 47:455–465

de Vernal A, Rochon A, Hillaire-Marcel C, Turon J-L, Guiot J (1993) Quantitative reconstruction of sea-surface conditions, seasonal extent of sea-ice cover and meltwater discharges in high latitude marine environments from dinoflagellate cycs assemblages. In: Peltier WR (ed). Ice in the Climate System. Springer-Verlag, Heidelberg, pp 611–621

Dean KG, McRoy CP, Ahlnas K, Springer A (1989) The plume of the Yukon river in relation to the oceanography of the Bering Sea. Remote Sensing of Environment 28:75–84

Dean KG, Stringer WJ, Ahlnäs K, Searcy C, Weingartner T (1994) The influence of river discharge on the thawing of sea ice, Mackenzie River Delta: albedo and temperature analyses. Pol Res 13:83–94

Degens ET, Kempe S, Richey JE (1991) Summary: Biogeochemistry of major world rivers. In: Degens ET, Kempe S, Richey JE (eds) Biogeochemistry of Major World Rivers. SCOPE 42, Wiley, Chichester, pp 323–347

Dell'Agnese DJ, Clark DL (1994) Siliceous microfossils from the warm Late Cretaceous and Early Cenozoic Arctic Ocean. J Paleont 68:31–47

Demaison GJ, Moore GT (1980) Anoxic environments and oil source bed genesis. Org Geochem. 2:9–31

Deming J, Rowe G, Cochran K, Walsh I, Bauerfeind E, Smith W, Daly K, Ashjian C, Yager P, Wallace D, Ritzrau W, Graf G (1995) Carbonflow in the NEW polynya: An interdisciplinary view. North East Water Polynya Symposium, Helsingør, Denmark 1–5 May 1995 [Abstracts]: 28

Deming JW, Baross JA (1993) Deep-sea smokers: Windows to a subsurface biosphere? Geochim Cosmochim Acta 57: 3219–3230

Denisenko NV, Rachor E, Denisenko SG (2003) Benthic Fauna in the southern part of the Kara Sea. In: Stein R, Fahl K, Fütterer DK, Galimov EM, Stepanets OV (eds), Siberian River Run-off in the Kara Sea: Characterisation, Quantification, Variability, and Environmental Significance, Proceedings in Marine Sciences Vol 6, Elsevier, Amsterdam, pp 213–236

Deser C (2000) On the teleconnectivity of the "Arctic Oscillation". Geophys Res Lett 27:779–782

Dethleff D, Loewe P, Weiel D, Nies H, Kuhlmann G, Bahe C, Tarasov G (1998) Winter expedition to the Southwestern Kara Sea – investigations on formation and transport of turbid sea-ice. Reps Pol Res 271, 40 pp

Dethleff D, Nürnberg D, Reimnitz E, Saarso M, Savchenko YP (1993) East Siberian Arctic Region Expedition '92: The Laptev Sea – its significance for Arctic sea ice formation and Transpolar sediment flux. Reps Pol Res 120:3–37

Dethleff D, Rachold V, Tintelnot T, Antonow M (2000) Sea-ice transport of riverine particles from the Laptev Sea to Fram Strait based on clay mineral studies. Int J Earth Sci 89:496–502

Deubel H (2000) Struktureigenschaften und Nahrungsbdarf der Zoobenthos-gemeinschaften im Bereich des Lomonossowrückens im Arktischen Ozean. Reps Pol Res 370, 147 pp

Deubel H, Engel M, Fetzer I, Gagaev S, Hirche HJ, Klages M, Larionov V, Lubin P, Lubina O, Nöthig EM, Okolodkov Y, Rachor E (2003) The southern Kara Sea ecosystem: phytoplankton, zooplankton and benthos communities influenced by river run-off. In: Stein R, Fahl K, Fütterer DK, Galimov EM, Stepanets OV (eds), Siberian River Run-off in the Kara Sea: Characterisation, Quantification, Variability, and Environmental Significance, Proceedings in Marine Sciences Vol 6, Elsevier, Amsterdam, pp 237–266

Deuser WG (1986) Seasonal and interannual variations in deep-water particle fluxes in the Sargasso Sea and their relationship to surface hydrography. Deep-Sea Res 33: 225–246

Deuser WG, Ross EH, Anderson RF (1981) Seasonality in the supply of sediment to the deep Sargasso Sea and implications for the rapid transfer of matter to the deep ocean. Deep-Sea Res 28A:495–505

Devol AH, Christensen JP (1993) Benthic fluxes and nitrogen cycling in sediments of the continental margin of the eastern North Pacific. J Mar Res 51: 345–372

Devol AH, Codispoti LA, Christensen JP (1996) Summer and winter denitrification rates in Arctic shelf sediments. Cont Shelf Res 17:1029–1050

Dibb JE (1996) Overview of field data on the deposition of aerosol-associated species to the surface snow of polar glaciers, particularly recent work in Greenland. In: Wolff EW, Bales RC (eds) Chemical Exchange Between the Atmosphere and Polar Snow. Springer, Berlin Heidelberg, pp 249–274

Dickson R, Yashayev I, Meincke J, Turrell B, Dye S, Holfort J (2002) Rapid freshening of the deep North Atlantic Ocean over the past four decades. Nature 416:832–837

Dickson RR (1999) All change in the Arctic. Nature 397: 389–391

Dickson RR, Osborn TJ, Hurrell JW, Meincke J, Blindheim J, Adlandsvik B et al. (2000) The Arctic Ocean response to the North Atlantic Oscillation. J Clim 13:2671–2696

Diel S, Tande KS (1992) Does the spawning of *Calanus finmarchicus* in high latitudes follow a reproducible pattern? Mar Biol 113:21–31

Dittmers K, Niessen F, Stein R (2003) Holocene sediment budget and sedimentary history of the Ob and Yenisei estuaries. In: Stein R, Fahl K, Fütterer DK, Galimov EM, Stepanets OV (eds), Siberian River Run-off in the Kara Sea: Characterisation, Quantification, Variability, and Environmental Significance, Proceedings in Marine Sciences Vol 6, Elsevier, Amsterdam, pp 457–484

Dixon RK, Brown S, Houghton A, Solomon AM, Trexler MC, Wisniewski J (1994) Carbon pools and flux of global forest ecosystems. Science 263:185–190

Dokken T, Hald M (1996) Rapid climatic shifts during isotope stages 2–4 in the polar North Atlantic. Geology 24:599–602

Doucette GJ, Fryxell GA (1983) *Thalassiosira antarctica*: Vegetative and resting stage chemical composition of an ice-related marine diatom. Mar Biol 78:1–6

Droop MR (1974) The nutrient status of algal cells in continuous culture. J Mar Biol Ass UK 54:825–855

Druffel ER, Williams PM, Bauer JE, Ertel JR (1992) Cycling of dissolved and particulate organic matter in the open ocean. J Geophys Res 97 (C10):15639–15659

Druzhkov NV, Druzhkova EI, Larionov VV, Tyuryakov AB (2001) Composition and vertical distribution of ice microbiota in northern part of the Barents Sea in beginning of winter. 1. Structure of microalgae community by means of light microscopical study. In: Lisitsin AP et al. (eds). An experience of complex oceanological researches in the Arctic. Nauchny Mir, Moscow, pp 325–343 (in Russian)

Duce RA, Duursma EK (1977) Input of organic matter to the ocean. Mar Chem 5:319–339

Duce RA, Liss PS, Merrill JT (1991) The atmospheric input of trace species to the World ocean. Glob Biogeoch Cycles 5/3:193–259

Ducklow HW (1999) The bacterial content of the oceanic euphotic zone FEMS. Microbiol Ecol 30:1–10

Ducklow HW, Steinberg DK, Buesseler KO (2001) Upperocean export and the biological pump. Oceanography 14: 50–58

Dunayev NN and Pavlidis JA (1988) A model of the Late Pleistocene glaciation of Eurasiatic Arctic Shelf. In: Kotlyakov VM and Sokolov VE (eds), Arctic Research – Advances and Prospects (2), Proceedings of the Conference of Arctic and Nordic Countries on Coordination of Research in the Arctic. Leningrad, Academy of Sciences of the USSR, December 1988, pp 70–72

Dunbar M (1969) The geographical position of the North Water. Arctic 22:438–441

Dunbar MJ (1968) Ecological development in polar regions. Prentice-Hall, Englewood Cliffs, pp 1–681

Dunton K (1992) Arctic biogeography: The paradox of the marine benthic fauna and flora. Trends in Ecology and Evolution 7:183–189.

Duplessy JC, Ivanova E, Murdmaa I, Paterne M, Labeyrie L (2001) Holocene paleoceanography of the northern Barents Sea and variations of the northward heat transport by the Atlantic Ocean. Boreas 30:2–16

Dyke AS, Dale JE, McNeeley RN (1996) Marine molluscs as indicators of environmental change in glaciated North American and Greenland during the last 18,000 years. Geographie physique et Quarternaire 50:125–184

Dyke AS, Hooper J, Savelle JM (1996) A history of sea ice in the Canadian Arctic Archipelago based on postglacial remains of the Bowhead Whale (*Balaena mysticetus*). Arctic 49:235–255

Eglinton G, Hamilton RJ (1967) Leaf epicuticular waxes. Science 156:1322–1323

Eglinton G and Hamilton RJ (1963) The distribution of alkanes. In: Swan T (ed), Chemical Plant Taxonomy. London, Academic Press, 187–208

Eicken H, Kolatschek J, Freitag J, Lindemann F, Kassens H, Dmitrenko I (2000) Identifying a major source area and constraints on entrainment for basin-scale sediment transport by Arctic sea ice. Geophys Res Lett 27:1919–1922

Eicken H, Krouse HR, Kadko D, Perovich DK (2002) Tracer studies of pathways and rates of meltwater transport through Arctic summer sea ice. J Geophys Res 107(C10): 10.1029/2000JC000583

Eicken H, Lensu M, Leppäranta M, Tucker WBI, Gow AJ, Salmela O (1995) Thickness, structure and properties of level summer multiyear ice in the Eurasian sector of the Arctic Ocean. J Geophys Res 100:22697–22710

Eicken H, Reimnitz E, Alexandrov V, Martin T, Kassens H, Viehoff T (1997) Sea-ice processes in the Laptev Sea and their importance for sediment export. Cont Shelf Res 17: 205–233

Eilertsen H-C, Hansen GA, Svendsen H, Hegseth EN (1993) The onset of spring phytoplankton bloom in the Barents Sea: Influence of changing light regime and other environmental factors. SPIE 2048:20–32

Eilertsen H-C, Taasen JP, Weslawski JM (1989) Phytoplankton studies in the fjords of West Spitzbergen: physical environment and production in spring and summer. J Plank Res 11:1245–1260

Eilertsen H-C, Taasen JP (1984) Investigations on the plankton community of Balsfjorden, Northern Norway. The phytoplankton 1976–1978. Environmental factors, dynamics of growth, and primary production. Sarsia 69:1–15

Ekwurzel B, Schlosser P, Mortlock RA, Fairbanks RG, Swift JH (2001) River runoff, sea ice meltwater, and Pacific water distribution and mean residence times in the Arctic Ocean. J Geophys Res 106:9075–9092

Elverhøi A, Andersen ES, Dokken T, Hebbeln D, Spielhagen RF, Svendsen JI, Sørflaten M, Rørnes A, Hald M, Forsberg CF (1995) The growth and decay of the Late Weichselian Ice Sheet in western Svalbard and adjacent areas based on provenance studies of marine sediments. Quat Res 44:303–316

Elverhøi A, Hooke RLB and Solheim A (1998) Late Cenozoic Erosion and Sediment Yield from the Svalbard-Barents Sea Region: Implications for Understanding Erosion of Glacierized Basins. Quat Sci Rev 17:209–241

Elverhoi A, Pfirman SL, Solheim A, Larsen BB (1989) Glaciomarine sedimentation in epicontinental seas exemplified by the Northern Barents Sea. Mar Geol 85:225–250

Elverhoi A, Solheim A (1983) The Barents Sea ice sheet – a sedimentological discussion. Pol Res 1(1):23–42

Elverhoi A, Andersen ES, Dokken T, Hebbeln D, Spielhagen RF, Svendsen JI, Sørflaten M, Rørnes A, Hald M and Forsberg CF (1995) The growth and decay of the Late Weichselian Ice Sheet in western Svalbard and adjacent areas based on provenance studies of marine sediments. Quat Res 44:303–316

Emerson S (1985) Organic carbon preservation in marine sediments. In: Sundquist ET, Broecker WS (eds) The Carbon Cycle and Atmospheric CO_2: Natural Variations Archean to Present. Geophysical Monograph Series, Vol 32, AGS, Washington DC, pp 78–87

Emerson S, Hedges JI (1988) Processes controlling the organic carbon content of open ocean sediments. Paleoanography 3:621–634

Emerson S, Stump C, Grootes PM, Stuiver M, Farwell GW, Schmidt FG (1987) Estimates of degradable organic carbon in deep-sea surface sediments from ^{14}C concentrations. Nature 329:51–53

English TS (1961) Some biological observations in the central North Polar Sea. Drift Station Alpha 1957–1958. Arctic Inst N Am Res Pap 13:8–80

Eppley RW (1972) Temperature and phytoplankton growth in the sea. Fish Bull NOAA 70:1063–1085

Escoubas J-M, Lomas M, LaRoche J, Falkowski, PG (1995) Light intensity regulation of *cab* gene transcription is signaled by the redox state of the plastoquinone pool. Proc Natl Acad Sci USA 92:10237–10241

Espitalié J, Laporte JL, Madec M, Marquis F, Leplat P, Paulet J, Boutefeu A (1977) Méthode rapide de caractérisation des roches mères, de leur potentiel pétrolier et de leur degre d'évolution. Rev Inst Franc Pétrol 32:23–42

Espitalié J, Makadi KS, Trichet J (1984) Role of mineral matrix during kerogen pyrolysis. Org Geochem 6:365–382

Fahl K, Cremer H, Erlenkeuser H, Hanssen H, Hölemann J, Kassens H, Knickmeier K, Kosobokova K, Kunz-Pirrung M, Lindemann F, Markhaseva E, Lischka S, Petryashov V, Piepenburg D, Schmid M, Spindler M, Stein R, Tuschling K (2001) Sources and pathways of organic carbon in the modern Laptev Sea (Arctic Ocean): Implications from biological, geochemical and geological data. Polarforschung 69 (1999):193–205

Fahl K, Kattner G (1993) Lipid content and fatty acid composition of algal communities in sea-ice and water from Weddell Sea (Antarctica). Pol Biol 13:405–409

Fahl K, Nöthig E, Stein R (1997) Organischer Kohlenstoff in der Laptewsee und auf dem Lomonossowrücken. Zweijahresbericht 1996/97, Alfred-Wegener-Institut Bremerhaven, pp 45–46

Fahl K, Stein R (1997) Modern organic carbon deposition in the Laptev Sea and the adjacent continental slope: Surface-water productivity vs terrigenous input. Org Geochem 26:379–390

Fahl K, Stein R (1999) Biomarkers as organic-carbon-source and environmental indicators in the Late Quaternary Arctic Ocean: Problems and perspectives. Mar Chem 63: 293–309

Fahl K, Stein R, Gaye-Haake B, Gebhardt C, Kodina LA, Unger D, Ittekkot V (2003) Biomarkers in surface sediments from the Ob and Yenisei estuaries and southern Kara Sea: Evidence for particulate organic carbon sources, pathways, and degradation. In: Stein R, Fahl K, Fütterer DK, Galimov EM, Stepanets OV (eds), Siberian River Run-off in the Kara Sea: Characterisation, Quantification, Variability, and Environmental Significance, Proceedings in Marine Sciences Vol 6, Elsevier, Amsterdam, pp 329–348

Fahrbach E, Meincke J, Østerhus S, Rohardt G, Schauer U, Tverberg V, Verduin J (2001) Direct measurements of volume transport though the Fram Straight. Pol Res 20: 217–224

Fairbanks RG (1989) A 17,000-year glacio-eustatic sea level record: influence of glacial melting rates on the Younger Dryas event and deep ocean-circulation. Nature 342: 637–642

Fairbanks RG (1990) The age and origin of the "Younger Dryas Climate Event" in Greenland ice cores. Paleoceanography 5:937–948

Falkowski PG, Barber RT, Smetacek V (1998) Biogeochemical controls and feedbacks on ocean primary production. Science 281:200–206

Falk-Petersen S, Pedersen G, Kwasniewski S, Hegseth EN, Hop H (1999) Spatial distribution and life-cycle timing of zooplankton in the marginal ice zone of the Barents Sea during the summer melt season in 1995. J Plankton Res 21:1249–1264

Falk-Petersen S, Sargent JR, Henderson J, Hegseth EN, Hop H, Okolodkov YB (1998) Lipids and fatty acids in ice algae and phytoplankton from the Marginal Ice zone in the Barents Sea. Pol Biol 20:41–47

Farrell JW, Pedersen TF, Calvert SE, Nielsen B (1995) Glacial-interglacial changes in nutrient utilization in the equatorial Pacific Ocean. Nature 377:514–517

Feder HM, Naidu AS (1991) The Chukchi Sea continental shelf: benthos-environmental interactions. Inst Mar Sci Univ Alaska Rep 92–1:1–250

Feder HM, Naidu AS, Baskaran M, Frost, K, Hammedi JM, Jewett SC, Johnson WR, Raymond J, Schell D (1991) Bering Strait-Hope Basin: habitat utilization and ecological characterization. Inst Mar Sci Univ Alaska Rep 92–2:1–457

Feder HM, Naidu AS, Jewett SC, Hameedi JM, Johnson WR, Whitledge TE (1994) The northeastern Chukchi Sea: benthos- environmental interactions. Mar Ecol Prog Ser 111:171–190

Feinberg LR, Dam HG (1998) Effects of diet on dimensions, density and sinking rates of fecal pellets of the copepod *Acartia tonsa*. Mar Ecol Prog Ser 175:87–96

Fennel W, Johannessen OM (1998) Wind forced oceanic responses near ice edges revisited. J Mar Syst 4:57–79

Fernandes MB, Sicre MA (2000) The importance of terrestrial organic carbon inputs on Kara Sea shelves as revealed by n-alkanes, OC and $\delta^{13}C$ values. Org Geochem 31:363–374

Field CB, Behrenfeld MJ, Randerson JT, Falkowski P (1998) Primary production of the biosphere: Integrating terrestrial and oceanic components. Science 281:237–240

Firth JV, Clark DL (1998) An early Maastrichtian organic-walled phytoplankton cyst assemblage from an organic-rich black mud in Core FL-533, Alpha Ridge: evidence for upwelling conditions in the Cretaceous Arctic Ocean. Mar. Micropal 34:1–27

Flato GM, Boer GJ (2001) Warming assymetry in climate change simulations. Geophys Res Lett 28:195–198

Forsyth D, Asudeh I, Green AG, Jackson HR (1986) Crustal structure of the northern Alpha Ridge beneath the Arctic Ocean. Nature 322:349–352

Fortier M (1999) Climatic and biological forcing of the vertical flux of biogenic carbon under the first-year ice in the Canadian Arctic in spring. PhD Thesis, Université Laval, Québec, Canada, 149 pp

Fortier M, Fortier L, Michel C, Legendre L (2002) Climatic and biological forcing of the vertical flux of biogenic particles under first-year sea ice in the Canadian Arctic. Mar Ecol Prog Ser 225:1–16

Francis JE (1988) A 50-million-year-old fossil forest from Strathcona Fiord, Ellesmere Island, Arctic Canada: evidence for a warm polar climate. Arct J Arct Inst North Am 41(4):314

Francis RC, Hare SR, Hollowed AB, Wooster WS (1998) Effects of interdecadal climate variability on the oceanic ecosystems of the NE Pacfic. Fisheries Oceanography 7:1–21

Francois R, Altabet MA (1992) Glacial to interglacial changes in surface nitrate utilization in the Indian sector of the Southern Ocean as recorded by sediment $\delta^{15}N$. Paleoceanography 7:589–603

Fransen A et al. (2001) The importance of shelf processes for the modification of chemical constituents in waters of the Eurasian Arctic Ocean: implications for carbon fluxes. Cont Shelf Res 21:225–242

Freitag J (1999) The hydraulic properties of Arctic sea ice – Implications for the small-scale particle transport (in German). Reps Pol Res 325, 150 pp

Friedrich MH (1997) Report about the characteristics of the NIOO-CEMO flume system – pilot study. NIOO Rapporten 6:1–62

Frost BW, Kishi MJ (1999) Ecosystem dynamics in the eastern and western gyres of the subarctic Pacific – a review of lower trophic level modelling. Prog Oceanogr 43: 317–333

Fry B, Sherr EB (1984) $\delta^{13}C$ measurements as indicators of carbon flow in marine and freshwater ecosystems. Contr Mar Sci 27:13–47

Fujishima Y, Ueda K, Maruo M, Nakayama E, Tokutome C, Hasegawa H, Matsui M, Sohrin Y (2001) Distribution of tracer bioelements in the Subarctic North Pacific Ocean and the Bering Sea (the R/V Hakuho Maru cruise KH-97-2). J Oceanogr 57:261–273

Fukuchi F, Hattori H, Sasaki H, Hoshiai T (1988) A phytoplankton bloom and associated processes observed with a long-term moored system in Antarctic waters. Mar Ecol Prog Ser 45: 279–288

Fukuchi M, Legendre L, Hoshiai T (1997) The Canada-Japan SARES project on first-year ice of Saroma-ko Lagoon (northern Hokkaido, Japan) and Resolute Passage (Canadian High Arctic). J Mar Syst 11:1–8

Fukuchi M, Sasaki H, Hattori H, Matsuda O, Tanimura A, Handa N, McRoy CP (1993) Temporal variability of particulate flux in the northern Bering Sea. Cont Shelf Res 13:693–704

Fulton RJ (1989) Foreword to Quaternary geology of Canada and Greenland. In: Fulton RJ (ed), Quaternary Geology of Canada and Greenland, Geol Soc Amer, The Geology of North America, K-1, pp 1–11

Fütterer DK (1992) ARCTIC'91: The Expedition ARK-VIII/3 of RV "Polarstern" in 1991. Reps Pol Res 107, 267 pp

Fütterer DK (1994) The Expedition ARCTIC'93 Leg ARK IX/4 of RV "Polarstern" 1993. Reps Pol Res 149, 244 pp

Gagosian RB, Peltzer ET, Zafiori OS (1981) Atmospheric transport of continentally derived lipids to the tropical North Pacific. Nature 291:321–324

Gagosian RB, Peltzer ET (1987) The importance of atmospheric input of terrestrial organic material to deep sea sediments. Org Geochem 10:661–669

Gakkel YY (1957) The science and developing of the Arctic. Sea Transport Press, Leningrad, 132 pp (in Russian)

Gakkel YY (1958) Destruction of Semenovsky Island. In: Problems of the Arctic. Sea Transport Press, Leningrad: 95–97 (in Russian)

Gakkel YY (1962) The exploration and development of polar lands. In: Harris CD (ed), Soviet geography; Accomplishments and tasks. New York, American Geographical Society Occasional Publication 1, pp 265–274.

Gard G (1986) Calcareous nannofossil biostratigraphy of late Quaternary Arctic sediments. Boreas 15:217–229

Gard G (1993) Late Quaternary coccoliths at the North Pole: Evidence of ice-free conditions and rapid sedimentation in the central Arctic Ocean. Geology 21:227–230

Gard G, Backman J (1990) Synthesis of Arctic and Sub-Arctic Coccolith Biochronology and History of North Atlantic Drift Water Influx during the last 500.000 Years. In: Bleil U, Thiede J (eds), Geological History of the Polar Oceans: Arctic versus Antarctic. NATO ASI Series C 308, Kluwer Academic Publishers, Dordrecht, pp 417–436

Gardner WD (2000) Sediment trap sampling in surface waters. In: Hanson RB, Ducklow HW, Field JG (eds). The changing ocean carbon cycle: A midterm synthesis of the Joint Global Ocean Flux Study. Cambridge University Press, pp 240–281

Gardner WD, Southard JB, Hollister CD (1985) Sedimentation, resuspension and chemistry of particles in the northwest Atlantic. Mar Geol 65:199–242

Gascard J-C, Watson A, Messias M-J, Olsson K, Johannessen T, Sumonsen K (2002) Long-lived vortices as a mode of deep-water ventilation in the Greenland Sea. Nature 416: 525–527

Gataullin VN, Polyak LV, Epstein OG, Romanyuk BF (1993) Glacigenic deposits of the Central Deep: a key to the Late Quaternary evolution of the Barents Sea. Boreas 22: 47–58

Gattuso JP, Frankignoulle M, Wollast R (1998) Carbon and carbonate metabolism in coastal aquatic ecosystems. Ann Rev Ecol Syst 29:405–434

Gauwet G, Sidorov I (1996) The biogeochemistry of Lena River: organic carbon and nutrients distribution Mar Chem 53:211–227

Gawarkiewicz G, Weingartner T, Chapman DC (1998) Sea-ice processes and water mass modification and transport over Arctic shelves. In: Brink KH, Robinson AR (eds) The Sea. Wiley & Sons, Inc., pp 171–190

Gebhardt AC, Lahajnar N, Gaye-Haake B, Fertzer I, Deubel H (2002) Sediment trap investigations in the Kara Sea. In: Stein R, Stepanets O (eds), Scientific Cruise report of the Kara Sea Expedition 2001 of RV "Akademik Boris Petrov": The German-Russian Project on Siberian River Run-off (SIRRO) and the EU Project "ESTABLISH", Reps Pol Mar Res 419:33–40

Geider RJ, MacIntyre HL, Kana TM (1996) A dynamic model of photoadaptation in phytoplankton. Limnol Oceanogr 41:1–15

Geology of the USSR (1970) The USSR Arctic islands (Vol 26) Nedra Moscow, 121 pp

Gianelli V, Thomas DN, Haas C, Kattner G, Kennedy H, Dieckmann GS (2001) Behavior of dissolved organic matter and inorganic nutrients during experimental sea-ice formation. Ann Glaciol 33:317–321

Gibson JAE, Trull T, Nichols PD, Summons RE, McMinn A (1999) Sedimentation of ^{13}C-rich organic matter from Antarctic sea-ice algae: A potential indicator of past sea-ice extent. Geology 27 (4):331–334

Gibson JAE, Vincent WF, Nieke B, Pienitz R (2000) Control of biological exposure to UV radiation in the Arctic Ocean: Comparison of the roles of ozone and riverine dissolved organic matter. Arctic 53:372–382

Gilstad M, Sakshaug E (1990) Growth rates of ten diatom species from the Barents Sea at different irradiances and day lengths. Mar Ecol Prog Ser 64:169–173

Giovando LF, Herlinveaux RH (1981) A discussion of factors influencing dispersion of pollutants in the Beaufort Sea. 81–4, Sidney, BC: Institute of Ocean Sciences, 198 pp

Gislason A, Astthorson OS (1998) Seasonal variations in biomass, abundance and composition of zooplankton in the subarctic waters north of Iceland. Pol Biol 20:85–94

Gloersen P, Campbell WJ (1991) Recent variations in Arctic and Antarctic sea-ice cover. Nature 352:33–36

Glud RN, Holby O, Hoffmann F, Canfield DE (1998) Benthic mineralization and exchange in Arctic sediment (Svalbard, Norway). Mar Ecol Progr Ser 173: 237–251

Glud RN, Kühl M, Wenzhöfer F, Rysgaard S (2002) Benthic diatoms of a high Arctic fjord (Young Sound, NE Greenland): importance for ecosystem primary production. Mar Ecol Prog Ser 238:15–29

Gobeil C, Macdonald RW, Smith JN, Beaudin L (2001) Lead contamination in Arctic basin sediments tracks Atlantic water flow pathways. Science 293:1301–1304.

Gobeil C, Macdonald RW, Sundby B (1997) Diagenetic separation of cadmium and manganese in suboxic continental margin sediments. Geochim Cosmochim Acta 61:4647–4654

Gobeil C, Paton D, McLaughlin FA, Macdonald RW, Paquette G, Clermont Y et al. (1991) Donnés géochimiques sur les eaux interstitielles et les sédiments de la mer de Beaufort, Mont Joli: Institut Maurice-Lamontagne, 101, 92 pp

Gobeil C, Sundby B, Macdonald RW, Smith JN (2001) Recent change in organic carbon flux to Arctic Ocean deep basins: Evidence from acid volatile sulfide, manganese and rhenium discord in sediments. Geophys Res Lett 28:1743–1746

Goericke R., Fry B (1994) Variations of marine plankton d^{13}C with latitude, temperature, and dissolved CO_2 in the world ocean. Glob Biogeochem Cycles 8:85–90

Goffart A, Catalano G, Hecq JH (2001) Factors controlling the distribution of diatoms and *Phaeocystis* in the Ross Sea. Deep-Sea Res II 48:161–175

Goldblatt RH, Mackas DL, Lewis AG (2001) Mesozooplankton community characteristics in the NE subarctic Pacific. Deep-Sea Res II 46:2619–2644

Goldman J (1999) Inorganic carbon availability and the growth of large marine diatoms. Mar Ecol Prog Ser 180: 81–91

Golikov AN (1990) Ecosystems of the New Siberian shoals and the fauna of the Laptev Sea and adjacent waters of the Arctic Ocean. Explorations of the fauna of the seas 37(45): Moscow: Nauka, (in Russian)

Goñi MA, Hedges JI (1992) Lignin dimers: structures, distribution, and potential geochemical applications. Geochim Cosmoschim Acta 56:4025–4043

Goñi MA, Ruttenberg KC, Eglinton TI (1997) Sources and contribution of terrigenous organic carbon to surface sediments in the Gulf of Mexico. Nature 389:275–278

Goñi MA, Ruttenberg KC, Eglinton TI (1998) A reassessment of the sources and importance of land-derived organic matter in surface sediments from the Gulf of Mexico. Org Geochem 62:3055–3075

Goñi MA, Yunker MB, Macdonald RW, Eglinton TI (2000) Distribution and sources of organic biomarkers in arctic sediments from the Mackenzie River and Beaufort Shelf. Mar Chem 71:23–51

Gordeev VV (2000) River input of water, sediment, major ions, nutrients and trace metals from Russian territory to the Arctic Ocean. In: Lewis EL (ed) Freshwater Budget of the Arctic Ocean. Kluwer, Dordrecht, pp 297–322

Gordeev VV, Dzhamalov RG, Zektser IS, Zhulidov AV, Brysgalo VN (1999) An assessment of groundwater input of nutrients into the coastal seas the Russian Arctic. Water Resources 38:1–6 (in Russian)

Gordeev VV, Martin JM, Sidorov IS, Sidorova MV (1996) A reassessment of the Eurasian river input of water, sediment, major elements, and nutrients to the Arctic Ocean. Am J Sci 296:664–691

Gordeev VV, Sidorov IS (1993) Concentrations of major elements and their outflow into the Laptev Sea (Russia). Mar Chem 43: 33–46

Gordeev VV, Tsirkunov VV (1998) River fluxes of dissolved and suspended substances. In: Kimstach V Meybeck M, Baroudy E (eds) A water quality assessment of the former Soviet Union. E and EN Spon, London and New York, pp 311–350

Gordon ES, Goñi MA, Roberts QN, Kineke GC, Allison MA (2001) Organic matter distribution and accumulation on the inner Louisiana shelf west of the Atchafalaya River. Cont. Shelf Res 21:1691–1721

Gorshkov SG (1983) World Ocean Atlas, Vol. 3, Arctic Ocean. Pergamon Press, 184 pp

Gorshkova TI (1970) Biogeochemistry of recent sediments in the seas of USSR. Abstr of Diss. VNIRO, Moscow pp 1–50 (in Russian)

Gorshkova TI (1974) Biogeochemistry of recent marine sediments and their biological significance. In: Bonitet of the World Ocean. Proc VNIRO. Pischevaya Promislennoste, Moscow, 1, pp 135–144 (in Russian)

Gosselin M, Levasseur M, Wheeler PA, Horner RA, Booth BC (1997) New measurements of phytoplankton and ice algal production in the Arctic Ocean. Deep-Sea Res II 44(8):1623–1644

Gough MA, Mantoura RFC, Preston M (1993) Terrestrial plant polymers in marine sediments. Geochim Cosmochim Acta 57:945–964

Gradinger R (1996) Occurrence of an algal bloom under Arctic pack ice. Mar Ecol Prog Ser 131:301–305

Gradinger R (1999) Vertical fine structure of the biomass and composition of algal communities in Arctic pack ice. Mar Biol 133:745–754

Graf G (1989) Benthic-pelagic coupling in a deep-sea benthic community. Nature 341:437–439

Graf G, Gerlach SA, Linke P, Queisser W, Ritzrau W, Schelz A, Thomsen L, Witte U (1995) Benthic-pelagic coupling in the Greenland-Norwegian Sea and its effect on the geological record. Geol Rundsch 84: 9–58

Grahl C, Boetius A, Nöthig E-M (1999) Pelagic-benthic coupling in the Laptev Sea affected by sea ice. In: Kassens H, Bauch HA, Dmitrenko I, Eicken H, Hubberten H-W, Melles M, Thiede J, Timokhov L (eds) Land-Ocean Systems in the Siberian Arctic: Dynamics and History. Springer, Berlin, pp 143–152

Gran HH (1902) Das Plankton des norwegischen Nordmeeres. Rep Norw Fish-Mar Invest 2(5):1–222

Gran HH (1904) Diatomaceae from ice-floes and plankton of the Arctic Ocean. Sci Res Norw North Polar Exp 4: 3–74

Gran HH (1929) Investigation of the production of plankton outside the Romsdalsfjord 1926–1927. Rapp. P-v Réun Cons Int Explor Mer 56:1–112

Gran HH (1931) On the conditions for the production of the plankton in the sea. Rapp. P-v Réun Cons Int Explor Mer 75:37–46

Grantz A, Clark DL, Phillips RL, Srivastava SP, Blome CD. Gray LB, Haga H, Mamet BL, McIntyre DJ, McNeil DH, Mickey MB, Mullen MW, Murchey BL, Ross CA, Stevens CH, Silberling NJ, Wall JH, Willard DA (1998) Phanerozoic stratigraphy of Northwind Ridge, magnetic anomalies in the Canada Basin, and the geometry and timing of rifting in the Amerasia Basin, Arctic Ocean. Geol Soc Amer Bull 110 (6):801–820

Grantz A, May SD, Taylor PT, and Lawner LA (1990) Canada Basin. In: Grantz, A, Johnson GL, and Sweeney JF (eds). The Arctic Ocean Region: The Geology of North America DNAG L, pp 379–402

Grantz A, Phillips RL, Jones GA (1999) Holocene pelagic and turbidite sedimentation rates in the Amerasian Basin, Arctic Ocean, from radiocarbon age-depth profiles in cores. In: Bruns P, Hass HC (eds) On the determination of sediment accumulation rates, GeoResearch Forum 5:209–222

Grantz A, Phillips RL, Mullen MW, Starratt SW, Jones GA, Naidu AS, Finney BP (1996) Character, paleoenvironment, rate of accumulation, and evidence for seismic triggering of Holocene turbidites, Canada Abyssal Plain, Arctic Ocean. Mar Geol 133:51–73

Grebmeier JM (1987) The ecology of benthic carbon cycling in the northern Bering and Chukchi Sea. PhD dissertation, Inst Marine Sci, Univ Alaska Fairbanks Alaska, pp 1–189

Grebmeier JM (1993) Studies on pelagic-benthic coupling extended onto the Soviet continental shelf in the Bering and Chukchi Seas. Cont Shelf Res 13:653–668

Grebmeier JM, Barry JP (1991) The influence of oceanographic processes on pelagic-benthic coupling in polar regions: A benthic perspective. J Mar Syst 2:495–518

Grebmeier JM, Cooper LW (1995) Influence of the St Lawrence Island Polynya upon the Bering Sea benthos. J Geophys Res 100:4439–4460

Grebmeier JM, Feder HM, McRoy CP (1989) Pelagic-benthic coupling on the shelf of the northern Bering and Chukchi Seas. II. Benthic community structure. Mar Ecol Prog Ser 51:253–268

Grebmeier JM, McRoy CP (1989) Pelagic-benthic coupling on the shelf of the northern Bering and Chukchi Seas. III. Benthic food supply and carbon cycling. Mar Ecol Prog Ser 53:79–91

Grebmeier JM, McRoy CP, Feder HM (1988) Pelagic-benthic coupling on the shelf of the northern Bering and Chukchi Seas. I. Food supply source and benthic biomass. Mar Ecol Prog Ser 48:57–67

Grebmeier JM, Smith WO Jr, Conover RJ (1995) Biological processes on Arctic continental shelves: ice-ocean biotic interactions. In: Smith OS Jr, Grebmeier JM (eds) Arctic oceanography: marginal ice zones and continental shelves. Am Geophys Union, Washington, DC, pp 231–261

Grigoriev MN (1993) Cryomorphogenesis of the Lena River mouth area. Permafrost Institute Press SB RAS, Yakutsk, 176 pp. (in Russian)

Grigoriev MN (1996) Rules of thermal abrasion and thermal denudation processes of the Arctic Sea shores (on an example of the Laptev Sea key sites). In: Materials of the first conference of Russian geocryologists Vol 1–3. Dynamic geocryology. Moscow Univ. Publ. House, Moscow: 504–511 (in Russian)

Grigoriev MN, Kunitsky VV (2000) The Ice Complex of the Yakutian Arctic coasts as a source of sediments on the shelf. In: Hydrometeorological and biogeochemical research in the Arctic. Proceedings of Arctic Regional Center (Vladivostok) Vol 2/1:109–116 (in Russian)

Grigoriev NF (1966) Permafrost in the Yakutian Coastal Zone. Nauka Press, Moscow, 180 pp (in Russian)

Groene T (1995) Biogenic production and consumption of dimethylsulphide (DMS) and dimethylsulfoniopropionate (DMSP) in the marine epipelagic zone: a review. J Mar Syst 6:191–209

Grønås S, Skeie P (1999) A case study of strong winds at an arctic front. Tellus 51A:865–879

Grootes PM, Stuiver M, White JWC, Johnson C, Jouzel J (1993) Comparison of oxygen isotope records from the GISP2 and GRIP Greenland ice cores. Nature 366:552–554

Grosswald MG, Hughes TJ (1999) The case for an ice shelf in the Pleistocene Arctic Ocean. Pol Geography 23:23–54

Guay CK, Klinkhammer GP, Falkner KK, Benner R, Coble PG, Whitledge TE, Black B, Bussell J, Wagner TA (1999) High-resolution measurements of dissolved organic carbon in the Arctic Ocean *by in situ* fiber-optic spectrometry. Geophys Res Lett 26:1007–1010

Guay CKH, Falkner KK, Muench RD, Mensch M, Frank M, Bayer R (2001) Wind-driven transport pathways for Eurasian Arctic river discharge. J Geophys Res 106:11469–11480

Gudkovich ZM (1961) Relation of the ice drift in the Arctic Basin to ice conditions in the Soviet Arctic seas (in Russian). Tr Okeanogr Kom Akad Nauk SSSR 11:14–21

Guest PS, Glendening JW, Davidson KL (1995) An observational and numerical study of wind stress variations within the marginal ice zone. J Geophys Res 100:10887– 10904

Gukov AY, Antonov M, Taldenkova EE, Schmid MK (2001) Peculiarities of the microscale (facial) benthos distribution at the fast ice edge in the Laptev Sea. Oceanology 41:395–402

Gulliksen B, Holte B (1992) List of the marine invertebrates and fish known from Norwegian territorial waters in Svalbard (in Norwegian). NPI Rapportserie, Oslo

Gurevich VI (1995) Recent sedimentogenesis and environment on the Arctic shelf of Western Eurasia. Meddelelser NR 131, Oslo, pp 1–92

Hald M, Dokken T, Hagen S (1996) Palaeoceanography on the European Arctic margin during the last deglaciation. In: Andrews JT, Austin WEN, Bergsten H, Jennings AE (eds). Late Quaternary palaeoceanography of the North Atlantic margins. Geol Soc Spec Pub III, The Geological Society, London, pp 275–289

Hald M, Dokken T, Mikalsen G (2001) Abrupt climatic change during the last interglacial-glacial cycle in the polar North Atlantic. Mar Geol 176:121–137

Hald M, Kolstad V, Polyak L, Forman SL, Herlihy FA, Ivanov G, Nescheretov A (1999) Late glacial and Holocene paleoceanography and sedimentary environments in the Saint Anna Trough, Eurasian Arctic Ocean margin. Palaeogeogr Palaeoclim Palaeoecol 146:229–249

Halldal P (1953) Phytoplankton investigations from weather ship M in the Norwegian Sea, 1948–49. Hvalråd Skr 38: 1–91

Halldal P, Halldal K (1973) Phytoplankton, chlorophyll, and submarine light conditions in Kings Bay, Spitsbergen, July 1971. Norw J Bot 20:99–108

Hamm C, Reigstad M, Riser CW, Mühlebach A, Wassmann P (2001) On the trophic fate of *Phaeocystis pouchetii*. VII. Sterols and fatty acids reveal sedimentation of *P. pouchetii*-derived organic matter via krill fecal strings. Mar Ecol Prog Ser 209:55–69

Hamm CE, Simson DA, Merkel R, Smetacek V (1999) Colonies of *Phaeocystis globosa* are protected by a thin but tough skin. Mar Ecol Prog Ser 187:101–111

Hansell DA, Carlson CA (1998) Deep-ocean gradients in the concentration of dissolved organic carbon. Nature 395: 263–266

Hansell DA, Whitledge TA, Goering JJ (1993) Patterns of nitrate utilization and new production over the Bering-Chukchi shelf. Cont Shelf Res 13:601–627

Hansell DA, Carlson CA (2001). Marine dissolved organic matter and the carbon cycle. Oceanography 14(4):41–49

Hansen B, Christiansen S, Pedersen G (1996) Plankton dynamics in the marginal ice zone of the central Barents Sea during spring: carbon flow and structure of the grazer food chain. Pol Biol 16:115–218

Hansen B, Turrell WR, Oesterhus S (2001) Decreasing overflow from the Nordic seas into the Atlantic Ocean through the Faroe Bank channel since 1950. Nature 411:927–930
Hansen-Bauer I (2002) Temperature and precipitation in Svalbard 1912–2050: Measurements and scenarios. Pol Res 38:225–232
Hare SR, Mantua NJ (2000) Empirical evidence for North Pacific regime shifts in 1977 and 1989. Prog Oceanog 47:103–145
Hargrave BT, Bodungen Bv, Conover RJ, Fraser AJ, Phillips G, Vass WP (1989) Seasonal changes in sedimentation of particulate matter and lipid content of zooplankton collected by sediment trap in the Arctic Ocean off Axel Heiberg Island. Pol Biol 9: 467–475
Hargrave BT, von Bodungen B, Stoffyn-Egli P, Mudi PJ (1993) Seasonal variability in particle sedimentation under permanent ice cover in the Arctic Ocean. Cont Shelf Res 14:279–293
Hargrave BT, Walsh ID, Murray DW (2002) Seasonal and spatial patterns in mass and organic matter sedimentation in the North Water Polynya. Deep-Sea Res II 49: 5227–5244
Harms IH, Karcher MJ, Dethleff D (2000) Modelling Siberian river runoff – implications for contaminant transport in the Arctic Ocean. J Mar Syst 27:95–115
Harper JR (1990) Morphology of the Canadian Beaufort Sea coast. Mar Geol 91:75–91
Harper JR, Reimer PD, Collins AD (1985) Beaufort Sea physical shore-zone analysis. Geological Survey of Canada Open File 1689, 105 pp
Harrison WG, Cota GF (1991) Primary production in polar waters: relation to nutrient availability. Pol Res 10:87–104
Hasle GR, Heimdal BR (1998) The net phytoplankton in Kongsfjorden, Svalbard, July 1988, with general remarks on species composition of arctic phytoplankton. Pol Res 17:31–52
Hattori H, Saito H (1997) Diel changes in vertical distribution and feeding activity of copepods in ice-covered Resolute Passage, Canadian Arctic, in spring 1992. J Mar Syst 11: 205–219
Hayes JM, Freeman KH, Popp BN, Hoham CH (1989) Compound-specific isotopic analyses: a novel tool for reconstruction of ancient biogeochmical processes. Adv Org Geochem 16:1115–1128
Head EJH, Harris LR, Campbell RW (2000) Investigations on the ecology of *Calanus* spp. in the Labrador Sea. I. Relationship between the phytoplankton bloom and reproduction and development of *Calanus finmarchicus* in spring. Mar Ecol Prog Ser 193:53–73
Hebbeln D (1991) Spätquartäre Stratigraphie und Paläozeanographie in der Fram-Straße. Berichte, Fachbereich Geowissenschaften, Universität Bremen (22), pp 1–174
Hebbeln D (1992) Weichselian glacial history of the Svalbard area: correlating the marine and terrestrial records. Boreas 21:295–304
Hebbeln D, Berner H (1993) Surface sediment distribution in the Fram Strait. Deep Sea Res I 40 (9):1731–1745
Hebbeln D, Dokken T, Andersen ES, Hald M, Elverhøi A (1994) Moisture supply for northern ice-sheet growth during the Last Glacial Maximum. Nature 370:357–359
Hebbeln D, Henrich R, Baumann KH (1998) Paleoceanography of the last interglacial/glacial cycle in the Polar North Atlantic. Quat Sci Rev 17 (1–3):125–153
Hebbeln D, Wefer G (1991) Effects of ice coverage and ice-rafted material on sedimentation in the Fram Strait. Nature 350:409–411
Hebbeln D, Wefer G (1997) Late Quaternary palaeoceanography in the Fram Strait. Palaeoceanography 12 (1): 65–78
Hedges JI (1992) Global biogeochemical cycles: progress and problems. Mar Chem 39:67–93
Hedges JI (2002) Why dissolved organics matter: A marine retrospective. In: Hansell D, Carlson C (eds) Biogeochemistry of marine dissolved organic matter. Academic Press, pp 1–33
Hedges JI, Clark WA, Cowie Gl (1988) Organic matter sources to the water column and surficial sediments of a marine bay. Limnol. Oceanogr 33:1116–1136
Hedges JI, Clark WA, Quay PD, Richey JE, Devol AH, Santos UDM (1986) Composition and fluxes of particulate organic material in the Amazon River. Limnol Oceanogr 31:717–738
Hedges JI, Keil R, Benner R (1997) What happens to terrestrial organic matter in the ocean? Org Geochem 27:195–212
Hedges JI, Keil RG (1995) Sedimentary organic matter preservation: an assessment and speculative synthesis. Mar Chem 49:81–115
Hedges JI, Mann DC (1979) The lignin geochemistry of marine sediments from the southern Washington coast. Geochim Cosmochim Acta 43:1809–1818
Hedges JI, Oades JM (1997) Comparative organic geochemistries of soils and marine sediments. Org Geochem 27: 319–361
Hedges JI, Stern JH (1984) Carbon and nitrogen determinations of carbonate-containing solids. Limnol Oceanogr 29:657–663
Heding SG (1942) Holothurioidea. Part II. Aspidochirota – Elasipoda – Dendrochirota. Danish INGOLF-Expedition 4:43–78
Heezen BC, Ewing M (1961) The Mid-Oceanic Ridge and its extension through the Arctic Basin. In: Raasch GO (ed). Geology of the Arctic. University of Toronto Press, pp 622–642
Hegseth EN (1989) Photoadaptation in marine diatoms. Pol Biol 9:479–486
Hegseth EN (1993) Sub-ice algal assemblages of the Barents Sea: Species composition, chemical composition, and growth rates. Pol Biol 12:485–496
Hegseth EN (1997) Phytoplankton of the Barents Sea – the end of a growth season. Pol Biol 17:235–241
Heimdahl BR (1983) Phytoplankton and nutrients in the waters northwest of Spitsbergen in the autumn 1979. J Plankton Res 5 (6): 901–918
Heiskanen AS, Keck A (1996) Distribution and sinking rates of phytoplankton, detritus, and particulate biogenic silicia in the Laptev Sea and Lena River (Arctic Siberia). Mar Chem 53:229–245
Hempel G (1985) On the biology of polar seas, particularly the Southern Ocean In: Gray JS, Christiansen MS (eds) Marine biology of polar regions and the effect of stress on marine organisms. Chichester, Wiley & Sons, pp 3–34
Henderson RJ, Hegseth EN, Park MT (1998) Seasonal variation in lipid and fatty acid composition of ice algae from the Barents Sea. Pol Biol 20:48–55
Henrichs SH, Reeburgh, WS (1987) Anaerobic mineralization of marine sediment organic matter: rates and the role of anaerobic processes in the oceanic carbon economy. Geomicrobiol J 5:191–237
Héquette A, Ruz M-H, Hill PR (1995) The effects of the Holocene sea level rise on the evolution of the southeastern coast of the Canadian Beaufort Sea. J Coast Res11:494–507

Herlihy FA, 1996. A marine sedimentary record of the last glacial cycle in the Franz Victoria and St. Anna Troughs, Northern Barents and Kara Seas, Russia. Unpublished MS Thesis, Columbus, Ohio State University, 196 pp

Herman Y (ed 1989) The Arctic Seas. Climatology, oceanography, geology, and biology. Van Nostrand Reinhold, New York, 888 pp

Herndl GJ, Obernosterer I (2001) UV-radiation and pelagic bacteria. In: Hessen DO (ed) Ecological studies. UV-radiation and Arctic ecosystems. Springer, Heidelberg

Hewes CD, Sakshaug E, Reid FMH, Holm-Hansen O (1990) Microbial autotrophic and heterotrophic eucaryotes in Antarctic waters: relationships between biomass and chlorophyll, adenosine triphosphate and particulate organic carbon. Mar Ecol Prog Ser 63:27–35

Hill P, Mudie PJ, Moran K, Blasco SM (1985) A sea-level curve for the Canadian Beaufort Shelf. Can J Earth Sci 22:1383–1393

Hill PR, Blasco SM, Harper JR, Fissel DB (1991) Sedimentation on the Canadian Beaufort Shelf. Cont Shelf Res 11: 821–842

Hill PR, Forbes DL, Dallimore SR, Morgan P (1986) Shoreface development in the Canadian Beaufort Sea. ACROSES Symposium on Cohesive Shores, Ottawa, NRC of Canada: 428–448

Hill PR, Héquette A, Ruz M-H (1993) Holocene sea-level history of the Canadian Beaufort shelf. Can J Earth Sci 30:103–108

Hill PR, Lewis PC, Desmarais S, Kauppaymuthoo V, Rais H (2001) The Mackenzie Delta: sedimentary processes and facies of a high-latitude, fine-grained delta. Sedimentology 48:1047–1078

Hill PR, Nadeau OC (1989) Storm-dominated sedimentation on the inner shelf of the Canadian Beaufort Sea. J Sed Petrol 59:455–468

Hirche H-J (1989) Egg production in the Arctic copepod *Calanus glacialis*: laboratory experiments. Mar Biol 103: 311–318

Hirche H-J (1997) Life cycle of the copepod *Calanus hyperboreus* in the Greenland Sea. Mar Biol 128:607–618

Hirche HJ, Baumann MEM, Kattner G, Gradinger R (1991) Plankton distribution and the impact of copepod grazing on primary production in Fram Strait, Greenland Sea. J Mar Sys 2:477–494

Hirche HJ, Kwasniewski S (1997) Production and development of Calanus species in the Northeast Water Polynya in relation to environmental conditions, J Mar Syst 10: 299–318

Hirche H-J, Niehoff B (1996) Reproduction of the Arctic copepod *Calanus hyperboreus* in the Greenland Sea – field and laboratory observations. Pol Biol 16:209–219

Hollerbach A, 1985. Grundlagen der organischen Geochemie. Springer, Berlin, 190 pp.

Holligan PM, Groom SB, Harbour DS (1993) What controls the distribution of the coccolithophore, *Emiliania huxleyi*, in the North Sea? Fish Oceanogr 2:175–183

Holloway G, Sou T (2001) Has arctic sea ice rapidly thinned? J Clim 15:1691–1701

Holmes MR, McClelland JW, Peterson BJ, Shiklomanov IA, Shiklomanov AI, Zhulidov AV, Gordeev VV, Bobrovitskaya NN (2002) A circumpolar perspective on fluvial sediment flux to the Arctic Ocean. Glob Biogeochem Cycles 16/4: 10.1029/2002GB001920

Holmes RM, Peterson BJ, Zhulidov AV, Gordeev VV, Makkaveev PN, Stunzhas PA, Kosmenko LS, Köhler GH, Shiklomanov AI (2001) Nutrient chemistry of the Ob and Yenisei Rivers, Siberia: results from June 2000 expedition and evaluation of long-term data sets. Mar Chem 75: 219–227

Holmes ML and Creager JS, 1974. Holocene history of the Laptev Sea continental shelf. In: Herman Y (ed), Marine Geology and Oceanography of the Arctic Seas. Springer-Verlag, New York, 211–230

Holte B, Dahle S, Gulliksen B, Naes K (1996) Some macrofaunal effects of local pollution and glacier induced sedimentation, with indicative chemical analyses, in the sediments of two Arctic fjords. Polar Biol 16:549–557

Hong, GH (1986) Fluxes, dynamics and chemistry of particulate matter and nutrient regeneration in the central basin of Boca de Quadra, southeast Alaska. PhD Thesis Univ Alaska Fairbanks. pp 1–225

Honjo S (1990) Particle fluxes and modern sedimentation in the polar ocean. In: Smith WO Jr (ed) Polar oceanography, part B: chemistry, biology, and geology. Academic Press, New York, pp 687–739

Hood DW, Calder JA (1981) The eastern Bering Sea Shelf: Oceanography and resources. Office of Marine Pollution Assessment, NOAA, Seattle, 1339 p (2 vols)

Hop H, Pearson T, Hegseth EN, Kovacs KM, Wiencke C, Kwasniewski S, Eiane K, Mehlum F, Gulliksen B, Wlodarska-Kowalczuk M, Lydersen C, Weslawski JM, Cochrane S, Gabrielsen GW, Leakey RJG, Lønne OJ, Zajaczkowksi M, Falk-Petersen D, Kendall M, Wängberg S-Å, Bischof K, Voronkov AY, Kovaltchouk NA, Wiktor J, Poltermann M, di Prisco G, Papucci C, Gerland S (2002) The marine ecosystem of Kongsfjorden, Svalbard. Pol Res 21:167–208.

Hop H, Poltermann M, Lønne O-J, Falk-Petersen S, Korsnes R, Budgell W (2000) Ice amphipod distribution relative to ice density and under-ice topography in the northern Barents Sea. Pol Biol 23:357–367

Hopkins DM (1979) Landscape and climate of Beringia during the late Pleistocene and Holocene time. In: Laughlin WS, Harper AB (eds) The First Americans: Origins, Affinities and Adaptations. Gustav Fisher, New York, pp 15–41

Hopkins TS (1991) The GIN Sea–A synthesis of its physical oceanography and literature review 1972–1985. Earth-Science Rev 30:175–318

Hoppe HG, Ducklow H, Karrasch B (1993) Evidence for dependency of bacterial growth on enzymatic hydrolysis of particulate organic matter in the mesopelagic ocean. Mar Ecol Prog Ser 93:277–283

Horner R (1984) Phytoplankton abundance, chorophyll *a*, and primary productivity in the western Beaufort Sea. In: Barnes PW, Schell DM, Reimnitz E (eds) The Alaskan Beufort Sea. Academic Press, New York, pp 295–310

Horner R, Alexander V (1972) Algal populations in Arctic sea ice: An investigation of heterotrophy. Limnol Oceanogr 17:454–458

Horner RA, Schrader GC (1982) Relative contributions of ice algae, phytoplankton and benthic microalgae to primary production in nearshore regions of the Beaufort Sea. Arctic 35:485–503

Horner, RA (1989) Arctic sea-ice biota. In: Herman Y (ed) The Arctic seas. Climatology, oceanography, geology, and biology. van Nostrand Reinhold, New York, pp 123–146

Houston AL, Deming JW (2002) Relationships between microbial extracellular enzymatic activity and suspended and sinking particulate organic matter: Seasonal transformations in the North Water. Deep-Sea Res II 49: 5211–5225

Hovland M (ed) (2001) The High-Arctic Drilling Challenge. Final Report of the Arctic's Role in Global Change Program Planning Group (APPG), Ocean Drilling Program, 38 pp

Huang WY, Meinschein (1976) Sterols as source indicators of organic material in sediments. Geochim Cosmochim Acta 40:323–330

Hufford GL (1974) On apparent upwelling in the southern Beaufort Sea. J Geophys Res 79:1305–1306

Huisman J, Weissing FJ (2001) Fundamental unpredictability in multispecies competition. Am Naturalist 157:488–494

Hulth S, Blackburn TH, Hall POJ (1994) Arctic sediments (Svalbard): consumption and microdistribution of oxygen. Mar Chem 46:293–316

Hulth S, Hall POJ, Blackburn TH, Landen A (1996) Arctic sediments (Svalbard): Pore water and solid phase distributions of C, N, P and Si. Pol Biol 16:447–462

Hulthe G, Hulth S, Hall POJ (1998) Effect of oxygen on degradation rate of refractory and labile organic matter in continental margin sediments. Geochim Cosmochim Acta 62:1319–1328

Hume JD, Schalk M, Hume PW (1972) Short-term climate changes and coastal erosion, Barrow, Alaska. Arctic 25: 271–278

Humphreys WF (1979) Production and respiration in animal populations. J Anim Ecol 48:427–453

Hunkins K (1968) Geomorphic provinces if the Arctic Ocean. In: Sater JE (ed) Arctic Drifting Stations. Arctic Institute of North America, pp 365–376

Hunkins KL (1961) Seismic Studies of the Arctic Ocean Floor, Geology of the Arctic 1:645–665

Hunt GL, Baduini CL, Brodeur RD, Coyle KO, Kachel NB, Napp JM et al. (1999) The Bering Sea in 1998: The second consecutive year of extreme weather-forced anomalies. EOS Transactions, AGU 80:561, 565–566

Hurrell JW (1995) Decadal trends in the North Atlantic Oscillation: regional temperatures and precipitation. Science 269:676–679

Hurrell JW (2002) North Atlantic Oscillation (NAO) Indices Information. http: //www.cgd.ucar.edu/ ~jhurrell /nao.html

Hygum BH, Rey C, Hansen BW (2000) Growth and development rates of *Calanus finmarchicus* nauplii during a diatom spring bloom. Mar Biol 136:1075–1085

IHO (International Hydrographic Office) (2001) S-23 "Limits of Ocean and Seas". Draft 4th Edition, October 2001

Ikävalko J, Gradinger R (1997) Flagellates and heliozoans in the Greenland Sea ice studied alive using light microscopy. Pol Biol 17:473–481

Ingvaldsen R, Loeng H, Asplin L (2002) Variability in the Atlantic inflow to the Barents Sea based on a one-year time series from moored current meters. Cont Shelf Res 22:505–519.

IPCC (2001) Climate Change 2001: The Scientific Basis. In: Houghton JT, Ding Y, Griggs DJ, Noguer M, van der Linden PJ and Xiaosu D (eds) Contribution of Working Group I to the Third Assessment Report of the Intergovernmental Panel on Climate Change (IPCC). Cambridge University Press, Cambridge, UK, 944 pp

Irigoyen X, Head R, Klenke U, Meyer-Harms B, Harbour D, Niehoff B, Hirche H-J, Harris R (1998) A high frequency time series at Weathership M, Norwegian Sea, during the 1997 spring bloom: feeding of adult female *Calanus finmarchicus*. Mar Ecol Prog Ser 172:127–137

Iseki K, Macdonald RW, Carmack EC (1987) Distribution of particulate matter in the southeastern Beaufort Sea in late summer. In: NIPR Symposium on Polar Biology, vol 1. National Institute of Polar Research: Tokyo, pp 35–46

ISHTAR (1988) ISHTAR cruise report, Institute of Marine Science. University of Alaska, Fairbanks

Ittekkot V (1988) Global trends in the nature of organic matter in river suspensions. Nature 332:436–438

Ittekkot V, Haake B, Bartsch M, Nair RR, Ramaswamy V (1992) Organic carbon removal in the sea: the continental connection. In: Summerhayes CP, Prell WL, Emeis KC (eds) Upwelling systems: Evolution since the early Miocene. Geol Soc Spec Pub, The Geological Society, London, pp 167–176

Ivanov GI, Nechsheretov AV, Ivanov VV (1999) Oceanographic Investigations in the St. Anna Trough, Kara Sea. Reps Pol Res 342:27–45

Ivanov GI, Shevchenko VP, Musatov EE, Ponomarenko TV, Svertilov AA, Neshcheretov AV (1997) Multidisciplinary research on the Svyataya Anna Trough (the Kara Sea). Oceanology 37:570–572

Ivanov YA, Aibulatov NA (1998) Multidisciplinary oceanological research in the Barents Sea in Cruise 11 of R/V Akademik Sergei Vavilov (August 27–October 15, 1997). Oceanology 38:575–579

Jackson HR, Mudie PJ, Blasco SM (1985) Initial geological report on CESAR. The Canadian expedition to study the Alpha Ridge. Geol Surv Canada, paper 84-22, 177 pp

Jackson HR, Oakey GN (1990) Sedimentary thickness map of the Arctic Ocean, Plate 5. In: Grantz A, Johnson GL and Sweeney JF (eds) The Arctic Ocean Region: The Geology of North America DNAG L

Jahnke RA (1989) Flux and recycling of bioactive substances in the surface sediments of the deep basins off southern California: Progress report, November 1987–July 1988. Rep US Dep Energy, pp 1–4

Jahnke RA (1996) The global ocean flux of particulate organic carbon: Areal distribution and magnitude. Glob Biogeochem Cycles 10:71–88

Jakobsson M (1999) First high-resolution chirp sonar profiles from the central Arctic Ocean reveal erosion of Lomonosov Ridge sediments. Mar Geol 158:111–123

Jakobsson M (2002) Hypsometry of the Arctic Ocean and its constituent seas. Geochem Geophys Geosys 3(5)

Jakobsson M, and IBCAO Editorial Board Members (2001) Improvement to the International Bathymetric Chart of the Arctic Ocean (IBCAO): Updating the Data Base and the Grid Model. EOS (Transactions, American Geophysical Union) 84(47) (Abstract)

Jakobsson M, Cherkis N, Woodward J, Coakley B, and Macnab R (2000) A new grid of Arctic bathymetry: A significant resource for scientists and mapmakers. EOS (Transactions, American Geophysical Union) 81(9): 89, 93, 96

Jakobsson M, Løvlie R, Arnold EM, Backman J, Polyak L, Knutsen JO and Musatov E (2001) Pleistocene stratigraphy and paleoenvironmental variation from Lomonosov Ridge sediments, central Arctic Ocean. Glob Planet Change 31(1–4):1–21

Jasper JP, Gagosian RB (1989) Alkenone molecular stratigraphy in an oceanic environment affected by glacial freshwater events. Paleoceanography 4:603–614

Jasper JP, Gagosian RB (1993) The relation between sedimentary organic carbon isotopic composition and organic biomarker compound concentrations. Geochim Cosmochim Acta 57:167–186

Jeffrey SW, Mantoura RFC, Wright SW (eds 1997) Phytoplankton pigments in oceanography: guidelines to modern methods. UNESCO, Paris, 661 pp

Jensen HM, Pedersen L, Burmeister A, Hansen BW (1999) Pelagic primary production during summer along 65 to 72∞N off West Greenland. Pol Biol 21:269–278

Johannessen OM, Bengtsson L, Miles MW, Kuzmina SI, Semenov VA, Alekseev GV, Nagurny AP, Zakharov VF, Bobylev LP, Pettersson LH, Hasselmann K, Cattle HP (2002) Arctic climate change - Observed and Modeled Temperature and Sea Ice. NERSC Technical report No 218

Johannessen OM, Miles M, Bjorgo E (1995) The Arctic's shrinking sea ice. Nature 376: 126–127

Johannessen OM, Shalina EV, Miles MW (1999) Satellite evidence for an Arctic sea ice cover in transformation. Science 286:1937–1939

Johannessen SC, Macdonald RW, Paton DW (2003) A sediment and organic carbon budget for the greater Strait of Georgia. Est Coast Shelf Sci 56:845–860

Johnsen G, Hegseth EN (1991) Photoadaptation of sea-ice microalgae in the Barents Sea. Pol Biol 11:179–184

Johnsen G, Sakshaug E (2000) Monitoring of harmful algal blooms along the Norwegian coast using bio-optical methods. S Afr J Mar Sci 22:309–321

Johnsen G, Sakshaug E, Vernet M (1992) Pigment composition, spectral characterization and photosynthetic parameters in *Chrysochromulina polylepis*. Mar Ecol Prog Ser 83:241–249

Johnsen G, Volent Z, Tangen K, Sakshaug E (1997) Time series of harmful and benign phytoplankton blooms in northwest European waters using the Seawatch buoy system. In: Kahru M, Brown C (eds). Monitoring algal blooms: New techniques for detecting large-scale environmental change. Landes Bioscience, New York, pp 115–143

Johnsen G. Eikrem W, Dalløkken R. Legrand C, Aure J, Skjoldal HR (1999) Eco-physiology, bio-optics and toxicity of the ichthyotoxic prymnesiophyte *Chrysochromulina leadbeateri*. J Phycol 35:1465–1476

Johnson GL (1969) Morphology of the Eurasian Arctic Basin. Pol Rec 14:619–628

Johnson GL, Grantz A, and Weber JR (1990) Bathymetry and Physiography. In: Grantz A, Johnson GL and Sweeney JF (eds) The Arctic Ocean Region: The Geology of North America DNAG L, pp 63–78

Johnson GL, Milligan DB (1967) Some geomorphological observations in the Kara Sea. Deep Sea Res 14:19–28

Johnson MA, Polyakov IV (2001) The Laptev Sea as a source for recent Arctic Ocean salinity changes. Geophys Res Lett 28:2017–2020

Johnson MA, Proshutinsky AY, Polyakov IV (1999) Atmospheric patterns forcing two regimes of Arctic circulation: A return to anticyclonic conditions? Geophys Res Lett 26: 1621–1624

Jokat W (ed) (1999) The Expedition ARKTIS-XV/2 of "Polarstern" in 1999. Reps Pol Res 368, 128 pp

Jokat W, Stein R, Rachor E, Schewe I (1999) Expedition gives Fresh view of Central Arctic Geology, EOS Transactions 80:465, 472–473

Jokat W, Uenzelmann-Neben G, Kristoffersen Y, Rasmussen TM (1992) Lomonosov Ridge: A double-sided continental margin. Geology 20:887–890

Jokat W, Weigelt E, Kristoffersen Y, Rasmussen TM (1995) New insights into the evolution of Lomonosov Ridge and the Eurasia Basin. Geophys J Int 122:378–392

Jones EP (2001) Circulation in the Arctic Ocean. Pol Res 20: 139–146

Jones GA, Darby DA, Bischof JF (1995) Holocene vs last glacial maximum sedimentation in the cental Arctic Ocean. In: Programs and Abstracts, 5th International Conference on Paleoceanography, Halifax Canada, pp 162–163

Jones GA, Keigwin LD (1988) Evidence from Fram Strait (78°N) for early deglaciation. Nature 336:56–59

Jones HLJ, Leadbeater BSC, Green JC (1994) Mixotrophy in haptophytes. In: Green JC, Leadbeater BSC (eds) The haptophytes. Systematics Association special vol. Clarendon Press, Oxford

Jorgensen BB (1978) A comparison of methods for the quantification of bacterial sulfate reduction in coastal marine sediments. Geomicrobiology 1:11–27

Jørgensen LL, Pearson TH, Anisimova NA, Gullikson B, Dahle S, Denisenko SG, Matishov GG (1999) Environmental influences on benthic fauna associations of the Kara Sea (Arctic Russia). Polar Biol 22:395–416

Jumars PA, Penry, DL, Baross, AJ, Perry MJ, Frost BW (1989) Closing the microbial loop: Dissolved carbon pathway to heterotrophic bacteria from incomplete ingestion, degestion and absorption in animals. Deep-Sea Res 36: 483– 495

Junge K, Krembs C, Deming J, Stierle A, Eicken H (2001) A microscopic approach to investigate bacteria under in-situ conditions in sea-ice samples. Ann Glaciol 33: 304–310

Juterzenka KV, Knickmeier K (1999) Chlorophyll *a* in water column and sea ice during the Laptev Sea freeze-up study in autumn 1995. In: Kassens H et al (eds) Land-ocean systems in the Siberian Arctic: Dynamics and history. Springer, Berlin, pp 153–160

Kähler P, Bauerfeind E (2001) Organic particles in a shallow sediment trap: substantial loss to the dissolved phase, Limnol Oceanogr 46: 719–723

Kaplin PA, Leont'ev OK, Luk'yanova SA, Nikiforov LG (1971) Coasts. Mysl' Press, Moscow, 479 pp (in Russian)

Kaplin PA, Selivanov AO (1999) Sea-level changes and Coasts of Russia: Past, Present, Future. GEOS Press, Moscow, 299 pp (in Russian)

Karasik, AM (1968) Magnetic anomalies of Gakkel Ridge and origin of the Eurasia subbasin of the Arctic Ocean. Geophysical methods of prospecting in the Arctic 5, Leningrad, pp 8–19

Karasik, AM (1974) The Euro-Asian Basin of the North Polar Ocean from the standpoint of plate tectonics. In: NIDRA, Problems of the Geology of the polar regions of the Earth, Leningrad, pp 24–31 (in Russian)

Kashkin NI (1964) On the winter deposits of phytoplankton in the littoral. Trudy Okeanol Inst Akad Nauk USSR 65: 49–57

Kassens H, Bauch H, Dmitrenko I, Eicken H, Hubberten HW, Melles M, Thiede J and Timokhov L (eds) (1999) Land-Ocean Systems in the Siberian: Dynamics and History. Springer-Verlag, Berlin, 711 pp

Kates K, Volcani BE (1966) Lipid components of diatoms. Biochem Biophys Acta 116:264–278

Kattner G (1999) Storage of dissolved inorganic nutrients in seawater: poisoning with mercuric chloride. Mar Chem 67:61–66

Kattner G, Budeus G (1997) Nutrient status of the North East Water Polynya. J Mar Syst 10:185–197

Kattner G, Gercken G, Eberlein K (1983) Development of lipids during a spring plankton bloom in the northern North Sea. Mar Chem 65:149–162

Kattner G, Lobbes JM, Fitznar HP, Engbrodt R, Lara RJ (1999) Tracing dissolved organic substances and nutrients of Lena River through the Laptev Sea (Arctic). Mar Chem 65: 25–39

Kawamura A (1967) Observations of phytoplankton in the Arctic Ocean in 1964. Inf Bull Plank Japan comm issue Dr Y Matsue 1967:71–89

Keil RG, Tsamakis E, Fuh CB, Giddings C, Hedges JI (1994) Mineralogical and textural controls on organic composition of coastal marine sediments: hydrographic separation using SPLITT fractionation. Geochim Cosmochim Acta 57:879–893

Kemp PF (1994) Microbial carbon utilization on the continental shelf and slope during the SEEP-II experiment. Deep-Sea Res II 41 (2/3):563–581

Kempema EW, Reimnitz E, Barnes PW (1989) Sea ice sediment entrainment and rafting in the Arctic. J Sed Pet 59: 308–317

Kendall MA (1996) Are Arctic soft-sediment macrobenthic communities impoverished? Polar Biol 16:393–399

Kiene RP (1992) Dynamics of methylsulfide and dimethylsulfoniopropionate in oceanic water samples. Mar Chem 37:29–52

Kierdorf C (2001) Organischer Kohlenstoff und Biomarker in Oberflächensedimenten vom Ostgrönländischen Kontinentalhang Unpublished Diploma Thesis. Geologisches Institut der Universität Köln/Alfred Wegener Institute for Polar and Marine Research, Bremerhaven, pp 1–76

Kiørboe T (1993) Turbulence, phytoplankton cell size, and the structure of pelagic food webs. Adv Mar Biol 29:2–72

Kirk JTO (1994) Light and photosynthesis in aquatic ecosystems. Cambridge Univ Press, 509 pp

Kirkegaard JB (1982) New records of abyssal benthic polychaetes from the Polar Sea. Steenstrupia 8:253–260

Kirst GO, Wiencke C (1995) Ecophysiology of polar algae. J Phycol 31:181–199

Kiselev VI, Ignat'ev AV, Kiyashko SI, Semiletov IP (2000) Terrigenous organic matter in deposits from the eastern area of the Arctic. DAN 371(2) 220–222

Kitchell JA, Clark DL (1982) Late Cretaceous-Paleogene paleogeography and paleocirculation; Evidence of North Polar upwelling, Paleogeogr Paleoclim Paleoecol 40:135– 165

Klages M, Deubel H, Rachor E (2003) Organic carbon consumption of the Kara Sea macrozoobenthos: a first assessment. In: Stein R, Fahl K, Fütterer DK, Galimov EM, Stepanets OV (eds), Siberian River Run-off in the Kara Sea: Characterisation, Quantification, Variability, and Environmental Significance, Proceedings in Marine Sciences Vol 6, Elsevier, Amsterdam, pp 267–280

Kleiber HP, Niessen F (1999) Late Pleistocene Paleoriver Channels on the Laptev Sea shelf – Implications from Sub-Bottom Profiling. Kassens H, Bauch H, Dmitrenko I, Eicken H, Hubberten HW, Melles M, Thiede J, and Timokhov L (eds), Land-Ocean Systems in the Siberian: Dynamics and History. Springer-Verlag, Berlin, pp 657–665

Kleiber HP, Niessen F (2000) Variations of continental discharge pattern in space and time – Implications from the Laptev Sea continental margin, Arctic Siberia. Int J Earth Sci 89:605–616

Klein B, LeBlanc B, Mei Z-P, Beret R, Michaud J, Mundy C-J, Quillfeldt C H von, Garneau M-É, Roy S, Gratton Y, Cochran JK, Bélanger S, Larouche P, Pakulski JD, Rivkin RB, Legendre L (2002) Phytoplankton biomass, production and potential export in the North Water. Deep-Sea Res II 49:4983–5002

Klenova MV (1960) Geology of the Barents Sea. Ac Sci of USSR, Moscow, pp 1–365 (in Russian)

Klenova MV (1961) Recent sedimentation in the Barents Sea. In: Recent sediments of the seas and the oceans. Ac Sci of USSR, Moscow, pp 419–436 (in Russian)

Kluyev YV (1970) Thermal abrasion of near shore zone of polar seas. News of All-Russian Geographic Society Vol 102/2:129–135 (in Russian)

Knauer GA, Martin JH, Bruland KW (1979) Fluxes of particulate carbon, nitrogen and phosphorus in the upper water column of the northeast Pacific. Deep-Sea Res 26: 97–108

Knies J (1994) Spätquartäre Sedimentation am Kontinentalhang nordwestlich Spitzbergens. Der letzte Glazial/Interglazial-Zyklus. Unpublished Diploma Thesis. Justus-Liebig-Universität, Gießen, pp 1–95

Knies J, Kleiber HP, Matthiessen J, Müller C, Nowaczyk N (2001) Marine ice-rafted debris records constrain maximum extent of Saalian and Weichselian ice sheets along the Eurasian margin. Glob Planet Change 31:45–64

Knies J, Müller C, Nowaczyk N, Stein R (2000) A multiproxy approach to reconstruct the environmental changes along the Eurasian continental margin over the last 160,000 years. Mar Geol 163:317–344

Knies J, Stein R (1998) New aspect of organic carbon deposition and its paleoceanographic implications along the northern Barents Sea margin during the last 30,000 years. Paleoceanography 13(4):384–394

Knies J, Vogt C, Stein R (1999) Late Quaternary growth and decay of the Svalbard/Barents Sea ice sheet and paleoceanographic evolution in the adjacent Arctic Ocean. Geo-Mar Lett 18:195–203

Kobari T, Ikeda T (2001) Life cycle of *Neocalanus flemingeri* (Crustacea: Copepoda) in the Oyashio region, western subarctic Pacific, with notes on its regional variations. Mar Ecol Prog Ser 209:243–255

Kodina LA (2001) The carbon isotope composition of phytoplankton along the Ob- Kara Sea transect in August-September 1999. Reps Pol Mar Res 393:157–160

Kodina LA, Tokarev VG, Vlasova LN, Korobeinik GS (2003) Contribution of biogenic methane to ikaite formation in the Kara Sea: evidence from the stable carbon isotope geochemistry. In: Stein R, Fahl K, Fütterer DK, Galimov EM, Stepanets OV (eds) Siberian River Run-off in the Kara Sea: Characterisation, Quantification, Variability, and Environmental Significance, Proceedings in Marine Sciences Vol 6, Elsevier, Amsterdam, pp 349–374

Kodina LA and Peresypkin VI, 2002. Stable carbon isotope ratio and lignin in surface sediments of the inner Kara Sea. Reps Pol Mar Res 419:134–142

Köhler H, Meon B, Gordeev VV, Spitzy A, Amon RMW (2003). Dissolved organic matter (DOM) in the rivers Ob and Yenisei and the adjacent Kara-Sea. In: Stein R, Fahl K, Fütterer DK, Galimov EM, Stepanets OV (eds) Siberian River Run-off in the Kara Sea: Characterisation, Quantification, Variability, and Environmental Significance, Proceedings in Marine Sciences Vol 6, Elsevier, Amsterdam, pp 281–308

Kohly A (1998) Diatom flux and species composition in the Greenland Sea and the Norwegian Sea in 1991–1992. Mar Geol 145:293–312

Koike I, Ogawa H, Nagata T, Fukuda R (2001) Silicate to nitrate ratio of the upper sub-arctic Pacific and the Bering Sea Basin in summer: Its implication for phytoplankton dynamics. J Oceanogr 57:253–260

Kolatschek JS (1998) Sea-ice dynamics and sediment transport in the Arctic: results from field measurements, remote sensing and modelling (in German). Ph D Thesis, University of Bremen, Germany

Korchagina YuI, Chetverikova OP (1976) Methods of studies into dispersed organic matter of the sedimentary rock. Nauka, Moscow, pp 1–229

Koreisha MM, Levantovskaya NP, Brushkov AV (1997) Cryogenic geosystems of the Kara coast and trends of their development. In: Fundamental investigations of the Earth Cryosphere. Nauka Press, Novosibirsk: 142–149 (in Russian)

Kosheleva VA, Yashin DS (1999) The bottom sediments of the Russia Arctic Seas. MNRRF, VNIIOkeangeologia, St-Petersburg, pp 1–286

Kosobokova KN, Hanssen H, Hirche H-J, Knickmeier K (1998) Composition and distribution of zooplankton in the Laptev Sea and adjacent Nansen Basin during summer, 1993. Pol Biol 1998 19:63–76

Kostka JE, Thamdrup B, Glud RN, Canfield DE (1999) Rates and pathways of carbon oxidation in permanently cold Arctic sediments. Mar Ecol Prog Ser 180:7–21

Kraus M, Matthiessen J, Stein R (2003) A high-resolution holocene marine pollen record from the northern Yenisei Estuary (southeastern Kara Sea) and paleoenvironmental implications. In: Stein R, Fahl K, Fütterer DK, Galimov EM, Stepanets OV (eds) Siberian River Run-off in the Kara Sea: Characterisation, Quantification, Variability, and Environmental Significance, Proceedings in Marine Sciences Vol 6, Elsevier, Amsterdam, pp 435–456

Krause G, Schauer U (eds) (2001) The Expeditions ARKTIS XVI/1 and ARKTIS XVI/2 of the Research Vessel "Polarstern" in 2000. Reps Pol Mar Res 389, 108 pp

Krishnamurthy RV, Machavaram M, Baskaran M, Brooks JM, Champ MA (2001). Organic carbon flow in the Ob, Yenisei rivers and Kara Sea of the Arctic Region. Mar Poll Bull 42:726–732

Kristiansen S, Farbrot T, Wheeler PA (1994) Nitrogen cycling in the Barents Sea – Seasonal dynamics of new and regenerated production in the marginal ice zone. Limnol Oceanogr 39:1630–1642

Kristiansen S, Lund BAA (1989) Nitrogen cycling in the Barents Sea – I. Uptake of nitrogen in the water column. Deep-Sea Res I 36:255–268

Kristoffersen Y (1990) Chapter 21: Eurasia Basin: in The Arctic ocean region, v. L, (eds) Grantz A, Johnson L and Sweeney JF, Boulder, Colorado, Geol Soc Amer, 365–378

Kröncke I (1994) Macrobenthic composition, abundance and biomass in the Arctic Ocean along a transect between Svalbard and the Makarov basin. Polar Biol 14:519–529

Kröncke I (1998) Macrofauna communities in the Amundsen Basin, at the Morris Jesup Rise and at the Yermak Plateau (Eurasian Arctic Ocean). Polar Biol 19:383–392

Kröncke I, Tan TL, Stein R (1994) High benthic bacteria standing stock in deep Arctic basins. Polar Biol 14:423–428

Kucheruk NV, Rybinkov PV, Sapoznikov FV (1999) Coastal benthic community of the Pechora Sea. Proc 12th Int Union of geodesy and geophysics General Assembly B:233

Kudelin BI (ed) (1975) A map of underground discharge in the USSR territory. Scale 1 : 2 500 000. Moscow-Leningrad, GUK (Glavnoe Upravlenie Kartografii) (in Russian)

Kudoh S, Robineau B, Suzuki Y, Fujiyoshi Y, Takahashi M (1997) Photosynthetic acclimation and the estimation of temperate ice algal primary production in Saroma-ko Lagoon, Japan. J Mar Syst 11:93–109

Kulikov NN, Lapina NN, Ivanov GI (1999) The history of investigation of bottom sediments of the western Arctic Shelf: the Barents and Kara seas. Reps Pol Res 342:3–10

Kunz-Pirrung M (1999) Distribution of aquatic palynomorphs in surface sediments from the Laptev Sea, Eastern Arctic Ocean. In: Kassens H, Bauch HA, Dmitrenko I, Eicken H, Hubberten HW, Melles M, Thiede J, Timikhov L (eds) Land-Ocean Systems in the Siberian Arctic: Dynamics and History. Springer, Berlin, pp 561–575

Kupriyanova EK, Jirkov IA (1997): Serpulidae (Annelida, Polychaeta) of the Arctic Ocean. Sarsia 82:203–236

Kupriyanova YK (1993a) Deep-water Serpulidae (Annelida, Polychaeta) from Kurile-Kamchatka trench 2. Genera *Bathyditrupa*, *Bathyvermilia*, and *Protis*. [Serpulidae (Annelida, Polychaeta)]. Zoologicheskiy Zhurnal 72: 21–28

Kupriyanova YK (1993b) Deep-water Serpulidae (Annelida, Polychaeta) from the Kuril-Kamchatka trench. 1. Genus *Hyalopomatus*. [Serpulidae (Annelida, Polychaeta)]. Zoologicheskiy Zhurnal 72:145–152

Kuptsov VM, Lisitzin AP (1996) Radiocarbon of Quaternary along shore and bottom deposits of the Lena and the Laptev Sea sediments. Mar Chem 53:301–312

Kuusk VV, Pronin AA, Radionov VF, Savchenko AV, Shevchenko VP, Smirnov VV, Vinogradova AA (1996) The short- and long range variations of dispersal and chemical composition of Arctic aerosols. In: Kulmala M (ed) Nucleation and Atmospheric Aerosols. Proceedings of 14-th International Conference. Helsinki, Finland, 26–30 August 1996. Pergamon Press, pp 546–549

Kuznetsov LL (1991) Production of microphytobenthos of the coastal zone of the Barents Sea. In: Production-destruction processes of pelagic zone of the Barents Sea shore. Apatity, pp 55–62 (in Russian)

Kuznetsov LL, Baitaz ON, Makarevich PR (1998) Structural-functional characteristics of the plankton community in the Baidaratskaya Bay by materials of the autumn 1991–1992 expeditions. In: Biology and Oceanography of the Kara and Barents sea (by Sevmorway). Apatiti, pp 88–95 (in Russian)

Lammers RB, Shiklomanov AI, Vörösmarty CJ, Fekete BM, Peterson BJ (2001) Assessment of contemporary Arctic river runoff based on observational discharge records. J Geophys Res 106:3321–3334

Lampitt RS (1985) Evidence for the seasonal deposition of detritus to the deep-sea floor and its subsequent resuspension. Deep-Sea Res 32:885–897

Lampitt RS, Antia AN (1997) Particle flux in deep-seas: regional characteristics and temporal variability. Deep-Sea Res I 44:1377–1403

Lampitt RS, Billett DSM, Rice AL (1986) Biomass of the invertebrate megabenthos from 500 to 4100m in the northeast Atlantic Ocean. Mar Bio 93:69–81

Langleben MP (1969) Albedo and degree of puddling of a melting cover of sea ice. J Glaciol 8:407–412

Langrock U, Stein R, Lipinski M, Brumsack H (2003a) Paleoenvironment and sea-level change in the early Cretaceous Barents Sea – implications from near-shore marine sapropels. Geo-Mar Lett, in press

Langrock U, Stein R, Lipinski M, Brumsack H (2003b) Late Jurassic to Early Cretaceous blak shale formation and paleoenvironment in high northern latitudes – examples from the Norwegian-Greenland-Seaway. Paleoceanography, in press

Lapina NN, Belov NA (1961) Features of sedimentation in the Arctic Ocean. In: Recent sediments of the seas and oceans. Ac Sci of USSR, Moscow, pp 86–97 (in Russian)

Lara RJ, Rachold V, Kattner G, Hubberten HW, Guggenberger G, Skoog A, Thomas DN. (1998) Dissolved organic matter and nutrients in the Lena River, Siberian Arctic: characteristics and distribution. Mar Chem 59:301–309

Larson RL (1991) Geological consequences of superplumes. Geology 19:963–966

Laws EA, Falkowski PG, Smith WO Jr, Ducklow H, McCarthy JJ (2000) Temperature effects on export production in the open ocean. Glob Biogeochem Cycles 14:1231–1246

Lawver LA, Müller RD (1994) Iceland hotspot track, Geology 22:311–314

Lawver LA, Müller RD, Srivastava SP, Roest W (1990) The opening of the Arctic Ocean. In: Bleil U, Thiede J (eds) Geological History of the Polar Oceans: Arctic versus Antarctic, NATO ASI Series C 308, pp 29–62

Lawver LA, Scotese CR (1990) A review of tectonic models for the evolution of the Canadian Basin. In: Grantz A, Johnson L and Sweeney JF (eds) The Arctic Ocean Region, The Geology of North America, Vol L, pp 593–618

Lawy EA, Popp BN, Bidigare RR, Kennicutt MC, Macko SA (1995) Dependence of phytoplankton carbon isotopic composition on growth rate and CO_{2aq}: theoretical considerations and experimental results. Geochim Cosmochim Acta 59:1131–1138

Leck C, Bigg EK, Covert DS (1996) Overview of the atmospheric research program during the International Arctic Ocean Expedition of 1991 (IAOE-91) and its scientific results. Tellus 48:131–155

Legendre L, Ackley SF, Dieckmann GS, Gulliksen B, Horner R, Hoshiai T, Melnikov IA, Reeburgh WS, Spindler M, Sullivan CW (1992) Ecology of sea ice biota. 2 Global significance. Pol Biol 12:429–444

Leith TL, Weiss HM. Mørk A, Arhus N, Elvebakk G, Embry AF, Brooks PW, Stewart KR, Pchelina TM, Bro EG, Verba ML, Danyushevskaya A, Borisov AV (1992) Mesozoic hydrocarbon source-rocks of the Arctic region. In: Vorren TO et al. (eds) Arctic Geology and Petroleum Potential. NPF Spec Publ 2:1–25

Leo RF, Barghorn ES (1970) Phenolic aldehydes: generation from fossil woods and carbonaceous sediments by oxidative degradation. Science 168:582–584

Levi BG (2000) The decreasing Arctic ice cover. Physics Today January 2000:19–20

Levitan MA, Arnold M, Bourtman MV, Ivanova EV, Marina MM (2000a) Link of Holocene marine and terrestrial records in the east Kara Sea. IMAGES Workshop "Correlation of post-glacial climatic events in the Arctic and Far East seas, IO RAS, Moscow, 1:15–16

Levitan MA, Dupsley J-C, Khusid TA, Belyaev NA, Bourtman MV (2000) Holocene sediments of the southern Novaya Zemlya Trough (the Pechora Sea) and brines history. IMAGES Workshop "Correlation of post-glacial climatic events in the Arctic and Far East seas". IORAS, Moscow 1:11–12

Levitan MA, Kuptsov VM, Romankevich EA, Kondratenko AV (2000) Some indication for late Quaternary Pechora River discharge: results of vibrocore studies in the southeastern Pechora Sea. Int J Earth Sci 89:533–540

Levitan MA, Khusid TA, Kuznetsov VM, Politova NV and Pavlova GA (1995) Sequence types of Upper Quaternary deposits of the Kara Sea. Oceanology 34:710–721

Lewis EL (2000) Introduction. In: Lewis EL (ed) The Freshwater Budget of the Arctic Ocean. NATO, ASI Series, pp xv–xxi

Lewis EL Ponton D, Legendre L, Leblanc B (1996) Springtime sensible heat, nutrients and phytoplankton in the Northwater Polynya, Canadian Arctic. Cont Shelf Res 16: 1775–1792

Lewis EL, Jones EP, Lemke P, Prowse TD, Wadhams P (2000) The Freshwater Budget of the Arctic Ocean. Kluwer Academic Publishers, London

Lindemann F (1994) Sonographische und sedimentologische Untersuchungen in der Laptevsee, sibirische Arktis. Master thesis, University of Kiel, 75 pp

Lindemann F (1998) Sedimente im Arktischen Meereis – Eintrag, Charakterisierung und Quantifizierung. Reps Pol Res 283, 124 pp

Lindemann RL (1942) The trophic-dynamic aspect of ecology. Ecology 23:399–418

Lipiatou E, Albaiges J (1994) Atmospheric deposition of hydrophobic organic chemicals in the northwestern Mediterranean Sea: comparison with the Rhone river input. Mar Chem 46:153–164

Lisitzin AP (1995) The marginal filter of the ocean. Oceanology 34:583–590

Lisitzin AP (1996) Oceanic Sedimentation. Lithology and Geochemistry. American Geophysical Union, Washington, DC, 400 pp

Lisitzin AP (ed) (1990) Organic matter of bottom deposits in polar zones of the World Ocean. Nedra Press, Leningrad, 280 pp (in Russian)

Lisitzin AP, Shevchenko VP, Burenkov VI (2000) Hydrooptics and suspended matter of Arctic seas. Atmos Oceanic Opt 13:61–71

Lisitzin AP, Shevchenko VP, Vinogradov ME, Severina OV, Vavilova VV, Mitskevich IN (1995) Particle fluxes in the Kara Sea and Ob and Yenisey estuaries. Oceanology 34: 683–693

Lisitzin AP, Vinogradov ME (1995) International high-latitude expedition in the Kara Sea (the 49th cruise of the R/V Dmitriy Mendeleev). Oceanology 34 (5):583–590

Liu KK, Iseki K, Chao SY (2000) Continental margin carbon fluxes. In: Hanson RB, Ducklow HW, Field JG (eds) The Changing Ocean Carbon Cycle, Cambridge University Press, Cambridge, pp 187–239

Lobbes JM, Fitznar HP, Kattner G (2000) Biogeochemical characteristics of dissolved and particulate organic matter in Russian rivers entering the Arctic Ocean. Geochim Cosmochim Acta 64:2973–2983

Loeng H (1989) Ecological features of the Barents Sea In: Rey L, Alexander V (eds) Proceedings of the 6th conference of the Comité Arctique International, Leiden, Brill, pp 327–365

Loeng H, Ozhigin V, Ådlandsvik B (1997) Water fluxes through the Barents Sea. ICES J Mar Sci 54: 310–317

Longhurst A, Sathyendranath S, Platt T, Caverhill C (1995) An estimate of global primary production in the ocean from satellite radiometer data. J Plankton Res 17:1245–1271

Lopatin GV (1952) Sediment load of the USSR Rivers. Geography, Moscow, 368 pp (in Russian)

Lopatin VN (1999) Global climate changes and economical activity in Russia. In: Ecological and socio-economic aspects of the development of Russia during global changes of environment and climate. GEOS Press, Moscow, pp 27–35 (in Russian)

Lopez P, Schmith T, Kaas E (2000) Sensitivity of the Northern Hemisphere circulation to North Atlantic SSTs in the ARPEGE Climate AGCM. Clim Dyn 16:535–547

Loughlin TR, Kiyotaka, O (eds) (1999) Dynamics of the Bering Sea. PICES/Univ Alaska Fairbanks, pp 1–825

Loughlin TR, Ohtani K (eds 1999) Dynamics of the Bering Sea. Univ Alaska Sea Grant AK-SG-99-03, Univ Alaska Fairbanks, 825 pp

Loughlin TR, Sukhanova IN, Sinclair EH, Herrero RC (1999) Summary of biology and ecosystem dynamics in the Bering Sea. In: Loughlin TR, Ohtani K (eds) Dynamics of the Bering Sea. Univ Alaska Sea Grant AK-SG-99-03, Univ Alaska Fairbanks, pp 387–407

Lozán JL, Graßl H, Hupfer P (2001) Climate of the 21st century: Changes and Risks Office: Wissenschaftliche Auswertungen, Hamburg, pp 1–449

Lubinski D, Korsun S, Polyak L, Forman SL, Lehman SJ, Herlihy FA, Miller GH (1996) The last deglaciation of the Franz Victoria Trough, northern Barents Sea. Boreas 25: 89–100

Lubinski DJ, Polyak L, Forman SL, 2001. Freshwater and Atlantic water inflows to the deep northern Barents and Kara seas since ca 13 ^{14}C ka: foraminifera and stable isotopes. Quat Sci Rev 20:1851–1879

Luchetta A, Lipizer M, Socal G (2000) Temporal evolution of primary production in the central Barents Sea. J Mar Syst 27:177–193

Lugwig W, Probst JL, Kempe S (1996) Predicting the oceanic input of organic carbon by continental erosion. Glob Biogeochem Cycles 10:23–41

Lukashin VN, Ljutsarev SV, Krasnyuk AD, Shevchenko VP, Rusakov VY (1999) Suspended particulate matter in the Ob and Yenisei estuaries. Reps Pol Res 300:155–178

Lukashin VN, Rusakov VY (1998) Sampling of aerosols, suspended matter, settled material and bottom sediments. Reps Pol Res 266:36–38

Lundberg L, Haugan P (1996) A Nordic Seas-Arctic Ocean carbon budget from volume flows and inorganic carbon data. Glob Biogeochem Cycles 10:493–510

Luther GW, Sundby B, Lewis BL, Brendel PJ, Silverberg N (1997) Interactions of manganese with the nitrogen cycle: Alternative pathways to dinitrogen. Geochim Cosmochim Acta 61:4043–4052

Macdonald RW (1996) Awakenings in the Arctic. Nature 380: 286–287

Macdonald RW (2000) Arctic estuaries and ice: a positive-negative estuarine couple. In: Lewis EL (ed) The Freshwater Budget of the Arctic Ocean. NATO, ASI Series, pp 383–407

Macdonald RW, Bewers JM (1996) Contaminants in the arctic marine environment: priorities for protection. ICES J Mar Sci 53:537–563

Macdonald RW, Carmack EC (1991) Age of Canada Basin deep waters: a way to estimate primary production for the Arctic Ocean. Science 254:1348–1350

Macdonald RW, Carmack EC, McLaughlin FA, Falkner KK, Swift JH (1999) Connections among ice, runoff and atmospheric forcing in the Beaufort Gyre. Geophys Res Lett 26:2223–2226

Macdonald RW, Carmack EC, Wallace DWR (1993) Tritium and radiocarbon dating of Canada Basin deep waters. Science 259:103–104

Macdonald RW, McLaughlin FA, Carmack EC (2002) Freshwater and its sources during the SHEBA drift in the Canada Basin of the Arctic Ocean. Deep-Sea Res I 49:1769–1785

Macdonald RW, Paton DW, Carmack EC, Omstedt A (1995) The freshwater budget and under-ice spreading of Mackenzie River water in the Canadian Beaufort Sea based on salinity and $^{18}O/^{16}O$ measurements in water and ice. J Geophys Res 100:895–919

Macdonald RW, Soloman SM, Cranston RE, Welch HE, Yunker MB, Gobeil C (1998) A sediment and organic carbon budget for the Canadian Beaufort Shelf. Mar Geol 144: 255–273

Macdonald RW, Thomas DJ (1991) Chemical interactions and sediments of the western Canadian Arctic Shelf. Cont Shelf Res 11:843–863

Macdonald RW, Wong CS, Erickson PE (1987) The distribution of nutrients in the southeastern Beaufort Sea: Implications for water circulation and primary production. J Geophys Res 92:2939–2952

Mackas DL, Tsuda A (1999) Mesozooplankton in the eastern and western subarctic Pacific: community structure, seasonal life histories, and interannual variability. Prog Oceanogr 43:335–363

Mackay JR (1963) Notes on shoreline recession along the coast of the Yukon territory. Arctic 16:195–197

Macklin SA (2001) Physical and biological time series originating from or used by fisheries oceanography coordinated investigations (FOCI) in the North Pacific Ocean and Bering Sea. In: Alexander V, Bychkov AS, Livingston P, McKinnell SM (eds) Proceedings of PICES/CoML/IPRC workshop on impact of climate variability on observation and prediction of ecosystem and biodiversity changes in the North Pacific. PICES Secretariat, Sidney, BC, pp 64–74

Magritsky DV (2001) Natural and anthropogenic changes of hydrological regime in the lower reaches and mouths of the largest rivers in the East Siberia. Ph D Thesis, Moscow State University, Moscow (in Russian)

Maier-Reimer E, Mikolajewicz U, Winguth A (1996) Future ocean uptake of CO_2 – Interaction between ocean circulation and biology. Climate Dynamics 12:711–721

Maita Y, Yanada M, Takahashi T (1999) Seasonal variation in the process of marine organism production based on downward fluxes of organic substances in the Bering Sea. In: Loughlin TR, Ohtani K (eds) Dynamics of the Bering Sea. Univ Alaska Sea Grant AK-SG-99-03, Univ Alaska Fairbanks, pp 341–352

Makarevich PR (1997) Phytoplankton of the Kara Sea. In: Matishov GG (ed) Plankton of the Seas of the Western Arctic Apatity, pp 51–65 (in Russian)

Makarov VN (1998) Resources of the trade algae. In: Trade and perspective for use algae and invertebrates of the Barents and White seas. Apatity, pp 257–276 (in Russian)

Makarov VN, Shoshina EV (1986) Algae-macrophytes of the Barents Sea. In: Life and conditions of its existence in the pelagic zone of the Barents Sea. Apatity, pp 52–66 (in Russian)

Maldonado MT, Price NM (1999) Utilization of iron bound to strong organic ligands by phytoplankton communities in the subarctic Pacific Ocean. Deep-Sea Res II 46:2447–2473

Maltseva AV (1980) Mean perennial discharge of organic substances from the territory of the USSR and its temporal variation. Hydrochemical Materials, Gidrometeoizdat, Leningrad 68:14–21 (in Russian)

Maltseva AV, Tarasov MN, Smirnov MP (1978) Estimates of the average multi-year influx of organic substances into the seas and oceans from the territory of the USSR. In: Mikhailov VN, Lupachev YuV (eds) Hydrology and hydrochemistry of sea and river-mouth areas. Tr. GOIN 142. Gidrometeoizdat, Moscow, pp 116–121 (in Russian)

Maltseva AV, Tarasov MN, Smirnov MP (1987) Organic matter discharge from the USSR territory. Hydrochem. Materials 102, Gidrometeoizdat, Leningrad, 130 pp (in Russian)

Manabe S, Stouffer RJ (1979) A CO_2 climate sensitivity study with a mathematical model for the global climate. Nature 282:491–492

Mangerud J, Astakhov VI, Svendsen JI (2002) The extent of the Barents-Kara ice-sheet during the Last Glacial Maximum. Quat Sci Rev 21:111–119

Mangerud J, Bolstad M, Elgersma A, Helliksen D, Landvik JY, Lønne I, Lycke AK, Salvigsen O, Sandahl T, Svendsen JI (1992) The last glacial maximum on Spitsbergen, Svalbard. Quat Res 38:1–31

Mangerud J, Dokken T, Hebbeln D, Heggen B, Ingólfsson Ó, Landvik JI, Mejdahl V, Svendsen JI, Vorren TO (1998) Fluctuations of the Svalbard-Barents Sea ice sheet during the last 150 000 years. Quat Sci Rev 17:11–42

Mangerud J, Gulliksen S (1975) Apparent radiocarbon ages of recent marine shells from Norway, Spitsbergen, and Arctic Canada. Quat Res 5:273–296

Mangerud J, Svendsen JI (1992) The Last Interglacial-Glacial Period on Spitsbergen, Svalbard. Quat Sci Rev 11:633–664

Manley O (1995) Branching of Atlantic Water within the Greenland-Spitsbergen passage: an estimate of recirculation. J Geophys Res 100 (C10):20 627–20 634

Manley TO, Smith WO (1994) Transport of biogenic matter to depth within the Greenland Sea. Geophys Monogr 85: 223–236

Margalef R (1978) Life-forms of phytoplankton as survival alternatives in an unstable environment. Oceanol Acta 1: 493–509

Martens CS, Klump JV (1984) Biogeochemical cycling in an organic rich coastal marine basin. 4. An organic carbon budget for sediments dominated by sulfate reduction and methanogenesis. Geochim Cosmochim Acta 48:1987–2004

Martin JH, Knauer GA, Karl DM, Broenkow WW (1987) VERTEX: carbon cycling in the northeast Pacific. Deep-Sea Res 34:267–285

Martin JM, Guan DM, Elbaz-Poulichet F, Thomas AJ, Gordeev VV (1993) Preliminary assessment of the distributions of some trace elements (As, Cd, Cu, Fe, Pb and Zn) in a pristine aquatic environment: the Lena River estuary (Russia). Mar Chem 43: 185–199

Maslanik JA, Serreze MC, Agnew T (1999) On the record reduction in 1998 western Arctic sea-ice cover. Geophys Res Lett 26:1905–1908

Maslanik JA, Serreze MC, Barry RG (1996) Recent decreases in Arctic summer ice cover and linkages to atmospheric circulation anomalies. Geophys Res Lett 23:1677–1680

Maslowski W, McLean J, Newton R, Schlosser P (1998) Modeling the interannual variability of the Arctic Ocean and sea ice circulation. Eos (supplement) 79:F414

Maslowski W, Newton B, Schlosser P, Semtner A, Martinson D (2000) Modeling recent climate variability in the Arctic Ocean. Geophys Res Lett 27:3743–3746

Matrai PA, Cooper DJ, Saltzman ES (1996) Frontal enhancement of dimethylsulfide concentrations across a Gulf Stream meander. J Mar Syst 7:1–8

Matrai PA, Vernet M (1997) Dynamics of the vernal bloom in the marginal ice zone of the Barents Sea: Dimethyl sulfide and dimethylsulfoniopropionate budgets. J Geophys Res 102(C10):22965–22979

Matsumoto E (1975) ^{234}Th-^{238}U radioactive disequilibrium in the surface layer of the ocean. Geochim Cosmochim Acta 39:205–212

Matthews JBL, Heimdal BR (1980) Pelagic productivity and food chains in fjord systems. In: Freeland HJ, Farmer DM, Levings CD (eds) Fjord Oceanography. Plenum, New York, pp 377–398

Matthews JBL, Hestad L, Bakke JLW (1978) Ecological studies in Korsfjorden, western Norway. The generations and stocks of *Calanus hyperboreus* and *C. finmarchicus* in 1971–1974. Oceanol Acta 1:277–184

Matthiessen J, Kunz-Pirrung M, Mudie PJ (2000) Freshwater chlorophycean algae in recent marine sediments of the Beaufort, Laptev and Kara Seas (Arctic Ocean) as indicators of river runoff. Int J Earth Sci 89:470–485

Matthiessen J, Stepanets OV (eds) (1998) Scientific Cruise Report of the Kara Sea Expedition of RV "Akademik Boris Petrov" in 1997. Reps Pol Res 266, 102 pp

Matthiessen J, Stepanets OV, Stein R, Fütterer DK, Galimov EM (eds) (1999) The Kara Sea Expedition of RV "Akademik Boris Petrov" in 1997: First Results of a joint Russian-German Pilot Study. Reps Pol Res 300, 239 pp

Matthiessen J, Knies J, Nowaczyk N and Stein R (2001) Late Quaternary dinoflagellate cyst ecostratigraphy along the Eurasian Continental Margin (Arctic Ocean): Variability of Atlantic water inflow in the last 150,000 years. Glob Plan Change 31/1-4: 65–86

Mauritzen C (1996a) Production of dense overflow waters feeding the North Atlantic across the Geenland-Scotland Ridge. Part 2: An inverse model. Deep-Sea Res I 43: 807–835

Mauritzen C (1996b) Production of dense overflow waters feeding the North Atlantic across the Greenland-Scotland ridge. Part 1: Evidence for a revised circulation scheme. Deep-Sea Res I 43:769–806

Mayer LM (1994) Surface area control of organic carbon accumulation in continental shelf sediments. Geochim Cosmochim Acta 58:1271–1284

Mayer M, Piepenburg D (1996) Epibenthic community patterns on the continental slope off East Greenland at 75°N. Mar Ecol Prog Ser 143:151–164

Mayer PA, Eadie BJ (1993) Sources, degradation and recycling of organic matter associated with sinking particles in Lake Michigan. Org Geochem 20:47–56

Maykut GA, Grenfell TC, 1975 The spectral distribution of light beneath first-year sea ice in the Arctic Ocean. Limnol Oceanogr 20:554–563

Maykut GA, Untersteiner N (1971) Some results from a time-dependent thermodynamic model of sea-ice. J Geophys Res 76:1550–1575

McCarthy M, Hedges J, Benner R (1996) Major biochemical composition of dissolved high molecular weight organic matter in seawater. Mar Chem 55:281–297

McCartney M (1997) Is the ocean at the helm? Nature 388: 520–522

McGhee R (1996) Ancient People of the Arctic. UBC Press in association with Canadian Museum of Civilization, Vancouver, pp 1–244

McKenna MC (1980) Eocene paleolatitude, climate, and mammals of Ellesmere Island. Palaeogeography, Palaeoclimatology, Palaeoecology 30(3–4):349–362

McLaren AS, Serreze MC, Barry RG (1987) Seasonal variations of sea ice motion in the Canada basin and their implications. Geophys Res Lett 14:1123–1126

McLaughlin FA, Carmack EC, Macdonald RW, Bishop JKB (1996) Physical and geochemical properties across the Atlantic/Pacific water mass boundary in the southern Canadian Basin. J Geophys Res 101 (C1):1183–1197

McPhee MG, Stanton TP, Morison JH, Martinson DG (1998) Freshening of the upper ocean in the Arctic: Is perennial sea ice disappearing? Geophys Res Lett 25:1729–1732

McRoy CP, Goering JJ, Shields WW (1972) Studies of primary production in the eastern Bering Sea. In: Takenouti AY et al (eds), Biological oceanography of the northern North Pacific Ocean, Motoda commemorative Vol. 3, Idemitsu Shoten, Tokyo, pp 199–216

Medlin LK, Barker GLA, Campbell L, Green JC, Hayes PK, Marie D, Wrieden S, Vaulot D (1996) Genetic characterization of *Emiliania huxleyi*. J Mar Syst 9:13–31

Medvedev VS (1972) Lithodynamics and sediment balance of the northern part of the White Sea. In: Evolutional processes and research methods of the sea coasts. Nauka Press, Moscow, pp 27–53 (in Russian)
Melle W, Skjoldal HR (1998) Reproduction and development of *Calanus finmarchicus*, *C. glacialis* and *C. hyperboreus* in the Barents Sea. Mar Ecol Prog Ser 169:211–228
Melling H (2000) Exchanges of freshwater through the shallow straits of the North American Arctic. In: Lewis EL, Jones EP, Lemke P, Prowse TD, Wadhams P (eds) The Freshwater Budget of the Arctic Ocean. Kluwer Academic Publishers, Netherlands, pp 479–502
Melling H, Lewis EL (1982) Shelf drainage flows in the Beaufort Sea and their effect on the Arctic Ocean pycnocline. Deep Sea Res 29:967–985
Melnikov IA (1989) Ecosystem of the arctic sea ice. Ac Sci of USSR, Moscow, pp 1–192 (in Russian)
Melnikov IA (1997) The Arctic ice ecosystem. Gordon and Breach Science Publishers, the Netherlands, 204 p
Melnikov IA, Kolosova EG, Welch HE, Zhitina LS (2002) Sea ice biological communities and nutrient dynamics in the Canadian Basin of the Arctic Ocean. Deep-Sea Research I 9:1623–1649
Menard HW and Smith SM (1966) Hypsometry of Ocean Basin Provinces. J Geophys Res 71:4305–4325.
Menzies RJ, George RY, Rowe GT (1973) Abyssal Environment and Ecology of the World Oceans, New York: Wiley-Interscience.
Metz JG, Roessler P, Facciotti D, Levering C, Dittrich F, Lassner M, Valentine R, Lardizabal K, Domergue F, Yamada A, Yazawa K, Knauf V, Browse (2001) Production of polyunsaturated fatty acids by polyketide synthases in both prokaryotes and eukaryotes. Science 293:290–293
Meybeck M (1982) Carbon, nitrogen, and phosphorous transport by world rivers. Am J Sci 282:401–450
Meybeck M (1993) C, N, P and S in rivers: from sources to global inputs. In: Wollast, R Mackenzie FT, Chou L (eds) Interactions of C, N, P and S. Biogeochemical Cycles and Global Change. Springer, Berlin, pp 163–194
Meyers PA (1994) Preservation of elemental and isotopic source identification of sedimentary organic matter. Chem Geol 144:289–302
Meyers PA (1997) Organic geochemical proxies of paleoceanographic, paleolimnologic, and paleoclimatic processes. Org Geochem 27:213–250
Meyers PA, Ishiwatari R (1993) Lacustrine organic geochemistry – an overview of indicators of organic matter sources and diagenesis in lake sediments. Org Geochem 20:867–900
Michaelis W, Ittekkot V, Degens ET (1986) River inputs into the oceans. In: Lasserre P, Martin JM (eds) Biogeochemical Processes at the Land-Sea Boundary. Elsevier, Amsterdam, pp 37–52
Michel C, Gosselin M, Nozais C (2002) Preferential sinking export of biogenic silica during the spring-summer in the North Water Polynya (northern Baffin Bay): temperature or biological control? J Geophys Res 107 (C7):3064, doi: 10.1029/2000JC000408
Michel C, Legendre L, Ingram RG, Gosselin M, Levasseur M (1996) Carbon budget of sea-ice algae in spring-evidence of a significant transfer to zooplankton grazers. J Geophys Res 101:18345–18360
Michel C, Legendre L, Taguchi S (1997) Coexistence of microalgal sedimentation and water column recycling in seasonally ice-covered ecosystem (Saroma-ko Lagoon, Sea of Okhotsk, Japan). J Mar Syst 11:133–148
Miller GH, Geirsdottir A, Koerner RM (2001) Climate implications of changing arctic sea ice. EOS Transactions, AGU 82:97, 103
Miller W, Zepp R (1995) Photochemical production of dissolved inorganic carbon from terrestrial organic matter: Significance to the oceanic organic carbon cycle. Geophys Res Lett 22:417–420
Miller WL, Moran MA, Sheldon WM, Zepp RG, Opsahl S (2002) Determination of apparent quantum yield spectra for the formation of biologically labile photoproducts. Limn Oceanogr 47:343–352
Milliman JD, Syvitski JPM (1992) Geomorphic/tectonic control of sediment discharge to the ocean: the importance of small mountain rivers. J Geol 100:525–544
Milne AR, Herlinveaux RH (1975) Crude oil in cold water. Fisheries and Oceans Canada, Sidney, BC, 119 pp
Minagawa M, Iseki K, Macdonald RW (1991) Carbon and nitrogen isotope composition of organic matter in the Arctic food webs. La mer 29:209–211
Minnet PJ, Bignami F, Böhm E, Budeus G, Galbraith PS, Gudmanson P, Hopkins TS, Ingram RG, Johnson MA, Niebauer HJ, Ramseier RO, Schneider W (1997) Summary of the formation and seasonal progression of the Northeast Water Polynya, J Mar Syst 10:79–85
Miyaka Y, Wada E (1967) The abundance ratio of $^{15}N/^{14}N$ in marine environments, Records of Oceanographic Works in Japan 9:32–53
Miyawaki O, Liu L, Nakamura K (1998) Effective partition constant of solute between ice and liquid phases in progressive freeze-concentration. J Food Sci 63:756–758
Mock T, Gradinger, R (1999) Determination of Arctic ice algal production with a new *in situ* incubation technique. Mar Ecol Prog Ser 177:15–26
Moisan TA, Olaizola M, Mitchell BG (1998) Xanthophyll cycling in *Phaeocystis antarctica*: changes in cellular fluorescence. Mar Ecol Prog Ser 169:113–121
Mopper K, Zhou X, Kieber RJ, Kieber DJ, Sikorski RJ, Jones RD (1991) Photochemical degradation of dissolved organic carbon and its impact on the oceanic carbon cycle. Nature 353:60–62
Moran SB, Ellis KM, Smith JN (1997) $^{234}Th/^{238}U$ disequilibria in the central Arctic Ocean: implications for particulate organic carbon export. Deep-Sea Res II 44:1593–1606
Moran SB, Smith JN (2000) ^{234}Th as a tracer of scavenging and particle export in the Beaufort Sea. Cont Shelf Res 20: 153–167
Moran SB, Weinstein SE, Edmonds HN, Smith JN, Kelly RP, Pilson MEQ, Harrison WG (2003) Does $^{234}Th/^{238}U$ disequilibrium provide an accurate record of the export flux of particulate organic carbon from the upper ocean? Limnol Oceanogr 48:1018–1029
Morison J, Aagaard K, Steele M (2000) Recent environmental changes in the Arctic: A review. Arctic 53:359–371
Morison J, Steele M, Andersen R (1998) Hydrography of the upper Arctic Ocean measured from the nuclear submarine USS Pargo. Deep-Sea Res I 45:15–38
Mørk A and Bjorøy M (1984) Mesozoic source-rocks on Svalbard. In: Spencer AM et al. (eds), Petroleum Geology of the North European Margin, Graham and Trotman, London, pp 371–382
Mountain DG (1974) Preliminary analysis of Beaufort Shelf circulation in summer. In: Reed JC, Sater JE (eds) The Coast and Shelf of the Beaufort Sea. Arctic Institute of North America, pp 27–43

Mueller-Lupp T, Bauch HA, Erlenkeuser H, Hefter J, Kassens H, Thiede J (2000) Changes in the deposition of terrestrial organic matter on the Laptev Sea shelf during the Holocene: evidence from stable carbon isotopes. Int J Earth Sci 89:563–568

Muggli DL, Harrison PJ (1997) Effects of iron on two oceanic phytoplankters grown in natural NE subarctic pacific seawater with no artificial chelators present. J Exp Mar Biol Ecol 212:225–237

Muggli DL, Lecourt M, Harrison PJ (1996) Effects of iron and nitrogen source on the sinking rate, physiology and metal composition of an oceanic diatom from the subarctic Pacific. Mar Ecol Prog Ser 132:215–227

Mulholland PJ, Watts JA (1982) Transport of organic carbon to the oceans by rivers of North America: a synthesis of existing data. Tellus 34:176–186

Mullen RE, Darby DA, Clark DL (1972) Significance of atmospheric dust and ice rafting for Arctic Ocean sediment. Bull Geol Soc Amer 83:205–212

Müller C (1995) Spätquartäre Sedimetationsprozesse in der östlichen Framstraße. Unpublished Diploma Thesis, Westfälische Wilhelms-Universität Münster, pp 1–84

Müller C (1999) Rekonstruktion der Paläo-Umweltbedingungen am Laptev-See-Kontinentalrand während der beiden letzten Glazial-/Interglazial-Zyklen anhand sedimentologischer und mineralogischer Untersuchungen. Reps Pol Res 328, 146 pp

Müller PJ (1977) C/N ratios in Pacific deep-sea sediments: effect of inorganic ammonium and organic nitrogen compounds sorbed by clays. Geochim Cosmochim Acta 41:765–776

Müller PJ, Suess E (1979) Productivity, sedimentation rate, and sedimentary organic matter in the oceans. I. – Organic matter preservation. Deep-Sea Res 26:1347–1362

Mumm N, Auel H, Hanssen H, Hagen W, Richter C, Hirche HJ (1998) Breaking the ice: large-scale distribution of mesozooplankton after a decade of Arctic and transpolar cruises. Pol Biol 20:189–197

Murata A, Takizawa T (2002) Impact of coccolithophorid bloom on the CO_2 system in surface waters of the eastern Bering Sea shelf. Geophys Res Lett 29(11): 10.1029/200GLO13906

Murdmaa I, Ivanova E, Duplessy J-C, Levitan M, Khusid T, Alekseeva T, Bourtman M, Belousov M, Serova V (2003) Facies system of the central and eastern Barents Sea since the last glaciation to Recent. Mar Geol, in press

Murdmaa IO, Ivanova EV (1999) Postglacial history of sedimentation in the shelf depressions of the Barents Sea. Lithology and mineral resources. N 6:576–595 (in Russian)

Murdmaa IO, Ivanova EV, Polyak LV (2000) High-resolution climatic record from Russkaya Gavan'Fjord, north-east Barents Sea. IMAGES Workshop "Correlation of postglacial climatic events in the Arctic and Far East seas". IO RAS, Moscow 1:13–14

Mutterlose J, Brumsack H, Flögel S, Hay W, Klein Ch, Langrock U, Lipinski M, Ricken W, Söding E, Stein R, Swientek O (2003) The Greenland-Norwegian Seaway: A key for understanding Late Jurassic to Early Cretaceous paleoenvironments. Paleoceanography 18(1) 10.1029/2001PA000625

Myhre AM, Thiede J, Firth JV et al. (1995) Proceedings Of the Ocean Drilling Program (Initial reports) 151, College Station, Texas, Ocean Drilling Program, pp 1–926

Myklestad S (1974) Production of carbohydrates by marine planktonic diatoms. I. Comparison of nine different species in culture. J Exp Mar Biol Ecol 15:261–274

Myklestad S, Haug A (1972) Production of carbohydrates by the marine diatom *Chaetoceros affinis* var. *willei* (Gran) Hustedt. I. Effect of the concentration of nutrients in the culture medium. J Exp Mar Biol Ecol 9:125–136

Mysak LA (2001) Patterns of Arctic circulation. Science 239: 1269–1270

Nagata T, Fukuda R, Fukuda H, Koike I (2001) Basin-scale geographic patterns of bacterioplankton biomass and production in the Subarctic Pacific, July–September 1997. J Oceanogr 57:301–313

Naidina OD, Bauch HA (1999) Distribution of pollen and spores in surface sediments of the Laptev Sea. Kassens H, Bauch H, Dmitrenko I, Eicken H, Hubberten HW, Melles M, Thiede J and Timokhov L (eds) (1999) Land-Ocean Systems in the Siberian: Dynamics and History. Springer-Verlag, Berlin, pp 577–585

Naidina OD, Bauch HA (2001) A holocene pollen record from the Laptev Sea shelf, northern Yakutia. Glob Planet Change 31:141–153

Naidu AS (1985) Organic carbon, nitrogen, and C/N ratios of deltaic sedimens, North Arctic Alaska. In: Degens ET, Kempe S, Herrera R (eds) Transport of carbon and minerals in major world rivers. SCOPE 58, Mitt Geol-Palaeontol Inst Univ Hamburg, pp 311–321

Naidu AS (1988) Marine surficial sediments, section 4. In: Bering, Chukchi and Beaufort Seas, coastal and ocean zones strategic assessment data atlas. US Department of Commerce, NOAA, Washington, DC

Naidu AS, Cooper LW, Finney BP, Macdonald RW, Alexander C, Semiletov IP (2000) Organic carbon isotope ratios ($\delta^{13}C$) of Amerasian continental shelf sediments. Int J Earth Sci 89:522–532

Naidu AS, Creager JS, Mowatt TC (1981) Clay mineral dispersal patterns in the northern Bering and Chukchi Seas. Mar Geol 47:1–15

Naidu AS, Finney BP, Baskaran M (1999) ^{210}Pb- and ^{137}Cs-based sediment accumulation rates in inner shelves and coastal lakes of subarctic and arctic Alaska: A synthesis. In: Bruns P, Hass HC (eds) On the Determiation of Sediment Accumulation Rates, GeoResearch Forum 5, Trans Tech Publications, Switzerland, pp 185–196

Naidu AS, Goering JJ, Kelley JJ, Venkatesan MI (2001) Historical changes in trace metals and hydrocarbons in the inner shelf sediments, Beaufort Sea: Prior and subsequent to petroleum-related industrial developments. OCS Study MMS 2001-061, Fairbanks, AK: Coastal Marine Institute

Naidu AS, Mowatt TC (1975) Depositional environments and sediment characteristics of the Colville and adjacent deltas, Northern Arctic Alaska. In: Broussard ML (ed) Deltas models for exploration. Houston Geological Society, Houston Texas, pp 283–309

Naidu AS, Mowatt TC, Rawlinson SE, Weiss HV (1984) Sediment characteristics of the lagoons of the Alaskan Beaufort Sea coast, and evolution of Simpson Lagoon. In: Barnes PW, Schell DM, Reimnitz E (eds) The Alaskan Beaufort Sea: ecosystems and environments. Academic Press, Orlando, pp 275–292

Naidu AS, Scalan RS, Feder HM, Goering JJ, Hameedi MJ, Parker PL, Behrens EW, Caughey ME, Jewett SC (1993) Stable organic carbon isotopes in sediments of the north Bering-south Chukchi seas, Alaskan-Soviet Arctic shelf. Cont Shelf Res 13:669–691

Nam SI, Stein R, Grobe H, Hubberten H (1995) Late Quaternary glacial-interglacial changes in sediment composition at the East Greenland continental margin and their paleoceanographic implications. Mar Geol 122: 243–262

Nanninga HJ, Tyrrell T (1996) Importance of light for the formation of algal blooms by *Emiliania huxleyi*. Mar Ecol Prog Ser 136:195–203

Nansen F (1897) Farthest North. Archibald Constable & Co, Whitehall Gardens, 510 pp

Nansen F (1902) The Norwegian North Polar Expedition 1893–1896: Scientific Results. London Longmans, Green, v. III

Nelson DM, Smith WO, Muench RD, Gordon LI, Sullivan CW and Husby DM (1989) Particulate matter and nutrient distribution in the ice-edge zone of the Weddell Sea: relationship to hydrography during late summer. Deep-Sea Res 36:191–209

Nesterova IL (1960) Chemical composition of suspended and dissolved substances in the Ob River. Geochemistry 4: 49–56 (in Russian)

Nichols PD, Jones GJ, de Leeuw, JW, Johns RB (1984) The fatty acid and sterol composition of two marine dinoflagellates. Phytochemistry 23 (5):1043–1047

Nichols PD, Palmisano AC, Rayner MS, Smith GA, White DC (1990) Occurrence of novel C_{30} sterols in Antarctic sea ice diatom communities during a spring bloom. Org Geochem 15:503–508

Niebauer HJ (1991) Bio-physical oceanographic interactions at the edge of the Arctic ice peak. Cont Shelf Res 1:49–98

Niebauer HJ (1998) Variability in Bering Sea ice cover as affected by a regime shift in the North Pacific in the period 1947–1996. J Geophys Res 103(C12):27717–27737

Niebauer HJ, Alexander V (1985) Oceanographic frontal structure and biological production at an ice edge. Cont Shelf Res 4:367–388

Niebauer HJ, Alexander V, Henrichs SM (1995) A time-series study of the spring blooms at the Bering Sea ice edge. I. Physical processes, chlorophyll, and nutrient chemistry. Cont Shelf Res 15:1859–1877

Niebauer HJ, Bond NA, Yakunin LP, Plotnikov VV (1999) An update of the climatology and sea ice of the Bering Sea. In: Loughlin TR, Ohtani K (eds) The physical oceanography of the Bering Sea. Univ Alaska Seagrant AK-SG-03, Univ Alaska Fairbanks, pp 29–59

Niehoff B, Klenke U, Hirche H-J, Irigoien X, Head R, Harris R (1999) A high frequency time series at Weathership M, Norwegian Sea, during the 1997 spring bloom: the reproductive biology of *Calanus finmarchicus*. Mar Ecol Prog Ser 176:81–92

Nielsen TG, Hansen B (1995) Plankton community structure and carbon cycling on the western coast of Greenland during and after the sedimentation of a diatom bloom. Mar Ecol Prog Ser 125:239–257

Niessen F, Dittmers K (2002) Geochirp and Elac sediment echograph profiling. Reps Pol Mar Res 419:64–73

Nihoul JCJ, Adam P, Brasseur P, Deleersnijder E, Djenidi S, Haus J (1993) Three dimensional general circulation model of the northern Bering Sea's summer ecohydrodynamcs. Cont Shelf Res 13:509–542

Nikiforov SL (1996) The main features of modern sediments in the southern part of the East-Siberian sea. Reps Pol Res 212:89–95

Noji TT, Børsheim KY, Rey F, Nortvedt R (1999) Dissolved organic carbon associated with sinking particles can be crucial for estimates of vertical carbon flux. Sarsia 84: 129–135

Noji TT, Miller LA, Skjelvan I, Falck E, Børsheim KY, Rey F, Urban-Rich J, Johannessen T (2001) Constraints on carbon drawdown and export in the Greenland Sea. In: Schäfer P, Ritzrau W, Schlüter M, Thiede J (eds) The Northern North Atlantic: a changing environment. Springer, Berlin, pp 39–52

Noji TT, Rey F, Miller L, Børsheim KY, Hirche H-J, Urban-Rich J (1996) Plankton dynamics and sedimentation. In: Wadhams P, Wilkinson JP, Wells SCS (eds) European Subpolar Ocean Programme (ESOP): sea ice – ocean interactions, Scott Polar Research Institute, Cambridge, England, pp 540–554

Nørgaard-Pedersen N, Spielhagen RF, Erlenkeuser H, Grootes PM, Heinemeier J, Knies J (2002) The Arctic Ocean during the Last Glacial maximum: Atlantic and Polar domains of surface water mass distribution and ice cover. Paleoceanography, in press

Nørgaard-Pedersen N, Spielhagen RF, Thiede J, Kassens H (1998) Central Arctic surface ocean environment during the past 80,000 yeras. Paleoceanography 13:193–204

Notholt H (1998) The implication of the "NorthEastWater" polynya on the sedimentation by NE-Greenland and Late Quaternary paleo-oceanic investigations. Alfred Wegener Institute for Polar and Marine Research, Bremerhaven. Reps Pol Res 275, 183 pp

Nowaczyk N, Baumann M (1992) Combined high-resolution magnetostratigraphy and nannofossil biostratigraphy for late Quaternary Arctic ocean sediments. Deep-Sea Res 39 (Suppl. 2) S567–S601

Nowaczyk N, Frederichs TW, Kassens H, Nørgaard-Petersen N, Spielhagen R, Stein R, Weiel D (2000) Sedimentation rates in the Makarov Basin, Central Arctic Ocean – A paleo- and rock magnetic approach. Paleoceanography 16:368–389

Nürnberg D, Wollenburg I, Dethleff D, Eicken H, Kassens H, Letzig T, Reimnitz E, Thiede J (1994) Sediments in Arctic sea ice – implications for entrainment, transport and release. Mar Geol 119:185–214

O'Leary MH (1988) Carbon isotopes in photosynthesis. Bioscience 38:328–336

Ogilvie AEJ, Junsson T (2000) The Iceberg in the Mist: Northern Research in Pursuit of a "Little Ice Age". Kluwer Academic Publishers, Dordrecht, pp 1–270

Ohman MD, Hirche H-J (2001) Density-dependent mortality in an oceanic copepod population. Nature 412:638–641

Olli K, Wexels Riser C, Wassmann P, Ratkova T, Arashkevich E (2001) Vertical export of biogenic matter, particulate nutrients and mesozooplankton faecal pellets off the NW coast of Galicia. Prog Oceanogr 51:443–466

Olli K, Wexels Riser C, Wassmann P, Ratkova T, Arashkevich E, Pasternak A (2002) Seasonal variation in vertical export of biogenic matter in the marginal ice zone and the central Barents Sea. J Mar Systems 38:189–204

Olsen A, Johannessen T, Rey F (2002) Inter-annual variability in the mechnaisms controlling the phytoplankton seasonal development at the entrance to the Barents Sea

Olsen A, Johannessen T, Rey F (2003) On the nature of the factors that control spring bloom development at the entrance to the Barents Sea and their interannual variability. Sarsia, in press

Omstedt A, Carmack EC, Macdonald RW (1994) Modeling the seasonal cycle of salinity in the Mackenzie shelf/estuary. J Geophys Res 99(C5):10011–10021

Opsahl S, Benner R (1997) Distribution and cycling of terrigenous dissolved organic matter in the ocean. Nature 386:480–482

Opsahl S, Benner R, Amon RMW (1999) Major flux of terrigenous dissoved organic matter through the Arctic Ocean. Limnol Oceanogr 44:2017–2023

O'Reilly JE, Busch DA (1984) Phytoplankton primary production on the northwest Atlantic shelf. Rapp P-v Réun Cons Int Explor Mer 183:255–268
Osterkamp TE, Gosink J P (1984) Observations and analyses of sediment-laden sea ice. In: Barnes PW, Schell DM, Reimnitz E (eds) The Alaskan Beaufort Sea: ecosystems and environments. Academic Press, Orlando, pp 73–93
Östlund GH, Possnert G, Swift JH (1987) Ventilation rate of the deep Arctic Ocean from Carbon 14 data. J Geophys Res 92(C4):3769–3777
Östlund HG (1982) The residence time of the freshwater component in the Arctic Ocean. J Geophys Res 87: 2035–2043
Østvedt OJ (1955) Zooplankton investigations from weathership "M" in the Norwegian Sea, 1948–49. Hvalråd Skr 40:1–93
Ottersen G, Ådlandsvik B, Loeng H (2000) Predicting the temperature of the Barents Sea. Fish Oceanogr 9:121–135
Overland JE, Bond NA, Adams JM (2001) North Pacific atmospheric and SST anomalies in 1997: Links to ENSO. Fisheries Oceanography 10:69–80
Overland JE, Salo SA, Kantha LH, Clayson AC (1999) Thermal stratification and mixing on the Bering Sea shelf. In: Loughlin TR, Ohtani K (eds) The dynamics of the Bering Sea. Univ Alaska Seagrant AK-SG-03, Univ Alaska Fairbanks, pp 129–146
Overland JE, Spillane MC, Hurlbert HE, Wallcraft AJ (1994) A numerical study of the Bering Sea Basin and exchange with the North Pacific. J Phys Oceanogr 24:736–758
Öztürk M, Steinnes E, Sakshaug E (2002) Iron speciation in the Trondheim Fjord from the perspective of iron limitation for phytoplankton. Est Coast Shelf Sci 55:197–212
Paasche E (1960) Phytoplankton distribution in the Norwegian Sea in June, 1954, related to hydrography and compared with primary production data. Fisk Skr Havunders 12(2):1–77
Pace ML, Knauer GA, Karl DM, Martin JH (1987) Primary production, new production and vertical flux in the eastern Pacific Ocean. Nature 325:803–804
Pacyna JM (1991) Chemical tracers of the origin of Arctic air pollution. In: Sturges WT (ed) Pollution of the Arctic Atmosphere. Elsevier, Berlin, pp 97–122
Painter T (1983) Carbohydrate origin of aquatic humus from peat. Carbohydrate Res 124:260–262
Pamatmat MM (1971) Oxygen consumption by the seabed. IV. shipboard and laboratory experiments. Limnol Oceanogr 16:536–550
Pamatmat MM (1973) Benthic community metabolism on the continental terrace and in the deep sea in the North Pacific. Int Revue ges Hydrobiol 58:345–368
Pamatmat MM, Banse K (1969) Oxygen consumption by the seabed. II. In situ measurements to a depth of 180 m. Limnol Oceanogr 14:250–259
Parkinson CL, Cavalieri DJ, Gloersen P, Zwally JH, Comiso JC (1999) Arctic sea ice extents, areas, and trends, 1978–1996. J Geophys Res 104(C9):20837–20856
Parrish DM (1987) Annual primary production in the Alaskan Arctic Ocean. University of Alaska, Fairbanks, Fairbanks, AK: MSc, 175 pp
Parsons TR, Maita Y, Lalli CM (1984) Manual of chemical and biological methods for seawater analysis. Pergamon Press, Oxford, 173 pp
Parsons TR, Webb DG, Dovey H, Haigh R, Lawrence M, Hopky G (1988) Production studies in the Mackenzie River-Beaufort Sea estuary. Pol Biol 8:235–239
Parsons TR, Webb DG, Rokeby BE, Lawrence M, Hopky GE, Chiperzak DB (1989) Autotrophic and heterotrophic production in the Mackenzie River/Beaufort Sea estuary. Pol Biol 9:261–266
Paul AZ, Menzies JR (1974) Benthic ecology of the high Arctic deep sea. Mar Bio 27:251–262
Pavlidis YuA, Ionin AS, Scherbakov FA (1988) Arctic shelf: Late Quaternary history as a base of evolution forecast. GEOS Press, Moscow, 187 pp (in Russian)
Pavlidis YuA, Ionin AS, Scherbakov FA, Dunaev NN, Nikiforov SL (1998) Arctic shelf: Late Quaternary history as basis for forecasting of its development. GEOS, Moscow, pp 1–187 (in Russian)
Pavlidis YuA, Scherbakov FA, Boyarskaya TD, Dunaev NN, Polyakova EI, Husid TA (1992) New data on stratification of Quaternary deposits and paleogeography of the southern part of the Barents Sea. Oceanology 32 (5):917–923 (in Russian)
Pavlov VK, Pfirman SL (1995) Hydrographic structure and variability of the Kara Sea: Implications for pollutant distribution. Deep-Sea Research II 42(6):1369–1390
Pavlov VK, Timohov LA, Baskakov GA, Kulakov MYu, Kurazhov VK, Pavlov PV, Pivovarov SV, Stanovy VV (1994) Hydrometeorological Regime of the Kara, Laptev and East-Siberian Seas. The Arctic and Antarctic Research Institute, St Petersburg, 179 pp
Pedersen OP, Tande KS, Slagstad D (2001) A model study of demography and spatial distribution of *Calanus finmarchicus* at the Norwegian coast. Deep-Sea Res II 48:567–587
Pedersen TF, Waters RD, Macdonald RW (1989) On the natural enrichment of cadmium and molybdenum in the sediments of Ucluelet Inlet, British Columbia. Sci Total Environ 79:125–139.
Peinert R, Antia A, Bauerfeind E, Haupt O, Krumbholz M, Peeken I, Bodungen Bv, Ramseier R, Voss M, Zeitzschel B (2001a) Particle flux variability in the Polar and Atlantic biogeochemical provinces of the Nordic Seas. In: Schäfer P, Ritzrau W, Schlüter M, Thiede J (eds) The Northern North Atlantic: a changing environment. Springer, Berlin, pp 53–68
Peinert R, Bauerfeind E., Gradinger R, Haupt O, Krumbholz M, Peeken I, Werner I, Zeitzschel B (2001b) Biogenic particle sources and vertical flux patterns in the seasonally ice-covered Greenland Sea. In: Schäfer P, Ritzrau W, Schlüter M, Thiede J (eds) The Northern North Atlantic: a changing environment. Springer, Berlin, pp 69–79
Peinert R, Bodungen Bv, Smetacek VS (1989) Food web structure and loss rate. In: Berger WH, Smetacek VS, Wefer G (eds) Productivity of the ocean: present and past. John Wiley & Sons Limited, pp 35–48
Pelletier BR (1975) Sediment dispersal in the Southern Beaufort Sea. Beaufort Sea Technical Report # 25a Beaufort Sea Project, 80 pp
Peña MA, Denman KL, Calvert SE, Thomson RE, Forbes JR (1999) The seasonal cycle in sinking particle fluxes off Vancouver Island, British Columbia. Deep-Sea Res II, 46: 2969–2992
Peresypkin VI, Petrova VI, Batova GI, Vetrov AA, Belyaev NA, Romankevich EA (2002) Organic geochemistry. In: The Pechora Sea. Experience of system researches. Nauchnii Mir, Moscow, in press (in Russian)
Peresypkin VI (2000) Geochemistry of lignin in aerosols and bottom sdiments of the oceanic tropics. The abstract of candidate's thesis (geology and mineralogy) IO RAN, Moscow, pp 1–32

Perry MJ, Boger J, English DC (1989) Primary production in Washington coastal waters. In: Hickey B, Landry M (eds) Coastal Oceanography of Washington and Oregon, Elsevier, New York, pp 117–138

Perry RK, Flemming HS, Weber HR, Kristoffersen Y, Hall JK, Grantz A and Johnson GL (1985) Bathymetry of the Arctic Ocean. Naval Research Laboratory, Washington, Scale 1:4, 704,075, 1 Sheet.

Pesant S, Legendre L, Gosselin M, Ashijan C, Booth B, Daly K, Fortier L, Hirche H-J, Michaud J, Smith REH, Smith S, Smith WO Jr (1998) Pathways of carbon cycling in the euphotic zone: the fate of large-sized phytoplankton in the Northeast Water Polynya, J Plankton Res 20: 1267– 1291

Pesant S, Legendre L, Gosselin M, Bauerfeind E, Budeus G (2002) Wind triggered events of phytoplankton downward flux in the Northeast water Polynya. J Mar Sys 31: 261–278

Peters KE (1986) Guidelines for evaluating petroleum source rocks using programmed pyrolysis. Amer Assoc Petrol Geol Bull 70:318–329

Peters KE, Moldowan JM (1993) The biomarker guide: Interpreting molecular fossils in petroleum and ancient sediments. Prentice Hall, pp 1–363

Petersen CH, Curtis MA (1980) Differences in energy flow through major components of subarctic, temperate and tropical marine shelf ecosystems Dana 1:53–64

Petit JR, Jouzel J, Raynout D, Barkov NI, Barnola J-M, Basile I, et al (1999) Climate and atmospheric history of the past 420,000 years from the Vostok ice core, Antarctica. Nature 399:429–436

Petrova VI (1999) Geochemistry of polycyclic aromatic hydrocarbons in the bottom sediments of the World Ocean. Abstr Diss. VNII Okeangeologiya, St-Petersburg, pp 1–30 (in Russian)

Petrova VI, Batova GI, Andreeva IA, Galishev MA (1999) Organic matter of the clay fraction in Holocene sediments: interaction between group, molecular, grain size and minerals composition. In: Abstracts of the XIII International Conference on Marine Geology, Moscow, 1: 300–302

Petrova VI, Batova GI, Galishev MA, Korotkova TA (2001) Organic matter in the bottom sediments of the East Siberia Sea – feature of distribution and genesis. In: Abstracts of the XIV International Conference on Marine Geology, Moscow, 2: 206–207

Petryashov VV, Sirenko BI, Golikov AA, Novoshilov AV, Rachor E, Piepenburg D, Schmid MK (1999) Macrobenthos in the Laptev Sea in relation to hydrology. In: Kassens H, Bauch HA, Dmitrenko I, Eiken H, Hubberten HW, Melles M, Thiede J, Timokhov L (eds) Land-Ocean Systems in the Siberian Arctic: Dynamics and history. Springer Verlag, Berlin, pp 169–180

Peulvé S, Sicre MA, Saliot A, De Leeuw JW, Baas M (1996) Molecular characterization of suspended and sedimentary organic matter in an Arctic Delta. Limnol Oceanogr 41(3):488–497

Pfannkuche O (1992) Organic carbon flux through the benthic community in the temperate abyssal Northeast Atlantic. In: Rowe GT, Pariente V (eds) Deep-sea Food Chains and the Global Carbon Cycle. Kluwer Academic Publishers 360:183–198

Pfannkuche O, Thiel H (1987) Meiobenthic standing stocks and benthic activity on the NE-Svalbard shelf and in the Nansen Nasin. Polar Biol 7:253–266

Pfirman S, Colony R, Nürnberg D, Eicken H, Rigor I (1997) Reconstructing the origin and trajectory of drifting Arctic sea ice. J Geophys Res 102(C6):12575–12586

Pfirman SL, Gascard JC, Wollenburg J, Mudie P, Abelmann A (1989b) Particle-laden Eurasian Arctic sea ice: observation from July and August 1987. Polar Res 7:59–66

Pfirman SL, Kögeler J, Anselme B (1995). Coastal environments of the western Kara and eastern Barents Seas. Deep-Sea Research II 42(6):1391–1412

Pfirman SL, Lange MA, Wollenburg I, Schlosser P (1990) Sea ice characteristics and the role of sediment inclusions in deep-sea deposition: Arctic – Antarctic comparisons. In: Bleil U, Thiede J Geological history of the Polar Oceans: Arctic versus Antarctic. Dordrecht, Kluwer Academic Publishers, pp 187–211

Pfirman SL, Wollenburg J, Thiede J, Lange MA (1989a) Lithogenic sediment on Arctic pack ice: potential aeolian flux and contribution to deep sea sediments. In: Leinen M, Sarnthein M (eds) Paleoclimatology and Paleometeorology: Modern and Past Patterns of Global Atmospheric Transport, NATO ASI Series C, Kluwer Academic Publishers 282:463–493

Phillips RL, Grantz A (1997) Quaternary history of sea ice and paleoclimate in the Amerasia Basin, Arctic Ocean, as recorded in the cyclical strata of Northwind Ridge. Geol Soc Amer Bull 109:1101–1115

Phillips RL, Grantz A (2001) Regional variations in provenance and abundance of ice-rafted clasts in Arctic Ocean sediments: implications for the configuration of late Quaternary oceanic and atmospheric circulation in the Arctic. Mar Geol 172:91–115

Phillips RL, Grantz A, Mullen MW (1990) Preliminary stratigraphy of piston cores from southeastern Northwind Ridge, Arctic Ocean. Open-File Report, US Geol Survey

Phoel WC (1982) A comparison of the seabed oxygen consumption rates of different benthic environments along the northeast United States continental shelf. Internat Council Explor of the Sea, CM 1982/E:23

Piepenburg D (1988) Zusammensetzung der Bodenfauna in der westlichen Framstrasse. Reps Pol Res 52:1–118

Piepenburg D (2000) Arctic brittle stars (Echinodermata: Ophiuroidea). Oceanogr Mar Biol Annu Rev 38:189–256

Piepenburg D, Blackburn TH, Dorrien CF von, Gutt J, Hall POJ, Hulth S, Kendall MA, Opalinski KW, Rachor E, Schmid MK (1995) Partitioning of benthic community respiration in the Arctic (northwestern Barents Sea). Mar Ecol Progr Ser 118:199–213

Piepenburg D, Brandt A, Juterzenka K von, Mayer M, Schnack K, Seiler D, Witte U, Spindler M (2001) Patterns and determinants of the distribution and structure of benthic faunal assemblages in the northern North Atlantic In: Schäfer P, Ritzrau W, Schlüter M, Thiede J (eds) The Northern North Atlantic: A Changing Environment. SpringerVerlag, Berlin, pp 179–198

Piepenburg D, Chernova NV, Dorrien CF v, Gutt J, Neyelov AV, Saldanha L, Schmid MK (1996) Megabenthic communities in the waters around Svalbard. Polar Biol 16:431–446

Piepenburg D, Schmid MK (1996) Brittle stars fauna (Echinodermata, Ophiuroidea) of the Arctic northwestern Barents Sea: composition, abundance, biomass and spatial distribution. Polar Biol 16:383–392

Piepenburg D, Schmid MK (1997) A photographic survey of the epibenthic megafauna of the Arctic Laptev Sea shelf: distribution, abundance, and estimates of biomass and organic carbon demand. Mar Ecol Prog Ser 147:63–75

Pilskaln C, Paduan JB, Chavez FP, Anderson RY, Berelson WM (1996) Carbon export and regeneration in the coastal upwelling system of Monterey Bay, central California. J Mar Res 54:1149–1178

Platt T, Harrison WG, Irwin B, Horne EP, Gallegos CL (1982) Photosynthesis and photoadaptation of marine phytoplankton in the Arctic. Deep-Sea Res I 38:29:1159–1170

Poltermann M, Hop H, Falk-Petersen S (2000) Life under Arctic sea ice – reproduction strategies of two sympagic (ice-associated) amphipod species, *Gammarus wilkitzskii* and *Apherusa glacialis*. Mar Biol 136:913–920

Polyak L, Edwards MH, Coakley BJ, Jakobsson M (2001) Ice shelves in the Pleistocene Artic Ocean inferred from glaciogenic deep-sea bedforms. Nature 410:453–457

Polyak L, Forman SL, Herlihy FA, Ivanov G, Krinitsky P (1997) Late Weichselian deglacial history of the Svyataya (Saint) Ann Trough, northern Kara Sea, Arctic Russia. Mar Geol 143:169–188

Polyak L, Gataullin V, Gainanov V, Gladysh V, Goremykin Y (2002) Kara Sea expedition yields insight into of LGM ice sheet. EOS 83:525, 529

Polyak L, Gataullin V, Okuneva O, Stelle V (2000) New constraints on the limits of the Barents-Kara ice sheet during the Last Glacial Maximum based on borehole stratigraphy from the Pechora Sea. Geology 28:611–614

Polyak L, Lehman SL, Gataullin V, Jull AJT (1995) Two-step deglaciation of the southeastern Barents Sea. Geology 23: 567–571

Polyak L, Solheim A (1994) Late- and postglacial environments in the northern Barents Sea west of Franz Josef Land. Pol Res 13:197–207

Polyakov IV, Johnson MA (2000) Arctic decadal and interdecadal variability. Geophys Res Lett 27:4097–4100

Pomeroy LR (1997) Primary production in the Arctic Ocean estimated from dissolved oxygen. J Mar Syst 10:1–8

Pomeroy LR, Wiebe WJ, Deibel D, Thompson RJ, Rowe GT, Pakulski JD (1991) Bacterial responses to temperature and substrate concentration during the Newfoundland spring bloom. Mar Ecol Prog Ser 75:143–159

Popov AV, Ivanov BV, Karelin ID (2000) Polynyas situated back of fast ice in the Laptev Sea and climatic change In: Hydrometeorological and biogechemical studies in the Arctic. Proceedings of The Arctic regional center, Vladivostok, 2 (1) 44–52

Popov BA, Sovershaev VA, Novikov VN (1988) The coastal zone of the Pechora-Kara region. In: Studies of geosystem stability in the North. Moscow Univ. Press. Moscow, pp 176–190 (in Russian)

Prahl FG, Carpenter R (1984) Hydrocarbons in Washington coastal sediments. Est Coast Shelf Sci 18:703–720

Prahl FG, Ertel JR, Goñi MA, Sparrow MA, Eversmeyer B (1994) Terrestrial organic-carbon contributions to sediments on the Washington margin. Geochim Cosmochim Acta 58 (14):3035–3048

Prahl FG, Muehlhausen LA (1989) Lipid biomarkers as geochemical tools for paleoceanographic study. In: Berger WH, Smetacek VS, Wefer G (eds) Productivity of the oceans: present and past. Wiley, New York, pp 271–289

Premuzic ET, Benkovitz CM, Gaffney JS, Walsh JJ (1982) The nature and distribution of organic matter in the surface sediments of world oceans and seas. Org Geochem 4:53–77

Proshutinsky A, Proshutinsky T, Weingartner T (1995) Northern Sea Route Reconnaissance Report: Appendix B, Climatology of environmental conditions affecting commercial navigation along the Northern Sea Route, US Army Corps of Engineers, Alaska District, Anchorage AK 194 pp

Proshutinsky AY, Johnson MA (1997) Two circulation regimes of the wind-driven Arctic Ocean. J Geophys Res 102(C6):12493–12514

Prowse TD, Flegg PO (2000) Arctic river flow: a review of contributing areas. In: Lewis EL (ed) Freshwater Budget of the Arctic Ocean. Kluwer, Dordrecht, pp 269–280

Quadfasel D, Gascard JC, Koltermann KP (1987) Large-scale oceanography in Fram Strait during the 1984 Marginal Ice Zone Experiment (MIZEX). J Geophys Res 92(C7): 6719–28

Rachold V (ed) (1999) Expeditions in Siberia in 1998. Reps Pol Res 316, 259 pp

Rachold V (ed) (2000) Expeditions in Siberia in 1999. Reps Pol Res 354, 269 pp

Rachold V and Hubberten HW (1999) Carbon isotope composition of particulate organic material in East Siberian rivers. In: Kassens H, Bauch H, Dmitrenko I, Eicken H, Hubberten HW, Melles M, Thiede J and Timokhov L (eds), Land-Ocean Systems in the Siberian: Dynamics and History. Springer, Berlin, pp 223–238

Rachold V, Grigoriev MN (eds.) (2001) Russian-German Cooperation System Laptev Sea 2000: The Expedition LENA 2000. Reps Pol Mar Res 388, 135 pp

Rachold V, Grigoriev MN, Are FE, Solomon S, Reimnitz E, Kassens H, Antonow M (2000) Coastal erosion vs riverine sediment discharge in the Arctic shelf seas. Int J Earth Sci 89:450–460

Rachold V, Grigoriev MN, Bauch HA (2002) An estimation of the sediment budget of the Laptev Sea during the last 5,000 years. Polarforschung 70 (2000):151–157

Ramseier RO, Bauerfeind E, Garrity C, Walsh ID (1997) Seasonal variability of sediment trap collections in the Northeast Water Polynya: Part 1. Sea-Ice parameters and particle flux. J Mar Sys 10: 359–369

Ramseier RO, Garrity C, Bauerfeind E, Peinert R (1999) Sea-ice impact on long term particle flux in the Greenland Sea's Is Odden-Nordbukta region during 1985–1996. J Geophys Res 104(C3): 5329–5343

Ramseier RO, Garrity C, Martin T (2001) An overview of sea-ice conditions in the Greenland Sea and the relationship of oceanic sedimentation to the ice regime. In: P Schäfer, W Ritzrau. M Schlüter and J Thiede (eds) The Northern North Atlantic: a changing environment. Springer, Berlin, pp 19–38

Rau GH, Riebesell U, Wolf-Gladrow D (1997) $CO_{2(aq)}$-dependent photosynthetic ^{13}C fractionation in the ocean: a model versus measurements. Glob Biogeochem Cycles 11:267–278

Rau GH, Sweeney RE, Kaplan IR (1982) Plankton ^{13}C:^{12}C ratio changes with latitude: differences between northern and southern oceans. Deep-Sea Research 29:1035–1039

Rau GH, Takahashi T, Des Marais DJ (1989) Latitudinal variations in plankton $\delta^{13}C$: implications for CO_2 and productivity in the past oceans. Nature 341:516–518

Rau GH, Sullivan CW, Gordon LI (1991) $\delta^{13}C$ and $\delta^{15}N$ variations in Wedell Sea particulate organic matter. Mar Chem 35: 355–369

Ravenschlag K, Sahm K, Knoblauch C, Jørgensen BB, Amann R (2000) Community structure, cellular rRNA content and activity of sulfate reducing bacteria in marine Arctic sediments. Appl Environm Microbiol 66:3590–3600

Razumov SO (2000) Rates of coastal thermoabrasion as a function of climate and morphological characteristics of the coast. Geomorphology 3: 88–94 (in Russian)

Rearic DM, Barnes PW, Reimnitz E (1990) Bulldozing and resuspension of shallow-shelf sediment by ice keels: Implications for arctic sediment transport trajectories. Mar Geol 91:133–147

Redfield AC, Ketchum BH, Richards FA (1963) The influence of organisms on the composition of sea water. In: Hill MN (ed) The Sea. Wiley, New York, pp 26–77

Reigstad M, Wexels Riser C, Øygarden S, Wassmann P, Rey F (2002) Variation in hydrography, nutrients and suspended biomass in the marginal ice zone and the central Barents Sea. J Mar Systems 38:9–29

Reimnitz E, Barnes PW (1982) Sea-ice influence on Arctic coastal retreat. Coastal Sediment 87:1578–1591

Reimnitz E, Barnes PW, Weber WS (1993) Particulate matter in pack ice of the Beaufort Gyre. J Glaciol 39:186–198

Reimnitz E, Dethleff D, Nürnberg D (1994) Contrasts in Arctic shelf sea-ice regimes and some implications: Beaufort Sea versus Laptev Sea. Mar Geol 119:215–225

Reimnitz E, Graves SM, Barnes PW (1988a) Beaufort Sea coastal erosion, sediment flux, shoreline evolution and the erosional shelf profile. US Geological Survey. To accompany Map I-1182–G, 22 pp

Reimnitz E, McCormick M, Bischof J, Darby DA (1998) Comparing sea-ice sediment load with Beaufort Sea shelf deposits: Is entrainment selective? J Sed Res 68:777– 787

Reimnitz E, McCormick M, McDougall K, Brouwers E (1993) Sediment export by ice rafting from a coastal polynya, Arctic Alaska, USA Arctic Alpine Res 25:83–98

Reinfelder JR, Kraepiel AML, Morel FMM (2000) Unicellular C_4 photosynthesis in a marine diatom. Nature 407:996– 999

Rey F (1991) Photosynthesis-irradiance relationships in natural phytoplankton populations of the Barents Sea. Pol Res 10:105–116

Rey F, Loeng H (1985) The influence of ice and hydrographic conditions on the development of stress on marine organisms. In: Gray JS, Christiansen ME (ed) Marine biology of polar regions and effect of stress on marine organisms. John Wiley & Sons Ltd, Chichester, pp 49–63

Rey F, Skjoldal HR (1987) Consumption of silicic acid below the euphotic zone by sedimenting diatom blooms in the Barents Sea. Mar Ecol Prog Ser 36: 307–312

Rice AL, Thurston MH, Bett BJ (1994) The IOSDL program – introduction and photographic evidence for the presence and absence of a seasonal input of phytodetritus at contrasting abyssal sites in the north-eastern Atlantic Deep-Sea Res II 41:1305–1320

Rich J, Gosselin M, Sherr E, Sherr B, Kirchman D (1997) High bacterial production, uptake and concentrations of dissolved organic matter in the Central Arctic Ocean. Deep-Sea Res II 44:1645–1663

Riebesell U, Wolf-Gladrow DA, Smetacek V (1993) Carbon dioxide limitation of marine phytoplankton growth-rates. Nature 361:249–251

Rigor IG (1992) Arctic ocean buoy program. ARCOS Newsletter 44:1–3

Rigor IG, Colony R (1997) Sea-ice production and transport of pollutants in the Laptev Sea, 1979–1993. Sci Tot Environm 202:89–110

Rigor IG, Colony RL, Martin S (2000) Variations in surface air temperature observations in the Arctic, 1979–97. J Clim 13:896–914

Rigor IG, Wallace JM, Colony R (2002) Response of sea ice to the Arctic Oscillation. J Clim 15:2648–2663

Roach AT, Aagaard K, Pease CH, Salo SA, Weingartner T, Pavlov V et al. (1995) Direct measurements of transport and water properties through the Bering Strait. J Geophys Res 100(C9):18,443–18,457

Robineau B, Legendre L, Kishino M, Kudoh S (1997) Horizontal heterogeneity of microbial biomass in the first-year sea ice of Saroma-ko Lagoon (Hokkaido, Japan). J Mar Syst 11:81–91

Rogers A, Billett D, Berger W, Flach E, Freiwald A, Gage J, Hebbeln D, Heip C, Pfannkuche O, Ramirez-Llodra E, Medlin L, Sibuet M, Soetaert K, Tendal O, Vanreusel A, Wlodarska-Kowalczuk M (2002) Life at the Edge: Achieving prediction from environmental variability and biological variety In: Wefer G, Billett D, Hebbeln D, Jørgensen BB, Schlüter M, vanWeering T (eds) Ocean margin systems. Springer, Berlin, pp 387–404

Romankevich EA (1984) Geochemistry of Organic Matter in the Ocean. Springer, Heidelberg, 334 pp

Romankevich EA, Artemiev VE (1985) Input of organic carbon into seas and oceans bordering the territory of the Soviet Union. In: Degens ET, Kempe S, Herrera R (eds) Transport of carbon and minerals in major world rivers. SCOPE 58, Mitt Geol-Palaeontol Inst Univ Hamburg: 450–459

Romankevich EA, Danyushevskaya AI, Belyaeva AN, Rusanov VP (1982) Biogeochemistry of organic matter of the Arctic Seas. Nauka, Moscow, pp 1–240

Romankevich EA, Vetrov AA (1997) Fluxes and masses of organic carbon in the Ocean. Geochemistry (9):945–952 (in Russian)

Romankevich EA, Vetrov AA (2001) Carbon cycle in the Russian Arctic seas. Moscow, Nauka, 302 pp (in Russian)

Romankevich EA, Vetrov AA, Vinogradov ME, Vedernikov VI (2000) Carbon cycling components in the Arctic Seas of Russia: Carbon fluxes from land, carbon in the bottom sediments, elements of balance. Oceanology 40:335–344

Romero N, Silverberg N, Roy S, Lovejoy C (2000) Sediment trap observations from the Gulf of St. Lawrence and the continental margin of eastern Canada. Deep-Sea Res II 47: 545–583

Romero-Wetzel MB, Gerlach SA (1991) Abundance, biomass, size-distribution and bioturbation potential of deep-sea macrozoobenthos on the Voering Plateau (1200–1500 m, Norwegian Sea). Rep Mar Res 33:247–265

Ronov AB (1993) Stratosphere or sedimentary cover of the Earth. Nauka Press. Moscow, 143 pp (In Russian)

Rosen H, Novakov T, Bodhaine BA (1981) Soot in the Arctic. Atmos Environ 15:1371–1374

Rossak BT, Kassens H, Lange H, Thiede J (1999) Clay mineral distribution in surface sediments of the Laptev Sea: Indicator for sediments provinces, dynamics and sources. In: Kassens H, Bauch H, Dmitrenko I, Eicken H, Hubberten HW, Melles M, Thiede J and Timokhov L (eds) Land-Ocean Systems in the Siberian: Dynamics and History. Springer, Berlin, pp 587–600

Rothrock DA (1978) Modeling sea ice features and processes. Symposium on dynamics of large ice masses 21–25 August 1978, Ottawa (unpubl ms)

Rothrock DA, Yu Y, Maykut GA (1999) Thinning of the Arctic sea-ice cover. Geophys Res Lett 26:3469–3472

Rowe GT, Boland GS, Escobar EGE, Cruz-Kaegi ME, Newton A, Piepenburg D, Walsh I, Demming J (1997) Sediment community biomass and respiration in the Northeast Water Polynya, Greenland: a numerical simulation of benthic lander and spade core data. J Mar Syst 10:497–515

Rowe GT, Boland GS, Phoel WC, Anderson RF, Biscaye PE (1994) Deep-sea respiration as an indication of lateral input of biogenic detritus from continental margins. Deep-Sea Res II 41:657–668

Rowe GT, Deming JW (1985) The role of bacteria in the turnover of organic carbon in deep-sea sediments. J Mar Res 43:925–950

Roy S et al. (2000) Importance of mesozooplankton feeding for the downward flux of biogenic carbon in the Gulf of St. Lawrence (Canada). Deep-Sea Res II, 47: 519–544

Rudels B, Friedrich HJ (2000) The transformations of Atlantic water in the Arctic Ocean and their significance for the freshwater budget. In: Lewis EL, Jones EP, Lemke P, Prowse TD, Wadhams P (eds) The Freshwater Budget of the Arctic Ocean. Kluwer Academic Publishers, Boston, pp 503–532

Rudels B, Jones EP, Anderson LG, Kattner G (1994) On the intermediate depth waters of the Arctic Ocean. In: OM Johannessen RDM, JE Overland (ed) The Polar Oceans and Their Role in Shaping the Global Environment. American Geophysical Union, Washington, DC, pp 33–46

Rudels B, Meyer R, Fahrbach E, Ivanov VV, Østerhus S, Quadfasel D, Schauer U, Tverberg V, Woodgate RA (2000) Water mass distribution in Fram Strait and over the Yermak Plateau in summer 1997. Annales Geophysicae 18: 687–705

Rudels B, Muench RD, Gunn J, Schauer U, Friedrich HJ (2000) Evolution of the Arctic Ocean boundary current north of the Siberian shelves. J Mar Syst 25:77–99

Rugh DJ, Muto MM, Moore SE, DeMaster DP (1999) Status review of the eastern North Pacific stock of gray whales. NMFS-AFSC-103: NOAA Technical Memorandum, NMFS-AFSC-103

Rühlemann C, Frank M, Hale W, Mangini A, Mulitza S, Müller P, Wefer G (1996) Late Quaternary productivity changes in the western equatorial Atlantic from ^{230}Th-normalized carbonate and organic carbon accumulation rates Mar Geol 135:127–152

Rullkötter J (2000) Organic matter: the driving force for early diagenesis. In: Schulz HD, Zabel M (eds) Marine Geochemistry. Springer, Berlin New York, pp 129–172

Rumohr H, Brey T, Ankar S (1987) A compilation of biometric conversion factors for benthic invertebrates of the Baltic Sea. Baltic Marine Biologists 9:1–56

Runge J, Therriault J-C, Legendre L, Ingram RG, Demers S (1991) Coupling between ice microalgal productivity and the pelagic, metazoan food web in the southeastern Hudson Bay: a synthesis of results. Pol Res 10:325–338

Ruttenberg KC, Goñi MA (1997) Phosphorus distribution, C:N:P ratios, and $\delta^{13}C_{OC}$ in arctic, temperate, and tropical coastal sediments: tools for characterizing bulk sedimentary organic matter. Mar Geol 139:123–145

Rysgaard S, Christensen PB, Sørensen MV, Funch P, Berg P (2000) Marine meiofauna, carbon and nitrogen mineralization in sandy and soft sediments of Disko Bay, West Greenland. Aquat Microb Ecol 21:59–71

Rysgaard S, Finster K, Dahlgaard H (1996) Primary production, nutrient dynamics and mineralisation in a northeastern Greenland fjord during the summer thaw. Pol Biol 16:497–506

Rysgaard S, Thamdrup B, Risgaard-Petersen N, Fossing H, Berg P, Christensen PB, Dalsgaard T (1998) Seasonal carbon and nutrient mineralization in a high-Arctic coastal marine sediment, Young Sound, Northeast Greenland. Mar Ecol Progr Ser 175:261–276

Saar RA (2000) The unbearable capriciousness of Bering. Science 287:1388–1389

Sahm K, Berninger U (1998) Abundance, vertical distribution, and community structure of benthic prokaryotes from permanently cold marine sediments (Svalbard, Arctic Ocean). Mar Ecol Progr Ser 165:71–80

Saito H, Hattori H (1997) Diel vertical migration and feeding rhythm of copepods under sea ice at Saroma-ko Lagoon. J Mar Syst 11:191–203

Saito K, Taniguchi A (1978) Phytoplankton communities in the Bering Sea and adjacent seas. II. Spring and summer communities in seasonally ice-covered areas. Astarte 11: 27–35

Sakshaug E (1972) Phytoplankton investigations in Trondheimsfjord, 1963–1966. K Norske Vidensk Selsk Skr 1972(1):1–56

Sakshaug E (1989) The physiological ecology of polar phytoplankton. In: Rey L, Alexander V (eds) Proc 6th Conf Com Arctique Intl, EJ Brill, Leiden, the Netherlands, pp 61–89

Sakshaug E, Andresen K (1986) Effect of light regime upon growth rate and chemical composition of a clone of *Skeletonema costatum* from the Trondheimsfjord, Norway. J Plank Res 8:619–638

Sakshaug E, Andresen K, Myklestad S, Olsen Y (1983) Nutrient status of phytoplankton communities in Norwegian waters (marine, brackish, and fresh) as revealed by their chemical composition. J Plank Res 5:175–196

Sakshaug E, Bjørge A, Gulliksen B, Loeng H, Mehlum F (1994) Structure, biomass distribution, and energetics of the pelagic ecosystem in the Barents Sea: A synopsis. Pol Biol 14:405–411

Sakshaug E, Bricaud A, Dandonneau Y, Falkowski PG, Kiefer DA, Legendre L, Morel A, Parslow J, Takahashi M (1997) Parameters of photosynthesis: definitions, theory and interpretation of results. J Plank Res 19:1637–1670

Sakshaug E, Holm-Hansen O (1977) Chemical composition of *Skeletonema costatum* (Grev.) Cleve and *Pavlova* (*Monochrysis*) *lutheri* (Droop)Green as a function of nitrate-, phosphate, and iron-limited growth. J Exp Mar Biol Ecol 29:1–34

Sakshaug E, Holm-Hansen O (1984) Factors governing pelagic production in polar oceans. In: Bolis O, Gilles R (eds) Marine phytoplankton and productivity. Lecture notes on coastal and estuarine studies 8, Springer, Berlin, pp 1–18

Sakshaug E, Holm-Hansen O (1986) Photoadaptation in Antarctic phytoplankton: variations in growth rate, chemical composition and P vs I curves. J Plank Res 8:459–473

Sakshaug E, Hopkins CCE, Øritsland NA (eds 1991) Proc Pro Mare Symp polar marine ecology, Trondheim, 12–16 May 1990. Pol Res 10(1–2):1–662

Sakshaug E, Myklestad S, Andresen K, Hegseth EN, Jørgensen L (1981) Phytoplankton off the Møre coast in 1975–1976: distribution, species composition, chemical composition and conditions for growth. In: Sætre R and Mork M (eds) The Norwegian Coastal Current. Univ Bergen, pp 688–711

Sakshaug E, Myklestad SM (1973) Studies of the phytoplankton ecology of the Trondheimsfjord. III. Dynamics of phytoplankton blooms in relation to environmental factors, bioassay experiments and parameters for the physiological state of the populations. J Exp Mar Biol Ecol 11:157–188

Sakshaug E, Rey F, Slagstad D (1995) Wind forcing of marine primary production in the northern atmospheric low-pressure belt. In: Skjoldal HR, Hopkins CEC, Erikstad KE, Leinaas, HP (eds) Ecology of fjords and Coastal Waters. Elsevier, Amsterdam, pp 15–25

Sakshaug E, Skjoldal HR (1989) Life at the ice edge. Ambio 18:60–67

Sakshaug E, Slagstad D (1991) Light and productivity of phytoplankton in polar marine ecosystems: a physiological view. Pol Res 10:69–85

Sakshaug E, Walsh JJ (2000) Marine biology: Biomass, productivity distributions and their variability in the Barents and Bering Seas. In: Nuttall M, Callaghan TV (eds) The Arctic: Environment, people, policy. Harwood, Amsterdam, pp 163–196

Sakshaug, E, Slagstad D (1992) Sea ice and wind: Effects on primary productivity in the Barents Sea. Atmos-Ocean 30:579–591

Saliot A, Cauwet G, Cahet G, Mazaudier D, Daumas R (1996) Microbial activities in the Lena River Delta and Laptev Sea. Mar Chem 53: 247–254

Salonen K, Sarvala J, Hakala I, Viljanen ML (1976) The relation of energy and organic carbon in aquatic invertebrates. Limnol Oceanogr 21:724–730

Sambrotto RN, Goering JJ (1983) Interannual variability of phytoplankton and zooplankton production on the southeast Bering Shelf. In: Wooster WS (ed) From year to year. Sea Grant Publ, Univ Washington, Seattle, pp 161–177

Sambrotto RN, Goering JJ, McRoy CP (1984) Large yearly production of phytoplankton in western Bering Strait. Science 225:1147–1150

Sambrotto RN, Niebauer HJ, Goering JJ, Iverson RL (1986) Relationships among vertical mixing, nitrate uptake, and phytoplankton growth during the spring bloom in the southeast Bering Sea middle shelf. Cont Shelf Res 5:161–199

Samoilovich YuG, Kagan LYa, Ivanova LV (1993) Quaternary deposits of the Barents Sea. Kola Sci Center RAS, Apatity pp 1–72 (in Russian)

Sampei M, Sasaki H, Hattori H, Kudoh S, Kashino Y, Fukuchi M (2002) Seasonal and spatial variability in the flux of biogenic particles in the North Water polynya in 1997–1998. Deep-Sea Res II 49: 5245–5257

Sandvik R, Dalløkken R, Sakshaug E (1996) Modelling primary production: Bio-optical parameters and data for validation. In: Wadhams P, Wilkinson JP, Wells SCS (eds) Sci Rep Eur Subpolar Ocean Progr (EC/MAST), Scott Polar Res Inst Cambridge, UK, pp 555–566

Sapozhnikov VV, Arzhanova NV, Zubarevitch VV (1993) Estimation of primary production in the ecosystem of the western Bering Sea. Russian J Aq Ecol 2:23–34

Sarmiento JL, Toggweiler JR (1984) A new model for the role of the oceans in determining atmospheric pCO_2. Nature 308:621–624

Sarnthein M, Pflaumann U, Ross R, Tiedemann R, Winn K (1992) Transfer functions to reconstruct ocean paleoproductivity: A comparison. In: Summerhayes CP, Prell WL, Emeis KC (eds), Upwelling systems. Evolution since the early Miocene. Geol Soc Spec Publ 64:411–427

Sarnthein M, Winn K and Zahn R (1987) Paleoproductivity of oceanic upwelling and the effect on atmospheric CO_2 and climatic change during deglaciation times. In: Berger WH and Labeyrie L (eds) Abrupt Climatic Change, Riedel Publ, Dordrecht, pp 311–337

Sasaki H, Hattori H, Matsuda O, Tanimura A, Fukuchi M (1993) Variability in the flux of rapidly sinking particles at the Chirikov Basin of the northern Bering Sea in summer 1988. La mer 31:73–82

Sasaki H, Taguchi S, Hattori H, Matsuda O, Tanimura A, Fukuchi M (1992) Variability in downward diatom flux in the northern Bering Sea in summer 1988. Porch. NIPR Symp. Pol Biol 5:166–167

Saupe SM, Schell DM, Griffiths WB (1989) Carbon-isotope ratio gradients in western arctic zooplankton. Mar Biol 103: 427–432

Sawyer VG, Semenov AD (1971) Photochemical methods of organic carbon determinations. Hydrochemical Materials 56:111–120 (in Russian)

Sayles FL, Martin WR, Deuser WG (1994) Response of benthic oxygen demand to particulate organic carbon supply in the deep sea near Bermuda. Nature 371:686–689

Schäfer P, Ritzrau W, Schlüter M, Thiede J (eds) (2001) The Northern North Atlantic: a changing environment. Springer, Berlin, 500pp

Schauer U, Muench R, Rudels B, Timokhov L (1997) Impact of eastern Arctic shelf waters on the Nansen Basin intermediate layers. J Geophys Res 102(C2):3371–3382

Schauer U, Rudels B, Jones EP, Anderson LG, Muench RD, Björk G et al. (2002) Confluence and redistribution of Atlantic water in the Nansen, Amundsen and Makarov basins. Annales Geophysicae 20:257–273

Scheffer F, Schachtschabel P (1984) Lehrbuch der Bodenkunde. Enke Verlag, Stuttgart, 442pp

Schell DM (1983) Carbon-13 and carbon-14 abundances in Alaskan aquatic organisms: Delayed production from peat in Arctic food webs. Science 219:1068–1071

Schell DM, Barnett BA, Vinette KA (1998) Carbon and nitrogen isotope ratios in zooplankton of the Bering, Chukchi and Beaufort Seas. Mar Ecol Prog Ser 162:11–23

Schell DM, Saupe SM, Haubenstock N (1989) Bowhead whale (Balaena mysticetus) growth and feeding as estimated by $\delta^{13}C$ techniques. Mar Biol 103:433–443

Schell DM, Saupe S (1989) Primary production, carbon energetics, and nutrient cycling. In: Truett JC (ed) Environmental characterization and biological utilization of the North Aleutian Shelf nearshore zone. MMS/NOAA OCSEAP Rep 60, pp 99–140

Schell DM, Zieman PJ (1983) Accumulation of peat carbon in the Alaska Arctic Coastal Plain and its role in biological productivity. In: PERMAFROST, Proceedings of the Fourth International Conference. National Academy Press, Washington DC: Fairbanks, Ak; 1983. pp 1105–1110

Schewe I (2001) Small-sized benthic organisms of the Alpha Ridge, central Arctic Ocean. Int Revue ges Hydrobiol 86: 317–335

Schewe I (2002) Leben unter permanenter Eisbedeckung: Effekte einer variablen Nahrungsverfügbarkeit auf die Ökologie und Verteilung kleinster benthischer Organismen in der Arktischen Tiefsee. Ph D thesis University Bremen, 165 pp

Schewe I, Soltwedel T (2003) Benthic response to ice-edge induced particle flux in the Arctic Ocean. Polar Biol, in press

Schlanger SO and Jenkyns HC (1976) Cretaceous oceanic anoxic events – causes and consequences. Geologie en Mijnbouw 55:179–184

Schlessinger WH, Melack JM (1981) Transport of organic carbon in the world rivers. Tellus 31:172–187

Schlitzer R (2000) Applying the adjoint method for biogeochemical modeling: export of particulate organic matter in the world ocean In: Kasibhatla P, Heimann M, Rayner P, Mahowald N, Prinn RG, Hartley DE (eds) Inverse methods in global biogeochemical cycles. Geophys Monogr Ser 114:107–124

Schlosser P, Bullister JL, Fine R, Jenkins WJ, Key R, Lupton J, Roether W, Smethie WM Jr (2001) Transformation and age of water masses. In: Siedler G, Church J, Gould J (eds) Ocean circulation and climate; observing and modelling the global ocean. Academic Press, San Diego, pp 431–452

Schlosser P, Kromer B, Ekwurzel B, Bönisch G, McNichol A, Schneider R et al. (1997) The first trans-Arctic ^{14}C section: comparison of the mean ages of the deep waters in the Eurasian and Canadian basins of the Arctic Ocean. Nuclear Instruments and Methods in Physics Research B 123:431–437

Schlosser P, Newton R, Ekwurzel B, Khatiwala S, Mortlock R, Fairbanks R (2002) Decrease of river runoff in the upper waters of the Eurasian Basin, Arctic Ocean, between 1991 and 1996: evidence from $\delta^{18}O$ data. Geophys Res Lett, 29: 10.1029/2001GL013135, 3-1-3-4

Schlosser P, Swift JH, Lewis D, Pfirman SL (1995) The role of large-scale Arctic Ocean circulation in the transport of contaminants. Deep Sea Res 42:1341–1367

Schlünz B, Schneider RR (2000) Transport of terrestrial organic carbon to the oceans by rivers: re-estimating flux and burial rates. Int J Earth Sci 88:752–763

Schlüter M, Pfannkuche O (2000) Benthic fluxes of nutrients and oxygen. Atlantic Data Base for Exchange Processes at the Deep Sea Floor. Final Report on a project of ADEPD. IOW. Warnemuende, pp 22–25

Schlüter M, Sauter E, Schäfer A, Ritztrau W (2000) Spatial budget of organic carbon flux to the seafloor of the northern North Atlantic (60°N–80°N). Glob Biogeochem Cycles 14:329–340

Schlüter M, Sauter EJ, Schulz-Bull D, Balzer W, Suess E (2001) Fluxes of organic carbon and biogenic silica reaching the seafloor: a comparison of High Northern and Southern latitudes of the Atlantic Ocean In: Schäfer P, Ritzrau W, Schlüter M, Thiede J (eds) The northern North Atlantic: A changing environment Springer Verlag, Berlin, pp 225–240

Schmid MK (1996) Zur Verbreitung und Respiration ökologisch wichtiger Bodentiere in den Gewässern um Svalbard (Arktis). Reps Pol Res 202, 93 pp

Schnack K (1998) Besiedlungsmuster der benthischen Makrofauna auf dem ostgrönlandischen Kontinentalhang. Reps Pol Res 294, 124 pp

Schneider W, Budéus G (1994) The North East water polynya (Greenland Sea). Polar Biol 14:1–9

Schneider W, Budéus G (1997) Summary of the Northeast Water Polynya formation and development (Greenland Sea). J Mar Syst 10:107–122

Schoster F, Levitan MA (2003) Scientific Cruise Report of the Joint Russian-German Kara-Sea Expedition of RV "Akademik Boris Petrov" in 2002. Reps Pol Mar Res 450, 109 pp

Schubert CJ (1995). Organischer Kohlenstoff in spätquartären Sedimenten des Arktischen Ozeans: Terrigener Eintrag und marine Produktivität. Reps Pol Res 177, 178 pp

Schubert CJ, Calvert SE (2001) Nitrogen and carbon isotopic composition of marine and terrestrial organic matter in Arctic Ocean sediments: implications for nutrient utilization and organic matter composition. Deep-Sea Res 48: 789–810

Schubert CJ, Stein R (1996) Deposition of organic carbon in Arctic Ocean sediments: Terrigenous supply vs marine productivity. Org Geochem 24:421–436

Schubert CJ, Stein R (1997) Lipid distribution in surface sediments from the eastern central Arctic Ocean. Mar Geol 138:11–25

Schubert CJ, Stein R, Calvert SE (2001) Tracking nutrient and productivity variations over the last deglaciation in the Arctic Ocean. Paleoceanography 16:199–211

Schultes S, Levasseur M, Michaud S, Cantin G, Wolfe G, Gosselin M, de Mora S (2000) Dynamics of dimethylsulfide production from dissolved dimethylsulfoniopropionate in the Labrador Sea. Mar Ecol Prog Ser 202:27–40

Schulz DJ, Calder JA (1976) Organic carbon $^{13}C/^{12}C$ variations in estuarine sediments. Geochim Cosmochim Acta 40: 381–385

Schwinghamer P, Hargrave B, Peer D, Hawkins CM (1986) Partitioning of production and respiration among size groups of organisms in an intertidal benthic community. Mar Ecol Prog Ser 31:131–142

Scott CL, Falk-Petersen S, Sargent JR, Hop H, Lønne O-J, Poltermann M (1999) Lipids and trophic interactions of ice fauna and pelagic zooplankton in the marginal ice zone of the Barents Sea. Pol Biol 21:65–70

Scully NM, Miller WL (2000) Spatial and temporal dynamics of colored dissolved organic matter in the north water polynya. Geophys Res.Lett 27:1009–1011

Seiler D (1999) Struktur und Kohlenstoffbedarf des Makrobenthos am Kontinentalhang Ostgrönlands. Reps Pol Res 307, 96 pp

Semenov AD (ed) (1977) Handbook of Chemical Analysis of Land Waters. Gidrometeoizdat, 541 pp (in Russian)

Semiletov IP (1999a) On aquatic sources and sinks of CO_2 and CH_4 in the Polar Regions. J Atmosph Sci 56(2): 286–306

Semiletov IP (1999b) Destruction of the coastal permafrost ground as an important factor in biogeochemistry of the Arctic Shelf waters. Transactions (Doklady) of Russian Academy of Sciences 368(6):679–682

Semiletov IP (2000) Greenhouse gases in the Arctic atmosphere: Terrestrial and marine sources of carbon dioxide and methane. In: Semiletov IP (ed) Hydrometeorological and Biogeochemical Research in the Arctic. Trudy Arctic Regional Center, 2, Vladivostok: 117–130 (in Russian)

Semiletov IP, Savelieva NI, Weller GE, Pipko II, Pugach SP, Gukov AY et al. (2000) The dispersion of Siberian river flows into coastal waters: meteorological, hydrological and hydrochemical aspects. In: Lewis EL, Jones EP, Lemke P, Prowse TD, Wadhams P (eds) The Freshwater Budget of the Arctic Ocean. Kluwer Academic Publishers, London, pp 323–366

Semyonov YuP (1965) Some peculiarities in the formation of bottom sediments of the East-Siberian and Chuckchee Seas Proceedings of NIIGA, Leningrad, 143:350–352

Serova VV, Gorbunova ZN (1997) Mineral composition of soils, aerosols, suspended matter and bottom sediments of the Lena River estuary and Laptev Sea. Oceanology 37(1)131–135

Serreze MC, Maslanik JA, Key JR, Kokaly RF (1995) Diagnosis of the record minimum in Arctic sea ice area during 1990 and associated snow cover extremes. Geophys Res Lett 22:2183–2186

Serreze MC, Walsh JE, Chapin FSI, Osterkamp T, Dyurgerov M, Romanovsky V et al. (2000) Observational evidence of recent change in the northern high-latitude environment. Clim Change 46:159–207

Shamov GI (1949) Discharge of suspended sediments in the USSR rivers. Trudy GGI 20 (74), Leningrad (in Russian)

Shapiro J (1961) Freezing-out, a safe technique for concentration of dilute solutions. Science 133:2063–2206

Sharma GD (1979) The Alaskan Shelf: hydrography, sedimentary and geochemical environment. Springer, Berlin Heidelberg New York, pp 1–498

Sherwood CR (2000) Numerical model of frazil ice and suspended sediment concentrations and formation of sediment laden ice in the Kara Sea. J Geophys Res 105(C6): 14061–14080

Shevchenko VP (2000) Aerosols – influence on accumulation of sediments and Arctic environment. Abstr Diss. IO RAS, Moscow, pp 1–32 (in Russian)

Shevchenko VP, Ivanov GI, Burovkin AA, Dzhinoridze RN, Zernova VV, Polyak LV, Shanin SS (1998) Sedimentary material flows in the St. Anna Trough and eastern Barents Sea. Doklady Earth Sciences (English Translation) 359A (3):400–403

Shevchenko VP, Ivanov GI, Zernova VV (1999) Vertical particle fluxes in the Saint Anna Trough and in the Eastern Barents Sea in August–September 1994. Reps Pol Res 342: 46–54

Shevchenko VP, Lisitzin AP, Kuptsov VM, Van Malderen H, Martin J-M, Van Grieken R, Huang WW (1999a) Composition of aerosols in the surface boundary layer of the atmosphere over the seas of the Western Russian Arctic. Oceanology 39/1:128–136

Shevchenko VP, Lisitzin AP, Polyakova EI (1999b) Abstracts of Reports at the XIII International Conference on Marine Geology, Moscow, GEOS 1:237–238

Shevchenko VP, Lisitzin AP, Stein R, Serova VV, Isaeva AB, Politova NV (1999c) The composition of the coarse fraction of aerosols in the marine boundary layer over the Laptev, Kara and Barents Seas. In: Kassens H, Bauch H, Dmitrenko I, Eicken H, Hubberten H-W, Melles M, Thiede J, Timokhov L (eds) Land-Ocean-Systems in the Siberian Arctic: Dynamics and History, Springer, Berlin, pp 53–59

Shevchenko VP, Lisitzin AP, Vinogradova AA, Smirnov VV, Serova VV, Stein R (2000) Arctic aerosols – results of ten-year investigations. Atmospheric and Oceanic Optics 13/6-7:510–533

Shevchenko VP, Lisitzin AP, Zernova VV, Lukashin VN, Politova NV, Rusakov VYu, Shanin SS (2000) Vertical particle fluxes in seas of the western Russian Arctic. In: Humanity and the World Ocean. Interdependence on the Dawn of the New Millennium. PACON 99 Proceedings. PACON International, p 239–249

Shiklomanov IA (1998) Comprehensive Assessment of the Freshwater Resources of the World: Assessment of eater Resources and Water Availability in the World. WMO, UNDP, UNED, FAO et al., WMO, Geneva, 88 pp

Shiklomanov IA, Shiklomanov AI, Lammers RB, Vörösmarty CJ, Peterson BJ, Fekete B (2000) The dynamics of river water inflow to the Arctic Ocean. In: Lewis EL, Jones EP, Lemke P, Prowse TD, P Wadhams (eds) The freshwater budget of the Arctic Ocean: Proceedings of the NATO advanced research workshop. Kluwer, Dortrecht, pp 281–296

Shiklomanov IA, Skakalsky BG (1994) Studying water, sediment and contaminant runoff of Siberian Rivers. Modern status and prospects. Arctic Research of the United States 8:295–306

Shindell DT, Miller RL, Schmidt GA, Pandolfo L (1999) Simulation of recent northern winter climate trends by greenhouse-gas forcing. Nature 399:452–455

Shiomoto A (1997) Size-fractionated chlorophyll a concentration in the Okhotsk Sea in October and November 1993, with special reference to the influence of dichothermal water. J Oceanogr 53:601–610

Shiomoto A (1999) Effect of nutrients on phytoplankton size in the Bering Sea Basin. In: Loughlin TR, Ohtani K (eds) The Physical Oceanography of the Bering Sea. Univ Alaska Seagrant AK-SG-03, Univ Alaska Fairbanks, pp 323–340

Sholkovitz ER (1976) Flocculation of dissolved organic and inorganic matter during the mixing of river water and seawater. Geochim Cosmochim Acta 40:831–845

Shuisky YuD (1983) Modern sediment balance in coastal zone. Abstract of Doctor Dissertation, Moscow State University, 41 pp (in Russian)

Shuisky YuD (1986) Research on the sediment balance in the coastal zone. Gidrometeoizdat Press, Leningrad, 240 pp (in Russian)

Shuisky YuD, Ogorodnikov VI (1981) Sedimentation conditions and main regularities of the formation of the granulometric composition of terrigenous deposits in the Chuckchee Sea. Lithology and minerals 2:11–25 (in Russian)

Shultz DJ, Calder JA (1976) Organic carbon $^{13}C/^{12}C$ variations in estuarine sediments. Geochim Cosmochim Acta 40: 381–385

Shvets VM (1973) Organic substances in under-ground waters. Nedra, Moscow, 191 pp (in Russian)

Sidorchuk AY, Panin AV (1996) Water supply from the Yana River basin since Late Pliocene. Terra Nostra 9:97

Sikes EL, Farrington JW, Keigwin LD (1991) Use of the alkenone unsaturation ratio U^{k}_{37} to determine past sea surface temperatures: core-top SST calibrations and methodology considerations. Earth Planet Sci Lett 104:36–47

Sime-Ngando T, Juniper SK, Demers S (1997) Ice-brine and planktonic microheterotrophs from Saroma-ko Lagoon, Hokkaido (Japan): quantitative importance and trophodynamics. J Mar Syst 11:149–161

Simó R, Perdós-Alió R, Mallin G, Grimalt JO (2000) Biological turnover of DMS, DMSP and DMSO in contrasting open-sea waters. Mar Ecol Prog Ser 203:1–11

Simoneit BRT (1977) Organic matter in eolian dusts over the Atlantic Ocean. Mar. Chem. 5: 443–464

Simoneit BRT, Elias VO (2000) Organic tracers from biomass burning in atmospheric particulate matter over the ocean. Mar Chem 69:301–312

Sirenko BI (2001) List of species of free-living invertebrates of Eurasian Arctic Seas and adjacent deep waters. Russian Academy of Science, Exploration of the Fauna of the Sea 51:1–129

Skjoldal HR, Hassel A, Rey F, Loeng H (1987) Spring phytoplankton development and zooplankton reproduction in the central Barents Sea in the period 1979–84. In: Loeng H (ed) Proc. 3d Soviet-Norwegian Symp Murmansk 1986, Inst Mar Res, Bergen, Norway, pp 59–89

Skjoldal HR, Rey F (1989) Pelagic production and variability of the Barents Sea ecosystem. In: Sherman K, Alexander LM (eds) Biomass yeilds and geography of large marine ecosystems. AAAS Selected Symposium 111, Washington DC, pp 241–286

Skoog A, Benner R (1997) Aldoses in various size fractions of marine organic matter: implications for carbon cycling. Limnol Oceanogr 42:1803–1813

Skopintsev BA, Krylova LP (1955) Organic substances input by the biggest rivers of the Soviet Union. Reports of the USSR Academy of Sciences (Doklady Akademii Nauk SSSR) 105 (4):770–773 (in Russian)

Slagstad D, Støle-Hansen K (1991) Dynamics of plankton growth in the Barents Sea: model studies. Pol Res 10: 173–186

Slagstad D, Wassmann P (1996) Climate change and carbon flux in the Barents Sea: 3-D simulations of ice-distribution, primary production and vertical export of particulate organic carbon. Memoirs of the National Institute of Polar Research, Tokyo, Special Issue 51:119–141

Smart CW, Gooday AJ (1997) Recent benthic foraminifera in the abyssal northeast Atlantic Ocean: Relation to phytodetrital inputs. J Foram Res 27:85–92

Smirnov MP, Tarasov MN, Maltseva AV (1978) Organic carbon discharge from the USSR territory to the seas of the Arctic Ocean. Hydrochemical Materials 66:62–76 (in Russian)

Smirnov MP, Tarasov MN, Maltseva AV (1988) Organic carbon discharge from the USSR territory and its temporal variability. Hydrochemical Materials 103:67–83 (in Russian)

Smirnov VV, Shevchenko VP, Stein R, Lisitzin AP, Savchenko AV, ARK XI/1 Polarstern Shipboard Scientific Party (1996) Aerosol size distribution over the Laptev Sea in July–September 1995: First results. Reps Pol Res 212:139–143

Smith D (1998) Recent increase in the length of the melt season of perennial Arctic sea ice. Geophys Res Lett 25: 655–658

Smith DG (1987) Late Paleozoic to Cenozoic reconstructions of the Arctic. In: Tailleur IL, Weimer P (eds) Alaskan North Slope Geology, Soc Econ Paleont Mineral, Spec Publ 50: 785–796

Smith JN, Ellis KM, Kilius LR (1998) ^{129}I and ^{137}Cs tracer measurements in the Arctic Ocean. Deep-Sea Res I 45: 959–984

Smith KL jr, Carlucci AF, Jahnke RA, Craven DB (1987) Organic carbon mineralization in the Santa Clara Basin: benthic boundary layer metabolism Deep-Sea Res II 34: 185–211

Smith OS jr, Grebmeier JM (eds) (1995) Arctic oceanography: marginal ice zones and continental shelves. Am Geophys Union, Washington, DC, pp 231–261

Smith RC, Baker KS (1981) Optical properties of the clearest natural waters (200–800 nm). Appl Optics 20:177–184

Smith REH, Gosselin M, Kudoh S, Robineau B, Taguchi S (1997) DOC and its relationship to algae in bottom ice communities. J Mar Syst 11:71–80

Smith SD, Muench RD, Pease CH (1990) Polynyas and leads: an overview of physical processes and environment. J Geophys Res 95:9461–9479

Smith SL (1991) Growth, development and distribution of the euphausiids *Thysanoessa raschi* (M. Sars) and *Thysanoessa inermis* (Krøyer) in the southeastern Bering Sea. Pol Res 10:461–478

Smith SL, Schnack-Schiel SB (1990) Polar zooplankton. In: Smith WO jr (ed) Polar oceanography part B. Academic Press New York, pp 527–598

Smith WO jr (1990 ed) Polar oceanography part B. Academic Press, New York, 406 pp

Smith WO jr, Codispoti L, Nelson DM, Manley T, Buskey EJ, Niebauer HJ, Cota GF (1991) Importance of *Phaeocystis* blooms in the high-latitude carbon cycle. Nature 352: 514–516

Smith WO jr, Gosselin M, Legendre L, Wallace D, Daly K, Kattner G (1997) New production in the Northeast Water Polynya. J Mar Syst 10:199–209

Smith WO jr, Sakshaug E (1990) Polar phytoplankton. In: Smith WO jr (ed) Polar oceanography part B. Academic Press New York, pp 477–525

Smith WO, Gordon LI (1997) Hyperproductivity of the Ross Sea (Antarctica) polynya during austral spring. Geophys Res Lett 24:233–236

Snelgrove PVR, Butman CA (1994) Animal-sediment relationships revisited: Cause versus effect. In: Butman CA, Ansell AD, Gibson RN, Barnes M (eds) Oceanography and Marine Biology: an Annual Review. UCL Press, London, 32, pp 111–177

Solheim A, Elverhoi A (1996) Surface sediments of the northwestern Barents Sea. Reps Pol Res 212:144–158

Solomon SM, Forbes DL (1994) Impacts of climate change on the Beaufort Sea coastal zone. In: Wells PG, Ricketts PJ (eds) Coastal Zone Canada '94, Cooperation in the Coastal Zone (vol 4), Halifax, NS, pp 1810–1822

Soltwedel T, Mokievsky V, Schewe I (2000) Benthic activity and biomass on the Yermak Plateau and in adjacent deep-sea regions northwest of Svålbard. Deep Sea Res I (47): 1761–1785

Soltwedel T, Schewe I (1998) Activity and biomass of the small benthic biota under permanent ice-coverage in the central Arctic Ocean. Polar Biol 19 (1):52–62

Søndergaard M le B, Williams PJ, Cauwet G, Riemann B, Robinson C, Terzic S, Woodward EMS, Worm J (2000) Net accumulation and flux of dissolved organic carbon and dissolved organic nitrogen in marine plankton communities. Limnol Oceanogr 45:1097–1111

Sorokin Y, Sorokin P (1999) Production in the Sea of Okhotsk. J Plank Res 21:201–230

Sorokin YuA, Sorokin PYu (1996) Plankton and primary production in the Lena River estuary and in the South-eastern Laptev Sea. Est Coast Shelf Sci. 43: 399–418

Sournia A (ed 1978) Phytoplankton manual. UNESCO, Paris, 337 pp

Sovershaev VA (1996) The research of the sea coasts in the cryolithozone with a view of rational economic developing. In: Materials of the first conference of Russian geocryologists Vol 1. Dynamic geocryology. Moscow Univ Publ House, Moscow, pp 494–503 (in Russian)

Sowers T, Bender M (1995) Climate records covering the last deglaciation. Science 269:210–214

Spielhagen R (2001) Enigmatic Arctic ice sheets. Nature 410:427–428

Spitzy A, Ittekkot V (1991) Dissolved and Particulate Organic Matter in Rivers. In: Mantoura RFC, Martin J-M, Wollast R (eds) Ocean Margin Processes in Global change, Wiley & Sons, New York, pp 5–17

Spitzy A, Leenheer J (1991) DOC in rivers. In: Degens ET, Kempe S, Richey JE (eds) Biogeochemistry of Major World Rivers. SCOPE 42, Wiley, Chichester

Springer AM (1998) Is it all climate change? Why marine bird and mammal populations fluctuate in the North Pacific. In: Holloway G, Muller P, Henderson D (eds) Biotic impacts of extratropical climate variability in the Pacific, Proceedings 'Aha Huliko'a Hawaiin Winter Workshop. University of Hawaii, Honalulu, pp 109–119

Springer AM, McRoy C, Flint MV (1996) The Bering Sea green belt: shelf edge processes and ecosystem production. Fish Oceanogr 5:205–223

Springer AM, McRoy CP (1993) The paradox of pelagic food webs in the northern Bering Sea-III. Patterns of primary production. Cont Shelf Res 13:575–599

Springer AM, McRoy CP, Turco KR (1989) The paradox of pelagic food webs in the northern Bering Sea-II. Zooplankton communities. Cont Shelf Res 9: 359–386

Springer AM, Roseneau DG (1985) Copepod-based food webs: auklets and oceanography in the Bering Sea. Mar Ecol Prog Ser 21:229–237

Stabeno P, Overland JE (2001) Bering Sea shifts toward an earlier spring transition. EOS Transactions AGU 82:317, 321

Stabeno PJ, Bond NA, Kachel NB, Salo SA, Schumacher JD (2001) On the temporal viability of the physical environment over the south-eastern Bering Sea. Fish Oceanogr 10:81–98

Stabeno PJ, Schumacher JD, Ohtani K (1999) The physical oceanography of the Bering Sea. In: Loughlin TR, Ohtani K (eds) Dynamics of the Bering Sea. Univ Alaska Seagrant AK-SG-03, Univ Alaska Fairbanks, pp 1–28

Stapleford LS, Smith REH (1996) The interactive effects of temperature and silicon limitation on the psychrophilic ice diatom *Pseudonitzschia seriata*. Pol Biol 16:589–594

Starmans A (1997) Vergleichende Untersuchungen zur Ökologie und Diversität des Mega-Epibenthos der Arktis und Antarktis. Reps Pol Res 250, 150 pp

Steele M, Boyd T (1998) Retreat of the cold halocline layer in the Arctic Ocean. J Geophys Res 103(C5):10419–10435

Steemann-Nielsen E (1935) The production of phytoplankton at the Faroe Isles, Iceland, East Greenland and in the waters around. Medd Komm Danm Fisk-Havunders 3(1): 1–93

Stefansson U, Thordardottir T, Olafsson J (1987) Comparison of seasonal oxygen cycles and primary production in the Faxafloi region, Southwest Iceland. Deep-Sea Res 34: 725–739

Stein R (1986) Surface-water paleo-productivity as inferred from sediments deposited in oxic and anoxic deep-water environments of the Mesozoic Atlantic Ocean. In: Degens ET et al. (eds) Biochemistry of Black Shales, Mitt Geol Paläont Inst Univ Hamburg, 60:55–70

Stein R (1990) Organic carbon content/sedimentation rate relationship and its paleoenvironmental significance for marine sediments. Geo-Mar Lett 10:37–44

Stein R (1991) Accumulation of organic carbon in marine sediments. Lecture Notes in Earth Sciences 34, Springer, Heidelberg, 217 pp

Stein R (1996) Organic-carbon and carbonate distribution in surface sediments from the Eastern Central Arctic Ocean and the Eurasian Continental Margin: sources and pathways. Reps Pol Res 212:243–267

Stein R (2001) Lithostratigraphy of gravity cores and correlation with sediment echograph profiles ("Akademik Boris Petrov" Kara Sea expeditions 1999 and 2000). Reps Pol Mar Res 393:120–140

Stein R (ed) (1998) Arctic Paleo-River Discharge (APARD) – A new research programme of the Arctic Ocean Science Board (AOSB). Reps Pol Res 279, 127 pp

Stein R (ed) (2000) Circum Arctic River Discharge and Its Geological Record. Int J Earth Sci 89:447–616

Stein R, Boucsein B, Fahl K, Garcia de Oteyza T, Knies J, Niessen F (2001) Accumulation of particulate organic carbon at the Eurasian continental margin during late Quarternary times: controlling mechanisms and paleoenvironmental signficance. Glob Planet Change 31:87–102

Stein R, Boucsein B, Hefter J, Kraus M, Lau N, Schoster F, Simstich J, Weiel D (2000) Marine Geology. In: Stein R, Stepanets OV (eds) Scientific Cruise Report of the Joint Russian-German Kara-Sea Expedition of RV "Akademik Boris Petrov" in 1999. Reps Pol Res 360:49–69

Stein R, Fahl K (2000) Holocene accumulation of organic carbon at the Laptev Sea continental margin (Arctic Ocean): sources, pathways, and sinks. Geo-Mar Lett 20: 27–36

Stein R, Fahl K (eds) (1997) Scientific Cruise Report of the Arctic Expedition ARK-XIII/2 of RV "Polarstern" in 1997. Alfred Wegener Institute for Polar and Marine Research, Bremerhaven. Reps Pol Res 255, 235 pp

Stein R, Fahl K, Dittmers K, Niessen F, Stepanets OV (2003a) Holocene siliciclastic and organic carbon fluxes in the Ob and Yenisei estuaries and the adjacent inner Kara Sea: Quantification, variability, and paleoenvironmental implications. In: Stein R, Fahl K, Fütterer DK, Galimov EM, Stepanets OV (eds) Siberian River Run-off in the Kara Sea: Characterisation, Quantification, Variability, and Environmental Significance, Proceedings in Marine Sciences Vol 6, Elsevier, Amsterdam, pp 401–434

Stein R, Fahl K, Fütterer DK, Galimov EM, Stepanets OV (eds) (2003b) Siberian River Run-off in the Kara Sea: Characterisation, Quantification, Variability, and Environmental Significance, Proceedings in Marine Sciences Vol 6, Elsevier, Amsterdam, 484 pp

Stein R, Fahl K, Ivanov GI, Levitan MA, Tarasov G (eds) (1999b) Modern and Late Quaternary Depositional Environment of the St. Anna Trough Area, Northern Kara Sea. Reps Pol Res 342, 244 pp

Stein R, Fahl K, Niessen F, Siebold M (1999a) Late Quarternary organic carbon and biomarker records from the Laptev Sea continental margin (Arctic Ocean): Implications for organic carbon flux and composition. In: Kassens H, Bauch HA, Dmitrenko I, Eicken H, Hubberten H-W, Melles M et al. (eds) Land-Ocean Systems in the Siberian Arctic: Dynamics and History. Springer, Berlin, pp 635–655

Stein R, Grobe H, Wahsner M (1994a) Organic carbon, carbonate, and clay mineral distributions in eastern central Arctic Ocean surface sediments. Mar Geol 119:269–285

Stein R, Korolev S (1994) Shelf-to-basin sediment transport in the eastern Arctic Ocean. Reps Pol Res 144:87–100

Stein R, Nam SI, Schubert C, Vogt C, Fütterer D, Heinemeier J (1994b) The last deglaciation event in the eastern central Arctic Ocean. Science 264:692–696

Stein R, Niessen F, Dittmers K, Levitan M, Schoster F, Simstich J, Steinke T, Stepanets OV (2002) Siberian river run-off and Late Quaternary glaciation in the southern Kara Sea: Preliminary results. Pol Res 21:315–322

Stein R, Nürnberg D (1995) Productivity proxies: Organic carbon and biogenic opal in surface sediments from the Laptev Sea shelf and the adjacent continental slope. Reps Pol Res 176:286–296

Stein R, Rullkötter J, Welte DH (1986) Accumulation of organic-carbon-rich sediments in the Late Jurassic and Cretaceous Atlantic Ocean – A synthesis. Chem Geol 56: 1–32

Stein R, Schubert C, Vogt C, Fütterer D (1994c) Stable isotope stratigraphy, sedimentation rates, and salinity changes in the Latest Pleistocene to Holocene eastern central Arctic Ocean. Mar Geol 119:333–355

Stein R, Stepanets OV (eds) (2000) Scientific Cruise Report of the Joint Russian-German Kara-Sea Expedition of RV "Akademik Boris Petrov" in 1999. Reps Pol Res 360, 141 pp

Stein R, Stepanets OV (eds) (2001) The German-Russian Project on Siberian River Run-off (SIRRO): Scientific Cruise Report of the Kara Sea Expedition "SIRRO 2000" of RV "Akademik Boris Petrov" and First Results. Reps Pol Mar Res 393, 287 pp

Stein R, Stepanets OV (eds) (2002) Scientific Cruise Report of the Kara Sea Expedition 2001 of RV "Akademik Boris Petrov": The German-Russian Project on Siberian River Run-off (SIRRO) and the EU-Project "ESTABLISH". Reps Pol Mar Res 419, 278 pp

Steinhauer MS, Boehm PD (1992) The composition and distribution of saturated and aromatic hydrocarbons in nearshore sediments, river sediments, and coastal peat of the Alaskan Beaufort Sea: implications for detecting anthropogenic hydrocarbon inputs. Mar Environ Res 33: 223–253

Stephantsev L, Shmelkov B (2000) Brief characteristics and results from measurements of hydrophysical structures of waters in the estuaries of Ob and Yenisei. Reps Pol Res 360:8–20

Stevenson FJ, Cheng CN (1972) Organic geochemistry of the Argentine Basin sediments: Carbon-nitrogen relationships and Quaternary correlations. Geochim Cosmochim Acta 36:653–671

Stewart RE (2000) The variable climate of the Mackenzie River Basin: Its water cycle and fresh water discharge. In: Lewis EL, Jones EP, Lemke P, Prowse TD, Wadhams P (eds) The Freshwater Budget of the Arctic Ocean. Kluwer Academic Publishers, Netherlands, pp 367–381
Stewart RE, Leighton HG, Marsh P, Moore GWK, Ritchie H, Rouse WR et al. (1998) The Mackenzie GEWEX study: The water and energy cycles of a major North American River Basin. Bull Amer Meteorol Soc 79:2665–2683
Stirling I (1997) The importance of polynyas, ice edges, and leads to marine mammals and birds. J Mar Syst 10:9–21
Stirling I (2002) Polar bears and seals in the eastern Beaufort Sea and Amundsen Gulf: A synthesis of population trends and ecological relationships over three decades. Arctic 55: 59–76
Stirling I, Andriashek D, Calvert W (1993) Habitat preferences of polar bears in the western Canadian Arctic in late winter and spring. Pol Rec 29:13–24
Stirling IS, Cleator H (1981) Polynyas in the Canadian Arctic. 46: Canadian Wildlife Service, 70 pp
Stockwell DA, Whitledge TE, Rho T, Stabeno PJ, Coyle KO, Napp JM et al. (2000) Field observations in the south-eastern Bering Sea during three years with extensive coccolithophorid blooms. EOS Transactions AGU 80: OS41
Stockwell DA, Whitledge TE, Zeeman SI, Coyle KO, Napp JN, Brodeur RD, Pinchuk AI, Hunt GL jr (2001) Anomalous conditions in the south-eastern Bering Sea, 1997: nutrients, phytoplankton and zooplankton. Fish Oceanogr 10: 99–116
Stowe LG, Teeri JA (1978) The geographic distribution of C4 species of the Dicotyledonae in relation to climate. Amer Nat 112: 609–623
Strass VH, Nöthig E-M (1996) Seasonal shifts in ice edge phytoplankton blooms in the Barents Sea related to the water column stability. Pol Biol 16:409–422
Strömberg JO (1989) Northern Svalbard Waters. In: Rey Alexander L & V (eds) Proceedings of the sixth conference of the Comité Arctique International, 13–15 May 1985: 402–426
Stuart V, Sathyendranath S, Head EJH, Platt T, Irwin B, Maass H (2000) Bio-optical characteristics of diatom and prymnesiophyte populations in the Labrador Sea. Mar Ecol Prog Ser 201:91–106
Stuiver M, Reimer PJ, Bard E, Beck JW, Burr GS, Hughen KA, Kromer B, McCormick G, van der Pflicht J, Spurk M (1998) INTCAL98 radiocarbon age calibration, 24,000–0 cal BP. Radiocarbon 40 (3):1041–1083
Subba Rao DV, Platt T (1984) Primary production of Arctic waters. Pol Biol 3:191–201
Suess E (1980) Particulate organic carbon flux in the oceans – surface productivity and oxygen utilisation. Nature 288:260–263
Sukhanova IN, Flint MV (2001) *Phaeocystis pouchetii* over the Eastern Bering Sea shelf. Oceanology 41:75–85
Sukhanova IN, Semina HJ, Venttsel MV (1999) Spatial distribution and temporal variability of phytoplankton in the Bering Sea. In: Loughlin TR, Ohtani K (eds) The physical oceanography of the Bering Sea. Univ Alaska Seagrant AK-SG-03, Univ Alaska Fairbanks, pp 453–483
Sullivan KA, Aller RC (1996) Diagenetic cycling of arsenic in Amazon shelf sediments. Geochim Cosmochim Acta 60: 1465–1477
Sültemeyer D (1998) Carbonic anhydrase in eukaryotic algae: characterization, regulation, and possible function during photosynthesis. Can J Bot 76:962–972
Sunda WG, Kieber DJ, Kiene RP, Huntsman S (2002) An antioxidant function for DMSP and DMS in marine algae. Nature 418:317–320
Sundby B, Gobeil C, Silverberg N, Mucci A (1992) The phosphorus cycle in coastal marine sediments. Limnol Oceanogr 37:1129–1145.
Suzdal'sky OV (1974) Lithodynamics of near-shore shallows of the White, Barents and Kara Seas. Sea Geology 3: 27–33 (in Russian)
Suzuki Y, Kudoh S, Takahashi M (1997) Photosynthetic and respiratory characteristics of an Arctic ice algal community living in low light and low temperature conditions. J Mar Syst 11:111–121
Svendsen J-I, Astakhov VI, Bolshiyanov DY, Demidov I, Dowdeswell JA, Gautallin V, Hjort. Ch, Hubberten HW, Larsen E, Mangerud J, Melles M, Möller P, Saarnisto M, Siegert MJ (1999) Maximum extent of the Eurasian ice sheets in the Barents and Kara Sea region during the Weichselian. Boreas 28:234–242
Svendsen JI, Elverhøi A, Mangerud J (1996) The retreat of the Barents Sea Ice Sheet on the western Svalbard margin. Boreas 25:244–256
Sverdrup HU (1953) On conditions for the vernal blooming of phytoplankton. J Cons Perm Int Explor Mer 18:287–295
Svindland KT, Vorren TO (2002). Late Cenozoic sedimentary environments in the Amundsen Basin, Arctic Ocean. Mar Geol 186:541–556
Sydnes LK (1991) Oil, water, ice and light. Pol Res 10:609–618
Syvertsen EE (1991) Ice algae in the Barents Sea: types of assemblages, origin, fate and role in the ice-edge bloom. Pol Res 10:277–288
Taguchi S, Smith REH (1997) Effect of nitrogen and silicate enrichment on photosynthate allocation by ice algae from Resolute Passage, Canadian Arctic. J Mar Syst 11:53–61
Takahashi T, Wanninkhof RH, Feely RA, Weiss RF, Chipman DW, Bates NR, Olafson J, Sabine CL, Sutherland SC (1999) Net air-sea Co_2 flux over the global oceans: An improved estimate based on the air-sea pCO_2 difference In: Nojiri Y (ed) Proceedings of the 2nd International Symposium on CO_2 in the Oceans. Tsukuba, Japan, pp 9–15
Tallqvist ME, Sundet JH (2000) Annual growth of the cockle *Clinocardium ciliatum* in the Norwegian Arctic (Svalbard area). Hydrobiologia 440:331–338
Tande KS (1991) *Calanus* in North Norwegian fjords and in the Barents Sea. Pol Res 10:389–407
Tande KS, Hopkins CCE (1981) Ecological investigations on the zooplankton community in Balsfjorden, northern Norway: the genital system in *Calanus finmarchicus* and the role of the gonad development in overwintering strategy. Mar Biol 63:159–164
Tande KS, Slagstad D (1990) Growth and production of the herbivorous copepod, *Calanus glacialis*, in the arctic waters of the Barents Sea. Mar Ecol Prog Ser 63:189–199
Tande KS, Slagstad D (1992) Regional and interannual variation in biomass and productivity of *Calanus finmarchicus*, in subarctic environments. Oceanol Acta 15:309–321
Taniguchi A (1999) Differences in the structure of the lower trophic level of pelagic ecosystems in the eastern and western subarctic Pacific. Limnol Oceanogr 43:289–315
Tarasov GA, Pogodina IA, Hasankaev VB, Kukina NA, Mityaev MV (2000) Processes of sedimentation on the glacial shelves. Kola Sci. Center RAS, Apatity, pp 1–473 (in Russian)
Tarduno JA, Brinkman DB, Renne PR, Cottrell RD, Scher H, Castillo P (1998) Evidence of extreme climatic warmth from Late Cretaceous Arctic vertebrates. Science 282: 2241–2244

Taylor GH, Teichmüller M, Davis A, Diessel CFK, Littke R, Roberts P (1998) Organic Petrology: A new handbook incorporation some revised parts of Stach's Textbook of Coal Petrology. Gebrüder Bornträger, Berlin, Stuttgart, 704 pp

Taylor J, Tranter M, Munhoven G (2002) Carbon cycling and burial in the glacially influenced Polar North Atlantic. Paleoceanography 17:10.1029/2001PA000644

Teeri JA (1988) Interaction of temperature and other environmental variables influencing plant distribution. Soc Experiment Biol 42:77–89

Teeri JA, Stowe LG (1976) Climatic pattern and the distribution of C4 grasses in North America. Oecologica 23:1–12

Telang SA, Pocklington R, Naidu AS, Romankevich EA, Gitelson II, Glachyshev MI (1991) Carbon and mineral transport in major North American, Russian Arctic, and Siberian rivers: the St. Lawrence, the Mackenzie, the Yukon, the Arctic Alaskan rivers, the Arctic basin rivers in the Soviet Union, and the Yenisei. In: Degens ET, Kempe S, Richey JE (eds) Biogeochemistry of Major World Rivers. SCOPE 42, Wiley, Chichester, pp 75–104

Thiede J (1986) Zur geologischen Geschichte der Polarmeere und ihrer Wassermasen. Universität Kiel, Mimeo

Thiede J (2002) POLARSTERN ARKTIS XVII/2 cruise report: AMORE 2001 (Arctic Mid-Ocean Ridge Expedition). Reps Pol Mar Res 421, 397 pp

Thiede J, Bauch HA, Hjort C, Mangerud J (eds) (2001) The late Quaternary stratigraphy and environments of northern Eurasia and the adjacent Arctic seas – new contributions from QUEEN. Glob Plan Change 31:1–474

Thiede J, Clark DL, Herman Y (1990) Late Mesozoic and Cenozoic paleoceanography of the northern polar oceans. In: Grantz A, Johnson L, Sweeney JF (eds) The Arctic Ocean Region, The Geology of North America, Vol L, pp 427–458

Thiel H (1975) The size structure of the deep-sea benthos. Int Revue ges Hydrobiol 60:575–606

Thingstad TF, Sakshaug E (1990) Control of phytoplankton growth in nutrient recycling ecosystems. Theory and terminology. Mar Ecol Prog Ser 63:261–272

Thomas DJ, Macdonald RW, Cornford AB (1986) Geochemical mass-balance calculations for the coastal Beaufort Sea, NWT, Canada. Rapp P-v Réun Cons Int Explor Mer 186:165–184

Thomas DN, Lara RJ, Eicken H, Kattner G, Skoog A (1995) Dissolved organic matter in Arctic multi-year sea ice during winter: major components and relationship to ice characteristics. Pol Biol 15: 477–483

Thomas, DN, Dieckmann GS (2002). Antarctic sea ice – a habitat for extremophiles. Science 295:641–644

Thomas, DR, Rothrock DA (1993) The Arctic Ocean ice balance: a Kalman smoother estimate. J Geophys Res 98 (C6):10053–10067

Thompson DWJ, Wallace JM (1998) The Arctic Oscillation signature in the wintertime geopotential height and temperature fields. Geophys Res Lett 25:1297–1300

Thomsen L, Graf G, Juterzenka K, Witte U (1995) An *in situ* experiment to investigate the depletion of seston above an interface feeder field on the continental slope of the western Barents Sea. Mar Ecol Prog Ser 123 295–300

Thomsen L, vanWeering T, Gust G (2002) Processes in the benthic boundary layer at the Iberian continental margin and their implication for carbon mineralization. Progr Oceanogr 52:315–329

Thomson J, Nixon S, Croudace IW, Pedersen TF, Brown L, Cook GT et al. (2001) Redox-sensitive element uptake in north-east Atlantic Ocean sediments (Benthic Boundary Layer Experiment sites). Earth Planet Sci Lett 184:535–547

Thurman EM (1985) Organic geochemistry of natural waters. Kluwer Academic Publisher Group, Dordrecht

Timokhov LA (1994) Regional characteristics of the Laptev and the East Siberian seas: climate, topography, ice phases, thermohaline regime, and circulation. Reps Pol Res 144: 15–31

Tissot B, Deroo G, Herbin JP (1979) Organic matter in Cretaceous sediments of the North Atlantic: Contribution to sedimentology and paleoceanography. In: Talwani M, et al. (eds) Deep Drilling Results in the Atlantic Ocean: Continental Margin and Paleoenvironment. Amer Geophys Union, M Ewing Ser, Vol 3, p 362–374

Tissot BP and Welte DH (1984) Petroleum Formation and Occurrence. Springer, Heidelberg, 699 pp

Toll EV (1897) Paleoglaciers of New-Siberian Islands, their relationship to the mammoth remains and to glacial epoch. In: Proceedings of the Russian Geographical Society, Vol 32 (1), St. Petersburg, Imperial Academy of Sciences Press, 139 pp (in Russian)

Toudal L, Valeur H (1996) Ice extent in the Greenland Sea 1978–1995. In: Wadhams P, Wilkinson JP, Wells SCS (eds) Sci Rep Eur Subpolar Ocean Progr (EC/MAST), Scott Polar Res Inst Cambridge, UK, pp 104–122

Tréguer P, Pondaven P (2000) Global change: Silica control of carbon dioxide. Nature 406:358–359

Tremblay J-E, Gratton Y, Fauchot J, Price NM (2002) Climatic and oceanic forcing of new, net, and diatom production in the North Water. Deep-Sea Res II 49:4927–4946

Trenberth KE, Hurrell JW (1994) Decadal atmosphere-ocean variations in the Pacific. Clim Dynam 9:303–319

Tseitlyn VB (1993) Relationship between primary production and the vertical flux of organic carbon in mesopelagic zone of the Ocean. Oceanology 33(2):189–192

Turner SM, Malin G, Liss PS, Harbour DS, Holligan PM (1988) The seasonal variation of dimethyl sulfide and dimethylsulfoniopropionate concentrations in nearshore waters. Limnol Oceanogr 33:364–375

Tyrrell T, Taylor AH (1996) A modelling study of *Emiliania huxleyi* in the NE Atlantic. J Mar Syst 9:83–112

Unger D, Neumann K, Fetzer I (2000) Sediment trap investigations in the Kara Sea. Reps Pol Res 360:28–35

Unger D, Schayen S, Gaye B (2001) Sediment trap investigations in the Kara Sea. Reps Pol Mar Res 393:29–33

Untersteiner N, Carmack E (1990) Arctic sea ice, freshwater, and global ocean climate: some possible connections. Appendix, WMO Working Group on Sea Ice and Climate, Rep of the 4th Session

Usachev PI (1961) Phytoplankton of the North Pole. Fish Res Bd Can Trans Ser 1285

Uspensky VA, Rodionova KF, Gorskaya AI (1966) Text-book on analysis of bituminous and dissipated organic matter of rocks. Nedra, Leningrad, pp 1–302 (in Russian)

Uye S (1996) Induction of reproductive failure in the planktonic copepod *Calanus pacificus* by diatoms. Mar Ecol Prog Ser 133:89–97

Vairavamurthy A, Andreae MO, Brooks JM (1986) Determination of acrylic acid in aqueous samples by Electron Capture Gas Chromatography after extraction with Tri-n-octylphosphine oxide and derivazation with Pentafluorobenzyl Bromide. Anal Chem 58:2684–2687

van Andel TH, Heath GH, Moore TC (1975) Cenozoic history and paleoceanography of the Central Equatorial Pacific, Mem Geol Soc Amer 143:1–134 van der Wal P, Kempers RS, Veldhuis MJW (1995) Production and downward flux of organic matter and calcite in a North Sea bloom of the coccolithophore *Emiliania huxleyi*. Mar Ecol Prog Ser 126:247–265

Van Wagoner NA, Robinson PT (1985) Petrology and Geochemistry of a CESAR bedrock sample; implication for the origin of the Alpha Ridge, Initial Geological Report on CESAR, Geol Surv Canada, 84–22, pp 47–58

Vanaverbeke J, Arbizu PM, Dahms HU, Schminke HK (1997) The metazoan meiobenthos along a depth gradient in the Arctic Laptev Sea with special attention to nematode communities. Pol Biol 18:391–401

Vanegas SA, Mysak LA (2000) Is there a dominant timescale of natural climate variability in the Arctic? J Clim 13: 3412–3434

Vanreusel A, Vincx M, Bett BJ, Rice AL (1995) Nematode biomass spectra at 2 abyssal sites in the NE Atlantic with a contratsing food supply. Int Revue ges Hydrobiol 80: 287–296

Varela DE, Harrison PJ (1999) Effect of ammonium on nitrate utilization by *Emiliania huxleyi*, a coccolithophore from the oceanic northeastern Pacific. Mar Ecol Prog Ser 186: 67–74

Vasiliev AA (1995) Thermoabrasion of coast in the Western Yamal. Abstracts of Int Conference on permafrost, Pushchino (in Russian)

Vedernikov VI, Demidov AB (1997) Primary production and chlorophyll in the Northeastern Norwegian Sea in July 1995. Oceanology 37(2):227–232

Vedernikov VI, Demidov AB, Sud'bin AI (1995) Primary production and chlorophyll in the Kara Sea in September 1993. Oceanology 34(5):630–640

Vedernikov VI, Gagarin VI (1998) Primary production and chlorophyll in the Barents Sea in September–October, 1997. Oceanology 38:642–649

Velikotsky MA (1998) Modern shore dynamics of Kolguev Island. In: Solomatin (ed) Dynamics of Russian Arctic shores. Moscow Univ Press, pp 93–101 (in Russian)

Velitchko AA, Dolukhanov PM, Rutter NW and Catto NR (eds) (1997) Quaternary of northern Eurasia: Late Pleistocene and Holocene landscapes, stratigraphy and environments. Quat Intern, 41/42, 1991 pp

Venkatesan MI, Kaplan IR, Ruth F (1983) Hydrocarbon geochemistry in surface sediments of Alaskan outer continental shelf: Part 1: C_{15+} hydrocarbons. Am Ass Petrol Geol Bull 67:831–840

Verity PG (2000) Grazing experiments and model simulations of the role of zooplankton in *Phaeocystis* food webs. J Plankton Res 43:317–343

Verity PG, Smayda TJ, Sakshaug E (1991) Photosynthesis, excretion and growth rates of *Phaeocystis* colonies and solitary cells. Pol Res 10:117–128

Vernet M, Matrai PA, Andreassen IJ (1998) Synthesis of particulate and extracellular carbon by phytoplankton at the marginal ice zone in the Barents Sea. J Geophys Res 103(C1):1023–1037

Vetter EW, Dayton PK (1998) Macrofaunal communities within and adjacent to a detritus-rich submarine canyon system. Deep-Sea Res II 45:25–54

Vetter YA, Deming J (1994) Extracellular enzyme activity in the Arctic Northeast Water polynya. Mar Ecol Progr Ser 114:23–34

Vézina AF, Demers S, Laurion I, Sime-Ngando T, Juniper SK, Devine L (1997) Carbon flows through the microbial food web of first-year ice in Resolute Passage (Canadian High Arctic). J Mar Syst 11:173–189

Vibe C (1967) Arctic Animals in Relation to Climatic Fluctuations. CA Reitzels Forlag, Copenhagen, 227 pp

Vidal J, Smith SL (1986) Biomass, growth, and development of populations of herbivorous zooplankton in the southeastern Bering Sea during spring. Deep-Sea Res I 33: 523–556

Vidal J, Whitledge TE (1982) Rates of metabolism of planktonic crustaceans as related to body weight and temperature of habitat. J Plank Res 1:77–83

Vinje T (1987) Dynamics and morphology of the Barents Sea ice fields. Proc 9th Int Conf POAC 263–268

Vinje T, Kvambek ÅS (1991) Barents Sea drift ice characteristics. Pol Res 10:59–68

Vinje T, Nyborg M, Kjrnli G (1996) Sea ice variation in the Greenland Sea during the nineteenth century. In: Wadhams P, Wilkinson JP, Wells SCS (eds) Science Reports of Euorpean Subpolar Ocean Progress (EC/ MAST). Scott Polar Research Institute, Cambridge, UK, pp 101–103

Vinje T (2001) Anomalies and trends of sea-ice extent and atmospheric circulation in the Nordic Seas during the period 1864–1998. J Clim 14:255–267

Vink GE, Jason Morgan W, Zhao WL (1984) Preferential Rifting of Continents: A Source of Displaced Terranes. J Geophys Res 89:10.072–10.076

Vinnikov KY, Robock A, Stouffer RJ, Walsh JE, Parkinson CL, Cavalieri DJ, Mitchell JFB, Garrett D, Zakharov VF (1999) Global warming and northern hemisphere sea ice extent. Science 286:1934–1937

Vinogradova NG (1979) The geographical distribution of the abyssal and hadal (ultra-abyssal) fauna in relation to the vertical zonation of the ocean. Sarsia 64:41–50

VNIIOkeangeologia (All Russia Research Institute for Geology and Mineral Resources of the World Ocean) (1999) Bottom relief of the Arctic Ocean. St. Petersburg, Department of Navigation and Oceanography, Russian Federation, scale 1:5,000,000, 1 Sheet

Vogelsang E, Sarnthein M, Pflaumann U (2001) $\delta^{18}O$ Stratigraphy, Chronology, and Sea Surface Temperatures of Atlantic sediment Records (GLAMAP-2000 Kiel). Institut für Geowissenschaften, Universität Kiel. Ber 13, pp 1–11

Vogt C (1997) Regional and temporal variations of mineral assemblages in Arctic Ocean sediments as climatic indicator during glacial/interglacial changes. Reps Pol Res 251, 309 pp

Vogt C, Knies J, Spielhagen RF, Stein R (2001) Detailed mineralogical evidence for two nearly identical glacial/deglacial cycles and Atlantic water advection to the Arctic during the last 90.000 years. Glob Planet Change 31 (1–4): 23–44

Vogt PR, Gardner J, Crane K (1999) The Norwegian-Barents-Svalbard (NBS) continental margin: Introducing a natural laboratory of mass wasting, hydrates, and ascent of sediment, pore water, and methane. Geo-Mar Lett 19: 2–21

Vogt PR, Taylor PT, Kovacs LC, Johnson GL (1979) Detailed aeromagnetic investigation of the Arctic Basin. J Geophys Res 84:1071–1089

Vogt PR, Taylor PT, Kovacs LC, Johnson GL (1982) The Canada Basin; Aeromagnetic constraints on structure and evolution, Tectonophysics 89:295–336

Volkman JK (1986) A review of sterol markers for marine and terrigenous organic matter. Org Geochem 9:83–99

Volkman JK, Barrett SM, Blackburn SI, Mansour MP, Sikes EL, Gelin F (1998) Microalgal biomarkers: a review of recent research developments. Org Geochem 29:1163–1179

Volkman JK, Barrett SM, Dunstan GA, Jeffrey SW (1993) Geochemical significance of the occurrence of dinosterol and other 4-methyl sterols in a marine diatom. Org Geochem 20:7–15

Volkman JK, Eglinton G, Corner EDS, Forsberg TEV (1980) Long-chain alkenes and alkenones in the marine coccolithophorid *Emiliania huxleyi*. Phytochem 19:2619–2622

von Quillfeldt CH (1997) Distribution of diatoms in the Northeast Water Polynya, Greenland. J Mar Syst 10:211–240

von Quillfeldt CH (2000) Common diatom species in arctic spring blooms: their distribution and abundance. Botanica Marina 43:499–516

Vörösmarty CJ, Hinzman LD, Peterson BJ, Bromwich DH, Hamilton LC, Morison J, Romanovsky VE, Sturm M, Webb RS (2001) The Hydrologic Cycle and its Role in the Arctic and Global Environmental Change: A Rationale and Strategy for Synthesis Study, Fairbanks, Alaska: Arctic Research Consortium of the U.S, pp 84

Voskresenskkii AI, Petrov LS (1985) Climate peculiarities. In: Arctic and Southern Oceans. Nauka, Leningrad: 45–64 (in Russian)

Wada E, Minagawa M, Mizutani H, Tsuji T, Imaizumi R, Karasawa K (1987) Biogeochemical studies on the transport of organic matter along the Otsuchi river watershed, Japan. Est Coast Shelf Sci 25:321–336

Wadhams P (1981) Sea-Ice Topography of the Arctic Ocean in the Region 70 degree W to 25 degree E. Phil Trans R Soc Lond 302:45–85

Wadhams P (1997) Variability of Arctic sea ice thickness – statistical significance and its relationship to heat flux. In: Stel JH, Behrens HWA, Borst JC, Droppert LJ, vd Meulen J (eds) Operational Oceanography. The Challenge for European Co-operation. Elsevier Science, pp 368– 384

Wadhams P (2000) Ice in the Ocean. Gordon and Breach Science Publishers, 351pp

Wagner T, Dupont L (1999) Terrestrial organic matter in marine sediments: analytical approaches and eolian-marine records of the central Equatorial Atlantic. In: G Fischer and G Wefer (eds) The use of Proxies in Paleoceanography: Examples from the South Atlantic. Springer, Heidelberg Berlin, pp 547–574

Wagner T, Henrich R (1994) Organo- and lithofacies of glacial-interglacial deposits in the Norwegian-Greenland Sea: Responses to paleoceanographic and paleoclimatic changes. Mar Geol 120: 335–364

Wahsner M, Müller C, Stein R, Ivanov G, Levitan M, Shelekova E, Tarasov G (1999) Clay mineral distributions in surface sediments from the Central Arctic Ocean and the Eurasian continental margin as indicator for source areas and transport pathways: A synthesis. Boreas 28: 215–233

Walker HJ (1974) The Colville River and the Beaufort Sea: some interactions. In: Reed JC, Sater JE (eds) The Coast and Shelf of the Beaufort Sea. Arctic Institute of North America, Arlington, Va, pp 513–540

Wallace DWR, Minett P, Hopkins TS (1995) Nutrients, oxygen, and inferred new production in the Northeast Water Polynya, 1992. J Geophys Res 100(C3):4323–4340

Wallace DWR, Moore RM, Jones EP (1987) Ventilation of the Arctic Ocean cold halocline: rates of diapycnal and isopycnal transport, oxygen utilization and primary production inferred using chlorofluoromethane distributions. Deep-Sea Res 34:1957–1979

Wallace JM, Thompson DWJ (2002) Annular modes and climate prediction. Physics Today February:28–33

Walsh JJ (1989) Arctic carbon sinks: present and future. Glob Biogeochem Cycles 3:393–411

Walsh JJ (1989) How much shelf production reaches the deep sea? In: Berger WH et al. (eds) Productivity of the Ocean: Past and Present, Life Sci Res Rep, Vol 44, Wiley & Sons, pp 175–192

Walsh JJ (1995) DOC storage in Arctic seas: the role of continental shelves. In: Smith WO jr, Grebmeier J (eds) Arctic oceanography: marginal ice zones and continental shelves. AGU Press, Washington DC, pp 203–230

Walsh JJ, Dieterle DA (1994) CO_2 cycling in the coastal ocean. I. A numerical analysis of the southeastern Bering Sea, with applications to the Chukchi Sea and the northern Gulf of Mexico. Prog Oceanogr 34:335–392

Walsh JJ, McRoy CP (1986) A tentative carbon (POC) budget for outer shelf and accumulation rate for continental slope, Bering Sea. Cont Shelf Res 5:259–288

Walsh JJ, McRoy CP, Coachman LK, Goering JJ, Nihoul JJ, Whitledge TE, Blackburn TH, Parker PL, Wirick CD, Shuert PG, Grebmeier JM, Springer AM, Tripp RD, Hansell DA, Djenidi S, Deleersnijder E, Henriksen K, Lund BA, Andersen P, Muller-Karger FE, Dean K (1989) Carbon and nitrogen cycling within the Bering/Chukchi seas: source regions for organic matter effecting AOU demands of the Arctic Ocean. Prog Oceanogr 22: 277–359

Walsh JJ, McRoy PC (1986) Ecosystem analysis in the southeastern Bering Sea. Cont Shelf Res 5:259–288

Walsh JJ, Rowe GT, Iverson RL, McRoy CP (1981) Biological export of shelf carbon is a sink of the global CO_2 cycle. Nature 291:196–201

Walsh JJ, Wirick CD, Pietrafesa LJ, Whitledge TE, Hoge FE, Swift RN (1988) High-frequency sampling of the 1984 spring bloom within the Mid-Atlantic Bight: Synoptic shipboard, aircraft, and in situ perspectives of the SEEP-I Experiment. Cont Shelf Res 8:529–563

Wassmann P (1984) Sedimentation and benthic mineralization of organic detritus in a Norwegian fjord. Mar Biol 83:83–94.

Wassmann P (1991) Dynamics of primary production and sedimentation in shallow fjords and polls of western Norway. Oceanogr Mar Biol Ann Rev 91:87–154

Wassmann P (1994) Significance of sedimentation for the termination of *Phaeocystis* blooms. J Mar Syst 5: 81–100

Wassmann P (1998) Retention versus export food chains: processes controlling sinking loss from marine pelagic systems. Hydrobiologia 363:29–57

Wassmann P (1999) Spring bloom development in the marginal ice zone and the Central Barents Sea. Mar Ecol 20:321–346

Wassmann P (2001) Vernal export and retention of biogenic matter in the north-eastern North Atlantic and adjacent Arctic Ocean: the role of the Norwegian Atlantic Current and topography. Mem Nat Inst Pol Res Spec Issue 54: 77–392

Wassmann P, Andreassen I, Reigstad M, Slagstad D (1996) Pelagic-benthic coupling in the Nordic Seas: The role of episodic events. PSZN I: Mar Ecol Ecol 17:447–471

Wassmann P, Olli K, Wexels Riser C, Svensen C (2003) Ecosystem function, biodiversity and vertical flux regulation in the twilight zone. In: Marine Science Frontiers for Europe, Springer, Heidelberg Berlin, in press

Wassmann P, Peinert R, Smetacek V (1991) Patterns of production and sedimentation in the boreal and polar Northeast Atlantic. Pol Res 10: 209–228

Wassmann P, Ratkova TN, Andreassen I, Vernet M, Pedersen G, Rey F (1999) Spring bloom development in the marginal ice zone and the central Barents Sea. PSZN I: Mar Ecol 20 (3/4):321–346

Wassmann P, Slagstad D (1993) Seasonal and annual dynamics of particulate carbon flux in the Barents Sea. Polar Biol 13:363–372

Wassmann P, Vernet M, Mitchell BG, Rey F (1990) Mass sedimentation of *Phaeocystis pouchetii* in the Barents Sea. Mar Ecol Prog Ser 66:183–195

Weatherhead EC, Morseth CM (1998) Climate change, ozone, and ultraviolet radiation. In: AMAP Assessment Report: Arctic Pollution Issues. Arctic Monitoring and Assessment Programme (AMAP), Oslo, Norway, 859

Weber JR (1983) Maps of the Arctic Basin Sea Floor: A History of Bathymetry and its Interpretation. Arctic 36 (2):121–142

Weber JR (1989) Physiography and Bathymetry of the Arctic Ocean Seafloor. In: Herman Y (ed) The Arctic Seas: Climatology, Oceanography, Geology, and Biology. Van-Nostrand Reinhold, New York, pp 797–828

Weber JR, Roots EF (1990) Historical background; Exploration, concepts, and observations. In: Grantz A, Johnson GL, Sweeney JF (eds) The Arctic Ocean Region: The Geology of North America DNAG L, pp 5–36

Weber ME, Niessen F, Kuhn G, Wiedicke M (1997) Calibration and application of marine sedimentary physical properties using a multi-sensor core logger. Mar Geol 136: 151–172

Weeks WF, Ackley SF (1986) The growth, structure and properties of sea ice. In: Untersteiner N (ed) The geophysics of sea ice. Martinus Nijhoff, Dordrecht (NATO ASI B146): 9–164

Weingartner T, Cavalieri DJ, Aagaard K, Sasaki Y. (1998) Circulation, dense water formation, and outflow on the northeast Chukchi Shelf. J Geophy Res 103(C4):7647–7661

Weingartner TJ, Danielson S., Saski Y, Pavlov V, Kulakov M (1999) The Siberian coastal current: A wind- and buoyancy-forced Arctic coastal current. J Geophy Res 104 (C12):29697–29713

Welch HE, Bergmann MA (1989) Seasonal development of ice algae and its prediction from environmental factors near Resolute, NWT, Canada. Can J Fish Aquat Sci 46: 1793–1804

Welch HE, Bergmann MA, Siferd TD, Martin KA, Curtis MF, Crawford RE, Conover RJ, Hop H (1992) Energy flow through the marine ecosystem of the Lancaster Sound region, Arctic Canada. Arctic 45:343–357

Weller G, Lange M (1999) Impacts of global climate change in the Arctic regions. An initial assessment. International Arctic Science Committee (IASC), Oslo, Norway, 30 pp

Weller G, Parker (1972) Radiation regime of the Arctic Basin – studies at T3, AIDJEX sites and Barrow. Tech Rep Geophys Inst Univ Alaska 2:4–11

Werner I, Arbizu PM (1999) The sub-ice fauna the Laptev Sea and the adjacent Arctic Ocean in summer 1995. Pol Biol 21:71–79

Werner, I (1997) Ecological studies on the Arctic under-ice habitat – Colonisation and processes at the ice-water interface. Berichte aus dem Sonderforschungsbereich 313, Univ Kiel, 70:1–167

Weslawski JM, Koszteyn J, Zajaczkowski M, Wiktor J, Kwasniewski S (1995) Fresh water in Svalbard fjord ecosystems. In: Skoldal HR, Hopkins C, Erikstad KE, Leinaas HP (eds) Ecology of fjords and coastal waters. Elsevier, Amsterdam, pp 229–242

Weslawski JM, Szymelfenig M, Zajaczkowski M, Keck A (1999) Influence of salinity and suspended matter on benthos of an Arctic tidal flat. J Mar Sci 56:194–202

Westerhausen L, Poynter J, Eglinton G, Erlenkeuser H, Sarnthein M (1993) Marine and terrestrial origin of organic matter in modern sediments of the equatorial East Atlantic: the delta ^{13}C and molecular record. Deep-Sea Res I 40:1087–1121

Wheeler PA (ed 1997) 1994 Arctic Ocean section. Deep-Sea Res II 44(8):1483–1757

Wheeler PA, Gosselin M, Sherr E, Thibault D, Kirchman DL, Benner R, Whitledge TE (1996) Active cycling of organic carbon in the central Arctic Ocean. Nature 380:697–699

Wheeler PA, Watkins JM, Hansing RL (1997) Nutrients, organic carbon and organic nitrogen in the upper water column of the Arctic Ocean: implications for the sources of dissolved organic carbon. Deep-Sea Res II 44: 1571– 1592

Whitehouse BG, Macdonald RW, Iseki K, Yunker MB, McLaughlin FA (1989) Organic carbon and colloids in the Mackenzie River and Beaufort Sea. Mar Chem 26:371–378

Whitledge TE, Blidiger RE, Zeeman SI, Sambrotto RN, Roscigno PF, Jensen PR, Brooks JM, Trees C, Veidt DM (1988) Biological measurements and related chemical features in Soviet and United States regions of the Bering Sea. Cont Shelf Res 8:1299–1319

Whitledge TE, Gorelbin MI, Cheryab SM (1992) Biogenic nutrient content. In: Nagel PA (ed) Results of the third joint US-USSR Bering and Chukchi Seas Expedition (BERPAC), summer 1988. USFWS, Washington DC, pp 39–49

Whitledge TE, Luchin VA (1999) Summary of chemical distributions and dynamics in the Bering Sea. In: Loughlin TR, Ohtani K (eds) The dynamics of the Bering Sea. Univ Alaska Seagrant AK-SG-03, Univ Alaska Fairbanks, pp 217–249

Whitledge TE, Reeburgh WS, Walsh JJ (1986) Seasonal inorganic nitrogen distributions and dynamics in the southeastern Bering Sea. Cont Shelf Res 5:109–132

Wilson JT (1963) Hypothesis of Earth's behavior. Nature 198: 925–929

Winther J-G, Gerland S, Ørbæk JB, Ivanov B, Zachek AS, Bezgreshnov AM (2001) Effects on spectral reflectance from snow ageing. Mem Natl Inst Polar Res Tokyo Spec issue 54:193–201

Wlodarska-Kowalczuk M, Szymelfenig M, Kotwicki L (1999) Macro- and meiobenthic fauna of the Yoldiabukta glacial bay (Isfjorden, Spitsbergen). Polish Polar Res 20:367–386

Wlodarska-Kowalczuk M, Weslawski JM, Gromisz S (1996) A comparison of the macrofaunal community structure and diversity in two arctic glacial bays – a "cold" one off Franz Josef Land and a "warm" one off Spitsbergen. Oceanologia 38:251–283

Wlodarska-Kowalczuk M, Weslawski JM, Kotwicki L (1998) Spitsbergen glacial bays macrobenthos – a comparative study. Polar Biol 20:66–73

Wolfe S Dallimore S, Solomon SM (1998) Coastal perma frost investigations along a rapidly eroding shoreline, Tuktoyaktuk, NWT. In: Lewkowicz AG, Allard M (eds) Proceedings of the Seventh International Permafrost Conference, 23–27 June, 1998, Yellowknife, NWT, Canada, pp 1125–1132

Wollast R (1991) The coastal organic carbon cycle: fluxes, sources, and sinks. In: Mantoura RFC, Martin JM, Wollast R (eds) Ocean margin processes in global change. Wiley, Chichester, pp 365–381

Wollenburg I (1993) Sediment transport by Arctic sea ice: The present-day lithogenic and biogenic particulate loading (in German). Reps Pol Res 127:1–159

Wollenburg JE, Spielhagen RF, Pirrung M, Knies J, Mackensen A (2002) High-resolution palaeoproductivity fluctuations since the Late Weichselian as indicated by benthic foraminifera in the marginal Arctic Ocean. Palaeogeogr Palaeoclimatol Palaeoecol, subm

Wong CS, Matear RJ (1999) Sporadic silicate limitation of phytoplankton productivity in the subarctic NE Pacific. Deep-Sea Res II 46:2539–2555

Woodgate R, Schauer U, Fahrbach E (1998) Moored current meters in Fram Strait at 79°N: Preliminary results with special emphasis on the West Spitsbergen current. AWI Berichte aus dem Fachbereich Physik, 28pp

Woodgate RA, Aagaard K, Muench RD, Gunn J, Björk G, Rudels B, Roach AT, Schauer U (2001) The Arctic Ocean boundary current along the Eurasian slope and adjacent Lomonosov Ridge: Water mass properties, transports and transformations from moored instruments. Deep-Sea Res I 48:1757–1792

Woodgate RA, Fahrbach E, Rohard G (1999) Structure and transports of the East Greenland Current at 75°N from moored current meters. J Geophys Res 104(C8): 18059–18072

Wyllie-Echeverria T, Wooster WS (1998) Year-to-year variations in Bering Sea ice cover and some consequences for fish distributions. Fish Oceanogr 7:159–170

Yackel JJ, Barber DG, Papakyriakou TN (2001) On the estimation of spring melt in the North Water Polynya using RADARSAT-1. Atmos-Ocean 39:195–208

Yashin DS (2000) Holocene sedimentation in the Arctic seas In: Geological-geophysical features of the litosphere of the Arctic Region. VNIIOkeangeologia, St. Petersburg 3: 57–67

Yayanos AA (1998) Empirical and theoretical aspects of life at high pressure in the deep-sea In: Horikoshi K, Grant WD (eds) Extremophiles: Microbial life in Extreme Environments. John Wiley & Sons, Chichester, pp 47–92

Yndestad H (1999) Earth nutation influence on the temperature regime of the Barents Sea. ICES J Mar Sci 56:381–387

Yunker MB, Backus SM, Graf Pannatier E, Jeffries DS, Macdonald RW (2002) Sources and significance of alkane and PAH hydrocarbons in Canadian arctic rivers. Est Coast Shelf Sci 55:1–31

Yunker MB, Cretney WJ (1995) Chlorinated dioxin trends between 1987 and 1993 for samples of crab hepatopancreas from pulp and paper mill and harbour sites in British Columbia, Can Tech Rep Fish Oc Sci 2082, Sidney: Institute of Ocean Sciences

Yunker MB, Macdonald RW (1995) Composition and origins of polycyclic aromatic hydrocarbons in the Mackenzie River and on the Beaufort Sea shelf. Arctic 48:118–129

Yunker MB, Macdonald RW, Cretney WJ, Fowler BR, McLaughlin FA (1993) Alkane, terpene, and polycyclic aromatic hydrocarbon geochemistry of the Mackenzie River and Mackenzie Shelf: Riverine contributions to Beaufort Sea coastal sediment. Geochim Cosmochim Acta 57:3041–3061

Yunker MB, Macdonald RW, Fowler BR, Cretney WJ, Dallimore SR, McLaughlin FA (1991) Geochemistry and fluxes of hydrocarbons to the Beaufort Sea shelf: A multivariate comparison of fluvial inputs and coastal erosion of peat using principal components analysis. Geochim Cosmochim Acta 55:255–273

Yunker MB, Macdonald RW, Veltkamp DJ, Cretney WJ (1995) Terrestrial and marine biomarkers in a seasonally ice-covered Arctic estuary – Integration of multivariate and biomarker approaches. Mar Chem 49:1–50

Yunker MB, McLaughlin FA, Fowler BR, Smyth TA, Cretney WJ, Macdonald RW et al. (1990) NOGAP B.6: Vol 7: Hydrocarbon determinations; Mackenzie River and Beaufort Sea shoreline peat samples. Canadian Data Report of Hydrography and Ocean Sciences 60:81

Zafiriou OC, Gagosian RB, Peltzer ET, Alford JB, Loder T (1985) Air-to-sea fluxes of lipids at Enewetak Atoll. J Geophys Res 90(D1):2409–2423

Zahn R, Markussen B, Thiede J (1985) Stable isotope data and depositional environments in the late Quaternary Arctic Ocean. Nature 314:433–435

Zdanowicz CM, Zielinski GA, Wake CP (1998) Characteristics of modern atmospheric dust deposition in snow on the Penny Ice Cap, Baffin Island, Arctic Canada. Tellus 50: 506–520

Zenkevitch L (1963) Biology of the seas of the USSR. George Allen and Unwin, London, 955 pp

Zenkovich VP (1962) Foundations of doctrine about the evolution of sea shores. AN USSR Press, Moscow, 710 pp (in Russian)

Zernova VV, Nöthig E-M, Shevchenko VP (2000) Vertical microalgae flux in the northern Laptev Sea. Oceanology 40:801–808

Zhang J, Rothrock D, Steele M (2000) Recent changes in the Arctic sea ice: The interplay between ice dynamics and thermodynamics. J Clim 13:3099–3114

Zheng Y, Schlosser P, Swift JH, Jones EP (1997) Oxygen utilization rates in the Nansen Basin, Arctic Ocean: implications for new production. Deep-Sea Res I 44:1923–1944

Zinchenko AG (2000) A new orographic scheme of the Arctic shelf of Russia. In: Geological-geophysical features of the lithosphere of the Arctic Region. VNIIOkeangeologia, St. Petersburg 3: 39–56

Zingone A, Chrétiennot-Dinet MJ, Lange M, Medlin L (1999) Morphological and genetic characterization of *Phaeocystis cordata* and *P. jahnii* (Prymnesiophyceae), two new species from the Mediterranean Sea. J Phycol 35:1322–1337

MIX
Papier aus verantwortungsvollen Quellen
Paper from responsible sources
FSC® C105338

If you have any concerns about our products,
you can contact us on
ProductSafety@springernature.com

In case Publisher is established outside the EU,
the EU authorized representative is:
Springer Nature Customer Service Center GmbH
Europaplatz 3, 69115 Heidelberg, Germany

Printed by Libri Plureos GmbH
in Hamburg, Germany